SCIENCE AND SANITY

AN INTRODUCTION TO NON-ARISTOTELIAN SYSTEMS AND GENERAL SEMANTICS

BY

ALFRED KORZYBSKI

AUTHOR OF *MANHOOD OF HUMANITY*
FOUNDER, INSTITUTE OF GENERAL SEMANTICS

FIFTH EDITION

With New Preface by
ROBERT P. PULA

International Non-Aristotelian Library
INSTITUTE OF GENERAL SEMANTICS
Englewood, New Jersey, USA

First Edition 1933
Second Edition 1941
Third Edition 1948
Fourth Edition 1958
Fifth Edition 1994

ISBN 0-937298-01-8

Library of Congress Catalog Card Number 58-6260

TO THE WORKS OF:

Aristotle	Cassius J. Keyser
Eric T. Bell	G. W. Leibnitz
Eugen Bleuler	J. Locke
Niels Bohr	Jacques Loeb
George Boole	H. A. Lorentz
Max Born	Ernst Mach
Louis de Broglie	J. C. Maxwell
Georg Cantor	Adolf Meyer
Ernst Cassirer	Hermann Minkowski
Charles M. Child	Isaac Newton
C. Darwin	Ivan Pavlov
René Descartes	Giuseppe Peano
P. A. M. Dirac	Max Planck
A. S. Eddington	Plato
Albert Einstein	H. Poincaré
Euclid	G. Y. Rainich
M. Faraday	G. F. B. Riemann
Sigmund Freud	Josiah Royce
Karl F. Gauss	Bertrand Russell
Thomas Graham	Ernest Rutherford
Arthur Haas	E. Schrödinger
Wm. R. Hamilton	C. S. Sherrington
Henry Head	Socrates
Werner Heisenberg	Arnold Sommerfeld
C. Judson Herrick	Oswald Veblen
E. V. Huntington	Wm. Alanson White
Smith Ely Jelliffe	Alfred N. Whitehead

Ludwig Wittgenstein

WHICH HAVE GREATLY INFLUENCED MY ENQUIRY,
THIS SYSTEM IS DEDICATED

"AT my alighting, I was surrounded with a crowd of people; but those who stood nearest seemed to be of better quality. They beheld me with all the marks and circumstances of wonder, neither, indeed, was I much in their debt; having never, till then, seen a race of mortals so singular in their shapes, habits, and countenances. Their heads were all reclined either to the right or the left; one of their eyes turned inward, and the other directly up to the zenith. Their outward garments were adorned with the figures of suns, moons, and stars, interwoven with those of fiddles, flutes, harps, trumpets, guitars, harpsicords, and many other instruments of music, unknown to us in Europe. I observed, here and there, many in the habit of servants, with a blown bladder fastened like a flail to the end of a short stick, which they carried in their hands. In each bladder was a small quantity of dried pease, or little pebbles (as I was afterwards informed). With these bladders they now and then flapped the mouths and ears of those who stood near them, of which practice I could not then conceive the meaning; it seems, the minds of these people are so taken up with intense speculations, that they neither can speak, nor attend to the discourses of others, without being roused by some external taction upon the organs of speech and hearing; for which reason, those persons, who are able to afford it always keep a flapper (the original is *climenole*) in their family, as one of their domestics, nor ever walk abroad, or make visits, without him. And the business of this officer is, when two or three more persons are in company, gently to strike with his bladder the mouth of him who is to speak, and the right ear of him or them to whom the speaker addresseth himself. This flapper is likewise employed diligently to attend his master in his walks, and, upon occasion, to give him a soft flap on his eyes, because he is always so wrapped up in cogitation that he is in manifest danger of falling down every precipice, and bouncing his head against every post; and in the streets, of jostling others, or being jostled himself, into the kennel.

It was necessary to give the reader this information, without which he would be at the same loss with me, to understand the proceedings of these people, as they conducted me up the stairs to the top of the island, and from thence to the royal palace. While we were ascending, they forgot several times what they were about, and left me to myself, till their memories were again roused by their flappers; for they appeared altogether unmoved by the sight of my foreign habit and countenance, and by the shouts of the vulgar, whose thoughts and minds were more disengaged.

. . . And although they are dextrous enough upon a piece of paper in the management of the rule, the pencil, and the divider, yet, in the common actions and behaviour of life, I have not seen a more clumsy, awkward, and unhandy people, nor so slow and perplexed in their conceptions upon all other subjects, except those of mathematics and music. They are very bad reasoners, and vehemently given to opposition, unless when they happen to be of the right opinion, which is seldom their case. Imagination, fancy, and invention they are wholly strangers to, nor have any words in their language by which those ideas can be expressed; the whole compass of their thoughts and mind being shut up within the two forementioned sciences."

JONATHAN SWIFT (*Gulliver's Travels*, A Voyage to Laputa)

CONTENTS

BOOK I

A GENERAL SURVEY OF NON-ARISTOTELIAN FACTORS

PART I

PRELIMINARIES

CONTENTS

PART II

GENERAL ON STRUCTURE

PART III

NON-ELEMENTALISTIC STRUCTURES

PART IV

STRUCTURAL FACTORS IN NON-ARISTOTELIAN LANGUAGES

PART V

ON THE NON-ARISTOTELIAN LANGUAGE CALLED MATHEMATICS

PART VI

ON THE FOUNDATION OF PSYCHOPHYSIOLOGY

BOOK II

A GENERAL INTRODUCTION TO NON-ARISTOTELIAN SYSTEMS AND GENERAL SEMANTICS

PART VII

ON THE MECHANISM OF TIME-BINDING

BOOK III

ADDITIONAL STRUCTURAL DATA ABOUT LANGUAGES AND THE EMPIRICAL WORLD

PART VIII

ON THE STRUCTURE OF MATHEMATICS

PART IX

ON THE SIMILARITY OF EMPIRICAL AND VERBAL STRUCTURES

PART X

ON THE STRUCTURE OF 'MATTER'

CONTENTS

A Note on Errata

We are grateful to sharp-eyed Hameed Khan and Stuart A. Mayper for noting some errors in *Science and Sanity*. In this Fifth Edition errors are identified with an asterisk in the margins of the text where they appear and are explained below:

Page	Correction
xxv	Line 12 "commensurate wth" should be: "with."
lxxxix	Quotation from "C. E. Coghill " should be **"G. E. Coghill."**
112	Name should be Ostwald.
112	"Half an acre" should be ~ 5.93 acres.
210	Tristram Shandy Paradox is stated incorrectly; while "*No day of his life would remain unwritten*," writing up each day augments the number of unwritten days by 364 or 365; the autobiography is never "*completed*."
211	The probability of an M-event for an audience of 2 in a town of 1,000 is 1 in 20 x 1/20.3; for 3, multiply again by 1/20.8; for 4, again by 1/21.2; of 5 again by 1/21.7, etc. (The probabilities depend on the size of the town; repeating 1/20 for each step assumes its size is infinite.) (It is not known whether the miscalculation was Korzybski's or whether he simply accepted it from Reference 203.)
318	Name should be Minkowski.
502	line should read "...semantic shock cannot produce a neurosis."
668	Formula for t′ has a misplaced parenthesis. It should read $t'=\beta(t-vx/c^2)$.
678	Formula should read: $$m_{\bar{N}}=\frac{m_0}{\sqrt{1-v^2/c^2}}$$
686	Hydrogen + ion does not have 2 atoms; H_2O splits into H^+ and OH^-.
703	Expressions (2), (3), (4) should have dots over **left-hand side** x_h's and over all q's **outside derivatives**. (No dots within the derivatives.) Third paragraph should have dots also over the q's in the first expression; no dots in the second expression. The final q_1 should also be dotted. Note that the superscript 2 after it is a **reference** (to Bibliography item 204), not an exponent.
771	Misprint in Reference 172. Should be "superstition."

PREFACE TO THE FIFTH EDITION, 1993

Six decades have passed since this book appeared in 1933. In the interim many thousands, in varying degrees of breadth and depth, have interacted with the formulations first presented herein. As of these hours of writing, people in every continent (save, perhaps, Antarctica) are studying and applying general semantics.

Those who have been attracted to and worked with Korzybski's formulations have largely come from the evaluationally energetic and self-selecting segment of our populations. They have tended to be leaders or those training to be leaders in a broad range of interests and disciplines. Through their efforts as teachers, managers, researchers, etc., Korzybski's formulations have explicitly and implicitly reached many thousands more.

Some aspects of general semantics have so permeated the (American) culture that behaviors derived from it are common; e.g., wagging fingers in the air to put 'quotes' around spoken terms which are deemed suspect. Original korzybskian terms are seen used without attribution, as if part of the general vocabulary; e.g., a paragraph-long explanation of "time-binding" appearing in a high school social studies text.

Some of Korzybski's coinages, particularly "neuro-linguistic," are now common coin and have extended the subset of English with the "neuro" prefix. (The *Random House Dictionary of the English Language*, 2nd Edition, p. 1291, mistakenly gives "1960-65" as the dates of origin for "neuro-linguistic" and the offspring terms, "neuro-linguistics" and "neuro-linguist.")

During the depth of the recent "Cold War" an interview about Korzybski's work was broadcast by Harry Maynard and Wladyslaw Marth to Poland over Radio Free Europe. A new study by a Polish scholar, Karol Janicki, credits Korzybski with being a precursor of what Janicki calls *Non-Essentialist Sociolinguistics.*[1]

Science and Sanity has by now spawned a whole library of works by other time-binders. Some of them have been listed in previous editions. Since the publication of the Fourth Edition, this 'parenting' has continued. Books, doctoral dissertations, masters theses, scholarly papers, essays and journal articles abound. The primary journals for on-going discussion and development of general semantics are the *General Semantics Bulletin*, published by the Institute of General Semantics, and *ETC.: A Review of General Semantics*, published by the International Society for General Semantics. Other journals, popular and scholarly, publish general semantics materials, pro and con.

For a sampling of books dealing with general semantics published since 1970, see the Bibliographic Note.[2] My choices do not reflect my evaluations of the books listed. They are included to indicate the continued growth, discussion of and influence of general semantics. Books critical of Korzybski/general semantics are included. They are tagged thus: (critique).

I consider Korzybski's *Collected Writings* 1920-1950 perhaps the most important publication in general semantics since *Science and Sanity*.[3] This book brings together all of Korzybski's known published writings other than his major books and a set of seminar lectures given at Olivet College in 1937. The 940 page book is a 'must' for anyone undertaking a serious study of general semantics.

Three works scheduled for publication in 1993 and 1994 include those of Susan Presby Kodish and Bruce I. Kodish, *Drive Yourself Sane! Using the Uncommon Sense of General Semantics*, Robert P. Pula, *Knowledge, Uncertainty and Courage: The Collected General Semantics Writings of Robert P. Pula*, and *General Semantics in Psychotherapy*, edited by Isabel Caro and Charlotte S. Read.

The Institute of General Semantics remains the primary center for training in general semantics. Seminar-workshops, weekend seminars and colloquia continue. An Advanced Studies and Teacher Certification Program for training leaders and teachers in general semantics has been established.

The distinguished annual Alfred Korzybski Memorial Lecture series, begun in 1952, continues to present highly regarded speakers whose work directly reflects or complements korzybskian orientations. These lecturers have been outstanding in the fields of anthropology, philosophy, physics, chemistry, physiology, embryology, medicine, neurology, surgery, education, sociology, linguistics, psychology, management, library science, law, ... etc.[4] Their participation bespeaks the growing regard for Korzybski's contributions and the importance of general semantics as a major twentieth century system.

The Index to this Fifth edition has been enhanced to facilitate general study and formulation searches. Grateful thanks are due to Bruce I. Kodish and Milton Dawes for their work in updating the Index, and to Bruce I. Kodish and Stuart Mayper for creating the Index of Diagrams. Grateful thanks also are due to Marjorie Zelner, Executive Secretary of the Institute of General Semantics, for her work as production editor of this Fifth Edition.

This briefly reviews the activities in the field of general semantics and shows that, as with any system that represents a challenge to its own culture, there remains much work to be done, but also that this work proceeds vigorously.

In the sixty years since Korzybski offered the first (not the last) non-aristotelian system in *Science and Sanity* public reaction has been both enthusiastic and critical. What about Korzybski's work continues to generate such interest and activity? If it were so, as some critics have asserted, that 'all' he did was to organize scattered insights, formulations and data into a system, that alone would have constituted a major achievement worthy of the gratitude of succeeding time-binders for centuries to come. Korzybski did that. He enunciated a system, incorporating aspects of but going beyond its predecessors, and proposed a methodology for making his system a living tool: *general semantics* (the name he selected), the first non-aristotelian system *applied* and made *teachable*.

In addition, I can list here the following selected formulations and *points of view, emphases*, etc., which I consider to be original with Korzybski.[5]

1. *Time-binding; time-binding ethics*. Rejecting both theological and zoological definitions, Korzybski adopted a natural science, operational approach and defined humans by what they can be observed *doing* which differentiates them from other classes of life; he defined them as the *time-binding class of life*, able to pass on knowledge from one generation to another over 'time.'[6]

Derived from this definition, which evaluates humans as a *naturally* cooperative class of life (the *mechanisms* of time-binding are descriptively social, cooperative), Korzybski postulated time-binding 'ethics' — modes of behavior, *choices* appropriate to time-binding organisms.

2. Korzybski recognized that *language (symbolizing in general) constitutes the basic tool of time-binding*. Others before him had noted that *language*, in the complex human sense, was one of the distinguishing features of humans. What Korzybski *fully* recognized was the central, *defining* role of language. No language, no time-binding. If so, then *structures of languages* must be determinative for time-binding.

3. The *neurologically* focused formulation of the *process of abstracting*. *No one* before Korzybski had so thoroughly and unflinchingly specified the *process* by which humans build and evolve theories, do their mundane evaluating, thrill to 'sunsets', etc. Korzybski's formulating of abstracting, particularly in the human realm, can constructively serve as a guide to on-going neuroscientific research.

4. As a function of the above, but deserving separate mention with the rigorously formulated notion of *orders* of abstracting, is the concurrent admonition that we should not confuse (identify) them. Given the hierarchical, sequential character of the nervous system (allowing, too, for horizontally related structures and parallel processing), it is inevitable that results along the way should manifest as (or 'at') differing orders or levels of abstracting. These *results* are inevitable. That they would be *formulated* at a given historical moment is *not* inevitable. Korzybski did it in the 1920's, publishing his descriptions in their mature form in 1933.

5. *Consciousness of abstracting*. If human organisms-as-a-whole-cum-nervous systems/brains *abstract* as claimed above and described herein (pp. 369-451 and *passim*), surely *consciousness* of these events must be crucial for

optimum human functioning. Animals (non-time-binders) abstract; but, so far as we know (1993), they do not *know* that they abstract. Indeed, many humans don't either — but they have the potentiality to do so. Korzybski recognized this, realized that *consciousness of abstracting* is essential for "fully functioning" humans, and made of it a primary goal of general semantics training.

6.   The *structure of language*.   Among Korzybski's most original formulations was the *multiordinal* character of many of the terms we most often use. He insisted that, with multiordinal terms, the 'meaning' is strictly a function of the order or level of abstraction at which the term is used and that its 'meaning' is so context-driven that it doesn't 'mean' *anything* definite until the context is specified or understood.

7.   *Structure as the only 'content' of knowledge*.   This represents the height/depth of non-elementalism; that what used to be designated as 'form' (structure) and 'content' are so intimately related as to be, practically speaking, fused, that structure and 'content' are functions of 'each other'. Further, and more deeply, all we *can* ever know expresses a set or sets of *relationships* and, most fundamentally, a *relationship* ('singular') between the 'known' and the knowing organism: the famous joint product of the observer-observed *structure*. "Structure is the only 'content' of knowledge" may qualify as Korzybski's deepest expression of anti-essentialism.   We can *not* know 'essences', *things in themselves*; all we can know is what we know as *abstracting nervous systems*. Although we can on-goingly know *more*, we can *not* 'transcend' ourselves as organisms that abstract.

8.   *Semantic reactions; semantic reactions as evaluations*.   Growing out of his awareness of the *transactive* character of human evaluating and wishing to correct for the elementalistic splitting involved in such terms as 'meaning', 'mental', 'concept', 'idea' and a legion of others, Korzybski consciously, deliberately formulated the term *semantic reaction*. It is central to his system.

9.   *The mathematical notion of function as applied to the brain-language continuum*.   Boldly grasping the neurophysiology of his day, Korzybski formulated what research increasingly finds: language is a function of (derives from, is invented by) brain; reciprocally, as a function of feedback mechanisms, brain is a function of (is modified by the electro-chemical structuring called) *language*.

10.   *Neuro-semantic environments as environments*.   The neuro-semantic environment constitutes a fundamental environmental issue unique to humans.

11.   Non-aristotelian system *as system*. Korzybski had non-aristotelian predecessors, as he well knew. What distinguishes his non-aristotelian stance is the *degree of formulational consciousness* he brought to it, and the

energetic courage with which he built it into a *system* — offered to his fellow humans as a better way to orient themselves.

12. *The Structural Differential as a model of the abstracting process and a summary of general semantics.* Korzybski realized the importance of *visualization* for human understanding. He knew, then, that to make some of the higher order, overarching *relationships* of his system accessible, *visible*, he must make a diagram, a model, a *map*, that people could *see* and *touch*. Thus the Structural Differential, a device for differentiating the structures of *abstracting*. As far as I know, this is the first *structurally appropriate* model of the abstracting process.

13. *Languages, formulational systems, etc., as maps and only maps of what they purport to represent.* This awareness led to the three premises (popularly expressed) of general semantics:

the map *is not* the territory

no map represents *all* of 'its' presumed territory

maps are self-reflexive, i.e., we can map our maps indefinitely. Also, every map is *at least*, whatever else it may claim to map, a map of the map-maker: her/his assumptions, skills, world-view, etc.

By 'maps' we should understand everything and anything that humans formulate — including this book and my present contributions, but also including (to take a few in alphabetical order), biology, Buddhism, Catholicism, chemistry, Evangelism, Freudianism, Hinduism, Islam, Judaism, Lutheranism, physics, Taoism, etc., etc., ...!

14. *Allness/non-allness as clear, to be dealt with, formulations.* If no map can represent *all* of 'its' presumed territory, we need to eschew habitual use of the term 'all' and its ancient philosophical correlates, *absolutes* of various kinds.

15. *Non-identity* and its derivatives, correlates, etc. At every turn in Korzybski's formulating we encounter his forthright challenge to the heart of aristotelianism — *and* its non-Western, equally essentialist counterparts. "Whatever you say a thing *is*, it *is not*." This rejection of the 'law of identity' ('everything is identical with itself') may be Korzybski's most controversial formulation. After all, Korzybski's treatment directly challenges the 'Laws of Thought', revered for over two thousand years in the West and, differently expressed, in non-Western cultures. Korzybski's challenge is thus *planetary*. We 'Westerners' can't (as some have tried) escape to the 'East'. Identifications, confusions of orders of abstracting, are common to all human nervous systems we know of.

16. *Extension of Cassius Keyser's "Logical Fate"*: from premises, conclusions follow, *inexorably*. Korzybski recognized that conclusions constitute *behaviors*, *consequences*, *doings*, and that these are not merely logical derivatives but *psycho-logical* inevitabilities. If we want to change

behaviors, we must first change the premises which gave birth to the behaviors. Korzybski's *strong version* of Keyser's restrictedly 'logical' formulation was first adumbrated in Korzybski's paper, "Fate and Freedom" of 1923 and received its full expression in the "Foreword" (with M. Kendig) to *A Theory of Meaning Analyzed* in 1942, both available in the *Collected Writings*. Both expressions well antedate Thomas Kuhn's "paradigm shifts" and, more pointedly than Kuhn, formulate the behavioral implications of logical and philosophical systems.

17. *The circularity of knowledge* (spiral-character-in-'time'). Korzybski noted that our most 'abstract' formulations are actually about non-verbal processes/events, and that how we formulate about these at a given date, how we talk to ourselves, through neural feedback mechanisms, relatively *determines* how we will subsequently abstract-formulate: healthfully if our abstracting is open, non-finalistic (non-absolute); pathologically if not.

18. *Electro-colloidal* (macro-molecular-biological) and related *processes*. Korzybski emphasized awareness of these as fundamental for understanding neuro-linguistic systems/organisms.

19. *Non-elementalism applied* to human organisms-as-a-whole-in-an-environment. Some of Korzybski's predecessors in the study of language and human error may have pointed to what he labeled 'elementalism' (verbally splitting what cannot be split empirically) as a linguistically-embedded human habit, but none I know of had so thoroughly *built against it* and recommended replacing it with *habitual non-elementalism*. Korzybski's practical insistence that adopting non-elementalistic *procedures* and *terms* would benefit the humans (including scientists) who adopt them is original and, for him, urgent.

20. Extensions of *logics* (plural) as subsets of non-aristotelian evaluating, including the limited usefulness (but *usefulness*) of aristotelian logic.[7]

21. Epistemology as centered in *neuro-linguistic, neuro-semantic* issues. Korzybski built squarely on the neuroscience of his day and affirmed the fundamental importance of epistemology (the study of how we know what we *say* we know) as the *sine qua non* for any sound system upon which to organize our interactions with our children, students, friends, lovers, bosses, trees, animals, government — the 'universe'. Becoming conscious of abstracting constitutes *applied* epistemology: *general semantics*.

22. The *recognition of and formulation of extensional and intensional orientations as orientations*. Here we see Korzybski at his most diagnostic and prognostic. Realizing that a person's epistemological-evaluational *style*, a person's *habitual way with evaluating determines* how life will go, he recommends adoption of an extensional orientation, with its emphasis on 'facts'. If a person is *over*-committed to verbal constructs, definitions, formulae, 'conventional wisdom', etc., that person may be so trapped in those

a priori decisions as to be unable to appropriately respond to new data from the non-verbal, not-yet-anticipated world. By definition, the *extensionally* oriented person, while remaining as articulate as any of her/his neighbors, is *habitually* open to new data, is *habitually* able to say, "I don't know; let's see." As an aid toward this more healthy orientation, Korzybski formulated the "extensional devices" explained in his Introduction to the Second Edition herein.

23. *Neuro-linguistic and neuro-semantic factors applied* to psychotherapeutic procedures and to the prevention of psycho-logical problems.

24. *Mathematics.* Korzybski's use of mathematical formulations and *point of view* qualifies as one of his most daring contributions.

25. *Science and mathematics as human behaviors.* Perhaps showing some korzybskian influence (much of it has come to be 'in the air'), writers on science and mathematics are increasingly addressing the human being who *does* science and/or mathematics. But Korzybski seems the first, to the *degree* that he did, to point to understanding these human behaviors as a necessary prerequisite or accompaniment to fully understanding science and mathematics as such. As Gaston Bachelard observed,

> The psychological and even physiological conditions of a
> non-Aristotelian logic have been resolutely faced in the
> great work of Count Alfred Korzybski, *Science and
> Sanity.*[8]

26. *Limitations of subject-predicate languages* (modes of representation) when employed without consciousness of abstracting. Korzybski addresses this central formulation fully in his book.

27. Insistence on *relative* 'invariance under transformation'. Korzybski was concerned that invariance of relations not be confused with 'invariance' of processes.

28. *General uncertainty* (all statements merely probable in varying degrees) as an inevitable derivative of korzybskian *abstracting, non-identity*, etc. Korzybski, drawing partly on his Polish milieu, anticipated and exceeded Heisenberg's mid-nineteen-twenties formulation of (restricted) uncertainty.

29. *The mechanism/machine-ism distinction.* This may seem too simple to list as an 'original' or even major point. Yet it is vital, indicating as it does Korzybski's strong commitment to finding out how something *works* as opposed to vague, 'spiritual' explanations.

Korzybski and some of his Institute successors who have worked to present *korzybskian* general semantics have sometimes met this resistance: "I'm not a machine!" People trained in the myriad 'intellectual', 'mystical' in varying degrees systems/traditions they bring to seminars often react as if they fear to lose their 'humanity' by being asked to consider the *mechanisms* which

XX PREFACE TO THE FIFTH EDITION

underpin or constitute their functioning. Korzybski took pains to explain that mechanism should not be confused with 'machine-ism'. His concern for investigations at this level was bracing and central to his approach.

30.   'Infinite'-valued evaluating and semantic *methods* of science (*not* 'content' of science or non-professional behavior of scientists at a given date) as *methods for sanity*. Thus the title of his book. General semanticists are *obliged to evaluate*, to analyze, criticize and sometimes reject the products of 'science' at a given date. The approach is scientific, not scientistic.

31.   *Predictability as the primary measure of the value of an epistemological formulation.* Korzybski was by no means an 'anti-aesthete'. He was deeply sensitive to (and knowledgeable about) music, married a portrait painter, read literature (Conrad was a favorite) including poetry, and even liked to relax with a good detective story. But he insisted that, for life issues, beauty or cleverness or mere consistency (logical coherence, etc.) were not enough.

Korzybski offered his non-aristotelian system with general semantics as its *modus operandi* as an on-going human acquisition, negentropic, ordering and self-correcting through and through, since it provides, self-reflexively, for its own reformulation, and assigns its users responsibility to do so should the need arise.

The above considerations have led me to the conclusion that Korzybski was not only a bold innovator, but also a brilliant synthesizer of available data into a coherent system. This system, when internalized and applied, can create a saner and more peaceful world, justifying the title of this book, *Science and Sanity*.

Robert P. Pula

September 1993

END NOTES

1. Karol, Janicki, *Toward Non-Essentialist Sociolinguistics*. Berlin and New York: Mouton de Gruyter, 1990.
2. Bibliographic Note: sample of books since 1970.
    J. Samuel Bois, *Breeds of Men: Toward the Adulthood of Humankind*, Harper and Row, 1970
    Lee Thayer, ed., *Communication: General Semantics Perspectives*, Spartan/Macmillan, 1970 (critique)
    William Youngren, *Semantics, Linguistics, and Criticism*, Random House, 1972 (critique)
    Kenneth G. Johnson, ed., *Research Designs in General Semantics*, Gordon and Breach, 1974

Donald E. Washburn and Dennis R. Smith, eds., *Coping With Increasing Complexity: Implications of General Semantics and General Systems Theory*, Gordon and Breach, 1974

Kenneth G. Johnson, *Lineamenti di Semantica Generale*, Roma: Editore Armando Armando, 1978

Ross Evans Paulson, *Language, Science, and Action: Korzybski's General Semantics — A Study in Comparative Intellectual History*, Greenwood Press, 1983 (critique)

Harold L. Drake, *General Semantics Views*, Millersville, PA: Millersville State College, 1983

Mary Morain, ed., *Bridging Worlds Through General Semantics*, International Society for General Semantics, 1984; and *Enriching Professional Skills Through General Semantics*, International Society for General Semantics, 1986

Gerard I. Nierenberg, *Workable Ethics*, Nierenberg and Zeif, 1987

Sanford I. Berman, ed., *Logic and General Semantics: Writings of Oliver Reiser and Others*, International Society for General Semantics, 1989

Karol Janicki, *Toward Non-Essentialist Sociolinguistics*, Berlin: Mouton de Gruyter, 1990

Kenneth G. Johnson, ed., *Thinking Creatically: A Systematic, Interdisciplinary Approach to Creative-Critical Thinking*, "Foreword" by Steve Allen, Institute of General Semantics, 1991

D. David Bourland, Jr. and Paul Dennithorne Johnston, eds., *To Be or Not: An E-Prime Anthology*, International Society for General Semantics, 1991

3. Korzybski, Alfred, *Collected Writings: 1920-1950*. Collected and arranged by M. Kendig. Final editing and preparation for printing by Charlotte Schuchardt Read, with the assistance of Robert Pula. Englewood, NJ: Institute of General Semantics, 1990.

4. The following have delivered lectures in the series (listed chronologically): William Vogt, M. F. Ashley Montagu, F. J. Roethlisberger, F. S. C. Northrop, Buckminster Fuller, Clyde Kluckhohn, Abraham Maslow, Russell Meyers (twice), Warren S. McCulloch, Robert R. Blake, Harold G. Cassidy, Henri Laborit, Joost A. M. Meerloo, Henry Lee Smith, Jr., Alvin M. Weinberg, Jacob Bronowski, Alastair M. Taylor, Lancelot Law White, Gregory Bateson, Henry Margenau, George Steiner, Harley C. Shands, Roger W. Wescott, Ben Bova, Elwood Murray, Don Fabun, Barbara Morgan, Thomas Sebeok, Robert R. Blake and Jane Srygley Mouton, Allen Walker Read, Karl H. Pribram, George F. F. Lombard, Richard W. Paul, Jerome Bruner, William V. Haney, Warren M. Robbins, Albert Ellis, Steve Allen, and William Lutz.   The following have

participated in colloquia in the series: William J. Fry, James A. Van Allen, Charles M. Pomerat, Jesse H. Shera, Allen Kent, Paul Ptacek, J. Samuel Bois, Elton S. Carter, Walter Probert, Kenneth G. Johnson and Neil Postman.

5. For Korzybski as a system-builder, see Dr. Stuart A. Mayper, "The Place of Aristotelian Logic in Non-Aristotelian Evaluating: Einstein, Korzybski and Popper," *General Semantics Bulletin* No. 47, 1980, pp. 106-110. For discussions of the continuing appropriateness of 'Korzybski's science', see Stuart A. Mayper, "Korzybski's Science and Today's Science," *General Semantics Bulletin*, No. 51, 1984, pp. 61-67; Barbara E. Wright, "The Heredity-Environment Continuum: Holistic Approaches at 'One Point in Time' and 'All Time'," *General Semantics Bulletin* No. 52, 1985, pp. 36-50; Russell Meyers, MD, "The Potentials of Neuro-Semantics for Modern Neuropsychology" (The 1985 Alfred Korzybski Memorial Lecture), *General Semantics Bulletin*, No. 54, 1989, pp. 13-59, and Jeffrey A. Mordkowitz, "Korzybski, Colloids and Molecular Biology: A View from 1985," *General Semantics Bulletin*, No. 55, 1990, pp. 86-89. For a detailed updating in the neurosciences by a non-general-semanticist which shows Korzybski's 1933 formulations as consistent with 1993 formulations, see "Part I" of Patricia Smith Churchland's *Neurophilosophy: Toward a Unified Science of the Mind/Brain*, MIT Press, 1986, pp. 14-235.)

6. Korzybski, Alfred, *Manhood of Humanity*.

7. See Dr. Mayper's paper: "The Place of Aristotelian Logic in Non-Aristotelian Evaluating...," listed above, and his earlier "Non-Aristotelian Foundations: Solid or Fluid?" *ETC.: A Review of General Semantics*, Vol. XVIII, No. 4, Feb., 1962, pp. 427-443.

8. Bachelard, Gaston, *The Philosophy of No*. Translated from the French by G. C. Waterson. New York: Orion Press, 1968, p. 108.

PREFACE TO THE FOURTH EDITION 1958

A quarter of a century has now elapsed since the first edition of Alfred Korzybski's principal work, *Science and Sanity,* appeared. The second edition was published in 1941 and the third was prepared in 1948, two years before the author's death. Although the second and third editions provided clarification and amplification of certain aspects of the non-aristotelian orientation originally proposed by the author, and while they cited important new data illustrating the rewards accruing to certain fields of human endeavor (e.g., psychotherapy) in consequence of the utilization of the orientations earnestly espoused by him, they represented no important departures from the first edition in respect of basic principles at theoretic and pragmatic levels. Nor, in serious retrospection, did any such appear to have been indicated.

Considering that the author himself, in applying the formulation of 'the self-reflexive map' to his own work, asserted on more than one occasion that perceptible revisions of his formulations must be anticipated and that such would very likely prove fairly compelling within a period estimated at twenty-five years, it comes as something of a surprise that as the 1958 reprinting of *Science and Sanity* goes to press no major alterations seem as yet to be required. In this modern world of rapid change—in which Man has acquired information regarding the intra- and extraorganic realms of his Universe at an unprecedented exponential rate; in which the Atomic Age has come into actual being; in which conquests of space that were but fanciful dreams only yesteryear have become astonishing realities; in which new specialties, bridging freely across the gaps of the unknown between conventional scientific disciplines, have sprung into life and become full-fledged within a matter of months; and in which far-seeing men of good will have organized their endeavors to unify the sciences, arts and humanitarian pursuits-at-large and appear as never before determined (despite recalcitrant and reactionary private interests) to implement a One World such as might befit the dignity of humanity in its manhood—the continuing substantiality of Korzybski's 1933 formulations must be regarded as a tribute to his vision and integrative genius. Now that we are able to stand a little apart from historical developments and view his life's work in some perspective, it can hardly be doubted that he grasped, as few had done before him and certainly none had so systematically and comprehensively treated, the abiding significance of linguistic habits and the communicative processes-in-general to all of Man's thinking-and-doing, from his loftiest metaphysical, epistemological and mathematical efforts to the most casual, trivial and mundane performances of his everyday living.

Like a skillful diagnostician, Korzybski penetrated deeply into the etiologic and pathologic substrates of what he perceived to be the more serious deterrents to current human endeavors and there succeeded in identifying certain grave strictures imposed upon Man's creative potentials and problem-solving proclivities by one of the *least suspected* of all possible agencies, namely, the academically-revered and ubiquitously-exercised aristotelian formulations of logic. This diagnostic act was, of course, the analogical equivalent of finding a positive Wasserman reaction in the blood serum of some long honored and beloved patriarch. Its disclosure promised and, in point of fact, proved to be no more popular. In this sense, Korzybski's position was wholly comparable to that of Copernicus and Galileo, who had been impelled by their private inquiries during the early Renaissance to challenge the popular ptolemaic cosmology and aristotelian mechanics of their day. It required an uncommon personal integrity, an unusual brand of courage and a plenum of physical energy to spell out the overt and covert effects produced by these widely-pervading, pathologic neuro-semantic processes in the community of humans. Korzybski was, as we now know, quite up to this formidable task.

To have made the diagnosis constituted·in itself an intellectual triumph. But Korzybski did more than this. His analyses enabled him to write effective prescriptions for both the prevention and treatment of the disorders he encountered round about and within the community of humans. These disorders, including cultural and institutional as well as personal misevaluations and delusions, he regarded as essentially those of inept semantic reactions. They were for him the unmistakable marks of unsanity, however 'normal' they might appear to be in a statistical sense.

The side-effects of Korzybski's formulations were hardly less significant than the prophylactic and therapeutic devices engendered by them. Among other things, they cast much needed light on the psychology of perception, child psychology, education, the cultural theories of modern anthropology, scientific method and operational ethics. As of the time of writing this introduction, a revolution in neurology, psychology, psychiatry and related disciplines, comparable in every way to that which broke upon the discipline of physics in the early years of the present century, appears both imminent and inevitable. The stirrings toward such appear to have been derived largely from a general semantic orientation, which has sufficiently influenced advanced investigators in these areas to make its impact apparent in their writings and in the character of their researches. The old dichotomies which have been for, lo! these many years the bed-rock terms of intellectual discourse, to which others referred and from which they derived their meanings—e.g., mental and physical, conscious and unconscious, thought and speech, structure and function, intellect and emotion, heredity and environment, organic and functional, reality and unreality, male and female, autonomic and cerebrospinal, pyramidal and extrapyramidal, motor and sensory, idiopathic and symptomatic, voluntary and involuntary, etc.—have exhibited visible signs of disintegration. New and operationally verifiable formulations are beginning to emerge in their place and field theory, commen- * see page xii surate wth that now being developed in the realm of nuclear physics, begins to accommodate data heretofore considered unconnected. Nowhere is it more apparent than in neurology, psychology and closely related disciplines that 'the word is not the thing.'

It would be a mistake, of course, for the reader to suppose that, because no major alterations in or additions to Korzybski's methodologic and applied formulations have appeared necessary up to the present, his students and others who find his views empathetic with their own embrace the inordinate faith that such will not eventually be required. Quite to the contrary. They are persuaded that modifications, major as well as minor, must come as newly acquired information necessitates; and they have deliberately provided for them. From its very inception, the discipline of general semantics has been such as to attract persons possessing high intellectual integrity, independence from orthodox commitments, and agnostic, disinterested and critical inclinations. On the whole, they have been persons little impressed with intellectual authority immanent within any individual or body of individuals. For them, authority reposes not in any omniscient or omnipresent messiah, but solely in the dependability of the predictive content of propositions made with reference to the non-verbal happenings in this universe. They apply this basic rubric as readily to korzybskian doctrine as to all other abstract formulations and theories and, like good scientists, they are prepared to cast them off precisely as soon as eventualities reveal them to be incompetent, i.e., lacking in reliable predictive content. This circumstance in itself should abrogate once and for all the feckless charges sometimes made by ill-informed critics that general semantics is but one more of a long succession of cults, having its divine master, its disciples, a bible, its own mumbo-jumbo and ceremonial rites. For, if there is any one denominator that can be regarded as common to all such cults, it is the self-sealing character of their dogmas, which *a priori* must stand as eternal verities, regardless of the advent of incompatible experiences. In antithetic contrast, general semanticists are fully sensible of the man-fashioned origin of general semantics and have taken pains to keep its structure open-ended. Far from being inclined to repel changes that appear to menace the make-up of general semantics, they actively antici-pate them and are prepared to foster those that seem to promise better predictions, better survival and better adaptation to the vicissitudes of this earthly habitat.

One cannot help but be aware, in 1958, that there is far less suspicion and misgiving among intellectuals concerning general semantics and gen-eral semanticists than prevailed ten and twenty years ago. Indeed, a certain receptivity is noticeable. The term 'semantics' itself is now fre-quently heard on the radio, TV and the public speaking platform and it appears almost as frequently in the public print. It has even found a recent 'spot' in a Hollywood movie and it gives some promise of becom-ing an integral part of our household jargon. This in no sense means that all such users of the term have familiarized themselves with the restricted meaning of the term 'semantics,' much less that they have inter-nalized the evaluative implications and guiding principles of action sub-sumed under general semantics. A comparable circumstance obtains, of course, in the layman's use of other terms, such as 'electronics.'

But more palpable gains than these can be counted. We have alluded to some of these as they bear on psychology, anthropology and the medical sciences. The years since the close of World War II have similarly witnessed the access of general semantics not only to academic curricula of the primary, secondary and collegiate levels of the North and South American continents, parts of Western Europe, Britain, Australia and Japan, but to the busy realms of commerce, industry and transportation: of military organization and civil administration; of law, engineering, soci-ology, economics and religion. These constitute no negligible extensions of general semantics into the world of 'practical' affairs. Large business enterprises, looking toward the improvement of intra- and extramural relations, more satisfying resolutions of the complicated problems that arise between labor and management, and the enhancement of service to their immediate constituents and fellow men in general have found it rewarding, in many instances, to reorganize their entire structure so as to assure the incorporation of general semantic formulations. Several organizations now in existence make it their sole business to advise and provide help in the implementation of such changes. The core of their prescriptions consists in the appropriate application of general semantics. It is becoming a routine for the high and intermediate level executives of certain industries, advertising agencies, banking establishments and the like to retreat for several days at a time while they receive intensive instruction and participate in seminar-workshops designed to indoctrinate them with the principles of general semantics. Comparable courses of instruction have been provided within recent years for the officers of the U.S. Air Academy, the traffic officers of the Chicago Police Department and the sales forces of several large pharmaceutical and biochemical houses. These innovations in business procedure entail, of course, enormous outlays of time, energy and money. They must in time pay perceptible dividends or suffer abandonment. That they are steadily on the increase appears to offer eloquent testimony of their effectiveness.

Other evidences of the growth and widening sphere of influence can be pointed to. Membership in the two major organizations concerned with the development, teaching and utilization of general semantics, namely, the Institute of General Semantics located at Lakeville, Connecticut and the International Society for General Semantics, with its central office at Chicago, has slowly but steadily increased over the years and, gratifyingly, has generally avoided the 'lunatic fringe' that appears ever ready to attach itself to convenient nuclei. The two current publications of these organizations, the *General Semantics Bulletin* and *ETC.: A Review of General Semantics,* continue to provide cogent original articles and synopses of progress in the field. Their subscription lists now include libraries scattered over the entire globe.

In 1949, the Third American Congress on General Semantics was held at the University of Denver. This turned out to be the last occasion at which Alfred Korzybski made a public appearance. During these stimulating sessions he had the satisfaction of hearing numerous reports of investigations by his former students and others who had profited roundly from their familiarity with the non-aristotelian formulations. Many of these papers, representing a wide and eclectic coverage of human interests, were subsequently published in the *General Semantics Bulletin.* Two additional conferences of national scope have been held in the interim—one in Chicago in 1951 and another in St. Louis in 1954. Another conference of international scope is planned for August, 1958. Meanwhile, numerous sectional conferences have been held in various cities each year and the number of courses sought and offered in general semantics is definitely on the increase.

All in all, then, a healthy state of affairs appears to prevail in respect of general semantics. The impact of Korzybski's work on Western culture is now unmistakable and there is every reason to be optimistic that his precepts will be read by ever-widening circles of serious students and that the latter, in their turn, must deeply influence generations of students yet to come. It remains to be seen what effects the regular implementation of these precepts will bring to mankind. Many of us are convinced that they will prove highly salutary.

RUSSELL MEYERS, M.D.

Division of Neurosurgery
College of Medicine
State University of Iowa
Iowa City, Iowa
October, 1957

BIBLIOGRAPHIC NOTE, 1958

In the second edition of this book, published in 1941, a short list of reprints and monographs available at the Institute of General Semantics was included—almost the only literature then on general semantics since *Science and Sanity* first appeared. (See page lxxxvi in this volume.) Since that date the number of books, papers, and reviews has grown profusely, and no attempt is made here to record them fully.

The principal articles in the field have appeared in *Papers From the Second American Congress on General Semantics,* M. Kendig, Editor (1943), in the *General Semantics Bulletin* and other materials published by the Institute for its Members, and in *ETC.: A Review of General Semantics,* the official organ of the International Society for General Semantics.

The *General Semantics Bulletin,* founded and edited by M. Kendig, is the official journal of the Institute of General Semantics, published since 1949 'for information and inter-communication among workers in the non-aristotelian discipline formulated by Alfred Korzybski.' It contains papers on many aspects of general semantics, theoretical and practical, as well as reports, discussions, news, book comments, etc. Numbers One & Two through Eighteen & Nineteen have been issued to date.

Some of the articles distributed by the Institute, 1947-1949, previous to the founding of the *Bulletin* are listed below:

KORZYBSKI, ALFRED, General Semantics: An Introduction to Non-aristotelian Systems, 1947.

———— 'Author's Note' introducing *Selections from Science and Sanity,* 1948.

———— Understanding of Human Potentialities, Key to Dealing with Soviet Union, 1948. Summary of an address, and introduction by Stuart Chase.

———— General Semantics: Toward a New General System of Evaluation and Predictability in Solving Human Problems. *American People's Encyclopedia,* 1949 edition.

ENGLISH, EARL, A General Semantics Course in the School of Journalism, University of Missouri, 1949.

KELLEY, DOUGLAS M., The Use of General Semantics and Korzybskian Principles as an Extensional Method of Group Psychotherapy in Traumatic Neuroses. *The Journal of Nervous and Mental Disease,* Vol. 114, No. 3, 1951.

LA BRANT, LOU, A Genetic Approach to Language, 1951.

LOOMIS, WILLIAM F., A Non-aristotelian Presentation of Embryology, Massachusetts Institute of Technology, 1949.

NEWTON, NORMAN T., A Non-aristotelian Approach to Design. Introductory lectures in a course, School of Architecture, Harvard University, 1948.

READ, ALLEN WALKER, An Account of the Word 'Semantics'. *WORD,* August 1948.

SKYNNER, ROBIN, Choice and Determinism. Distributed to Institute Members 1948; published in *Rationalist Annual* (England), 1949.

Also by Korzybski and others, not included elsewhere:

POLLOCK, THOMAS C., A Theory of Meaning Analyzed: Critique of I. A. Richards' Theory of Language, and SPAULDING, J. GORDON, Elemental-ism: The Effect of an Implicit Postulate of Identity on I. A. Richards' Theory of Poetic Value. With a supplementary paper by ALLEN WALKER READ, The Lexicographer and General Semantics. Foreword by ALFRED KORZYBSKI and M. KENDIG. General Semantics Monographs No. III, 1942.

KORZYBSKI, ALFRED, The Role of Language in the Perceptual Processes, Chapter 7 in *Perception: An Approach to Personality* by Robert R. Blake and Glenn V. Ramsey, Ronald Press, 1951.

*   *   *

For books, published and in preparation, see International Non-aristotelian Library list opposite the title page.

Current lists of publications in the field of general semantics are available from the Institute of General Semantics, 163 Engle Street, Englewood, NJ 07631. Telephone 201-568-0551. Fax 201-569-1793.

PREFACE TO THE THIRD EDITION 1948

> If thinkers will only be persuaded to lay aside their prejudices and apply themselves to studying the evidences . . . I shall be fully content to await the final decision. (402)   CHARLES S. PEIRCE

> For the mass of mankind . . . if it is their highest impulse to be intellectual slaves, then slaves they ought to remain. (402)   CHARLES S. PEIRCE

In spite of the fact that since 1933 a great many new discoveries in sciences have been made, to be analysed in a separate publication, the fundamental *methodological* issues which led even to the release of nuclear energy remain unaltered, and so this third edition requires no revision of the text.

Soon after the publication of the second edition in 1941, the Second American Congress on General Semantics was held at the University of Denver. The papers presented there have been compiled and edited by M. Kendig[1] and show applications in a wide variety of fields. A third congress, international in scope, is being planned for 1949. Students of our work who have made applications in their fields of interest are invited to submit papers to the Institute. The rapid spread of interest, by now on all continents, has indicated the need for the new methods set forth here, and many study groups have been formed here and abroad.

As the center for training in these non-aristotelian methods, the Institute of General Semantics was incorporated in Chicago in 1938. In the summer of 1946 the Institute moved to Lakeville, Connecticut, where its original program is being carried on.

I must stress that I give no panaceas, but experience shows that when the methods of general semantics are *applied,* the results are usually beneficial, whether in law, medicine, business, etc., education on all levels, or personal inter-relationships, be they in family, national, or international fields. If they are not applied, but merely talked about, no results can be expected. Perhaps the most telling applications were those on the battlefields of World War II, as reported by members of the armed forces, including psychiatrists on all fronts, and especially by Dr. Douglas M. Kelley,* formerly Lieutenant Colonel in the Medical Corps, who reports in part as follows:

> General semantics, as a modern scientific method, offers techniques which are of extreme value both in the prevention and cure of such [pathological] reactive patterns. In my experience with over seven thousand cases in the European Theater of Operations, these basic principles

---

* Chief Consultant in Clinical Psychology and Assistant Consultant in Psychiatry to the European Theater of Operations; also Chief Psychiatrist in charge of the prisoners at Nuremberg. Author of *22 Cells in Nuremberg,* Greenberg, New York, 1947.

were daily employed as methods of group psychotherapy and as methods of psychiatric prevention. It is obvious that the earlier the case is treated the better the prognosis, and consequently hundreds of battalion-aid surgeons were trained in principles of general semantics. These principles were applied (as individual therapies and as group therapies) at every treatment level from the forward area to the rear-most echelon, in front-line aid stations, in exhaustion centers and in general hospitals. That they were employed with success is demonstrated by the fact that psychiatric evacuations from the European Theater were held to a minimum.[2]

The origin of this work was a new functional definition of 'man', as formulated in 1921,[3] based on an analysis of uniquely human *potentialities;* namely, that each generation may begin where the former left off. This characteristic I called the 'time-binding' capacity. Here the reactions of humans are not split verbally and elementalistically into separate 'body', 'mind', 'emotions', 'intellect', 'intuitions', etc., but are treated from an organism-as-a-whole-in-an-environment (external and *internal*) point of view. This parallels the Einstein-Minkowski space-time integration in physics, and both are necessitated by the modern evolution of sciences.

This new definition of 'man', which is neither zoological nor mythological, but functional and extensional (factual), requires a complete revision of what we know about humans. If we would judge human reactions by statistical data of psychiatric patients, or many other special groups, our understanding of 'human nature' must be completely twisted. Both the zoological and mythological assumptions must limit human society to animalistic biological, instead of time-binding psycho-biological evaluations, which involve socio-cultural responsibilities and thus may mark a new period of human development.*

In *Manhood of Humanity* I stressed the *general* human unique characteristic of time-binding, which potentially applies to all humans, leaving no place for race prejudices. The structure of science is interwoven with Asiatic influences, which through Africa and Spain spread over the continent of Europe, where it was further developed. Through the discovery of factors of sanity in physico-mathematical *methods,* science

---

* Some readers do not like what I said about Spengler. It is perhaps because they did not read carefully. Spengler, the mathematician and historian, dealt with the spasms of periods of human evolution which paralleled the development of science and mathematics, and his erudition must be acknowledged. In my honest judgment, he gave 'a great description of the *childhood of humanity*', which he himself did not outgrow. In 1920 Sir Auckland Geddes said, 'In Europe, we know that an age is dying.' And in 1941 I wrote, 'The terrors and horrors we are witnessing in the East and the West are the deathbed agonies of that passing epoch.' With Spengler's limitations, no wonder the Nazis joined hands with him. They made good death-bedfellows, demonstrating empirically the 'Decline of the West',

and sanity became linked in a structurally non-aristotelian methodology, which became the foundation of a *science of man.*

We learned from anthropology that the degrees of socio-cultural developments of different civilizations depend on their capacity to produce higher and higher abstractions, which eventually culminate in a *general consciousness of abstracting,* the very key to further human evolution, and the thesis of this book. As Whitehead justly said, 'A civilisation which cannot burst through its current abstractions is doomed to sterility after a very limited period of progress.'

In mankind's cultural evolution its current abstractions became codified here and there into systems, for instance the aristotelian system, our main concern here. Such systematizations are important, for, as the *Talmud* says, 'Teaching without a system makes learning difficult.' In analysing the aristotelian codifications, I had to deal with the two-valued, 'either-or' type of orientations. I admit it baffled me for many years, that practically all humans, the lowest primitives not excluded, who never heard of Greek philosophers, have some sort of 'either-or' type of evaluations. Then I made the obvious 'discovery' that our relations to the world outside and inside our skins often happen to be, *on the gross level,* two-valued. For instance, we deal with day *or* night, land *or* water, etc. On the living level we have life *or* death, our heart beats *or* not, we breathe *or* suffocate, are hot *or* cold, etc. Similar relations occur on higher levels. Thus, we have induction *or* deduction, materialism *or* idealism, capitalism *or* communism, democrat *or* republican, etc. And so on endlessly on all levels.

In living, many issues are not so sharp, and therefore a *system which posits the general sharpness of 'either-or', and so objectifies 'kind',* is unduly limited; it must be revised and made more flexible in terms of 'degree'. This requires a physico-mathematical 'way of thinking', which a *non-*aristotelian system supplies.

Lately the words 'semantics' and 'semantic' have become widely used, and generally misused, even by important writers, thus leading to hopeless confusion. 'Semantics' is a name for an important branch of philology, as complex as life itself, couched in appropriate philological terms, and as such has no direct application to life problems. The 'significs' of Lady Welby was closer to life, but gave no techniques for application, and so did not relate linguistic structures to the structures of non-verbal levels by which we actually live. In modern times, with their growing complexities, a theory of *values,* with extensional tech-

---

by exposing in the dramatic finale the prevailing outworn, unrevised, now pathological, doctrines to which unfortunately most of the politicians of the world still subscribe.

xxxiv   PREFACE TO THE THIRD EDITION

niques for educational guidance and self-guidance, became imperative. Such a theory, the first to my knowledge, required a modern scientific approach, and this was found in physico-mathematical methods (space-time) and the foundations of mathematics. It originated in 1921 in *Manhood of Humanity*, was formulated in a methodological outline in my papers in 1924, 1925, and 1926, and in 1933 it culminated in the present volume.

My work was developed entirely independently of 'semantics', 'significs', 'semiotic', 'semasiology', etc., although I know today and respect the works of the corresponding investigators in those fields, who explicitly state they do not deal with a general theory of values. Those works do not touch my field, and as my work progressed it has become obvious that a theory of 'meaning' is impossible (page xv ff.), and 'significs', etc., are unworkable. Had I not become acquainted with those accomplishments shortly before publication of this book, I would have labelled my work by another name, but the system would have remained fundamentally unaltered. The original manuscript did not contain the word 'semantics' or 'semantic', but when I had to select some terms, from a time-binding point of view and in consideration of the efforts of others, I introduced the term *'General Semantics'* for the *modus operandi* of this first non-aristotelian system. This seemed appropriate for historical continuity. A theory of evaluation appeared to follow naturally in an evolutionary sense from 1) 'meaning' to 2) 'significance' to 3) *evaluation*. *General* Semantics turned out to be an empirical natural science of non-elementalistic evaluation, which takes into account the living individual, not divorcing him from his reactions altogether, nor from his neuro-linguistic and neuro-semantic environments, but allocating him in a *plenum* of some values, no matter what.

The present theory of values involves a clear-cut, workable discipline, limited to its premises, a fact which is often disregarded by some readers and writers. They seem also often unaware of the core of the inherent difficulties in these age-old problems, and the solutions available through changing not the language, but the *structure* of language, achieved by the habitual use of the extensional devices in our evaluational reactions.

For instance, in *Ten Eventful Years,* an *Encyclopaedia Britannica* publication, appears an article on 'Semantics, General Semantics', which considerably increases the current confusions concerning these subjects. It is not even mentioned that 'semantics' is a branch of philology, nor is there any clarifying discrimination made between the noun 'semantics' and the adjective 'semantic'. Moreover it has many misstatements and even falsifications of my work and the work of others, and some statements make no sense. Fortunately there is another popular publication, the *American People's Encyclopedia,* which is publishing a reliable article on *general* semantics.

It is not generally realized that with human progress, the complexities and difficulties in the world increase following an exponential function of 'time', with indefinitely accelerating accelerations. I am deeply convinced that these problems cannot be solved at all unless we boldly search for and revise our antiquated notions about the 'nature of man' and apply modern extensional methods toward their solution.

Fortunately at present we have an international body, the United Nations Educational, Scientific and Cultural Organization,[4] which with its vast funds, has the services of the best men in the world, and a splendid program. It is true that they are very handicapped by dependence on translations, which seldom convey the same implications in different languages. Yet this need not be a handicap, for the methods of exact sciences disregard national boundaries, and so the extensional methods and devices of general semantics can be applied to all existing languages, with deep psycho-logical effects on the participants and through them on their countrymen. Thus the world would gain an international common denominator for inter-communication, mutual understanding, and eventual agreement. I would suggest that students of general semantics write on this subject. The activities of this international body after all affect all of us.

We *need not* blind ourselves with the old dogma that 'human nature cannot be changed', for we find that it *can be changed*. We must begin to realize our potentialities as humans, then we may approach the future with some hope. We may feel with Galileo, as he stamped his foot on the ground after recanting the Copernican theory before the Holy Inquisition, *'Eppur si muove!'* The evolution of our human development may be retarded, but it cannot be stopped.

A. K.

Lakeville, Connecticut
October, 1947

REFERENCES

1. KENDIG, M., Editor. *Papers from the Second American Congress on General Semantics.* Institute of General Semantics, 1943.
2. KORZYBSKI, A. A Veteran's Re-Adjustment and Extensional Methods. *ETC.: A Review of General Semantics,* Vol. III, No. 4. See also
   SAUNDERS, CAPTAIN JAMES, USN (Ret.). Memorandum. *Training of Officers for the Naval Service: Hearings Before the Committee on Naval Affairs, United States Senate,* June 12 and 13, 1946, pp. 55-57. U. S. Government Printing Office, Washington, D. C., 1946.
3. KORZYBSKI, A. *Manhood of Humanity: The Science and Art of Human Engineering.* E. P. Dutton, New York, 1921; 2nd ed., Institute of General Semantics, Lakeville, Conn., Distributors, 1950.
4. HUXLEY, JULIAN. *UNESCO: Its Purpose and Its Philosophy.* Public Affairs Press, Washington, D. C., 1947.

INTRODUCTION TO THE SECOND EDITION
1941

There is what may perhaps be called the method of optimism, which leads us either willfully or instinctively to shut our eyes to the possibility of evil. Thus the optimist who treats a problem in algebra or analytic geometry will say, if he stops to reflect on what he is doing: 'I know that I have no right to divide by zero; but there are so many other values which the expression by which I am dividing might have that I will assume that the Evil One has not thrown a zero in my denominator this time.'

MAXIME BÔCHER*

God may forgive you your sins,
but your nervous system won't.

OLD MAXIM.

When in perplexity, read on.

OLD MAXIM.

Section A.   Recent developments and the founding of the Institute of General Semantics.

Science and Sanity: An Introduction to Non-aristotelian Systems and General Semantics, first published in October, 1933, was intended to be a textbook showing how in modern scientific methods we can find factors of sanity, to be tested empirically.   Although a great many scientific discoveries have been made since the first publication, it did not seem necessary to revise the text for this second edition because the methodological data given, important for our purpose, have not changed.   However, the list of books in preparation for the Non-aristotelian Library has been revised, and in this introduction I indicate some new developments in general semantics and include a short new bibliography, supplementing the bibliography of 619 titles given on page 767 ff.

In 1935 I began to conduct seminar courses in general semantics in schools, colleges and universities, and before various groups of educators,** scientists, and physicians, including psychiatrists.   In the same year a group of students of *Science and Sanity* organized the First Amer-

---

* *Congress of Arts and Science,* St. Louis, 1904, Vol. I, p. 472.

** I use the word 'educator' in its standard English sense; namely, 'one who or that which educates'. I use 'educate' in the sense of: 'to rear . . . bring up from childhood, so as to form habits, manners, mental and physical aptitudes . . . To provide schooling for . . . train generally . . . train so as to develop some special aptitude, taste, or disposition.' Etc. (*The Shorter Oxford English Dictionary on Historical Principles,* Oxford, at the Clarendon Press, 1933.)   In this sense any teacher from nursery school through university professors are 'educators'.   From a *life point of view* this would include even parents, nurses, etc.

ican Congress on General Semantics at the Washington College of Education at Ellensburg, where a number of papers from various fields were presented. The present writer delivered three addresses on the application of general semantics to education and medicine, which are printed in the proceedings of the Congress.* The Second American Congress on General Semantics will be held at the University of Denver in August, 1941. This Congress is organized by Professor Elwood Murray of the University and M. Kendig, Educational Director of the Institute.

In 1938 the Institute of General Semantics was incorporated in Chicago for neuro-linguistic, neuro-epistemologic, scientific research and education. Since that date, as director of the Institute, my major efforts have been concentrated on further research and co-ordination of rapidly accumulating empirical data, along with the conduct of seminar courses to train in the new extensional methods for application in personal adjustment, and in the respective special fields of the students. At present several universities are offering accredited courses in general semantics, and in a number of other universities and colleges general semantics is incorporated in the presentation of other courses.

From scientific necessity this book was written inductively; the seminar courses are presented deductively, and so the two complement each other. The seminars include much illustrative empirical material accumulated in the five years of application of the system by my co-workers and myself, together with the pertinent, factual, newest findings of other sciences.

The non-aristotelian system presented here has turned out to be a strictly *empirical* science, as predicted, with results which have greatly surpassed even my expectations. General semantics is not any 'philosophy', or 'psychology', or 'logic', in the ordinary sense. It is a new extensional discipline which explains and trains us how to use our nervous systems most efficiently. It is not a medical science, but like bacteriology, it is indispensable for medicine in general, and for psychiatry, mental hygiene, and education in particular. In brief, it is the formulation of a new non-aristotelian system** of orientation which affects every

---

* Distributed by the Institute of General Semantics, Lakeville, Connecticut.

** The terms 'era', 'epoch' and 'system' will frequently appear here, and to avoid confusion it may be advisable to indicate in what sense these terms are used. 1) *Era:* 'A date or an event, which begins a new period in the history of anything; an important date. . . . A period marked by the prevalence of some particular state of things.' Etc. 2) *Epoch:* '. . . a period of history defined by the prevalence of some particular state of things. . . . A period . . . in the history of a process.' Etc. 3) *System:* 'A set or assemblage of things connected, associated, or interdependent, so

branch of science and life.   The separate issues involved are not entirely new; their methodological formulation *as a system* which is workable, teachable and so elementary that it can be applied by children, is entirely new.

The experience of my co-workers, mostly educators and psychiatrists, and my own, shows that about ninety per cent of those who train them-selves seriously in the new extensional methods definitely benefit in various degrees, and in ways so varied as to be unpredictable.

Theory and empirical results show that these new methods involve psychosomatic factors which help the balancing and integration of the functions of the nervous system, while the prevalent and traditional in-tensional methods of evaluation tend to disintegrate these functions.   The nervous mechanisms involved work automatically one way or another, harmfully or beneficially, depending on the methods with which we utilize them.   This has not been fully realized before.

The new methods eliminate or alleviate different semantogenic block-ages; many 'emotional disturbances', including even some neuroses and psychoses; various learning, reading, or speech difficulties, etc.; and general maladjustments in professional and/or personal lives.   These difficulties result to a large extent from the failure to use 'intelligence' adequately so as to bring about proper evaluation.

It is well known that many psychosomatic symptoms such as some heart, digestive, respiratory, and 'sex' disorders, some chronic joint dis-eases, arthritis, dental caries, migraines, skin diseases, alcoholism, etc., to mention a few, have a semantogenic, and therefore neuro-semantic and neuro-linguistic origin.   In general semantic training we do not go into the medical angle as such.   We eliminate the harmful semantogenic factors, and in most cases the corresponding symptoms disappear—pro-vided the student is willing to work at himself seriously.

Section B.   Some difficulties to be surmounted.

1. THE ATTITUDES OF 'PHILOSOPHERS', ETC.

'Philosophers', 'psychologists', 'logicians', mathematicians, etc., are somehow unable to comprehend that their work is the product of the

---

as to form a complex unity; a whole composed of parts in orderly arrangement accord-ing to some scheme or plan. . . . A set of principles, etc.; a scheme, method.   The set of corollated principles . . . or statements belonging to some department of knowledge . . . a department of knowledge . . . considered as an organized whole; a comprehensive body of doctrines, conclusions . . . An organized scheme or plan of action; an orderly or regular method of procedure. . . . A formal, definite or established scheme or method . . . systematic form or order.' Etc. (*The Shorter Oxford English Dictionary on Historical Principles.*)

working of *their own nervous systems*. For most of them it is only detached verbalism such as we often find in hospitals for 'mentally' ill. For
instance, a very gifted, well-minded mathematician and professor of 'philosophy' wrote to me: 'I do not, however, think that neuro-psychology is
relevant to the analysis of the nature of meaning. . . . I do not believe
in confusing logic with neuro-psychology'. These professionals would be
shocked if they would study the many volumes of verbal rationalizations
by patients in hospitals. They would find very quickly that the words
interplay with the other words somehow, but they have very little, if any,
connection with the facts, and that is one reason why the patients are confined. Why speculate on academic verbal definitions instead of investigating facts in such hospitals, where patients also pay no attention to the
functioning of their own nervous systems? Even a gramophone record
undergoes some physical changes before words or noises can be 'stored'
and/or reproduced. Is it so very difficult to understand that the extremely
sensitive and highly complex human nervous system also undergoes some
electro-colloidal changes before words, evaluations, etc., are stored, produced, or reproduced? In the work of general semantics we deal with
the *living neuro*-semantic and *neuro*-linguistic reactions, not mere detached verbal chatter in the abstract. In our experience we have found
that even seriously maladjusted persons benefit considerably if we can
succeed in making them 'think' about themselves in neurological electro-
colloidal terms (see chapter IX).

Most 'philosophers' who reviewed this book made particularly shocking performances. Average intelligent readers can understand this book,
as they usually have some contact with life. It is not so with those who
indulge in mere verbalism. I can give here only a classical example of
some 'philosophical' performances. A reviewer in the *Journal of Philosophy*, February 1, 1934, writes:

'Except for his stimulating discussion of the mathematical infinite
(p. 206) and his hints on the nature of theory (p. 253), he contributes
nothing to the clarification of meanings by definite analyses of special
problems. Indeed, he only adds to the confusion when he declares that
hypotheses contrary to the fact are meaningless (e.g., p. 168) ; if his views
were correct, science would come to an end. His theory of meaning, like
his theory of social causation, is very naïve, to say the least.'

I suggest that the reader verify whether on page 168 there is such a
statement, or even a hint at such a notion, which I could not possibly have.
Besides, I do not give any theories of 'meaning' or of 'social causation'!

Most 'philosophers', 'logicians', and even mathematicians look at this
non-aristotelian system of *evaluation* as some system of formal non-aris-

totelian 'logic', which is not the case. They are somehow not able to take the natural science point of view that all science, mathematics, 'logic', 'philosophy', etc., are the product of the functioning of the human nervous system, involving some sort of internal orientations, or evaluations, which are not necessarily formalized. The analysis of such *living reactions* is the sole object of general semantics as a natural *empirical science.*

These 'philosophers', etc., seem unaware, to give a single example, that by teaching and preaching 'identity', which is empirically non-existent in this actual world, they are *neurologically* training future generations in the pathological identifications found in the 'mentally' ill or maladjusted. As explained on page 409, and also Chapter XXVI, whatever we may *say* an object '*is*', *it is not,* because the statement is verbal, and the facts are not.

It is pathetic, if not tragic, that society should invest millions of dollars to support such specialists who train future generations in maladjustments just because they disregard the unavoidable neuro-linguistic and neuro-semantic effects of their teachings on the lives of their pupils.

Most scientists and educators are either entirely innocent of these problems, or indifferent and passive, or even negativistic. Like some animals that can outwit humans because of their keen observations, the cunning, often pathological, thoroughly ignorant present day totalitarian leaders are not unaware of the academic shortcomings based on inertia, verbalism, etc., and openly proceed to utilize these human nervous weaknesses destructively, with very telling results. Nazism, wars *of* and *on* nerves, wars of *verbal distortion,* etc., with their following disasters are in 1941 only too obvious examples. I will return to this subject at the end of this introduction.

The terms 'philosophy', 'system', etc., as ordinarily used, stand for too broad generalizations. Different 'philosophies' represent nothing but *methods of evaluation,* which may lead to empirical mis-evaluation if science and empirical facts are disregarded. Different systems may be very broad and general, such as, say, the aristotelian system (A) (see Fig. 1), within which is a more limited and less general system such as 'christianity' (C), within which is, for instance, the leibnitzian system (L), and within which there are individual, personal systems (P). Every Smith$_1$ has an individual interpretation of broader systems, and so in actuality has *a*

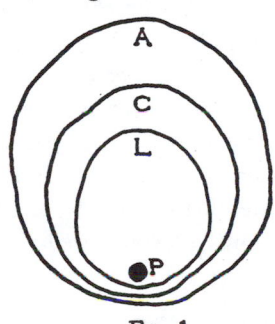

Fig. 1

system of his own. As a rule, personal systems are a part of, and influenced by, larger systems, which in turn are influenced by still more general systems. Such problems can be handled at present only by the methods of general semantics and by topological methods.*

'Mental' illness and every form of maladjustment are to be considered as mis-evaluations, involving some 'philosophies', public or individual, one within the other, as usual. 'Philosophers', etc., who wish to become aware of such dangers, would do well to study the verbalizations and mis-evaluations of the 'mentally' ill in hospitals.

2. PERPLEXITIES IN THEORIES OF 'MEANING'

There is a fundamental confusion between the notion of the older 'semantics' as connected with a theory of *verbal* 'meaning' and words defined by words, and the present theory of 'general semantics' where we deal only with *neuro*-semantic and *neuro*-linguistic *living reactions* of $Smith_1$, $Smith_2$, etc., as their reactions to neuro-semantic and neuro-linguistic *environments as environment.*

The present day theories of 'meaning' are extremely confused and difficult, ultimately hopeless, and probably harmful to the sanity of the human race. Of late in the United States some members of the progressive education movement have written much on 'referents' and 'operational' methods, in the abstract, based on verbalism. Let us consider some facts, and how the theories of referents and operational methods fit *human evaluations.* Here is, for instance, $Smith_1$ who, through family, social, economic, political, etc., conditions has become 'insane'. $Smith_1$ finally, in ordinary parlance, kills $Smith_2$. From a human point of view it is a very complex and tragic situation. Let us account for it in terms of referents and operations. The body and the heart of $Smith_2$, the hand of $Smith_1$, the knife, etc., are perfectly good referents. The grabbing of the knife by $Smith_1$ and plunging it in the heart of $Smith_2$, the falling down on the ground by $Smith_2$ and the kicking of his legs are perfectly good operations. However, where is *human evaluation?* Where is concern with 'sanity' and 'insanity'? Here we deal with some of the deepest human and social tragedies which, in this case, involve not only the killing of $Smith_2$ by $Smith_1$, but the sick, unhappy, twisted life of $Smith_1$, affecting all his life connections, and with which we must be concerned if we are to be *human beings* and different from apes.

---

* Lewin, Kurt. *Principles of Topological Psychology.* McGraw-Hill, New York, 1936.

Such an example is of course extreme and over-simplified, although it illustrates the principles. However, officially teaching such *methods* which are *inadequate* to handle evaluation, and so human values, has a definite sinister effect, among others, on the 'sex' life of the students. Many of them are taught to orient themselves generally by referents and operations only; and so mere physiological performance is often *identified* with mature love life, etc., and is a causative factor in the wide-spread marital unhappiness, promiscuity and other lowerings of human cultural and ethical standards.

Thus, theories of 'meaning' or still worse, 'meaning of meaning', based on 'referents' and 'operational' methods are thoroughly *inadequate* to account for human values, yet they do affect the nervous systems of humans. We must, therefore, work out a *theory of evaluation* which is based on the optimum electro-colloidal action and reaction of the nervous system.

There is no doubt that a civilized society needs some mature 'morals', 'ethics', etc. In a general theory of evaluation and sanity we must consider seriously such problems, if we are to be sane humans at all. Theory and practice show that healthy, well-balanced people are naturally 'moral' and 'ethical', unless their educations have twisted their types of evaluations. In general semantics we do not 'preach' 'morality' or 'ethics' *as such,* but we train students in consciousness of abstracting, consciousness of the multiordinal mechanisms of evaluation, *relational* orientations, etc., which bring about cortico-thalamic integration, and then as a result 'morality', 'ethics', awareness of social responsibilities, etc., follow automatically. Unfortunately our educational systems are unaware of, or even negativistic toward, such *neuro*-semantic and *neuro*-linguistic issues. These are sad observations to be made about our present educational systems.

May I suggest that readers consult *Apes, Men and Morons* and *Why Men Behave Like Apes* by Earnest A. Hooton; *The Mentality of Apes* by W. Köhler, *The Social Life of Apes and Monkeys* by S. Zuckerman, and many other studies of this kind. They might then more clearly understand how the aristotelian type of education leads to the humanly harmful, gross, macroscopic, brutalizing, biological, animalistic types of orientations which are shown today to be *humanly inadequate*. These breed such 'führers' as different Hitlers, Mussolinis, Stalins, etc., whether in political, financial, industrial, scientific, medical,* educational, or even publishing, etc., fields, fancying that they represent 'all' of the *human*

---

* See Carrel's *Man The Unknown.*

world! Such delusions must ultimately be destructive to *human* culture, and responsible for the tragic 'cultural lag', stressed so much today by social anthropologists.

Existing theories of 'meaning' of any school do not take into consideration that any definition of words by words must be based ultimately on *undefined terms*. To the best of my knowledge this problem is not considered at all in present day educational systems, outside of some sciences, and so the existing theories run in a vicious circle, just like a dog chasing his tail, and are bound to be ineffective, if not harmful.

As Professor Keyser aptly formulates the problem: 'If he contend, as sometimes he will contend, that he has defined all his terms and proved all his propositions, then either he is a performer of logical miracles or he is an ass; and, as you know, logical miracles are impossible.'*

Similarly the theorists in the 'theory of meaning' as described above disregard the *inadequacy* for human orientation of the subject-predicate form of representation. I must refer the reader to my chapter on relations, page 188 ff., for further information.

In principle, a type of orientation which restricts formally everything to subject-predicate forms of representation can account only for symmetrical relations, and we may beat in the bush about 'meaning'; in principle, however, a theory of evaluation is then impossible. *Evaluation* must be based on asymmetrical relations such as 'more' or 'less', etc., which cannot be dealt with at all adequately if restricted formally to subject-predicate forms of representation, that harmfully affect our orientations.

What I have said here is correct in principle; however, in practice, in the neuro-semantic and neuro-linguistic development of the white race we had to invent, by living necessity, some asymmetrical relations such as 'more' or 'less', etc. The difficulty lies in the fact that these methods of escape from a subject-predicate grammatical structure of language were used only haphazardly, and not formulated generally into a workable system based on asymmetrical relations, which would be *teachable*.

Similarly with the problem of intensional orientation by verbal definitions and extensional orientation by facts (see p. 173); there is also confusion about it. 'Pure' extension is humanly impossible; 'pure' intension is possible, and is often found in hospitals for 'mentally' ill, and in some chairs of 'philosophy'. These issues and problems are seriously confusing to the average person because they have not been formulated before in a methodological system.

---

* Keyser, Cassius J. *Mathematical Philosophy.* E. P. Dutton, New York, 1922, p. 152.

3. *INADEQUACY OF FORMS OF REPRESENTATION AND THEIR STRUCTURAL REVISION*

It is not generally realized what serious difficulties an *inadequate, unduly limited form of representation or theory* brings about. This is well known in science. Thus, for instance, the euclidean and newtonian systems cannot deal successfully with electricity and so it was imperative to produce non-euclidean and non-newtonian systems, which do apply to the sub-microscopic electrical levels and also to the macroscopic gross levels. Similarly in life, the two-valued aristotelian system could not deal adequately with the electro-colloidal sub-microscopic levels of the functioning of our nervous systems, on which sanity depends. Thus the formulation of the present infinite-valued non-aristotelian system became also an imperative necessity.

I must stress that as the older systems are only special limitations of the new more general 'non' systems (see p. 97), it would be incorrect to interpret a 'non' system as an 'anti' system.

Such a non-aristotelian system is long overdue. It was retarded because of persecution by the church and other influential bodies, the general belief that 'Aristotle said the last word', etc., and particularly because of the inherent difficulties of such a revision.

The problem of inadequacy in the forms of representation has handicapped science and life a great deal until relatively adequate systems were produced. In life the situation is much more aggravated, for if our orientations and evaluations are inadequate, our predictability is impaired, and we feel with the poet Housman, 'I, a stranger and afraid, in a world I never made'. If we have a more adequate or proper evaluation, we would have more correct predictability, etc., (see p. 58 ff. and p. 750 ff.). We would then feel, 'We are *not* strangers, and *not* afraid, in this *human mess* you and I have made'.

Another of the main difficulties is that a language or a system of a given structure can be somewhat altered from within, but cannot be *revised structurally* without going *outside* the former system. For instance, all the attempts to revise the structure of the euclidean and newtonian systems from within were ineffective. Those who revised these systems structurally had to go outside the systems first, after which they were able to produce different, independent, new systems. Only then did an effective evaluation of the former systems become possible.

Similarly the aristotelian, two-valued, intensional system can be revised structurally and evaluated properly only by building independently a non-aristotelian, infinite-valued, extensional system. This verifies the contention of Bertrand Russell made in 1922 that there is a 'possibility' that 'every language has . . . a structure concerning which, *in the language,* nothing can be said, but that there may be another language dealing with the structure of the first language, and having itself a new structure'.* What Russell calls a 'possibility' becomes a fact once a system of *different structure* is built. Then the issues become clear.

Russell limits himself to the structure of a language, and disregards the fact that this limitation is artificial, and that any language involves structural assumptions which build up a *system of orientations* that may be racial, national, personal, etc.

4. *IDENTIFICATIONS AND MIS-EVALUATIONS*

The problem of *general* identification is a major problem which does not seem to be understood at all even by specialists. Psychiatrists know professionally the tragic consequences of identifications in their patients. But what even psychiatrists do not realize is that identifications in daily life are extremely frequent and bring about every kind of difficulties.

As a matter of fact we live in a world in which non-identity is as entirely general as gravitation, and so *every identification* is bound to be in some degree a mis-evaluation. In a four-dimensional world where 'every geometrical point has a date', even an 'electron' at different dates is not identical with itself, because the sub-microscopic processes actually going on in this world cannot empirically be stopped but only transformed. We can, however, through extensional and four-dimensional methods translate the dynamic into the static and the static into the dynamic, and so establish a similarity of structure between language and facts, which was impossible by aristotelian methods. Unfortunately even some modern physicists are unable to understand these simple facts.

To communicate to my classes what I want to convey to my readers here, the following procedure has been useful. In my seminars I pick a young woman student and pre-arrange with her a demonstration about which the class knows nothing. During the lecture she is called to the platform and I hand her a box of matches which she takes carelessly and drops on the desk. That is the only 'crime' she has committed. Then I begin to call her names, etc., with a display of anger, waving my fists in front of her face, and finally with a big gesture, I slap her face gently. Seeing this 'slap', as a rule ninety per cent of the students recoil and shiver; ten per cent show no overt reactions. The latter have seen what they have seen, but they *delayed their evaluations.* Then I explain to the students that their recoil and shiver was an *organismal evaluation* very

---

* Wittgenstein, Ludwig. *Tractatus Logico-Philosophicus,* with an introduction by Bertrand Russell. Harcourt, Brace, New York, 1922, p. 23.

harmful in principle, because they *identified* the seen facts with their judgements, creeds, dogmas, etc. Thus their reactions were entirely unjustified, as what they have *seen* turned out to be merely a scientific demonstration of the mechanism of identification, which identification I expected.

Such identifications are very common. The late Dr. Joshua Rosett, formerly Professor of Neurology in Columbia University, and Scientific Director, Brain Research Foundation, New York, gives an example from his own experience. 'A vivid picture on the cinema screen represented a boy and a girl pulling down hay from a stack for bedding. I sneezed—from the dust of the hay shown on the screen.'*

The problem of identification in values is neurologically strictly connected with the pathological reversal of the natural order of evaluation, which is found in different degrees in the maladjusted, neurotics, psychotics, and even in some 'normal' persons. Thus, the supposedly innocent 'shiver' and the sneezing in the examples above, or the attack of hay fever when *paper* roses are shown (see p. 128), etc., may as well in other cases end in a sudden death or in a neurosis or psychosis. The neurological mechanisms are similar, involving identifications in values of different orders of abstractions, and therefore the very common reversal of the natural order of evaluation.

In the evolution of the human race and language there was a natural order of evaluation established; namely, the life facts came first and labels (words) next in importance. Today, from childhood up, we inculcate words and language first, and the facts they represent come next in value, another pathologically reversed order, by which we are unconsciously being trained to identify words with 'facts'. Even in medicine we much too often evaluate by the definitions of 'diseases' instead of dealing with an individual sick patient, whose illness seldom fits textbook definitions.

The foregoing considerations deal directly with aristotelian orientations by intension, or verbal definitions, where verbiage comes first in importance, and facts next. By non-aristotelian methods we train in the natural order; namely, that first order empirical facts are more important than definitions or verbiage. It should be noticed that the average child is born extensional, and then his evaluations are distorted as the result of intensional training by parents, teachers, etc., who are unaware of the heavy neurological consequences.

These are key problems involved in the passing from aristotelian to non-aristotelian orientations, which affect our future personal, national

---

* Rosett, Joshua. *The Mechanism of Thought, Imagery, and Hallucination.* Columbia University Press, New York, 1939, p. 212.

and international adjustments. For a detailed discussion the reader is referred to this text, see index under the terms 'identification', 'order', 'natural order', etc.

5. METHODS OF THE MAGICIAN

Another very serious difficulty arises due to the fact that our knowledge of the world and ourselves involves unavoidable factors of deception and self-deception. A scientific study of magic with its methods of psycho-logical deception is most revealing, as it shows the mechanisms by which we are continually and unknowingly being deceived in science and daily life.* The stock in trade of the magician to fool the public consists of methods of misdirection, of mis-evaluation, half-truths, etc., used to play on the ordinary associations and implications, habits of hasty generalizations, etc., of the audience, thus leading to misinterpretations, identifications, lack of predictability, etc. These general, and so common, psycho-logical mechanisms are very deep, and to a large extent are connected with the aristotelian type of intensional, subject-predicate orientations, which ultimately may become harmful.

For maximum adjustment, and therefore sanity, we need neurological methods to prevent and counteract these heretofore unavoidable old deceptions and self-deceptions. In a non-aristotelian system these difficulties are recognized and empirical methods are discovered to eliminate them step by step. Such methods of prevention and counteraction culminate in training in consciousness of abstracting (see Chapters XXVI, XXVII, XXIX and p. 499 ff.).

I must stress that as far as we humans are concerned, we cannot possibly be entirely ignorant about ourselves; we may have only *false knowledge* or *half-truths*. It is psychiatrically known that in many instances false knowledge, particularly about ourselves, breeds maladjustments, often of a serious character, just because it is based fundamentally on self-deception. In the meantime we react and act *'as if'* our half-truths or false knowledge were 'all there is to be known'. Thus we are bound to be bewildered, confused, obsessed with fears, etc., because of mistakes due to our mis-evaluations, when we orient ourselves by verbal structures which do not fit facts.

Section C. Revolutions and evolutions.

One of the gravest difficulties facing the world today is the passing from one historical era to another. Such passings, as history shows, have

---

* Kelley, Douglas M. Conjuring as an Asset to Occupational Therapy. *Occupational Therapy and Rehabilitation*. Vol. 19, No. 2, April, 1940.

always been painful, and pregnant with consequences. To illustrate: the transition from papal control to non-papal control, passing through murderous religious persecutions and slaughters, including the devastating Thirty Years' War, etc.; from French royalism to republicanism, passing through the ferocious French Revolution and Commune; from czarism to state capitalism, passing through the latest bloody Russian Revolution and a period of so-called 'communism'. Now we are witnessing the struggles of 'democracies' with 'totalitarian states', passing as yet through the recent ruthless Spanish War, second World War, etc., etc.

Similarly we can give illustrations of painful transitions from one system to another from the history of science, which were also accompanied by bewilderment and labour: for instance, the passing from the ptolemaic to the copernican, from euclidean to non-euclidean, newtonian to non-newtonian (einsteinian), etc., systems.

In all these transitions it took one or more generations before the upheaval subsided and an adjustment was made to the new conditions.

No matter how painful and disturbing these transitions were, they were still changes and revisions *within* the then most general, intensional aristotelian system. This system was imposed on the white race by the 'church fathers'. Its strength and influence was due to its academically rationalized general verbal formulations which were set forth in textbooks, and thus became teachable. From the beginning the aristotelian system as formulated was inadequate and many attempts at corrections were made. The white race was impressed by the church that 'Aristotle spake', and there was nothing more to be said. In fact, attempts to revise this system were prohibited even up to very recent times. Just the same, new facts which would not fit the aristotelian and church patterns were accumulating and so new methods, languages of special structure, etc., were required.

Perhaps an illustration from the history of mathematics will help. For more than 2,000 years by necessity mathematicians differentiated and integrated in some clumsy fashion in order to solve individual problems. But only after the formulation of a *general theory* by Newton and by Leibnitz did the *general method* become teachable and communicable as a general practical discipline (see p. 574) which provided the foundations for future developments in mathematics.

The aristotelian system had been formulated in a very rationalized way. Non-aristotelian attempts have been and are being made continually in limited areas. The difficulty was that no methodological general theory based on the new developments of life and science had been formulated until general semantics and a general, extensional, teachable and communicable, non-aristotelian system was produced.   The main difficulties ahead are neuro-semantic and neuro-linguistic because for more than 2,000 years our nervous systems have been canalized in the inadequate, intensional, often delusional, aristotelian orientations, which are reflected even in the *structure of the language* we habitually use.

It may be helpful to indicate some historical facts of the development of our orientations since Socrates (469–399 B.C.).   Socrates was the son of a *sculptor* and himself did *some work with the chisel and his hands*. He became an important founder of a school of 'philosophy'.   In brief, this school had very high standards for science, seeking the application of the science of the time to life, so that it became what may be called a 'school of wisdom'.

One of his students, Plato (427–347 B.C.), who came from an *aristocratic family,* became the founder of a different school, called the 'Academy', and the 'father' of what may be called 'mathematical philosophy'. Unlike his teacher, he began, in his 'Doctrine of Ideas', to verbally split humans into 'body' *and* 'mind', as if they could be so split in living beings. He built a system of *'immaterialism'* or *'idealism'*.

Aristotle (384–322 B.C.), the son of a *physician,* was the student of Plato, and particularly interested in *biology, other natural sciences,* etc. He founded the most influential of the three schools, which is called by his name.   He was undoubtedly one of the most gifted men mankind has ever known.   As usual in such cases, the study of one branch of knowledge leads to another, so Aristotle was led to the study of 'logic', *linguistic structure,* etc., about which he produced scholarly treatises or textbooks, ultimately formulating the most complete system of his time. Because of the completeness of the system, backed by powerful influences, it has moulded our orientations and evaluations up to the present.   The man on the street, our education, medicine and even sciences, are still in the clutches of the system of Aristotle, a system *inadequate* for 1941 yet perhaps satisfactory 2,300 years ago, when conditions of life were relatively so simple, when orientations were on the macroscopic level only, and knowledge of scientific facts was practically nil (see p. 371 ff.).

In Aristotle's system *as applied,* the split becomes complete and institutionalized, with jails for the 'animal' and churches for the 'soul'.   Now we begin to realize how pernicious and retarding for civilization that split is.   For instance, only since Einstein and Minkowski do we begin to understand that 'space' *and* 'time' cannot be split empirically, otherwise we create for ourselves delusional worlds.   Only since their work has modern sub-microscopic physics with all its accomplishments become possible.

Similarly, and tragically, this applies to medicine. Until recently we have had a split medicine. One branch, general medicine, was interested in the 'body' (soma) ; the other was interested in the 'soul' ('psyche'). The net result was that general medicine was a glorified form of veterinary science, while psychiatry remained metaphysical.* However, it has been found empirically that a great many 'physical' ailments are of a semantogenic origin. Only a few years ago general physicians began to understand that they cannot deal with humans without knowing something about psychiatry, and psychosomatic medicine began to be formulated. I cannot go into further detail here, except to mention that this is another constructive step away from the aristotelian system, which as applied trains us in artificial, verbal splits.

If we train in methods which in principle lead to splitting the personality, we obviously train or prepare the ground for dementia praecox or schizophrenia, which very often involves a split personality. At any rate, it does not seem to be advisable for sanity, and so proper evaluation of 'facts' and 'reality', to train our children in delusional methods. Personally, the author is always profoundly shocked that parents, who after all care for their children, can tolerate educators, physicians, scientists, etc., who train their children in such pernicious and hopelessly antiquated methods. I also always wonder whether educators, physicians, scientists and other professionals realize what harm they can do by disregarding factors of sanity, or by ignoring them.

It is pitiful to watch how even some of the most outstanding scientists in the world are unable to understand what a passing from one system to another means. Thus, for example, an Encyclopedia of Unified Science was projected. A number of very scholarly treatises were published in it, and yet because the difficulties were not faced squarely the authors are missing the point that neuro-semantic and neuro-linguistic mechanisms are involved and that we are passing from one system to another.

One of the tremendous obstacles in the revision of the aristotelian system is exactly the excellence of the work of Aristotle based on the very few scientific facts known 2,300 years ago. The aim of his work *circa* 350 B.C. was to formulate the *essential nature of science* (350 B.C.) and the forms and laws of science. His immediate goal was entirely *methodological* (350 B.C.), and he aimed to formulate a *general method* for 'all' scientific work. He was even expounding the theory of symmetrical relations, the relation of the general to the particular, etc. In his days these orientations were by necessity two-valued and 'objective'; hence

---

*Korzybski, Alfred. Neuro-semantic and Neuro-linguistic Mechanisms of Extensionalization. *American Journal of Psychiatry*, Vol. 93, No. 1, July, 1936.

follows his whole system, then more or less satisfactory on macroscopic levels.   A modern revision of the aristotelian system or the building of a non-aristotelian system involves, or is based on, similar aims; namely, the formulation of a *general method* not only for scientific work, but also life, as we know it *today* (1941).

Modern scientific developments show that what we label 'objects' or 'objective' are mere nervous constructs inside of our skulls which our nervous systems have abstracted electro-colloidally from the actual world of electronic processes on the sub-microscopic level.   And so we have to face a complete methodological departure from two-valued, 'objective' orientations to *general, infinite-valued, process orientations,* as necessitated by scientific discoveries for at least the past sixty years.

The aim of the work of Aristotle and the work of the non-aristotelians is similar, except for the date of our human development and the advance of science.   The problem is whether we shall deal with science and scientific methods of 350 B.C. or of 1941 A.C.   In general semantics, in building up a non-aristotelian system, the aims of Aristotle are preserved yet scientific methods are brought up to date.

Section D. A non-aristotelian revision.

In an attempt to convey the magnitude of the task we are now confronting, I can do no better than to summarize roughly in the following tabulation some of the more outstanding points of difference between the aristotelian system as it shapes our lives today, *and is lived by;* and a scientific, non-aristotelian system, as it will, perhaps, guide our lives sometime in the future.

OLD ARISTOTELIAN ORIENTATIONS (circa 350 B.C.)	NEW GENERAL SEMANTIC NON-ARISTO-TELIAN ORIENTATIONS (1941 A.C.)
1. Subject-predicate methods	Relational methods
2. Symmetrical relations, inadequate for proper *evaluation*	Asymmetrical relations, indispensable for proper *evaluation*
3. *Static, 'objective', 'permanent', 'sub-stance', 'solid matter', etc.,* orientations	*Dynamic,* ever-changing, etc., electronic *process* orientations
4. 'Properties' of 'substance', 'attributes', 'qualities' of 'matter,' etc.	Relative invariance of function, dynamic structure, etc.
5. Two-valued, 'either-or', inflexible, dogmatic orientations	Infinite-valued flexibility, degree orientations
6. Static, finalistic *'allness';* finite number of characteristics attitudes	Dynamic *non-allness;* infinite number of characteristics attitudes

Aristotelian Orientations	Non-Aristotelian Orientations
7. *By definition* 'absolute sameness in "all" respects' ('identity')	*Empirical* non-identity, a natural law as universal as gravitation
8. Two-valued 'certainty', etc.	Infinite-valued maximum probability
9. Static absolutism	Dynamic relativism
10. *By definition* 'absolute emptiness', 'absolute space', etc.	*Empirical* fullness of electro-magnetic, gravitational, etc., fields
11. *By definition* 'absolute time'	*Empirical* space-time
12. *By definition* 'absolute simultaneity'	*Empirical* relative simultaneity
13. Additive ('and'), linear	Functional, non-linear
14. $(3+1)$–dimensional 'space' *and* 'time'	4-dimensional space-time
15. Euclidean system	Non-euclidean systems
16. Newtonian system	Einsteinian or non-newtonian systems
17. 'Sense' data predominant	Inferential data as fundamental new factors
18. Macroscopic and microscopic levels	*Sub*-microscopic levels
19. Methods of magic (self-deception)	Elimination of self-deception
20. Fibers, neurons, etc., 'objective' orientations	Electro-colloidal *process* orientations
21. Eventual 'organism-as-a-whole', disregarding environmental factors	Organism-as-a-whole-*in-environments*, introducing new unavoidable factors
22. Elementalistic *structure* of language and orientations	Non-elementalistic *structure* of language and orientations
23. 'Emotion' *and* 'intellect', etc.	Semantic reactions
24. 'Body' *and* 'mind', etc.	Psychosomatic integration
25. Tendency to split 'personality'	Integrating 'personality'
26. Handicapping nervous integration	Producing automatically thalamo-cortical integration
27. Intensional *structure* of language and orientations, perpetuating:	Extensional *structure* of language and orientations, producing:
28. Identifications in value: a) of electronic, electro-colloidal, etc., stages of processes with the silent, non-verbal, 'objective' levels b) of individuals, situations, etc. c) of orders of abstractions	Consciousness of abstracting Extensional devices
29. Pathologically reversed order of evaluation	Natural order of evaluation
30. Conducive to neuro-semantic tension	Producing neuro-semantic relaxation

Aristotelian Orientations	Non-Aristotelian Orientations
31. Injurious psychosomatic effects	Beneficial psychosomatic effects
32. Influencing toward un-sanity	Influencing toward sanity
33. 'Action at a distance', metaphysical false-to-fact orientations	'Action by contact', neuro-physiological scientific orientations
34. Two-valued causality, and so consequent 'final causation'	Infinite-valued causality, where the 'final causation' hypothesis is not needed
35. Mathematics derived from 'logic', with resulting verbal paradoxes	'Logic' derived from mathematics, eliminating verbal paradoxes
36. Avoiding empirical paradoxes	Facing empirical paradoxes
37. Adjusting empirical facts to verbal patterns	Adjusting verbal patterns to empirical facts
38. Primitive static 'science' (religions)	Modern dynamic 'religions' (science)
39. Anthropomorphic	Non-anthropomorphic
40. *Non-similarity of structure* between language and facts	*Similarity of structure* between language and facts
41. Improper evaluations, resulting in:	Proper evaluations, tested by:
42. Impaired predictability	Maximum predictability
43. *Disregarded*	Undefined terms
44. *Disregarded*	Self-reflexiveness of language
45. *Disregarded*	Multiordinal mechanisms and terms
46. *Disregarded*	$\frac{\text{Over}}{\text{Under}}$ defined character of terms
47. *Disregarded*	Inferential *terms* as *terms*
48. *Disregarded*	Neuro-linguistic environments considered as environment
49. *Disregarded*	Neuro-semantic environments considered as environment
50. *Disregarded*	Decisive, automatic effect of the *structure of language* on types of evaluation, and so neuro-semantic reactions
51. Elementalistic, verbal, intensional 'meaning', or still worse, 'meaning of meaning'	Non-elementalistic, extensional, by fact *evaluations*
52. Antiquated	Modern, 1941

THE NEW NON-ARISTOTELIAN ORIENTATIONS DIFFER AS
MUCH FROM THE ARISTOTELIAN AS THE ARISTOTELIAN
DIFFER FROM THE PRIMITIVE TYPES OF EVALUATION.

The old orientations are being perpetuated, as a rule unknowingly, through the aristotelian structure of our language, our institutions, etc. The new orientations are simpler than the old because they are closer to empirical facts, and so are even more easily absorbed by children—provided parents, teachers, etc., are themselves aware of the new methods and so can give the children guidance.

The difficulties we are now facing, with the many important new factors introduced in a non-aristotelian system, listed roughly in the tabulation, cannot be evaluated effectively unless we understand the rôle that *new factors* play in our generalizations.

Section E. New factors: the havoc they play with our generalizations.

In mathematics and science we use extensively the method of interpolation. In building curves we do not have all the points or data. We have a number of them and then connect the points with a smooth curve.

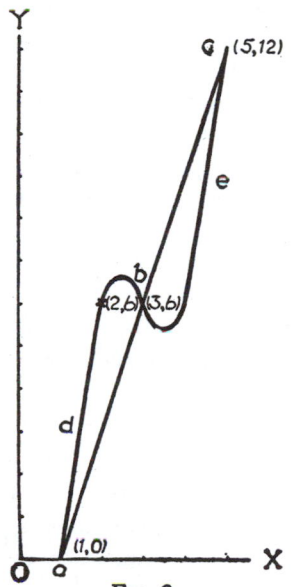

FIG. 2

The equation of that curve is given on the basis of the actual data at hand. The nervous processes which are involved in interpolations and building up equations are also involved in producing *ordinary generalizations* in daily life; that is, we interpolate from the data we have and then generalize in words instead of equations. It is well known that sometimes when a new datum is discovered it transforms the curve entirely, with a corresponding change in the equation (generalization).

Fig. 2 as an illustration will make this clearer. If we measure the experimental points (1,0), (3,6), (5,12), we would find them to lie on the line *abc* with the equation $y = 3x - 3$, and we might conclude therefrom that further similar experiments would confirm the linearity of the relationship being studied. But if a further analysis yields the point (2,6), the simplest curve fitting these data is now the curve *adbec*, expressed by the equation $y = x^3 - 9x^2 + 26x - 18$, which is different and much more complex than before, because it is a cubic equation instead of a linear equation.*

---

*I am indebted for this example to Dr. A. S. Householder, University of Chicago.

It is not generally recognized what havoc the discovery of a single new, important, structural factor may play with our generalizations. In science and ordinary life we are coming across such new factors quite often, and we have to change our equations or generalizations, and so our standards of evaluation, if we do not want to build up delusional situations for ourselves.

As an example I can suggest here the work of Professor W. Burridge,* who in his physiological investigations introduced the new unavoidable factor of the electro-colloidal structure of life. In this case it does not matter whether the particular colloidal theory suggested by Burridge is correct or not. The fact that he introduced an important new structural factor leads to entirely different interpretations, generalizations, etc., although the first order empirical facts remain. Such an introduction requires a complete revision of the generalizations of biology, physiology, neurology, etc., and therefore even medicine and psychiatry. Incidentally, psychosomatic results become at least intelligible.

Other examples may be given, such as the work of Professor William F. Petersen,** who introduced the new factor of weather into medicine; or of Freud, who introduced the 'unconscious', etc.; or of Lorentz, Einstein and others, who introduced the finite velocity of light into the newtonian system, etc., etc. As is well known, the introduction of these new factors revolutionized constructively the older theories.

The scientific requirements of a new theory are very exacting. A new theory must account for the known facts and predict new facts following the new generalizations, which in turn depend upon the new factors or structural assumptions introduced. The predicted new facts must then be verified empirically.

In general semantics we introduce a number of new unavoidable structural factors; among others, our neuro-semantic (neuro-evaluational) and neuro-linguistic *environments as environment*. Such introductions also require a radical revision of what we know, and have wide applications in daily life, as well as in sciences, including the foundation of mathematics (see chapters XIV, XV, XVIII, and XIX) and physics (see chapter XVII). These new factors should particularly interest parents, educators, medical men, psychiatrists, and other specialists.

The introduction of new factors may at first produce seeming diffi-

---

* Dean of the Medical Faculty and Principal of King George's Medical College, Lucknow, India.
** Professor of Pathology, University of Illinois, College of Medicine.

culties because of the unfamiliarity of a new terminology which embodies the new structural assumptions, and because of the necessity of a re-canalization of our neuro-linguistic habits, etc. Yet after the new orientations are acquired, the new issues become much simpler than the older, because they are better understood (see p. 97).

In at least one historical case, it was the omission of an unnecessary artificial assumption that brought about a transformation of the whole system. I speak here about euclidean geometry, which assumes the equal distance of parallels, and the non-euclidean geometries, which eliminated this equal distance postulate as unnecessary. The results were very striking. Thus, in the euclidean system we build curves out of little bits of 'straight lines'. We do the opposite in the newer geometries—we start with curves, shortest distances, etc., not 'straight lines' (as no one knows what that means), and build up 'straight lines' as the limit of an arc of a circle with an 'infinite radius' (see p. 590).

Further explanations are given in the text, but I hope that I have conveyed to the reader the fundamental character of these problems and some of the difficulties encountered at first when new structural factors are introduced. Even the elimination of a postulate may be translated into an introduction of a new negative factor. This translation is important in life, although it may be unimportant in technical mathematics. In science as well as in life we deal all the time with this kind of problems, and when they are not understood structurally, we are only plunged into paradoxes and bewilderment, and potential maladjustment.

Section F. Non-aristotelian methods.

1. NEUROLOGICAL MECHANISMS OF EXTENSIONALIZATION

There is an especially broad generalization, already referred to, which empirically indicates a fundamental difference between the traditional, aristotelian, intensional orientations, and the new non-aristotelian extensional orientations, and in many ways summarizes the radical differences between the two systems. This is the problem of *intension* (spelled with an *s*) and *extension*. Aristotle, and his followers even today, recognized the difference between intension and extension. However, they considered the problem *in the abstract*, never applying it to human *living reactions* as *living reactions*, which can be predominantly intensional or predominantly extensional. The interested reader is advised to consult any textbook on 'logic' concerning 'intension' and 'extension', as well as the material given in this text (see index).

The difference can be illustrated briefly by giving examples of 'definitions'. Thus a 'definition' by intension is given in terms of aristotelian 'properties'. For instance, we may verbally 'define' 'man' as a 'featherless biped', 'rational animal', and what not, which really makes no difference, because no listing of 'properties' could possibly cover 'all' the characteristics of Smith₁, Smith₂, etc., and their inter-relations.

By extension 'man' is 'defined' by exhibiting a class of individuals made up of Smith₁, Smith₂, etc.

On the surface this difference may appear unimportant; not so in *living life applications*. The deeper problems of neurological mechanisms enter here. If we orient ourselves predominantly by intension or verbal definitions, our orientations depend mostly on the cortical region. If we orient ourselves by extension or facts, this type of orientation by necessity follows the natural order of evaluation, and involves thalamic factors, introducing automatically cortically delayed reactions. In other words, orientations by intension tend to train our nervous systems in a split between the functions of the cortical and thalamic regions; orientations by extension involve the integration of cortico-thalamic functions.

Orientations by extension induce an *automatic* delay of reactions, which *automatically* stimulates the cortical region and regulates and protects the reactions of the usually over-stimulated thalamic region.

What was said here is elementary from the point of view of neurology. The difficulty is that this little bit of neurological knowledge is not applied in practice. Neurologists, psychiatrists, etc., have treated these problems in an 'abstract', 'academic', detached way only, somehow, entirely unaware that living human reactions depend on the working of the human nervous system, from which dependence there is no escape. No wonder 'philosophers', 'logicians', mathematicians, etc., disregard the working of their nervous systems if even neurologists and psychiatrists still orient themselves by verbal fictions in the '*abstract*'.

If we investigate, it seems appalling how little of the vast knowledge we have is actually applied. Even the ancient Persians showed their understanding of the difference between *learning* and *applying* in their proverb: 'He who learns and learns and yet *does* not what he knows, is one who plows and plows yet never sows'. In this new modern non-aristotelian system we have not only to '*know*' elementary facts of modern science, including neuro-linguistic and neuro-semantic researches, but also to *apply* them. In fact, the whole passage from the aristotelian to non-aristotelian systems depends on this change of attitude from intension to extension, from macroscopic to sub-microscopic orientations, from 'objective' to process orientations, from subject-predicate to relational evaluations, etc. This is a laborious process and months of self-discipline are required for adults before these new methods can be applied generally; children as a rule have no difficulties.

If we stop to reflect, however, it seems obvious that those who are trained in two-valued, macroscopic, *'objective'*, aristotelian orientations only, are thoroughly unable to have modern, electro-colloidal, sub-microscopic, infinite-valued, *process* orientations in life, which can be acquired only by training in non-aristotelian methods.

It is sad indeed to deal with even young scientists in the colloidal and quantum fields who, after taking off their aprons in the laboratory, relapse immediately into the two-valued, prevalent aristotelian orientations, thus ceasing to be scientific 1941. In many ways these scientists are worse off than the 'man on the street', because of the artificially *accentuated* split between their scientific and their life orientations. Although they work in an infinite-valued, non-aristotelian field, even they need special training to become conscious of how to apply their own scientific non-aristotelian methods to life problems.

Empirically the consequences of training in the new methods are astonishingly far-reaching. This is easily understood after reflection, because the integrating of the functions of the cortical and thalamic regions brings about better functioning of glands, organs, etc. Although general semantics is not a medical science, we can understand why the non-aristotelian extensional thalamo-cortical methods bring about a great deal of stabilization and even psychosomatic consequences, as the empirical results achieved by my psychiatric co-workers and myself indicate.

2. *NEURO-SEMANTIC RELAXATION*

The optimum working of the nervous system depends, among other things, on 'normal' blood pressure, which is predominantly a thalamic function, supplying the nervous system with necessary blood circulation. As both affective, or 'emotional', responses and blood pressure are neurologically closely connected, it is fundamental for 'emotional' balance to have 'normal' blood pressure, and *vice versa*.

In general semantics we utilize what I call 'neuro-semantic relaxation', which, as attested by physicians, usually brings about 'normal' blood pressure; that is, it lowers abnormally high pressure and raises abnormally low pressure, thus regulating the essential *blood circulation*, and so blood supply. The standards for 'normal' are given in statistical averages and are not accurate for the given individual, and at different times. These conditions and beneficial consequences are strictly empirical, and must be taken into account, regardless of the fact that the present scientific theories on this subject are not yet clear. It must be

realized that for the 'normal' working of the nervous system we must
have a proper blood circulation, which may be affected by the *tension* of
blood vessels, and is also connected with 'emotional' *tension*. We are
never aware of this particular steady kind of 'emotional' tension, which
involves hidden fears, anxieties, uncertainties, frustrations, etc., and
through the nervous mechanisms of projection colour harmfully our attitudes toward the world and life in general. Such conditions result in
defensiveness, which is no defense, but a wasteful, useless drain on the
limited nervous capacities.

Some details of the mechanisms and techniques involved, as they
affect, among others, so-called 'speech difficulties', (stuttering, etc.) are
given by Professor Wendell Johnson, University of Iowa, in his *Language and Speech Hygiene: An Application of General Semantics,* published as the first monograph of the Institute of General Semantics.
More details concerning neuro-semantic relaxation will be presented in
professional papers.

3. EXTENSIONAL DEVICES AND SOME APPLICATIONS

To achieve extensionalization we utilize what I call 'extensional
devices':

1) Indexes
2) Dates ⎫ Working Devices
3) Etc. (*et cetera*) ⎬
4) Quotes ⎫ Safety Devices
5) Hyphens ⎭

It should be noticed that in a four-dimensional world dating is only a
particular temporal index by which we can deal effectively with space-
time. In non-aristotelian orientations these extensional devices should
be used habitually and permanently, with a slight motion of the hands
to indicate absolute individuals, events, situations, etc., which change at
different dates, also different orders of abstraction, etc. Thus thalamic
factors become involved, *without which the coveted thalamo-cortical integration cannot be accomplished.*

I may add that all existing psychotherapy, no matter of what school,
is based on the partial and particular extensionalization of a given patient,
depending on the good luck and personal skill of the psychiatrist. Unfortunately these specialists are in the main unaware of what is said here,
and of the existence of a theory of sanity which gives general, simple, and
workable thalamo-cortical methods for extensionalization, and so thalamo-
cortical integration.

A few illustrations of the wide practical applications of some of the
devices may be given here. In many instances serious maladjustments

follow when 'hate' absorbs the whole of the affective energy of the given individual.   In such extreme cases 'hate' exhausts the *limited* affective energy.   No energy is left for positive feelings and the picture is often that of a dementia praecox, etc.   Thus an individual 'hates' a *generalization* 'mother', 'father', etc., and so by identification 'hates' 'all mothers', 'all fathers', etc., in fact, hates the whole fabric of human society, and becomes a neurotic or even a psychotic.   Obviously, it is useless to preach 'love' for those who have hurt and have done the harm.   Just the opposite; as a preliminary step, by *indexing* we *allocate* or *limit* the 'hate' to the individual Smith$_1$, instead of a 'hate' for a generalization which spreads over the world.   In actual cases we can watch how this allocation or limitation of 'hate' from a generalization to an individual helps the given person.   The more they 'hate' the individual Smith$_1$ instead of a generalization, the more positive affective energy is liberated, and the more 'human' and 'normal' they become.   It is a long struggle, but so far empirically invariably successful, provided the student is willing to work persistently at himself.

But even this indexed *individualized* 'hate' is not desirable, and we eliminate it rather simply by *dating*.   Obviously Smith$_1^{1920}$ *is not* Smith$_1^{1940}$ and most of the time hurt$_1^{1920}$ would not be a 'hurt' in 1940. With such types of orientations the individual becomes adjusted, and serious improvements in family and social relationships follow, because the student has trained himself in a general method for handling his own problems.

Similar mechanisms of generalization through identifications are involved in morbid and other generalized *fears* which are so disastrous for daily adjustment.   Because *thalamic factors are involved,* these difficulties are helped greatly or eliminated by a similar use of the extensional devices to individualize and then date the allocated fears.

What a heavy price we may sometimes pay for the disregard of extensional devices in connection with the structure of language, can be illustrated no better than by the life history and work of Dr. Sigmund Freud. In his writings Freud ascribed *one* intensional undifferentiated 'sex' even to infants, which revolted public opinion.   If Freud would have used the extensional devices he would not have gotten into such detrimental professional and other difficulties.   He would not have used the fiction 'sex' without indexes, dates and quotes, and he would have explained that an infant has a ticklish organ which could be labelled 'sex$_0^0$' at birth, 'sex$_1^1$' at the age of one, 'sex$_2^2$' at the age of two, etc.   These are obviously different in life, but the differences are hidden by the one abstract definitional term 'sex', and made obvious only by the extensional techniques.

Let us be frank about it. The intensional abstract 'sex' labels a fiction. By extension or facts, 'sex' varies with every individual not only with age (dates), but in relation to endless other factors, and can be handled adequately only by the use of extensional devices.

4. IMPLICATIONS OF THE STRUCTURE OF LANGUAGE

In what is said above we were already dealing with the change from an intensional to an extensional *structure* of language, and so orientation. We can investigate a step further, and find that the aristotelian structure of language is in the main *elementalistic*, implying, through structure, a split or separation of what in actuality cannot be separated. For instance, we can verbally split 'body' *and* 'mind', 'emotion' *and* 'intellect', 'space' *and* 'time', etc., which as a matter of fact cannot be separated empirically, and can be split only verbally.   These elementalistic, splitting, structural characteristics of language have been firmly rooted in us through the aristotelian training. It built for us a *fictitious animistic world* not much more advanced than that of the primitives, a world in which under present conditions an optimum adjustment is in principle impossible.

In a non-aristotelian system we do not use elementalistic terminology to represent facts which are non-elementalistic. We use terms like 'semantic reactions', 'psychosomatic', 'space-time', etc., which eliminate the verbally implied splits, and consequent mis-evaluations. In the beginning of my seminars when I am explaining space-time, students often react by saying, 'Oh, you mean "space" *and* "time"'. This translation would abolish the whole modern advances of physics, because of the structural implications of a delusional verbal split. Similarly the habitual use of the non-elementalistic term 'semantic reactions' eliminates metaphysical and verbal speculations on such elementalistic fictions as 'emotion' *and* 'intellect', etc., considered as separate entities.

Unfortunately these considerations of structural implications have been entirely disregarded in daily life even by scientists, often befuddling issues very seriously. Thus, the term 'concept' is widely used, and the users are not conscious that this term has elementalistic implications of 'mind' or 'intellect' taken *separately,* which then become verbal fictions. The actual facts, however, can be simply expressed with correct structural implications. What is called 'concept' amounts to nothing more or less than a verbal *formulation,* a term which eliminates the false-to-fact implications. Students of general semantics are strongly advised never to use the elementalistic term 'concept', but the non-elementalistic 'formulation' instead. We could eventually berate and ridicule people for their old neuro-linguistic habits, but in our work we take the *neurological attitude* and realize the difficulties of linguistic habits and neurological recanalization. From this point of view we only face understandingly the inherent difficulties. I can even now hear the reactions of some of my readers, 'I fully agree with you, and I believe it is a very fine *concept*'! And so it goes.

From the above it becomes obvious that without changing the language itself, which is practically impossible, we can easily change the *structure* of language to one free from false-to-fact implications. This change is feasible.

Another example may make issues clearer. Thus the intensional verbal definition of 'man' or 'chair', etc., brings to our consciousness *similarities,* and, so to say, drives the *differences* into the 'unconscious'. In a world of processes and non-identity it follows that no individual, 'object', event, etc., can be the 'same' from one moment to the next. And so individualizing (indexes) and temporal devices (dates), etc., should be used *conjointly*. Thus, obviously $chair_1^{1600}$ is not the 'same' as $chair_1^{1940}$, nor is $Smith_1^{Monday}$ the 'same' as $Smith_1^{Tuesday}$. Orientations in such extensional terms bring to our consciousness not only similarities but also differences. Through training in the consciousness of abstracting we become aware that characteristics are left out in the process of abstracting by our nervous systems, and so we become conscious of the possibility that new factors may arise at any time which would necessitate a change in our generalizations.

Once more we can get a bit of wisdom from mathematical method. I believe it was the great mathematician Sylvester who said that 'in mathematics we look for similarities in differences and differences in similarities', which statement should apply to our whole life orientation. This is made uniquely applicable to life by the new non-aristotelian extensional structure of language and so orientations.

The reader will find in this work the use of certain terms which, although they are standard English words, are not habitually used. The terms used here have been carefully selected and tested, and found to be more similar to the structure of the actual facts. The power of terminology, because of its *structural implications,* is well known in science, but is entirely disregarded in our daily neuro-linguistic habits.

It is shocking to realize that even such scholarly aristotelians as the Jesuits, and other devotees, are unable *or unwilling* to comprehend the obvious structural modern neuro-semantic and neuro-linguistic facts. When confronted with them they hide behind a verbal smoke screen of medieval terms such as 'nominalism', 'realism', etc., which in modern science are hopelessly antiquated, useless, confusing, and so eventually harmful. Their attitude even today is that all those problems were settled and disposed of by different monks in the Middle Ages. Modern researches reveal that nothing of the sort was settled or disposed of, and that a new, up-to-date revision is necessary to eliminate the false knowledge from which present day tragedies follow automatically. The reader is referred to the Encyclopaedia Britannica under such terms as 'nominalism', 'realism', and related terms.

Section G. $\frac{Over}{Under}$defined terms.*

As was explained before, for a revision of a system we must first get outside of the system. Only after producing a non-aristotelian extensional system can the aristotelian intensional structure of our traditional system and language be properly evaluated.

Here we introduce a most important technical term which describes a fundamental characteristic of a correct attitude toward language; namely, that most terms are '$\frac{over}{under}$defined'. They are over-defined (over-limited) by intension, or verbal definition, because of our *belief* in the definition; and are hopelessly under-defined by extension or facts, when generalizations become merely hypothetical. For instance, the euclidean parallels with their equal distance are over-defined by intension and under-defined by extension, as 'equal distance' is unnecessary and also is denied by facts. Similarly the newtonian equations are over-defined (over-limited) by intension, while under-defined by extension, which includes the necessary finite velocity of a signal (Lorentz-Einstein).

From these two examples alone we may see how heavy the problem is, as the discovery of a new important factor makes it obvious that most generalizations must be $\frac{over}{under}$defined, depending upon whether our attitude is intensional or extensional. Unfortunately only those who have studied psychiatry and/or general semantics can fully comprehend the difficulties involved. Different maladjusted, neurotics, psychotics, etc., orient themselves by intension most of the time. This means they evaluate by over-definition, just because they *believe* in their limited verbal-

---

* The term 'over-defined class' was introduced to the best of my knowledge by Dr. A. S. Householder. This term is inadequate for our purpose, as it disregards the problems of intension and extension, which represent different types of evaluation. Besides, the term 'class' is very ambiguous. In science and life we deal mostly with $\frac{over}{under}$defined *terms,* as will be explained.

isms, and not by extensional facts, which make us conscious of under-definition.

To make this fundamental difficulty clearer I will use a rather trivial illustration. The dictionaries define 'house' as a 'building for human habitation or occupation', etc. Let us imagine that we buy a house; this buying is an extensional activity, usually with some consequences. If we orient ourselves by intension we are really buying a definition, although we may even inspect the house, which may appear desirable, etc. Then suppose we move into the house with our furniture and the whole house collapses because termites have destroyed all the wood, leaving only a shell, perhaps satisfying to the eye. Does the verbal definition of the house correspond to the extensional facts? Of course not. It becomes obvious then that by intension the term 'house' was over-defined, or over-limited, while by extension, or actual facts, it was hopelessly under-defined, as many important characteristics were left out. In no dictionary definition of a 'house' is the possibility of termites mentioned.

'Philosophers', etc., and 'philosophizing' laymen, if they ever will be able to face facts and verbal paradoxes, will have a merry time arguing back and forth about the above human and neuro-linguistic situation because they know nothing about psychiatry and *empirical* data of general semantics. Without serious neuro-linguistic study, including the 'philosophical treatises' of 'mentally' ill in hospitals, they will not be able to understand why, by intension or *belief* in verbal definitions, most terms are hopelessly over-defined, while by extension they are hopelessly under-defined. Their analysis of intensional 'over-definitions' will be *extensional by necessity,* and they will have great difficulties in realizing the very important fact that we deal for the most part *only* with $\frac{\text{over}}{\text{under}}$defined terms.

I must stress again that this difficulty is not inherent in our language as such, but depends exclusively on our *attitude* toward the *use* of language.

The ignorance of 'philosophers', etc., about *neuro*-semantic and *neuro*-linguistic issues is not only appalling, but positively harmful to sanity, civilization and culture. To justify their own existence in civilization they should have investigated such problems professionally long ago, and incorporated them in their work. Even the present world tragedies are one of the results of their intensional delusional *neuro*-semantic and *neuro*-linguistic detachment. Present day totalitarianisms were built by the dumping on the human nervous systems of such terms as 'communism,' 'bolshevism,' etc., which induced corresponding fearful signal reactions (see chapter XXI) of the ruling classes, resulting in their imbecilic and suicidal behaviour.   The ruling classes welcomed in many ways the totalitarians as an eventual safe-guard of their personal selfish interests.   The extensional results are that the dreaded 'communists' and 'bolshevists' have united with the totalitarians, and today, 1940, the 'communists' are as 'imperialistic' as any czar has ever been.

To give another example of $\frac{over}{under}$defined terms, it may be helpful to cite a paradox formulated by the mathematician Frege in connection with linguistic difficulties underlying mathematical foundations.

In a village there was only one barber, who shaved only those who did not shave themselves.   The question arises whether the barber shaves himself or not.   If we say 'yes,' then he did not shave himself; if we say 'no,' then he shaved himself.   In daily life we deal all the time with such paradoxes, which if not clarified result only in bewilderment.

The *term* 'barber' *as a term*, since it omits the living human being, is a label for a fiction, because there is no such thing as a 'barber' without a living human being.   By extension the given specialist in shaving, $Smith_1$, is not so simple.   He is peppered with complex chain-indexes and dates.   Thus, $Smith_{11}$ may be by profession a barber, $Smith_{12}$ may be a father, $Smith_{13}$ may be a member of the village council, and anyway $Smith_{1n}$ is a living person who has his own life and personality outside his profession, and ultimately he has to shave himself if he does not want a beard, verbalism or no verbalism.   Obviously the *term* 'barber' is over-defined, over-limited, by intension, and is under-defined by extension.

One of my co-workers, commenting on this paradox, suggested that the barber may be a woman and have no beard; or, the barber may be a beardless hermaphrodite or eunuch; or, the barber may have a full beard. Thus, we have traditionally assumed, in analyzing this old paradox, that the barber was a man with a beard which was somehow shaved.

The difficulties of this $\frac{over}{under}$defined terms situation affect not only our daily lives, but science as well.   For example, H_2O is by intension or definition over-defined; by extension or in practice we do not deal with 'pure' H_2O, which is only a symbol on paper, because actually unavoidable impurities are always present.

Similarly let us consider '*blood transfusion*'.   In the beginning we used the term 'blood transfusion' as over-defined; by extension it turned out to be under-defined, because different bloods have different characteristics, and often blood of one type killed the patient who had blood of another type.

Here I will list a few of the many heavy terms we use in science and daily life which are the cause of endless verbal bickering and confusion, because of our lack of realization of their $\frac{over}{under}$-defined character, depending uniquely on our attitudes.   Terms such as *variation* in biology and anthropology, *learning, frustration, education, needs, intelligence, instincts, genius, teacher, leadership, love, hate, fear, sex, man, woman, infantilism, maladjustment, dementia praecox, personality, democracy, totalitarianism, dollar, god, gold, war, peace, aggression, neutral, jew, number, velocity,* etc., etc., can serve as illustrations.

One psychoanalyst suggests *ego* and *super-ego;* another writes: 'I could quote you a considerable part of psychoanalytic terms'.   An epistomologist says, '*Meaning* is a forbidden term in my courses. . . .   In linguistics the terms *phoneme, word, sentence* are mazes of confusion. . . . *Philosophy* is in as bad a situation.   *Metaphysics* is even worse.' To quote a prominent anthropologist: '$\frac{Over}{Under}$-definition is notably common in the field of so-called social anthropology in which students attempt to disregard the human organism and deal with human affairs as discrete phenomena'.   For example, '*culture* may be technology, morals, philosophy, or a wooden leg—all most vaguely formulated. . . . When some change in the anatomy and physiology of the organism is attributed to *environment,* the latter term is not broken down into climate, rainfall, food supply, etc.   *Social environment* may be arts, industries, law, morals, religion, familial institutions, tradition, etc.'

The following comment by a mathematician shows the generality of this problem: 'A term would seem to be extensionally under-defined so long as we cannot in practice exhaust its instances by enumeration.   But this much is true of just about every term of the kind traditionally known as "general concrete"; e.g. *house, dime, star, neurone.*'

A journalist suggests: 'As an example recently come to our attention I would mention those magic words *Monroe Doctrine.*   Even when Mr. Hull discusses it, as he does as nearly correctly as anyone "in the know", he omits some real facts, such as the economic implications of overturning the international *status quo* in this hemisphere.   But when Japan and/or Germany (high order abstractions as used here) refer to Asiatic and/or European Monroe Doctrines, the meaning of the original words has been completely metamorphosed through $\frac{over}{under}$-definition.   The American accepted meaning includes no actual control of those falling within the doctrine's sphere, whereas Japan and Germany mean an actual hegemony in their respective spheres.   The relationship between ours and theirs is therefore a vast confusion of terms.

'Then consider the *incidents* growing out of *insults* in the international fields.   What is an insult?   It is usually pure verbalism with great affective characteristics manipulated to sway others as the swayer directs. To bring it into the domestic field, call a *Republican* (what is that?) a *New Dealer* (again, what is that?) and the fur begins to fly.'

A leading moving picture executive says that actors have frequent verbal arguments about what is *funny*.   The only thing to do is to try it before an audience.   'If it makes them . . . laugh, it may be termed *funny*.   If it fails to make them laugh, it is *not funny*.'   In the meantime, 'your audience may tell you that the subject in dispute is neither funny nor not-funny.   It is merely boring.'

There is no need to give further examples here, as practically the whole dictionary could be quoted.   In my enquiry concerning $\frac{over}{under}$-defined terms in many fields I got a number of answers which were very fundamental, which I gratefully acknowledge.   Some replies were to the effect that 'I would gladly give you examples such as you ask for, but I do not think I have any that would be new to you', which shows their understanding of the problem.   Yet the most extensional answer was given by that brilliant jurist, Dr. Robert M. Hutchins, who sent to me his Convocation Address of June, 1940 with a letter, which he has kindly given me permission to quote, as follows: 'I am afraid you will feel that all the words I use are examples of the errors you are attacking.   Here is my last Convocation Address, with a sample in every line.'   Such a judgement is profoundly justified whenever language is utilized.   This address is a splendid piece of work, and it implies the intuitive recognition of the fundamental neuro-linguistic difficulties we are up against.

But an intuitive grasp by exceptional persons does not make that recognition *teachable* in general education.   We need crisp, general *methodological formulations* which will make people aware of the rôle the structure of language plays in affecting our types of reactions.   For instance, our language may be elementalistic or non-elementalistic, intensional or extensional, in structure, etc.   We discover also the fundamental multiordinal character of the most important terms we have, the $\frac{over}{under}$-defined character of most of our terms, etc.

As the difficulties mentioned here are inherent in our neuro-semantic and neuro-linguistic mechanisms, which control our reactions, the only possible safe-guard against the dangers of hopeless bewilderment, fears, anxieties, etc., is the *consciousness of the mechanisms*.   Certainly 'philosophers', 'logicians', psychiatrists, educators, etc., should be aware of these problems, and introduce this consciousness even in elementary education and in psychotherapy.

The problem of $\frac{over}{under}$ defined terms is very difficult to explain briefly. It is discussed more fully in two of my papers presented before professional societies.*

Section H. The passing of the old aristotelian epoch.

1. 'MAGINOT LINE MENTALITIES'

Present day scientific researches and historical world developments show there is no doubt that the old aristotelian epoch of human evolution is dying. The terrors and horrors we are witnessing in the East and the West are the deathbed agonies of that passing epoch, and not the beginning of a new system. The changes of historical periods in human development are often accompanied by the disorganization, and sometimes acute suffering, of mankind, and the price is bound to be paid by one or more generations.

I doubt if in the whole of human history there is a more accentuated illustration than the tragic and sudden collapse, in the summer of 1940, of the French government and army, and eventually of French culture and 'democracy'. The degree of stupidity, treachery, graft, dishonesty, ignorance, and ultimately decadence, etc., the French plutocrats and politicians, and so-called 'intelligentsia' displayed is unprecedented, particularly because of the fine historical record the French have had. We test the freshness or deterioration of fishes by smelling the head end, and as we know at the date of this writing, the head ends of the French 'democracy' have a putrid odor. This deterioration affected the French military men, who once were the finest in the world, and so the collapse was complete. I can give no better, no more pitiful, no more shocking illustration of the collapse of the old system.

The 'Maginot line mentality' will become a historical classic, and will be applied quite appropriately to other than military fields. It means a thoughtless, self-deceptive, etc., 'security' in antiquated systems as matched against modern methods of 1940. Well, the French Marianna felt secure from the front and was taken from behind by the German army men, who traditionally pay no attention to such 'details'.

---

* (a) $\frac{Over}{Under}$ defined Terms, 1939, the third of a trilogy of papers presented before annual meetings of the American Mathematical Society on General Semantics: I. Extensionalization in Mathematics, Mathematical Physics and General Education, 1935; II. Thalamic Symbolism and Mathematics, 1938. Institute of General Semantics, Lakeville, Connecticut.

(b) General Semantics, Psychiatry, Psychotherapy and Prevention, presented before the annual meeting of the American Psychiatric Association, 1940.

Dealing with those tragic and painful collapses in civilizations, and eventually passing to another spasm of civilization, what interests us most in considering the problems of sanity, is the newest, psychopathological methods of destroying sanity, not merely the organized orgies of murder, rape, arson, looting, drugging, and destruction under different dictators, mikados, etc.

I mention the 'mikado' especially here as a tragic human example of the effect of $\frac{over}{under}$defined terms, which in *life application* sway the history of mankind.   By definition and/or creed the mikado is supposed to be some sort of a 'god', etc.   By extension or facts, the best we know, he is probably a sort of a nice, supposedly educated, collegian.   He has a wife and makes babies, but he is told about Japanese people, the behaviour of Japanese troops in China, etc., only as much as the ruling clique in Japan allows him to know.   If he would be allowed to know what 'his' soldiers, and so his representatives, are *actually* doing in China with their *governmentally organized* murder, rape, looting, drugging, etc., I doubt if he, as a 'nice collegian', would approve it.   However, if he would try to do something about it, he probably would be 'liquidated' by the ruling clique.   From a historical, civilization, human point of view he must be adjudged responsible, as the head of his government, for what the ruling clique and the Japanese army do in China in his name.

This applies to many other 'rulers', who seldom know what is going on extensionally because they rely on the use of $\frac{over}{under}$defined terms in the reports of those who are in *actual* control.   Ignorance in high places cannot humanly be an excuse.

Imagine a British empire tolerating so long a Chamberlain in the government, or the endless petty, befuddling, deluding, etc., bickerings of political partisanship, which are good enough to wreck any system of 'democracy' (in practice another $\frac{over}{under}$defined term).

It seems, however, there is at least one point the totalitarian and 'democratic' politicians have in common, best expressed by Kipling:

> ' 'Ow the loot!
> Bloomin' loot!
> That's the thing to make the boys git up an' shoot!
> It's the same with dogs an' men,
> If you'd make 'em come again
> Clap 'em forward with a Loo! loo! Lulu! Loot!
> Whoopee! Tear 'im, puppy! Loo! loo!
> Lulu! Loot! loot! loot!'

2. *WARS* OF *AND* ON *NERVES*

It was explained already how the introduction of new factors is bound to change our generalizations and therefore evaluations. But this somehow is disregarded by most rulers and politicians who are on the defensive, while those who are on the offensive introduce new psycho-logical factors to confuse the old generalizations, as a rule successfully. Politicians, gangsters, military men, etc., without any understanding of the depth of destructiveness to the human nervous systems, utilize these methods quite successfully. Magicians have studied those methods professionally, but they utilize them for entertainment, not for destruction.

These destructive methods are the bases of the 'war *of* nerves', and the 'war *on* nerves', etc., to the point of using '*screaming*' bombs, verbal distortion, the 'psychology' of deception, etc. These methods can be counteracted *only* when governments who feel their responsibility not only to the ruling classes, but also to the *people* of their nations, will employ experts in neuro-psychiatry, anthropology, general semantics, etc., for guidance, if the present world neurosis is to be checked.

There are persistent reports that the Nazi government is utilizing a staff of psycho-logical experts for *destructive* purposes. Other totalitarian governments ape their successfully worked out and tested methods. The 'democratic' governments in this present fundamental *nerve contest* appear a tragic joke of ignorance, inefficiency, etc. In practice this amounts to betrayal, because they fail to recognize the overwhelming importance and vulnerability of the human nervous system, and do not utilize such experts in a *constructive* way. The 'scream' of a bomb, for instance, is much more destructive to the 'enemy' than the destruction by the bomb itself, which may kill a few people at the cost of at least $100,000 per corpse, while the 'scream' alone brings demoralizing terror to hundreds, if not thousands of people. It is certainly an expertly calculated and efficient 'war *on human* nerves'. But what can be done if ignorant 'democratic' governments refuse to live up to their duties?

Humanity, civilizations, cultures, etc., are ultimately based on the constructive use of neuro-semantic and neuro-linguistic mechanisms present in every one of us. Many pathological Nazi leaders utilize these constructive mechanisms in civilization for destructive selfish purposes. Under experts they have turned against mankind the essential assets of mankind. The beginning was 'mental' illness of a few leaders, based on hates, fears, revenges, etc. Later this destructive task was passed on to governmental psycho-logical experts, to build up methods to tear down human *neuro*-semantic and *neuro*-linguistic mechanisms, quite successfully because of the abysmal ignorance of modern scientific issues exhibited by the political verbalists and enchanters of other nations.

One of the most effective of these methods is the use of *pathological verbal distortion* such as is found among the 'mentally' ill. For instance, a paranoiac may believe 'honestly' that he is persecuted, become dominated by 'hate', etc., and ultimately may kill to 'defend' himself. Unfortunately at present only psychiatrists, familiar with verbal distortions and 'rationalizations' of patients in hospitals, can fully understand these problems.

A 'mentally' ill person is not necessarily a 'genius,' but it is well known to psychiatrists that some 'mentally' ill are often very cunning and will outwit any doctor or nurse. At present the people of the world do not realize that they are being trained in psychopathological uses of their nervous systems, and a future generation or two will become semantically crippled because trained in such distortions.

The violation, through ignorance and/or *un*-sanity, of the similarity of structure in the map-territory relationship (see p. 58 ff. and p. 750 ff.), and/or deliberate, professionally planned distortion of it, abolishes predictability, proper evaluation, trust, etc. This results only in breeding fears, anxieties, hates, etc., which disorganize individuals and even nations. There must be a correspondence and similarity of structure between language and facts, and so consequent thalamo-cortical integration, if we are to survive as a sane 'civilized' race.

In a few years history will judge these dying spasms of the aristotelian system, a system which was the best of its kind 2,300 years ago, as formulated by a great man under the conditions of the very few scientific facts known at that date. It is not so today, 1941. Most of the knowledge of scientific facts and methods of Aristotle are obsolete today, and in the main harmful, like the 'Maginot line' orientation.

By necessity the aristotelian system was based on macroscopic or animal, 'sense', levels, which even now predominantly guide the masses. It could take into consideration 'sense' data, etc., but cannot deal adequately with 1941 cultural as well as sanity conditions which, as we know today, are resultants of sub-microscopic, electro-colloidal processes.

In a non-aristotelian system we are stressing the differences between the animal reflex, automatic *signal* reactions, which do not involve 'thinking', human 'intelligence', etc., and human *symbol* reactions, with their flexibility, based on conscious evaluations, etc. These differences could hardly be conveyed better than by studying *The Rape of the Masses; The Psychology of Totalitarian Propaganda,* by Dr. Serge Chakotin, (Alliance Book Corporation, New York, 1940). A former student of

Professor Pavlov, Dr. Chakotin bases his analysis of totalitarian methods on Pavlov's fundamental researches of conditional reactions in dogs.

3. *HITLER AND PSYCHO-LOGICAL FACTORS IN HIS LIFE*

The groping dissatisfaction with the old system was so general that only a catalyst was needed to precipitate the crisis.   This catalyst was found in the son of Alois Schicklgruber (also spelled Schücklgruber) who later changed his name to 'Hitler'.   There was a history of illegitimacy in the family.   Rudolf Olden in his biography of Hitler says, 'Hitler has given the simplest and clearest picture possible of conditions in his father's home.   But we have only to look at the facts to see that, far from being simple, the married life of his father was unusual and tempestuous. Three wives, seven children, one divorce, one birth before marriage, two shortly after the wedding, one wife fourteen years older than himself and another twenty-three years younger—that is saying a good deal for a Customs officer.'

There were other important circumstances in Adolf Hitler's life which were influential and found their fulfillment in totalitarian systems. (*a*) He was born from a peasant stock, by tradition prepared to carry a heavy load of work with persistency.   (*b*) He was baptized in the Catholic Church, an institution well known to have totalitarian orientations, and which up to this day in principle proclaims authority over 'all' the Catholics in the world.   Having absorbed that totalitarian orientation from childhood up, which applies also to Mussolini, Stalin, etc., it was simple for those so trained to switch to *state totalitarianism,* where such leaders could find a 'lebensraum' for themselves as individuals, thus enhancing their own 'egos', and incidentally filling their pockets.   No one who has actually studied the public appearances of various totalitarian 'führers' can miss the utter similarity between their reactions and the reactions of the mobs to them.   They act like little 'gods on wheels', and the mobs react with unreasoned, blind, fanatical subjection, which the führers and their aides know how to manufacture.

(*c*) Hitler was born into Austrian bureaucracy, one of the most inefficient, dishonest, hypocritical, etc., bureaucracies in the world, permeated with the Hapsburg motto, 'Divide et impera'.   The older Schicklgruber wanted his son also to become a Hapsburg bureaucrat.   Schicklgruber, Jr. had a natural repulsion for them, and so deliberately boycotted any education, to disqualify himself for such a fate.   This lack of education ostracized him from the class of so-called 'intelligentsia', to which a Hapsburg bureaucrat eventually belonged.   Through living necessities he had to become a plain labour hand, yet because of his paranoia tendencies, delusions of grandeur based on unhealthy worship of historical 'heroes', etc., he was also not acceptable to the plain workers, who are generally sane and do not look at life as a Wagnerian opera. So in reality he found that he was not acceptable anywhere, belonged nowhere, a misfit everywhere, until he adhered to totalitarianism as a 'religion' which he and his closest associates modified to suit the Prussian character, selected by them as a standard of German perfection, to be imposed on the rest of the world.

(d) When he joined the German army with its orderly efficiency, etc., he found an ideal for himself as an escape from Hapsburg decadence. No matter how he hated the Hapsburg polite perfidy, he was too much of an Austrian not to utilize to the limit the Hapsburg methods. Ultimately through this combination of methods he 'out-Prussianed' the Prussians, whose particular arrogant, brutal methods were never approved and often disliked throughout the world and even in Germany.

I give these data as partial explanations of how through life and other circumstances the whole life of Hitler, as well as his political program, was based on hate, revenge and destruction of what he feared and hated as a person, driven by his delusions of persecution and grandeur. It was only natural in his 'chosen people' delusion that he should hate and try to destroy other 'chosen people'; obviously there is no place in this world for two or more 'chosen people'. The absurdity of Hitler's ignorant anthropological theories has been definitely established by science and history, and in fact are not taken seriously by many of the informed Nazi leaders themselves.

Some such analysis of a few of the more important factors in Hitler's life indicates how his 'mental' illness developed, involving 'inferiority' and 'persecution' complexes, etc., and explains why for his own comfort he surrounded himself personally with mostly psychopathological people, although their psychiatric classifications may be different.

Very soon psychiatric treatises will be written on the 'Jehovah complex' of Schicklgruber, Jr., etc. Perhaps the following quotations will illustrate how the 'Jehovah', as recorded in *Exodus* 19 and 20, is being copied today:

'Now therefore, if ye will obey my voice indeed, and keep my covenant, then ye shall be mine own possession from among all peoples: for all the earth is mine: and ye shall be unto me a kingdom of priests, and a holy nation. These are the words which thou shalt speak unto the children of Israel [Nazis].'

Or, 'I am Jehovah thy God, who brought thee . . . out of the house of bondage [England].'

Or, 'for I Jehovah thy God am a jealous God, visiting the iniquity of the fathers upon the children, upon the third and upon the fourth generation of them that hate me, and showing mercy unto thousands of them that love me and keep my commandments.'

Or, 'An altar of earth thou shalt make unto me, and shalt sacrifice thereon thy burnt-offerings, and thy peace-offerings, thy sheep, and thine oxen: in every place where I record my name I will come unto thee and I will bless thee.'   Etc., etc.

These suggestions are given only to indicate how psychiatrists can help future historians.

4. EDUCATION FOR INTELLIGENCE AND DEMOCRACY

It may become clearer why I speak of a dying, aristotelian, two-valued system by giving examples of how this type of evaluation is at the foundation of present day confusions and terrors.   Thus, for instance, the Nazi militant delusion of 'chosen people' gives us an excellent illustration of a two-valued, 'either-or' orientation.   The two-valued semantic twisting of 'real neutrality' is another significant example.   This distortion has kept the 'neutrals' in terrors, disorganizing their national and political life to the point of complete collapse, which today is a historical fact. The Nazi two-valued evaluation of 'neutrality' was: *either* be 'really neutral' and endorse and fight for the Nazis, *or* be 'not really neutral' and not help them.   According to this orientation a 'really neutral' Belgium, Holland, Denmark, Norway, etc., should fight against England, France, etc., to prove that they are 'really neutral'!

A similar analysis applies to the 'aggression' of China *against* Japan, Czechoslovakia *against* Germany, Poland *against* Germany, Poland *against* Russia, Finland *against* Russia, Greece *against* Italy, etc., and so on endlessly, which shows only the pathological application of the two-valued, 'either-or' patterns *in action*.   This analysis applies also to the first World War and the 'war guilt'.   In a non-aristotelian orientation we ask for actual facts, and do not accept mere verbalism.   Who invaded whom?   The historical facts are simple.   We know by now *who invaded whom,* and never mind verbal definitions.

When analysed *from a non-aristotelian point of view,* such orientations appear pathologically twisted.   Yet they produced results, as history shows.   It is not accidental that some years ago Hitler in one of his speeches took a definite stand for the prevailing aristotelianism, two-valued orientations, etc., and against modern science, which naturally develops in a non-aristotelian direction.   Quite soon whole volumes will be written on this subject; here it is possible only to indicate the main *methodological* issues involved.

Dr. Irving J. Lee in his article, 'General Semantics and Public Speaking', *Quarterly Journal of Speech,* December, 1940, formulates a fundamental contrast between the types of 'rhetorics' of Aristotle and Hitler, and the non-aristotelian type of communication found in general semantics which is based on proper evaluation, made possible by thalamo-cortical integration.

We should not make the mistake of fancying that Hitler, etc., or the mikado are building a new non-aristotelian system, and a future new saner civilization. It is only a rebellion *within* the old 'either-or' system, a changing from one scheme of selfishness, greed and force to another cabal of selfishness, greed, and brute force, this time unavoidably lowering human cultural standards by training future generations in pathological abuses of neuro-semantic and neuro-linguistic mechanisms, emasculating and misusing science, etc.

A non-aristotelian system must include considerations of neuro-semantic and neuro-linguistic environments as environment. Introductions of such new factors necessitate a complete revision of all known doctrines, pet creeds, etc., and make possible the building of a *science of man,* which under the old aristotelian conditions was impossible. The tabulation given here indicates some of the many older fictitious factors which have been eliminated as false to facts and destructive, while new, constructive factors have been introduced. This by necessity requires the utilization of more adequate methods and techniques by which we can cope with a new world.

The new, non-aristotelian types of evaluations are forthcoming in every field of human endeavour, in science and/or life, necessitated by the urgencies of modern conditions. The main problem today is to formulate *general methods* by which these many separate attempts can be unified into a general system of evaluation, which can become communicable to children and, with more difficulty, even to adults. History shows that whenever older methods prove their inefficiency new methods are produced which meet the new conditions more effectively. But the difficulties involved must first be clearly *formulated* before methods and techniques can be devised with which we can deal with them more successfully.

It seems unnecessary to enlarge on the present day world tragedies because many excellent volumes have already been written and are continuing to accumulate, psychiatric evaluations included. I must stress, however, that no writer I know of has ever understood the depth of the pending transition from the aristotelian system to an already formulated non-aristotelian system. This transition is much deeper than the change from merely one aristotelian 'ism' to another.

We argue so much today about 'democracy' versus 'totalitarianism'. Democracy presupposes intelligence of the masses;* totalitarianism does *not* to the same degree. But a 'democracy' without intelligence of the masses under modern conditions can be a worse human mess than any dictatorship could be.** Certainly present day education, while it may cram students' heads with some data, without giving them any *adequate methodological synthesis* and extensional working methods, does not train in 'intelligence' and how to become adjusted to life, and so does not work toward 'democracy'. Experiments show that even a root can learn a lesson (see p. 120), and animals can learn by trial and error. But we humans after these millions of years should have learned how to utilize the 'intelligence' which we supposedly have, with some predictability, etc., and use it *constructively,* not *destructively,* as, for example, the Nazis are doing under the guidance of specialists.

In general semantics we believe that some such thing as healthy human intelligence is possible, and so somehow we believe in the eventual possibility of 'democracy'. We work, therefore, at methods which could be embodied even in elementary education to develop the coveted thalamo-cortical integration, and so sane intelligence. Naturally in our work *prevention* is the main aim, and this can be accomplished only through education, and as far as the present is concerned, through *re*-education, and *re*-training of the human nervous system.

Section I. Constructive suggestions.

As far back as 1933, on page 485 ff. of the present book, I drew attention to the human dangers of the abuse of neuro-semantic and neuro-linguistic mechanisms, with suggestions for preventive measures. In September, 1939, I advanced further constructive suggestions to some leading governments, urging the employment of permanent boards of neuro-psychiatrists, psycho-logicians, and other specialists, to counteract similar dangers in connection with the present world crises. I received only *two* polite acknowledgements of my letters. But both forewarnings of 1933 and 1939 have been disregarded in practice, even by specialists, with known disastrous results.

---

* Mumford, Lewis. *Men Must Act.* Harcourt, Brace, New York, 1939.
** Consult, for example, comments of Supreme Court justices about the impossibility of 'justice' when juries are made up of individuals of *low grade* 'mentality', etc.

In the meantime the more far-sighted Nazi government employed a staff of specialists working at methods to *disorganize* the nervous functioning of their adversaries which, as facts show, have worked very successfully and devastatingly upon the unlucky citizens whom the short-sighted, unscientific, etc., governments never guided toward the proper use of their nervous systems, or safe-guarded from the abuses.*

Perhaps at present, 1941, after some irreparable harm has been done, the governments of the world will awaken and realize that the proper functioning of the nervous systems of their citizens is in many ways more important than any gun, battleship or aeroplane, etc., could possibly be, as there must be a Smith$_1$ behind the gun!

No matter who is finally 'victorious' in the present world struggle, no matter which way we look at it, the return to the old conditions is impossible. A complete neuro-semantic and neuro-linguistic revision is inevitable, and this revision is bound to lead away from aristotelianism. For this revision we are preparing the foundations in the formulations of general semantics. Before any lasting adjustments in the future social, economic, political, ethical, etc., fields are accomplished we have to be able to *evaluate properly and talk sense.* Otherwise the situation is hopeless.

Obviously, regardless of what the 'politicians' may say, in every country we necessarily have some kind of guidance by the government and executive power, no matter in what direction. Even 'complete lack of guidance' must be considered guidance of some sort, in the direction, say, of 'rugged individualism', etc., which, if carried to the limit, becomes the unworkable ideal of anarchy. In practical life such attitudes ultimately engender animal competition instead of human co-operation, and the very opposite of what we consider as the social feeling imperative for 'democracy'.

The real question is whether the existing governments are informed enough about human neurological problems, sanity, etc., and are intelligent enough, honest enough, etc., to guide and advise their people *constructively and efficiently* in constantly emerging *neurological situations* such as occur in home and school lives, in national and international affairs, etc. Unfortunately the answer is in the negative. At present there is no such government I know of. The Nazi government, on the other hand, has mobilized the psycho-logical knowledge available to them for *destructive* purposes, which must be *professionally counteracted* by the rest of the governments of the civilized world, if sanity is to prevail.

---

*Taylor, Edmond. *The Strategy of Terror.* Houghton-Mifflin, Boston, 1940.

Depending on science for more and better killing *machines* is certainly not the solution for *human* problems, culture and civilization. Without being sentimental, in a human civilization humans matter more than machines, or symbols such as a 'dollar', a 'pound sterling', a 'pound of flesh', a 'scalp', etc., or such verbal generalizations as 'liberty', 'equality', etc. The *living reactions* of Smith$_1$ are more important than the verbalisms of Smith$_1$, who nevertheless can shake the air with his verbal tricks, as many of us too often do, affecting the nervous systems of others.

At present the totalitarians have exploited neuro-semantic and neuro-linguistic mechanisms to their destructive limit, the best they knew how, to date. Counteraction, reconstruction, and/or prevention are impossible unless such mechanisms are utilized *constructively* under the guidance of governmental specialists in the fields of anthropology, neuro-psychiatry, general semantics, etc., who would understand the language of their fellow workers in related scientific fields, and would be FREE TO DEVOTE THEIR ENTIRE TIME AND EFFORTS TO THIS TASK, AND TO FURTHER INVESTIGATIONS.

Although practically all civilized states employ psychiatrists in their governmental hospitals for 'mentally' ill, these physicians are necessarily preoccupied with their patients and cannot undertake the special duties of the board I suggest. Such a board would require the full time and attention of its members, as they would be called upon for consultation by various other governmental departments such as interior, state, labour, commerce, health, army, navy, etc., and so special studies and co-ordinating knowledge in related branches of science would be essential.

It seems extremely short-sighted in 1941 that governments should employ permanently specialists in chemistry, physics, engineering, etc.; other specialists who advise how to eliminate lice from poultry, raise pigs, conserve wild life, etc.—and yet have no *permanent* consulting board of specialists who would advise how to conserve and prevent the abuse of human nervous systems. Even a Chamberlain would have intelligence and/or honesty enough to pass a problem of a 'magnetic mine' to physicists and engineers, and not to party politicians, who know nothing about such mechanisms, but would nevertheless be ready to debate 'politically' on the subject.

For example, if consulted, such a suggested body of governmental specialists would have studied *Mein Kampf* and various speeches of Hitler, Goebbels, etc., as a part of their duties, long ago, and would have advised their governments that psychopathological people are getting in control of world affairs and that their words cannot be trusted at all.

There would have been no 'appeasements', etc., and other measures would have been taken to cope with the depth of the problems involved.

It seems that the suggestions made on page 485 ff., although necessary, are not sufficient at the date of this writing, and the latest suggestions become imperative to safe-guard our future.

CONCLUSION

To summarize, under present world conditions the rôle of governments is becoming more and more difficult and important. With all modern complexities it is impossible for governmental men to be specialists in every field of science, and therefore they must depend on professional experts *attached to the government,* not only in the fields of chemistry, engineering, physics, agriculture, etc., which they already utilize; but also in anthropology, neuro-psychiatry, general semantics, and related professions. Otherwise the governments will indefinitely play the rôle of the blind leading the blind. It is unreasonable to wait ten or twenty years to learn by bitter experience how short-sighted and incompetent our governments have been. Why not utilize some human intelligence, proper evaluation, etc., toward which extensional methods lead, and thereby have some *predictability.* This is definitely an imperative, immediate need.

We should not delude ourselves. Once the psychopathological misuses of neuro-semantic and neuro-linguistic mechanisms have been so successfully introduced, they will remain with us unless reconstructive and preventive governmental measures are undertaken by experts, at once.

The conditions of the world are such today that private scientific undertakings and even professional opinions of scientific societies, or international congresses, etc., are bound to be ineffective. Only governmental interest, backing, financing, etc., can organize and enforce a serious movement for sanity, the more so since scientists, physicians, educators, and other professionals do not have the necessary time, money, authority, or even initiative to carry forward concerted plans. We have learned this group wisdom by now in the case of smallpox vaccination, control of epidemics, etc., and I venture to suggest that only such group wisdom will be effective as far as the health of our nervous systems is concerned. In terms of money certainly it would be economical to spend for *preventive* and *permanent* measures an amount even less than the cost of a single aeroplane which is made today and shot down tomorrow.

It must be sadly admitted that even professionals, no matter how prominent they may be in their narrow specialties, as individuals or specialized groups are at present scientifically unequipped to deal with such large and complex problems as the passing from one system of orientation to another, because those whose duty it was to integrate methodologically the vast knowledge at hand, have failed.   Such conditions can be remedied only by diversified methodological investigations, co-operation, and *concerted action* of specialists in different fields, which no private undertaking can organize effectively.   The reader is referred to page 558 ff. and also to my 'Science of Man'.*

There can be little doubt that self-seeking politicians, to cover up their own tracks, will be against such scientific sanity guidance, but enlightened public opinion will sooner or later force the issues to the only possible intelligent solution.

The prevalent and constantly increasing general deterioration of human values is an unavoidable consequence of the crippling misuse of *neuro*-linguistic and *neuro*-semantic mechanisms.   In general semantics we are concerned with the *sanity* of the race, including particularly methods of prevention; eliminating from home, elementary, and higher education inadequate aristotelian types of evaluation, which too often lead to the *un-sanity* of the race, and building up for the first time a positive theory of sanity, as a workable non-aristotelian *system*.

The task ahead is gigantic if we are to avoid more personal, national, and even international tragedies based on unpredictability, insecurity, fears, anxieties, etc., which are steadily disorganizing the functioning of the human nervous system.   Only when we face these facts fearlessly and intelligently may we save for future civilizations whatever there is left to save, and build from the ruins of a dying epoch a new and saner society.

I seriously appeal to scientists, educators, medical men, especially psychopathologists, parents, and other forward-looking citizens to investigate and co-operate in urging the governments to carry out their duty to guide the people scientifically, as suggested here.

A non-aristotelian re-orientation is inevitable; the only problem today is when, and at what cost.

A. K.

CHICAGO, MARCH, 1941.

* Korzybski, A.   The Science of Man.   *Amer. Jour. of Psychiatry.*   May, 1937

ACKNOWLEDGEMENTS

I gratefully acknowledge the valuable comments of the following in connection with $\frac{over}{under}$ defined terms: Franz Alexander, M.D., Professor Leonard Bloomfield, Douglas Gordon Campbell, M.D., Professor Morris R. Cohen, C. B. Congdon, M.D., Professor S. I. Hayakawa, Professor Earnest A. Hooton, Doctor Robert M. Hutchins, N. E. Ischlondsky, M.D., Professor Wendell Johnson, Professor Kurt Lewin, Professors H. G. and L. R. Lieber, Mr. Robert Lord, Jules H. Masserman, M.D., Mr. H. L. Mencken, Professor Charles W. Morris, the late Professor Raymond Pearl, Professor W. V. Quine, Professor Oliver Reiser, Professor Bertrand Russell, Doctor Eugene Randolph Smith, Mr. A. Ranger Tyler, and many students and friends too numerous to list here.

I wish to express my warm appreciation to my students and secretaries, Miss Charlotte Schuchardt and Miss Pearl Johnecheck, for the constructive help they have given me in the preparation of this introduction. The drawings on pages xiv and xxviii were made by Miss Johnecheck. I am also genuinely indebted to Miss M. Kendig, Educational Director of the Institute, and to Doctor Irving J. Lee of Northwestern University, for their important criticism and co-operation.

SPECIAL ACKNOWLEDGEMENT

On behalf of students of general semantics who have attended seminars at the Institute, and on my own behalf, I want to express my deep gratitude to Cornelius Crane, whose vision, interest, and financing made possible the founding of the Institute in 1938. The widespread influence and rapid development of the work of the Institute in this world turmoil became a living reality because of Mr. Crane's generous contributions during the first two and a half years of our pioneer effort. Forces of destruction are working steadily, and Mr. Crane should be credited with helping to organize constructive efforts.

I also gratefully acknowledge the contributions of the other students who are now helping to support the work of the Institute.

ALFRED KORZYBSKI

SUPPLEMENTARY BIBLIOGRAPHY
TO THE SECOND EDITION

The following bibliography is only illustrative of points made in the introduction to the second edition. A number of volumes listed here give extensive bibliographies in their fields. For instance, the book of Dr. Dunbar has 130 pages of bibliography which cover 2,358 items. Some of the most important and latest empirical data on electrical brain-waves, electro-physiology, conditional reactions in humans, electro-colloidal processes of the nervous system, experimental neuroses and psychoses in animals, the reactions of apes, data on human psychotherapy, the methods of deception and sensory misdirection as utilized by magicians, etc., are given mostly in technical journals and monographs, and the interested reader may find them in libraries.

This applies also to the many applications of the methods of general semantics in education, mental hygiene, speech difficulties, etc., carried on in universities and colleges, as well as applications in the practice of physicians, including psychiatrists; these are in preparation, or printed at present only by professional journals or by the Institute of General Semantics (see special list).

I list also some new pertinent, professional publications such as *Psychosomatic Medicine, Journal of Symbolic Logic, Encyclopedia of Unified Science*, etc., without listing the titles of the individual contributions. It is suggested that the interested reader, and particularly educators, medical men, etc., become acquainted with such material, or at least know that it does exist. The reader is also referred to the foreword to the bibliography given on page 767, and the titles which follow.

In science and life a great deal depends on proper evaluation, tested by predictability, which depends in turn on the similarity of structure between territory-map or fact-language. Thus, we have to know scientific facts, as well as the intricacies and difficulties of language and its structure. Fortunately there is a weekly *Science News Letter*, published by Science Service, Washington, D. C., giving brief, authoritative, non-technical *factual* summaries of progress in science, mathematics, medicine, etc., including sources, which every specialist as well as intelligent layman should know.

1. ADLER, MORTIMER. *How to Read a Book.* Simon & Schuster, New York, 1940.
2. ARENSBERG, CONRAD M. See Chapple.
3. ARNOLD, THURMAN. *The Symbols of Government.* Yale Univ. Press, New Haven, 1935.
4. *The Folklore of Capitalism.* Yale Univ. Press, New Haven, 1937.
5. *The Bottlenecks of Business.* Reynal, Hitchcock, New York, 1940.
6. BORN, MAX. *The Restless Universe.* Harper & Bros., New York, London, 1936.
7. BURRIDGE, W. *Excitability, A Cardiac Study.* Oxford Univ. Press, London, New York, 1932.
8. *A New Physiology of Sensation.* Oxford Univ. Press, London, New York, 1932.
9. *A New Physiological Psychology.* Arnold & Co., London, Baltimore, 1933.
10. *Alcohol and Anaesthesia.* Williams & Norgate, London, 1934.
11. CARNAP, R. *The Unity of Science.* K. Paul, Trench, Trubner & Co., London, 1934.
12. *Philosophy and Logical Syntax.* K. Paul, Trench, Trubner & Co., London, 1935.
13. *The Logical Syntax of Language.* Harcourt, Brace, New York, 1937.
14. *Foundations of Logic and Mathematics.* Univ. of Chicago Press, 1939.
15. CARREL, ALEXIS. *Man the Unknown.* Harper & Bros., New York, 1935.
16. CHAPPLE, ELIOT D. *Measuring Human Relations; An Introduction to the Study of the Interaction of Individuals.* With the collaboration of Conrad M. Arensberg. Genetic Psychology Monographs. Feb., 1940.
17. CHASE, STUART. *The Tyranny of Words.* Harcourt, Brace, New York, 1938.
18. DUNBAR, H. F. *Emotions and Bodily Changes.* Columbia Univ. Press, New York, 1938, 2nd ed. Extensive bibliography of 2,358 titles.

19. EINSTEIN, A., and INFELD, L. *The Evolution of Physics; the Growth of Ideas from Early Concepts to Relativity and Quanta.* Simon & Schuster, New York, 1938.
20. ESSER, P. H. Waan als Meerwaardige Term. *Psychiatrische en Neurologische Bladen,* 1939, No. 4, Bennebroek, Holland.
21. Psycho-logie en Semantiek. *Nederl. Tijdschrift voor Psychologie.* Vol. 8, 1940. Zutphen, Holland.
22. ESSER, P. H., and KRANS, R. L. Korzybski's Wetenschap van 'Den Mensch'. *Mensch en Maatschappij,* 1940, No. 2. Amsterdam, Holland.
23. FRANK, JEROME. *Law and the Modern Mind.* Tudor Publ. Co., New York, 1935.
24. GOLDBERG, ISAAC. *The Wonder of Worlds; An Introduction to Language for Everyman.* Appleton-Century, New York, 1938.
25. GRAY, LOUIS H. *Foundations of Language.* Macmillan, New York, 1939.
26. HEIDEN, K. *Hitler; A Biography.* Knopf, New York, 1936.
27. HOGBEN, L. *Genetic Principles in Medicine and Social Science.* Knopf, New York, 1932.
28. *Mathematics for the Million.* Norton, New York, 1937.
29. *The Retreat from Reason.* Random House, New York, 1937.
30. *Science for the Citizen.* Knopf, New York, 1938.
31. *Dangerous Thoughts.* Norton, New York, 1940.
32. *Principles of Animal Biology.* Norton, New York, 1940.
33. HOOTON, E. A. *Up From the Ape.* Macmillan, New York, 1931.
34. *Apes, Men and Morons.* Putnam's, New York, 1937.
35. An Anthropologist Looks at Medicine. *Science.* March 20, 1936.
36. *Twilight of Man.* Putnam's, New York, 1939.
37. *Why Men Behave Like Apes and Vice Versa.* Princeton Univ. Press, 1940.
38. HORNEY, K. *The Neurotic Personality of Our Time.* Norton, New York, 1937.
39. INFELD, L. See Einstein.
40. *International Encyclopedia of Unified Science.* Otto Neurath, Editor-in-chief. Vols. I and II. Foundations of the Unity of Science. Univ. of Chicago Press, 1939.
41. ISCHLONDSKY, N. E. *Neuropsyche und Hirnrinde,* 2 vol. German. Under the titles: I. *The Conditional Reflex and Its Importance in Biology, Medicine, Psychology and Pedagogics;* II. *Physiological Foundations of Deep Psychology, with Special Application to Psychoanalysis.* Urban & Schwarzenberg, Berlin and Vienna, 1930.
42. KASNER, EDWARD, and NEWMAN, JAMES. *Mathematics and the Imagination.* Simon & Schuster, New York, 1940.
43. KELLEY, DOUGLAS M. Conjuring as an Asset to Occupational Therapy. *Occupational Therapy and Rehabilitation.* Vol. 19, No. 2, April, 1940.
44. KOPEL, D. See Witty.
45. KRANS, R. L. See Esser.
46. *Language in General Education.* A Report of the Committee on the Function of English in General Education for the Commission on Secondary School Curriculum of the Prog. Educ. Asso. Appleton-Century, New York, 1940.
47. LEWIN, K. *Principles of Topological Psychology.* McGraw-Hill, New York, 1936.
48. LEWIS, NOLAN D. C. *Research in Dementia Praecox.* Natl. Comm. for Mental Hygiene, New York, 1936.
49. LUDECKE, KURT G. W. *I Knew Hitler.* Scribners, New York, 1938.
50. LUNDBERG, G. A. *Foundations of Sociology.* Macmillan, New York, 1939.
51. MACKAYE, J. *The Logic of Language.* Dartmouth Coll. Publs., Hanover, N. H., 1939.
52. MALINOWSKI, B. *Coral Gardens and Their Magic; A Study of the Methods of Tilling the Soil and of Agricultural Rites in the Trobriand Islands,* 2 vol. I. *Introduction;* II. *An Ethnographic Theory of Language and some Practical Corollaries.* Allen & Unwin, London, 1935.
53. *The Foundations of Faith and Morals; An Anthropological Analysis of*

Primitive Beliefs and Conduct with Special Reference to the Fundamental Problems of Religion and Ethics. Univ. of Oxford Press, London, New York, 1936.
54. MEYER, ADOLF. Mental Health. *Science.* Sept. 27, 1940.
55. MUMFORD, L. *The Culture of Cities.* Harcourt, Brace, New York, 1938.
56. *Men Must Act.* Harcourt, Brace, New York, 1939.
57. *Faith for Living.* Harcourt, Brace, New York, 1940.
58. MUNCIE, W. *Psychobiology and Psychiatry.* With a Foreword by Adolf Meyer. Mosby, St. Louis, 1939.
59. NEWMAN, JAMES. See Kasner.
60. NISSEN, H. W. See Yerkes.
61. OLDEN, R. *Hitler.* Covici, Friede, New York, 1936.
62. PERKINS, F. THEODORE. See Wheeler.
63. PETERSEN, WILLIAM F. *The Patient and the Weather,* 4 vol. Edwards Bros., Ann Arbor, Mich., 1938.
64. PITKIN, W. B. *Escape From Fear.* Doubleday, Doran, New York, 1940.
65. PRESCOTT, DANIEL A. *Emotion and the Educative Process.* Amer. Council on Educ., Washington, D. C., 1938.
66. *Psychosomatic Medicine.* Published *quarterly* by Comm. on Problems of Neurotic Behavior, Natl. Research Council, Washington, D. C.
67. QUINE, W. V. *Mathematical Logic.* Norton, New York, 1940.
68. RAUSCHNING, H. *The Revolution of Nihilism.* Alliance Book Corp., 1939.
69. *The Voice of Destruction (Hitler Speaks).* Putnam's, New York, 1940.
70. RICHARDS, I. A. *Interpretation in Teaching.* Harcourt, Brace, New York, 1938.
71. ROSETT, J. *The Mechanism of Thought, Imagery, and Hallucination.* Columbia Univ. Press, New York, 1939.
72. RYAN, CARSON W. *Mental Health Through Education.* Commonwealth Fund, New York, 1938.
73. SAPIR, E. *Totality.* Language Monograph of Linguistic Soc. of Amer. Waverly Press, Baltimore, 1930.
74. *The Expression of the Ending Point Relation.* Language Monograph of Linguistic Soc. of Amer., 1932.
75. SARGENT, PORTER. *Human Affairs.* Porter Sargent, Boston, 1938.
76. *Education; A Realistic Appraisal.* Porter Sargent, Boston, 1939.
77. *What Makes Lives.* Porter Sargent, Boston, 1940.
78. SCHIFERL, MAX. *An Introduction to Interpretation.* Stanford Language Arts Investigation, Interpretation Series I. Stanford Univ. Press, 1939.
79. SMITH, GEDDES. See Stevenson.
80. STEVENSON, GEORGE S., and SMITH, GEDDES. *Child Guidance Clinics; One Quarter Century of Development.* Commonwealth Fund, New York, 1934.
81. SULLIVAN, LAWRENCE. *The Dead Hand of Bureaucracy.* Bobbs-Merrill, New York, 1940.
82. TAYLOR, EDMOND. *The Strategy of Terror.* Houghton-Mifflin, Boston, 1940.
83. URBAN, WILBUR M. *Language and Reality; The Philosophy of Language and the Principles of Symbolism.* Macmillan, New York, 1939.
84. WHEELER, RAYMOND H., and PERKINS, F. THEODORE. *Principles of Mental Development.* Crowell Co., New York, 1932.
85. WILLIAMS, JESSE F. *A Textbook of Anatomy and Physiology.* Saunders, Philadelphia, 1939, 6th ed.
86. WITTY, PAUL, and KOPEL, DAVID. *Reading and the Educative Process.* Ginn & Co., Boston, 1939.
87. WOODGER, J. H. *The Axiomatic Method in Biology.* Cambridge Univ. Press, London, 1937.
88. YERKES, ROBERT M., AND NISSEN, HENRY W. Pre-linguistic Sign Behavior in Chimpanzee. *Science.* June 23, 1939.
89. ZUCKERMAN, S. *The Social Life of Monkeys and Apes.* Harcourt, Brace, New York, 1932.

The following items in the bibliography have either been omitted by inadvertence, or they appeared after the numbering of the bibliography was completed.

90. BARNARD, R. H. General Semantics and the Controversial Phases of Speech. *Quar. Jour. of Speech.* Dec., 1940.
91. CHAKOTIN, S. *The Rape of the Masses; The Psychology of Totalitarian Political Propaganda.* Alliance Book Corporation, New York, 1940.
92. HITLER, ADOLF. *Mein Kampf.* Stackpole Sons, New York, 1939.
93. LEE, IRVING. General Semantics and Public Speaking. *Quar. Jour. of Speech.* Dec., 1940.
94. The Adult in Courses in Speech. Accepted for publication. *College English* 1941.
95. LIEBER, H. G. and L. R. *Non-Euclidean Geometry or Three Moons in Mathesis.* Galois Institute of Mathematics, Long Island Univ., Brooklyn, New York, 1931.
96. *Galois and the Theory of Groups; A Bright Star in Mathesis.* Galois Institute of Mathematics, 1932.
97. *The Einstein Theory of Relativity,* Part I. Galois Inst. of Mathematics, 1936.
98. *Psychiatry; Jour. of the Biology and the Pathology of Interpersonal Relations.* Wm. A. White Psychiatric Foundation, Washington, D. C.
99. REISER, O. L. *The Promise of Scientific Humanism.* Oskar Piest, New York, 1940.
100. ROBINSON, EDWARD S. *Law and the Lawyers.* Macmillan, New York, 1935.

LIST OF REPRINTS AND MONOGRAPHS

INSTITUTE OF GENERAL SEMANTICS—1941

1. BARRETT, L. G. General Semantics and Dentistry. *Harvard Dental Record,* June, 1938.
2. Evaluational Disorders and Caries; Semantogenic Symptoms. *Jour. of Amer. Dental Asso.* Nov., 1939.
3. BREWER, JOSEPH. Education and the Modern World. Convocation Address, Olivet College, Sept., 1937. Reproduced from No. 21.
4. BURRIDGE, W. A New Colloido-Physiological Psycho-Logics. Reproduced from No. 21.
5. CAMPBELL, D. G. General Semantics; Implications of Linguistic Revision for Theoretical and Clinical Neuro-Psychiatry. *Amer. Jour. of Psychiatry.* Jan., 1937.
6. Neuro-Linguistic and Neuro-Semantic Factors of Child Development. Address, Chicago Pediatric Soc. Jan., 1938. Reproduced from No. 21.
7. General Semantics in Education, Counseling, and Therapy. *Natl. Educ. Asso. Proc.* 1939.
8. General Semantics and Schizophrenic Reactions; Neuro-Linguistic and Neuro-Semantic Mechanisms of Pathogenesis and Their Implications for Prevention and Therapy. Presented before Amer. Psychiatric Asso., Chicago. May, 1939. To be published.
9. See *Congdon.*
10. CONGDON, C. B., and CAMPBELL, D. G. A Preliminary Report on the Psychotherapeutic Application of General Semantics. March, 1937. Reproduced from No. 21.
11. DEVEREUX, G. A Sociological Theory of Schizophrenia. *Psychoanalytic Rev.* July, 1939.
12. HAYAKAWA, S. I. General Semantics and Propaganda. Presented before the Natl. Council of Teachers of English, St. Louis, Mo. Nov., 1938. *Pub. Opinion Quar.* April, 1939.
13. *Language in Action.* Experimental second edition of a text for Freshman English Courses. An application of the principles of General Semantics which provides an orientation towards language based upon modern linguistic, scientific, and literary theory. Institute of General Semantics. Chicago, 1940. Final text published by Harcourt, Brace, New York, 1941.

14. HERRICK, C. JUDSON. A Neurologist Makes Up His Mind. The Mellon Lecture, Univ. of Pittsburgh, School of Medicine, May, 1939. *Scientific Monthly.* Aug., 1939.
15. JOHNSON, WENDELL. *Language and Speech Hygiene; An Application of General Semantics.* Outline of a Course, Iowa Univ. General Semantics Monographs, No. I. Institute of General Semantics, Chicago, 1939.
16. KENDIG, M. Language Re-Orientation of High School Curriculum and Scientific Control of Neuro-Linguistic Mechanisms for Better Mental Health and Scholastic Achievement. Presented before Educ. Section, A.A.A.S., St. Louis, Dec., 1935. Reproduced from No. 21.
17. Book Reviews for Students of General Semantics; First Series. *The Psychiatric Exchange of the Ill. State Institutions.* March, 1939.
18. Comments on the Controversy over the 'Nature and Constancy of the I.Q. as a Measure of Potential Growth'. *Educational Method.* Jan., 1940.
19. KEYSER, CASSIUS J. Mathematics and the Science of Semantics. *Scripta Mathematica.* May, 1934.
20. KORZYBSKI, A. Preface to First Edition, *Science and Sanity,* 1933. Separately published. Institute of General Semantics, Chicago.
21. *General Semantics: Papers from the First American Congress for General Semantics,* 1935. With an introductory 'Outline of General Semantics' by Alfred Korzybski and other related contributions. Bibliography. Collected and arranged by Hansell Baugh. (Author index Korzybski) Arrow Editions, New York, 1938. Distributed also by the Institute of General Semantics, Chicago.
22. Outline of General Semantics; The Application of Some Methods of Exact Sciences to the Solution of Human Problems and Educational Training for General Sanity. Presented before First American Congress for General Semantics, Ellensburg, Wash., 1935. Reproduced from No. 21.
23. Neuro-Semantic and Neuro-Linguistic Mechanisms of Extensionalization; General Semantics as a Natural Experimental Science. Presented before the Psychology Section, A.A.A.S., St. Louis, Dec., 1935. *Amer. Jour. of Psychiatry.* July, 1936.
24. The Science of Man. *Amer. Jour. of Psychiatry.* May, 1937.
25. *General Semantics; Extensionalization in Mathematics, Mathematical Physics and General Education.* Three papers presented before annual meetings of the Amer. Math. Soc., 1935, 1938, 1939. With an introductory Outline of General Semantics. General Semantics Monographs No. II. Institute of General Semantics, Chicago, 1941.
26. *A Memorandum* on the Institute of General Semantics. A preliminary report, 1940.
27. General Semantics, Psychiatry, Psychotherapy and Prevention. Paper presented before the Amer. Psychiatric Asso., May, 1940. *Amer. Jour. of Psychiatry.* Sept., 1941.
28. Introduction to Second Edition, *Science and Sanity,* 1941. Separately published. Institute of General Semantics, Chicago.
29. MICHIE, S. A New General Language Curriculum for the Eighth Grade. *Modern Language Jour.* Feb., 1938.
30. SEMMELMEYER, M. The Application of General Semantics to a Program for Reading Readiness. Paper presented before the Third Annual Conference on Reading, University of Chicago, June, 1940. Institute of General Semantics, Chicago. An abridgement of this paper is published in the proceedings of the conference, *Reading and Pupil Development,* under the title, 'Promoting Readiness for Reading and for Growth in the Interpretation of Meaning'. Suppl. Educ. Monographs, No. 51. October, 1940. Univ. of Chicago Press.
31. WEINBERG, A. M. General Semantics and the Teaching of Physics. *Amer. Physics Teacher.* April, 1939.
32. WEYL, HERMANN. The Mathematical Way of Thinking. *Studies in the History of Science.* Univ. of Pa. Press, 1941.
Latest list of publications available from Institute of General Semantics, Lakeville, Connecticut.

PREFACE TO THE FIRST EDITION 1933

It is difficult for a philosopher to realise that anyone really is confining his discussion within the limits that I have set before you. The boundary is set up just where he is beginning to get excited. (573)   A. N. WHITEHEAD

That all debunkers must add new boshes of their own to supply the vacua created by the annihilation of the old, is probably a law of nature. (22)

E. T. BELL

Teaching without a system makes learning difficult.   *The Talmud*

The layman, the 'practical' man, the man in the street, says, What is that to me? The answer is positive and weighty. Our life is entirely dependent on the established doctrines of ethics, sociology, political economy, government, law, medical science, etc. This affects everyone consciously or unconsciously, the man in the street in the first place, because he is the most defenseless. (280)   A. K.

When new turns in behaviour cease to appear in the life of the individual its behaviour ceases to be intelligent. (106)   C. E. COGHILL   * see page xii

> 'Tis a lesson you should heed,
>   Try again;
> If at first you don't succeed,
>   Try again;
> Then your courage should appear,
> For if you will persevere
> You will conquer, never fear,
>   Try again.
>
> WILLIAM EDWARD HICKSON.

The main portions of the present work have already been presented in the form of lectures before different universities, technological institutes, teachers' and physicians' associations, and other scientific bodies. The general outline was presented for the first time before the International Mathematical Congress in Toronto in 1924, and published in the form of a booklet. A further elaboration of the system was read before the Washington (D. C.) Society for Nervous and Mental Diseases in 1925, and the Washington (D. C.) Psychopathological Society in 1926, and later published. A fuller draft was presented before the Congrès des mathématiciens des pays Slaves, in Warsaw, Poland, in 1929. A special and novel aspect of the subject, in connection with the conditional reflexes of Pavlov, was outlined before the First International Congress of Mental Hygiene, Washington, D.

C., in 1930. Other aspects were discussed before the American Mathematical Society, October 25, 1930, and the Mathematical Section of The American Association for the Advancement of Science, December 28, 1931. The latter paper is printed as Supplement III in this volume.

The general character of the present work is perhaps best indicated by the two following analogies. It is well known that for the working of any machine some lubricant is needed. Without expressing any judgement about the present 'machine age', we have to admit that technically it is very advanced, and that without this advancement many scientific investigations necessitating very refined instruments would be impossible. Let us assume that mankind never had at its disposal a clean lubricant, but that existing lubricants always contained emery sand, the presence of which escaped our notice. Under such conditions, existing technical developments, with all their consequences, would be impossible. Any machine would last only a few weeks or months instead of many years, making the prices of machines and the cost of their utilization entirely prohibitive. Technical development would thus be retarded for many centuries. Let us now assume that somebody were to discover a simple means for the elimination of emery from the lubricants; at once the present technical developments would become possible, and be gradually accomplished.

Something similar has occurred in our human affairs. Technically we are very advanced, but the elementalistic premises underlying our human relations, practically since Aristotle, have not changed at all. The present investigation reveals that in the functioning of our nervous systems a special harmful factor is involved, a 'lubricant with emery' so to speak, which retards the development of sane human relations and prevents general sanity. It turns out that in the structure of our languages, methods, 'habits of thought', orientations, etc., we preserve delusional, psychopathological factors. These are in no way inevitable, as will be shown, but can be easily eliminated by special training, therapeutic in effect, and consequently of educational preventive value. This 'emery' in the nervous system I call identification. It involves deeply rooted 'principles' which are invariably false to facts and so our orientations based on them cannot lead to adjustment and sanity.

A medical analogy here suggests itself. We find a peculiar parallel between identification and infectious diseases. History proves that under primitive conditions infectious diseases cannot be controlled. They spread rapidly, sometimes killing off more than half of the affected population. The infectious agent may be transmitted either directly, or through rats, insects, etc. With the advance of science, we are able to control the disease, and various important preventive methods, such as sanitation, vaccination, etc., are at our disposal.

Identification appears also as something 'infectious', for it is transmitted directly or indirectly from parents and teachers to the child by the mechanism and structure of language, by established and inherited 'habits of thought', by rules for life-orientation, etc. There are also large numbers of men and women who make a profession of spreading the disease. Identification makes general sanity and complete adjustment impossible. Training in non-identity plays a therapeutic role with adults. The degree of recovery depends on many factors, such as the age of the individual, the severity of the 'infection', the diligence in training in non-identity, etc. With children the training in non-identity is extremely simple. It plays the role both of sanitation and of the equally simple and effective preventive vaccination.

As in infectious diseases, certain individuals, although living in infected territory, are somehow immune to this disease. Others are hopelessly susceptible.

The present work is written on the level of the average intelligent layman, because before we can train children in non-identity by preventive education, parents and teachers must have a handbook for their own guidance. It is not claimed that a millenium is at hand, far from it; yet it seems imperative that the *neuro*-psycho-logical factors which make general sanity impossible should be eliminated.

I have prefaced the parts of the work and the chapters with a large number of important quotations. I have done so to make the reader aware that, on the one hand, there is already afloat in the 'universe of discourse' a great deal of genuine knowledge and wisdom, and that, on the other hand, this wisdom is not generally applied and, to a large extent, cannot be applied as long as we fail to build a simple system based on the complete elimination of the pathological factors.

A system, in the present sense, represents a complex whole of co-ordinated doctrines resulting in methodological rules and principles of procedure which affect the orientation by which we act and live. Any system involves an enormous number of assumptions, presuppositions, etc., which, in the main, are not obvious but operate unconsciously. As such, they are extremely dangerous, because should it happen that some of these unconscious presuppositions are false to facts, our whole life orientation would be vitiated by these unconscious delusional factors, with the necessary result of harmful behaviour and maladjustment.   No system has ever been fully investigated as to its underlying unconscious presuppositions.   Every system is expressed in some language of some structure, which is based in turn on silent presuppositions, and ultimately reflects and reinforces those presuppositions on and in the system.   This connection is very close and allows us to investigate a system to a large extent by a linguistic structural analysis.

The system by which the white race lives, suffers, 'prospers', starves, and dies today is not in a strict sense an aristotelian system. Aristotle had far too much of the sense of actualities for that.   It represents, however, a system formulated by those who, for nearly two thousand years since Aristotle, have controlled our knowledge and methods of orientations, and who, for purposes of their own, selected what today appears as the worst from Aristotle and the worst from Plato and, with their own additions, imposed this composite system upon us.   In this they were greatly aided by the structure of language and psycho-logical habits, which from the primitive down to this very day have affected all of us consciously or unconsciously, and have introduced serious difficulties even in science and in mathematics.

Our rulers: politicians, 'diplomats', bankers, priests of every description, economists, lawyers, etc., and the majority of teachers remain at present largely or entirely ignorant of modern science, scientific methods, structural linguistic and semantic issues of 1933, and they also lack an essential historical and anthropological background, without which a sane orientation is impossible.*   This ignorance is often wilful as they mostly refuse, with various excuses, to read modern works dealing with such problems.   As a result a conflict is created and maintained between the advance of science affecting conditions of actual life and the orientations of our rulers, which often remain antiquated by centuries, or one or two thousand years.   The present world conditions are in chaos; psycho-logically there exists a state of helplessness— hopelessness, often resulting in the feelings of insecurity, bitterness, etc., and we have lately witnessed psychopathological mass outbursts, similar to those of the dark ages.   Few of us at present realize that, as long as such ignorance of our rulers prevails, *no solution of our human problems is possible.*

---

* The literature of these subjects is very large and impossible to give here or in my bibliography; but as primers I may as well suggest numbers 299, 334, 492, 558, 589 in my bibliography.   These books in turn give further references.

The distinctly novel issue in a non-aristotelian system seems to be that in a human class of life elementary methodological and structural ignorance about the world and ourselves, as revealed by science, is bound to introduce delusional factors, for no one can be free from some conscious or unconscious structural assumptions. The real and only problem therefore seems to be whether our structural assumptions in 1933 are primitive or of the 1933 issue. The older 'popularization of science' is not the solution, it often does harm. The progress of science is due in the main to scientific methods and linguistic revisions, and so the new facts discovered by such methods cannot be properly utilized by antiquated psycho-logical orientations and languages. Such utilization often results only in bewilderment and lack of balance. Before we can adjust ourselves to the new conditions of life, created in the main by science, we must first of all revise our grossly antiquated methods of orientation. Then only shall we be able to adjust ourselves properly to the new facts.

Investigations show that the essential scientific structural data of 1933 about the world and ourselves are extremely simple, simpler even than any of the structural fancies of the primitives. We usually have sense enough to fit our shoes to our feet, but not sense enough to revise our older methods of orientation to fit the facts. The elimination of primitive identifications, which is easily accomplished once we take it seriously, produces the necessary psycho-logical change toward sanity.

'Human nature' is not an elementalistic product of heredity alone, or of environment alone, but represents a very complex organism-as-a-whole end-result of the enviro-genetic manifold. It seems obvious, once stated, that in a human class of life, the linguistic, structural, and semantic issues represent powerful and ever present environmental factors, which constitute most important components of all our problems. 'Human nature' *can be changed*, once we know how. Experience and experiments show that this 'change of human nature', which under verbal elementalism was supposed to be impossible, can be accomplished in most cases in a few months, if we attack this problem by the non-elementalistic, *neuro*-psycho-logical, special non-identity technique.

If the ignorance and identifications of our rulers could be eliminated a variety of delusional factors through home and school educational and other powerful agencies would cease to be imposed and enforced upon us, and the revision of our systems would be encouraged, rather than hampered. Effective solutions of our problems would then appear spontaneously and in simple forms; our 'shoes' would fit our 'feet' and we could 'walk through life' in comfort, instead of enduring the present sufferings.

Since our existing systems appear to be in many respects unworkable and involve psychopathological factors owing in the main to certain presuppositions of the aristotelian system, and also for brevity's sake, I call the whole operating systemic complex 'aristotelian'. The outline of a new and modern system built after the rejection of the delusional factors I call 'non-aristotelian'. To avoid misunderstandings I wish to acknowledge explicitly my profound admiration for the extraordinary genius of Aristotle, particularly in consideration of the period in which he lived. Nevertheless, the twisting of his system and the imposed immobility of this twisted system, as enforced for nearly two thousand years by the controlling groups, often under threats of torture and death, have led and can only lead to more disasters. From what we know about Aristotle, there is little doubt that, if alive, he would not tolerate such twistings and artificial immobility of the system usually ascribed to him.

The connection between the study of psychiatry and the study of mathematics and the foundations of mathematics is very instructive. In the development of civilization and science we find that some disciplines, for instance, the very young science of psychiatry, have progressed rapidly. Other disciplines such as mathematics, physics, etc., until recently progressed slowly, mainly on account of certain dogmas and prejudices. Of late some of these prejudices have been eliminated, and since then the progress of these sciences has become extremely rapid. Still other disciplines such as 'psychology', the traditional 'philosophy', sociology, political economy, ethics, etc., have developed their principles very little in nearly two thousand years notwithstanding a wealth of accumulated new data.

Many reasons are responsible for this curious state of affairs, but I will suggest only three, in the order of their importance. (1) First of all, the last mentioned slowly developing disciplines are the closest to us humans, and a primitive man, or an entirely ignorant person 'knows all about' these most complex problems in existence. This 'know it all' general tendency produces an environmental, psychological, linguistic, etc., manifold, filled with identifications which produce dogmas, prejudices, misunderstandings, fears, and what not, making an impersonal, impartial scientific approach next to impossible. (2) Few of us realize the unbelievable traps, some of them of a psychopathological character, which the structure of our ordinary language sets before us. These also make any scientific approach or agreement on vital points impossible. We grope by animalistic trial and error, and by equally animalistic strife, wars, revolutions, etc. These first two points apply practically to all of us, and introduce great difficulties even into mathematics. (3) One of the main reasons why psychiatry has advanced so rapidly in such a short period in con-tradistinction to 'psychology', is that it studies relatively simple and relatively singled-out symptoms. But as these symptoms are not iso-lated, and represent the reactions of the organism-as-a-whole, their partial study yields glimpses of the general and fundamental mecha-nisms. If we study mathematics and mathematical sciences as forms of human behaviour, we study also simplified and singled-out human reactions of the type: 'one and one make two', 'two and one make three', etc., and we also get glimpses of general mechanisms. In psy-chiatry we study simplified psycho-logical reactions at their worst; in mathematics and mathematical sciences we study simplified psycho-logical reactions at their best. When both types of reactions are studied conjointly, most unexpected and very far-reaching results follow which deeply affect every known phase of human life and activity, science included. The results of such widely separated studies do not conflict, but supplement each other, elucidating very clearly a general mechanism which operates in all of us. Psychiatrical studies help us most unexpectedly in the solution of mathematical paradoxes; and mathematical studies help us to solve very important problems in psychotherapy and in prevention of psycho-logical disorders.

History shows that the advancement of science and civilization involves, first, an accumulation of observations; second, a preliminary formulation of some kind of 'principles' (which always involve some unconscious assumptions); and, finally, as the numbers of observations increase, it leads to the revision and usually the rejection of unjusti-fied, or false to facts 'principles', which ultimately are found to repre-sent only postulates. Because of the cumulative and non-elementalistic character of human knowledge, a mere challenge to a 'principle' does not carry us far. For expediency, assumptions underlying a system have (1) to be discovered, (2) tested, (3) eventually challenged, (4) eventually rejected, and (5) a *system*, free from the eventually objectionable postulates, has to be built.

Examples of this abound in every field, but the histories of the non-euclidean and non-newtonian systems supply the simplest and most obvious illustrations. For instance, the fifth postulate of Euclid did not satisfy even his contemporaries, but these challenges were ineffective for more than two thousand years. Only in the nineteenth century was the fifth postulate eliminated and non-euclidean systems built without it. The appearance of such systems marked a profound revolution in human orientations. In the twentieth century the much more important 'principles' underlying our notions about the physical world, such as 'absolute simultaneity', 'continuity' of atomic processes, 'certainty' of our experiments and conclusions, etc., were challenged, and systems were then built without them. As a result, we now have the magnificent non-newtonian physics and world outlooks, based on the work of Einstein and the quantum pioneers.

Finally, for the first time in our history, some of the most important 'principles' of all principles, this time in the 'mental world', were challenged by mathematicians. For instance the universal validity of the so-called 'logical law of the excluded third' was questioned. Unfortunately, as yet, no full-fledged systems based on this challenge have been formulated, and so it remains largely inoperative, although the possibilities of some non-aristotelian, though elementalistic and unsatisfactory 'logics', are made obvious.

Further researches revealed that the *generality* of the 'law of the excluded third' is not an independent postulate, but that it is only an elementalistic consequence of a deeper, invariably false to facts principle of 'identity', often unconscious and consequently particularly pernicious. Identity is defined as 'absolute sameness in all respects', and it is this 'all' which makes identity impossible. If we eliminate this 'all' from the definition, then the word 'absolute' loses its meaning, we have 'sameness in some respects', but we have no 'identity', and only 'similarity', 'equivalence', 'equality', etc. If we consider that all we deal with represents constantly changing sub-microscopic, interrelated processes which are not, and cannot be 'identical with themselves', the old dictum that 'everything is identical with itself' becomes in 1933 a principle invariably false to facts.

Someone may say, 'Granted, but why fuss so much about it?' My answer would be, 'Identification is found in all known primitive peoples; in all known forms of "mental" ills; and in the great majority of personal, national, and international maladjustments. It is important, therefore, to eliminate such a harmful factor from our prevailing systems.' Certainly no one would care to contaminate his child with a dangerous germ, once it is known that the given factor is dangerous. Furthermore, the results of a complete elimination of identity are so far-reaching and beneficial for the daily life of everyone, and for science,*

* While correcting the proofs of this Preface, I read a telegraphic press report from London by Science Service, that Professor Max Born, by the applica-

that such 'fussing' is not only justified, but becomes one of the primary tasks before us. Anyone who will study the present work will be easily convinced by observations of human difficulties in life, and science, that the majority of these difficulties arise from necessary false evaluations, in consequence of the unconscious false to facts identifications.

The present work therefore formulates a system, called non-aristotelian, which is based on the complete rejection of identity and its derivatives, and shows what very simple yet powerful structural factors of sanity can be found in science. The experimental development of science and civilization invariably involves more and more refined discriminations. Each refinement means the elimination of some identifications somewhere, but many still remain in a partial and mostly unconscious form. The non-aristotelian system formulates the general problem of non-identity, and gives childishly simple non-elementalistic means for a complete and conscious elimination of identification, and other delusional or psychopathological factors in all known fields of human endeavours, in science, education, and all known phases of private, national, and international life. This work, in its application to education and psychotherapy, has been experimental for more than six years.

The volume is divided into three main divisions. Book I gives a general survey of non-aristotelian structural factors discovered by science, which are essential in a textbook. Only such data are selected, interpreted and evaluated as are necessary for a full mastery of the system. Book II presents a general introduction to non-aristotelian systems and general semantics free from identity, and gives a technique for the elimination of delusional factors from our psychological reactions. Book III gives additional structural data about languages, and also an outline of the essential structural characteristics of the empirical world, but only such as are pertinent for training in the non-aristotelian discipline.

Following each quotation prefacing each part and chapter, the number in parenthesis indicates the number of the book in the bibliography from which the quotation is taken.

---

tion of the *non-elementalistic* methods of Einstein, has succeeded in making a major contribution to the formulation of a unified field theory which now includes the quantum mechanics. Should this announcement be verified in its scientific aspects, our understanding of the structure of 'matter', 'electron', etc., would be greatly advanced, and would involve of course most important practical applications. For the semantic aspects of these problems, see pp. 378, 386 f., 541, 667, 698–701, and Chapter XXXIX.

I have tried to avoid footnotes as much as possible. The small numbers after some words in the text refer to the Notes on p. 763 ff., where the references to the bibliography are given.

Book II is largely self-contained and therefore can be read independently of the others, after the reader has become acquainted with the short tables of abbreviations given on pp. 15 and 16, and with Chapters II and IV. I believe, however, that for the best results the book should be read consecutively without stopping at passages which at first are not entirely clear, and read at least twice. On the second reading, passages which at first were not clear will become obvious because, in such a wide system, the beginning presupposes the end, and vice versa.

The discovery of such important and entirely general delusional factors in the older systems leads to a far-reaching revision of all existing disciplines. Because of modern complexities of knowledge this revision can only be accomplished by the activities of specialists working together as a group, and unified by one principle of non-identity, which necessitates a structural treatment.

To facilitate this most urgent need, and to present the results of this work to the public at reasonable prices, an International Non-aristotelian Library has been organized, to be printed and distributed by The Science Press Printing Company, Lancaster, Pennsylvania, U. S. A., and Grand Central Terminal, New York City.

It is also intended to organize an International Non-aristotelian Society with branches in connection with all institutions of learning throughout the world, where co-operative scientific work for the elimination of identity can be carried out, as this work is beyond the capacities of any one man.

Since the scope of the Library and Societies is international, I have accepted, in the main, the Oxford spelling and rules, which are a happy medium between the English used in the United States of America and that of the rest of the world. In certain instances I had to utilize some forms of expressions which are not entirely customary, but these slight deviations were forced upon me by the character of the subject, the need for clarity, and the necessity for cautiousness in generalizations. The revision of the manuscript and reading of the proofs in connection with other editorial and publishing duties has been a very laborious task for one man, and I only hope that not too many mistakes have been overlooked. Corrections and suggestions from the readers are invited.

The International Non-aristotelian Library is a non-commercial, scientific venture, and the interest and help of scientists, teachers, and those who are not indifferent to the advancement of science, civilization, sanity, peace, and to the improvement of social, economic, international, etc., conditions, will be greatly appreciated.

From one point of view, this enquiry has been independent; from another, much material has been adapted. In some instances it is impossible to give specific credit to an author, particularly in a text-book, and it is simpler and fairer to state that the works of Professors H. F. Biggs, G. Birtwistle, E. Bleuler, R. Bonola, M. Born, P. W. Bridgman, E. Cassirer, C. M. Child, A. S. Eddington, A. Einstein, A. Haas, H. Head, L. V. Heilbrunn, C. J. Herrick, S. E. Jelliffe, C. J. Keyser, C. I. Lewis, J. Loeb, H. Minkowski, W. F. Osgood, H. Piéron, G. Y. Rainich, B. Russell, C. S. Sherrington, L. Silberstein, A. Sommerfeld, E. H. Starling, A. V. Vasiliev, H. Weyl, W. A. White, A. N. Whitehead, E. B. Wilson, L. Wittgenstein and J. W. Young have been constantly consulted.

Although I have had no opportunity to use directly the fundamental researches of Doctor Henry Head on aphasia, and particularly on semantic aphasia, my whole work has been seriously influenced by his great contributions. Doctor Head's work, in connection with a non-elementalistic analysis, makes obvious the close connection between: (1) identifications; (2) structural ignorance; (3) lack of proper evaluations in general, and of the full significance of words and phrases in particular; and (4) the corresponding necessary, at least colloidal lesions of the nervous system.

I am under heavy obligations to Professors: E. T. Bell, P. W. Bridgman, C. M. Child, B. F. Dostal, M. H. Fischer, R. R. Gates (London), C. Judson Herrick, H. S. Jennings, R. J. Kennedy, R. S. Lillie, B. Malinowski (London), R. Pearl, G. Y. Rainich, Bertrand Russell (London), M. Tramer (Bern), W. M. Wheeler, H. B. Williams, W. H. Wilmer; and Doctors: C. B. Bridges, D. G. Fairchild, W. H. Gantt, P. S. Graven, E. L. Hardy, J. A. P. Millet, P. Weiss, W. A. White, Mr. C. K. Ogden (London), and Miss C. L. Williams, for reading the manuscript and/or the proofs as a whole, or in part, and for their invaluable criticism, and suggestions.

I also owe much to Doctor C. B. Bridges and Professor W. M. Wheeler, not only for their important criticisms and constructive suggestions, but also for their painstaking editorial corrections and interest.

Needless to say, I assume full responsibility for the following pages, the more, that I did not always follow the suggestions made.

I wish to express my deep appreciation to Doctor W. A. White and the staff of St. Elizabeth's Hospital, Washington, D. C., who, during my two years of study in the hospital, gave me every assistance to facilitate my research work there. I am indebted to Doctor P. S. Graven for supplying me with his as yet unpublished experimental clinical material, which was very useful to me.

Three important terms have been suggested to me; namely, 'envirogenetic' by Doctor C. B. Bridges, 'actional' by Professor P. W. Bridgman, and 'un-sane' by Doctor P. S. Graven, which debt I gladly acknowledge.

I am also deeply grateful to Professor R. D. Carmichael for writing Supplement I for this book on the Theory of Einstein, and to Doctor P. Weiss for his kind permission to reprint as Supplement II his article on the Theory of Types.

I warmly appreciate the kindness of those authors who gave me their permission to utilize their works.

During my twelve years of research work in the present subject and preparation of this volume I have been assisted by a number of persons, to whom I wish to express my appreciation. My particular appreciation is extended to my secretary, Miss Lily E. MaDan who, besides her regular work, made the drawings for the book; to Miss Eunice E. Winters for her genuine assistance in reading the proofs and compiling the bibliography; and to Mr. Harvey W. Culp for the difficult reading of the physico-mathematical proofs and the equally difficult preparation of the index.

The technical efficiency in all departments of the Science Press Printing Company, and the zealous and courteous co-operation of its compositors and officials, have considerably facilitated the publication of this book, and it is my pleasant duty to extend my thanks to them.

My heaviest obligations are to my wife, formerly Mira Edgerly. This work was difficult, very laborious, and often ungrateful, which involved the renouncing of the life of 'normal' human beings, and we abandoned much which is supposed to make 'life worth living'. Without her whole-hearted and steady support, and her relentless encouragement, I neither would have formulated the present system nor written the book which embodies it. If this book proves of any value, Mira Edgerly is in fact more to be thanked than the author. Without her interest, no non-aristotelian system, nor theory of sanity would have been produced in 1933.

A. K.

New York, August, 1933.

ACKNOWLEDGEMENT

The author and the publishers gratefully acknowledge the following permissions to make use of copyright material in this work:

Messrs. G. Allen and Unwin, London, for permission to quote from the works of Bertrand Russell.

The publishers of the Archives of Neurology and Psychiatry, for permission to quote from a paper by W. A. White.

Messrs. Blackie and Son, London and Glasgow, for permission to quote from the works of E. Schrödinger.

Messrs. Gebrüder Borntraeger, Berlin, for permission to utilize material from the work of L. V. Heilbrunn.

The Cambridge University Press, Cambridge, England, for permission to quote from the works of G. Birtwistle, A. S. Eddington, A. N. Whitehead, A. N. Whitehead and Bertrand Russell.

The Chemical Catalog Company, New York, for permission to quote from *Colloid Chemistry,* edited by J. Alexander.

Messrs. J. and A. Churchill, London, for permission to quote from the work of E. H. Starling.

Messrs. Constable and Company, London, for permission to utilize material from the works of A. Haas.

Messrs. Doubleday, Doran and Company, Garden City and New York, for permission to quote from the work of J. Collins.

Messrs. E. P. Dutton and Company, New York, for permission to quote from the works of C. J. Keyser.

The Franklin Institute, Philadelphia, Pa., for permission to quote from the paper of G. Y. Rainich printed in the Journal of the Franklin Institute.

Messrs. Harcourt, Brace and Company, New York, for permission to quote from the work of W. M. Wheeler.

Messrs. Henry Holt and Company, New York, for permission to quote from the works of C. M. Child, A. Einstein, and C. J. Herrick.

Messrs. Kegan Paul, Trench, Trubner and Company, London, and Messrs. Harcourt, Brace and Company, New York, for permission to quote from the works of C. K. Ogden and I. A. Richards, H. Piéron, and Bertrand Russell.

Messrs. Macmillan and Company, London and New York, for permission to quote from the works of L. Couturat, W. S. Jevons, J. Royce, and S. P. Thompson.

The Macmillan Company, New York and London, for permission to quote from the works of E. Bleuler, M. Bôcher, P. W. Bridgman, A. S. Eddington, E. V. McCollum, W. F. Osgood, and A. N. Whitehead.

Messrs. Methuen and Company, London, and Messrs. Dodd, Mead and Company, New York, for permission to quote from the works of A. Einstein, H. Minkowski, and H. Weyl.

Messrs. Methuen and Company, London, and Messrs. E. P. Dutton and Company, New York, for permission to quote from the works of M. Born, A. Haas, A. Sommerfeld, and H. Weyl.

The publishers of 'Mind', Cambridge, England, for permission to reprint the article of P. Weiss.

Sir John Murray, London, and Messrs. P. Blakiston's Son and Company, Philadelphia, Pa., for permission to quote from the work of W. D. Halliburton.

The Nervous and Mental Disease Publishing Company, Washington, D. C., for permission to quote from the works of S. E. Jelliffe and W. A. White.

Messrs. W. W. Norton and Company, New York, for permission to quote from the works of H. S. Jennings.

The Open Court Publishing Company, Chicago, Ill., for permission to quote from the works of R. Bonola, R. D. Carmichael, Bertrand Russell and J. B. Shaw.

The Oxford University Press, London and New York, for a ten dollar permission to quote from the works of I. P. Pavlov and H. F. Biggs.

The publishers of Physical Review, New York, for permission to quote from the paper of C. Eckart.

The Princeton University Press, Princeton, New Jersey, for permission to quote from the work of A. Einstein.

Messrs. G. P. Putnam's Sons, New York and London, for permission to utilize the works of J. Loeb.

The W. B. Saunders Company, Philadelphia, Pa., for permission to quote from the works of W. T. Bovie, A. Church and F. Peterson, and C. J. Herrick.

The Science Press, New York and Lancaster, Pa., for permission to quote from the work of H. Poincaré.

Messrs. Charles Scribner's Sons, New York, for permission to quote from the works of G. Santayana.

The University of California Press, Berkeley, Calif., for permission to quote from the work of C. I. Lewis.

The University of Chicago Press, Chicago, Ill., for permission to quote from the works of W. Heisenberg, C. J. Herrick, R. S. Lillie, and J. Loeb.

The publishers of the University of Washington Chapbooks, Seattle, Wash., for permission to quote from the work of E. T. Bell.

The Williams and Wilkins Company, Baltimore, Md., for permission to quote from the work of E. T. Bell.

In several instances I have quoted a few lines from other publications, without asking special permission. I now wish to express my gratitude to these respective publishers.

In all instances the sources, from which the quotations and the material used were taken, are explicitly indicated in the text of the book.

BOOK I

A General Survey of

Non-Aristotelian Factors

Allow me to express now, once and for all, my deep respect for the work of the experimenter and for his fight to wring significant facts from an inflexible Nature, who says so distinctly "No" and so indistinctly "Yes" to our theories. (550)   HERMANN WEYL

The firm determination to submit to experiment is not enough; there are still dangerous hypotheses; first, and above all, those which are tacit and unconscious. Since we make them without knowing it, we are powerless to abandon them. (417)   H. POINCARÉ

The empiricist . . . thinks he believes only what he sees, but he is much better at believing than at seeing. (461)   G. SANTAYANA

For a Latin, truth can be expressed only by equations; it must obey laws simple, logical, symmetric and fitted to satisfy minds in love with mathematical elegance.
The Anglo-Saxon to depict a phenomenon will first be engrossed in making a *model*, and he will make it with common materials, such as our crude, unaided senses show us them. . . . He concludes from the body to the atom.
Both therefore make hypotheses, and this indeed is necessary, since no scientist has ever been able to get on without them. The essential thing is never to make them unconsciously. (417)   H. POINCARÉ

If a distinction is to be made between men and monkeys, it is largely measurable by the quantity of the subconscious which a higher order of being makes conscious. That man really lives who brings the greatest fraction of his daily experience into the realm of the conscious.*
                                    MARTIN H. FISCHER

The thought of the painter, the musician, the geometrician, the tradesman, and the philosopher may take very different forms; still more so the thought of the uncultivated man, which remains rudimentary and revolves for ever in the same circles. (411)   HENRI PIÉRON

One need only open the eyes to see that the conquests of industry which have enriched so many practical men would never have seen the light, if these practical men alone had existed and if they had not been preceded by unselfish devotees who died poor, who never thought of utility, and yet had a guide far other than caprice. (417)   H. POINCARÉ

---

* Spinal Cord Education. *Ill. Med. Jour.* Dec., 1928.

1

The men most disdainful of theory get from it, without suspecting it, their daily bread; deprived of this food, progress would quickly cease, and we should soon congeal into the immobility of old China. (417) H. POINCARÉ

If one wishes to obtain a definite answer from Nature one must attack the question from a more general and less selfish point of view. (415)
                                                                                    M. PLANCK

In particular—if we use the word intelligence as a synonym for mental activity, as is often done—we must differentiate between the primitive forms of sensory intelligence, with their ill-developed symbolism beyond which backward children cannot advance, . . . and the forms of verbal intelligence created by social education, abstract and conceptual forms. (411)
                                                                            HENRI PIÉRON

A civilisation which cannot burst through its current abstractions is doomed to sterility after a very limited period of progress. (575)
                                                                    A. N. WHITEHEAD
. . . almost any idea which jogs you out of your current abstractions may be better than nothing. (575)   A. N. WHITEHEAD

That is precisely what common sense is for, to be jarred into uncommon sense. One of the chief services which mathematics has rendered the human race in the past century is to put 'common sense' where it belongs, on the topmost shelf next to the dusty canister labeled 'discarded nonsense.' (23)
                                                                            E. T. BELL

If you have had your attention directed to the novelties in thought in your own lifetime, you will have observed that almost all really new ideas have a certain aspect of foolishness when they are first produced. (575)
                                                                    A. N. WHITEHEAD

To know how to criticize is good, to know how to create is better. (417)
                                                                            H. POINCARÉ

The explanatory crisis which now confronts us in relativity and quantum phenomena is but a repetition of what has occurred many times in the past. . . . Every kitten is confronted with such a crisis at the end of nine days. (55)
                                                                            P. W. BRIDGMAN

The concept does not exist for the physicist until he has the possibility of discovering whether or not it is fulfilled in an actual case. . . . As long as this requirement is not satisfied, I allow myself to be deceived as a physicist (and of course the same applies if I am not a physicist), when I imagine that I am able to attach a meaning to the statement of simultaneity. (I would ask the reader not to proceed farther until he is fully convinced on this point.) (150)   A. EINSTEIN

Einstein, in thus analyzing what is involved in making a judgment of simultaneity, and in seizing on the act of the observer as the essence of the situation, is actually adopting a new point of view as to what the concepts of physics should be, namely, the operational view . . . if we had adopted the operational point of view, we would, before the discovery of the actual physical facts, have seen that simultaneity is essentially a relative concept, and would have left room in our thinking for the discovery of such effects as were later found. (55)   P. W. BRIDGMAN

Let any one examine in operational terms any popular present-day discussion of religious or moral questions to realize the magnitude of the reformation awaiting us. Wherever we temporize or compromise in applying our theories of conduct to practical life we may suspect a failure of operational thinking. (55) P. W. BRIDGMAN

I believe that many of the questions asked about social and philosophical subjects will be found to be meaningless when examined from the point of view of operations. It would doubtless conduce greatly to clarity of thought if the operational mode of thinking were adopted in all fields of inquiry as well as in the physical. Just as in the physical domain, so in other domains, one is making a significant statement about his subject in stating that a certain question is meaningless. (55) P. W. BRIDGMAN

There is a sharp disagreement among competent men as to what can be proved and what cannot be proved, as well as an irreconcilable divergence of opinion as to what is sense and what is nonsense. (22) E. T. BELL

Notice the word "nonsense" above. It was their inability, among other things, to define this word . . . that brought to grief the heroic attempt of Russell and Whitehead to put mathematical reasoning on a firm basis. (22) E. T. BELL

The structure of all linguistic material is inextricably mixed up with, and dependent upon, the course of the activity in which the utterances are embedded. (332) B. MALINOWSKI

To sum up, we can say that the fundamental grammatical categories, universal to all human languages, can be understood only with reference to the pragmatic Weltanshauung of primitive man, and that, through the use of Language, the barbarous primitive categories must have deeply influenced the later philosophies of mankind. (332) B. MALINOWSKI

Since no two events are identical, every atom, molecule, organism, personality, and society is an emergent and, at least to some extent, a novelty. And these emergents are concatenated in such a way as to form vast ramifying systems, only certain ideal sections of which seem to have elicited the attention of philosophers, owing to their avowedly anthropocentric and anthropodoxic interests. (555) WILLIAM MORTON WHEELER

The words *is* and *is not*, which imply the agreement or disagreement of two ideas, must exist, explicitly or implicitly, in every assertion. (354) AUGUSTUS DE MORGAN

The little word *is* has its tragedies; it marries and identifies different things with the greatest innocence; and yet no two are ever identical, and if therein lies the charm of wedding them and calling them one, therein too lies the danger. Whenever I use the word *is*, except in sheer tautology, I deeply misuse it; and when I discover my error, the world seems to fall asunder and the members of my family no longer know one another. (461) G. SANTAYANA

The complete attempt to deal with the term *is* would go to the form and matter of every thing in *existence*, at least, if not to the possible form and matter of all that does not exist, but might. As far as it could be done, it would give the grand Cyclopaedia, and its yearly supplement would be the history of the human race for the time. (354) AUGUSTUS DE MORGAN

Consciousness is the feeling of negation: in the perception of 'the stone as grey,' such feeling is in barest germ; in the perception of 'the stone as not grey,' such feeling is full development. Thus the negative perception is the triumph of consciousness. (578) A. N. WHITEHEAD

But, if we designate as intelligence, quantitatively, the totality of mental functioning, it is evident that the suppression of verbal thought involves a defect, relatively very important among cultivated individuals leading a complex social life: the uneducated person from this point of view is a defective. (411)   HENRI PIÉRON

The philosophy of the common man is an old wife that gives him no pleasure, yet he cannot live without her, and resents any aspersions that strangers may cast on her character. (461)   G. SANTAYANA

It is terrible to see how a single unclear idea, a single formula without meaning, lurking in a young man's head, will sometimes act like an obstruction of inert matter in an artery, hindering the nutrition of the brain, and condemning its victim to pine away in the fullness of his intellectual vigor and in the midst of intellectual plenty. (402)   CHARLES S. PEIRCE

PART I

PRELIMINARIES

Corpus Errorum Biologicorum

. . . .

But exactly the distinctive work of science is the modification, the reconstruction, the abandonment of old ideas; the construction of new ones on the basis of observation. This however is a distressing operation, and many refuse to undergo it; even many whose work is the practice of scientific investigation. The old ideas persist along with the new observations; they form the basis—often unconsciously—for many of the conclusions that are drawn.

This is what has occurred in the study of heredity. A burden of concepts and definitions has come down from pre-experimental days; the pouring of the new wine of experimental knowledge into these has resulted in confusion. And this confusion is worse confounded by the strange and strong propensity of workers in heredity to flout and deny and despise the observations of the workers in environmental action; the equally strange and strong propensity of students of environmental effects to flout and deny and despise the work on inheritance. If one accepts the affirmative results of both sets, untroubled by their negations, untroubled by definitions that have come from the past, there results a simple, consistent and useful body of knowledge; though with less pretentious claims than are set forth by either single set.

Our first fallacy springs from the situation just described. It is:

I.   The fallacy of non-experimental judgments, in matters of heredity and development. . . .

Our second general fallacy is one that appears in the interpretation of observational and experimental results; it underlies most of the special fallacies seen in genetic biology. This is the fallacy that Morley in his life of Gladstone asserts to be the greatest affliction of politicians; it is indeed a common plague of humanity. It is:

II.   The fallacy of attributing to one cause what is due to many causes. . . .

III.   The fallacy of concluding that because one factor plays a role, another does not; the fallacy of drawing negative conclusions from positive observations. . . .

IV.   The fallacy that the characteristics of organisms are divisible into two distinct classes; one due to heredity, the other to environment. . . .

VII.   The fallacy of basing conclusions on implied premises that when explicitly stated are rejected. . . .
Many premises influencing reasoning are of this hidden, unconscious type. Such ghostly premises largely affect biological reasoning on the topics here dealt with; they underlie several of the fallacies already stated, and several to come. . . .

VIII.   The fallacy that showing a characteristic to be hereditary proves that it is not alterable by the environment. . . .

5

IX. The fallacy that showing a characteristic to be altered by the environment proves that it is not hereditary. . . .

It appears indeed probable, from the present state of knowledge and the trend of discovery, that the following sweeping statements will ultimately turn out to be justified:—

(1) All characteristics of organisms may be altered by changing the genes; provided we can learn how to change the proper genes.

(2) All characteristics may be altered by changing the environmental conditions under which the organism develops; provided that we learn what conditions to change and how to change them.

(3) Any kind of change of characteristics that can be induced by altering genes, can likewise be induced (if we know how) by altering conditions. (This statement is open to more doubt than the other two; but it is likely eventually to be found correct.) . . .

X. The fallacy that since all human characteristics are hereditary, heredity is all-important in human affairs, environment therefore unimportant. . . .

XI. The fallacy that since all important human characteristics are environmental, therefore environment is all-important, heredity unimportant, in human affairs. (247)   H. S. JENNINGS

CHAPTER I

AIMS, MEANS AND CONSEQUENCES OF A
NON-ARISTOTELIAN REVISION

The process of intellectualism is not the subject I wish to treat: I wish to speak of science, and about it there is no doubt; by definition, so to speak, it will be intellectualistic or it will not be at all. Precisely the question is, whether it will be. (417)   H. POINCARÉ

The aim of science is to seek the simplest explanations of complex facts. . . . Seek simplicity and distrust it. (573)   A. N. WHITEHEAD

The present enquiry originated in my attempt to build a science of man. The first task was to define man scientifically in non-elementalistic, functional terms. I accomplished that in my book *Manhood of Humanity* (New York, E. P. Dutton & Co.), and in it I called the special characteristic which sharply distinguishes man from animal the time-binding characteristic.

In the present volume I undertake the investigation of the mechanism of time-binding. The results are quite unexpected. We discover that there is a sharp difference between the nervous reactions of animal and man, and that judging by this criterion, nearly all of us, even now, copy* animals in our nervous responses, which copying leads to the general state of un-sanity reflected in our private and public lives, institutions and systems. By this discovery the whole situation is radically changed. If we copy animals in our nervous responses through the lack of knowledge of what the appropriate responses of the human nervous system should be, we can stop doing so, both individually and collectively, and we are thus led to the formulation of a first positive theory of sanity.

The old dictum that we 'are' animals leaves us hopeless, but if we merely *copy* animals in our nervous responses, we can stop it, and the hopeless becomes very hopeful, provided we can discover a *physiological* difference in these reactions. Thus we are provided with a definite and promising program for an investigation.

Such an investigation is undertaken in the present volume.

The result of this enquiry turned out to be a non-aristotelian system, the first to be formulated, as far as I know, and the first to express the very scientific tendency of our epoch, which produced the non-euclidean and non-newtonian (Einstein's and the newer quantum theories) systems. It seems that these three, the non-aristotelian, non-euclidean and non-newtonian systems are as much interwoven and interdependent

---

*The use of the term 'copy' is explained in Chapter II.

as were the corresponding older systems. The aristotelian and the non-aristotelian systems are the more general, the others being only special and technical consequences arising from them.

Both the aristotelian and the non-aristotelian systems affect our lives deeply, because of psycho-logical factors and the immediacy of their application. Each is the expression of the psycho-logical tendencies of its period. Each in its period must produce in the younger generations a psycho-logical background which makes the understanding of its appropriate disciplines 'natural' and simple. In an aristotelian human world the euclidean and newtonian systems are 'natural', while the youth educated in the non-aristotelian habits will find the non-euclidean and non-newtonian systems simpler, more 'natural', and the older systems 'unthinkable'.

The functioning of the human nervous system is a more generalized affair than that of the animal, with more possibilities. The latter is a special case of the former, but not vice versa. John Smith, through ignorance of the mechanism, may use his nervous system as a Fido; but Fido cannot copy Smith. Hence, the danger for Smith, but not for Fido. Fido has many of his own difficulties for survival, but, at least, he has no *self-imposed* conditions, mostly silly and harmful, such as Smith has ignorantly imposed on himself and others. The field covered by this enquiry is very wide and involves unexpectedly special suggestive contributions in diverse branches of science. To list a few for orientation:

1. The formulation of General Semantics, resulting from a General Theory of Time-binding, supplies the scientists and the laymen with a general modern method of orientation, which eliminates the older psychological blockages and reveals the mechanisms of adjustment;

2. The departure from aristotelianism will allow biologists, physiologists, etc., and particularly medical men to 'think' in modern colloidal and quantum terms, instead of the inadequate, antiquated chemical and physiological terms. Medicine may then become a science in the 1933 sense;

3. In psychiatry it indicates on colloidal grounds the solution of the 'body-mind' problem;

4. It shows clearly that desirable human characteristics have a definite *psychophysiological* mechanism which, up till now, has been misused, to the detriment of all of us;

5. It gives the first definition of 'consciousness' in simpler physico-chemical terms;

6. A general theory of sanity leads to a general theory of psychotherapy, including all such existing medical schools, as they all deal with

disturbances of the *semantic reactions* (psycho-logical responses to words and other stimuli in connection with their *meanings*);

7. It formulates a physiological foundation for 'mental' hygiene, which turns out to be a most general preventive *psychophysiological* experimental method;

8. It shows the *psychophysiological* foundation of the childhood of humanity as indicated by the *infantilism* in our present private, public, and international lives;

9. In biology it gives a semantic and structural solution of the 'organism-as-a-whole' problem;

10. In physiology and neurology it reformulates to human levels the Pavlov theory of conditional reflexes, suggesting a new scientific field of *psychophysiology* for experiments;

11. In epistemology and semantics it establishes a definite non-elementalistic theory of meanings based not only on definitions but also on *undefined* terms;

12. It introduces a new development and use of 'structure';

13. It establishes structure as the only possible content of knowledge;

14. It discovers the *multiordinality* of the most important terms we have, thus removing the psycho-logical blockage of semantic origin and helping the average man or scientist to become a 'genius', etc.;

15. It formulates a new and physiological theory of mathematical types of extreme simplicity and very wide application;

16. It offers a non-aristotelian solution of the problem of mathematical 'infinity';

17. It offers a new non-aristotelian, semantic (from Greek, to signify) definition of *mathematics* and *number,* which clarifies the mysteries about the seemingly uncanny importance of number and measurement and throws a new light on the role, structural significance, meaning, and methods of mathematics and its teaching;

18. In physics, the enquiry explains some fundamental, but as yet disregarded, semantic aspects of physics in general, and of Einstein's and the new quantum theories in particular;

19. It resolves simply the problem of 'indeterminism' of the newer quantum mechanics, etc.

I realize that the thoughtful reader may be staggered by such a partial list. I am in full sympathy with him in this. I also was staggered.

As this enquiry claims to be scientific, in the 1933 sense, I must explain how, in spite of great difficulties and handicaps, I was able to accomplish the work. As my work progressed, it turned out to be 'speaking about speaking'. Now all scientific works in all fields are written or spoken, and so are ultimately verbal. In order to speak about speaking, in any satisfactory and fundamental 1933 sense, I had to become acquainted with the special languages used by scientists in all fields. I did not realize beforehand what a very serious undertaking this was. It took many years and much hard labour to accomplish it, but, once accomplished, the rest was simple. Scientists do not differ from the rest of us. They usually disregard entirely structural, linguistic, and semantic issues, simply because no one has, as yet, formulated these problems or shown their importance. The structural revision of their language led automatically to new results and new suggestions, and hence the surprising list.

The present enquiry is limited and partial, but because it deals with linguistic and semantic issues and their *physiological* and psycho-logical aspects, it is, as far as it goes, *unusually general*. I found that, in writing, it is extremely difficult and impracticable always to state explicitly the limitations of a statement. It seems most practical to say here that, in general, *all statements here made are limited* by further considerations of the actualities of an analysed problem.

Thus, for instance, a 'theory of sanity' deals with the most important semantic issues from limited semantic aspects, and has nothing to do with forms of 'insanity' arising from different organic, or toxic, or other disturbances, which remain as serious as ever. The statements made cover just as much as further investigations will allow them to cover—and no more.

The reader should be warned against undue generalizations, as they may be unjustified. It is impossible, at this stage, to foresee all the ramifications of the present work. The verbal issues, which correspond roughly to the older 'mental' issues, seem to pervade all *human* problems to some extent, and so the field of application and influence of any such enquiry must be very large. Most of the results of the present work involve factors of unusual security of conclusion, though they may violate canons of our 'philosophical' creeds.

The explanation is astonishingly simple and easily verified. The present non-aristotelian system is based on fundamental *negative* premises; namely, the complete denial of 'identity', which denial *cannot be denied* without imposing the burden of impossible proof on the person who denies the denial. If we start, for instance, with a statement that 'a word is *not* the object spoken about', and some one tries to deny that, he would have to produce an actual physical object which would *be the word,*—impossible of performance, even in asylums for the 'mentally' ill. Hence my security, often 'blasphemously cheerful', as one of my friends calls it.

This general denial of the 'is' of identity gives the main fundamental non-aristotelian premise, which necessitates a structural treatment. The status of negative premises is much more important and secure to start with than that of the positive 'is' of identity, found in the aristotelian system, but easily shown to be false to fact, and involving important delusional factors.

Any map or language, to be of maximum usefulness, should, in structure, be similar to the structure of the empirical world. Likewise, from the point of view of a theory of sanity, any system or language should, in structure, be similar to the structure of our nervous system. It is easily shown that the aristotelian system differs structurally from these minimal requirements, and that the non-aristotelian system is in accordance with them.

This fact turns out to be of *psychophysiological* importance. The above considerations, and others impossible to mention in this chapter, have suggested to me the form and structure of the whole work. I have spared no effort to make the presentation as connected, simple, and, particularly, as *workable* as I could. As I deal with structure, and similarity of structure, of languages and the empirical world, a definite selection of topics is immediately suggested. I must give enough structural data about languages in general, and enough structural data about the empirical world, and then select, or, if necessary, build, my terminology and system of similar structure.

The reader should not be afraid if some parts of the book look technical and mathematical. In reality, they are not so. Speaking of the language called mathematics, from a structural point of view, I have had to illustrate what was said, and the few symbols or diagrams are used only for that purpose. Many of the structural points are of genuine importance and interest to professional scientists, teachers, and others, who seldom, if ever, deal with such structural, linguistic, and semantic problems as are here analysed. The layman who will read the book diligently and repeatedly, without skipping any part of it, will get at least a *feeling* or vague notion that *such problems do exist,* which will produce a very important psycho-logical effect or release from the old animalistic unconditionality of responses, whether or not he feels that he has 'understood' them fully.

My earnest suggestion, backed by experience, to the reader is to read the book through several times, but not to dwell on points which are not clear to him. At each reading the issues will become clearer, until they will become entirely his own. Superficial reading of the book is to be positively discouraged, as it will prove to be so much time wasted. Experience teaches me that the number of semantic maladjustments, particularly among the white-collar class, is very large. At present, I do not know any case where a thorough *training* in such a non-aristotelian semantic discipline would not give very serious means for better adjustment. It will quiet down affective, semantic disturbances, sharpen orientation, judgement, the power of observation, and so forth; it will eliminate different psycho-logical blockages, help to overcome the very annoying and common 'inferiority' feelings; it will assist the outgrowing of the adult *infantile state,* which is a nervous deficiency practically always connected with some pathological sex-reactions or lack of normal and healthy impulses.

After all, we should not be surprised at this. Language, as such, represents the highest and latest physiological and neurological function of an organism. It is unique with Smith and of uniquely *human* circular structure, to use a logical term—or of spiral structure, to use a four-dimensional or a physico-chemical-aspect term. Before we can use the semantic nervous apparatus properly, we must first know how to use it, and so formulate this use.

In these processes an 'effect' becomes a *causative* factor for future effects, influencing them in a manner particularly subtle, variable, flexible, and of an endless number of possibilities. 'Knowing', if taken as an end-product, must be considered also as a causative psychophysiological factor of the next stage of the semantic response. The disregard of this mechanism is potentially of serious danger, particularly in the rearing of children, as it trains them in unanalysed linguistic habits, the more so since the human nervous system is not complete at birth. This structural and functional circularity introduces real difficulties in our analysis, disregarded or neglected in the aristotelian system. Human life, in its difference from animal life, involves many more factors and is inherently of different and more complex structure. Before we can be fully human and, therefore, 'sane', as a 'normal' human being should be, we must first know how to handle our nervous responses—a circular affair.

A non-aristotelian system must not disregard this *human-natural-history structural fact* of the inherent circularity of all physiological functions which in any form involve human 'knowing'. A non-aristotelian system differs essentially in structure from its predecessor, which, by necessity, through the lack of knowledge characteristic of its epoch, disregarded these structural semantic issues and so was constructed on cruder *animalistic patterns.*

The difficulty in passing from the old system to another of different structure is not in the non-aristotelian system as such, which is really simpler and more in accord with common sense; but the serious difficulty lies rather in the older habits of speech and nervous responses, and in the older semantic reactions which must be overcome. These difficulties are, perhaps, more serious than is generally realized. Only those who have experienced the passing from euclidean to non-euclidean, and from newtonian to non-newtonian systems can fully appreciate this semantic difficulty; as a rule, it takes a new generation to do it painlessly and with entire success. This applies to the general public, but is not an excuse for scientists, educators, and others who are entrusted with the education of, or who otherwise influence, the semantic reactions of children. If any reader realizes his difficulties and *seriously wants* to overcome them, another suggestion may be given. A structural diagram in the present work, called the Structural Differential, shows the structural difference between the world of animal and the world of man. This structural difference is not yet fully realized; neither is its semantic importance understood, as it has never been formulated in a simple way before; yet the permanent and instinctive realization of these structural differences is unconditionally necessary for the mastering of the present theory of sanity. This diagram, indeed, involves all the psychophysiological factors necessary for the transition from the old semantic reactions to the new, and it gives in a way a *structural summary* of the whole non-aristotelian system. As the diagram is based on the denial of the 'is' of identity, its use is practically indispensable; it has been made in relief and in printed forms, to be kept on the wall or the desk as a permanent visual structural and semantic reminder. Without actual handling, pointing the finger or waving the hand at it, seeing the *order,* and so on, it is practically impossible, or very difficult, to become *trained,* or to explain the present system to ourselves or others, because the foundation of all 'knowing' is structural, and the Structural Differential actually shows this structural difference between the world of animal and the world of man.

One of the best ways for grown-up persons to train themselves in the present theory of sanity is to try to explain it to others, repeatedly pointing to the Structural Differential. In my experience, those who have disregarded this advice have always made very slow progress, and have never got the full semantic benefit of their efforts. As regards the verbal side of the training, it is as important to use exclusively the terms

given in this book, which are non-aristotelian and non-elementalistic, as
it is to *abandon entirely* the 'is' of identity and some of the elementalistic primitive terms.

The reader should be warned from the beginning of a very fundamental semantic innovation; namely, of the discovery of the *multiordinality* of the most important terms we have. This leads to a conscious
use of these terms in the multiordinal, extremely flexible, full-of-con-
ditionality sense. Terms like 'yes', 'no', 'true', 'false', 'fact', 'reality',
'cause,' 'effect', 'agreement', 'disagreement', proposition', 'number', 'relation', 'order', 'structure', 'abstraction', 'characteristic', 'love', 'hate',
'doubt', etc., are such that if they can be applied to a statement they can
also be applied to a statement about the first statement, and so, ultimately,
to all statements, no matter what their order of abstraction is. Terms of
such a character I call *multiordinal terms*. The main characteristic of
these terms consists of the fact that on different levels of orders of abstractions they may have different meanings, with the result that they
have no general meaning; for their meanings are determined solely by the
given context, which establishes the different orders of abstractions.
Psycho-logically, in the realization of the multiordinality of the most important terms, we have paved the way for the specifically *human* full
conditionality of our semantic responses. This allows us great freedom
in the handling of multiordinal terms and eliminates very serious psycho-
logical fixities and blockages, which analysis shows to be animalistic in
their nature, and, consequently, pathological for man. Once the reader
understands this multiordinal characteristic, this semantic freedom does
not result in confusion.

Accidentally, our vocabulary is enormously enriched without becoming cumbersome, and is made very exact. Thus a 'yes' may have an
indefinite number of meanings, depending on the context to which it is
applied. Such a blank 'yes' represents, in reality, 'yes$_\infty$', but this includes
'yes$_1$', 'yes$_2$', 'yes$_3$', etc., all of which are, or may be, different. All speculations about such terms *in general*—as, for instance, 'what a fact or
reality is?'—are futile, and, in general, illegitimate, as the only correct
answer is that 'the terms are multiordinal and devoid of meaning outside
of a context'. This settles many knotty epistemological and semantic
questions, and gives us a most powerful method for promoting human
mutual freedom of expression, thus eliminating misunderstandings and
blockages and ultimately leading to agreement.

I suspect that without the discovery of the multiordinality of terms
the present work could not have been written, as I needed a more flexible
language, a larger vocabulary, and yet I had to avoid confusion. With

the introduction of the multiordinality of terms, which is a *natural* but, as yet, an unnoticed fact, our ordinary vocabulary is enormously enriched; in fact, the number of words in such a vocabulary *natural for man* is infinite. The multiordinality of terms is the fundamental mechanism of the *full conditionality* of *human* semantic reactions; it eliminates an unbelievable number of the old animalistic blockages, and is fundamental for sanity.

A number of statements in the present work have definite meanings for different specialists, often running entirely counter to the accepted scientific creeds. As they followed naturally from the context, I inserted them for the specialist, without warning, for which I have to apologize to the general reader, although they will be useful to him also.

To make issues sharper, some words will be repeated so often that I abbreviate them as follows:

Abbreviation	Stands for	Abbreviation	Stands for
A	aristotelian	$\bar{N}$	non-newtonian
$\bar{A}$	non-aristotelian	*el*	elementalistic
E	euclidean	*non-el*	non-elementalistic
$\bar{E}$	non-euclidean	*m.o* or *(m.o)*	multiordinal
N	newtonian	*s.r* or *(s.r)*	semantic reactions, both singular and plural

In some instances, for special emphasis, the words will be spelled in full.

A $\bar{A}$-system, being extensional, requires the enumeration of long lists of names, which, in principle, cannot be exhausted. Under such conditions, I have to list a few representatives followed by an 'etc.', or its equivalents. As the extensional method is characteristic of a $\bar{A}$ treatment, the expression 'etc.' occurs so often as to necessitate a special $\bar{A}$ extensional punctuation whenever the period does not indicate another abbreviation, as follows:

Abbreviation	Stands for	Abbreviation	Stands for
.,	etc.,	.:	etc.:
,.	,etc.	.?	etc.?
.;	etc.;	.!	etc.!

This book is intended as a handbook, and I have avoided referring the reader to other books, but have given as much of structural data as I deemed useful for a general orientation. In a work of such wide scope and novelty, it seemed desirable to give a general outline rather than to elaborate in detail on some particular points, so that this work is not exhaustive in any field; nor, at present, can it be.

The notes at the end of the book are given for the purposes (among others) of indicating sources of information, as an acknowledgement, and to facilitate the work of the future student. As much as I could, I have avoided direct quotations from other authors, because usually it has seemed more expedient to change the wording slightly. In many instances, I have followed the original wordings very closely, always giving the proper credit.

I have not avoided repetitions, because I have found, through sad experience, that many times, when I was reproached for a repetition, the hearer or reader was disregarding quite happily and unconsciously the said 'repetition', as if he had never heard it before. For such a work as the present one, the standard literary habits—'avoid repetitions', 'let the reader discover it for himself'., are extremely detrimental to the understanding of a few fundamental issues and to the acquiring of $\bar{A}$ habits and new $s.r.$ To facilitate the student's task, I had no other choice than to write as I did.

In 1933, scientific opinion is divided as to whether we need more science or less science. Some prominent men even suggest that scientists should take a vacation and let the rest of mankind catch up with their achievements. There seems no doubt that the discrepancy between human adjustments and the advances of science is becoming alarming. Is, then, such a suggestion justified?

The answer depends on the *assumptions* underlying such opinions. If humans, as such, have reached the limit of their nervous development, and if the scientific study of man, as man, should positively disclose these limitations, then such a conclusion would be justified. But is this the case?

The present investigation shows most emphatically that this is not the case. All sciences have progressed exclusively because they have succeeded in establishing their own A languages. For instance, a science of thermodynamics could not have been built on the terms of 'cold' and 'warm'. Another language, one of relations and structure, was needed; and, once this was produced, a science was born and progress secured. Could *modern* mathematics be built on the Roman notation for numbers —I, II, III, IV, V.? No, it could not. The simplest and most child-like arithmetic was so difficult as to require an expert; and all progress was very effectively hampered by the symbolism adopted. History shows that only since the unknown Hindu discovered the most revolutionary and modern principle of *positional notation*—1, 10, 100, 1000., modern mathematics has become possible. Every child today is more skilful in his arithmetics than the experts of those days. Incidentally, let us notice that positional notation has a definite *structure*.

Have we ever attempted anything similar in the study of man? As-a-whole? In all his activities? Again, most emphatically, No! We have never studied man-as-a-whole scientifically. If we make an attempt, such as the present one, for instance, we discover the astonishing, yet simple, fact that, even now, we all copy animals in our nervous responses, although these can be brought to the human level if the difference in the mechanism of responses is discovered and formulated.

Once this is understood, we must face another necessity. To abolish the discrepancy between the advancement of science and the power of adjustment of man, we must first establish the science of man-as-a-whole, embracing *all* his activities, science, mathematics and 'mental' ills *not* excluded. Such an analysis would help us to discover the above-mentioned difference in responses, and the $s.r$ in man would acquire new significance.

If the present work has accomplished nothing more than to suggest such possibilities, I am satisfied. Others, I hope, will succeed where I may have failed. Under such conditions, the only feasible resort is to produce a science of man, and thus have not less, but more, science, ultimately covering all fields of human endeavour, and thereby putting a stop to the animalistic nervous reactions, so vicious in their effects on man.

At present, nowhere in the world are there such *psychophysiological* researches being made. There are large sums of money invested in different well-established institutions for scientific research, for 'mental' hygiene, for international peace, and so forth, but not for what is possibly the most important of all lines of research; namely, a general

2

science of man in all aspects of his behaviour, science, mathematics, and 'mental' ills included.

It is to be hoped that, in the not-too-distant future, some individuals and universities will awaken to the fact that language is a fundamental *psychophysiological* function of man, and that a scientific investigation of man in *all* his activities, is a necessary, pressing, very promising, and practical undertaking. Then, perhaps, special chairs will be established in universities, and some such researches in *semantic reactions* and *sanity* will command as much interest and public encouragement as other scientific investigations.

I, personally, have no doubt that this would mark the beginning of a new era, *the scientific era,* in which all human desirable characteristics would be released from the present animalistic, psychophysiological, *A* semantic blockages, and that sanity would prevail.

That this is not a dream, and that such nervous mechanisms producing blockages do exist, has been demonstrated by Pavlov on his dogs, by all psychotherapy, and the experiments now being made on the elimination of the disturbances of the *s.r.* The abundance of geniuses among younger physicists, since the einsteinian structural revolution and semantic release, is also an important empirical evidence that different man-made verbal systems can stimulate or hamper the functioning of the human nervous system.

What has been said here has very solid structural, neurological foundations. For our purpose, we may consider a rough structural difference between the nervous systems of man and animal. Briefly, we can distinguish in the brain two kinds of nervous fibres, the radiating projection fibres and the tangential correlation and association fibres. With the increase of complexities and modifiability of the behaviour, we find an increased number and more complex interrelations of association fibres. The main difference, for instance, between the brain of a man and the brain of a higher ape is found not in the projection apparatus, but in the association paths, which are enormously enlarged, more numerous, and more complex in man than in any animal. Obviously, if these association paths are blocked to the passage of nervous impulses by some psychophysiological process, the reactions of the individual must be of a lower order, and such blockage must give the effect of the given individual's being organically deficient, and must, therefore, result in animalistic behaviour.

The present investigation discloses that the *s.r* may assume very diversified forms, one of which is the production of very powerful psychophysiological blockages. These, when once we understand their mechanism, can be eliminated by proper education and training in appropriate *s.r.*

CHAPTER II

TERMINOLOGY AND MEANINGS

The representation of mental phenomena in the form of reactions, conditioned reflexes, Bechterew's 'psycho-reflexes,' leads to a truly physiological schematization. . . . (411)   HENRI PIÉRON

Now I claim that the Ethnographer's perspective is the one relevant and real for the formation of fundamental linguistic conceptions and for the study of the life of languages, whereas the Philologist's point of view is fictitious and irrelevant. . . . To define Meaning, to explain the essential grammatical and lexical characters of language on the material furnished by the study of dead languages, is nothing short of preposterous in the light of our argument. (332)   B. MALINOWSKI

If he contend, as sometimes he will contend, that he has defined all his terms and proved all his propositions, then either he is a performer of logical miracles or he is an ass; and, as you know, logical miracles are impossible. (264)   CASSIUS J. KEYSER

Finally, in semantic aphasia, the full significance of words and phrases is lost. Separately, each word or each detail of a drawing can be understood, but the general significance escapes; an act is executed upon command, though the purpose of it is not understood. Reading and writing are possible as well as numeration, the correct use of numbers; but the appreciation of arithmetical processes is defective. . . . A general conception cannot be formulated, but details can be enumerated. (411)
HENRI PIÉRON
Moreover, the aphasic patient in his mode of life, in his acts and in all his behaviour may seem biologically and socially normal. But he has nevertheless suffered an unquestionable loss, for he no longer has any chance of undergoing further modifications of social origin, and of reacting in his turn as a factor in evolution and progress. (411)   HENRI PIÉRON

Particularly it expresses that most important step in the treatment, the passing over from a mere intellectual acceptance of the facts of the analysis, whether in interpretation of the underlying complexes or in recognition of the task to be accepted, to an emotional appreciation and appropriation of the same. *Intellectual acceptance can work no cure* but may prove seriously misleading to the patient who is attempting to grasp the situation and to the beginner in analysis as well. (241)   SMITH ELY JELLIFFE

Section A. On semantic reactions.

The term *semantic reaction* is fundamental for the present work and *non-elementalistic systems*. The term 'semantic' is derived from the Greek *semantikos,* 'significant', from *semainein* 'to signify', 'to mean', and was introduced into literature by Michel Bréal in his *Essai de Sémantique.* The term has been variously used in a more or less general or restricted sense by different writers. Of late, this term has been used by the Polish School of Mathematicians, and particularly L. Chwistek (see Supplement III), A. F. Bentley[1], and has been given a medical application by Henry Head[2] in the study of different forms of Aphasias. 'Aphasia', from the Greek *aphasia,* 'speechlessness', is used to describe disorders in comprehension or expression of written and spoken language which result from lesions of the brain. Disturbances of the semantic reactions in connection with faulty education and ignorance must be considered in 1933 as sub-microscopic colloidal lesions.

Among the many subdivisions of the symbolic disturbance, we find *semantic aphasia,* to be described (after Head) as the want of recognition or the full significance or intention of words and phrases, combined with the loss of power of appreciating the 'ultimate or non-verbal meaning of words and phrases' to be investigated presently, and the failure to recognize the intention or goal of actions imposed upon the patient.

The problems of 'meaning' are very complex and too little investigated, but it seems that 'psychologists' and 'philosophers' are not entirely in sympathy with the attitude of the neurologists. It is necessary to show that in a $\bar{A}$-system, which involves a new theory of meanings based on *non-el* semantics, the neurological attitude toward 'meaning' is the only structurally correct and most useful one.

The explanation is quite simple. We start with the negative $\bar{A}$ premise that words are *not* the un-speakable objective level, such as the actual objects outside of our skin *and* our personal feelings inside our skin. It follows that the only link between the objective and the verbal world is exclusively structural, necessitating the conclusion that the only content of all 'knowledge' is structural. Now structure can be considered as a complex of relations, and ultimately as multi-dimensional order.

From this point of view, all language can be considered as names either for un-speakable entities on the objective level, be it things or feelings, or as *names for relations.* In fact, even objects, as such, *could* be considered as relations between the sub-microscopic events and the human nervous system. If we enquire what the last relations represent, we find that an object represents an abstraction of low order produced by our nervous system as the result of the sub-microscopic events acting as stimuli upon the nervous system. If the objects represent abstractions of some order, then, obviously, when we come to the enquiry as to language, we find that words are still higher abstractions from objects. Under such conditions, a theory of 'meaning' looms up naturally. If the objects, as well as words, represent abstractions of different order, an individual, A, cannot know what B abstracts, unless B tells him, and so the 'meaning' of a word *must* be given by a definition. This would lead to the dictionary meanings of words, provided we could define all our words. But this is impossible. If we were to attempt to do so, we should soon find that our vocabulary was exhausted, and we should reach a set of terms which could not be any further defined, from lack of words. We thus see that all linguistic schemes, if analysed far enough, would depend on a set of undefined terms. If we enquire about the 'meaning' of a word, we find that it depends on the 'meaning' of other words used in defining it, and that the eventual new relations posited between them ultimately depend *on the m.o meanings of the undefined terms,* which, at a given period, cannot be elucidated any further.

Naturally, any fundamental theory of 'meaning' cannot avoid this issue, which must be crucial. Here a semantic experiment suggests itself. I have performed this experiment repeatedly on myself and others, invariably with similar results. Imagine that we are engaged in a friendly serious discussion with some one, and that we decide to enquire into the meanings of words. For this special experiment, it is not necessary to be very exacting, as this would enormously and unnecessarily complicate the experiment. It is useful to have a piece of paper and a pencil to keep a record of the progress.

We begin by asking the 'meaning' of every word uttered, being satisfied for this purpose with the roughest definitions; then we ask the 'meaning' of the words used in the definitions, and this process is continued usually for no more than ten to fifteen minutes, until the victim begins to speak in circles—as, for instance, defining 'space' by 'length' and 'length' by 'space'. When this stage is reached, we have come usually to the *undefined* terms of a given individual. If we still press, no matter how gently, for definitions, a most interesting fact occurs. Sooner or later, signs of *affective disturbances* appear. Often the face reddens; there is a bodily restlessness; sweat appears—symptoms quite similar to those seen in a schoolboy who has forgotten his lesson, which he 'knows but cannot tell'. If the partner in the experiment is capable of self-observation, he invariably finds that he feels an internal *affective pressure,* connected, perhaps, with the rush of blood to the brain and probably best expressed in some such words as 'what he "knows" but cannot tell', or the like. Here we have reached the bottom and the foundation of all *non-elementalistic meanings*—the meanings of *undefined terms,* which we 'know' somehow, but cannot tell. In fact, we have reached the un-speakable level. This 'knowledge' is supplied by the lower nerve centres; it represents affective first order effects, and is interwoven and interlocked with other affective states, such as those called 'wishes', 'intentions', 'intuitions', 'evaluation', and many others. It should be noticed that these first order effects have an objective character, as they are un-speakable—are *not* words.

'Meaning' must be considered as a multiordinal term, as it applies to all levels of abstractions, and so has no general content. We can only speak legitimately of 'meanings' in the plural. Perhaps, we can speak of the meanings of meanings, although I suspect that the latter would represent the un-speakable first order effect, the affective, personal raw material, out of which our ordinary meanings are built.

The above explains structurally why most of our 'thinking' is to such a large extent 'wishful' and is so strongly coloured by affective factors. Creative scientists know very well from observation of themselves, that all creative work starts as a 'feeling', 'inclination', 'suspicion', 'intuition', 'hunch', or some other un-speakable affective state, which only at a later date, after a sort of nursing, takes the shape of a verbal expression, worked out later in a rationalized, coherent, linguistic scheme called a theory. In mathematics we have some astonishing examples of intuitively proclaimed theorems, which, at a later date, have been proven to be true, although the original proof was false.

The above explanation, as well as the neurological attitude toward 'meaning', as expressed by Head, is *non-elementalistic*. We have not illegitimately split organismal processes into 'intellect' and 'emotions'. These processes, or the reactions of the organism-as-a-whole, can be contemplated at different neurological stages in terms of order, but must never be split or treated as separate entities. This attitude is amply justified structurally and empirically in daily and scientific life. For instance, we may assume that educated Anglo-Saxons are familiar with the Oxford Dictionary, although it must be admitted that they are handicapped in the knowledge of their language by being born into it; yet we know from experience how words which have one standard definition carry different meanings to, and produce different affective individual reactions on, different individuals. Past experiences, the knowledge., of different individuals are different, and so the *evaluation* (affective) of the terms is different. We are accustomed to such expressions as 'it means nothing to me', even in cases when the dictionary wording is accepted; or 'it means a great deal to me', and similar expressions which indicate that the meanings of meanings are somehow closely related to, or perhaps represent, the first order un-speakable affective states or reactions.

Since 'knowledge', then, is not the first order un-speakable objective level, whether an object, a feeling.; structure, and so relations, becomes the only possible content of 'knowledge' *and of meanings.* On the lowest level of our analysis, when we explore the objective level (the un-speakable feelings in this case), we must try to define every 'meaning' as a conscious feeling of actual, or assumed, or wished., *relations* which pertain to first order objective entities, psycho-logical included, and which can be evaluated by personal, varied, and racial—again un-speakable first order—psychophysiological effects. Because relations can be defined as multi-dimensional order, both of which terms are *non-el,* applying to 'senses' and 'mind', after *naming* the un-speakable entities, all experience can be *described* in terms of relations or multi-dimensional order. The meanings of meanings, in a given case, in a given individual at a given moment., represent composite, affective psycho-logical configurations of all relations pertaining to the case, coloured by past experiences, state of health, mood of the moment, and other contingencies.

If we consistently apply the organism-as-a-whole principle to any psycho-logical analysis, we must conjointly contemplate at least both aspects, the 'emotional' and the 'intellectual', and so *deliberately ascribe* 'emotional' factors to any 'intellectual' manifestation, and 'intellectual' factors to any 'emotional' occurrence. That is why, on human levels, the *el* term 'psychological' must be abolished and a new term *psycho-logical* introduced, in order that we may construct a science.

From what has been said, we see that not only the structure of the world is such that it is made up of absolute individuals, but that mean-ings in general, and the meanings of meanings in particular—the last representing probably the un-speakable first order effects—also share, in common with ordinary objects, the absolute individuality of the objective level.

The above explains why, by the inherent structure of the world, life, and the human nervous system, human relations are so enormously complex and difficult; and why we should leave no stone unturned to discover beneath the varying phenomena more and more general and invariant foundations on which human understanding and agreement may be based. In mathematics we find the only model in which we can study the invariance of relations under transformations, and hence the need for future psycho-logicians to study mathematics.

It follows from these considerations that any psycho-logical occur-rence has a number of aspects, an 'affective', and an 'intellectual', a physiological, a colloidal, and what not. For the science of psychophysiol-ogy, resulting in a theory of sanity, the above four aspects are of most importance. As our actual lives are lived on objective, un-speakable levels, and not on verbal levels, it appears, as a problem of evaluation, that the objective level, including, of course, our un-speakable feelings, 'emotions'., is the most important, and that the verbal levels are only auxiliary, sometimes useful, but at present often harmful, because of the disregard of the *s.r.* The role of the auxiliary verbal levels is only ful- filled if these verbal processes are translated back into first order effects. Thus, through verbal intercourse, in the main, scientists discover useful first order abstractions (objective), and by verbal intercourse again, *culture* is built; but this only when the verbal processes affect the un- speakable psycho-logical manifestations, such as our feelings, 'emotions',.

Some extraordinary parrot could be taught to repeat all the verbal 'wisdom' of the world; but, if he survived at all, he would be just a par- rot. The repeated noises would not have affected his first order effects— his affects—these noises would 'mean' nothing to him.

Meanings, and the meanings of meanings, with their inseparable affective components, give us, therefore, not only the *non-elementalistic* foundation on which all civilization and culture depends, but a study of the *non-el* mechanisms of meanings, through psychophysiology and general semantics, gives us, also, powerful physiological means to achieve a host of desirable, and to eliminate a large number of undesirable, psycho-logical manifestations.

The physiological mechanism is extremely simple and necessitates a breaking away from the older elementalism. But it is usually very diffi- cult for any given individual to break away from this older elementalism, as it involves the established *s.r*, and to be effective is, by necessity, a little laborious. The working tool of psychophysiology is found in the *semantic reaction.* This can be described as the psycho-logical reaction of a given individual to words and language and other symbols and events *in connection with their meanings,* and the psycho-logical reac- tions, which *become meanings and relational configurations* the moment the given individual begins to analyse them or somebody else does that for him. It is of great importance to realize that the term 'semantic' is *non-elementalistic,* as it involves conjointly the 'emotional' as well as the 'intellectual' factors.

From the *non-el* point of view, any affect, or impulse, or even human instinct, when made conscious acquires *non-el* meanings, and becomes ultimately a psycho-logical configuration of desirable or undesir- able to the individual relations, thus revealing a workable *non-el* mechan- ism. Psychotherapy, by making the unconscious conscious, and by ver- balization, attempts to discover meanings of which the patient was not aware. If the attempt is successful and the individual meanings are revealed, these are usually found to belong to an immature period of evaluation in the patient's life. They are then consciously revised and rejected, and the given patient either improves or is entirely relieved. The condition for a successful treatment seems to be that the *processes should be managed in a non-elementalistic* way. Mere verbal formalism is not enough, because the full *non-elementalistic* meanings to the patient are not divulged; consequently, in such a case, the *s.r* are not affected, and the treatment is a failure.

The *non-el* study of the *s.r* becomes an extremely general scientific discipline. The study of relations, and therefore order, reveals to us the mechanism of *non-el* meanings; and, in the application of an ordinal *physiological* discipline, we gain psychophysiological means by which powerfully to affect, reverse, or even annul, undesirable *s.r*. In psychophysiology we find a *non-el* physiological theory of meanings and sanity.

From the present point of view, all affective and psycho-logical responses to words and other stimuli *involving meanings* are to be considered as *s.r*. What the relation between such responses and a corresponding persistent psycho-logical *state* may be, is at present not clear, although a number of facts of observation seem to suggest that the re-education of the *s.r* results often in a beneficial change in some of these states. But further investigation in this field is needed.

The realization of this difference is important in practice, because most of the psycho-logical manifestations may appear as evoked by some event, and so are to be called responses or reflexes. Such a response, when lasting, should be called a given *state,* perhaps a semantic state, but not a semantic reflex. The term, 'semantic reaction', will be used as covering both semantic reflexes and states. In the present work, we are interested in *s.r*, from a psychophysiological, theoretical and experimental point of view, which include the corresponding states.

If, for instance, a statement or any event evokes some individual's attention, or one train of associations in preference to another, or envy, or anger, or fear, or prejudice., we would have to speak of all such responses on psycho-logical levels as *s.r*. A stimulus was present, and a response followed; so that, by definition, we should speak of a reaction. As the active factor in the stimulus was the individual meanings to the given person, and his response had meanings to him as a first order effect, the reaction must be called a *semantic* reaction.

The present work is written entirely from the *s.r* point of view; and so the treatment of the material, and the language used, imply, in general, a psycho-logical response to a stimulus in connection with meanings, this response being expressed by a number of such words as 'implies', 'follows', 'becomes', 'evokes', 'results', 'feels', 'reacts', 'evaluates', and many others. All data taken from science are selected, and only those which directly enter as factors in *s.r* are given in an elementary outline. The meanings to the individual are dependent, through the influence of the environment, education, languages and their structure, and other factors, on racial meanings called science, which, to a large extent, because of the structural and relational character of science, become physiological semantic factors of the reactions. In fact, science, mathematics, 'logic'., may be considered from a *non-elementalistic* point of view as *generalized* results of *s.r* acceptable to the majority of informed and not heavily pathological individuals.

To facilitate the writing and the reading of the work, I am compelled to use definite devices. As in case of structure, multiordinal terms, so in the case of *s.r*, I often employ an ordinary form of expression and use the words 'structur*al*', 'multiordin*al*', 'semantic', as adjectives, or 'structurally', 'semantically'., as adverbs, always implying the full meanings, that under such and such conditions of a given stimulus, the given *s.r* would be such and such. In many instances, the letters *s.r* or (*s.r*) will be inserted to remind the reader that we deal with semantic reactions or the psycho-logical reactions in connection with the meanings of the problems analysed. It is not only useful, but perhaps essential, that the reader should stop in such places and try to evoke in himself the given *s.r*. The present work leads to new *s.r* which are beneficial to every one of us and fundamental for sanity. The casual reading of the present book is not enough. Any one who wants the full or partial benefit of the joint labours of the author and the reader must, even in the reading, begin to re-train his *s.r*.

As the organism works as-a-whole, and as the training is psycho-physiological in terms of order, reversing the reversed pathological order., organism-as-a-whole means *must* be employed. For this purpose, the Structural Differential has been developed. The reader will later understand that it is practically impossible to achieve, without its help, the maximum beneficial semantic results.

From a *non-el* point of view, which makes illegitimate any *el* verbal splitting of 'emotions' and 'intellect'., these processes must be analysed in terms of order, indicating the stages of the psycho-neural process-as-a-whole. Empirically, there is a difference between an 'emotion' which becomes 'rationalized' and 'emotions' invoked or produced by 'ideas'. The order is different in each instance, and if, in a given nervous system, at a given moment, or under some special conditions, the lower or higher nerve centres work defectively, the nervous reactions are not well balanced and the manifestations acquire a one-sided character. The other aspect is not abolished, but is simply less prominent or less effective. Thus, in morons, imbeciles, and in many forms of infantilism, the 'thinking' is very 'emotional' and of a low grade; in so-called 'moral imbeciles', and perhaps in 'schizophrenia', the 'thinking' may be seemingly 'normal', yet it does not affect the 'feelings', which are deficient.

From the *non-el* semantic *human* point of view, any affect only gains meanings when it is conscious; or, in other words, when an actual or assumed set of relations is present. In an ideally balanced and efficient human nervous system, the 'emotions' would be translated into 'ideas', and 'ideas' translated into 'emotions', *with equal facility*. In other words, the $s.r$ of a given individual would be under full control and capable of being educated, influenced, transformed quickly and efficiently —the very reverse of the present situation. The present enquiry shows that the lack of psychophysiological methods for training and lack of analysis and understanding of the factors involved, are responsible for this deplorable situation.

The above processes are quite obvious on *racial* grounds, if we study science and mathematics from the semantic point of view. With very few exceptions, we only fail individually. For instance, a Euclid and a Newton had 'hunches', 'intuitions'.; then they rationalized and verbalized them and so affected the rest of us and established the 'natural' feeling for E geometries, N mechanics,. When new $\bar{E}$ or $\bar{N}$ systems were produced, many of the older scientists could 'understand' them, could even master the new symbolic technique; yet their 'feelings'., were seldom affected. They 'thought' in the new way, but they continued to 'feel' in the old; their $s.r$ did not follow fully the transformation of their 'ideas', and this produced a split personality.

Any fundamentally new system involves new $s.r$; and this is the main difficulty which besets us when we try to master a new system. We must re-educate, or change, our older $s.r$. As a rule, the younger generation, which began with the new $s.r$, has no such difficulties with the new systems. Just the opposite—the older $s.r$ become as difficult or impossible to them as the new were to the older generation. To both generations, with their corresponding $s.r$, the non-habitual $s.r$ are 'new', no matter what their historical order and how difficult or how simple they are. However, there is an important difference. The newer systems, as, for instance, the $\bar{E}$, $\bar{N}$, and the present corresponding $\bar{A}$-system, are *more general:* which means that the newer systems include the older as particular cases, so that the younger generation has $s.r$ which are *more flexible, more conditional,* with a broader outlook., semantic conditions absent in the older systems.

The problems connected with the *s.r* are not new, because these are inherent in man, no matter on what low or primitive level or on what high level of development he may be; but, up to the undertaking of the present analysis, the problems of *s.r* were not formulated, their psychophysiological mechanisms were not discovered, and so, to the detriment of all of us, we have had no workable educational means by which to handle them effectively.

That is why the passing from one era to another is usually so difficult and so painful. The new involves new *s.r*, while, as a rule, the older generations have enforced their systems, and, through them, by means of controlled education and linguistic structure and habits, the old *s.r*. This the younger generation, *always* having more racial experience, cannot accept, so that revolutions, scientific or otherwise, happen, and, when successful, the new systems are imposed on the older generation without the older generation's changing their *s.r*. All of which is painful to all concerned. The next generation after such a 'revolution' does not have similar difficulties, because from childhood they are trained in the new *s.r*, and all appears as 'natural' to them, and the older as 'unthinkable', 'silly',.

As a descriptive fact, the present stage of human development is such that with a very few exceptions our nervous systems do not work properly in accordance with their survival structure. In other words, although we have the potentialities for correct functioning in our nervous system, because of the neglect of the physiological control-mechanism of our *s.r*, we have semantic blockages in our reactions, and the more beneficial manifestations are very effectively prevented.

The present analysis divulges a powerful mechanism for the control and education of *s.r*; and, by means of proper evaluation, a great many undesirable manifestations on the psycho-logical level can be very efficiently transformed into highly desirable ones. In dealing with such a fundamental experimental issue as the *s.r*, which have been with us since the dawn of mankind, it is impossible to say new things all the time. Very often the issues involved become 'common sense'; but what is the use, in practice, of this 'common sense', if it is seldom, if ever, applied, and in fact cannot be applied because of the older lack of workable psychophysiological formulations? For instance, what could be simpler or more 'common sense' than the $\bar{A}$ premise that an object is *not* words; yet, to my knowledge, no one *fully* applies this, or has *fully acquired* the corresponding *s.r*. Without first acquiring this new *s.r*, it is impossible to discover this error and corresponding *s.r* in others; but as soon as we have trained ourselves, it becomes so obvious that it is impossible to miss it. We shall see, later, that the older $s.r$ were due to the lack of structural investigations, to the old structure of language, to the lack of consciousness of abstracting, to the low order conditionality of our conditional reactions (the semantic included), and a long list of other important factors. All scientific discoveries involve $s.r$, and so, once formulated, and the new reactions acquired, the discoveries become 'common sense', and we often wonder why these discoveries were so slow in coming in spite of their 'obviousness'. These explanations are given because they also involve some $s.r$; and we must warn the reader that such evaluations ($s.r$) : 'Oh, a platitude!', 'A baby knows that', are very effective $s.r$ *to prevent* the acquisition of the new reactions. This is why the 'discovery of the obvious' is often so difficult; it involves very many of semantic factors of new evaluation and meanings.

A fuller evaluation is only reached at present on racial grounds in two or more generations, and never on individual grounds; which, of course, for *personal generalized adjustment* and happiness, is very detrimental. Similarly, only in the study of racial achievements called science and mathematics can we discover the appropriate $s.r$ and the nervous mechanism of these so varied, so flexible, and so fundamental reactions.

In fact, without a structural formulation and a $\bar{A}$ revision based on the study of science and mathematics, it is impossible to discover, to control, or to educate these $s.r$. For this reason it was necessary to analyse the semantic factors in connection with brief and elementary considerations taken from modern science. But, when all is said and done, and the important semantic factors discovered, the whole issue becomes extremely simple, and easily applied, even by persons without much education. In fact, because the objective levels are *not* words, the only possible aim of science is to discover *structure,* which, when formulated, is *always simple* and easily understood by everyone, with the exception, of course, of very pathological individuals. We have already seen that structure is to be considered as a configuration of relations, and that relations appear as the essential factors in meanings, and so of $s.r$. The present enquiry, because structural, reveals vital factors of $s.r$. The consequences are extremely simple, yet very important. We see that by a simple *structural re-education* of the $s.r$, which in the great mass of people are still on the level of copying animals in their nervous reactions, we powerfully affect the $s.r$, and so are able to impart very simply, to all, in the most elementary education of the $s.r$ of the child, *cultural* results at present sometimes acquired unconsciously and painfully in university education.

The above considerations have forced upon me the structure of the present work and the selection and presentation of the material. Of course, the reader can skip many parts and at once plunge into Part VII, and discover that it is all 'childishly simple', 'obvious' and 'common sense'. Such a reader or a critic with this particular *s.r* would miss the point, which can be verified as an *experimental* fact in the meantime, that in spite of its seeming simplicity, *no one,* not even the greatest genius, *fully* applies these 'platitudes' outside of his special work, which *s.r*, in his limited field, represent the semantic components *that make up his genius.*

The full acquisition of the new *s.r* requires special training; but, when acquired, it solves for a given individual, without any outside interference, all important human problems I know of. It imparts to him some of the *s.r* of so-called 'genius', and thus enlarges his so-called 'intelligence'.

The problems of the structure of a given language are of extreme, and as yet unrealized, semantic importance. Thus, for instance, the whole Einstein theory, or any other fundamental scientific theory, must be considered as the building of a new language of similar structure to the empirical facts known at a given date. In 1933, the general tendency of science, as made particularly obvious in the works of J. Loeb, C. M. Child, psychiatry, the Einstein theory, the new quantum mechanics., and the present work, is to build languages which take into consideration the many important invariant relations, a condition made possible only by the use of *non-el* languages. In my case, I must construct a *non-el* language in which 'senses' and 'mind', 'emotions' and 'intellect'., are no longer to be verbally split, because a language in which they *are* split is not similar in structure to the known empirical facts, and all speculations in such an *el* language must be misleading.

This *non-el* language involves a new *non-el* theory of meanings, as just explained. The term 'semantic', 'semantically', 'semantic reactions', 'semantic states'., are *non-el*, as they involve both 'emotions' and 'intellect', since they depend on 'meanings', 'evaluation', 'significance', and the like, based on structure, relations, and ultimately multi-dimensional order. All these terms apply equally to 'senses' and to 'mind', to 'emotions' and to 'intellect'—they are not artificially split.

It is important to preserve the *non-el* or organism-as-a-whole attitude and terminology throughout, because these represent most important factors in our *s.r.* Sometimes it is necessary to emphasize the origin, or the relative importance, of a given aspect of the impulse or reaction, or to translate for the reader a language not entirely familiar to him into one to which he is more accustomed. In such cases, I use the old *el* terms in quotation marks to indicate that I do not eliminate or disregard the other aspects—a disregard which otherwise would be implied by the use of the old terms.

The term psycho-logical will always be used either with a hyphen to indicate its *non-el* character, or in quotation marks, without a hyphen, when we refer to the old elementalism. Similarly, with the terms psycho-logics, psycho-logicians, for 'psychology' and 'pyschologist'. The terms 'mental ills', 'mental hygiene' are unfortunate ones, since they are used by the majority as *el*. Psychiatrists, it is true, use them in the organism-as-a-whole sense to include 'emotions'. Because of the great semantic influence of the structure of language on the masses of mankind, leading, as it does, through lack of better understanding and *evaluation* to *speculation on terms,* it seems advisable to abandon completely terms which imply to the *many* the suggested elementalism, although these terms are used in a proper *non-el* way by the *few*.

If specialists, to satisfy their *s.r,* disregard these issues and persist in the use of *el* terms, or use such expressions as 'man is an animal' and the like, they misunderstand the importance of semantic factors. Through lack of appreciation or of proper evaluation of the problems involved, they *artificially* and most effectively prevent the rest of us from following their work without being led astray by the inappropriate structure of their language. The harm done through such practices is quite serious, and, at present, mostly disregarded. For this reason, I either use quotation marks on the terms 'mental', 'mental' ills, 'mental' hygiene., or else I use the terms psycho-logical, semantic ills, psycho-logical or semantic hygiene,. The above two terms are not only *non-el* but also have an important advantage of being international. The terms 'affects', 'affective' are little used outside of scientific literature, where they are used mostly in the *non-el* ordinal sense. I use them in a similar way, without quotation marks.

All the issues involved in the present work are, of necessity, interconnected. Thus, order leads to relations, relations to structure, and these, in turn, to *non-el* meanings and evaluations, which are the fundamental factors of all psycho-logical states and responses, called more specifically semantic reactions, states, and reflexes. The reader should be careful to remain at all times aware of these connections and implications. Whenever we find order, or relations, or structure, in the outside world, or in our nervous system, these terms, because of their *non-el* character, imply similar order, relations, and structure in our psycho-logical processes, thus establishing meanings, proper evaluations., ultimately leading toward appropriate *s.r.* The reverse applies also. Whenever we speak of *s.r*, *non-el* meanings, structure, relations, and, finally, order, are implied.

The use of *non-el* languages is seriously beneficial, as it is structurally more correct and establishes *s.r* which are more appropriate, more flexible, or of higher order conditionality, a necessity for the optimum working of the human nervous system,—all of which results follow automatically from the structure of the language used.

A *non-el*, structurally correct, but non-formulated, attitude is a private benefit. Once it is formulated in a *non-el* language, it becomes a public benefit, as it induces in others the *non-el* attitudes, thus transforming the former *s.r*. In this way, a 'feeling' has been translated structurally into language; which, in turn, through structure, involves other people's attitudes and 'feelings', and so their *s.r*.

The whole process is extremely simple, elementary, and automatic; yet, before we acquire the new *s.r*, we find difficulties because of the fundamental novelty of these reactions. Any persistent student will acquire them easily, provided he does not expect too rapid a progress. The subject matter of the present analysis is closely related to the 'feelings' of everybody; yet the difficulties in acquiring the new reactions are similar to those the older scientists found in acquiring the *s.r* necessary for mastering the $\bar{E}$ and $\bar{N}$ systems.

In physics, we often need 'space-like' or 'time-like' intervals, although the *non-el* implications of the term 'interval' remain. Similarly, in our problems when we are interested in the 'emotion-like' or 'mind-like' aspects of the *non-el s.r*, we shall indicate the special aspects by using the old terms in quotation marks. This method prevents wasteful and futile speculations on *el* terms, and serves as a reminder that the other aspects are present, although in a given discussion we do not deem them to be important. The above has, by itself, very far-reaching semantic influence on our reactions.

From what has already been said, it is clear that the terminology of semantic reactions., covers in a *non-el* way all psycho-logical reactions which were formerly covered by *el* terms of 'emotions' and 'intellect', the reactions themselves always being on the objective levels and un-speakable. As *s.r* can always be analysed into terms of meanings and evaluation, and the latter into terms of structure, relations, and multi-dimensional order, which involves physiological factors, the term 'semantic' ultimately appears as a physiological or rather psychophysiological term. It suggests workable and simple educational methods which will be explained later. The reader should notice that the use of a language of a new structure has led to new results, which, in turn, directly affect our *s.r.*

An important point should be stressed; namely, that the issues are fundamentally simple, because they are similar in structure to the structure of human 'knowledge' and to the nervous structure on which so-called 'human nature' depends. Because of this similarity, it is unconditionally necessary to become fully acquainted with the new terms of new structure, and to use them habitually. Only then will the beneficial results follow. All languages have some structure; and so all languages involve automatically the, of necessity, interconnected *s.r.* Any one who tries to translate the new language into the old while 'thinking' in the older terms is confronted with an inherent neurological difficulty and involves himself in a hopeless confusion of his own doing. The reader must be warned against making this mistake.

In the present work, I have tried to realize fully my duties toward my reader; and I am certain that the reader who will read the book diligently and repeatedly will be repaid for his labours. The realization that some problems *do exist,* even if we do not fully appreciate or understand them, has very serious semantic influence on all of us. Realizing my responsibilities toward the reader, I have not spared difficult labour in order to bring these semantic facts to his attention. I seriously suggest that no reader ought to disregard Parts VIII, IX, and X, but that he should become at least acquainted with the existence of the problems there discussed. If this is conscientiously done, many beneficial *s.r* will appear sooner or later.

The present system is an interconnected whole: the beginning implies the end, and the end implies the beginning. Because of this characteristic, the book should be read *at least twice,* and preferably oftener. I wish positively to discourage any reader who intends to give it merely a superficial reading.

The problems of *s.r* have not, so far, been analysed at all from the point of view of structure, and the present enquiry is, as far as my knowledge goes, the first in existence. The problems of meanings are vast, extremely important, and very little analysed. The interested reader will find some material in the excellent critical review of the problems of meanings in Ogden's and Richards' *The Meaning of Meaning,* in some parts of Baldwin's *Thought and Things or Genetic Logic,* and in Lady Welby's article in the *Encyclopaedia Britannica* on *Significs.* In these three studies, a partial literature of the subject is given.

The present work involves issues taken from many and diverse branches of knowledge which have not hitherto been seen to be connected.

What is of importance is that the issues presented should be sound *in the main,* even if not perfect in details, which often have no bearing on the subject. Specialists in the fields here analysed should pass their professional judgement as to the soundness of their *special parts* of the system. They do not need even to be enthusiastic, it is enough if they approve it. The main issue is the building of a $\bar{A}$-system, which *co-ordinates* many disconnected fields of knowledge on the basis of structure, from the special point of view of *non-el s.r.* If these results have been accomplished, the author is satisfied.

Section B. On the un-speakable objective level.

The term 'un-speakable' expresses exactly that which we have up to now practically entirely disregarded; namely, that an object or feeling, say, our toothache, is *not* verbal, is *not* words. Whatever we may say will not be the objective level, which remains fundamentally un-speakable. Thus, we can sit on the object called 'a chair', but we cannot sit on the noise we made or the name we applied to that object. It is of utmost importance for the present $\bar{A}$-system not to confuse the verbal level with the objective level, the more so that all our immediate and direct 'mental' and 'emotional' reactions, and all *s.r*, states, and reflexes, belong to the un-speakable objective levels, as these are *not* words. This fact is of great, but unrealized, importance for the training of appropriate *s.r.* We can train these reactions simply and effectively by 'silence on the objective levels', using familiar *objects* called 'a chair' or 'a pencil', and this training automatically affects our 'emotions', 'feelings', as well as other psycho-logical immediate responses difficult to reach, which are also *not* words. We can train simply and effectively the *s.r* inside our skins by training on purely objective and familiar grounds outside our skins, avoiding unnecessary psycho-logical difficulties, yet achieving the desired semantic results. The term 'un-speakable' is used in its strict English meaning. The objective level is *not* words, can *not* be reached by words alone, and has nothing to do with 'good' or 'bad'; neither can it be understood as 'non-expressible by words' or 'not to be described by words', because the terms 'expressible' or 'described' already presuppose words and symbols. Something, therefore, which we call 'a chair' or 'a toothache' may be *expressed* or *described* by words; yet, the situation is not altered, because the given description or expression will *not* be the actual objective level which we call 'a chair' or 'a toothache'.

Semantically, this problem is genuinely crucial. Any one who misses that—and it is unfortunately easily missed—will miss one of the most important psycho-logical factors in all *s.r* underlying sanity. This omission is facilitated greatly by the older systems, habits of thought, older *s.r*, and, above all, by the primitive *structure* of our *A* language and the use of the 'is' of identity. Thus, for instance, we *handle* what we call a pencil. Whatever we *handle* is un-speakable; yet we *say* 'this *is* a pencil', which statement is unconditionally false to facts, because the object appears as an absolute individual and *is not* words. Thus our *s.r* are at *once trained in delusional values*, which must be pathological.

I shall never forget a dramatic moment in my experience. I had a very helpful and friendly contact prolonged over a number of years with a very eminent scientist. After many discussions, I asked if some of the special points of my work were clear to him. His answer was, 'Yes, it is all right, and so on, *but*, how can you expect me to follow your work all through, if I still do not know what an object *is*?' It was a genuine shock to me. The use of the little word 'is' as an identity term applied to the objective level had paralysed most effectively a great deal of hard and prolonged work. Yet, the semantic blockage which prevented him from acquiring the new *s.r* is so simple as to seem trifling, in spite of the semantic harm done. The definite answer may be expressed as follows: 'Say whatever you choose *about* the object, and whatever you might say *is not* it.' Or, in other words: 'Whatever you might *say* the object "is", well it *is not*.' This negative statement is *final*, because it is *negative*.

I have enlarged upon this subject because of its crucial semantic importance. Whoever misses this point is missing one of the most vital factors of practically all *s.r* leading toward sanity. The above is easily verified. In my experience I have never met any one, even among scientists, who would *fully* apply this childish 'wisdom' as an *instinctive* 'feeling' and factor in all his *s.r*. I want also to show the reader the extreme simplicity of a *Ā*-system based on the denial of the 'is' of identity, and to forewarn him against very real difficulties induced by the primitive structure of our language and the *s.r* connected with it. Our actual lives are lived entirely on the objective levels, including the un-speakable 'feelings', 'emotions'., the verbal levels being only *auxiliary*, and effective only if they are translated back into first order un-speakable effects, such as an object, an action, a 'feeling'., all on the silent and un-speakable objective levels. In all cases of which I know at present, where the retraining of our *s.r* has had beneficial effects, the results were obtained when this 'silence on the objective levels' has been attained, which affects all our psycho-logical reactions and regulates them to the benefit of the organism and of his survival adaptation.

Section C. On 'copying' in our nervous reactions.

The selection of the term 'copying' was forced upon me after much meditation. Its standard meaning implies 'reproduction after a model', applicable even to mechanical processes, and although it does not exclude, it does *not* necessarily include conscious copying. It is not generally realized to what an important extent copying plays its role in higher animals and man.

Some characteristics are inborn, some are acquired. Long ago, Spalding made experiments with birds. Newly hatched birds were enclosed in small boxes which did not allow them to stretch their wings or to see other birds fly. At the period when usually flying begins, they were released and began to fly at once with great skill, showing that flying in birds is an inborn function. Other experiments were made by Scott to find out if the characteristic song of the oriole was inborn or acquired. When orioles, after being hatched, were kept away from their parents, at a given period they began to sing; but the peculiar melody of their songs was different from the songs of their parents. Thus, singing is an inborn characteristic, but the special melody is due to copying parents, and so is acquired.[3]

In our human reactions, speech in general is an inborn characteristic, but what special language or what special *structure* of language we acquire is due to environment and copying—much too often to unconscious and, therefore, uncritical copying. As to the copying of animals in our nervous reactions, this is quite a simple problem. Self-analysis, which is rather a difficult affair, necessitating a serious and efficient 'mentality', was impossible in the primitive stage. Copying parents in many respects began long before the appearance of man, who has naturally continued this practice until the present day. The results, therefore, are intimately connected with reactions of a pre-human stage, transmitted from generation to generation. But for our present purpose, the most important form of the copying of animals was, and is, the copying of the comparative unconditionality of their conditional reflexes, or lower order conditionality; the animalistic identification or confusion of orders of abstractions, and the lack of consciousness of abstracting, which, while natural, normal, and necessary with animals, becomes a source of endless semantic disturbances for humans. More about copying will be explained as we proceed.

It should be noticed also that because of the structure of the nervous system and the history of its development, the more an organism became 'conscious', the more this copying became a neurological necessity, as exemplified in parrots and apes. With man, owing to the lack of consciousness of abstracting, his copying capacities became also much more pronounced and often harmful. Even the primitive man and the child are 'intelligent' enough to observe and copy, but not informed enough in the racial experiences usually called science, which, for him, are non-existent, to discriminate between the reactions on the 'psychological' levels of animals and the typical responses which man with his more complex nervous system should have. Only an analysis of *structure* and *semantic reactions,* resulting in consciousness of abstracting, can free us from this unconscious copying of animals, which, let us repeat, must be pathological for man, because it eliminates a most vital regulating factor in human nervous and *s.r*, and so vitiates the whole process. This factor is not simply additive, so that, when it is introduced and *superimposed* on any response of the human nervous system allowing such superimposition, the whole reaction is *fundamentally changed* in a beneficial way.

CHAPTER III

INTRODUCTION

And so far as the actual, fundamental, biological structure of our society is concerned and notwithstanding its stupendous growth in size and all the tinkering to which it has been subjected, we are still in much the same infantile stage. But if the ants are not despondent because they have failed to produce a new social invention or convention in 65 million years, why should we be discouraged because some of our institutions and castes have not been able to evolve a new idea in the past fifty centuries? (553)

WILLIAM MORTON WHEELER

The ancient who desired to illustrate illustrious virtue throughout the empire, first ordered well their own state. Wishing to order well their own state, they first regulated their families. Wishing to regulate their families, they first cultivated their own persons. Wishing to cultivate their persons, they first rectified their heart. Wishing to rectify their hearts, they first sought to be sincere in their thoughts. Wishing to be sincere in their thoughts, they extended their knowledge to the utmost; and this extension of knowledge lay in the investigation of things. Things being investigated, knowledge became complete. Their knowledge being complete, their thoughts were sincere. Their thoughts being sincere, their hearts were then rectified. Their heart being rectified, their persons were cultivated. Their persons being cultivated, their families were regulated. Their family being regulated, their states were rightly governed. Their states being rightly governed, their whole empire was made tranquil and happy. From the emperor down to the mass of the people, all must consider the cultivation of the person, the root of every thing besides.   CONFUCIUS

My service at the front during the World War and an intimate knowledge of life-conditions in Europe and the United States of America have convinced me that a scientific revision of *all* our notions about ourselves is needed. Investigation disclosed that all disciplines dealing with the affairs of man either do not have a definition of man, or, if they do, that it is formulated in metaphysical, *el,* subject-predicate languages, which are unscientific and ultimately semantically harmful.

As we have, at present, no general science of man embracing *all* his functions, language, mathematics, science and 'mental' ills included, I believed that to originate such a science would be useful. This task I began in my *Manhood of Humanity,* and have continued in the present volume. The selection of a name for such a science is difficult. The only really appropriate name, 'Anthropology', is already pre-empted to cover a most fundamental and sound discipline, without which even modern psychiatry would be impossible. This name, at present, is used in a *restricted* sense to signify the animalistic natural history of man, disregarding the fact that the *natural history of man* must include factors non-existent in the animal world, but which are his *natural* functions,

38

such as language and its structure, the building of his institutions, laws, doctrines, science, mathematics., which condition his environment, his *s.r*, which, in their turn, influence and determine his development.

We see that the 'natural history' of animals differs greatly in structure from a future scientific *'natural history'* of man, a structural difference very seldom fully realized. I propose, then, to call the very valuable existing Anthropology the *Restricted Anthropology,* and to call the generalized science of man *General Anthropology,* so as to include *all* his natural functions, of which those covered by the Restricted Anthropology would be a part.

Such a definite General Anthropology would be very different from the existing restricted one. It would include all disciplines of human interest from a special anthropological and semantic point of view. Very often an anthropological discipline—for instance, anthropological psycho-logics, anthropological sociology, law, history, or 'philosophy'— would turn out to be a *comparative* discipline. Such, by necessity, would have to use a language of four-dimensional structure, which would necessitate, as a preliminary, a fundamental revision of the structure of the language they use—a semantic factor which, up to now, has been largely neglected.

It is to be frankly admitted that the present enquiry has led to some quite unexpected and startling results. In my *Manhood of Humanity,* I defined man *functionally* as a time-binder, a definition based on a *non-el* functional observation that the human class of life differs from animals in the fact that, in the rough, each generation of humans, at least potentially, can start where the former generation left off—a definition which, in the language of this particular structure, *is sharp,* and corresponds to empirical facts. We should notice, also, that in the case of primitive tribes which apparently have not progressed at all for many thousands of years, we always find, among other reasons, some special doctrines or creeds, which proclaim very efficiently, often by killing off individuals (who always are responsible for progress in general), that any progress or departure from 'time honoured' habits or prejudices 'is a mortal sin' or what not. Even in our own case we are not free from such semantic tendencies. Only the other day, historically speaking, the 'holy inquisition' burned or silenced scientists. The discovery of the microscope and telescope, for instance, was delayed for a long time because the inventor, in fear of priestly persecution, was afraid to write his scientific discoveries in plain language. He had to write them in cipher—a fact discovered only a few years ago. Those afflicted with diseases can easily realize where our science in general, and medical science in particular, *might* be today if not for the holy zeal of powerful enemies of science who vehemently and ruthlessly sponsored ignorance, old *s.r*, and so disease.

In some countries, even at present, science is persecuted, and the attempt is made to starve scientists, a device often quite as effective as burning at the stake, of which the Tennessee trial and others are examples. But, in spite of all these primitive semantic tendencies, which, unfortunately, are often very efficient, the general time-binding characteristic of man remains unaltered, although its rate can be slowed down by the ignorance of those who control our symbolism—words, money,.

The failure to understand these problems is due to the fundamental fact that, until now, we have had no scientific functional *non-el* definition of man; neither have we originated any scientific enquiry into the inherent 'nature of man', which is impossible if we disregard *s.r*. We should remember that in this commercialized era we offer large incomes to those who preach with great zeal how 'evil' 'human nature is', and who tell us how, without their services, all kinds of dreadful things will happen to a given individual.

In the light of modern enquiry, the above issues come to a very sharp focus. Either these apostles *do* know that what they promise has only delusional value, yet they want to retain their incomes, or else they live in *delusional* worlds, and a sane mankind should take care of them. In either case, they are *unfit* to be any longer entrusted with the care of the further development of culture and the future of mankind. Sooner or later, we must meet this semantic situation squarely, as too many factors of human sanity are at stake.

In my *Manhood of Humanity,* it is shown that the canons of what we call 'civilization' or 'civilizations' are based on animalistic generalizations taken from the obvious facts of the lives of cows, horses, dogs, pigs., and applied to man. Of course, such generalizations started with *insufficient data*. The generalizations had to be primitive, superficial; and when applied in practice, periodical collapses were certain to follow. No bridge would stand or could even be built, if we tried to apply rules of surfaces to volumes. The rules or generalizations in the two cases are different, and so the results of such primitive semantic confusion must be disastrous to all of us.

The present enquiry began with the investigation of the characteristic difference between animal and man; namely, the mechanism of time-binding. This analysis, because of the different structure of the language used, had to be carried on independently and anew. The results are, in many instances, new, unexpected even by myself, and they show unmistakably that, to a large extent, even now we nearly all copy animals in our nervous processes. Investigation further shows that such nervous reactions *in man* lead to non-survival, pathological states of general *infantilism,* infantile private and public behaviour, infantile institutions, infantile 'civilizations' founded on strife, fights, brute competitions., these being supposedly the 'natural' expression of 'human nature', as different commercialists and their assistants, the militarists and priests, would have us believe.

As always in human affairs, in contrast to those of animals, the issues are circular. Our rulers, who rule our symbols, and so rule a symbolic class of life, impose their own infantilism on our institutions, educational methods, and doctrines. This leads to nervous maladjustment of the incoming generations which, being born into, are forced to develop under the un-natural (for man) semantic conditions imposed on them. In turn, they produce leaders afflicted with the old animalistic limitations. The vicious circle is completed; it results in a general state of human un-sanity, reflected again in our institutions. And so it goes, on and on.

At first, such a discovery is shocking. On second consideration, however, it seems natural that the human race, being relatively so recent, and having passed through different low levels of development, should misunderstand structurally their human status, should misuse their nervous system,. The present work, which began as the 'Manhood of Humanity' turned out to be the 'Adulthood of Humanity', for it discloses a *psychophysiological* mechanism of infantilism, and so points toward its prevention and to adulthood.

The term 'infantilism' is often used in psychiatry. No one who has had any experience with the 'mentally' ill, and studied them, can miss the fact that they always exhibit some infantile symptoms. It is also known that an adult, otherwise considered 'normal', but who exhibits marked infantile semantic characteristics, cannot be a fully adjusted individual, and usually wrecks his own and other persons' lives.

In the present investigation, we have discovered and formulated a definite psychophysiological mechanism which is to be found in all cases of 'mental' ills, infantilism, and the so-called 'normal' man. The differences between such neural disturbances in different individuals turn out to vary only in degree, and as they resemble closely the nervous responses of animals, which are necessarily regressive for man, we must conclude that, generally, we do not use our nervous system properly, and that we have not, as yet, entirely emerged from a very primitive semantic stage of development, in spite of our technical achievements.

Indeed, experience shows that the more technically developed a nation or race is, the more cruel, ruthless, predatory, and commercialized its systems tend to become. These tendencies, in turn, colour and vitiate international, national, capital-labour, and even family, relations.

Is, then, the application of science at fault? No, the real difficulty lies in the fact that different primitive, animalistic, un-revised doctrines and creeds with corresponding *s.r* have *not* advanced in an equal ratio with the technical achievements. When we analyse these creeds semantically, we find them to be based on structural assumptions which are false to facts, but which are strictly connected with the unrevised structure of the primitive language, all of which is the more dangerous because it works unconsciously.

When we study comparatively the nervous responses of animals and man, the above issues become quite clear, and we discover a definite psychophysiological mechanism which marks this difference. That the above has not been already formulated in a workable way is obviously due to the fact that the *structure* of the old language successfully prevented the discovery of these differences, and, indeed, has been largely responsible for these human semantic disturbances. Similarly, in the present $\bar{A}$-system, the language of a new and modern structure, as exemplified by terms such as 'time-binding', 'orders of abstractions', 'multi-ordinal terms', 'semantic reactions'., led automatically to the disclosure of the mechanism, pointing the way toward the means of control of a special therapeutic and preventive character.

The net results are, in the meantime, very promising. Investigation shows that, in general, the issues raised are mostly *linguistic,* and that, in particular, they are based on the analysis of the *structure* of languages in connection with *s.r.* All statements, therefore, which are made in this work are about empirical facts, language and its structure. We deal with an obvious and well-known inherent psychophysiological function of the human organism, and, therefore, all statements can be readily verified or eventually corrected and refined, allowing easy application, and automatically eliminating primitive mythologies and *s.r.*

After all is done and said, one can only wonder why such a simple fact that language represents a very important, unique, and inherent psychophysiological function of the human organism has been neglected so long.

The answer seems to be that: (1) the daily language is structurally extremely complex; (2) it is humanly impossible to analyse its structure by using the language of A structure, so that before anything can be done at all in this field a $\bar{A}$-system must first be formulated; (3) there is a general innocence on the part of nearly all specialists, a very few mathematicians excepted, of the structural and semantic role of the simplest—although still inadequate—$\bar{A}$ language called mathematics; (4) all these issues involve most powerful unconscious factors which work automatically against any revision, and (5) the building of a $\bar{A}$-system in 1933 is an extremely laborious enterprise, to say the least, and, in all probability, really beyond the power of any single man to complete.

The last point is quite important; and, although I have no intention to apologize or present any alibis, because any thoughtful reader will understand it, I must explain, nevertheless, briefly why the present work has probably fallen far short of what it eventually could be.

In the days of Aristotle, we knew extremely little of science in the 1933 sense. Aristotle, in his writings, formulated for us a whole scientific program, which we followed until very lately. Whoever, in 1933, attempts to build a $\bar{A}$-system, must, by *internal necessity,* connected with the problems of the *structure of language,* do something similar. Obviously, in 1933, with the overwhelming number of most diversified facts known to science, the question is no more to sketch a scientific program for the future, but to build a system which, at least in structure, is similar to the structure of the known facts from all branches of knowledge.

Let me repeat: the necessity is internal, and connected with the *structure* of language as such, involving new *s.r,* and so no one can avoid it, as this whole work shows in detail.

Now such structural adjustment requires an immense amount of study of diverse empirical facts, and then it must depend on new generalizations, concerned in the main with structure. Many statements of scientists, when even accepted as reliable, still have to be translated into a special language in which structural issues are made quite obvious, divulging factors in *s.r.* This is a very serious difficulty, particularly when many branches of knowledge are drawn upon, as each uses its own special language; so that such a unitary translation in terms of structure imposes a serious burden on the memory of the translator, and often little details escape attention in the implications of the translation, although they may be well known to the translator. As this is probably the chief difficulty, it is in this field that the main corrections will have to be made.

I admit that I started this enquiry without fully realizing its inherent difficulties, or whither it would eventually lead. The more I advanced, the more special knowledge was required. I had to go to sources, and, in a way, partially specialize in many branches of science which were never connected before. The progress was extremely slow; in fact, it has taken ten years to write this book, but I had to go through the necessary preliminaries or abandon the whole enterprise.

Now I present the results of this work to the public. It is the best I can do, although I fully realize its limitations and imperfections. The unexpected drama of such an enterprise is found, in that a $\bar{A}$-system, like its predecessor, involves full-fledged structural metaphysics of some sort, to be explained later.

The A-system involved primitive structural metaphysics; a $\bar{A}$-system, to be of any semantic value at all, must start with the structural metaphysics or structural assumptions as given by science 1933. The first step in building such a system is to study science 1933 and mathematics, and so to know these structural data (and assumptions where we lack data). Such a study is very laborious, slow, and even ungrateful, because the issues with which we are concerned are *structural*. Thus, years of patient and sometimes painful labour often result merely in a very few and brief, but important, sentences.

The active, and only very lately relaxed, persecution of those scholars who dared to attempt the revision of Aristotle has been very effective in keeping the primitive *s.r*. There are in this field practically no important works of a critical character, and this fact, naturally, made my own work more difficult.

It appears that, during the last few years, most of the physiological functions of the human organism have been investigated, with the exception of psychophysiological *semantic reactions* and their disturbances from the present point of view.

The study of aphasia is rather recent, and that of *semantic aphasia* still more so. Only since the World War has much new knowledge been accumulated in this field. With the 1933 scientific outlook, macroscopic structure becomes a function of the sub-microscopic dynamic structure, and the considerations of colloidal structure and disturbances become extremely important. We must, therefore, enlarge the study of semantic aphasia as connected with macroscopic lesions, to include semantic phasic (not only a-phasic) sub-microscopic functional disturbances connected with *order*, natural survival order and its pathological reversal, the disturbances of the *multiordinal semantic reactions*.

It is known that 'mental' ills or difficulties often disturb physiological functions of the human organism, and vice versa. Something similar appears to be true about the last and little-investigated *s.r*. In this case, more and special difficulties arise, because of the fact that these particular reactions are strictly connected with different 'emotional' or affective responses, which are due to the knowledge (or lack of knowledge) of their mechanism. They are circular, as all functions connected with knowledge are. This difficulty is very serious and closely connected with the *structure of language,* disclosing also a most important fact, that languages *may* have structure. This subject could not have been even suggested by the A-system; nor could it have been analysed by A means.

The most encouraging feature of this work is the fact that it is *experimental,* and that in cases where it has been applied, it worked remarkably well. It appears that all desirable human characteristics, high 'mentality' included, have a definite psychophysiological mechanism, easily understood and easily trained. One should not expect this training to be more quickly acquired than the mastering of spelling, or driving a car, or typewriting. Practice shows that it requires approximately as much diligence and persistence as learning to spell or to typewrite. The results accomplished in the field of 'mental' health, widened horizons, and unlimited possibilities for personal and public adjustment, justify so small a price. This applies to adults, but in a different way is true for children. From an educational point of view, it is as simple or as complex to train children in the improper use of an important physiological function, such as language, as to train them in the proper use of the nervous system and appropriate *s.r.* In fact, the new is simpler and easier, if we start with it, because it is in accordance with 'human nature'. In theory, it plays a most important role in the prevention of many future eventual break-downs, which the old misuse of the function was certain to involve.

The problems of a $\overline{A}$-system are, to the best of my knowledge, novel. They are of two kinds: (1) scientific, leading to a theoretical, general structural revision of all systems, and (2) purely practical, such as can be grasped and applied by any individual who will spend the time and effort necessary to master this system and acquire the corresponding *s.r.*

The results are far-reaching. They help any individual to solve his problems by himself, to his own and others' satisfaction. They also build up an *affective* semantic foundation for personal as well as for national and international agreement and adjustment.

Some of the results are quite unexpected. For instance, it is shown that the older systems, with their linguistic methods of handling our nervous system, led inevitably to 'universal disagreement'. In individual life this led to pathological conflicts with ourselves; in private life, to

family strifes and unhappiness, and so to nervous disturbances; in
national life, to political strifes, revolutions.; in international affairs, to
mutual misunderstanding, suspicion, impossibility to agree, wars, World
Wars, 'trade wars'., ultimately ending in slaughter, general unemployment, and an unnecessarily great amount of general unrest, worry, confusion, and suffering in different degrees for all, helping again to disturb the proper working of the human nervous system.

The subject of this work is ultimately 'speaking about speaking'.
As all human institutions depend upon speaking—even the World War
could not have been staged without speaking—and as all science is ultimately verbal, such an analysis must cover a large field. In such an
attempt, therefore,. we must first understand the speaking of scientists,
of different specialists., and so must get acquainted with their languages,
and what they are talking *about*. This is the semantic reason why I have
had to explain to the reader many simple, yet necessary, scientific structural issues.

The present book is written on the level of the average intelligent
reader, and any such reader will get the full benefit for his labour provided he is willing to put into it the necessary work and persistence.
Perhaps a word of warning is necessary. My own personal experience
is that, when once we have acquired a bad habit, let us say in making
errors in typewriting, this bad habit is very difficult to eliminate. A
similar remark applies to the old habits of speech, and of the semantic
responses connected with them. A re-education is simple in principle,
but it requires a great deal of persistent effort to overcome undesirable
s.r. My experience convinces me that the self-satisfied, the 'happy'
person, who has no problems at all, if there is such a person, should
not attempt to read this book. He will waste just so much effort. But I
can confidently promise that any one who has any problems to solve,—
be they personal difficulties with himself, his family, or his associates—
the scientist, the teacher, or the professional person who wants to become
more efficient in his own work, will be amply repaid for spending the
necessary 'time' in mastering the linguistic, and so neurological problems
involved in such a structural semantic re-education.

This investigation has turned out to be a general introduction to a
theory of sanity, the first ever made, as far as I know. When applied, it
genuinely works; but, of course, we have to apply it fully. Mere lip-
service will not do; because while superficial agreement is quite easy,
yet physiologically, on a deeper level, we continue to follow the older
harmful *s.r*. In such a case nothing is actually changed; the old neurologically harmful animalistic responses persist. For this the author and

the present work must not be held responsible; the fault will lie in the disregard of fundamental conditions by the reader or the students.

From an educational point of view, these problems are particularly important. If teachers disregard the structural linguistic semantic issues, they disregard a most powerful and effective educational method. If they train in structurally and physiologically harmful A habits, after this mechanism has been disclosed, such teachers, to my understanding, do not honestly perform their very serious social obligations. Ignorance is no excuse when once we know that ignorance is the only possible excuse.

The present $\bar{A}$-system is far from perfect. Such a work as this has, of necessity, to be altered with the years, as the structure of the language used has to be continually adjusted to newly discovered empirical structural data. But the present enquiry at least shows that in the researches in the linguistic and structural semantic fields there are un-dreamed-of possibilities of tremendous power, and the circularity of human knowledge is made to work with an increasing acceleration in a constructive way toward the adulthood of man.

In Chapter I, I have given a tentative list of some results following from the present work. Among them we found a new and *semantic definition of number* and mathematics, to be explained in Chapter XVIII. This has very far-reaching consequences, because the existing definition of number is A, in terms of classes, and makes the importance of mathematics still more mysterious. With the discovery that the only content of knowledge is structural, understood as a complex of relations and multiordinal and multi-dimensional order, and that the structure of the nervous system is such that only in mathematics do we find a *language of similar structure,* the importance of mathematics, considered as a language, becomes of fundamental semantic significance for a theory of sanity. But to show that, and to be able to apply this fact in practice, we must clarify or rather eliminate the mystery which surrounds number and measurement. The semantic definition of number is given in terms of relations, and so number and measurement become the most potent factors for supplying us with information about structure, which we know already gives the only content of knowledge.

At this point, I feel it essential to refer to the unique and astonishing work of Oswald Spengler, *The Decline of the West.* This work is a product of such unusual scholarship and breadth of vision that, in many instances, the details do not matter. Its method, its scope, and the complete novelty of its general point of view, combined with such tremendous erudition, are of main importance. The work is labeled by the author as a 'philosophy of history', or a morphology of history, or morphology of cultures. The word 'morphology' is used as implying the study of forms, and the term 'form' appears very frequently throughout the book.

The main contention of this work—and it is an entirely justified observation—is that the behaviour of the organisms called humans is such that, at different periods, they have produced definite aggregates of achievements, which we dissect and label 'science', 'mathematics', architecture', 'sculpture', 'music'., and that at any given period all these achievements are interconnected by a psycho-logical necessity. To this statement I would add: the *structure of languages* of a given period which affect the *s.r* should not be disregarded.

Spengler is a mathematician, an extremely well and generally informed mathematician at that, with a great vision. He surveys these aggregates as definite units and shows the necessary psycho-logical connections between all the achievements and the evolution of the notion of number. It does not matter whether all his connections are always beyond criticism. That some such connections do exist, seems beyond doubt. All human achievements have been accomplished in some definite period, and they have been accomplished at a definite period only because of the necessary psycho-logical attitude and *s.r* of that period.

With regard to the method followed in this work of Spengler, we must notice, first of all, that the attitude of the work is frankly anthropological, in the sense of *General Anthropology;* namely, as the *natural history of man, not* disregarding man's *natural* behaviour, such as building up sciences, mathematics, arts and institutions, and creating new environments, which again influence his development. Morphology means 'study of forms', which carries *static* implications. Taken from the dynamic point of view 1933, when we know that the dynamic unit, out of which the world and ourselves appear to be built, is found in the dynamic atom of 'action'; his 'form' becomes four-dimensional *dynamic structure,* the equivalent of 'function'; and then the whole outlook of Spengler becomes a *structural* enquiry into the world of man, including all his activities.

This 'form', or rudimentary structural point of view, or feeling, or inclination, or tendency, or *s.r*—call it what you choose—Spengler, the mathematician and historian, acquires from the deep study of mathematics considered as a *form of human behaviour;* which, in turn, is a part of his behaviour when he was planning and accomplishing his work, a natural expression of the strivings of *his* epoch, which is also our own. In my own work, I have attempted to formulate these vague strivings of our epoch in the form of a general *semantic* psychophysiological theory.

From this point of view, his achievement is momentous, a great description of the childhood of humanity. In spite of the title, there is nothing pessimistic about it, although most of his readers understood it in that way. 'The Decline of the West' implies the birth of a new era, perhaps the adulthood of humanity. There is no doubt in 1933 that the collapses of the older systems which we witness are probably irrevocable. Sir Auckland Geddes, the British Ambassador to the United States of America, foresaw them when he said in 1920: 'In Europe, we know that an age is dying. Here it would be easy to miss the signs of coming changes, but I have little doubt that it will come. A realization of the *aimlessness* of life lived to labour and to die, having achieved nothing but avoidance of starvation, and the birth of children also doomed to the weary treadmill, has seized the minds of millions.'

In 1932 Ambassador Mellon, of the United States of America to Great Britain, said:

'Part of our difficulty arises because we look on the present industrial economic crisis as merely a sporadic illness of the body politic due to conditions in some particular country or section of the world which can be cured by applications of some magic formula. Greater difficulty arises because we who are left over from the last century continue to look on the last decade as merely a prolongation of all that is gone before. We insist upon trying to make life flow in the same channels as before the war, whereas in the years since the war ended are, in reality, the beginning of a new era, not the end of the old.'

To this statement of Ambassador Mellon, the newspapers comment as follows:

'This is an important utterance, since, as far as we know, it is the first admission from the ruling forces in this country that the present panic is not "just another panic".'

No doubt, a period of human development has ended. The only sensible way is to look forward to a full understanding of the next phase, get hold of this understanding, keep it under conscious and scientific control, and *avoid this time,* perhaps for the first time in human history, the unnecessary decay, bewilderment, apathy, individual and mass suffering in a *human* life-period, animalistically believed, up to now, to be unavoidable in the passing of an era. Instead of being swept down by animalistic resistance to the *humanly* unavoidable change, we must analyse, understand, and so keep conscious control of one change to another, and, as yet, always higher state of human culture.

This is no place to analyse these issues in detail. Volumes have already been written about the work of Spengler. More volumes will

4

be necessary to analyse the issues raised, and not always solved, in this present volume.

I want to make clear only that words are *not* the things spoken about, and that there is *no* such thing as an object in absolute isolation. These assertions are negative and experimental and cannot be successfully denied by any one, except by producing positive evidence, which is impossible.

We must realize that *structure, and structure alone,* is *the* only link between languages and the empirical world. Starting with undeniable negative premises, we can always translate them into positive terms, but such translation has a new and hitherto unprecedented security. In the era which is passing, positive premises were supposed to be important, and we did not know that a whole $\bar{A}$-system can be built on negative premises. The new era will have to revaluate these data, and build its systems on *negative premises,* which are of much greater security. *A priori,* we cannot know if such systems can be built at all, for in this field the only possible 'proof' is actual performance and the exhibition of a sample.

This has been attempted in my work, and so the possibility of such systems becomes a fact on record.

In the new era, the role of mathematics considered as a form of human behaviour and as a language will come to the fore. Means can be found, as exemplified in the present volume, to impart mathematical structure to language without any technicalities. It is enough to understand the above-mentioned negative premises and the role of structure, and to produce systems from this angle.

The role of mathematics has been and, in general, is still misunderstood, perhaps because of the very unsatisfactory definition of number. Even Spengler asserts that 'If mathematics were a mere science, like astronomy or mineralogy, it would be possible to define their object. This, man is not, and never has been, able to do.' The facts are that mathematicians have been prone to impress us with some religious awe of mathematics; meanwhile, by definition, whatever has symbols and propositions must be considered a *language.* All mystery vanishes in this field, and the only question is as to what kind of language mathematics represents. From the structural point of view, the answer is simple and obvious. Mathematics, although in daily life it appears as a most insufficient language, seems to be *the only language* ever produced by man, which, *in structure, is similar,* or the most similar known, to the structure of the world *and of our nervous system.* To be more explicit, let me say at once that in this 'similarity of structure' we find *the only* positive 'knowledge' of 1933, and, perhaps, of any date.

So far, I am in full agreement with the great work of Spengler. More than that, *The Decline of the West* may be considered as a preliminary and preparatory survey of the great cultural spasms which have rocked mankind. It may be instructive, in the meantime, to point out some differences, and the eventual difference of conclusions, between the present work and the work of Spengler.

The difference, to start with, is in language. Spengler announces his work as 'intuitive and depictive through and through, which seeks to present objects and relations illustratively instead of offering an army of ranked concepts'.

My own aim is not merely descriptive, but structural and analytical, and so I must use a different language, helping to discover semantic psychophysiological *mechanisms* of the events of which Spengler is giving us a very exceptional *picture*.

Spengler missed two points: that mathematics must be considered a language; and that the connection which he asserts between the mathematics of each period and other achievements is more general than he suspects and applies to the inherent *structure* of languages, in general, and the *s.r* of each period, in particular.

Although his analysis is in effect $\bar{A}$, yet the $\bar{A}$ issues are not formulated clearly or consciously; nor does he mention that the present-day accepted definition of number in terms of classes is still A. It is true that he considers 'forms' and the use of number as relations, but he does not emphasize that number has not been formally defined as a relation, which is essential in a $\bar{A}$-system. His 'form' is still static and not dynamic structure; nor did he discover that the only possible content of knowledge is structural, a fact which is *the* semantic factor responsible for 'cultures', and 'periods', and everything else in human development.

This present work is in great sympathy with the momentous work of Spengler; but, as it culminates in a $\bar{A}$-system, it goes further than his, and is more workable and more practical, confirming the general contention of Spengler, that cultures have their periods of growth and development and that, so far, without conscious human intention, they are superseded by others.

We should notice once more in this respect that the issues we deal with, whenever *human* psycho-logical reactions are involved, are circular, as distinguished from animal reactions. Human structures, in language or in stone, reflect the psycho-logical status, feelings, intuitions, structural metaphysics, and other *s.r*, of their makers and periods; and, vice versa, once these structural strivings and tendencies are formulated as such, they help to quicken and transform one period into the next one.

To the best of my knowledge, the general analysis of this most fundamental structural semantic problem of human knowledge is here formulated for the first time. It will allow, as a result, the making of human progress conscious; it will enable us to control it and so to make it uninterrupted by the painful and wasteful semantic periods of hopelessness and helplessness so characteristic of the older periods of transition.

If it is an historical fact and also a psycho-logical fact that a time-binding class of life has to have periods of development, let us have them! Let us investigate the mechanism of time-binding, of *s.r*, which are the dynamic factors of those changes and developments! Let us direct that development consciously, and this will lead to the elimination of unnecessary and painful panics, unrest, and the often bloody bursting of those animalistic barriers self-imposed on the dynamic class of life called 'man'.

When all is said and done, one cannot but see, at least as far as the white race is concerned, that a change from an A to a $\bar{A}$-system must be momentous. Such a change will mark the difference between a period when the mystery of 'human knowledge' was not solved and a period when it has been solved. This inherent human circular characteristic has, so far, been neurologically abused. We have not known how to handle our nervous structure. We have imposed upon and hampered human development by animalistic methods. The solving of the problems of the content of 'human knowledge' will open a new era of man as a *man,* leading toward proper handling of his capacities, and an era scientific in all respects, not merely in a few exact sciences. Psychologically, it will be an era of sanity, and, therefore, of human general adjustment, agreement, and co-operation. The dreams of Leibnitz will become sober reality.

The only determined attempt made, so far, to deal with the symbolic problems whose importance is emphasized in the present work is that of the Orthological Institute (10 King's Parade, Cambridge, England). This research organization, founded by C. K. Ogden, editor of 'The International Library of Psychology', is concerned with the influence of language on 'thought' in all its bearings; and it is to be hoped that further endowment may be forthcoming to enable the scope of its enquiries to be extended to include *structure, non-elementalistic semantic reactions,* and *non-aristotelian systems.* Reference to International Languages or a Universal Language will be found in the notes.[1]

PART II

GENERAL ON STRUCTURE

The relativity theory of physics reduces everything to relations; that is to say, it is structure, not material, which counts. The structure cannot be built up without material; but the nature of the material is of no importance. (147)   A. S. EDDINGTON

Structure and function are mutually related. Function produces structure and structure modifies and determines the character of function. (90)
                                                    CHARLES M. CHILD

These difficulties suggest to my mind some such possibility as this: that every language has, as Mr. Wittgenstein says, a structure concerning which, *in the language*, nothing can be said, but that there may be another language dealing with the structure of the first language, and having itself a new structure, and that to this hierarchy of languages there may be no limit. Mr. Wittgenstein would of course reply that his whole theory is applicable unchanged to the totality of such languages. The only retort would be to deny that there is any such totality. (456)   BERTRAND RUSSELL

CHAPTER IV

ON STRUCTURE

No satisfactory justification has ever been given for connecting in any way the consequences of mathematical reasoning with the physical world.
(22)   E. T. BELL

Any student of science, or of the history of science, can hardly miss two very important tendencies which pervade the work of those who have accomplished most in this field. The first tendency is to base science more and more on experiments; the other is toward greater and more critical verbal rigour. The one tendency is to devise more and better instruments, and train the experimenters; the other is to invent better verbal forms, better forms of representation and of theories, so as to present a more coherent account of the experimental facts.

The second tendency has an importance equal to that of the first; a number of isolated facts does not produce a science any more than a heap of bricks produces a house. The isolated facts must be put in order and brought into mutual structural relations in the form of some theory. Then, only, do we have a science, something to start from, to analyse, ponder on, criticize, and improve. Before this something can be criticized and *improved,* it must first be produced, so the investigator who discovers some fact, or who formulates some scientific theory, does not often waste his time. Even his errors may be useful, because they may stimulate other scientists to investigate and improve.

Scientists found long ago that the common language in daily use is of little value in science. This language gives us a form of representation of very old structure in which we find it impossible to give a full, coherent account of ourselves or of the world around us. Each science has to build a special terminology adapted to its own special purposes. This problem of a suitable language is of serious importance. Too little do we realize what a hindrance a language of antiquated structure is. Such a language does not help, but actually prevents, correct analysis through the semantic habits and structural implications embodied in it. The last may be of great antiquity and bound up, by necessity, with primitive-made structural implications, or, as we say, metaphysics, involving primitive *s.r.*

The above explains why the popularization of science is such a difficult and, perhaps, even a semantically dangerous problem. We attempt to translate a creative and correct language which has a structure

55

similar to the structure of the experimental facts into a language of different structure, entirely foreign to the world around us and ourselves. Although the popularization of science will probably remain an impossible task, it remains desirable that the *results* of science should be made accessible to the layman, if means could be found which do not, by necessity, involve misleading accounts. It seems that such methods are at hand and these involve *structural* and semantic considerations.

The term 'structure' is frequently used in modern scientific literature, but, to the best of my knowledge, only Bertrand Russell and Wittgenstein have devoted serious attention to this problem, and much remains to be done. These two authors have analysed or spoken about the structure of propositions, but similar notions can be generalized to languages considered as-a-whole. To be able to consider the structure of one language of a definite structure, we must produce another language of a *different* structure in which the structure of the first can be analysed. This procedure seems to be new when actually performed, although it has been foreseen by Russell.[1] If we produce a $\bar{A}$-system based on 'relations', 'order', 'structure'., we shall be able to discuss profitably the A-system, which does not allow asymmetrical relations, and so cannot be analysed by A means.

The dictionary meaning of 'structure' is given somewhat as follows: Structure: Manner in which a building or organism or other complete whole is constructed, supporting framework or whole of the essential parts of something (the structure of a house, machine, animal, organ, poem, sentence; sentence of loose structure, its structure is ingenious; ornament should emphasize and not disguise the lines of structure),. The implications of the term 'structure' are clear, even from its daily sense. To have 'structure' we must have a complex of ordered and interrelated parts.

'Structure' is analysed in *Principia Mathematica* and is also simply explained in Russell's more popular works.[2] The *Tractatus* of Wittgenstein is built on structural considerations, although not much is explained about structure, for the author apparently assumes the reader's acquaintance with the works of Russell.[3]

One of the fundamental functions of 'mental' processes is to distinguish. We distinguish objects by certain characteristics, which are usually expressed by adjectives. If, by a higher order abstraction, we consider individual objects, not in some perfectly *fictitious* 'isolation', but as they appear empirically, as members of some aggregate or collection of objects, we find characteristics which belong to the collection and not to an 'isolated' object. Such characteristics as arise from the fact that the object belongs to a collection are called 'relations'.

In such collections, we have the possibility of *ordering* the objects, and so, for instance, we may discover a relation that one object is 'before' or 'after' the other, or that A is the father of B. There are many ways in which we can order a collection, and many relations which we can find. It is important to notice that 'order' and 'relations' are, for the most part, empirically present and that, therefore, this language is fit to represent the facts as we know them. The structure of the actual world is such that it is *impossible* entirely to isolate an object. An *A* subject-predicate language, with its tendency to treat objects as in isolation and to have no place for relations (impossible in complete 'isolation'), obviously has a structure not similar to the structure of the world, in which we deal *only* with collections, of which the members are related.

Obviously, under such empirical conditions, only a language originating in the analysis of collections, and, therefore, a language of 'relations', 'order'., would have a *similar structure* to the world around us. From the use of a subject-predicate form of language alone, many of our fallacious anti-social and 'individualistic' metaphysics and *s.r* follow, which we will not analyse here, except to mention that their structural implications follow the structure of the language they use.

If we carry the analysis a step further, we can find relations between relations, as, for instance, the *similarity of relations*. We follow the definition of Russell. Two relations are said to be similar if there is a *one-one* correspondence between the terms of their fields, which is such that, whenever two terms have the relation P, their correlates have the relation Q, and vice versa. For example, two series are similar when their terms can be correlated without change of order, an accurate map is similar to the territory it represents, a book spelt phonetically is similar to the sounds when read , .[4]

When two relations are similar, we say that they have a *similar structure,* which is defined as the class of all relations similar to the given relation.

We see that the terms 'collection', 'aggregate', 'class', 'order', 'relations', 'structure' are interconnected, each implying the others. If we decide to face empirical 'reality' boldly, we must accept the Einstein-Minkowski four-dimensional language, for 'space' and 'time' *cannot be separated empirically,* and so we must have a language of *similar structure* and consider the facts of the world as series of *interrelated ordered events,* to which, as above explained, we must ascribe 'structure'. Einstein's theory, in contrast to Newton's theory, gives us such a *language*, *similar in structure* to the empirical facts as revealed by science 1933 *and common experience*.

The above definitions are not entirely satisfactory for our purpose. To begin with, let us give an illustration, and indicate in what direction some reformulation could be made.

Let us take some actual territory in which cities appear in the following order: Paris, Dresden, Warsaw, when taken from the West to the East. If we were to build a *map* of this territory and place Paris *between* Dresden and Warsaw thus:

Actual territory   *————————*————————*
                   Paris   Dresden   Warsaw

Map   *————————*————————*
                   Dresden   Paris   Warsaw

we should say that the map was wrong, or that it was an incorrect map, or that the map has a *different structure* from the territory. If, speaking roughly, we should try, in our travels, to orient ourselves by such a map, we should find it misleading. It would lead us astray, and we might waste a great deal of unnecessary effort. In some cases, even, a map of wrong structure would bring actual suffering and disaster, as, for instance, in a war, or in the case of an urgent call for a physician.

Two important characteristics of maps should be noticed. A map *is not* the territory it represents, but, if correct, it has a *similar structure* to the territory, which accounts for its usefulness. If the map could be ideally correct, it would include, in a reduced scale, the map of the map; the map of the map, of the map; and so on, endlessly, a fact first noticed by Royce.

If we reflect upon our languages, we find that at best they must be considered *only as maps*. A word *is not* the object it represents; and languages exhibit also this peculiar self-reflexiveness, that we can analyse languages by linguistic means. This self-reflexiveness of languages introduces serious complexities, which can only be solved by the theory of multiordinality, given in Part VII. The disregard of these complexities is tragically disastrous in daily life and science.

It has been mentioned already that the known definitions of structure are not entirely satisfactory. The terms 'relation', 'order', 'structure' are interconnected by implication. At present, we usually consider order as a kind of relation. With the new four-dimensional notions taken from mathematics and physics, it may be possible to treat relations and structure as a form of *multi-dimensional order*. Perhaps, theoretically, such a change is not so important, but, from a practical, applied, educational, and semantic point of view, it seems very vital. Order seems *neurologically simpler* and more fundamental than relation. It is a characteristic of the empirical world which we recognize directly by our lower nervous centres ('senses'), and with which we can deal with great accuracy by our higher nervous centres ('thinking'). This term seems most distinctly of the organism-as-a-whole character, applicable both to the activities of the higher, as well as lower, nervous centres, and so *structurally* it must be fundamental.

The rest of this volume is devoted to showing that the common, A-system and language which we inherited from our primitive ancestors *differ entirely in structure* from the well-known and established 1933 structure of the world, ourselves and our nervous systems included. Such antiquated map-language, by necessity, must lead us to semantic disasters, as it imposes and reflects its *unnatural* structure on the structure of our doctrines and institutions. Obviously, under such *linguistic* conditions, a science of man was impossible; differing in structure from our nervous system, such language must also disorganize the functioning of the latter and lead us away from sanity.

This once understood, we shall see clearly that researches into the structure of language and the adjustment of this structure to the structure of the world and ourselves, as given by science at each date, must lead to new languages, new doctrines, institutions., and, in fine, may result in a new and saner civilization, involving new *s.r* which may be called the scientific era.

The introduction of a few new, and the rejection of some old, terms suggests desirable structural changes, and adjusts the structure of the language-map to the known structure of the world, ourselves, and the nervous system, and so leads us to new *s.r* and a theory of sanity.

As words *are not* the objects which they represent, *structure, and structure alone,* becomes the only link which connects our verbal processes with the empirical data. To achieve adjustment and sanity and the conditions which follow from them, we must study structural characteristics of this world *first,* and, then only, build languages of similar structure, instead of habitually ascribing to the world the primitive structure of our language. All our doctrines, institutions., depend on verbal arguments. If these arguments are conducted in a language of wrong and unnatural structure, our doctrines and institutions must reflect that linguistic structure and so become unnatural, and inevitably lead to disasters.

That languages, as such, all have some structure or other is a new and, perhaps, unexpected notion. Moreover, every language having a structure, by the very nature of language, reflects in its own structure that of the world as assumed by those who evolved the language. In other words, we read unconsciously into the world the structure of the language we use. The guessing and ascribing of a fanciful, mostly primitive-assumed, structure to the world is precisely what 'philosophy' and 'metaphysics' do. The empirical search for world-structure and the building of new languages (theories), of necessary, or similar, structure, is, on the contrary, what science does. Any one who will reflect upon these structural peculiarities of language cannot miss the semantic point that the scientific method uses the only correct language-method. It develops in the *natural order,* while metaphysics of every description uses the reversed, and ultimately a pathological, order.

Since Einstein and the newer quantum mechanics, it has become increasingly evident that the only content of 'knowing' is of a *structural* character; and the present theory attempts a formulation of this fact in a generalized way. If we build a $\bar{A}$-system by the aid of new terms and of methods excluded by the A-system, and stop some of our primitive habits of 'thought' and *s.r,* as, for instance, the confusion of order of abstractions, reverse the reversed order, and so introduce the natural order in our analysis, we shall then find that all human 'knowing' exhibits a structure similar to scientific knowledge, and appears as the *'knowing' of structure.* But, in order to arrive at these results, we must depart completely from the older systems, and must abandon permanently the use of the 'is' of identity.

It would seem that the overwhelming importance *for mankind* of systems based on 'relations', 'order', 'structure'., depends on the fact that such terms allow of an exact and 'logical' treatment, as two relations of similar structure have all their logical characteristics in common. It becomes obvious that, as in the A-system we could not deal in such terms, higher rationality and adjustment were impossible. It is not the human 'mind' and its 'finiteness' which is to be blamed, but a primitive language, with a structure foreign to this world, which has wrought havoc with our doctrines and institutions.

The use of the term 'structure' does not represent special difficulties when once we understand its origin and its meanings. The main difficulty is found in the old A habits of speech, which do not allow the use of structure, as, indeed, this notion has no place in a complete A subject-predicativism.

Let us repeat once more the two crucial *negative* premises as established firmly by *all* human experience: (1) Words *are not* the things we are speaking about; and (2) There *is no* such thing as an object in absolute isolation.

These two most important *negative* statements cannot be denied If any one chooses to deny them, the burden of the proof falls on him. He has to establish what he affirms, which is obviously impossible. We see that it is safe to start with such solid *negative* premises, translate them into positive language, and build a $\bar{A}$-system.

If words *are not* things, or maps *are not* the actual territory, then, obviously, the only possible link between the objective world and the linguistic world is found in *structure, and structure alone*. The only usefulness of a map or a language depends on the *similarity of structure* between the empirical world and the map-languages. If the structure is not similar, then the traveller or speaker is led astray, which, in serious human life-problems, must become always eminently harmful,. If the structures *are similar,* then the empirical world becomes 'rational' to a potentially rational being, which means no more than that verbal, or map-predicted characteristics, which follow up the linguistic or map-structure, are applicable to the empirical world.

In fact, in structure we find the mystery of rationality, adjustment., and we find that the whole content of knowledge is exclusively structural. If we want to be rational and to understand anything at all, we must look for structure, relations, and, ultimately, multi-dimensional order, all of which was impossible in a broader sense in the A-system, as will be explained later on.

Having come to such important *positive* results, starting with undeniable *negative* premises, it is interesting to investigate whether these results are *always* possible, or if there are limitations. The second *negative* premise; namely, that there *is no* such thing as an object in absolute isolation, gives us the answer. If there *is no* such thing as an absolutely isolated object, then, at least, we have two objects, and we shall *always* discover some relation between them, depending on our interest, ingenuity, and what not. Obviously, for a man to speak about anything at all, *always* presupposes *two* objects at least; namely, the object spoken about and the speaker, and so a *relation* between the two is always present. Even in delusions, illusions, and hallucinations, the situation is not changed; because our immediate feelings are also un-speakable and *not* words.

The semantic importance of the above should not be minimized. If we deal with organisms which possess an inherent activity, such as eating, breathing., and if we should *attempt to build for them conditions* where such activity would be impossible or hampered, these *imposed* conditions would lead to degeneration or death.

Similarly with 'rationality'. Once we find in this world at least potentially rational organisms, we should not *impose* on them conditions which hamper or prevent the exercise of such an important and inherent function. The present analysis shows that, under the all-pervading aristotelianism in daily life, asymmetrical relations, and thus structure and order, have been impossible, and so we have been *linguistically* prevented from supplying the potentially 'rational' being with the means for rationality. This resulted in a semi-human so-called 'civilization', based on our copying animals in our nervous process, which, by necessity, involves us in arrested development or regression, and, in general, disturbances of some sort.

Under such conditions, which, after all, may be considered as firmly established, because this investigation is based on undeniable *negative* premises, there is no way out but to carry the analysis through, and to build up a $\bar{A}$-system based on *negative* fundamental premises or the denial of the 'is' of identity with which rationality will be possible.

Perhaps an illustration will make it clearer, the more that the old subject-predicate language rather conceals structure. If we take a statement, 'This blade of grass is green', and analyse it only as a statement, superficially, we can hardly see how any structure could be implied by it. This statement may be analysed into substantives, adjectives, verb.; yet this would not say much about its structure. But if we notice that these words can also make a question, 'Is this blade of grass green?', we begin to realize that the *order* of the words plays an important role in some languages connected with the meanings, and so we can immediately speak of the structure of the sentence. Further analysis would disclose that the sentence under consideration has the subject-predicate form or structure.

If we went to the objective, silent, un-speakable level, and analysed this objective blade of grass, we should discover various structural characteristics in the blade; but these are not involved in the statement under consideration, and it would be illegitimate to speak about them. However, we can carry our analysis in another direction. If we carry it far enough, we shall discover a very intricate, yet definite, relation or complex of relations between the objective blade of grass and the observer. Rays of light impinge upon the blade, are reflected from it, fall on the retina of our eye, and produce within our skins the feeling of 'green'., an extremely complex process which has some definite structure.

We see, thus, that any statement referring to anything objective in this world can always be analysed into terms of relations and structure, and that it involves also definite structural assumptions. More than that, as the only possible content of knowledge and science is structural, whether we like it or not, to *know* anything we must search for structure, or posit some structure. Every statement can also be analysed until we come to definite structural issues. This applies, however, with certainty only to significant statements, and, perhaps, not to the various noises which we can make with our mouth with the semblance of words, but which are meaningless, as they are not symbols for anything. It must be added that in the older systems we did not discriminate between words (symbols) and noises (not symbols). In a $\bar{A}$-system such a discrimination is essential.

The structure of the world is, in principle, *unknown;* and the only aim of knowledge and science is to discover this structure. The structure of languages is potentially *known,* if we pay attention to it. Our only possible procedure in advancing our knowledge is to match our verbal structures, often called theories, with empirical structures, and see if our verbal predictions are fulfilled empirically or not, thus indicating that the two structures are either similar or dissimilar.

We see, thus, that in the investigation of structure we find not only means for rationality and for adjustment, and so sanity, but also a most important tool for exploring this world and scientific advance.

From the educational point of view, also, the results of such an investigation seem to be unusually important, because they are extremely simple, *automatic* in their working, and can be applied universally in elementary education. As the issue is merely one of linguistic structure, it is enough to train children to abandon the 'is' of identity, in the habitual use of a *few new terms,* and to warn them repeatedly against the use of some terms of antiquated structure. We shall thus eliminate the pre-human and primitive semantic factors included in the structure of a primitive language. The moralizing and combating of primitive-made metaphysics is not effectual; but the habitual use of a language of modern structure, free from identity, produces semantic results where the old failed. Let us repeat again, a most important point, that the new desirable semantic results follow as *automatically* as the old undesirable ones followed.

It should be noticed that terms such as 'collection', 'fact', 'reality', 'function', 'relation', 'order', 'structure', 'characteristics', 'problem'., must be considered as *multiordinal terms* (see Part VII), and so, in general ∞-valued and ambiguous. They become specific and one-valued only in a given context, or when the orders of abstractions are distinguished.

In the following enquiry an attempt to build a science of man, or a *non-aristotelian system,* or a theory of sanity, is made, and it will be necessary to introduce a few terms of new structure and to abide by them.

Let me be entirely frank about it: the main issues are found in the *structure* of language, and readers who are interested in this work will facilitate their task if they make themselves familiar with these new terms and use them habitually. This work will then appear simple, and often self-evident. For those other readers who insist on translating the new terms with *new structural implications* into their old habitual language, and choose to retain *the old terms* with *old structural implications* and old *s.r.*, this work will not appear simple.

Examples illustrating what has just been said abound; here I shall mention only that the $\bar{E}$ geometries, the new revision of mathematics originated by Brouwer and Weyl, the Einstein theory, and the newer quantum mechanics., have similar main aims; namely, to produce *non-el* statements which are structurally closer to the empirical facts than the older theories, and to reject those unwarranted structural assumptions which vitiated the old theories. The reader should not be surprised to learn that these new theories are not a passing whim of scientists, but represent lasting advances *in method*. Whether these attempts at restatements are finally found to be valid or not, they remain steps in the right direction.

It is quite natural that with the advance of experimental science some generalizations should appear to be established from the facts at hand. Occasionally, such generalizations, when further analysed, are found to contain serious structural, epistemological and methodological implications and difficulties. In the present work one of these empirical generalizations becomes of unusual importance, so important, indeed, that Part III of this work is devoted to it. Here, however, it is only possible to mention it, and to show some rather unexpected consequences which it entails.

That generalization states: that *any* organism must be treated as-a-whole; in other words, that the organism is not an algebraic sum, a *linear* function of its elements, but always *more* than that. It is seemingly little realized, at present, that this simple and innocent-looking statement involves a full structural revision of our language, because that language, of great pre-scientific antiquity, is *elementalistic,* and so singularly inadequate to express *non-elementalistic* notions. Such a point of view involves profound structural, methodological, and semantic changes, vaguely anticipated, but never formulated in a definite theory. The problems of structure, 'more', and 'non-additivity' are very important and impossible to analyse in the old way.

If this generalization be accepted—and on experimental, structural, and epistemological grounds we cannot deny its complete structural justification—some odd consequences follow; that is to say, odd, as long as we are not accustomed to them. For instance, we see that 'emotion' and 'intellect' cannot be divided, that this division structurally violates the organism-as-a-whole generalization. We must, then, choose between the two: we must either abandon the organism-as-a-whole principle, or abandon accepted speculations couched in *el* verbal terms which create insoluble *verbal* puzzles. Something similar could be said about the distinction of 'body' versus 'soul', and other verbal splittings which have hampered sane advance in the understanding of ourselves, and have filled for thousands of years the libraries and tribunes of the world with hollow reverberations.

The solution of these problems lies in the field of structural, symbolic, linguistic, and semantic research, as well as in the fields of physics, chemistry, biology, psychiatry., because from their very nature these problems are structural.

5

CHAPTER V

GENERAL LINGUISTIC

... to be an abstraction does not mean that an entity is nothing. It merely means that its existence is only one factor of a more concrete element of nature. (573)   A. N. WHITEHEAD

In my opinion the answer to this question is briefly, this:—As far as the laws of mathematics refer to reality, they are not certain; and as far as they are certain, they do not refer to reality. (151)   A. EINSTEIN

Thus it would seem that, wherever we infer from perceptions, it is only structure that we can validly infer; and structure is what can be expressed by mathematical logic, which includes mathematics. (457)
                                                        BERTRAND RUSSELL

The current accounts of perception are the stronghold of modern metaphysical difficulties. They have their origin in the same misunderstanding which led to the incubus of the substance-quality categories. The Greeks looked at a stone, and perceived that it was grey. The Greeks were ignorant of modern physics; but modern philosophers discuss perception in terms of categories derived from the Greeks. (574)   A. N. WHITEHEAD

To the biochemist, biophysicist, biologist, and physiological psychologist, however, life and mind are so amazingly complex and comprise so many heterogeneous processes that their blanket designation as two emergent levels cannot seem very illuminating, and to the observer who contemplates the profuse and unabated emergence of idiots, morons, lunatics, criminals, and parasites in our midst, Alexander's prospect of the emergence of deity is about as imminent as the Greek kalends. (555)
                                                WILLIAM MORTON WHEELER

In speaking of linguistic researches, I do not mean only an analysis of printed 'canned chatter', as Clarence Day would call it, but I mean the behaviour, the performance, *s.r* of living Smiths and Browns and the connections between the noises uttered by them and their behaviour. No satisfactory analysis has been made, and the reason seems to be in the fact that each existing language really represents a conglomeration of *different* languages with different structures and is, therefore, extremely complex as long as structure is disregarded. That 'linguists', 'psychologists', 'logicians'., were, and usually are, very innocent of *mathematics,* a type of language of the greatest simplicity and perfection, with a clearcut structure, similar to the structure of the world, seems to be responsible for this helplessness. Without a study of mathematics, the adjustment of structure seems impossible.

We should not be surprised to find that mathematics must be considered a language. By definition, whatever has symbols and propositions is called a language, a form of representation for this something-going-on which we call the world and which is admittedly *not words*. Several interesting statements can be made about mathematics considered as a language. First of all, mathematics appears as a form of human behaviour, as genuine a human activity as eating or walking, a function in which the human nervous system plays a very serious part. Second, from an empirical point of view a curious question arises: why, of all forms of human behaviour, has mathematizing proved to be *at each historical period* the most excellent human activity, producing results of such enormous importance and unexpected validity as not to be comparable with any other musings of man? Briefly, it may be said that the secret of this importance and the validity of mathematics lie in the mathematical *method* and structure, which the mathematizing Smith, Brown, and Jones have used—we may even say, were *forced* to use. It is not necessary to assume that the mathematicians were 'superior' men. We will see later that mathematics is not a very superior activity of the 'human mind', but it is perhaps the *easiest*, or simplest activity; and, therefore, it has been possible to produce a structurally perfect product of this simple kind.

The understanding and proper evaluation of what has been said about the structure and method of mathematics will play a serious semantic role all through this work, and, therefore, it becomes necessary to enlarge upon the subject. We shall have to divide the abstractions we make into two classes: (1) objective or physical abstractions, which include our daily-life notions; and (2) mathematical abstractions, at present taken from pure mathematics, in a restricted sense, and later generalized. As an example of a mathematical abstraction, we may take a mathematical circle. A circle is defined as the locus of all points in a plane at equal distance from a point called the centre. If we enquire whether or not there is such an actual thing as a circle, some readers may be surprised to find that a mathematical circle must be considered a pure fiction, having nowhere any objective existence. In our definition of a mathematical circle, *all particulars* were included, and whatever we may find about this mathematical circle later on will be strictly dependent on this definition, and no new characteristics, not already included in the definition, will ever appear. We see, here, that *mathematical abstractions are characterized by the fact that they have all particulars included.*

If, on the other hand, we draw an objective 'circle' on a blackboard or on a piece of paper, simple reflection will show that what we have drawn is *not* a mathematical circle, but a *ring*. It has colour, temperature, thickness of our chalk or pencil mark ,. When we draw a 'circle', it is no longer a mathematical circle with *all particulars included in the definition,* but it becomes a *physical ring* in which *new characteristics* appear not listed in our definition.

From the above observations, very important consequences follow. Mathematizing represents a very simple and easy human activity, because it deals with fictitious entities with all particulars included, and we proceed by remembering. The structure of mathematics, because of this over-simplicity, yet structural similarity with the external world, makes it possible for man to build verbal systems of remarkable validity.

Physical or daily-life abstractions differ considerably from mathematical abstractions. Let us take any actual object; for instance, what we call a pencil. Now, we may describe or 'define' a 'pencil' in as great detail as we please, yet it is impossible to include *all* the characteristics which we may discover in this actual objective pencil. If the reader will try to give a 'complete' description or a 'perfect' definition of any actual physical object, so as to include 'all' particulars, he will be convinced that this task is humanly impossible. These would have to describe, not only the numerous rough, macroscopic characteristics, but also the microscopic details, the chemical composition and changes, sub-microscopic characteristics and the endlessly changing relationship of this objective something which we have called pencil to the rest of the universe., an inexhaustible array of characteristics which could never be terminated. In general, physical abstractions, including daily-life abstractions are such that *particulars are left out*—we proceed by a process of forgetting. In other words, no description or 'definition' will ever include all particulars.

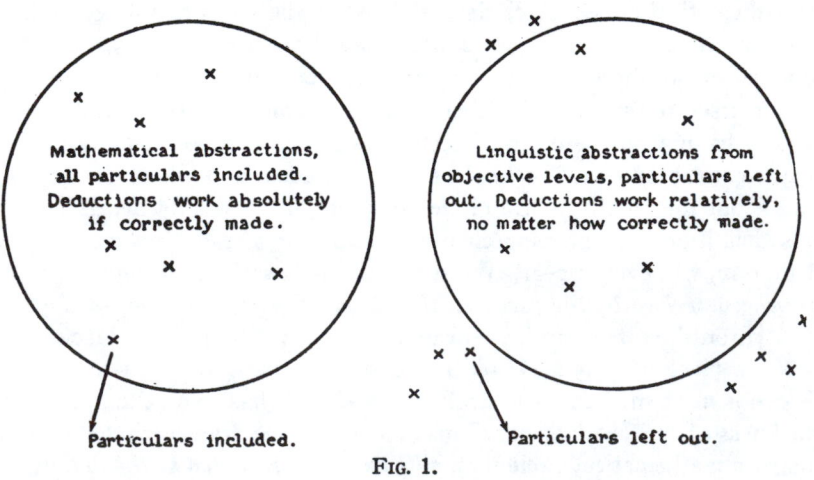

Fig. 1.

Only and exclusively in mathematics does deduction, if correct, work absolutely, for no particulars are left out which may later be discovered and force us to modify our deductions.

Not so in abstracting from physical objects. Here, particulars are left out; we proceed by forgetting, our deductions work only relatively, and must be revised continuously whenever new particulars are discovered. In mathematics, however, we build for ourselves a fictitious and *over-simplified* verbal world, with abstractions which have all particulars included. If we compare mathematics, taken as a language, with our daily language, we see readily that in both verbal activities we are building for ourselves forms of representation for this something-going-on, which is *not* words.

Considered as a language, mathematics appears as a language of the highest perfection, but at its lowest development. Perfect, because the structure of mathematics makes it possible to be so (all characteristics included, and no physical content), and because it is a language of *relations* which are also found in this world. At the lowest development, because we can speak in it as yet about very little and that in a very narrow, restricted field, and with limited aspects.

Our other languages would appear, then, as the other extreme, as the highest mathematics, but also at their lowest development—highest mathematics, because in them we can speak about everything; at their lowest development because they are still A and not based on asymmetrical relations. Between the two languages there exists as yet a large unbridged structural gap. The bridging of this gap is the problem of the workers of the future. Some will work in the direction of inventing new mathematical methods and systems, bringing mathematics closer in scope and adaptability to ordinary language (for instance, the tensor calculus, the theory of groups, the theory of sets, the algebra of states and observables,.). Others will undertake linguistic researches designed to bring ordinary language closer to mathematics (for instance, the present work). When the two forms of representation meet on relational grounds, we shall probably have a simple language of mathematical structure, and mathematics, as such, might then even become obsolete.

It is not desirable that the reader should be under the impression that all mathematical 'thinking' is low-grade 'thinking'. The mathematicians who discover or invent new *methods* for relating and structures are the biggest 'mental' giants we have had, or ever shall have. Only the technical interplay of symbols, to find out some new possible combination, can be considered as low-grade 'thinking'.

From what has now been said, it is probably already obvious that if any one wants to work scientifically on problems of such enormous complexity that they have so far defied analysis, he would be helped enormously if he would train his *s.r* in the simplest forms of correct 'thinking'; that is, become acquainted with mathematical methods. The continued application of this relational method should finally throw some light on the greatest complexities, such as life and man. In contrast to enormous advances in all technical fields, our knowledge of 'human nature' has advanced very little beyond what primitives knew about themselves. We have tried to analyse the most baffling phenomena while disregarding structural peculiarities of languages and thus failing to provide sufficient fundamental training in new *s.r*. In practically all universities at present, the mathematical requirements, even for scientists, are extremely low, much lower, indeed, than is necessary for the progress of these scientists themselves. Only those who specialize in mathematics receive an advanced training, but, even with them, little attention is devoted to *method* and *structure* of languages *as such*. Until lately, mathematicians themselves were not without responsibility for this. They treated mathematics as some kind of 'eternal verity', and made a sort of religion out of it; forgetting, or not knowing, that these 'eternal verities' last only so long as the nervous systems of Smiths and Browns are not altered. Besides, many, even now, disclaim any possible connection between mathematics and human affairs. Some of them seem, indeed, in their religious zeal, to try to make their subjects as difficult, unattractive, and mysterious as possible, to overawe the student. Fortunately, a strong reaction against such an attitude is beginning to take place among the members of the younger mathematical generation. This is a very hopeful sign, as there is little doubt that, without the help of professional mathematicians who will understand the general importance of *structure* and *mathematical methods,* we shall not be able to solve our human problems in time to prevent quite serious break-downs, since these solutions ultimately depend on structural and semantic considerations.

The moment we abandon the older theological attitude toward mathematics, and summon the courage to consider it as a form of human behaviour and the expression of *generalized s.r*, some quite interesting problems loom up. Terms like 'logic' or 'psychology' are applied in many different senses, but, among others, they are used as labels for certain disciplines called sciences. 'Logic' is defined as the 'science of the laws of thought'. Obviously, then, to produce 'logic' we should have to study *all* forms of human behaviour connected directly with mentation; we should have to study not only the mentations in the daily life of the average Smiths, Browns., but we should have to study the mentations of Joneses and Whites when they use their 'mind' at its best; namely, when they mathematize, scientize., and we should also have to study the mentations of those whom we call 'insane', when they use their 'mind' at its worst. It is not our aim to give a detailed list of these forms of human behaviour which we should study, since all should be studied. It is enough for our purpose to emphasize the two main omissions; namely, the study of mathematics and the study of 'insanity'.

As a similar reasoning applies to 'psychology', we must sadly admit that we have as yet no general theory which deserves the name of 'logic' or psycho-logics. What has passed under the name of 'logic', for instance, is not 'logic' according to its own definition, but represents a philosophical grammar of a primitive-made language, of a structure different from the structure of the world, unfit for serious use. If we try to apply the rules of the old 'logic', we find ourselves blocked by ridiculous impasses. So, naturally, we discover that we have no use for such a 'logic'.

It follows also that any one who has any serious intention of becoming a 'logician' or a psycho-logician must, first of all, be a thorough mathematician and must also study 'insanity'. Only with such preparation is there any possibility of becoming a psycho-logician or semantician. Sometimes it is useful to stop deceiving ourselves; and it is deceiving ourselves if we claim to be studying *human* psycho-logics, or *human* 'logic', when we are generalizing only from those forms of human behaviour which we have in common with the animals and neglect other forms, especially the most characteristic forms of human behaviour, such as mathematics, science, and 'insanity'. If, as psycho-logicians, we want to be 'behaviourists', it is clear that we must study *all* known forms of human behaviour. But it seems never to have occurred to the 'behaviourists' that mathematics and 'insanity' are very characteristic forms of human behaviour.

Some readers may be puzzled by my calling the daily forms of representation we use 'primitive-made'. Let me illustrate what I mean by a classical example. For more than two thousand years the famous paradox of Zeno has puzzled 'philosophers', without any solution, and only in our own day has it been solved by mathematicians. The paradox reads: Achilles was supposed to be a very swift runner, and in a race with a tortoise, which was given the benefit of starting first, Achilles could never overtake his slow competitor, because, the argument runs, before he could overtake the tortoise he would have to halve the distance between them, and again halve the remaining half, and so on. No matter how long this might last, there still would be some distance to halve, and so it was concluded he could never pass the tortoise. Now any child knows that this conclusion is not true; yet the *verbal* argument for the untrue conclusion remained, in the hands of 'philosophers' and 'logicians', perfectly valid for more than two thousand years. This instance throws light on the stage of development which we have reached and of which we often boast.

Having, then, no scientific *general* theory of 'logic' and psycho-logics to guide us, the task of an enquiry like the present is very much handicapped. We must merely go ahead groping and pioneering; and this is always a difficult, blundering task.

It is indeed very important that not only the scientists but also the intelligent public, as a whole, should understand that at present we have no general theory which may be called 'logic' or psycho-logics. Perhaps an illustration will help to bring home this really shocking state of affairs. Imagine, for example, that we should try to study dinosaurs exhaustively. The standard methods of study would centre about the actual fossil remains when such are available; but, in the case of those extinct forms of which the fossil remains are very meagre, or entirely lacking, much information is obtained from the study of the tracks which have been left on the mud flats that have become rocks. It seems undeniable that such a study of fossil tracks would contribute a large share to the formulation of any 'general theory' of the characteristics of dinosaurs. We could go further and say that no 'general theory' could be complete if such study were entirely neglected.

Now, that is precisely the situation in which 'psychologists' and 'logicians' find themselves; they have made many studies, and gathered some facts, but they have entirely disregarded as yet these unique and peculiar black tracks which the mathematicians and others have left on white paper when they mathematized or scientized. The old 'psychological' generalizations were made from insufficient data, in spite of the fact that *sufficient* data; namely, these black marks on white paper, exist, and have existed for a long time. But these marks the 'psychologists' and 'logicians' were not able properly to read, analyse, and interpret.

Under such circumstances, it should not be surprising to find that, in the study of animals, we have vitiated our researches by reading into the animals our own activities, and that we have vitiated our own understanding of ourselves by faulty generalizations from a few data taken mostly from those activities which we have in common with animals. Thus we measure ourselves by animalistic standards. This error is mainly due to the ignorance of mathematical method and the disregard of structural problems by those who deal with human affairs. Indeed, as I have already shown in my *Manhood of Humanity,* what we call 'civilization' rests upon faulty generalizations taken from the lives of cows, horses, cats, dogs, pigs., and self-imposed upon Smith and Brown.

The main thesis of this $\bar{A}$-system is that *as yet we all (with extremely few exceptions) copy animals in our nervous processes,* and that practically *all* human difficulties, 'mental' ills of all degrees included, have this characteristic as a component. I am glad to be able to report that a number of experiments undertaken with 'mentally' or nervously ill individuals have shown decided benefit in cases where it proved possible to re-educate them to appropriate human *s.r.*

Here, perhaps, it may be advisable to interpolate a short explanation. When we deal with human affairs and man, we sometimes use the term 'ought', which is very often used arbitrarily, dogmatically, and absolutistically, and so its use has become discredited. In many quarters, this term is very unpopular, and, it must be admitted, justly so. My use of it is that of the engineer, who undertakes to study a machine entirely unknown to him—let us say, a motorcycle. He would study and analyse *its structure,* and, finally, would give a verdict that with such a structure, under certain circumstances, this machine *ought* to work in a particular way.

In the present volume, this engineering attitude is preserved. We shall investigate the structure of human knowledge, and we shall conclude that with such a structure it should work in this particular way. In the motorcycle example, the proof of the correctness of the reasoning of the engineer would be to fill the tank with gasoline and make the motorcycle go. In our analogous task, we have to *apply* the information we get and see if it works. In the experiments mentioned above, the $\bar{A}$-system actually has worked, and so there is some hope that it is correct. Further investigations will, of course, add to, or modify, the details, but this is true of all theories.

Another reason why a non-mathematician cannot study psychological phenomena adequately is that mathematics is the only science which has *no physical content* and, therefore, when we study the performances of Smiths and Browns when they mathematize, we study the *only* available working of 'pure mind'. Moreover, mathematics is the only language which at present has a structure similar to that of the world and the nervous system. It must be obvious that from such a study we should learn more than by the study of any other 'mental' activity. In some quarters it is believed, I think erroneously, that 'psychology' and 'logic' have no 'physical content'. 'Psychology' and 'logic' have a very definite content—Smith, Brown.,—and we should treat these disciplines in relation to the living organism. Quite probably, when the above issues are fully realized, these specialists, future psycho-logicians and semanticians, will begin to study mathematical methods and pay attention to structure, and a number of mathematicians, in their turn, will become psycho-logicians, psychiatrists, semanticians,. When this happens, we may expect marked advance in these lines of endeavour.*

In the course of this book, it will be shown that the structure of human knowledge precludes any serious study of 'mental' problems without a thorough mathematical training. We shall take for granted all the partial light thrown on man by existing disciplines and shall make some *observations* from the study of the *neglected* forms of human behaviour, such as mathematics, exact sciences, and 'insanity', and with these new data re-formulate, in the rough, all available data at hand in 1933.

At the present early stage of our enquiry, we must, of necessity, be often vague. Before we give the new data, it is impossible to speak in a more definite way. Besides, in such a general survey, we shall have to use what I call *multiordinal* terms. At present, all the most humanly important and interesting terms are multiordinal, and no one can evade the use of such terms. Multiordinality is inherent in the structure of 'human knowledge'. This multiordinal mechanism gives the key to many seemingly insoluble contradictions, and explains why we have scarcely progressed at all in the solution of many human affairs.

The main characteristic of these multiordinal terms is found in that they have *different meanings* in general, depending on the order of abstractions. Without the level of abstraction being specified, a *m.o* term is only ambiguous; its use involves shifting meanings, variables, and therefore generates, not propositions, but propositional functions. It may not be an exaggeration to say that the larger number of human tragedies, private, social, racial., are intimately connected with the non-realization of this multiordinality of the most important terms we use.

A similar confusion between orders of abstractions is to be found in all forms of 'insanity', from the mildest, which afflicts practically

---

*There are already signs that the more serious workers, as, for instance, the Gestalt 'psychologists', begin to feel their handicaps. Others, as yet, do not seem to realize the hopelessness of their endeavours—as best exemplified by the American school of Behaviourists, who seem to think that the splendid name they have selected will solve their problems. It would be very interesting to see the Behaviourists *deny* that the writing of a mathematical treatise, or of some new theory of quantum mechanics represents a form of *human behaviour* which they should study. Some day they must face the fact that they have neglected to consider a great many forms of human behaviour—the *most characteristic* forms at that—and that, therefore, they could not produce an adequate theory of the nature of the 'human mind'.

every one of us, to the most pronounced and violent. Indeed, the discovery of this mechanism leads conversely to a *theory of sanity*. Imperfect as this theory of sanity probably is, it opens a wide field of possibilities which I myself, at this stage, am unable fully to appreciate.

There seems one thing certain, at present; namely, that the old theories and methods tended strongly to produce morons and 'insane' persons, while 'geniuses' were only born in spite of these handicaps. Perhaps in the future we shall be able to produce 'geniuses', while morons and 'insane' persons will be born only in spite of our precautions. If this should actually prove to be true, and the experimental results seem to give some hope in this direction, this world would then become quite a different place in which to live.

CHAPTER VI

ON SYMBOLISM

Philosophers have worried themselves about remote consequences, and the inductive formulations of science. They should confine attention to the rush of immediate transition. Their explanations would then be seen in their native absurdity. (578)   A. N. WHITEHEAD

It is often said experiments must be made without a preconceived idea. That is impossible. Not only would it make all experiment barren, but that would be attempted which could not be done. Every one carries in his mind his own conception of the world, of which he can not so easily rid himself. We must, for instance, use language; and our language is made up only of preconceived ideas and can not be otherwise. Only these are unconscious preconceived ideas, a thousand times more dangerous than the others. (417)   H. POINCARÉ

. . . the patriotic archbishop of Canterbury, found it advisable—' "
"Found *what?*" said the Duck.
"Found *it*," the Mouse replied, rather crossly: "of course you know what 'it' means."
"I know what 'it' means well enough, when *I* find a thing," said the Duck: "it's generally a frog, or a worm."*   LEWIS CARROLL

. . . psychiatry works specifically on the social organ of man itself— the person's assets and behavior, that which we must adjust before we can expect the individual to make proper use of most of our help.**
                                                              ADOLF MEYER

Perhaps, as has often been said, the trouble with people is not so much with their ignorance as it is with their knowing so many things that are not so. . . . So that it is always important to find out about these fears, and if they are based upon the knowledge of something that is not so, they may perhaps be corrected. (568)   WILLIAM A. WHITE

The affairs of man are conducted by our own, man-made rules and according to man-made theories. Man's achievements rest upon the use of symbols. For this reason, we must consider ourselves as a symbolic, semantic class of life, and those who rule the symbols, rule us. Now, the term 'symbol' applies to a variety of things, words and money included. A piece of paper, called a dollar or a pound, has very little value if the other fellow refuses to take it; so we see that money must be considered as a symbol for human agreement, as well as deeds to property, stocks, bonds,. The *reality* behind the money-symbol is doctrinal, 'mental', and one of the most precious characteristics of mankind. But it must be used properly; that is, with the proper understanding of

---

*Alice in Wonderland.
**Historical Sketch and Outlook of Psychiatric Social Work. *Hosp. Soc. Serv.* V, 1922, 221.

its structure and ways of functioning. It constitutes a grave danger when misused.

When we say 'our rulers', we mean those who are engaged in the manipulation of symbols. There is no escape from the fact that they do, and that they always will, rule mankind, because we constitute a symbolic class of life, and we cannot cease from being so, except by regressing to the animal level.

The hope for the future consists in the understanding of this fact; namely, that we shall always be ruled by those who rule symbols, which will lead to scientific researches in the field of symbolism and *s.r.* We would then *demand* that our rulers should be *enlightened* and *carefully selected*. Paradoxical as it may seem, such researches as the present work attempts, will ultimately do more for the stabilization of human affairs than legions of policemen with machine guns, and bombs, and jails, and asylums for the maladjusted.

A complete list of our rulers is difficult to give; yet, a few classes of them are quite obvious. Bankers, priests, lawyers and politicians constitute one class and work together. They do not *produce* any values, but manipulate values produced by others, and often pass signs for no values at all. Scientists and teachers also comprise a ruling class. They produce the main values mankind has, but, at present, they do not realize this. They are, in the main, themselves ruled by the cunning methods of the first class.

In this analysis the 'philosophers' have been omitted. This is because they require a special treatment. As an historical fact, many 'philosophers' have played an important and, to be frank, sinister role in history. At the bottom of any historical trend, we find a certain 'philosophy', a structural implication cleverly formulated by some 'philosopher'. The reader of this work will later find that most 'philosophers' gamble on multiordinal and *el* terms, which have *no definite single (one-valued) meaning*, and so, by cleverness in twisting, can be made to appear to mean anything desired. It is now no mystery that some quite influential 'philosophers' were 'mentally' ill. Some 'mentally' ill persons are tremendously clever in the manipulation of words and can sometimes deceive even trained specialists. Among the clever concoctions which appear in history as 'philosophic' systems, we can find flatly opposing doctrines. Therefore, it has not been difficult at any period for the rulers to select a cleverly constructed doctrine perfectly fitting the ends they desired.

One of the main characteristics of such 'philosophers' is found in the delusion of grandeur, the 'Jehovah complex'. Their problems have appeared to them to be above criticism or assistance by other human beings, and the correct procedure known only to super-men like themselves. So quite naturally they have usually refused to make enquiries. They have refused even to be informed about scientific researches carried on outside the realms of their 'philosophy'. Because of this ignorance, they have, in the main, not even suspected the importance of the problems of structure.

In all fairness, it must be said that not all so-called 'philosophy' represents an episode of semantic illness, and that a few 'philosophers' really do important work. This applies to the so-called 'critical philosophy' and to the *theory of knowledge* or *epistemology*. This class of workers I call epistemologists, to avoid the disagreeable implications of the term 'philosopher'. Unfortunately, epistemological researches are most difficult, owing mainly to the lack of scientific psycho-logics, general semantics, and investigations of structure and *s.r.* We find only a very few men doing this work, which, in the main, is still little known and unapplied. It must be granted that their works do not make easy reading. They do not command headlines; nor are they aided and stimulated by public interest and help.

It must be emphasized again that as long as we remain humans (which means a symbolic class of life), the rulers of symbols will rule us, and that no amount of revolution will ever change this. But what mankind has a right to ask—and the sooner the better—is that our rulers should not be so shamelessly ignorant and, therefore, pathological in their reactions. If a psychiatrical and scientific enquiry were to be made upon our rulers, mankind would be appalled at the disclosures.

We have been speaking about 'symbols', but we have not yet discovered any general theory concerning symbols and symbolism. Usually, we take terms lightly and never 'think' what kind of implication and *s.r* one single important term may involve. 'Symbol' is one of those important terms, weighty in meanings. If we use the term 'food', for instance, the presupposition is that we take for granted the existence of living beings able to eat; and, similarly, the term 'symbol' implies the existence of intelligent beings. The solution of the problem of symbolism, therefore, presupposes the solution of the problem of 'intelligence' and structure. So, we see that the issues are not only serious and difficult, but, also, that we must investigate a semantic field in which very little has been done.

In the rough, a symbol is defined as a sign which stands for something. Any sign is not necessarily a symbol. If it stands for something, it becomes a symbol for this something. If it does not stand for something, then it becomes not a symbol *but* a *meaningless* sign. This applies to words just as it does to bank cheques. If one has a zero balance in the bank, but still has a cheque-book and issues a cheque, he issues a sign but not a symbol, because it does not stand for anything. The penalty for such use of these particular signs as symbols is usually jailing. This analogy applies to the oral noises we make, which occasionally become symbols and at other times do not; as yet, no penalty is exacted for such a fraud.

Before a noise., may become a symbol, something must exist for the symbol to symbolize. So the first problem of symbolism should be to investigate the problem of 'existence'. To define 'existence', we have to state the standards by which we judge existence. At present, the use of this term is not uniform and is largely a matter of convenience. Of late, mathematicians have discovered a great deal about this term. For our present purposes, we may accept two kinds of existence: (1) the physical existence, roughly connected with our 'senses' and persistence, and (2) 'logical' existence. The new researches in the foundations of mathematics, originated by Brouwer and Weyl, seem to lead to a curtailment of the meaning of 'logical' existence in quite a sound direction; but we may provisionally accept the most general meaning, as introduced by Poincaré. He defines 'logical' existence as a statement free from self-contradictions. Thus, we may say that a 'thought' to be a 'thought' must not be self-contradictory. A self-contradictory statement is meaningless; we can argue either way without reaching any valid results. We say, then, that a self-contradictory statement has no 'logical' existence. As an example, let us take a statement about a square circle. This is called a contradiction in terms, a non-sense, a meaningless statement, which has no 'logical' existence. Let us label this 'word salad' by a special noise—let us say, 'blah-blah'. Will such a noise become a word, a symbol? Obviously not—it stands for nothing; it remains a mere noise., no matter if volumes should be written about it.

It is extremely important, semantically, to notice that not all the noises., we humans make should be considered as symbols or valid words. Such empty noises., can occur not only in direct 'statements', but also in 'questions'. Quite obviously, 'questions' which employ noises., instead of words, are not significant questions. They ask nothing, and cannot be answered. They are, perhaps, best treated by 'mental' pathologists as symptoms of delusions, illusions, or hallucinations. In asylums the noises., patients make are predominantly meaningless, as far as the external world is concerned, but *become symbols in the illness of the patient.*

A complicated and difficult problem is found in connection with those symbols which have meaning in one context and have no meaning in another context. Here we approach the question of the application of 'correct symbolism to facts'. We will not now enlarge upon this subject, but will only give, in a different wording, an illustration borrowed from Einstein. Let us take anything; for example, a pencil. Let us assume that this physical object has a temperature of 60 degrees. Then the 'question' may be asked: 'What is the temperature of an "electron" which goes to make up this pencil?' Different people, many scientists and mathematicians included, would say: '60 degrees'; or any other number. And, finally, some would say: 'I do not know'. All these answers have one common characteristic; namely, that they are senseless; for they try to answer a meaningless question. Even the answer, 'I do not know', does not escape this classification, as there *is nothing to know about a meaningless question.* The only correct answer is to explain that the 'question' has no meaning. This is an example of a symbol which has no application to an 'electron'. Temperature by *definition* is the vibration of molecules (atoms are considered as mon-atomic molecules) ; so to have temperature at all, we must have at least two molecules. Thus, when we take one molecule and split it into atoms and electrons, the term 'temperature' does not apply by definition to an electron at all. Although the term 'temperature' represents a perfectly good symbol in one context, it becomes a meaningless noise in another. The reader should not miss the plausibility of such gambling on words, for there is a very real semantic danger in it.

In the study of symbolism, it is unwise to disregard the knowledge we gather from psychiatry. The so-called 'mentally' ill have often a very obvious and well-known semantic mechanism of projection. They project their own feelings, moods, and other structural implications on the outside world, and so build up delusions, illusions, and hallucinations, believing that what is going on *in* them is going on *outside* of them. Usually, it is impossible to convince the patient of this error, for his whole illness is found in the semantic disturbance which leads to such projections.

In daily life we find endless examples of such semantic projections, of differing affective intensity, which projections invariably lead to consequences more or less grave. The structure of such affective projections will be dealt with extensively later on. Here we need only point out that the problems of 'existence' are serious, and that any one who claims that something 'exists' outside of his skin must show it. Otherwise, the 'existence' is found only inside of his skin—a psycho-logical state of affairs which becomes pathological the moment he projects it on the outside world. If one should claim that the term 'unicorn' is a symbol, he must state what this symbol stands for. It might be said that 'unicorn', as a symbol, stands for a *fanciful* animal in heraldry, a statement which happens to be true. In such a sense the term 'unicorn' becomes a symbol for a fancy, and rightly belongs to psycho-logics, which deals with human fancies, but does not belong to zoology, which deals with actual animals. But if one should believe firmly and intensely that the 'unicorn' represents an actual animal which has an external existence, he would be either mistaken or ignorant, and could be convinced or enlightened; or, if not, he would be seriously ill. We see that in this case, as in many others, all depends to what 'ology' our semantic impulses assign some 'existence'. If we assign the 'unicorn' to psycho-logics, or to heraldry, such an assignment is correct, and no semantic harm is done; but if we assign a 'unicorn' to zoology; that is to say, if we believe that a 'unicorn' has an objective and not a fictitious existence, this *s.r* might be either a mistake, or ignorance, and, in such a case, it could be corrected; otherwise, it becomes a semantic illness. If, in spite of all contrary evidence, or of the lack of all positive evidence, we hold persistently to the belief, then the affective components of our *s.r* are so strong that they are beyond normal control. Usually a person holding such affective beliefs is seriously ill, and, therefore, no amount of evidence can convince him.

We see, then, that it is not a matter of indifference to what 'ology' we assign terms, and some assignments may be of a pathological character, if they identify psycho-logical entities with the outside world. Life is full of such dramatic identifications, and it would be a great step forward in semantic hygiene if some 'ologies'—e.g., demonologies of different brands, should be abolished as such, and their subject-matter transferred to another 'ology'; namely, to psycho-logics, where it belongs.

The projection mechanism is one fraught with serious dangers, and it is very dangerous to develop it. The danger is greatest in childhood, when silly teachings help to develop this semantic mechanism, and so to affect, in a pathological way, the physically undeveloped nervous system of the human child. Here we meet an important fact which will become prominent later—that ignorance, identification, and pathological delusions, illusions, and hallucinations, are dangerously akin, and differentiated *only* by the 'emotional' background or intensity.

An important aspect of the problem of existence can be made clear by some examples. Let us recall that a noise or written sign, to become a symbol, must stand for *something*. Let us imagine that you, my

6

reader, and myself are engaged in an argument. Before us, on the table, lies something which we usually call a box of matches: you argue that there are matches in this box; I say that there are no matches in it. Our argument can be settled. We open the box and look, and both become convinced. It must be noticed that in our argument we used *words,* because they stood for something; so when we began to argue, the argument could be solved to our mutual satisfaction, since there was a *third* factor, the object, which corresponded to the symbol used, and this settled the dispute. A third factor was present, and agreement became possible. Let us take another example. Let us try to settle the problem: 'Is blah-blah a case of tra-tra?' Let us assume that you say 'yes', and that I say 'no'. Can we reach any agreement? It is a real tragedy, of which life is full, that such an argument cannot be solved at all. We used noises, not words. There was *no third* factor for which these noises stood as symbols, and so we could argue endlessly without any possibility of agreement. That the noises may have stood for some *semantic disturbance* is quite a different problem, and in such a case a psycho-pathologist should be consulted, but arguments should stop. The reader will have no difficulty in gathering from daily life other examples, many of them of highly tragic character.

We see that we can reach, even here, an important conclusion; namely, that, first of all, we must distinguish between words, symbols which symbolize something, and noises, not symbols, which have no meaning (unless with a pathological meaning for the physician); and, second, that if we use words (symbols for something), all disputes can be solved sooner or later. But, in cases in which we use noises as if they were words, such disputes can *never* be settled. Arguments about the 'truth' or 'falsehood' of statements containing noises are useless, as the terms 'truth' or 'falsehood' do not apply to them. There is one characteristic about noises which is very hopeful. If we use words, symbols, not-noises, the problems may be complicated and difficult; we may have to wait for a long time for a solution; but we know that a solution will be forthcoming. In cases where we make noises, and treat them as words, and this fact is exposed, then the 'problems' are correctly recognized at once as 'no-problems', and such solutions remain valid. Thus, we see that one of the obvious origins of human disagreement lies in the use of noises for words, and so, after all, this important root of human dissension might be abolished by proper education of *s.r* within a single generation. Indeed, researches in symbolism and *s.r* hold great possibilities. We should not be surprised that we find meaningless noises in the foundation of many old 'philosophies', and that from them arise most of the old 'philosophical' fights and arguments. Bitterness and tragedies follow, because many 'problems' become 'no-problems', and the discussion leads nowhere. But, as material for psychiatrical studies, these old debates may be scientifically considered, to the great benefit of our understanding.

We have already mentioned the analogy between the noises we make when these noises do not symbolize anything which exists, and the worthless 'cheques' we give when our bank balance is zero. This analogy could be enlarged and compared with the sale of gold bricks, or any other commercial deal in which we try to make the other fellow accept something by a representation which is contrary to fact. But we do not realize that when we make noises which are not words, because they are not symbols, and give them to the other fellow as if they were to be considered as words or symbols, we commit a similar kind of action. In the concise *Oxford Dictionary of Current English,* there is a word, 'fraud', the definition of which it will be useful for us to consider. Its standard definition reads: 'Fraud, n. Deceitfulness (rare), criminal deception, *use of false representations.* (in Law, . . .) ; dishonest artifice or trick (*pious* fraud, deception intended to benefit deceived, and especially to strengthen religious belief) ; person or thing not fulfilling expectation or description.'* Commercialism has taken good care to prevent one kind of symbolic fraud, as in the instances of spurious cheques and selling gold bricks or passing counterfeit money. But, as yet, we have not become intelligent enough to realize that another most important and similar kind of fraud is continually going on. So, up to the present, we have done nothing about it.

No reflecting reader can deny that the passing off, on an unsuspecting listener, of noises for words, or symbols, must be classified as a fraud, or that we pass to the other fellow contagious semantic disturbances. This brief remark shows, at once, what serious ethical and social results would follow from investigation of correct symbolism.

On one side, as we have already seen, and as will become increasingly evident as we proceed, our *sanity* is connected with correct symbolism. And, naturally, with the increase of sanity, our 'moral' and 'ethical' standards would rise. It seems useless to preach metaphysical 'ethics' and 'morals' if we have no standards for sanity. A fundamentally *un*-sound person cannot, in spite of any amount of preaching, be either 'moral' or 'ethical'. It is well known that even the most good-natured person becomes grouchy or irritable when ill, and his other

---

*The first italics are mine.—A. K.

semantic characteristics change in a similar way. The abuse of symbolism is like the abuse of food or drink: it makes people ill, and so their reactions become deranged.

But, besides the moral and ethical gains to be obtained from the use of correct symbolism, our economic system, which is based on symbolism and which, with ignorant commercialism ruling, has mostly degenerated into an abuse of symbolism (secrecy, conspiracy, advertisements, bluff, 'live-wire agents'.,), would also gain enormously and become stable. Such an application of correct symbolism would conserve a tremendous amount of nervous energy now wasted in worries, uncertainties., which we are all the time piling upon ourselves, as if bent upon testing our endurance. We ought not to wonder that we break down individually and socially. Indeed, if we do not become more intelligent in this field, we shall inevitably break down racially.

The semantic problems of correct symbolism underlie *all human* life. Incorrect symbolism, similarly, has also tremendous semantic ramifications and is bound to undermine any possibility of our building a structurally *human* civilization. Bridges cannot be built and be expected to endure if the cubic masses of their anchorages and abutments are built according to formulae applying to *surfaces*. These formulae are structurally different, and their confusion with the formulae of volumes must lead to disasters. Similarly, we cannot apply generalizations taken from cows, dogs, and other animals to man, and expect the resultant social structures to endure.

Of late, the problems of meaninglessness are beginning to intrigue a number of writers, who, however, treat the subject without the realization of the multiordinal, ∞-valued, and *non-el* character of meanings. They assume that 'meaningless' has or may have a definite general content or unique, one-valued 'meaning'. What has been already said on *non-el* meanings, and the example of the unicorn given above, establish a most important semantic issue; namely, that what is 'meaningless' in a given context on one level of analysis, may become full of sinister meanings on another level when it becomes a symbol *for a semantic disturbance*. This realization, in itself, is a most fundamental semantic factor of our reactions, without which the solution of the problems of sanity becomes extremely difficult, if at all possible.

CHAPTER VII

LINGUISTIC REVISION

This would appear to put at least part of the Theory of Demonstration in a category with the efforts of beginners in Geometry: To prove that A equals B: let A equal B; therefore A equals B. (22) E. T. BELL

To what final conclusions are we then led respecting the nature and extent of the scholastic logic? I think to the following: that it is not a science, but a collection of scientific truths, too incomplete to form a system of themselves, and not sufficiently fundamental to serve as the foundation upon which a perfect system may rest. (44) GEORGE BOOLE

. . . the subject-predicate habits of thought . . . had been impressed on the European mind by the overemphasis on Aristotle's logic during the long mediaeval period. In reference to this twist of mind, probably Aristotle was not an Aristotelian. (578) A. N. WHITEHEAD

The Euclidean space alone is one which at the same time is free of electricity and of gravitation. (551) HERMANN WEYL

To imagine that Newton's great scientific reputation is tossing up and down in these latter-day revolutions is to confuse science with omniscience. (149) A. S. EDDINGTON

This latter objection was sanctioned by Newton, who was not a strict Newtonian. (457) BERTRAND RUSSELL

The evil produced by the Aristotelian 'primary substance' is exactly this habit of metaphysical emphasis upon the 'subject-predicate' form of proposition. (578) A. N. WHITEHEAD

The belief or unconscious conviction that all propositions are of the subject-predicate form—in other words, that every fact consists in some thing having some quality—has rendered most philosophers incapable of giving any account of the world of science and daily life. (453) BERTRAND RUSSELL

The alternative philosophic position must commence with denouncing the whole idea of 'Subject qualified by predicate' as a trap set for philosophers by the syntax of language. (574) A. N. WHITEHEAD

And a well-made language is no indifferent thing; not to go beyond physics, the unknown man who invented the word *heat* devoted many generations to error. Heat has been treated as a substance, simply because it was designated by a substantive, and it has been thought indestructible. (417) H. POINCARÉ

Aristotle was almost entirely concerned with establishing what had been conceived already or of refuting error, but not with solving the problem of the discovery of truth. Now and then, in reading his organon, one feels that he has almost sensed the nature of this problem, only to find that he lapses immediately into a discussion of the logic of demonstration. He thinks of confirming truth rather than of finding it. (82) R. D. CARMICHAEL

85

It is necessary here to give a short account of the great scientific revolution which started some years ago, but which is still going on with very beneficial results. This scientific revolution started in geometry, and, in a deeper sense, is carried on by geometry. Until the work of Gauss, Lobatchevski, Bolyai, Riemann., the E geometry, being *unique*, was believed to be *the* geometry of *the* 'space'. The moment a second geometry was produced, just as good, self-consistent, yet contradictory to the old one, *the* geometry became *a* geometry. None was unique. One absolute was dead. Until Einstein (roughly), *the* universe of Newton was, for us, *the* universe. With Einstein, it became *a* universe. Something similar happened to man.* A new 'man' was produced, just as good, certainly contradictory to the old one. *The* man became *a* man, otherwise a 'conceptual construction', one among the infinity of possible ones.

It is not difficult to see that in all these advances there is a common characteristic, which can be put simply in that it consists in a little change from a 'the' into an 'a'. Some people insist upon sentences in one-syllable words; here we could indeed satisfy them! The change, no doubt, can be expressed by the exchange of one syllable for another. But the problems, in spite of this apparent simplicity, are quite important; and the rest of this volume will be devoted to the examination of this change and of what it structurally involves.

In mentioning the above names, a very important one was omitted, that of Aristotle. I merely mentioned these names as representative of certain trends. Otherwise, of course, it would have been necessary to mention additional names, including sometimes those of their predecessors and the followers who have carried their work further. It would have been particularly necessary in the case of Aristotle, who was not only a most gifted man, but who, also, because of the character of his work, has semantically affected perhaps the largest number of people ever influenced by a single man; and so his work has undergone a most marked elaboration. Because of this, his name, in this book, will usually stand for the body of doctrines known as aristotelianism. It is important to keep this in consideration, because it is becoming more and more evident that the work of Aristotle and his followers has had an unprecedented influence upon the development of the Aryan race, and so the study of aristotelianism may help us to understand ourselves. In using the name of the founder of the school as a synonym for the school itself,

*See my *Manhood of Humanity, The Science and Art of Human Engineering.*

we make our statements less cumbersome. Some of the statements may not be true about the founder of the school; yet they remain true about the school.

Aristotle (384-322 B.C.) was born in Stagira, Greece. He was the son of a physician and had marked predilection for natural history and a distinct dislike for mathematics. Plato, who is considered the 'father of mathematicians', was his teacher. Early in his career Aristotle reacted strongly against the mathematical philosophy of his teacher, and began to build up his own system, which had a strongly biological bias and character. Psycho-logically, Aristotle was a typical extrovert, who projects all his internal processes on the outside world and objectifies them: so his reaction against Plato, the typical introvert, for whom 'reality' was all inside, was a natural and rather an inevitable consequence. The struggle between these two giants was typical of the two *extreme* tendencies which we find in practically all of us, as they represent two most diverse, and yet fundamental psycho-logical tendencies. In 1933 we know that either of these extremes in our make-up is undesirable and un-sound, in science as well as in life. In science, the extreme extroverts have introduced what might be called gross empiricism, which, as such, is a mere *el* fiction—practically a delusion. For no 'facts' are ever free from 'doctrines': so whoever fancies he can free himself from 'doctrines', as expressed in the structure of the language he uses., simply cherishes a delusion, usually with strong affective components. The extreme introverts, on the other hand, originated what might be called the 'idealistic philosophies', which in their turn become *el* delusions. We should not overlook the fact that both these tendencies are *el* and structurally fallacious. Belief in the separate existence of *el*, and, therefore, fictitious, entities must be considered as a structurally un-sound *s.r* and accounts in a large degree for many bitter fights in science and life.

In asylums, these two tendencies are sometimes very obvious. The extreme extrovert is found mostly among the paranoiacs; the extreme introvert among cases of schizophrenia (dementia praecox). Between the two extremes we find all possible shades and degrees represented in daily life as well as in asylums. Both extreme tendencies involve harmful *s.r*, because both produce delusions of some elementalism which, as such, is always *fictitious* and *impossible*. 'Mentally' ill are often characterized by *s.r* involving this capacity for building for themselves fictitious worlds in which they can find refuge from actual life. If we, who live outside of asylums, *act* as if we lived in a fictitious world—that is to say, if we are consistent with our beliefs—we cannot adjust ourselves to actual conditions, and so fall into many *avoidable* semantic difficulties.

But the so-called normal person practically never abides by his beliefs, and when his beliefs are building for him a fictitious world, he saves his neck by *not* abiding by them. A so-called 'insane' person *acts* upon his beliefs, and so cannot adjust himself to a world which is quite different from his fancy.

Let me repeat that the nervous system of the human child is not physically finished at birth: and, therefore, it is easy to give it quite harmful semantic twists, by wrong doctrines. To eliminate the vicious and fictitious *el* outlook and *s.r*, it is of paramount importance to try to educate a child to be neither an extreme extrovert nor an extreme introvert, but a balanced extroverted-introvert.

In psychotherapy, the attempt is often made to re-educate these tendencies. The physician usually tries to make an extrovert more introverted, and an introvert more extroverted. In case of success, the patient either recovers altogether or improves considerably.

In practice there is a considerable difference between the re-education of an extrovert to an introvert and that of an introvert to an extrovert. We have already seen that the balanced person should be both. In daily *el* language, the introvert is 'all thought' and has not much use for the external world, while the extrovert is 'all senses' and has little use for 'thought'. It often happens that it is easier to re-educate an introvert, because at least he 'thinks', but difficult to re-educate an extrovert, as he has not cultivated his capacity to 'think'. He may be a remarkable player on words, but all his verbal plays, though clever, are shallow.

Now we shall be able to understand why Aristotle, the extrovert, and his doctrines have appealed, and still appeal, to those who can 'think' but feebly. The fact that the fuller linguistic system of the extrovert Aristotle was accepted in preference to the work of the introvert, Plato, is of serious semantic consequence to us. It is evident that mankind, in its evolution, had to pass through a low period of development; but this fact is not the only reason why the *A* doctrines have had such a tremendous influence upon the Aryan race. The reason is much more deeply rooted and pernicious. In his day, over two thousand years ago, Aristotle inherited a structurally primitive-made *language*. He, as well as the enormous majority of us at present, never realized that what is going on outside of our skins is certainly *not* words. We never 'think' about this distinction, but we all take over semantically from our parents and associates their habitual forms of representation involving structure as *the* language in which to talk about this world, not knowing, or else forgetting, that a language to be fit to represent this world should at least have the *structure* of this world.

Let me illustrate this by a structural example: let us take a man-made green leaf. We see that in it *green colour was added*. Now let us take a natural green leaf. We see that the green colour was *not added* to it, but that the natural green leaf must be considered a process, a *functional* affair which *became* green without anybody's adding green colour. In the old savage mythologies, there were always demons in *human* shape, who actually made everything with *their hands*. This primitive mythology built up a 'plus' or additive language which attributed to the world an anthropomorphic structure. This false notion of the world's structure was, in turn, reflected in the language. It was a subject-predicate, 'plus' language, and not as it should be, to fit the structure of the world, a *functional* language.

Here we come across a tremendous fact; namely, that a language, any language, has at its bottom certain metaphysics, which ascribe, consciously or unconsciously, some sort of structure to this world. Our old mythologies ascribed an anthropomorphic structure to the world, and, of course, under such a delusion, the primitives built up a language to picture such a world and gave it a subject-predicate form. This subject-predicate form also was closely related to our 'senses', taken in a very *el*, primitive form.

This 'plus' tendency not only shaped our language, but even in mathematics and in physics we are still much more at home with linear ('plus') equations. Only since Einstein have we begun to work seriously at new forms of representation which are no longer expressed by linear (or 'plus') equations. At present, we have serious difficulties in this field. It must be admitted that linear equations are much simpler than non-linear equations. I will explain later that the notion of two-valued causality is strictly connected with this linearity or *additivity*.

Neither Aristotle nor his immediate followers realized or could realize what has been said here. They took the structure of the primitive-made language for granted, and went ahead formulating a philosophical grammar of this primitive language, which grammar—to our great semantic detriment—they called 'logic', defining it as the 'laws of thought'. Because of this formulation in a general theory, we are accustomed even today to inflict this 'philosophical grammar' of primitive language upon our children, and so from childhood up imprison them unconsciously by *the structure* of the language and the so-called 'logic', in an anthropomorphic, structurally primitive universe.

Investigation shows that three great names in our history have been very closely interconnected: Aristotle, who formulated a general theory of a primitive language, a kind of 'philosophical grammar' of this language, and called it 'logic'; Euclid, who built the first nearly autonomous 'logical' system, which we call 'geometry'; and, finally, Newton, who rounded up these structural systems by formulating the foundations of macroscopic mechanics. These three systems happen to have one underlying structural metaphysics, in spite of the fact that Newton corrected some of the most glaring errors of Aristotle. Such first systems are never structurally satisfactory, and, in time, it was found that these systems contained unjustified structural assumptions which their followers tried to evade. It was natural that the innovators should meet with a strong resistance, as these old systems had become so elaborated as to impress the 'thoughtless' with their finality. So the revisions went very slowly and very shyly. In the case of Aristotle, revision was still more difficult because the current religious 'philosophies' of the Western world were inextricably bound up with the A-system. The religious leaders took a strong stand, and as late as the seventeenth century threatened death to the critics of Aristotle.

Even today a revision of Aristotle is extremely difficult, for these three systems have a tremendous semantic hold upon us. Many semantic factors have contributed to this hold. First, they were established by men who were really very gifted. Second, they were not wise epigrams but were genuine systems with definite structure, and, as such, extremely difficult to replace. Obviously, it was not enough to pick some weak spot in one of these systems; the new system-builder would have to replace the old structure by an equally full-fledged structure, and this was a very laborious and difficult task. Third, these systems were strictly united by one structural metaphysics and $s.r$; they collaborated with each other, and gave each other assistance. Finally, the interdependence of these systems rested to a large degree on the structure of the primitive language, upon which Aristotle had legislated, and which was accepted by practically all Aryans, and so was inherently bound up with our daily habits of speech and $s.r$. Together, these four factors constituted a tremendous power, working against any attempts at revision.

We do not realize what tremendous power the structure of an habitual language has. It is not an exaggeration to say that it enslaves us through the mechanism of $s.r$ and that the structure which a language exhibits, and impresses upon us unconsciously, is *automatically projected* upon the world around us. This semantic power is indeed so unbelievable that I do not know any one, even among well-trained scientists, who, after having admitted some argument as correct, does not the next minute deny or disregard (usually unconsciously) practically every word he had admitted, being carried away again by the structural implications of the old language and his *s.r*.

This linguistic slavery makes criticism very difficult, for the majority of critics with their *s.r* defend unconsciously structural and linguistic implications, instead of analysing open-mindedly the structure of the facts at hand. All our advances are going very slowly, very painfully and haltingly, because the new work in science, the Einstein and the new quantum theories included, is all of a *non-el* structure, while our daily languages are *el* and absolutistic and twist pathologically our habits of 'thought' and *s.r*. No help is forthcoming from the so-called 'psychologists'. Not to keep the reader guessing too long, let me say here— although this will be explained at length later on—that the main achievement of Einstein was precisely in the fact that he refused to divide *verbally* 'space' *and* 'time', which experimentally cannot be so divided. This was accomplished by the help of the mathematician Minkowski, who invented a language of *new* structure; namely, the four-dimensional 'space-time', in which to talk about events. This device made the Einstein General Theory possible, and affected the new quantum theories. In the present work, in order to be able to talk about the organism-as-a-whole, we must introduce this *non-el* principle as fundamental and apply it.

The first science to break the traditional structural ring was geometry. Full-fledged $\bar{E}$ systems were built. Following these $\bar{E}$ systems, $\bar{N}$ systems were built (Einstein, quantum), and the 'time' is ripe to build a $\bar{A}$-system, which the present writer originated in his *Manhood of Humanity,* and which is formulated as a structural outline of a general theory in the present volume.

As soon as this new $\bar{A}$-system was definitely formulated, a most curious, natural, and yet unexpected result became apparent; namely, that the three new systems, the $\bar{A}$, $\bar{E}$, and the $\bar{N}$ have *also* one underlying structure and metaphysics. This fact adds to the importance of the situation. All these three new systems have been produced independently. They express between them the structural and semantic urge and longing of all modern science. Their mutual interdependence, mutual structure, mutual metaphysics, mutual method are helpful, for when the vital nature of the issues at hand is clearly seen, it will be found expedient to *start* from this interdependence as a basis, although, historically speaking, it was not a factor in the production of these systems.

This does not seem to be clearly understood by all scientists. I have read, for instance, scientific papers in which Einstein is reproached that he did not *start* with $\bar{E}$ geometries, but only at a later stage incorporated them into his system. This argument, of course, is not against

Einstein but for Einstein. Similar remarks could be made about this present work; and again this would not be an argument against this work, but for it. All these new systems represent methodological and structural advances, and will have played their semantic roles even if some day they should be dismissed and systems of different structure take their place.

Historically, attempts in the direction of a $\bar{A}$ discipline have been very numerous. Indeed, the invention of any new important term of a non-subject-predicate character, or of a functional character, was, in itself, an attempt in the $\bar{A}$ direction. All sciences have had to abandon the common vocabularies and build their own terminologies, many of which are also $\bar{A}$. Although all these attempts have been made, and have quite often been successful in their fields, to the best of my knowledge, they were not made consciously. The term accepted here; namely, 'non-aristotelian' is very useful, not only because it is appropriate and illustrates very well what we have to contend with, but also because it places the emphasis properly and makes us conscious of the structural issues. The fact that the three new non-systems have as much in common as the older three had, recommends and justifies the use of the term. The new problem which looms up; namely, the validity or non-validity of the A law of the excluded third, leads automatically to the non-chrysippian and $\bar{A}$ ∞-valued 'logics', which merge with the theory of probability.[1] According to the accepted use, it is enough to build a system differing from an older system by *one single postulate,* to justify (for instance) the name 'non-euclidean'.

The scope of this particular chapter does not permit me to enlarge upon this difficult and important problem as to the differences between the A and $\bar{A}$ systems, but for orientation, a short list of structural differences is given here; all of which involves new semantic factors.

The primitive form of representation which Aristotle inherited, together with its structural implications and his 'philosophical grammar', which was called 'logic', are strictly interconnected, so much so that one leads to the other.

In the present $\bar{A}$-system, I reject Aristotle's assumed structure, usually called 'metaphysics' (*circa* 350 B.C.), and accept modern science (1933) as my 'metaphysics'.

I reject the following structurally and semantically important aspects of the A-system, which I shall call postulates, and which underlie the A-system-function:

1) The postulate of uniqueness of subject-predicate representation

2) The two-valued *el* 'logic', as expressed in the law of 'excluded third'.

3) The necessary confusion through the lack of discrimination between the 'is' of identity, which I reject completely, and the 'is' of predication, the 'is' of existence, and the 'is' used as an auxiliary verb.

4) The elementalism, as exemplified by the assumed sharp division of 'senses' *and* 'mind', 'percept' *and* 'concept', 'emotions' *and* 'intellect', .

5) The *el* theory of 'meaning'.

6) The *el* postulate of two-valued 'cause-effect'.

7) The *el* theory of definitions, which disregards the undefined terms.

8) The three-dimensional theory of propositions and language.

9) The assumption of the cosmic validity of grammar.

10) The preference for intensional methods.

11) The additive and *el* definition of 'man'.

This list is not complete but sufficient for my purpose and for orientation.

I reject the use of the 'is' of identity entirely, because identity is never found in this world, and devise methods to make such a rejection possible.

I base the $\bar{A}$-system-function and system all through on negative *'is not'*, premises which cannot be denied without the production of impossible data, and so accept 'difference', 'differentiation'., as fundamental.

I accept relations, structure, and order as fundamental.

I accept the many-valued, more general, structurally more correct 'logic of probability' of Łukasiewicz and Tarski, which in my *non-el* system becomes infinite-valued (∞-valued) semantics.*

I accept functional representation whenever possible.

I introduce the *principle of non-elementalism* and apply it all through, which leads to: (*a*) A *non-el* theory of *meanings;* (*b*) A *non-el* theory of *definitions* based on *undefined terms;* (*c*) A *psycho-physiological* theory of *semantic reactions.*

I accept the absolute individuality of events on the un-speakable objective levels, which necessitates the conclusion that *all statements* about them are only probable in various degrees, introducing a *general principle of uncertainty* in *all statements.*

I accept 'logical existence' as fundamental.

I introduce differential and four-dimensional methods.

---

*I use the term infinite-, or ∞-valued in the sense of Cantor as a *variable finite*.

I accept the propositional function of Russell.

I accept the doctrinal function of Keyser, and generalize the system function of Sheffer.

I introduce the four-dimensional theory of propositions and language.

I establish the *multiordinality* of terms.

I introduce and apply psychophysiological considerations of *non-el* orders of abstractions.

I expand the two-term 'cause-effect' relation into an ∞-valued causality.

I accept the ∞-valued determinism of maximum probability instead of the less general two-valued one.

I base the $\bar{A}$-system on extensional methods, which necessitates the introduction of a new punctuation indicating the 'etc.' in a great many statements.

I define 'man' in *non-el* and functional terms.

This list is also not complete and is given for orientation and justification of the name of a *non-aristotelian system*.

In the rough, all science is developing in the $\bar{A}$ direction. The more it succeeds in overcoming the old structural implications of speech, and the more successful it is in building new vocabularies, the further and more rapidly it will progress.

Our human relations at present are still mostly based on the A-system-function. The issues are definite. Either we shall have a science of man, and, therefore, have to part company with the structural implications of our old language and corresponding *s.r*—and this means we shall have to build up a new terminology, which is $\bar{A}$ in structure, and use different methods;—or we shall remain in A semantic clutches, use A language and methods, involving older *s.r*, and have no science of man. As I am engaged in building up a science of man, all departures I am forced to make from accepted methods are necessary semantic preliminaries to the building of my system and need no apology.

It is no exaggeration to say that the A, E, and N systems have one most interesting structural and semantic characteristic in common; namely, that they have a few unjustified 'infinities' too many. The modern $\bar{E}$, $\bar{N}$, and, finally, $\bar{A}$ systems, after analysis, eliminate these unjustified notions. New systems arise, quite different from the old ones, which again have this structural characteristic in common, that they have a few 'infinities' less—an important semantic factor, especially in the $\bar{A}$-system, as it helps to eliminate our older delusional mythologies. In the mathematical reconstruction of Brouwer, Weyl, and the Polish School, a similar tendency is apparent, leading to revision of the mathematical notions of infinity. For instance, the E-system involves several structural 'infinity' assumptions. In it, a line has infinite length; the space constant is infinite; and the natural unit of length is also infinite. In the N-system, the velocity of light is assumed unconsciously to be infinite, a structural assumption false to facts. The A-system involves also false to facts infinity assumptions, explained later. It is extremely interesting to note that in any system a similar result follows from the introduction of these different 'infinities'; namely, when such an 'infinity' is introduced in the denominator, it makes the whole expression vanish. When, in the observation of actual facts, we *miss* some characteristic entirely; for instance, order, it leads to the introduction of some 'infinity' somewhere. In other words, faulty, insufficient observation leads to the introduction somewhere in our systems of some fanciful 'infinities'.

I must emphasize again the semantic difficulties which beset us, in the formation of a new and $\bar{A}$-system, mainly because of the lack of scientific *non-el* psycho-logics and general semantics. Having no general theories to guide us in our researches, we must select some other devices. We can survey those achievements of mankind which have proved to be the most beneficial and of most lasting value, *study* their structure and try to train ourselves, and our *s.r*, in repeating the psycho-logical processes and methods which have made them. In this way, we are led to the study of the structure of mathematics and science, and acquire the habit of rigorous and critical 'thought' and acquire new *s.r*. Naturally, such a method is wasteful; it would be simpler to have general *non-el* theories, which I have proposed to call general semantics and psycho-logics, replacing the older *el* 'logic' and 'psychology', and study these short, structurally correct, ready-made formulations to train our *s.r* rather than to study the actual performance of scientists and mathematicians, and formulate these generalizations for ourselves. But, until the present work, this could not be done.

For these reasons, we shall have to make, in the following chapters, a short survey of different scientific achievements without going into technical details, but giving enough of these details to indicate structure and its bearing on *s.r*. Every thing given will be strictly of an elementary character, and the intelligent reader will find no special difficulties in following the survey.

The selection of suitable material presented a very serious problem. I consulted with many friends and used my best judgement, backed by some experience. An important factor was the class of readers for whom this book is written. Sooner or later a new branch of science must be—

and will be—established for the pursuit of this $\bar{A}$ enquiry; so the future student and teacher must have at least an outline of the main problems. It seemed more advisable to outline main issues relevant to the subject, than to work out some of them in more detail. A great deal of new scientific literature on structure and $s.r$ must be produced by mathematicians, psychiatrists, linguists, psychophysiologists,. In this field, experience has taught me that very little has been done and that much of what has been done cannot be accepted without a *non-el* revision. It seems to be more convenient that the reader shall not be referred to too many books, and more expedient that the writer should not take too much for granted; so most of the structural and semantic informations which are necessary for an intelligent reading are given, together with additional references for students who wish to go deeper into the subject.

The reader will find that the *non-el* principle has been emphasized. In the meantime, in the writing I have had to use some *el* terms. In such cases, I used the old terms in quotation marks. The reason for this is that before the full general theory is developed, it is impossible to do otherwise. Besides, even if organism-as-a-whole terms were used from the beginning, this also would not be entirely adequate; for the organism-as-a-whole cannot and should not be structurally separated from its environment; and so the terms should be enlarged to cover, by implication, the environment.

Later we shall see that all languages have some characteristics similar to mathematical languages. For instance, the A word, 'apple', as it has no individual subscripts or date, is *not* a name for a definite object or stage of a process *which are all different,* but a name *for a definition,* which, in principle, is one-valued, while the objective processes are ∞-valued. If this mechanism is not clearly understood, we are bound in dealing with actual ∞-valued stages of processes, to identify the ∞ values into one or a few values. The above considerations necessitate a *non-el* new theory of meanings in accordance with the structure of the world and our nervous system.

The distinction between mathematical and physical languages is structurally most important, although once identification is entirely eliminated, we discover that all possible characteristics found in this world are due to *structure*, and so can be expressed in terms of structure, relations, and multi-dimensional order.

Several similar difficulties will appear later on, all having a similar general characteristic; namely, that we seem to reach an impasse, from which there is no way out. Yet escape can be found, not by solution in the old way, but by reformulating the problem so as to make a solution possible. This method is of extreme usefulness in mathematics, and seemingly can be applied to life also.

If we compare the three systems of Aristotle, Euclid, and Newton, designated, $A, E, N,$ respectively, in Fig. 1, with the non-aristotelian, non-euclidean and non-newtonian systems, designated $\bar{A}, \bar{E}, \bar{N},$ a very important fact should be noticed; namely, that the $\bar{A}\bar{E}\bar{N}$ trilogy is *more general* than AEN. This fact has far-reaching semantic and practical consequences and perhaps can be best explained by the aid of a diagram. We see that the $\bar{A}\bar{E}\bar{N}$ trilogy includes the AEN trilogy as a particular case, from which it follows that all those readers who are already re-educated to the new $\bar{A}\bar{E}\bar{N}$ s.r, have less difficulty in understanding the older $AEN,$ simply because the older systems are only

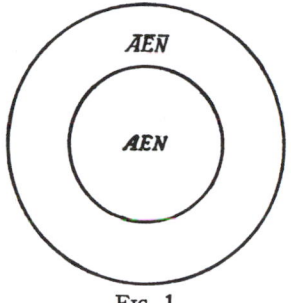

FIG. 1

particular cases of the new $\bar{A}\bar{E}\bar{N}$. But this is not so with those readers who still have the old AEN s.r; they have to enlarge their limited point of view, grasp more than they knew before, and so will have serious semantic difficulties for a while, and, perhaps, become impatient or even angry. With the understanding of this larger generality of the new $\bar{A}\bar{E}\bar{N},$ perhaps a great deal of this semantic futile unpleasantness can be eliminated.

I know of no better example to illustrate this than to refer the reader to a little elementary book, *Debate on the Theory of Relativity,* published by the Open Court Co., Chicago.[2] It is really interesting to watch how good-natured the einsteinists are as compared with the newtonians. This book is suggested because it is elementary, extremely instructive, and very well worth reading. But the whole literature of euclideanism, non-euclideanism, newtonianism and non-newtonianism gives ample proof of the above statements. What kind of verbal flowers the aristotelians will throw to the non-aristotelians remains to be seen; but some verbal and semantic uproar can be expected.

It should be expected that this widening of horizons can only be attained, after all, with difficulty, because it requires an alteration of habitual reactions, from one-, two-, and three-valued to ∞-valued new s.r—usually not easy to achieve. But there seems little doubt that the future depends on it, and so we shall not be able to escape it indefinitely.

As we usually fail to make allowances for the 'emotional' aspects of 'intellectual' pursuits, let me once more point to the fact that even purely 'intellectual' achievements have their 'emotional' components and these

7

are included in the *non-el s.r* It seems that broader ∞-valued under-standing has beneficial effect on our *s.r,* a result which should be expected, if, as at present, we have no reason to doubt that the organism-as-a-whole is a dependable structural *non-el* generalization.

PART III

NON-ELEMENTALISTIC STRUCTURES

The history of human thought may be roughly divided into three periods, each period has gradually evolved from its predecessor. The beginning of one period overlaps the other. As a base for my classification I shall take the relationship between the observer and the observed. . . .

The first period may be called the Greek, or Metaphysical, or Pre-Scientific Period. In this period the observer was everything, the observed did not matter.

The second period may be called the Classical or Semi-Scientific—still reigning in most fields—where the observer was almost nothing and the only thing that mattered was the observed. This tendency gave rise to that which we may call *gross* empiricism and *gross* materialism.

The third period may be called the Mathematical, or Scientific Period. . . . *In this period mankind will understand (some understand it already) that all that man can know is a joint phenomenon of the observer and the observed. . . .*

Someone may ask, How about "intuitions," "emotions," etc.? The answer is simple and positive. It is a fallacy of the old schools to divide man into parcels, elements; all human faculties consist of an inter-connected whole . . . (280)   A. K.

The organism is inexplicable without environment. Every characteristic of it has some relation to environmental factors. And particularly the organism as a whole, *i. e.*, the unity and order, the physiological differences, relations and harmonies between its parts, are entirely meaningless except in relation to an external world. (92)   CHARLES M. CHILD

In reality it is the brain as a whole which is the centre of association, and the association is the very raison d'être of the nervous system as a whole. (411)   HENRI PIÉRON

The views of space and time which I wish to lay before you have sprung from the soil of experimental physics, and therein lies their strength. They are radical. Henceforth space by itself, and time by itself, are doomed to fade away into mere shadows, and only a kind of union of the two will preserve an independent reality. (352)   H. MINKOWSKI

This assumption is not permissible in atomic physics; the interaction between observer and object causes uncontrollable and large changes in the system being observed, because of the discontinuous changes characteristic of atomic processes. (215)   W. HEISENBERG

Well, this is one of the characteristics by which we recognize the facts which yield great results. They are those which allow of these happy innovations of language. The crude fact then is often of no great interest; we may point it out many times without having rendered great services to science. It takes value only when a wiser thinker perceives the relation for which it stands, and symbolizes it by a word. (417)   H. POINCARÉ

99

CHAPTER VIII

GENERAL EPISTEMOLOGICAL

The physiological gradient is a case of protoplasmic memory since it represents the persistence of the effects of environmental action. The establishment of a gradient in a protoplasm may be regarded as a process of learning.   CHARLES M. CHILD

In what has already been said, we have emphasized repeatedly the 'organism-as-a-whole' principle. The principle is structural, involving most important semantic factors and so deserves a more detailed consideration.

Since the days of Aristotle, more than two thousand years ago, this principle has been often emphasized, often belittled, but, withal, seldom applied. That all we know about life and organisms seems to justify such a principle seems obvious.

The arguments of those experimentalists who belittle or object to such a principle seem to be all of a similar type, and are, perhaps, best expressed by Professor H. S. Jennings, who, in his friendly review of Ritter's book on the *Organismal Conception of Life,* concludes that such an 'organismal conception' is quite justified, but is entirely sterile and does not help laboratory workers.

It must be granted that at the date when the book of Ritter and the review of Jennings were written such a statement was *seemingly* justified. The principle is usually treated as a rough generalization from experience and is not analysed further; the *structural,* epistemological, psycho-logical and semantic consequences were not known, and so the laboratory workers actually did not realize that they *have much help.*

As we have already seen, the main semantic issues were, and are, *structural.* How can we apply the organism-as-a-whole principle if we insist on keeping an *el* language and attitude? Naturally, if the principle is *not* applied, it is futile to look for semantic consequences of a non-applied principle. But once the principle *is* applied, a new language has to be built, of different structure and, *therefore, new implications,* which suggest a long series of new experiments.

A new and structurally different theory may be summarized in a single term—as, for instance, 'tropism' or 'dynamic gradient', a fact which not only revolutionizes our knowledge but which leads also to entirely new experiments and further knowledge. Experiments, as such, always give relational, structural data, that, under such and such conditions, such and such results follow. The *non-el* attitude and language, as opposed to the old elementalism, is a part of a broader and more fundamental semantic problem; namely, *similarity of structure* between language and the external world. Such similarity leads to similarity of 'logical' relations, predictability, and so forth, and, in general, to the understanding of the structure of the world and *new s.r.*

There are many examples of such organism-as-a-whole terms, but for the present we will mention only the terms 'tropism' in the generalized sense of Loeb, and the 'dynamic or physiological gradients' of Professor Child. The term 'tropism' means the response of the organism-as-a-whole to special external stimuli. For instance, the term 'heliotropism' is applied in cases when the organism responds to the influence of light; 'chemotropism', when it reacts to chemical stimuli; 'galvanotropism', when the organism responds to galvanic (electrical) stimulation, .

The term 'dynamic or physiological gradient' is the foundation of the $\bar{A}$ biological system of Professor Child. Because of its importance, I shall explain the meaning of this term in some detail.[1]

All protoplasm exhibits empirically a structural characteristic which may be called 'irritability', which appears as a reaction of living protoplasm to external dynamic influences. That 'irritability', as a structural characteristic, becomes obvious when we consider that structurally disintegrated protoplasm is colloidally inactive and becomes 'dead'. Many of the most important characteristics of living protoplasm are strictly bound up with structural integrity.

This 'irritability' occurs in a *structural plenum* and is transmitted to other regions of the protoplasm with differing yet *finite* velocities, and not in 'no time', as Alice would say. Let us imagine a non-differentiated, except for the limiting surface (A), and living bit of protoplasm. This limiting surface represents that part of the protoplasm which is in direct contact with the environment. If the external dynamic factor (S) excites this living bit of protoplasm at a point (B), this stimulus will be the strongest at (B), and it will spread to the further removed portions of (A) in a diminishing gradient. If the decrement is not too sharp, the stimulus will reach the furthest regions of (A); namely, (C), (D), (E), (F), .

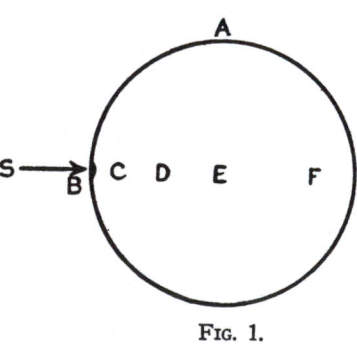

FIG. 1.

The presence or absence of the decrement or its steepness and the intensity of the excitation during transmission depends on the specific character of the protoplasm, and varies from individual to individual, and in different regions and under different conditions varies in one individual.

Thus, we see that a living cell has a necessary relationship with the environment and with external energies because of its limiting surface. The difference between the 'inside of the skin' and the 'outside of the skin' establishes the organism-as-a-whole. The interplay between the inside and the outside is *structural* and supplies the energies which activate the organism. The membrane formation is mostly not dependent upon the constitution of any particular protoplasm, but is rather a general reaction of all protoplasm to environmental influences.

The evidence we have seems to show that in all protoplasm in which we find no specialized conducting paths a certain decrement appears, so that the effectiveness of transmission is limited. In a primitive non-differentiated protoplasm different points further removed from (B) will show different degrees of excitatory changes decreasing from (B). At a certain point the transmission may cease altogether.

The result, then, becomes an excitation-transmission gradient of greater or lesser length, the different levels of which represent various degrees or intensities of excitation.

The primary region of excitation (B) is physiologically more affected and dominant over the other regions to which the excitation is transmitted, because it has more effect upon them than they have upon it. The effect of such conditions gives rise to a temporary structural organismal pattern. The region of primary excitation (B) becomes the dominant region, and the other regions become subordinate to it.

The potentiality for the excitation and the transmission was structurally present in the protoplasm, but this could not produce the pattern which resulted from the external excitation. We see that the action of the external factor was necessary for the realization of the definite physiological pattern whose potentialities were in the protoplasm.

These new excitation-transmission patterns exhibit all the characteristics of new structural patterns in the protoplasmic mass. They determine localized differences at different points, (C), (D), (E),. These differences and *relations* with the dominant region (B) constitute a physiological axis with (B) at one pole. This new pattern constitutes a new structural integration, which is a joint phenomenon of the potentialities of the protoplasm and the environmental action. This relation is of a functional and not merely of a 'plus' character. Child shows that the physiological axes in their simpler forms are similar to, if not the result of, such excitation-transmission gradients.

For the organism to work as-a-whole, some sort of integrating pattern is necessary. The behaviour of the organism-as-a-whole results, first, from patterns already present, and, second, from the possibilities of further development and integration in response to particular external factors. The physiological gradients give such means.

The development of our nervous system is strictly connected with the above principles discovered by Child.[2] In axiate animals and man the chief aggregation of nervous tissues is localized in the apical (head) end, which region is characterized primarily by a higher rate of metabolism in the early stages. Physiological gradients originate as simple protoplasmic reactions to external stimuli, and so the nervous system originates in protoplasmic behaviour. Nerves then become simply *structuralized* and permanent physiological gradients, and so exert a physiological dominance over other tissues.

From an epistemological point of view, we should notice that the gradients are primarily quantitative and that we do not need specific factors to determine them. Any factor that will determine a more or less persistent quantitative differential in the protoplasm ought to be effective. The above theory is structurally supported by a large number of experiments. For instance, we can experimentally obliterate or determine new gradients.[3] The organism appears in this new light as a behaviour reaction-pattern, and substantiates the old saying that the function builds the organ. Not only should the organism be treated as-a-whole, but it is impossible to isolate the organism from its environment. A functional interrelationship is established between the two.

This theory appears, also, to be fundamental for psychiatry and for psycho-logics, for it establishes the head as a dominant region on the base of an experimentally proven higher rate of metabolism. From Child's point of view, as suggested by Dr. William A. White, the main dynamic gradient, the central nervous axis, gives the *structuralized* evidence of the degree of *correlation* of the other organs and of the degree that the body is under the control of the head-end of this gradient. The failure to keep in touch with this centre of control leads to the disintegration of the individual[4]. The head-end is also the most modifiable point in the axis of control, a conclusion which is of the utmost significance in psychotherapy. It is known that the metabolism of organs can be affected by 'psychic' stimuli, and it is only one step further to understand, as White says, why we may have other structuralized functions, such as structuralized anti-social feelings, structuralized greed, structuralized hate., facts which are observed daily in ordinary life and in asylums. From the point of view of the theory of Child, the nervous system appears not only as a structuralized conducting gradient, but it also explains how specific conducting tissues could have evolved from non-specific living protoplasm. It is important to notice the dominance which the primary region of excitation exerts over the others, since, with the great complexities of the human brain, we understand better why so-called 'mental' and semantic issues, which are phylogenetically the youngest, are of such importance.

In our daily life we deal with different people, some of whom are seriously ill 'mentally' and who, under favorable conditions, would be under medical attendance. The majority of us—some specialists consider it to run even as high as ninety per cent of the whole population—would be better off if taken care of by some psychiatrist, or, at least, if under consultation from time to time.

Owing to old religious prejudices, often unconscious, it is still believed that those 'mentally' ill are either obsessed by 'demons' or are being punished for some 'evil',. The majority even of enlightened people have a kind of semantic horror or fright at 'mental' ills, not realizing that under the *animalistic conditions* which prevail at present in our theories, 'ethical', social, economic., those only with the least human traits are favored, while those most human cannot stand such animalistic conditions and often break down. It is not a novelty that a moron cannot be 'insane'. A moron lacks something; only the more gifted individuals, the more human (as compared with animals), break down. I know of many psychiatrists who say that 'it takes a "good mind" to be "insane" '.

Now, 'mental' and semantic excitation, which phylogenetically appeared so recently, naturally plays, in many instances, a dominant part, a fact which science, until very lately, has completely disregarded. The present theory makes it quite obvious that with animalistic theories in existence, and un-sanity (lack of consciousness of abstracting, confusion of orders of abstractions resulting from identifications.,) practically universally operating in every one of us, a seriously unbalanced race must be produced.

There can be no doubt that the consistent application of a *non-el* language in the analysis of animal behaviour has suggested new experiments and that, as a result, the use of such terms had its influence on laboratory workers. It does not matter to what extent these terms, or the theories which they represented, were 'right' or 'wrong', they were terms of the *non-el* type, and they expressed in one term entirely *struc-*

turally new and far-reaching theories. In testing these theories, new series of experiments were required. Even when the new experiments were devised to verify the older experiments, again the laboratory workers got direct benefit of the structurally new terms. But these benefits were largely unconscious, and so biologists *could* believe in the older days that they had no laboratory benefits from the use of such terms; however, this belief is now entirely unjustified.

Since the *non-el* principle is not only a structurally justified empirical generalization, but also involves for its application the structural rebuilding of our language and old theories, the semantic issues are far-reaching and of great practical value.

The application of the principle means the rejection of the old elementalism which results and leads to identifications and to blinding semantic disturbances, which, in turn, prevent clear vision and unbiased creative freedom.

According to the modern theory of materials, as given in Part X, the mutual interdependence, the mutual action and reaction of everything in this world upon everything else appears as a *structural* fact and a necessity, and so *el* languages cannot be expected to lead to satisfactory semantic solutions. We should not be surprised to find that the struggle against identification and elementalism appears at some stage in every science.

Some of the most prominent examples of this tendency outside of biology, psychiatry., can be found in modern physics. From a structural point of view the whole theory of Einstein is nothing else than an attempt to reformulate physics on a *structurally new non-el* and $\bar{A}$ foundation—an exact structural parallel of the biological organism-as-a-whole principle.

Einstein realized that the empirical structure of 'space' and 'time' with which the physicist and the average man deals is such that it cannot be empirically divided, and that we actually deal with a blend which we have split only elementalistically and verbally into these fictitious entities. He decided to build a verbal system closer in structure to the facts of experience and, with the help of the mathematician Minkowski, he formulated a system of new structure which employed a *non-el language* of space-time. As we know from physics and astronomy, this *non-el* language *suggested experiments,* and so it had beneficial laboratory application. But, in fact, the influence goes still deeper, as the present work will show, for such structural advances carry with them profound psycho-logical, semantic effects. Although, at present, these beneficial influences operate unconsciously, they, nevertheless, tend to counteract the *el* and absolutistic semantic effects of identification.

It is interesting to note that the Einstein theory, because structural, has had the effect upon the younger physicists of a semantic release from the old structural elementalism and has prepared the semantic ground for the crop of young geniuses which has sprung up lately in the quantum field. It was found that the *el* 'absolute' division of the 'observer' and the 'observed' was false to facts, because every observation in this field disturbs the observed. The elimination of this elementalism in the quantum field led to the most revolutionary restricted 'uncertainty principle' of Heisenberg, which, without abolishing determinism, requires the transforming of the two-valued A 'logic' into the ∞-valued semantics of probability. Again, this advance in quantum formulations has suggested new experiments.

The $\bar{A}$-system, as originated by the writer in his *Manhood of Humanity* and other writings, is also the result of the structurally *non-el* tendency. In *Manhood of Humanity*, I introduced a *non-el* term, 'time-binding', by which is meant *all* the factors *which as-a-whole* make man a man, and which differentiate him from animals. In carrying the system further in the present book, I reject the structurally *el* separation involved in such terms as 'senses' and 'mind'., and introduce, instead, *non-el* terms, such as 'different orders of abstractions'., where 'mind' and 'senses'., are no longer divided. Curiously enough, even in such a field, the method has suggested experiments, and so again the new language has laboratory importance.

What has been said above about the organism-as-a-whole, and illustrated by particular cases, seems to show a general characteristic of all our abstracting capacities. We usually disregard, or fail to appreciate, the fact that a single structurally important new term might lead to the re-postulation of the whole structure of the language in the given field. In science we search for structure; so any structurally new term is useful, because, when tested, it always gives structural information, whether positive or negative. In our human affairs, it is different. All our human institutions follow the structure of the language used; but we never 'think' of that, and, when the silly institutions do not work, we blame it all on 'human nature', without any scientific justification.

Poincaré, in one of his essays, speaks about the harmful effect which the term 'heat' had on physics. Grammatically, the term 'heat' is classified as a substantive, and so physics was labouring for centuries looking for some 'substance' which would correspond to the substantive name 'heat'. We know by now that there is no such thing, but that 'heat' must be considered as a manifestation of 'energy'. If we choose to carry this analysis further, we should find that 'energy' is also not a very satisfactory term, but that 'action', perhaps, is more fundamental.

In dealing with ourselves and the world around us, we must take into account the structural fact that everything in this world is strictly interrelated with everything else, and so we must make efforts to discard primitive *el* terms, which imply structurally a *non-existing isolation*.

The moment this is realized, we shall have to treat the *non-el* principle seriously. As the new terms have, also, their *non-el* implications, such terms throw new light on old problems.

In practice, it is difficult, at first, to avoid the use of old terms. When we want to digest fully a new and important work based on new structural terms and acquire corresponding *s.r*, the best way to train oneself in the use of the new terms is by gradually dropping the old terms. If we *have* to use the old terms, then we should train ourselves to be aware of their *insufficiency* and of their *fallacious* structural *implications,* and so be free from the old *s.r*.

The use of the new terms should be deliberate. We should put the problem to ourselves somewhat as follows: The old language is structurally, and, therefore, by implication, semantically unsatisfactory; the new terms seem to correspond closer to facts; let us test the new terms. Are the new terms always structurally satisfactory? Probably not, but in science experiments check predictions, and so new structural issues become clarified.

We have been speaking about new and old terms quite simply, yet the issues are not so simple. The invention of a single structurally new term always involves new structural and relational notions, which, again, involve *s.r*. For instance, if we study any event, and in that study use the terms 'tropism', or 'dynamic gradient', or 'time-binding', or 'order of abstractions', or 'space-time', or 'wave-packets'., we must use all structural and semantic implications the terms involve.

Using the first four terms, we are bound to treat the organism-as-a-whole, for the terms are not *el*. They are not based on the notion of, nor do they postulate, some fictitious 'isolated' elements. In using space-time, we introduce the individuality of events, as every 'point of space' carries with itself a *date,* which, by necessity, makes every 'point' in space-time unique and individual. In using the term 'wave-packet', we re-interpret the older objectified and, perhaps, fictitious 'electron',.

The consistent and permanent use of such terms naturally involves, structurally, a new world-outlook, new *s.r*, more justified by our scientific and daily experience. But the greatest gain is usually in getting away from primitive structural notions and metaphysics, with their vicious semantic disturbances. In creative work, semantic *limitations* hamper a clear understanding, and prevent scientists from inventing or formulating better, simpler, and more effective theories of different structure.

As soon as we possess 'knowledge', then we shall 'know' all that there is to be known. By definition, there cannot be any *unknowable*. There is a place for the unknown structure. The unknown is rather extensive, partly because science has been, and still is, persecuted, as has already been pointed out.

The so-called 'unknowable' was the semantic result of identification, of a semantic unbalance, which posits for knowledge something 'beyond' knowledge. But has such a postulation any meanings outside of psychopathology? Of course not, as it starts with a self-contradictory assumption, which, being senseless, must lead to senseless results.

We have dwelt on the problems of the structure of terms at such length, because they are generally disregarded, but they are, for semantic purposes, fundamental. The reader will get the main benefit of this book and will receive help in understanding modern scientific issues if he becomes entirely convinced of the seriousness of structural and semantic problems.

Terms are artifices of humans which are necessary to economize effort in the field of 'experience' and experimentation. They are useful in reducing the actual amount of experience necessary, by allowing verbal experimentation. The human rate of progress is swifter than that of the animals, and this is due mainly to the fact that we can summarize and transmit past experiences to the young generation in a degree far more effective than that of the animals. We have also extra-neural means for recording experiences, which the animals lack entirely.

That such verbal experimentation is possible at all is conditioned by the fact that languages have *structure,* and that our knowledge of the world is *structural* knowledge. Let us repeat once more that if two relations have similar structure, all of their 'logical' characteristics are similar; therefore, once structure is discovered, such a process of verbal experimentation becomes extremely effective, and an accelerating cultural device. The use of an antiquated language in our human affairs, in addition to other drawbacks, prevents our being more intelligent in those affairs.

The natural order of investigation is indicated thus: (1) Empirical search for structure in the sciences; (2) Once this structure is discovered, at each date, the structure of our language is adjusted to it and our new *s.r* trained. Historically, we have partially followed the reversed, and ultimately pre-human, and so pathological, order. Without science, and with extremely meagre and primitive knowledge of the structure of the world, we have produced grunts and languages of primitive false structure, reflecting, of necessity, its implications as to the assumed structure of the world. We have made out of it primitive dogmas which are still in full sway and embodied in the structure of the old language. This is also the reason why, outside of technical achievements, we are still on such primitive levels. It is easy to understand why experimental science is of such importance and why theoretical (verbal) predictions must be tested experimentally. The above also gives a deeper and a new justification for what is called 'pragmatism'.

Experiments constitute a search for relations and structure in the empirical world. Theories produce languages of some structure. If the two structures are similar, the 'theories work'; otherwise, they do not, and suggest further search and structural adjustments.

It should be mentioned, perhaps, that the main epistemological principle which has led to the writing of the works of the present author was a definite inclination to abandon identification and the resulting structurally unsatisfactory *el* language in general use, and to produce a *non-el* system, which, in structure, would be similar to the world around us, ourselves and our nervous system included. This structural novelty was the foundation on which the $\bar{A}$-system has been gradually built.

CHAPTER IX

COLLOIDAL BEHAVIOUR

In fact, to-day colloids may be regarded as an important, perhaps the most important connecting link between the organic and the inorganic world. (7)   WOLFGANG PAULI

In our researches, let us follow the natural order and give a brief structural account of what we know, empirically, about the medium in which life is found; namely, about the colloids. The following few elementary particulars show the empirical importance of structure, and so are fundamental in the present work.

At present, physicians are usually too innocent of psychiatry, and psychiatrists, although they often complain about this innocence of their colleagues, seldom, if ever, themselves pay any attention to the colloidal structure of life; and their arguments about the 'body-mind' problem are still scientifically incomplete and unconvincing, though the 'body-mind' problem has been present with us for thousands of years. It is a very important semantic problem, and, as yet, not solved scientifically, although there is a simple solution of it to be found in the *colloidal* structure of life.

The reader should not ascribe any uniqueness of the 'cause-effect' character to the statements which follow, as they may not be true when generalized. Colloidal science is young and little known. Science has accumulated a maze of facts, but we do not have, as yet, a general theory of colloidal behaviour. Statements, therefore, should not be unduly generalized.

We shall only indicate a few structural and relational connections important for our purpose.

When we take a piece of some material and subdivide it into smaller pieces, we cannot carry on this process indefinitely. At some stage of this process the bits become so small that they cannot be seen with the most powerful microscope. At a further stage, we should reach a limit of the subdivision that the particles can undergo without losing their chemical character. Such a limit is called the molecule.* The smallest particle visible in the microscope is still about one thousand times larger than the largest molecule. So we see that between the molecule and the smallest visible particle there is a wide range of sizes.

---

*This statement is only approximate, because there is evidence that chemical characteristics change as the molecule is approached.

* see page xii Findlay calls these the 'twilight zone of matter'; and it was Oswald, I believe, who called it the 'world of neglected dimensions'.

This 'world of neglected dimensions' is of particular interest to us, because in this range of subdivision or smallness we find very peculiar forms of behaviour—life included—which are called 'colloidal behaviour'.

The term 'colloid' was proposed in 1861 by Thomas Graham to describe the distinction between the behaviour of those materials which readily crystallize and diffuse through animal membranes and those which form 'amorphous' or gelatinous masses and do not diffuse readily or at all through animal membranes. Graham called the first class 'crystalloids' and the second 'colloids', from the Greek word for glue.

In the beginning colloids were regarded as special 'substances', but it was found that this point of view was not correct. For instance, NaCl may behave in solution either as a crystalloid or as a colloid; so we began to speak about the *colloidal state*. Of late, even this term became unsatisfactory and is often supplanted by the term *'colloidal behaviour'*.

In general, a colloid may be described as a 'system' consisting of two or more 'phases'. The commonest represent emulsions or suspensions of fine particles in a gaseous, liquid, or other medium, the size of the particles grading from those barely visible microscopically to those of molecular dimensions. These particles may be either homogeneous solids, or liquids, or solutions themselves of a small percentage of the medium in an otherwise homogeneous complex. Such solutions have one characteristic in common; namely, that the suspended materials may remain almost indefinitely in suspension, because the tendency to settle, due to gravity, is counteracted by some other factor tending to keep the particles suspended. In the main, colloidal behaviour is not dependent upon the physical state or chemistry of the finely subdivided materials or of the medium. We find colloidal behaviour exhibited not only by colloidal suspensions and emulsions where solid particles or liquid droplets are in a liquid medium, but also when solid particles are dispersed in gaseous medium (smokes), or liquid droplets in gaseous media (mists),.

Materials which exhibit this special colloidal behaviour are always in a very fine state of subdivision, so that the ratio of *surface exposed* to *volume of material* is very large. A sphere containing only 10 cubic centimetres, if composed of fine particles 0.00000025 cm. in diameter, would have a total area of all the surfaces of the particles nearly equal * see page xii to half an acre.[1] It is easy to understand that under such *structural* conditions the *surface forces* become important and play a prominent role in colloidal behaviour.

The smaller the colloidal particles, the closer we come to molecular and atomic sizes. Since we know atoms represent electrical *structures,* we should not be surprised to find that, in colloids, surface energies and electrical charges become of fundamental importance, as by necessity all surfaces are made up of electrical charges. The surface energies operating in finely grained and dispersed systems are large, and in their tendency for a minimum, every two particles or drops tend to become one; because, while the mass is not altered by this change, the surface of one larger particle or drop is less than the surface of two smaller ones—an elementary geometrical fact. Electrical charges have the well-known characteristic that like repels like and attracts the unlike. In colloids, the effect of these factors is of a fundamental, yet opposite, character. The surface energies tend to unite the particles, to coagulate, flocculate or precipitate them. In the meanwhile the electrical charges tend to preserve the state of suspension by repelling the particles from each other. On the predominance or intensity of one or other of these factors, the instability or the stability of a suspension depends.

In general, if 'time' limits are not taken into consideration, colloids are *unstable* complexes, in which continuous transformation takes place, which is induced by light, heat, electric fields, electronic discharges, and other forms of energy. These transformations result in a great variation of the characteristics of the system. The dispersed phase alters its characteristics and the system begins to coagulate, reaching a *stable* state when the coagulation is complete. This process of transformation of the characteristics of the system which define the colloid, and which ends in coagulation, is called the 'ageing' of the colloid. With the coagulation complete, the system loses its colloidal behaviour—it is 'dead'. Both of these terms apply to inorganic as well as to organic systems.

Some of the coagulating processes are partial and reversible, and take the form of change in viscosity; some are not. Some are slow; some extremely rapid, particularly when produced by external agencies which alter the colloidal equilibrium.

From what has been said already, it is obvious that colloids, particularly in organisms, are extremely sensitive and complex structures with enormous possibilities as to degree of stability, reversibility., and allow a wide range of variation of behaviour. When we speak of 'chemistry', we are concerned with a science which deals with certain materials which preserve or alter certain of their characteristics. In 'physics', we go beyond the obvious characteristics and try to discover the *structure* underlying these characteristics. Modern researches show clearly that atoms have a very complex structure and that the macroscopic character-

8

istics are directly connected with sub-microscopic structure. If we can alter this structure, we usually can alter also the chemical or other characteristics. As the processes in colloids are largely structural and physical, anything which tends to have a structural effect usually also disturbs the colloidal equilibrium, and then different macroscopic effects appear. As these changes occur as series of interrelated events, the best way is to consider colloidal behaviour as a physico-electro-chemical occurrence. But once the word 'physical' enters, structural implications are involved. This explains also why all known forms of radiant energy, being structures, can affect or alter colloidal structures, and so have marked effect on colloids.

As all life is found in the colloidal form and has many characteristics found also in inorganic colloids, it appears that colloids supply us with the most important known link between the inorganic and the organic. This fact also suggests entirely new fields for the study of the living cells and of the *optimum conditions for their development, sanity included*.

Many writers are not agreed as to the use of the terms 'film', 'membrane', and the like. Empirically discovered structure shows clearly, however, that we deal with surfaces and *surface energies* and that a 'surface tension film' behaves as a membrane. In the present work, we accept the obvious fact that organized systems are film-partitioned systems.

One of the most baffling problems has been the peculiar periodicity or rhythmicity which we find in life. Lately, Lillie and others have shown that this rhythmicity could not be explained by purely physical nor purely chemical means, but that it is satisfactorily explained when treated as a *physico-electro-chemical structural occurrence*. The famous experiments of Lillie, who used an iron wire immersed in nitric acid and reproduced, experimentally, a beautiful periodicity resembling closely some of the activities of protoplasm and the nervous system, show conclusively that both the living and the non-living systems depend for their rhythmic behaviour on the chemically alterable film, which divides the electrically conducting phases. In the iron wire and nitric acid experiment, the metal and the acid represent the two phases, and between the two there is found a thin film of oxide. In protoplasmic structures, such as a nerve fibre, the internal protoplasm and the surrounding medium are the two phases, separated by a surface film of modified plasm membrane. In both systems, the electromotive characteristics of the surfaces are determined by the character of the film.[2]

That living organisms are film-bounded and partitioned systems accounts also for irritability. It appears that irritability manifests itself as sensitiveness to electrical currents. These currents seem to depend on polarizability or resistance to the passage of ions, owing to the presence of semi-permeable boundary films or surfaces enclosing or partitioning the system. It is obvious that we are here dealing with complex *structures* which are intimately connected with the characteristics of life. Living protoplasm is electrically sensitive only as long as its structure is intact. With death, semi-permeability and polarizability are lost, together with electrical sensitivity.

One of the baffling peculiarities of organisms is the rapidity with which the chemical and metabolic processes spread. Indeed, it is impossible to explain this by the transportation of material. All evidence shows that electrical and, perhaps, other energy factors play an important role; and that this activity again depends on the presence of surfaces of protoplasmic structures with electrode-like characteristics which form circuits.

The great importance of the electrical charges of the colloidal particles arises out of the fact that they prevent particles from coalescing; and when these charges are neutralized, the particles tend to form larger aggregates and settle out of the solution. Because of these charges, when an electrical current is sent through a colloidal solution, the differently charged particles wander to one or the other electrode. This process is called cataphoresis. There is an important difference in behaviour in inorganic and organic colloids under the influence of electrical currents, and this is due to the difference in structure. In inorganic colloids, an electrical current does not coagulate the whole, but only that portion of it in the immediate vicinity of the electrodes. Not so in living protoplasm. Even a weak current usually coagulates the entire protoplasm, because the inter-cellular films probably play the role of electrodes and so the entire protoplasm structurally represents the 'immediate vicinity' of the electrodes. Similarly, structure also accounts for the extremely rapid spread of some effects upon the whole of the organism.

Electrical phenomena in living tissue are mainly of two more or less distinct characters. The first include electromotive energy which produces electrical currents in nerve tissue, the membrane potentials,. The second are called, by Freundlich, electrokinetic, and include cataphoresis, agglutination,. There is much evidence that the mechanical work of the muscles, the secretory action of the glands, and the electrical work of the nerve cells are closely connected with the colloidal structure of these tissues. This would explain why *any factor* (semantic reactions included) capable of altering the colloidal structure of the living protoplasm must have a marked effect on the behaviour and welfare of the organism.

Experiments show that there are four main factors which are able to disturb the colloidal equilibrium: (1) Physical, as, for instance, X-rays, radium, light, ultra-violet rays, cathode rays.; (2) Mechanical, such as friction, puncture.; (3) Chemical, such as tar, paraffin, arsenic.; and, finally, (4) Biological, such as microbes, parasites, spermatozoa,. *In man, another (fifth) potent factor; namely, the semantic reactions,* enters, but about this factor, I shall speak later.

For our purpose, the effects produced by the physical factors, because obviously structural, are of main interest, and we shall, therefore, summarize some of the experimental structural results. Electrical currents of different strength and duration, as well as acids of different concentration, or addition of metallic salts, which produce marked acidity, usually coagulate the protoplasm, these effects being structurally interrelated. Slow coagulation involves changes in viscosity, all of which, under certain conditions, may be reversible.[3] When cells are active, their fluidity often changes in a sharp and rapid manner.[4]

Fat solvents are called surface-active materials; when diluted, they decrease protoplasmic viscosity; but more concentrated solutions produce increased viscosity or coagulation.[5] The anaesthetics, which always are fat solvents and surface-active materials, are very instructive in their action for our purpose, as they affect very diversified types of protoplasm similarly, this similarity of action being due to the similarity of colloidal structure. Thus, ether of equal concentration will make a man unconscious, will prevent the movement of a fish and the wriggling of a worm, or stop the activity of a plant cell, without permanently injuring the cells.[6] In fact, the action of all drugs is based on their effect upon the colloidal equilibrium, without which action a drug would not be effective. It is well known that various acids or alkalis always change the electrical resistance of the protoplasm.[7]

The working of the organism involves mostly a structural and very important 'vicious circle', which makes the character of colloidal changes *non-additive*. If, for instance, the heart, for any reason, slows down the circulation, this produces an accumulation of carbonic acid in the blood, which again increases the viscosity of the blood and so throws more work on the already weakened heart.[8] Under such structural conditions, the results may accumulate very rapidly, even at a rate which can be expressed as an exponential function of higher degree.

Different regions of the organism have different charges; but, in the main, an injured, or excited, or cooler part is electro-negative (which is connected with acid formation), and the electro-positive particles rush to those parts and supply the material for whatever physiological need there may be.[9]

The effects of different forms of radiant energy on colloids and protoplasm are being extensively studied, and the results are very startling. The different forms of radiant energy differ in wave-length, frequency.,—that is to say, generally in *structure*,—and, as such, may produce structural effects on colloids and organisms, which effects may appear on the gross macroscopic level in many different forms.

Electrical currents, for instance, retard reversibly the growth of roots, may activate some eggs into larval stages without fertilization., which makes it possible to understand why, in some cases, a mere puncturing of the egg may disturb the equilibrium and produce the effects of fertilization.[10]

The X-, or Röntgen-rays have been shown to accelerate 150 times the process of mutation. Muller, in his experiments with several thousand cultures of the fruit fly, has established the above ratio of induced mutations, which become hereditary.[11] 'Cosmic rays' in the form of radiation from the earth, in tunnels, for instance, show similar results, except that mutation occurs only twice as often as under the usual laboratory conditions. Under the influence of X-rays, mice change their colour of hair; gray mice become white, and white ones darker. Sometimes further additional bodily changes appear; as, for instance, one or no kidneys, abnormal eyes or legs, occur more often than under ordinary conditions. Some animals lose their power of reproduction, although the body is not obviously changed. Plants respond also to the X-ray treatment. They grow faster, flower more, and produce new forms more readily. In humans the effect of X-ray irradiation has often proven disastrous to the health of experimenters. There are even data that the irradiation of pregnant mothers may result in deformation of the head and limbs of the unborn child and, in one-third of the cases, feeble-mindedness of the children has resulted.[12]

Ultra-violet rays also show a marked effect. In some instances, they slow down or stop the streaming of protoplasm, because of increased viscosity or coagulation; plants grow slowly or rapidly; certain valuable ingredients in plants are increased; certain animals, as, for instance, small crustacea or bacteria are killed; eggs of Nereis (a kind of sea worm), which usually have 28 chromosomes, after irradiation have 70; certain bone malformations in children are cured; the toxin in the blood serum of pernicious anaemia patients is destroyed,.[13] In this respect, we should notice again that ultra-violet irradiation produces curative effects like those of cod-liver oil, which shows that the effect of both factors is ultimately colloidal and structural.

Extensive experimentation with cathode rays is very recent, but already we have a most astonishing array of structural facts. Moist air is converted into nitric acid, synthetic rubber is produced rapidly, the milk from rubber trees is made solid and insoluble without the use of sulphur, liquid forms of bakelite are solidified without heating, linseed oil becomes dry to the touch in three hours and hard in six hours, certain materials, like cholesterol, yeast, starch, cottonseed oil, after exposure for thirty seconds, heal rickets, and similar unexpected results. What are usually called 'vitamins' do not only represent 'special substances', but become structurally active factors; and this is why ultra-violet rays may produce results like those of some 'substance'. It seems that in 'vitamins' the surface activities are important; the parallelism shown by von Hahn between the surface activities of different materials and the Funk table of vitamin content is quite suggestive. Some data seem to show that, in some instances, surface-active materials, such as coffee or alcohol, produce beneficial surface activities similar to the 'vitamins'.[14]

The above short list gives only an approximate picture of the overwhelming importance of the roles which structure in general, and colloids in particular, play in our lives. We see about us many human types. Some are delicate, some are heavy-set, some flabby, some puffy, all of which indicates a difference in their colloidal structure. Paired with these physical colloidal states are also nervous, 'mental', and other characteristics, which vary from weak and nervous to the extreme limitation of nervous activities, as in idiocy, which is a negation of activity.

It is curious that in all illnesses, whether 'physical' or 'mental', the symptoms are very few, and fundamentally of a standard type. In 'physical' illness we find the following common characteristics: fever, chills, headaches, convulsions, vomiting, diarrhoea,. In 'mental' ills, identifications, illusions, delusions, and hallucinations—in general, the reversed pathological order—are found. It is not difficult to understand the reason. Because of the general colloidal background of life, different disturbances of colloidal equilibrium *should* produce similar symptoms. In fact, many of these symptoms have been reproduced experimentally by injecting inert precipitates incapable of chemical reactions, which have induced artificial colloidal disturbances. Thus, if the serum from an epileptic patient is injected into a guinea pig, it results in an attack of convulsions, often ending in death. But, if the guinea pig is previously made immune by an injection of some colloid which accustoms the nerve-endings to the colloidal flocculation, then, for a few hours following, we can, with impunity, introduce into the circulation otherwise fatal doses of epileptic serum. Epileptic serum can also be made immune by filtration, or by strong centrifugation, or by long standing, which frees it from colloidal precipitates.[15]

Death through blood transfusion or the injection of *any* colloid into the circulation has also, in the main, similar symptoms, regardless of the chemical character of the colloid, indicating once more the importance and fundamental character of structure.[16]

That illnesses are *somehow* connected with colloidal disturbances (note the wording of this statement) becomes quite obvious when we consider catarrhal diseases, inflammations, swellings, tumours, cancer, blood thrombi., which involve colloidal injuries, resulting in extreme cases in complete coagulation or fluidification, the variation between 'gel' and 'sol' appearing in a most diversified manner.[17]   Other illnesses are connected with precipitation or deposits of various materials. Gout, for instance, results from a morbid deposit of uric acid, and different concretions, such as the 'stones', are very often found in different fluids of the organism. We have, thus, concretions in the intestines, the bile, the urine, the pancreas, the salivary glands; lime deposits in old softened tissues, 'rice bodies' in the joints, 'brain sand', .[18]

In bacterial diseases, the micro-organisms rapidly produce acids and bases which tend to destroy the colloidal equilibrium. Lately, it has been found that even tuberculosis is more than a mere chapter in bacteriology. All the main tubercular symptoms can be reproduced, experimentally, by means of colloidal disturbances without the intervention of a single bacterium.[19] This would explain also why, in some instances, psychotherapy is effective in diseases with tubercular symptoms.[20]

By structural necessity, every expression of cellular activity involves some sort of colloidal behaviour; and any factor disturbing the colloidal structure must be disturbing to the welfare of the organism. Vice versa, a factor which is beneficial to the organism must reach and affect the colloids.

After this brief account of the structural peculiarities of the domain in which life is found, we can understand the baffling 'body-mind' problem. We do not yet know as many details as we could wish, but these will accumulate the moment a general solution is clearly formulated. It is a well-established *experimental* fact that all nervous and 'mental' activities are connected with, or actually generate, electrical currents, which of late are scrupulously studied by the aid of an instrument called the psychogalvanometer.[21] It is not suggested that electrical currents are the only ones which are involved. There may be many different forms of radiant energy produced or effective, which we have not yet the instruments to record. Experiments suggest such a possibility. Thus, for instance, the apex of a certain rapidly growing vegetable or animal tissue emits some sort of invisible radiation which stimulates the growth of living tissue with which it is not in contact. The tip of a turnip or onion root, if placed at right angles to another root, at a distance of a quarter of an inch, so stimulates the growth of the latter that the increase of the number of cells, on its side nearest the point of stimulation, is as high as seventy per cent. These radiations accelerate the growth of some bacteria. Other examples could be given.[22]

A classical example of the effect left on protoplasm by energetic factors is given by Bovie.[23]

As yet, we have not assumed that the protoplasm of plants also shows lasting structural and functional results of stimulation, some sort of 'learning' or 'habit-formation' characteristics. But such is the case; and further experimentation along these lines will help greatly to understand the mechanism of 'mental' processes in ourselves.

If we take the seed of a plant, for instance, of a squash, and keep it in a moist tropism chamber in the dark, it will grow a root. When the root is about one inch long, we begin our experiment. Originally, under the influence of gravitation, the root grows vertically downwards (A). If we rotate the tropism chamber 90° so that the root is horizontal (B), the root will soon bend downwards under the influence of positive geotropism. But the bending does not occur at once. There is a latent period—in the case of the squash seed, about ten minutes—after which pause the root is bent downwards. When we have determined this latent period for a given seedling, we then rotate the chamber to the positions (B), (C), (A), (B), (C)., just within the 'time' limit before the bending would occur. We repeat such procedure several times. When we set the root again in its vertical downward position (A), we notice that the root, without any more changes of position, will wag backwards and forward with the period as was used in the experiment. This unexpected behaviour will last for several days. It shows that the alternating stimulus of gravitation, as applied to the root, has produced some structural changes in the protoplasm which persist for a comparatively long period after the stimulus has ceased to act. It be-

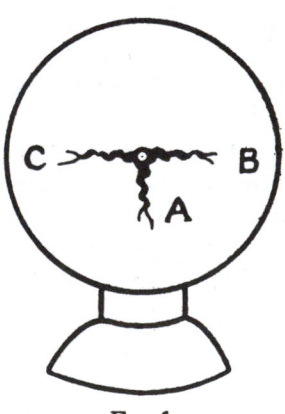

Fig. 1

comes obvious that teachability and the structural tendency for forming engrams is a general characteristic of protoplasm.

All the examples given above show clearly that *structure* in general, and of colloids in particular, gives us a satisfactory basis for the understanding of the *equivalence* between occurrences which belonged formerly to 'chemistry' and those classified as 'physical', and ultimately between these and those we call 'mental'. Structure, and structure alone, gives not only the *unique* content of what we call 'knowledge', but also the bridge between the different classes of occurrences—a fact which, as yet, has not been fully understood.

To sum up: It is known that colloidal behaviour is exhibited by materials of very fine subdivision, the 'world of neglected dimensions', which involves surface activities and electrical characters of manifold and complex structure, and therefore the flexibility of gross macroscopic characteristics. It is well known that all life-processes, 'feelings', 'emotions', 'thought', semantic reactions, and so forth, involve *at least* electrical currents. As electrical currents and other forms of energy are able to affect the colloidal structure on which our physical characteristics depend, obviously 'feelings', 'emotions', 'thought'.; in general, *s.r*, which are connected with manifestations of energy, will also have some effect on our bodies, and vice versa. Colloidal structure supplies us with an extremely flexible mechanism with endless possibilities.

When we analyse the known empirical facts from a structural point of view, we find not only the equivalence which was mentioned before, but we must, also, legitimately consider the so-called 'mental', 'emotional', and other semantic and nervous occurrences in connection with manifestations of energy which have a powerful influence on the colloidal behaviour, and so ultimately on the behaviour of our organisms as-a-whole. Under such environmental conditions, we must take into account all energies which have been discovered, *semantic reactions not excluded,* as all such energies have structural effect. As language is one of the expressions of one of these energies, we ought to find it quite natural that the structure of language finds its reflection in the structure of the environmental conditions which are dependent on it.

Until lately, the disregard of colloidal science and of structure in general has greatly retarded advance in biology, psychiatry, and other sciences. Biology, for instance, has mostly studied 'life' where none existed; namely, in death. If we study corpses, we study death, not life, and life is a function of living cells. The living cell is semi-fluid, and all the forces which act in colloidal solutions and constitute colloidal behaviour are acting because they *can act,* while a dead cell is *coagulated* and so a different set of energies is operating there.[24]

Should we wonder that life, being a form of colloidal behaviour on microscopic and sub-microscopic levels, conditioned by little colloidal 'wholes', and structures separated from their environment by surfaces, preserves a similar character on macroscopic levels? We should, instead, be surprised if this did not turn out to be the case.

CHAPTER X

THE 'ORGANISM-AS-A-WHOLE'

... in hypnotized children real colours and suggested colours are blended to form the complementary colour. (189)   W. HORSLEY GANTT

Section A. Illustrations from biology.

Because of the semantic importance of the structural non-elemental-istic principle, and the weighty, yet in the beginning odd, consequences which follow the consistent application of this principle in practice, we will give a short account of some other experimental structural facts taken from widely separated fields.

A worm, a marine planarian, called a *Thysanozoön (Brochii)*, is common in the bay of Naples. If we put a normal *Thysanozoön* on its back, it soon will right itself. When the brain of the worm has been removed, under similar conditions of the experiment, the worm will right itself, but *more slowly*. In this case, we see a general tendency of the organism-as-a-whole; the nervous system only facilitated a quicker action. If we cut the worm partly in two, so that the longitudinal nerves are severed, but a thin piece of tissue keeps the two parts together, the two parts move in a co-ordinated way, as if not cut. The organism still works as-a-whole, although the conditions seem not favourable.[1]

If we cut a fresh-water planarian (*Planaria torva*) in two, trans-versely, the posterior part, which has no brain, moves about as well as the anterior part, which has the brain. If we try to find the effect of light on the part devoid of brain and eyes, we see that the effect of light is not changed, and that the posterior part crawls away from light into dark corners as a normal animal would, except that the action takes place at a slower rate. In normal animals, the reaction usually begins in about one minute after the exposure; in the brainless part, it takes nearly five minutes of exposure.[2]

How chemical conditions affect the activities of the organism-as-a-whole can be well illustrated by the following examples. In a jellyfish, we can increase or decrease the locomotor activities by simply changing the chemical constitution of the water. If we increase the number of Na ions in the sea-water, the rhythmical contractions increase and the animal becomes restless. If we increase the number of Ca ions, the con-tractions decrease. In a similar way, we can change the orientation toward light in a number of marine animals by changing the constitution of the medium. The larvae of *Polygordius,* which usually go away from light into dark corners, can be compelled to go toward light by two methods: either by lowering the temperature of the sea-water, or else by increasing the concentration of the salts in the sea-water. This behaviour can be reversed by raising the temperature or lowering the concentration of the salts.[3]

An extremely instructive group of experiments has been performed in artificial fertilization of the eggs of a large number of marine animals, such as starfish, molluscs, and others.

Under usual conditions, these eggs cannot develop unless a spermatozoon enters the egg, which results in a thickening of the membrane called the 'fertilization membrane'. Experiments show that such a transformation can be produced artificially in an unfertilized egg, with resulting 'fertilization', by several artificial means, as, for instance, by the treatment of the eggs with special chemicals, and, in some instances, by merely puncturing the egg with a needle. The late Jacques Loeb succeeded in producing in this way parthenogenetic frogs, which lived a normal life.[4]

Under normal conditions, the eggs of different sea animals can be fertilized only by their proper sperm. But, if we raise the alkalinity of the sea-water slightly, we find that the eggs can be fertilized by different sperms, often of widely separated kinds of animals.[5] If we put unfertilized eggs of a sea-urchin into sea-water which contains a trace of saponin, we find that the eggs acquire the characteristic 'membrane of fertilization'. If the eggs are taken out, washed carefully and put back into sea-water, they develop into larvae.[6] The change in the chemical constitution of sea-water will also often produce twins from one egg Change in temperature may change the colour of butterflies,.[7]

A very large class of such organism-as-a-whole reactions is given in the works of Professor C. M. Child on regeneration. I suggest these works, not only because they are particularly interesting, even to the layman, but mainly because Professor Child has formulated a $\bar{A}$ biological system, the importance of which is becoming paramount, and is beginning to be applied even in psychiatry by Dr. Wm. A. White and others.

We find the characteristic of profiting by past experiences and acquiring negative reactions very low in the scale of life. Thus, even infusoria, which ingest a grain of carmine, soon learn to refuse it.[8] Most interesting experiments were performed on worms by Yerkes in 1912 and verified repeatedly. Yerkes built a T-shaped maze. In one arm (C) he placed a piece of sand-paper (S), beyond which there was an electrical device (E) which could give an electrical shock. The animal used for experimenting was an earthworm. The worm was admitted through the entrance (A). If he selected his way through (B), he got out without disagreeable consequences. If he selected (C), he received, first, a fair warning through the sand-paper (S) and, if this was not enough, he received an electrical shock at (E). After a number of experiences, the worm learned his lesson and avoided the path (C). After this habit was acquired, the five anterior segments of the worm were cut off. The beheaded worm retained the habit, although it reacted

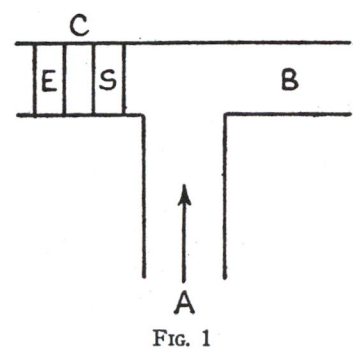

FIG. 1

more slowly. During the following two months, the worm grew a new brain and the habit disappeared. When trained again, he partially reacquired the above habit. Further experiments established that normal worms acquire the avoiding habit in approximately two hundred trials; and when the electrical device was put in the other arm, the worm learned how to reverse his habit in about sixty-five trials. Once the habit was acquired, the removal of the brain did not alter it. Worms with removed brains were also able to acquire a similar habit. Since the brain of an earthworm is a very small part of his whole nervous system, it has only a small dominance, and the neuro-muscular habits are acquired by the whole system and not simply by the brain. But, when a new brain began to operate, its dominance was seemingly sufficient to eliminate the habit.[9]

Experiments of McCracken with silkworm moths have shown that a beheaded moth can live as long as a normal one. It can be mated and will lay the normal number of fertile eggs arranged in the usual way. But it will not lay eggs spontaneously, and cannot select the proper kind of leaves on which to deposit them. If the head and the thorax were cut off, the females were unable to mate and their life was shortened to about five days. If mated before the operation, they would still lay eggs when stimulated.

In these more complicated cases, the brain is necessary for the more complicated behaviour, as, for instance, the selection of a mulberry leaf.[10] Although the organism works as-a-whole, the differentiation and relative importance (domination) of different organs becomes more accentuated, the higher we go in the scale of life.

Section B. Illustrations from nutrition experiments.

We find striking illustrations of the *non-el* principle in the study of 'vitamins'. A few years ago it was discovered that certain widely spread and pernicious diseases were due to deficiencies of some factors in diet. These factors, which normally are present in very minute amounts, were called 'vitamins' by the Polish biologist, Funk. The most important vitamin-deficiency diseases are called Rickets, Scurvy, Beri-Beri, and Pellagra. In all these cases, it is important to notice that the lack of a minute amount of some factor may have the most varied, pronounced, and seemingly unrelated consequences. The symptoms can now be produced deliberately on experimental animals, by diets free from the particular 'vitamins' and can also be cured at will by feeding them with the proper 'vitamins'.[11]

Rickets appears essentially as a disease of infancy or childhood. In mild cases, the disease may only be discovered after the death of the adult. In these cases, the lesions have not become pronounced enough during life to attract attention.

The diagnosis usually depends on manifestations in the bones, but rickets affects the whole organism and not merely the skeleton. The children are nervous and irritable, but apathetic. They sleep poorly and perspire excessively. The muscles become wasted and weak. Often a secondary anaemia occurs. The children sit, stand, and walk later than usual ; the teeth appear later in life and decay sooner. The bones usually become much affected. Areas of softening appear in the long bones, which become bent. In more severe cases, the bones may even become fractured and the head of the bone may separate from the shaft. The general resistance of the children to other diseases is lowered and mortality increases.

Cod liver oil *or* sunshine usually effects a cure. We should notice the little word 'or', for quite different 'causes' produce similar 'effects'— an example illustrating that in life 'cause' and 'effect' do not correspond in a one-to-one relation, but in a many-to-one relation.

Experiments have shown that not less than three primary dietary factors are concerned with the development of skeletal tissue. These are phosphorus, calcium, and at least one organic compound which is known as antirachitic vitamin. The work of Professor E. V. McCollum and his co-workers seems to show an interesting point ; namely, that the ratio between the concentrations of calcium and of phosphorus in the food may be more important than the absolute amounts of these substances.

Scurvy develops gradually. The patient loses weight, appears anae-
mic, pale, weak, and short of breath. The gums become swollen, bleed
easily, and often develop ulcers. The teeth loosen and may fall out.
Hemorrhages between the mucous membranes and the skin often occur.
Blue-black spots in the skin are very easily produced, or even occur
spontaneously. The ankles become swollen, and, in severe cases, the skin
becomes hard. Nervous symptoms of a varied character appear, some of
which are due to the rupture of blood vessels. In later stages of the disease, delirium and convulsions may occur. Autopsy reveals significant
data; namely, hemorrhages and fragility of the bones. Scurvy appears
also as a deficiency disease, produced mainly by the lack in food of the
so-called 'anti-scorbutic vitamin'.

Beri-Beri labels a form of inflammation of the peripheral nerves,
the nerves of motion and sensation being equally affected. In the beginning of the disease, the patient feels fatigue, depression, and stiffness of
the legs. We distinguish two forms, the wet and the dry. In the dry
form, wasting, anaesthesia and paralysis are the chief manifestations.
The most marked manifestation in the wet form is the accumulation of
serum in the cellular tissue affecting the trunks, limbs and extremities.
Usually, in both forms, there appear tenderness of the calf muscles and
a tingling or burning in the feet, legs, and arms. The mortality is high.

Pellagra involves the nervous system, the digestive tract, and skin.
Normally, one of the first symptoms to appear is soreness and inflammation of the mouth. Symmetrical redness of the skin occurs on parts of
the body. The nervous symptoms become more pronounced as the disease advances. The spinal cord is particularly involved, but the central
nervous system is also often affected.

Speaking about 'vitamins' and how their absence affects the organism-as-a-whole, we should mention that sterility in females may be connected with lack of vitamins. Astonishing experiments by Professor
McCollum showed that such diverse phenomena as loss of weight, premature old age, high infant mortality., are largely due to diet, and that
even such fundamental instincts as the motherly instinct are also affected.
The normally nourished rat very seldom destroys its young and, as a rule,
rats are good mothers. If we put such a mother rat on an abundant diet
that is deficient in some vitamins, the mother reacts quite differently
toward her young and destroys them soon after their birth. This characteristic has been controlled experimentally, and reversed at will by
proper diets. Nervousness and irritability in rats can also be controlled
experimentally by means of the vitamins they receive, or lack in food.

Section C. Illustrations from 'mental' and nervous diseases.

Simple and striking examples of what the *non-el* principle means can also be given from psychiatry.

White quotes the report of Prince that a patient was subject to severe attacks of hay fever when exposed to roses. On one occasion, a bunch of roses was unexpectedly produced from behind a screen. The patient started a severe attack with all the usual symptoms, lachrymation, congestion of the mucosa., although the roses were *made of paper*. This interesting case shows clearly how 'mental' factors (the belief that the roses were genuine) produce a series of reactions involving sensory, motor, vasomotor disturbances, and secretory disturbances of a definitely 'physical' character.[12]

Migraine labels a disturbance in the tension of blood vessels (vasomotor), which is due to a great variety of possible stimuli acting on the vegetative nervous system. In some instances, the stimuli may be purely physical, as severe blows, falls, fast movements, sudden alteration in temperature, of pressure.; or they may be chemical, and due to nicotine, alcohol, morphine, or to some endocrinal disturbances (adrenals, thyroid), toxins,. They may be of a purely somatic reflex character, due to fatigue, tumour formations,. They may also be of a semantic character, due to anger, fear, disappointment, worry, and other semantic states, which may act by disturbing the metabolism.

Migraine appears usually as a periodical abnormal state, in which the patient suffers from an oppressive pain in the head which gradually passes from heaviness and dullness to splitting intensity. Often characteristic visual signs also appear. The patient sees dark spots in the visual field, flying specks, and may become even partially blind. Chilliness, depression, sensory disturbances, particularly in the stomach, with vomiting, are often present. An attack may last a few hours or even several days.[13]

Cretinism labels a physical and 'mental' disturbance due mainly to the loss or diminution of the function of the thyroid gland. The patient (child) falls behind in his physical development, which often results in dwarfism, except for the skull, which grows larger in proportion to the rest of the body. The bone defects give rise to widely separated eyes, pug nose,. The bony tissue becomes unusually hard, and there is also a marked dental deficiency. The neck is usually thick and short, the abdomen puffy, the navel sunken. The hair line begins low on the forehead, the nose is sunken, the eyelids swollen, the face puffy, the tongue protruding. The liver is usually enlarged, respiration is slow, and changes in the blood can be detected. The nervous system is affected: we also find defects in sensory and motor nerve structure. On the 'mental' level, we find different degrees of stupidity, 'mental' weakness (morons), imbecility, and even idiocy. Smell, eyesight, and hearing are often poor, speech disturbed, so that we often find the patients deaf and dumb. The patients have an unsteady gait, with wobbling of the head. Over-activity of the thyroid gland results in the well-known goitre.

Hyperpituitarism results in acromegaly, characterized by the gradual enlargement of the bones of the nose, jaw, hands and feet, gigantism, often connected with profound disturbances. Hypopituitarism, or deficiency of the pituitary hormones, gives rise to a group of diseases characterized by a progressive accumulation of fat, and is connected with other abnormalities and disturbances.

From the field of the psychoneuroses, I shall mention only hysteria. It is very interesting to note that the many and various physical and somatic symptoms are of a purely semantic origin. The symptoms of hysteria are many and very complex, but they group themselves mainly in disturbances of motion and 'sensation'. We find every kind of paralysis and anaesthesia. Paralysis of the limbs is frequent; anaesthesia may be distributed in many ways, involving the superficies or the various 'sense-organs'. It is interesting to note that the distribution of these symptoms does not follow the anatomical areas of nerve distribution, but shows a symbolic (psycho-logical) grouping. The disturbances of motility are usually in the form of paralysis. Tremors, muscular debility, fatigability, involuntary muscular twitching, tics and spasms are often hysterical in origin. Speech is often involved; sometimes patients can only whisper, although their vocal organs are healthy. Stuttering is often hysterical, and analysis shows that the words which give difficulties usually have special semantic significance for the patient. Respiratory disturbances of an asthmatic character and disturbances of the gastro-intestinal tract are also often hysterical.

It should be emphasized that since non-elementalism has a physico-chemical structural base in colloidal behaviour, all life and all organisms give ample material for illustration. We have given here only a very few examples, selected mainly because of their simple and spectacular empirical character, but not generally too well known. Empirical data show clearly that the most diversified factors, acting as partial stimuli, ultimately affect or result in the response of the whole.

The handling of such empirical, structural, fundamental problems involves serious structural, linguistic and semantic difficulties which have to be solved *entirely* by adjusting the structure of the language used. But

9

such adjustment requires a full understanding of the structural issues at hand and a fundamental structural departure from *A* methods and *means*. These structural issues and means of departures from *A* methods are explained in the following chapters.

To sum up: The *non-el* principle formulates a structural character inherently found in the structure of the world, ourselves, and our nervous system on all levels; the knowledge and application of which is unconditionally necessary for adjustment on all levels, and, therefore, in humans, for *sanity*.

As 'knowledge', 'understanding', and such functions are *solely* relational, and, therefore, structural, the unconditional and inherent condition for adjustment on all human levels depends on building languages of similar structure to the experimental facts. Once this is accomplished, all the former desirable semantic consequences follow *automatically*.

For simplicity, we have considered only examples of the 'organism-as-a-whole', but, as a matter of fact, such a detached consideration cannot be considered entirely satisfactory, as, *structurally,* every organism depends on its environment; and, therefore, in building our languages, we ought to coin terms which also involve the latter by implication. Fortunately, this condition does not involve us in serious difficulties, when once identity is eliminated and the fundamental problems of structure are grasped. Indeed, the terms which we have already used, or which will be used as we proceed, are all of such a *non-el* structure as to involve the environment by implication.

In dealing with 'Smith', the difficulties are particularly serious because his nervous system is the most complex known. It is stratified four-dimensionally (in space-time), and the dominance of some centres introduces prodigious and manifold interrelations non-existent in nervous systems of simpler structure; and we still have to learn how to handle the former. Fortunately, mathematical methods and psychiatry explain a good deal about this question, and give us the desired means to apply what we have learned.

Obviously, to 'know' something is quite different from the *habitual* application of what we have learned. This semantic difference is particularly acute in the case of language, as it involves *structural* implications which work *unconsciously*. It is not enough to 'understand' and 'know' the content of the present work; one must *train* oneself in *the use* of the new terms. Then only can he expect the maximum semantic results.

PART IV

STRUCTURAL FACTORS IN NON-ARISTOTELIAN LANGUAGES

Without objects conceived as unique individuals, we can have *no Classes*. Without classes we can, as we have seen, define *no Relations*, without relations we can have *no Order*. *But to be reasonable is to conceive of order-systems, real or ideal. Therefore, we have an absolute logical need to conceive of individual objects as the elements of our ideal order systems.* This postulate is the condition of defining clearly any theoretical conception whatever. The further metaphysical aspects of the concept of an individual we may here ignore. *To conceive of individual objects is a necessary presupposition of all orderly activity.* (449)   JOSIAH ROYCE

The connections shown by these particular examples hold in general: given a transformation, you have a function and a relation; given a function, you have a relation and a transformation; given a relation, you have a transformation and a function: *one* thing—*three* aspects; and the fact is exceedingly interesting and weighty. (264)   CASSIUS J. KEYSER

It can, you see, be said, with the same approximation to truth, that the whole of science, including mathematics, consists in the study of transformations or in the study of relations. (264)   CASSIUS J. KEYSER

Science is never merely knowledge; it is orderly knowledge. (449)
                                                                    JOSIAH ROYCE

Philosophers have, as a rule, failed to notice more than two types of sentence, exemplified by the two statements "this is yellow" and "buttercups are yellow." They mistakenly suppose that these two were one and the same type, and also that all propositions were of this type. The former error was exposed by Frege and Peano; the latter was found to make the explanation of order impossible. Consequently the traditional view that all propositions ascribe a predicate to a subject collapsed, and with it the metaphysical systems which were based upon it, consciously or unconsciously. (457)   BERTRAND RUSSELL

Interesting analyses by Van Woerkom have shown a general incapacity in aphasics for grasping relations, realizing ordered syntheses, etc.; all of them are operations which are based, in the normal individual, on the use of verbal symbolization. When confronted by groups of figures or of geometrical forms, the aphasic, even though he may perceive them correctly, is unable to analyse or to order the elements, to grasp their succession . . .
(411)   HENRI PIÉRON

CHAPTER XI

ON FUNCTION

The whole science of mathematics rests upon the notion of function, that is to say, of dependence between two or more magnitudes, whose study constitutes the principal object of analysis.   C. E. PICARD

Every one is familiar with the *ordinary* notion of a function—with the notion, that is, of the lawful dependence of one or more variable things upon other variable things, as the area of a rectangle upon the lengths of its sides, as the distance traveled upon the rate of going, as the volume of a gas upon temperature and pressure, as the prosperity of a throat specialist upon the moisture of the climate, as the attraction of material particles upon their distance asunder, as prohibitionary zeal upon intellectual distinction and moral elevation, as rate of chemical change upon the amount or the mass of the substance involved, as the turbulence of labor upon the lust of capital, and so on and on without end. (264)   CASSIUS J. KEYSER

The infinite which it superficially gets rid of is concealed in the notion of "any," which is but one of the protean disguises of mathematical generality. (22)   E. T. BELL

The famous mathematician, Heaviside, mentions the definition of quaternions given by an American schoolgirl. She defined quaternions as 'an ancient religious ceremony'. Unfortunately, the attitude of many mathematicians justified such a definition. The present work departs widely from this religious attitude and treats mathematics simply as a most important and unique form of human behaviour. There is nothing sacred about any single verbal formulation, and even those that now seem most fundamental should be held subject to structural revision if need should arise. The few mathematicians who have produced epoch-making innovations in mathematical method had this behaviouristic attitude *unconsciously,* as will be shown later. The majority of mathematicians take mathematics as a clear-cut entity, 'by itself'. This is due, first, to a confusion of orders of abstractions and to identification, as will be explained later; and, second, to its seeming simplicity. In reality, such an attitude introduces quite unexpected complications, leading to mathematical revolutions, which are always bewildering. The mathematical revolutions occur only because of this *over-simplified,* and thus fallacious, attitude of the mathematicians toward their work. Had all mathematicians the semantic freedom of those who make the mathematical 'revolutions', there would be *no* mathematical 'revolutions', but an extremely swift and constructive progress. To re-educate the *s.r* of such mathematicians, the problem of the psycho-logics of mathematics must receive more attention. This means that some mathematicians must

133

become psycho-logicians also, or that psycho-logicians must study mathematics.

For, let us take a formula which exemplifies mathematics at its best; namely, one and one make two $(1 + 1 = 2)$. We see clearly that this human product involves a threefold relation: between the man who made it,

$$(A) \quad \diagup 1 + 1 = 2 \diagdown$$
$$\text{Brown} \longleftrightarrow \text{Smith}$$

let us say, Smith, and the black-on-white marks (A), between these marks and Brown, and between Brown and Smith. This last relationship is the only *important* one. The marks (A) are only auxiliary and are *meaningless by 'themselves'*. They would never occur if there were no Smiths to make them, and would be of no value if there were no understanding Browns to use and to appreciate them. It is true that when we take into account this threefold relation the analysis becomes more difficult and must involve a revision of the foundations of mathematics. Although it is impossible to attempt in this book a deeper analysis of these problems in a general way, yet this behaviouristic attitude follows the rejection of the 'is' of identity, and is applied all through this work.

The notion of 'function' has played a very great role in the development of modern science, and is structurally and semantically fundamental. This notion was apparently first introduced into mathematical literature by Descartes. Leibnitz introduced the term. The notion of a 'function' is based on that of a *variable*. In mathematics, a variable is used as an ∞-valued symbol that can represent *any one* of a series of numerical elements.

It is useful to enlarge the mathematical meaning of a variable to include any ∞-valued symbol of which the value is not determined. The various determinations which may be assigned to the variable we call the *value* of the variable. It is important to realize that a mathematical variable does not vary or change in itself, but can take *any* value within its range. If a particular value is selected for a variable, then this value, and, therefore, the variable, becomes fixed—a one-valued constant. In the use of these terms, we should take into account the behaviour of the mathematizer. His 'x' is like a container, into which he may pour any or many liquids; but once the selection has been made, the content of the container is one or a constant. So 'change' is not inherent in a variable; it is due only to the volition of the mathematizer, who can change one value for another. Thus, the value changes by quanta, in definite lots, according to the pleasure of the operator. This quantum character of the variable has serious structural and semantic consequences, which will become clearer further on. It allows us, without stretching our definitions, to apply the new vocabulary to any problem whatsoever. It is in structural accord with the trend of the quantum theory, and, therefore, with the *structure* of this world, as we know it at present.

The notion of a variable originated in mathematics, and, in the beginning, dealt only with numbers. Now numbers, when given, represent, structurally, a manifold or aggregate which is *not* supposed to change. So, when we consider a variable, we should 'think' *not* of a changing entity, but of *any* element we choose out of our perfectly constant collection (when given). Let me repeat that the notion of *change* enters in, only in connection with the volition and the *s.r* of the one who operates these unchanging entities. The notion of a variable is taken always in an extensional ∞-valued sense, to be explained later, as it always implies structurally a collection of many individuals, out of which collection a selection of one can be made. The notion of a variable is general and, in principle, ∞-valued; a constant is a special one-valued case of a variable in which the collection contains a single element, making alternative selection impossible.

Variables are usually symbolized by the end letters of the alphabet, x, y, z,. The supply is increased as desired by the use of indices; for instance, x', y', z'.; x'', y'', z''.; or x_1, y_1, z_1.; x_2, y_2, z_2,. This gives a flexible means of denoting numerous individuals, and so manufacturing them indefinitely, as the extensional method of mathematics requires. Another method, introduced not long ago, has proven useful in dealing with a definite selection of variables in a simplified manner. One letter or one equation can be used instead of many. The variable sign x is modified by another letter which may have different values, in a given range; for instance, x_i, x_k,. The modifying letter i or k can take the serial values; let us say i or $k = 1, 2, 3$,. Since the one symbol x_k stands for the array of many *different* variables x_1, x_2, x_3., statements can be greatly simplified, and yet preserve structurally the *extensional* individuality.

It is important that the non-mathematical reader should become acquainted with the above methods and notations, as they involve a profound and far-reaching structural and *psycho-logical* attitude, useful to *everybody,* involving most fundamental *s.r.*

The *extensional* method means dealing structurally with many *definite individuals;* as, for instance, with 1, 2, 3., a series in which each individual has a special and *unique* name or symbol. This extensional method is structurally the *only* one by which we may expect to acquire $\bar{A}$ ∞-valued *s.r.* In a strict sense, the problems in life and the sciences do not differ structurally from this mathematical problem. In life and science, one deals with many, actual, unique individuals, and all *speaking* is using abstractions of a very high order (abstractions from abstraction from abstraction,.). So, whenever we speak, the individual is never completely covered, and some characteristics are left out.

A rough definition of a function is simple: y is said to be a function of x, if, when x is given, y is determined. Let us start with a simple mathematical illustration: $y = x + 3$. If we select the value 1 for x our $y = 1 + 3 = 4$. If we select $x = 2$, then $y = 2 + 3 = 5$,. Let us take a more complicated example; for instance: $y = x^2 - x + 2$. We see that for $x = 1$, $y = 1 - 1 + 2 = 2$; for $x = 2$, $y = 4 - 2 + 2 = 4$; for $x = 3$, $y = 9 - 3 + 2 = 8$,.

In general, y is determined when we fulfill all the indicated *operations* upon the variable x, and so get the final results of these operations. In symbols, $y = f(x)$, which is read, y equals function of x, or y equals f of x.

In our example, we may call x the independent variable, meaning that it is the one to which we may assign any value at our pleasure, if not limited by the conditions of our problem, and y would then be the dependent variable, which means that its value is no longer dependent on our pleasure, but is determined by the selection of the value of x. The terms dependent and independent variables are not absolute, for the dependence is mutual, and we could select either variable as the independent one, according to our wishes.

The notion of a 'function' has been generalized by Bertrand Russell to the very important notion of a 'propositional function'.[1] For my purpose, a rough definition will be sufficient. By a propositional function, I mean an ∞-*valued* statement, containing one or more variables, such that when single values are assigned to these variables the expression becomes, in principle, a *one-valued* proposition. The ∞-valued character of propositional functions seems essential, because we may have a one-valued descriptive function with variables, or a one-valued expression formulating a semantic relational law expressed in variable terms., yet these would be propositions. Thus, the ∞-*valued* statement, 'x is black', would exemplify a propositional function; but the one-valued relation 'if x is more than y, and y is more than z, then x is more than z' exemplifies a proposition. This extended *m.o* notion of a propositional function becomes of crucial importance in a $\bar{A}$-system, because most of our speaking is conducted in ∞-valued languages to which we mostly and delusionally ascribe single values, entirely preventing proper evaluation.

An important characteristic of a propositional function, for instance, 'x is black', is that such a statement is neither true nor false, but ambiguous. It is useless to discuss the truth or falsehood of propositional functions, since the terms true or false cannot be applied to them. But if a definite, single value is assigned to the variable x, then the propositional function becomes a proposition which may be true or false. For instance, if we assign to x the value 'coal', and say 'coal is black', the ∞-valued propositional function has become a one-valued true proposition. If we should assign to x the value 'milk', and say 'milk is black', this also would make a proposition, but, in this case, false. If we should assign to x the value 'blah-blah', and say 'blah-blah is black', such a statement may be considered as meaningless, since it contains sounds which have *no* meaning; or we *may* say, 'the statement blah-blah is *not* black but meaningless', and, therefore, the proposition 'blah-blah is black', is *not* meaningless but false.

We should notice—a fact disregarded in the *Principia Mathematica* —that there is no hard and fast rule by which we can distinguish between meaningless and false statements in general, but that such discrimination depends on many factors in each specific case. A propositional function, 'x is black', cannot be its own argument: for instance, if we substitute the whole propositional function, 'x is black', for the variable x in the original propositional function, and then consider the expression, 'x is black, is black', which Whitehead and Russell classify as *meaningless,* this expression is *not* necessarily meaningless, but *may* be considered *false.* For, the statement, 'x is black', is defined as a *propositional function,* and, therefore, the statement, 'x is black, is black', *may* be considered *false.*

The problems of 'meaning' and 'meaningless' are of great semantic importance in daily life, but, as yet, little has been done, and little research made, to establish or discover valid criteria. To prove a given statement false is often laborious, and sometimes impossible to do so, because of the undeveloped state of knowledge in that field. But with meaningless *verbal* forms, when their meaninglessness is exposed in a given case, the non-sense is exploded for good.

From this point of view, it is desirable to investigate more fully the mechanism of our symbolism, so as to be able to distinguish between statements which are false and verbal forms which have no meanings. The reader should recall what was said about the term 'unicorn', used as a symbol in heraldry and, eventually, in 'psychology', since it stands for a human *fancy,* but, in zoology, it becomes a noise and not a symbol, since it does not stand for any actual animal whatsoever.

A very curious semantic characteristic is shared in common by a propositional function and a statement containing meaningless noises; namely, that neither of them can be true or false. In the old A way all sounds man made, which could be written down and looked like words, were considered words; and so every 'question' was expected to have an answer. When spell-marks (noises which can be spelled) were put together in a specified way, each combination was supposed to say something, and this statement was supposed to be true or false. We see clearly that this view is not correct, that, in addition to words, we make noises (spell-marks) which may have the appearance of being words, but should *not* be considered as words, as they say nothing in a given context. Propositional functions, also, cannot be classified under the simple two opposites of true and false.

The above facts have immense semantic importance, as they are directly connected with the possibility of human agreement and adjustment. For upon statements which are neither true nor false we can always disagree, if we insist in applying criteria which have no application in such cases.

In *human* life the semantic problems of 'meaninglessness' are fundamental for sanity, because the evaluation of noises, which do not constitute symbols in a given context as symbols in that context, must, of necessity, involve delusions or other morbid manifestations.

The solution of this problem is simple. Any noises or signs, when used semantically as symbols, *always* represent *some symbolism,* but we must find out to what field the given symbolism applies. We find only three possible fields. If we apply a symbol belonging to one field to another field, it has very often no meaning in this latter. In the following considerations, the theory of errors is disregarded.

A symbol may stand for: (1) Events outside our skin, or inside our skin in the fields belonging to physics, chemistry, physiology, . (2) Psycho-logical events inside our skin, or, in other words, for *s.r* which may be considered 'sane', covering a field belonging to psycho-logics. (3) Semantic disturbances covering a pathological field belonging to psychiatry.

As the above divisions, together with their interconnections, cover the field of human symbolism, which, in 1933, have become, or are rapidly becoming, *experimental* sciences, it appears obvious that older 'metaphysics' of every description become illegitimate, affording only a very fertile field for study in psychiatry.

Because of *structural* and the above *symbolic* considerations based on $\bar{A}$ negative, non-identity premises, these conclusions appear as *final;* and, perhaps, for the first time bring to a focus the age-long problem of the subject-matter, character, value, and, in general, the status of the older 'metaphysics' in human economy. From the *non-el,* structural, and semantic point of view, the problems with which the older 'metaphysics' and 'philosophy' dealt, should be divided into two quite definite groups. One would include 'epistemology', or the theory of knowledge, which would ultimately merge with scientific and *non-el* psycho-logics, based on general semantics, structure, relations, multi-dimensional order, *and* the quantum mechanics of a given date; and the rest would represent semantic disturbances, to be studied by a generalized up-to-date psychiatry.

Obviously, considerations of structure, symbolism, sanity., involve the solutions of such weighty problems as those of 'fact', 'reality', 'true', 'false'., which are completely solved only by the consciousness of abstracting, the multiordinality of terms.,—in general, a $\bar{A}$-system.

Let me repeat the rough definition of a propositional function— as an ∞-valued statement containing variables and characterized by the fact that it is ambiguous, neither true nor false.

How about the terms we deal with in life? Are they all used as one-valued terms for constants of some sort, or do we have terms which are inherently ∞-valued or variable? How about terms like 'mankind', 'science', 'mathematics', 'man', 'education', 'ethics', 'politics', 'religion', 'sanity', 'insanity', 'iron', 'wood', 'apple', 'object', and a host of other terms? Are they labels for one-valued constants or labels for ∞-valued stages of processes. Fortunately, here we have no doubt.

We see that a large majority of the terms we use are names for ∞-valued stages of processes with a *changing content.* When such terms are used, they generally carry different or many contents. The terms represent ∞-valued variables, and so the statements represent ∞-valued propositional functions, not one-valued propositions, and, therefore, in principle, are neither true nor false, but ambiguous.

Obviously, before such propositional functions can become propositions, and be true or false, single values must be assigned to the variables by some method. Here we must select, at least, the use of co-ordinates. In the above cases, the 'time' co-ordinate is sufficient. Obviously, 'science 1933' is quite different from 'science 1800' or 'science 300 b.c.'.

The objection may be made that it would be difficult to establish means by which the use of co-ordinates could be made workable. It seems that this might involve us in complex difficulties. But, no matter how simple or how complex the means we devise, the details are *immaterial*, and, therefore, we can accept the roughest and simplest; let us say, the year, and usually no spatial co-ordinates. The invaluable semantic effect of such an innovation is *structural*, one-, versus ∞-valued, *psycho-logical* and methodological, and affects deeply our *s.r.*

From time immemorial, some men were supposed to deal in one-valued 'eternal verities'. We called such men 'philosophers' or 'metaphysicians'. But they seldom realized that all their 'eternal verities' consisted only of *words,* and words which, for the most part, belonged to a primitive language, reflecting in its structure the assumed structure of the world of remote antiquity. Besides, they did not realize that these 'eternal verities' last only so long as the human nervous system is not altered. Under the influence of these 'philosophers', two-valued 'logic', and confusion of orders of abstractions, nearly all of us contracted a firmly rooted predilection for 'general' statements—'universals', as they were called—which, in most cases, inherently involved the semantic one-valued conviction of validity for all 'time' to come.

If we use our statements with a date, let us say 'science 1933', such statements have a profoundly modified structural and psycho-logical character, different from the old general legislative semantic mood. A statement concerning 'science 1933', whether correct or not, has no element of semantic conviction concerning 1934.

We see, further, that a statement about 'science 1933' might be quite a definite statement, and that if the person is properly informed, it probably would be true. Here we come in contact with the structure of one of those human semantic impasses which we have pointed out. We humans, through old habits, and because of the inherent structure of human knowledge, have a tendency to make static, definite, and, in a way, absolutistic one-valued statements. But when we fight absolutism, we quite often establish, instead, some other dogma equally silly and harmful. For instance, an active atheist is psycho-logically as unsound as a rabid theist.

A similar remark applies to practically all these opposites we are constantly establishing or fighting for or against. The present structure of human knowledge is such, as will be shown later, that we tend to make definite statements, static and one-valued in character, which, when we take into account the present pre-, and $\bar{A}$ one-, two-, three-valued affective components, inevitably become absolutistic and dogmatic and extremely harmful.

It is a genuine and fundamental semantic impasse. These static statements are very harmful, and yet they cannot be abolished, for the present. There are even weighty reasons why, without the formulation and application of ∞-valued semantics, it is not possible (1933) to abolish them. What can be done under such structural circumstances? Give up hope, or endeavour to invent methods which cover the discrepancy in a satisfactory (1933) way? The analysis of the psycho-logics of the mathematical propositional function and $\overline{A}$ semantics gives us a most satisfactory structural solution, necessitating, among others, a four-dimensional theory of propositions.

We see (1933) that we can make definite and *static* statements, and yet make them semantically *harmless*. Here we have an example of abolishing one of the old A tacitly-assumed 'infinities'. The old 'general' statements were supposed to be true for 'all time'; in quantitative language it would mean for 'infinite numbers of years'. When we use the date, we reject the fanciful tacit A 'infinity' of years of validity, and *limit* the validity of our statement by the date we affix to it. Any reader who becomes accustomed to the use of this structural device will see what a tremendous semantic difference it makes psycho-logically.

But the above does not exhaust the question structurally. We have seen that when we speak about ∞-valued processes, and stages of processes, we use variables in our statements, and so our statements are not propositions but propositional functions which are not true or false, but are ambiguous. But, by assigning single values to the variables, we make propositions, which might be true or false; and so investigation and agreement become possible, as we then have something definite to talk about.

A fundamental structural issue arises in this connection; namely, that in doing this (assigning single values to the variables), our attitude has automatically changed to an extensional one. By using our statements with a date, we deal with definite issues, on record, which we can study, analyse, evaluate., and so we make our statements of an extensional character, with all cards on the table, so to say, at a given date. Under such extensional and limited conditions, our statements then become, eventually, propositions, and, therefore, true or false, depending on the amount of information the maker of the statements possesses. We see that this criterion, though difficult, is feasible, and makes agreement possible.

A structural remark concerning the A-system may not be amiss here. In the A-system the 'universal' proposition (which is usually a propositional function) always implies *existence*. In A 'logic', when it is said that 'all A's are B', it is assumed that there are A's. It is obvious that always assuming existence leaves no place for non-existence; and this is why the old statements were supposed to be true or false. In practical life, collections of noises (spell-marks) which look like words, but which are not, are often not suspected of being meaningless, and action based on them may consequently entail unexplicable disaster. In our lives, most of our miseries do not originate in the field where the terms 'true' and 'false' apply, but in the field where they *do not apply;* namely, in the immense region of propositional functions and meaninglessness, where agreement must fail.

Besides, this sweeping and unjustified structural assumption makes the A-system *less general.* To the statement, 'all A's are B', the mathematician adds 'there may or may not be A's'. This is obviously *more general.* The old pair of opposites, true and false, may be enlarged to three possibilities—statements which might be true, or false, and verbal forms which have the appearance of being statements and yet have no meaning, since the noises used were spell-marks, *not symbols* for anything with actual or 'logical' existence.

Again a $\bar{A}$-system shares with the $\bar{E}$ and $\bar{N}$ systems a useful and important methodological and structural innovation; namely, it limits the validity of its statements, with weighty semantic beneficial consequences, as it tends from the beginning to eliminate undue, and often intense, dogmatism, categorism, and absolutism. This, on a printed page, perhaps, looks rather unimportant, but when *applied,* it leads to a fundamental and structurally beneficial alteration in our semantic *attitudes* and behaviour.

In the present work, each statement is merely the best the author can make in 1933. Each statement is given *definitely,* but with the semantic *limitation* that it is based on the information available to the author in 1933. The author has spared no labour in endeavouring to ascertain the state of knowledge as it exists in the fields from which his material is drawn. Some of this information may be incorrect, or wrongly interpreted. Such errors will come to light and be corrected as the years proceed.

A great source of difficulty and of possible objections is that science is, at present, so specialized that it is impossible for one man to know all fields, and that, therefore, the use of a term such as 'science 1933', might be fundamentally unsound. This objection should not be lightly dismissed, as it is serious. Yet it can, I believe, be answered satisfactorily. At this early stage of our enquiry, a large number of the facts of knowledge does not affect my investigation; therefore, it has not proved impossible to keep sufficiently well informed on the points which are covered. Also, the further scientific theories advance, the simpler they become. For instance, books on physics are simpler and less voluminous now than twenty years ago. Something similar could be said about mathematics. The general outlook is simpler.

The main interest of the author at this stage of his work is structural and semantic, rather than technical, and so he has only had to know enough of the technique of different sciences to be able to understand sufficiently their *structure* and *method*. Revolutionary structural and methodological advances are few in the history of mankind; and so it is possible, though not easy, to follow them up in 1933.

But the main point is that the affixing of the date has very far-reaching structural methodological and, therefore, psycho-logical semantic consequences. For instance, it changes propositional function into propositions, converts semantically one-valued intensional methods into ∞-valued extensional methods, introduces four-dimensional methods., and so the 'date' method is to be recommended on these *structural and semantic grounds alone*. As it is beneficial to affix the date in 1933, we affix the date 1933, not to give the impression that from a technical point of view I am familiar with the results of all branches of science at that date, but to indicate that no advance in *structure* and *method* of 1933 has been disregarded. It will become obvious later in this book, when additional data have been taken into consideration, and a new summary and new abstractions made, that the result is a surprising simplification, which can be clearly understood by laymen as well as by scientists. With the help of the generalizations of new structure and ∞-valued semantics, it will be easier to follow the advance of science, because we shall then have a better outlook on the psycho-logics of science as-a-whole.

It will become clear, too, that to provide for a further elaboration of this work in the future, the establishment of a special branch of research in $\bar{A}$-systems must become a *group* activity; for, as I have been painfully aware, the production of even this outline of that branch of research has overstrained the powers of one man.

The most cheering part of this work is, perhaps, the practical results which this investigation has accomplished, combined with the simplicity of means employed. One of the dangers into which the reader is liable to fall is to ascribe too much generality to the work, to forget the limitations and, perhaps, one-sidedness which underlie it. The limitation and the generality of this theory lie in the fact that if we symbolize our human problems ($H = f(x_1, x_2, x_3, x_4, x_5, \ldots x_n)$) as a function of an enormous number of variables, the present theory deals only with a few of these variables, let us say x_1 (say, structure), x_2 (say, evaluation)., but these variables have been found, up to the present, in *all* our experience and all our equations.

A most important extension of the notion of 'function' and 'propositional function' has been further accomplished by Cassius J. Keyser, who, in 1913, in his discussion of the multiple interpretations of postulate systems, introduced the notion of the 'doctrinal function'. Since, the doctrinal function has been discussed at length by Keyser in his *Mathematical Philosophy* and his other writings, by Carmichael[2], and others. Let us recall that a propositional function is defined as an ∞-valued statement, containing one or more variables, such that when single values are assigned to these variables the expression becomes a one-valued proposition. A manifold of interrelated propositional functions, usually called postulates, with all the consequences following from them, usually called theorems, has been termed by Keyser a *doctrinal function*. A doctrinal function, thus, has no specific content, as it deals with variables, but establishes *definite relations* between these variables. In principle, we can assign many single values to the variable terms and so generate many doctrines from *one* doctrinal function. In an ∞-valued $\bar{A}$-system which eliminates identity and is based on structure, doctrinal functions become of an extraordinary importance.

In an ∞-valued world of absolute individuals on objective levels, our statements can always be formulated in a way that makes obvious the use of ∞-valued terms (variables) and so the postulates can always be expressed by propositional function. As postulates establish relations or multi-dimensional order, a set of postulates which defines a doctrinal function gives, also *uniquely*, the *linguistic structure*. As a rule, the builders of doctrines do not start with sets of postulates which would explicitly involve variables, but they build their doctrine around some specific content or one special respective value for the variables, and so the *structure* of a doctrine, outside of some mathematical disciplines, has never been explicitly given. If we trace a given *doctrine with specific content* to its *doctrinal function without content,* but variable terms, then, only, do we obtain a set of postulates which gives us the *linguistic structure*. Briefly, to find the structure of a doctrine, we must formulate the doctrinal function of which the given doctrine is only a special interpretation. In non-mathematical disciplines, where doctrines are not traced down to a set of postulates, we have no means of knowing their structure, or whether *two different* doctrines originated from *one* doctrinal function, or from *two*. In other words, we have no simple means of ascertaining whether the two different doctrines have similar or different structure. Under aristotelianism, these differentiations were impossible, and so the problems of linguistic structure, propositional and doctrinal functions., were neglected, except in the recent work of mathematicians. The entirely general semantic influence of these structural conditions becomes obvious when we realize that, no matter whether or not our doctrines are traced down to their doctrinal functions, our semantic processes and all 'thinking' follow *automatically* and, by necessity, the conscious or unconscious postulates, assumptions., which are given (or made conscious) *exclusively* by the doctrinal function.

The terms 'proposition', 'function', 'propositional function', 'doctrinal function'., are multiordinal, allowing many orders, and, in a given analysis, the different orders should be denoted by subscripts to allow a differentiation between them. When we deal with more complex doctrines, we find that in structures they represent higher order doctrines, or a higher whole, the constituents of which represent lower order doctrines. Similarly, with doctrinal functions, if we take any *system,* an analysis will discover that it is a whole of related doctrinal functions. As this situation is the most frequent, and as 'thinking', in general, represents a process of relating into higher order relational entities which are later *treated as complex wholes,* it is useful to have a term which would symbolize doctrinal functions of higher order, which are made up of doctrinal functions of lower orders. We could preserve the terminology of 'higher' and 'lower' order; but as these conditions are always found in all *systems,* it seems more expedient to call the higher body of interrelated doctrinal functions, which ultimately produce a system—a *system-function.* At present, the term 'system function' has been already coined by Doctor H. M. Sheffer[3]; but, to my knowledge, Sheffer uses his 'system function' as an equivalent for the 'doctrinal function' of Keyser. For the reasons given above, it seems advisable to limit the term 'doctrinal function' to the use as introduced by Keyser, and to enlarge the meaning of Sheffer's term 'system function' to the use suggested in the present work, this natural and wider meaning to be indicated by the insertion of a hyphen.

In a $\bar{A}$-system, when we realize that we live, act., in accordance with *non-el s.r,* always involving integrated 'emotions' and 'intellect' and, therefore, some explicit or implicit postulates which, by structural necessity, utilize variable, multiordinal and ∞-valued terms, we must recognize that *we live and act by some system-functions* which consist of doctrinal functions. The above issues are not only of an academic interest, as, without mastering all the issues emphasized in the present work, it is

10

impossible to analyse the extremely complex difficulties in which, as a matter of fact, we are immersed.

At present, the doctrinal functions and the system-functions have not been worked out, and even in mathematics, where these notions originated, we speak too little about them. But in mathematics, as the general tendency is to bring all mathematical disciplines to a postulational base, and these postulates always involve multiordinal and ∞-valued terms, we actually produce doctrinal or system-functions, as the case may be. In this way, we find the *structure* of a given doctrine or system, and so are able to compare the structures of different, and sometimes very complex, verbal schemes. Similar structure-finding methods must be applied some day to all other, at present, non-mathematical disciplines. The main difficulty, in the search for structure, was the absence of a clear formulation of the issues involved and the need for a $\bar{A}$-system, so as to be able to *compare two systems,* the comparison of which helps further structural discovery. It is not claimed that either the A or $\bar{A}$ system-functions have been formulated here, but it seems that, in the presence or absence of identification, we find a fundamental postulate which, once formulated, suggests a comparison with experience. As we discover that 'identity' is invariably false to facts, this A postulate must be rejected from any future $\bar{A}$-system.

It happens that any new and revolutionary doctrine or system is always based on a new doctrinal or system-function which establishes its new structure with a new set of relations. Thus, any new doctrine or system, when traced to its postulates, allows us to verify and scrutinize the initial postulates and to find out if they correspond to experience,.

A few examples will make it clearer. Cartesian analytical geometry is based on *one* system-function, having one system-structure, although we may have indefinitely many different cartesian co-ordinates. The vector and the tensor systems also depend on two different system-functions, different from the cartesian; they have three different structures. Intertranslations are possible, but only when the fundamental postulates do not conflict. Thus, the tensor language gives us invariant and intrinsic relations, and these can be translated into the cartesian relations. It seems certain, however, although I am not aware that this has been done, that the indefinitely many extrinsic characteristics which we can manufacture in the cartesian system, cannot be translated into the tensor language, which does not admit extrinsic characteristics.

Similar relations are found between other doctrines and systems, once their respective structural characteristics are discovered by the formulating of their respective functions, which, by the explicit or implicit postulates, determine their structure.

Thus all existing schools of psychotherapy, prior to 1933, result from *one* system-function which underlies *implicitly* the system originated by Freud.[4] The *particular* freudian doctrine is only one of the indefinitely many variants of *similar system-structure*, which can be manufactured from the one system-function underlying the particular freudian system. In other words, it is of no importance what 'complex' we emphasize or manufacture, the *structural principles* which underlie this new freudian and revolutionary system-function remain unchanged. From this point of view, all existing schools of psychotherapy could be called 'cartesian', because, although they all have *one* general system-structure, yet they allow indefinitely many particular variations. The present $\bar{A}$-system suggests that the 'cartesian' school of psychotherapy is still largely A, *el* and fundamentally of one structure.

The present system involves a different system-function of different structure, rejecting identity, discovering the 'structural unconscious', establishing psychophysiology,. The mutual translatability follows the rules of general semantic principles or conditions which apply also to mathematics; namely, that a $\bar{A}$-system, being based on relations; on the elimination of identity; on structure., gives us only intrinsic characteristics and might be called the 'tensor' school of psychotherapy. This system allows all the intrinsic characteristics discovered, no matter by whom, but has no place for the indefinitely many, quite consistent, yet irrelevant metaphysical, extrinsic characteristics, which we can manufacture at will.

Without the realization of the structural foundations emphasized in the present system, it is practically impossible not to confuse linguistic structural issues, which lead inevitably to semantic blockages. When we deal with doctrines or systems of *different* structure, each of which involves different doctrinal or system-functions, it is of the utmost importance to keep them at first *strictly separated;* to work out each system by itself, and only after this is accomplished can we carry out an independent investigation as to the ways they *mutually intertranslate.* Let me again repeat, that the mixing of different languages of different structures is fatal for clear 'thinking'. Only when a system is traced to its system-function, and the many implications worked out in their *un-mixed* form, can we make a further *independent* investigation of the ways in which the different systems intertranslate. As a general rule, every new scientific system eliminates a great deal of spurious metaphysics from the older systems. In practice, the issues are extremely simple if one decides to follow the general rule; namely, either completely to reject or completely to accept *provisionally,* at a given date, a new system; use *exclusively* the structurally new terms; perform our semantic operations exclusively in these terms; compare the conclusions with experience; perform *new* experiments which the structurally new terminology suggests; and only then, as an independent enquiry, investigate how one system translates into the other. In those translations, which correspond to the transformation of frames of reference in mathematics, we find the most important *invariant* characteristics or relations which survive this translation. If a characteristic appears in all formulations, it is a sign that this characteristic is *intrinsic,* belongs to the subject of our analysis, and is not accidental and irrelevant, belonging only to the accidental structure of the language we use. Once these invariant, intrinsic characteristics are discovered, and there is no way to discover them except by reformulating the problems in different languages (in mathematics we speak about the transformation of frames of reference), we then know that we have discovered invariant relations, which survive transformation of different forms of representations, and so realize that we are dealing with something genuinely important, *independent* from the structure of the language we use.

History shows that the discovery of isolated, though interesting, facts has had less influence on the progress of science than the discovery of *new system-functions* which produce *new* linguistic *structures and new methods.* In our own lifetime, some of the most revolutionary of these advances in structural adjustment and method have been accomplished. The work of Einstein, the revision of mathematical foundations, the new quantum mechanics, colloidal science, and advances in psychiatry, are perhaps structurally and semantically the most important. There seems no escape from admitting that no modern man can be really intelligent in 1933 if he knows nothing about these structural scientific revolutions. It is true that, because these advances are so recent, they are still represented in very technical terms; their system-functions have not been formulated, and so the deeper structural, epistemological and semantic simple aspects have not been worked out. These aspects are of enormous human importance. But they must be represented without such an abundance of dry technicalities, which are only a means, and not an end, in search for structure.

A scientist may be very much up to date in his line of work, let us say, in biology; but his physico-mathematical structural knowledge may be somewhere in the eighteenth or nineteenth century and his epistemology, metaphysics, and structure of language of 300 B.C. This classification by years gives a fairly good picture of his semantic status. Indeed, we can fortell quite often what kind of reaction such a man will exhibit.

This functional, propositional-function, and system-function structural attitude is in accord with the methods developed by psychiatry. In psychiatry, 'mental' phenomena are considered, in some instances, from the point of view of arrested development; in others, as regression to older and more primitive levels. With this attitude and understanding, we cannot ignore this peculiar intermixing of different personalities in one man when different aspects of him exhibit *s.r* of different ages and epochs of the development of mankind. In this connection should be mentioned the problem of the multiple personalities which often occur in the 'mentally' ill. Such splitting of personality is invariably a serious semantic symptom, and a person who exhibits different ages in his semantic development, as, for instance, 1933 in some respects, sixteenth century in others, and 300, or even 5000, B.C. in still others, cannot be a well co-ordinated individual. If we teach our children, whose nervous systems are *not* physically finished at birth, doctrines structurally belonging to entirely different epochs of human development, we ought not to wonder that semantic harm is done. Our efforts should be to co-ordinate and integrate the individual, help the *nervous* system, and not split the individual semantically and so disorganize the nervous system.

It is necessary to remember that the organism works as-a-whole. In the old days we had a comforting delusion that science was a purely 'intellectual' affair. This was an *el* creed which was structurally false to facts. It would probably be below the dignity of an older mathematician to analyse the 'emotional' values of some piece of mathematical work, as, for instance, of the 'propositional function'. But such a mathematician probably never heard of psychogalvanic experiments, and how his 'emotional curve' becomes expressive when he is solving some mathematical problem.

In 1933, we are not allowed to follow the older, seemingly easier, and simpler paths. In our discussion, we have tried to analyse the problems at hand as ∞-valued manifestations of human behaviour. We were analysing the doings of Smith, Brown., and the semantic components which enter into these forms of behaviour must be especially emphasized, emphasized because they were neglected. In well-balanced persons, all psycho-logical aspects should be represented and should work harmoniously. In a theory of sanity, this semantic balance and co-ordination should be our first aim, and we should, therefore, take particular care of the neglected aspects. The *non-el* point of view makes us postulate a permanent connection and interdependence between all psycho-logical aspects. Most human difficulties, and 'mental' ills, are of *non-el* affective origin, extremely difficult to control or regulate by *el* means. Yet, we now see that purely technical scientific discoveries, because structural, have unsuspected and far-reaching beneficial *affective* semantic components. Perhaps, instead of keeping such discoveries for the few 'highbrows', who never use them fully, we could introduce them as structural, semantic and *linguistic* devices into elementary schools, with highly bene-ficial psycho-logical results. There is really no difficulty in explaining what has been said here about structure to children and training them in appropriate *s.r.* The effect of doing so, on sanity, would be profound and lasting.

CHAPTER XII

ON ORDER

The fundamental importance of the subject of order may be inferred from the fact that all the concepts required in geometry can be expressed in terms of the concept of order alone. (237) E. V. HUNTINGTON

Dimensions, in geometry, are a development of order. The conception of a *limit*, which underlies all higher mathematics, is a serial conception. There are parts of mathematics which do not depend upon the notion of order, but they are very few in comparison with the parts in which this notion is involved. (455) BERTRAND RUSSELL

The notion of continuity depends upon that of *order*, since continuity is merely a particular type of order. (454) BERTRAND RUSSELL

Logistic may be defined as *the science which deals with types of order as such.* (300) C. I. LEWIS

The branch of physics which is called Elementary Geometry was long ago delivered into the hands of mathematicians for the purposes of instruction. But, while mathematicians are often quite competent in their knowledge of the abstract structure of the subject, they are rarely so in their grasp of its physical meaning. (529) OSWALD VEBLEN

We often think that when we have completed our study of *one* we know all about *two*, because "two" is "one and one". We forget that we have still to make a study of "and". Secondary physics is the study of "and"— that is to say, of organisation. (149) A. S. EDDINGTON

. . . the geometry of paths can be regarded as a generalization both of the earliest part of elementary geometry and of some of the most refined of physical theories. The study of the projective, the affine and the metric geometry of paths ought to result in a comprehensive idea of what types of physical theory it is possible to construct along the lines which have been successful in the past. (529) OSWALD VEBLEN

What I wish to emphasize now is the need of logistic studies which will make it possible to say more definitely than is yet possible in this field what is assumed, what is proved, and how the group of theorems and definitions hang together. (529) OSWALD VEBLEN

Memory, in fact, is nothing but the reinforcement and facilitation of the passage of the nervous impulse along certain paths. (411) HENRI PIÉRON

But before dealing with the brain, it is well to distinguish a second characteristic of nervous organization which renders it an organization in levels. (411) HENRI PIÉRON

This affective repercussion seems to take place at the penultimate stage of the nervous system and governs complicated reflexes or instinctive reactions. (411) HENRI PIÉRON

Furthermore, there are even symbols of symbols, evocative of images only in the second degree, by means of primary stations of the co-ordination centres. (411) HENRI PIÉRON

In this way it is seen that the order in which a given group of stimuli taking part in a stimulatory compound are arranged, and the pauses between them are the factors which determine the final result of the stimulation, and therefore most probably the form of the reaction, and we know already that different intensities of the same stimulus can be differentiated very accurately, one definite intensity being connected with excitation and another with inhibition. (394)   I. P. PAVLOV

Whoever studies Leibniz, Lambert and Castillon cannot fail to be convinced that a consistent calculus of concepts in intension is either immensely difficult or, as Couturat has said, impossible. (300)   C. I. LEWIS
The relation between intensions and extensions is *unsymmetrical*, not symmetrical as the medieval logicians would have it. (300)   C. I. LEWIS
The old "law" of formal logic, that if α is contained in β in extension, then β is contained in α in intension, and vice versa, is *false*. The connection between extension and intension is by no means so simple as that. (300)   C. I. LEWIS

I do not suggest explicit confusions of this sort, but only that traditional elementary logic, taught in youth, is an almost fatal barrier to clear thinking in later years, unless much time is spent in acquiring a new technique. (457)   BERTRAND RUSSELL

Section A. Undefined terms.

We can now introduce a structural *non-el* term which underlies not only all existing mathematics, but also the present work. This *bridging* term has equal importance in science and in our daily life; and applies equally to 'senses' and 'mind'. The term in question is 'order', in the sense of 'betweenness'. If we say that a, b, and c are in the order a, b, c, we mean that b is *between* a and c, and we say, further, that a, b, c, has a different order from c, b, a, or b, a, c,. .The main importance of numbers in mathematics is in the fact that they have a definite *order*. In mathematics, we are much concerned with the fact that numbers represent a definite ordered series or progression, 1, 2, 3, 4,.

In the present system, the term 'order' is accepted as *undefined*. It is clear that we cannot *define* all our terms. If we start to define all our terms, we must, by necessity, soon come to a set of terms which we cannot define any more because we will have no more terms with which to define them. We see that the structure of *any* language, mathematical or daily, is such that we must start implicitly or explicitly with undefined terms. This point is of grave consequence. In this work, following mathematics, I explicitly start with undefined important terms.

When we use a series of names for objects, 'Smith, Brown, Jones'., we say *nothing*. We do not produce a proposition. But if we say 'Smith kicks Brown', we have introduced the term 'kicks', which is not a name for an object, but is a term of an entirely different character. Let us call it a 'relation-word'. If we analyse this term, 'kicks', further, we will find that we can define it by considering the leg (objective) of Smith (objective), some part of the anatomy of Brown (objective), and, finally, Brown (objective). We must use a further set of terms that describes how the leg of Smith 'moves' through an 'infinity' of 'places' in an 'infinity' of 'instants' of 'time', 'continuously' until it reaches Brown.

When a donkey kicks a donkey, there may be a broken leg; but that is, practically, the only consequence. Not so when Smith kicks Brown. Should Brown happen to be a royal or a business man in Nicaragua or Mexico, this might be considered 'a mortal offense of a great sovereign nation to another great sovereign nation', a war might follow and many non-royals or non-business men might die. When a *symbolic* class of life enters the arena, semantic complications may arise not existing with animals.

In the relation terms, the statement, 'Smith kicks Brown', has introduced still further symbolic complications. It involves a full-fledged metaphysics, as expressed in the terms 'moves', 'infinity', 'space', 'time', 'continuity', and what not. It must be emphasized that *all* human statements, savage or not, involve a structural metaphysics.

These relational terms should be elucidated to the utmost. Until lately, the 'philosophers', in their 'Jehovah complex', usually said to the scientists: 'Hands off; those are superior problems with which only we chosen ones can deal'. As a matter of history, 'philosophers' have not produced achievements of any value in the structural line. But the 'mere' scientists, mainly mathematicians and mathematical physicists, have taken care of these problems with extremely important structural (1933) results. In the solutions of these semantic problems, the term 'order' became paramount.

Perhaps this example of an analysis of the statement, 'Smith kicks Brown', shows the justice of the contention of this work that no man can be 'intelligent' if he is not acquainted with these new works and their structural elucidations.

We see that no statement made by man, whether savage or civilized, is free from some kind of structural metaphysics involving *s.r.* We see also that when we explicitly start with *undefined* words, these undefined words have to be taken on faith. They represent some kind of implicit creed, or metaphysics, or structural assumptions. We meet here with a tremendously beneficial semantic effect of modern methods, in that we deliberately state our undefined terms. We thus divulge our creeds and metaphysics. In this way, we do not blind the reader or student. We invite criticism, elaboration, verification, evaluation., and so accelerate progress and make it easier for others to work out issues. Compare the statement of Newton, *'Hypotheses non fingo'* (I do not make hypotheses), in his *Philosophiae Naturalis Principia Mathematica,* when he proceeded to coin some very doubtful hypotheses; and such works as produced by Peano, Whitehead, Russell, and others, in which not only all assumptions are stated explicitly, *but even the assumptions,* given in single *undefined* terms, are listed. It is not assumed here that even Peano, Whitehead, Russell, and the others have fulfilled this program entirely. It is quite probable that not all of their assumptions are stated explicitly. However, a very serious and revolutionary beginning has been made in this direction. We have still far to go, for at present even mathematicians disregard the threefold relational character of mathematics, and, by a semantic confusion of orders of abstraction, make structurally *el* assumptions false to the facts of 1933; namely, that mathematics exists 'by itself', detached from the producers, Smith and others. This procedure reminds one of the old N 'I do not make hypotheses', proffered just at the moment he begins to legislate about the structure of the universe and to postulate his 'absolute space' and 'absolute time' 'without reference to any external object whatsoever'. This, of course, was structurally unascertainable, and so was a mere figment of his imagination, inside of his skin, and may become a pathological semantic projection when externalized by *affective* pressure.

That we must all start with undefined terms, representing blind creeds which cannot be elucidated further *at a given moment,* may fill the hearts of some metaphysicians with joy. 'Here', they might say, 'we have the goods on the scientists; they criticize us and reject our theories, and yet they admit that they also must start with blind creeds. Now we have full justification for assuming whatever we want to.' But this joy would be short-lived for any reasonably sane individual. In mathematics, we deliberately assume the minimum, and not the maximum, as in metaphysics. The undefined terms selected for use are the *simplest* of our experience; for instance, 'order' (betweenness). Also, in experimental sciences, we assume the least possible. We demand from a *scientific* theory, according to the standards of 1933, that it should account for all relevant facts known in 1933 and should serve as a basis for the *prediction* of new facts, which can be checked by new experiments. If metaphysicians and 'philosophers' would comply with such scientific standards, their theories would be scientific. But their old theories would have to be abandoned and their new theories would become branches of science. Under such structural circumstances, there is no possibility of going *outside* of science, as we can enlarge the bounds of science without known limits, in search for structure.

It must be pointed out that no set of undefined terms is ultimate. A set remains undefined only until some genius points out simpler and more general or structurally more satisfactory undefined terms, or can reduce the number of such terms. Which set we accept is determined, in the main, by pragmatic, practical, and structural reasons. Out of two systems which have many characteristics of usefulness., in common, we would and should select the one which assumes least, is the simplest, and carries the furthest. Such changes from one set to another, when scientific, are usually epoch-making, as exemplified in mathematics.

It is important to realize that this semantic attitude signalizes a new epoch in the development of science. In scientific literature of the old days, we had a habit of demanding, 'define your terms'. The new 1933 standards of science really should be, *'state your undefined terms'*. In other words, 'lay on the table your metaphysics, your assumed structure, and then only proceed to define your terms in terms of these *undefined terms'*. This has been done completely, or approximately so, only in mathematics. Yet, probably no one will deny that the new requirements of science (1933), no matter how laborious, are really desirable, and constitute an advance in method, in accordance with the structure of human knowledge.

In the present work, this method will be employed practically all through. Of course, names for objects may be accepted without enquiry. So we have already a large vocabulary at our disposal. But names alone do not give propositions. We need *relation-words,* and it is here where our undefined terms become important. Up to this stage of the present work, I have accepted, without over-full explanation, the vocabularies made by the linguists of exact science, whom we usually call mathematicians. There is an enormous benefit in doing so, because, no matter how imperfect the mathematical vocabulary may be, it is an extensive and developed linguistic system of similar structure to the world around us and to our nervous system 1933. (See Part V.)

Some of the most important undefined terms which play a marked role in this work are 'order' (in the sense explained), 'relation', and 'difference', although we could define relation in terms of multi-dimensional order. There is a remarkable structural characteristic of these terms; namely, that they are *non-el*, and that they apply to 'senses' as well as to 'mind'. It is, perhaps, well to suggest that, in future works, the terms selected should be of the *non-el* type. Since these terms apply equally to 'senses' and to 'mind', we see that in *such* terms we may attempt to give a 'coherent' account of what we experience. The expression 'coherent' implies 'mind', and 'experience' implies 'senses'. It is amusing to watch this peculiar circularity of human knowledge, many instances of which will appear later on. Thus, there was great difficulty in expressing organism-as-a-whole notions; we had to grope about in establishing the beginnings of a suitable vocabulary before we could approach problems which were antecedent in order.

It is necessary to notice a rather curious structural similarity between the $\bar{N}$ and $\bar{A}$ systems. In both cases, we deal with certain velocities about which we know positively that they are *finite*. The velocity of light in the N-system was assumed to be infinite, although we know it is not so, and so 'simultaneity' had absolute meaning. The $\bar{N}$ systems introduced the finite velocity of light by ordering events, which happens to be true to facts, and thus 'simultaneity' lost its absolute character. Likewise in the $\bar{A}$-system and language, the velocity of nervous impulses was assumed to be infinite, to spread 'instantaneously'. And so we had most perplexing 'philosophical' rigmaroles about 'emotions', 'intellect'., taken as independent separate entities. When we introduce explicitly the finite velocity of nervous impulses (on the average, 120 metres per second in the human nerves), we are able to reach a perfectly clear understanding, *in terms of order,* of the spread of impulses. Some 'infinite velocity' does not involve *order*. Conversely, by considering the order of events, we introduce finite velocities. We shall see later that 'infinite velocity' is *meaningless* and so all actual happenings can be ordered. The above is an important factor in our *s.r.*

Let us give a rough example. Assume that Smith has had a bad dinner. Some nervous impulse, originating from the bad dinner, starts going. At this stage, we may call it an 'undifferentiated' nervous impulse. It travels with *finite* velocity, reaches the brain-stem and the approximately central part of his brain, which we call the thalamus; is affected by them and is no longer 'undifferentiated' but becomes, let us say, 'affective'. In the cortex, it is affected again by the lessons of past experiences. It returns again to the lower centres and becomes, let us say, 'emotion'; and then anything might happen, from sudden death to a glorious poem.

The reader must be warned that this example is rough and over-simplified. Impulses are reinforced and 'inhibited' from a complex chain of nervous interconnections. But what I wish to show by this example, is that, by accepting the *finite* velocity of nerve currents, in terms of *order,* we can build up a definite vocabulary to deal, not only with the 'organism-as-a-whole', but also with the different stages of the process. This is important because, without some such *ordinal* scheme, it is structurally impossible to evade enormous verbal and semantic difficulties which lead to great confusion.

In the analysis of the above example, only the structural and *methodological* aspects are emphasized. No attempt is made to legislate for neurologists or to instruct them how they should define and use their terms.

Section B. Order and the nervous system.

We know that, structurally, not all parts of the nervous system are of equal phylogenetic age. The ventral part of the brain, the thalamus (in the rough), which is of most interest in this connection, is older than the cerebral cortex. By the term 'thalamus' I denote all the sub-divisions and most important appendages which we need not mention by their technical names. In man, both the thalamus and the cerebral cortex are much enlarged and have a complex structural cyclic inter-connection. The cerebral cortex is a term applied to a superficial layer of grey nervous tissue covering the cerebral hemispheres. It is called the 'new brain' by Edinger. The higher correlation centres in the cere-bral hemispheres can act only through the agency of the lower centres, the brain-stem, and the thalamus. In other words, the cerebral cortex, the functioning of which is connected chiefly with the higher associa-tions, is of such structure that no nervous impulse can enter it without first passing through the lower centres of the ventral parts of the brain and brain-stem.

In the lower vertebrates, which lack the cerebral cortex, the sub-cortical mechanisms are adequate for all simple exigencies of life and simple association processes, these sub-cortical mechanisms being older phylogenetically than the cerebral cortex, yet younger than some still more ventral parts.

The brush-like connections between the nerves are called synapses, and although a nerve-fibre seems to be capable of transmitting nerve impulses in both directions, the nervous impulse can seemingly pass the synapse in only one direction; so a nervous polarity is established whenever synapses are present.[1]

The following diagram and explanations are taken from Professor Herrick's *Introduction to Neurology,* pp. 60, 61, 62, 63, 69, 70. In the quotations I retain the spelling but change Herrick's Fig. 18 to my Fig. 1; all but one word of the italics are mine.*

---

*I quote here from the Second Edition of the *Introduction to Neurology* an account of the classical theory of reflex circuits which is quite satisfactory for my purpose. In his later work (see my bibliography), Professor Herrick forcibly points out the limitations of the reflex theories as *partial patterns,* as opposed to the activity of the organism-as-a-whole. In the Fifth Edition (1931), the chapter on reflex circuits has been entirely rewritten, and the *non-el* attitude is expressed very clearly. I am much indebted to Professor Herrick for drawing my attention to this rewritten chapter.

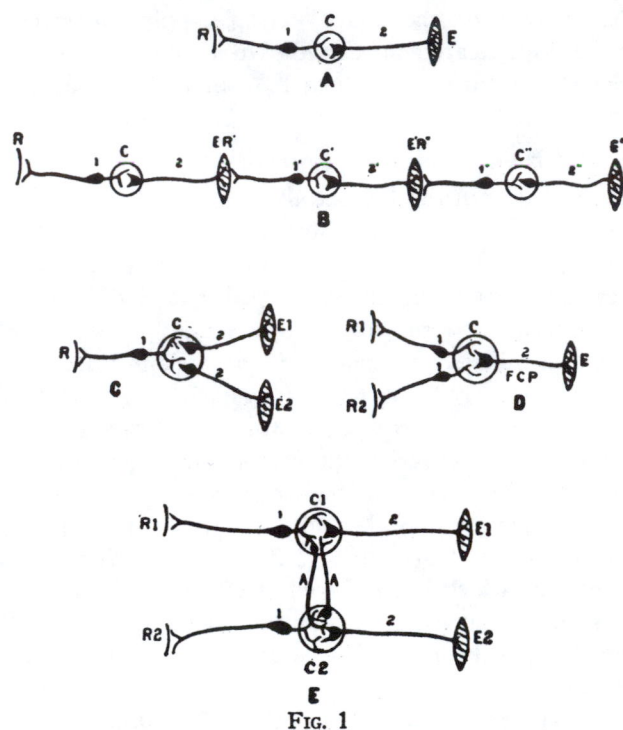

FIG. 1

Diagrams representing the relations of neurons in five types of reflex arcs: A, simple reflex arc; B, chain reflex; C, a complex system illustrating allied and antagonistic reflexes and physiological resolution; D, a complete system illustrating allied and antagonistic reflexes with a final common path; E, a complex system illustrating the mechanism of physiological association A,A, association neurons; $C,C',C'',C1$, and $C2$, centres (adjustors); E,E',E'', $E1$, and $E2$, effectors; FCP, final common path; $R,R',R'',R1$, and $R2$, receptors.

'The structure of the simple reflex circuit is diagrammatically illustrated in Fig. 1, A. The receptor (R) may be a simple terminal expansion of the sensory nerve-fiber or a very complex sense organ. The effector (E) may be a muscle or a gland. The cell body of the afferent neuron (1) may lie within the center (C) or outside, as in the diagram. . . . A simple reflex act involving the use of so elementary a mechanism as has just been described is probably never performed by an adult vertebrate. The nervous impulse somewhere in its course always comes into relation with other reflex paths, and in this way complications in the response are introduced. . . .

'Separate reflex circuits may be so compounded as to give the so-called chain reflex (Fig. 1, B). Here the response of the first reflex serves as the *stimulus* for the second, and so on in series. The units of these chain reflexes are usually not simple reflexes as diagrammed, but *complex* elements of the types next to be described. . . . The chain reflex . . . is a very common and a very important type. Most of the ordinary acts in the routine of daily life employ it in one form or another, the *completion of one stage* of the process serving as the *stimulus* for the *initiation of the next.* . . .

'Figure 1, C illustrates another method of compounding reflexes so that the stimulation of a single sense organ may excite either or both of two responses. If the two effectors, $E1$ and $E2$, can coöperate in the performance of an adaptive response, the case is similar to that of Fig. 1, A, with the possibility of a more complex type of reaction. This is an allied reflex. If, however, the two effectors produce antagonistic movements, so that both cannot act at the same time, the result is a physiological dilemma. Either no reaction at all results, or there is a sort of physiological resolution (sometimes called physiological choice), one motor pathway being taken to the exclusion of the other. Which path will be chosen in a given case may be determined by the physiological state of the organs. If, for instance, one motor system, $E2$, is greatly fatigued and the other rested, the threshold of $E2$ will be raised and the motor discharge will pass to $E1$.

'Figure 1, D illustrates the converse case, where two receptors discharge into a single center, which, in turn, by means of a final common path (FCP) excites a single effector (E). If the two receptors upon stimulation normally call forth the same response, they will reinforce each other if simultaneously stimulated, the response will be strengthened, and we have another type of allied reflex. But there are cases in which the stimulation of $R1$ and $R2$ (Fig. 1, D) would naturally call forth antagonistic reflexes. Here, if they are simultaneously stimulated, a physiological dilemma will again arise which can be resolved only by one or the other afferent system getting control of the final common path.

'Figure 1, E illustrates still another form of combination of reflexes. Here there are connecting tracts (A, A) between the two centers so arranged that stimulation of either of the two receptors $(R1$ and $R2)$ may call forth a response in either one of two effectors $(E1$ and $E2)$. These responses may be allied or antagonistic, and much more complicated reflexes are here possible than in any of the preceding cases. . . .

'It must be kept in mind that in higher vertebrates all parts of the nervous system are bound together by connecting tracts (internuncial pathways). . . . These manifold connections are so elaborate that every part of the nervous system is in nervous connection with every other part, directly or indirectly. This is illustrated by the way in which the digestive functions (which normally are quite autonomous, the nervous control not going beyond the sympathetic system, . . .) may be disturbed by mental processes whose primary seat may be in the association centers of the cerebral cortex; and also by the way in which strychnin-poisoning seems to lower the neural resistance everywhere, so that a very slight stimulus may serve to throw the whole body into convulsions. . . .

'Our picture of the reflex act in a higher animal will, then, include a view of the whole nervous system in a state of neural tension. The stimulus disturbs the equilibrium at a definite point (the receptor), and the wave of nervous discharge thus set up irradiates through the complex lines determined by the neural connections of the receptor. If the stimulus is weak and the reflex path is simple and well insulated, a simple response may follow immediately. Under other conditions the nervous discharge may be inhibited before it reaches any effector, or it may irradiate widely, producing a very complex reflex pattern. . . .

'The mechanism of the reflex should not be regarded as an open channel through which energy admitted at the receptive end-organ is transmitted to be discharged into the effector organ. It is rather a *complex* apparatus, containing reserves of potential energy which can be released upon the application of an adequate stimulus in accordance with a pattern determined by the inherent *structure* of the apparatus itself. In other words, the nervous *discharge* [italics of Professor Herrick] is *not* a mere transmission of the energy of the stimulus, but it implies *active* consumption of material and release of energy (*metabolism*) within both the nerve centers and the nerve-fibers. The energy acting upon the effector organ may, therefore, be different in both kind and amount from that applied to the receptive end-organ. The response likewise involves the liberation of the latent energy of the effector (*muscle or gland*), the nervous impulse serving merely to release the trigger which discharges this reserve energy.'

It is necessary to warn the reader that the human nervous system is structurally of inconceivable complexity. It is estimated that there are in the human brain about twelve thousand millions of nerve cells or neurons, and more than half of these are in the cerebral cortex. Most probably, the majority of the neurons of the cerebral cortex are directly or indirectly connected with every cortical field. Were we to consider a million cortical nerve cells connected one with another in groups of only two neurons each, and compute the possible combinations, we would find the number of possible interneuronic connection-patterns to be represented by $10^{2,783,000}$. What such a number is like is hardly possible to imagine. For comparison, it may be said that probably the whole visible sidereal universe does not contain more than 10^{66} atoms![2]

Our present knowledge of the nervous system is limited as regards its complexities and possibilities, but we know many structural facts which seem to be well established. One of these is that the human nervous system is more complex than that of any animal. Another is that the human cortex is of later origin than, and in a way an outgrowth from, the more central parts of the brain (which establishes a structure of *levels*). A third is that the interconnection of the parts of the nervous system is *cyclic*. A fourth is that the velocity of nerve currents is *finite*. The last fact is of serious structural importance, and, as a rule, disregarded.

Section C. Structure, relations, and multi-dimensional order.

In such an ordered cyclic chain, the nerve impulses reach and traverse the different levels with *finite* velocity and so, in each case, in a *definite order*. 'Intelligence' becomes a manifestation of life of the organism-as-a-whole, structurally impossible in some fictitious 'isolation'. To 'be' means *to be related*. To be related involves multi-dimensional *order* and results in *structure*.

'Survival', 'adaptation', 'response', 'habit formation', 'orientation', 'learning', 'selection', 'evaluation', 'intelligence', 'semantic reactions', and all similar terms involve structurally an ordered, interrelated structural complex in which and by which we live and function. To 'comprehend', to 'understand', to 'know', to be 'intelligent'., in the pre-human as well as the human way, means the most useful survival adjustment to such an ordered, interrelated structure as the world and ourselves.

In this vocabulary, 'structure' is the highest abstraction, as it involves a whole, taken as-a-whole, made up of *interrelated parts,* the relations of which can be defined in still simpler terms of order. 'Knowing', in its broadest as well as in its narrow human sense, is conditioned by structure, and so consists of *structural* knowings. All empirical structures involve relations, and the last depend on multi-dimensional order. A language of *order*, therefore, is the simplest form of language, yet in structure it is similar to the structure of the world and ourselves. Such a language is bound to be useful for adaptation and, therefore, sanity; it results in the understanding of the structural, relational, multi-dimensional order in the environment on all levels.

We must stress the structural fact that the introduction of *order* as a fundamental term abolishes some fanciful and semantically very harm-

11

ful 'infinities'. If an impulse could travel in 'no time' or with 'infinite velocity', which is a *structural impossibility* in this world, such an impulse would reach different places 'instantaneously', and so there would be *no order involved*. But, as soon as the actual order in which impulses reach their destination is found, 'infinite velocities' are abolished. We shall show later that 'infinite velocity' is a meaningless noise; here we stress only the point that it is a structural impossibility, as structure involves relations and orders, and order could not exist in a world where 'infinite velocities' were possible.

Conversely, if in our analysis we disregard order, we are bound to disregard relations and structure and to introduce, by necessity, some fanciful 'infinite velocities'. Any one who treats 'mind' in 'isolation' makes a structurally false assumption, and, by necessity, unconsciously ascribes some meaningless 'infinite velocity' to the nerve currents.

We have dwelt upon this subject at such length because of its general structural and semantic importance. The first step towards understanding the theory of Einstein is to be entirely convinced on the above points. Newton's disregard of order introduced an unconscious false to facts assumption of the 'infinite velocity' of light, which fatalistically leads to an objectification called 'absolute time', 'absolute simultaneity', and so introduces a terminology of inappropriate structure. A similar remark applies to arguments about 'mind' in an objectified, 'isolated' way. These arguments disregard the *order* in which the nervous impulses spread and so, by necessity, introduce a silent false to facts assumption of the 'infinite velocity' of nerve currents.

On empirical structural grounds, we know neurological and general facts on two levels. (1) Macroscopically, we have a structure in levels, stratified, so to say, with complexities arising from the general colloidal physico-chemical structure of the organism-as-a-whole. (2) The general sub-microscopic, atomic, and sub-atomic structure of all materials simply gives us the persistence of the macroscopic characteristics as the relative *invariance of function,* due to dynamic equilibrium, and ultimately reflected and conditioned by this *sub-microscopic structure* of all materials. Under such actual structural conditions, terms like 'substance', 'material'., and 'function', 'energy', 'action'., become interconnected—largely a problem of preference or necessity of selecting the level with which we want to deal.

On sub-microscopic levels, 'iron', or anything else, means only a persistence for a limited 'time' of certain gross characteristics, representing a process (structurally a four-dimensional notion involving 'time') which becomes a question of structure. With the 1933 known unit of the world called an 'electron', which appears as an *'energy'* factor, the relative persistence or invariance of dynamic sub-microscopic structure gives us, on macroscopic levels, an average, or statistical, persistence of gross macroscopic characteristics, which we label 'iron'.

The above should be thoroughly understood and digested. As a rule, we all identify orders and levels of abstractions and so have difficulty in keeping them separate verbally (and, therefore, 'conceptually'). Thorough structural understanding helps us greatly to acquire these new and beneficial *s.r*.

Under such structural *empirical* conditions, a language of order, which implies relations and structure, as enlarged to the order of abstractions or level of consideration, largely volitional, becomes the only language which, in *structure,* is similar to the structure of the world, ourselves included, and so, of necessity, will afford the maximum of semantic benefits.

It should be understood that, on structural grounds, terms like 'substance' and 'function' become, in 1933, perfectly *interchangeable,* depending on the order of abstraction. 'Substance', for instance, on the macroscopic level becomes 'invariance of function' on the sub-microscopic level. It follows that what we know about the macroscopic ('anatomical') structure can be quite legitimately enlarged by what we know of *function* (structure on different levels). This interchangeability and complementary value of evidence is conditioned by structural considerations, and the fact that 'structure' is multiordinal. On gross anatomical grounds, we know a great deal about this structure of the nervous system. Because of experimental difficulties, very little is known of the structural sub-microscopic happenings, yet we can speak about them with benefit in *functional* terms; as, for instance, of 'activation', 'facilitation', 'resistance', 'psychogenetic effects', 'diffusion', 'permeability', the older 'inhibition',.

In such a cyclic chain as the nervous system, there is, as far as energy is concerned, no last stage of the process. If there is no motor reaction or other reflex, then there is a semantic or associative reaction with 'inhibitory' or activating consequences, which are functionally equivalent to a motor reaction. At each stage, a 'terminal' receptor is a *reacting* organ in the chain.[3]

We know quite well from psychiatry how nervous energy may deviate from constructive and useful channels into destructive and harmful channels. The energy is not lost, but misdirected or misapplied. For instance, an 'emotional' shock may make some people release their energy into useful channels, such as concentrated efforts in some direction, which would have been impossible without this shock; but, in others, an 'emotional' shock leads to the building up of morbid 'mental' or physical symptoms.

Since the nervous structure is cyclic in most of its parts, as well as as-a-whole, and since these cycles are directly or indirectly interconnected, mutual interaction of those cycles may produce most elaborate behaviour patterns, which may be spoken of, in their manifestations, in terms of *order*. As each more important nerve centre has incoming and outgoing nerve-fibres, the activation, or reinforcement, or diffusion of nerve currents may sometimes manifest itself in our *s.r* as *reversal of order* in some aspects. Neurologically, considered on the sub-microscopic level, it would only be a case of activation, or of diffusion, or of 'inhibition'., probably *never* a problem of reversed order in the actual nerve currents.

The semantic manifestations of order and reversed order are of *crucial importance,* for we are able to *train* the individual to different orders or reversals of orders. This procedure neurologically involves activations, enforcements, diffusions, 'inhibitions', resistances, and all the other types of nervous activities, which without the formulation of psychophysiology were all *most inaccessible to* direct training. The structural fact that order and reversal of order in semantic manifestation, which all are on the un-speakable objective level, have such intimate and profound connection with fundamental nervous processes, such as activation, enforcement, diffusion, permeability, 'inhibition', resistance., gives us tremendous new powers of an educational character in building up *sanity,* and supplies methods and means to affect, direct, and train nervous activities and *s.r,* which we were *not able formerly to train* psychophysiologically. Perhaps one of the main values of the present work is the discovery of physiological means, to be given presently, for training the human nervous system in 'sanity'.

The reader should be aware, when we speak of order and reversal of order, that we mean the order and reversal of order in the un-speakable *s.r*; but the neurological mechanism is of a different character, as already explained. Our analysis of the simple semantic manifestations involve *evaluation* and so *order*, permitting a most complex re-education and re-training of the nervous system, which were entirely beyond our reach with the older methods.

Experimental evidence seems to corroborate what has been suggested here, and the analysis, in terms of order, seems to have serious practical neurological significance, owing to *similarity of structure,* resulting in *evaluation*, and so appropriate *s.r.*

For our analysis in terms of order, we start with the simplest imaginable nervous cycle; but it must be explicitly understood that such simple cycles actually never exist, and that our diagrams have value only as picturing the cyclic *order*, without complications. Let us repeat that the introduction of an analysis in terms of *order* or *reversal of order* in *the manifestation* involves, under educational influence, various other and *different actual nervous activities,* a class of activities which hitherto have always evaded our educational influences. For structural purpose, it is sufficient to make use of the distinction between lower and higher centres (a rough and ready distinction) and to consider the lower centres generally in connection with the thalamus and brain-stem (perhaps also other sub-cortical layers), and the higher centres generally in connection with the cortex. This lack of precision is intentional, for we need only sufficient structural stratification to illustrate *order,* and it seems advisable to assume only the well-established *minimum* of structure.

We have already mentioned the absolute individuality of the organism and, as a matter of fact, of everything else on objective levels. The reader need have no metaphysical shivers about such extreme individuality on the un-speakable levels. In our human economy, we need both similarities and differences, but we have as yet, in our *A*-system, chiefly concentrated our attention and training on similarities, disregarding differences. In this work, we start structurally *closer to nature* with un-speakable levels, and make *differences* fundamental, similarities appearing only at a later stage (order) *as a result of higher abstractions.* In simple words, we obtain similarities by disregarding differences, by a process of abstracting. In a world of only absolute differences, without similarities, recognition, and, therefore, 'intelligence', would be impossible.

It is possible to demonstrate how 'intelligence' and abstracting both started together and are due to the physico-chemical structure of protoplasm. All living material, usually called protoplasm, has, in some degree, the nervous functions of irritability, conductivity, integration,. It is obvious that a stimulus S does not affect the little piece of primitive protoplasm A 'all over and at once' (infinite velocity), but that it affects it first in a definite spot B, that the wave of excitation spreads, with finite velocity and usually in a diminishing gradient, to the more remote portions of A. We notice also that the *effect* of the stimulus S on A is not *identical* with the stimulus itself. A falling stone is *not* identical with the pain we feel when the stone falls on our foot. Neither do our feelings furnish a full report as to the characteristics of the stone, its internal structure, chemistry,. So we see that the bit of protoplasm is affected only *partially,* and in a *specific* way, by the stimulus. Under physico-chemical conditions, as they exist in life, there is no place for any 'allness'. In life, we deal structurally only with 'non-allness'; and so the term, '*abstracting* in different orders', seems to be structurally and uniquely appropriate for describing the effects of external stimuli on living protoplasm. 'Intelligence' of any kind is connected with the abstracting (non-allness) which is characteristic of all protoplasmic response. Similarities are perceived only as differences become blurred, and, therefore, the process is one of abstracting.

The important novelty in my treatment is in the structural fact that I treat the term 'abstracting' in the *non-el* way. We find that all living protoplasm 'abstracts'. So I make the term abstracting fundamental, and I give it a wide range of meanings to correspond to the facts of life by introducing abstractions of *different orders*. Such a treatment has great structural advantages, which will be explained in Part VII.

As our main interest is in 'Smith$_n$', we will speak mostly of him, although the language we use is structurally appropriate for characterizing all life. 'Abstracting' becomes now a physiological term with structural, actional, physico-chemical, and *non-el* implications.

Accidentally, some light is thus thrown on the problem of 'evolution'. In *actual* objective life, each new cell is different from its parent cell, and each offspring is *different* from its parents. Similarities appear only as a result of the action of our nervous system, which does not register absolute differences. Therefore, we register similarities, which evaporate when our means of investigation become more subtle. Similarities are read *into nature* by our nervous system, and so are structurally less fundamental than differences. Less fundamental, but no less important, as life and 'intelligence' would be totally impossible without *abstracting*. It becomes clear that the problem which has so excited the *s.r* of the people of the United States of America and added so much to the merriment of mankind, 'Is evolution a "fact" or a "theory"?', is simply silly. Father and son are never identical—that surely is a structural 'fact'—so there is no need to worry about still higher abstractions, like 'man' and 'monkey'. That the fanatical and ignorant attack on the theory of evolution should have occurred may be pathetic, but need concern us little, as such ignorant attacks are always liable to occur. But that biologists should offer 'defences', based on the confusion of orders of abstractions, and that 'philosophers' should have failed to see this simple dependence is rather sad. The problems of 'evolution' are verbal and have nothing to do with life as such, which is made up all through of *different* individuals, 'similarity' being structurally a manufactured article, produced by the nervous system of the observer.

In my own practice, I have become painfully aware of a similar discrepancy in the learned *s.r* of some older professors of biology, who quite often try to inform me that 'Life is overlapping', and that 'no sharp distinction between "man" and "animal" can be made'. They forget or do not know that, structurally, actual 'life' is composed of *absolute individuals,* each one *unique* and different from all other individuals. Each individual should have its individual name, based on mathematical extensional methods; for instance, x_1, x_2, x_3, ... x_n; of Smith$_1$, Smith$_2$., or Fido$_1$, Fido$_2$,. 'Man' and 'animal' are only labels for verbal fictions and are not labels for an actual living individual. It is obvious that as these verbal fictions, 'man' and 'animal', are not the living individual, their 'overlapping' or 'not overlapping' depends only on *our ingenuity,* our power of observation and abstraction, and our capacity of coining *non-el* functional definitions.

Let us see how adaption might work in practice. Let us consider two or three caterpillars, which we may name C_1, C_2, C_3, since each of them is an absolute individual and different from the others. Let us assume that C_1 is positively heliotropic, which means that he is compelled to go toward light; that C_2 is negatively heliotropic, which means that he would tend to go away from light; and that C_3 is non-heliotropic, which means that light would have no effect on him of a directional character. At a certain age, C_1 would crawl up the tree near which he was born and so reach the leaves, eat them, and, after eating them, would be able to complete his development. C_2, and probably C_3, would die, as they would not crawl up the tree toward sufficient food. Thus, we see that among the indefinite number of possible individual make-ups of C_k ($k = 1, 2, 3, ... n$), each one being different, only those which were positively heliotropic would survive *under the conditions of this earth,* and all the rest would die. The positively heliotropic would propagate and their positive heliotropism might be perpetuated, the negatively heliotropic and non-heliotropic becoming extinct. This would only occur, however, in a world in which trees have roots in the ground and leaves on their parts toward the sun. In a world where the trees grew with their roots toward the sun and leaves in the ground, the reverse would happen; namely, the negatively heliotropic and non-heliotropic would survive, and the positively heliotropic would die out. We can not foretell whether, in such a world, there would be caterpillars; so this is an hypothetical example.

Experiments made with such caterpillars have shown that the positively heliotropic ones crawl toward the sun, even upon a plant which has been turned over, with the roots toward the sun. They crawl a*way* *from food,* and die. We see that the external *environmental* conditions determine which characteristics survive, and so we reach the notion of adjustment.

The practical result of these conditions is that the indefinite number of individual variations, although they undoubtedly exist, seldom come to our attention, as those variations which do not fit their environmental conditions become extinct; their variations do not become hereditary, and consequently we can seldom find them outside of a laboratory.

This shows, also, the permanent connection and interdependence of the facts of nature. The structural fact that our trees grow with their roots in the ground and their leaves upward is not an independent fact; it has something to do with the structure of the world and the position and the effect of the sun,. So the fact that we have positively heliotropic caterpillars of a special kind, and not negatively heliotropic ones, has something to do with the structure of the rest of the world.

To illustrate this interconnection and interdependence of nature still more clearly, let me suggest an hypothetical question. How would conditions, as they are on this earth, compare with those which would obtain, if it were, let us say, one mile greater in diameter? Some try to guess the answer; yet this question cannot be answered at all. The diameter of this earth is strictly dependent on all the structural conditions which prevail in this world. Since it is impossible to know what kind of a universe it would be, in which this earth could be different from what we know, it is, of course, equally impossible to foresee whether on such a fictitious earth, in such a fictitious universe, there would even be life at all. Because the structure of the world is such as we know it, our sun, our earth, our trees, our caterpillars, and, finally, ourselves have their structure and characteristics. We do not need to enter here into the problems of determinism versus indeterminism, as these problems are purely verbal, depending on our orders of abstractions, the 'logic' we accept, and so, finally, on our pleasure, as is explained more in detail later on, and could not be solved satisfactorily in an A, *el* system, with its two-valued 'logic'.

According to the evidence at hand in 1933, 'Smith$_n$' appears among the latest inhabitants of this earth, and subject to the general test of survival, as already explained. The few thousand years during which there had been any 'Smith$_n$' are too short a period to test, with any certitude, his capacity for survival. We know of many species of animals, and also races of man, of which very little trace has been left. What we know about their history is mostly through a few fossils, which are kept in museums.

The external world is full of devastating energies and of stimuli too strong for some organisms to endure. We know that only those organisms have survived which could successfully either protect themselves from over-stimulations or else were under protective circumstances. If we look over the series of surviving individuals, paying special attention to the higher animals and man, we find that the nervous system has, besides the task of conducting excitation., the task of so-called 'inhibition'. Response to stimuli, by survival, proved its usefulness. But to diminish the response to some stimuli or avoid stimuli, proved also to be useful, again by survival. It is known that the upper or latest layers of the nervous system are mostly such *protective* layers, to prevent immediate responses to stimulation. With the development of the nervous system from the simplest to the most complex, we see an increase in behaviour of a modifiable or individually adjustable type. In terms of *order,* and using the old language, 'senses' came first in order and 'mind' next, in all their forms and degrees.

If we speak in neurological terms, we may say that the present nervous structure is such that the entering nerve currents have a natural direction, established by survival; namely, they traverse the brain-stem and the thalamus first, the sub-cortical layers next, then the cerebral cortex, and return, transformed, by various paths. Experience and science in 1933 are showing that this is the order established under a heavy toll of destruction and non-survival in a system of adjustment, and so should be considered the 'natural' order, because of its survival value.

We all know in practice about a 'sensation', and a 'mental picture' or 'idea'. As 'sensations' were often very deceptive and, therefore, did not always lead to survival, a nervous system which somehow retained vestiges, or 'memories', of former 'sensations' and could recombine them, shift them., proved of higher survival value, and so 'intelligence' evolved, from the lowest to the highest degrees.

Experience and experiments show that the natural order was 'sensation' first, 'idea' next; the 'sensation' being an abstraction of some order, and the 'idea' already an abstraction from an abstraction or an abstraction of higher order.

Experience shows again that among *humans,* this order in manifestations is sometimes *reversed;* namely, that some individuals have 'idea' first; namely, some vestiges of memories, and 'sensations' next, without any external reason for the 'sensations'. Such individuals are considered 'mentally' ill; in legal terms, they are called 'insane'. They 'see', where there is nothing to see; they 'hear', where there is nothing to hear; they are paralysed, where there is no reason to be paralysed; they have pains,

when there is no reason to have pains, and so on, endlessly. Their survival value, if not taken care of, is usually nil. This reversal of order,
but in a mild degree, is extremely common at present among all of us
and underlies mainly all human misfortunes and un-sanity.

This reversal of order in its mild form is involved in identification
or the confusion of orders of abstractions; namely, when we act as if an
'idea' were an 'experience' of our 'senses', thus duplicating in a mild
form the mechanism of the 'mentally' ill. This implies nervous disturbances, since we *violate the natural (survival) order of the activities of the
nervous system*. The mechanism of *projection* is also connected with
this *reversal of order*. This reversal transforms the external world into
a quite different and fictitious entity. Both ignorance and the old metaphysics tend to produce these undesirable nervous effects of *reversed
order and so non-survival evaluation*. If we use the nervous system in a
way which is against its survival structure, we must expect non-survival.
Human history is short, but already we have astonishing records of
extinction.

That such reversal of order in the manifestations of the functioning
of the nervous system must be extremely harmful, becomes evident when
we consider that ·in such a case the upper layers of our nervous system
(the cortex) not only do not protect us from over-stimulation originating in the external world and inside us, but actually contribute to the
over-stimulation by producing fanciful, yet very real, irritants. Experiments on some patients have shown how they benefit *physically* when
their internal energy is liberated from fighting phantoms and so can be
redirected to fight the colloidal disturbances. Such examples could be
cited endlessly from practically every field of medicine and life. This
problem of 'reversal of order' is not only very important semantically,
but also very complex, and it will be analysed further on.

The reader should not miss the fact that an analysis in terms of
order throws a new light on old problems, and so the scientific benefit of
the use of such a term is shown. But this is not all; the use of the term
order has brought us to the point where we can see far-reaching practical applications of the knowledge we already possess, and of which we
have not so far made any systematic use.

We know that the activity of the nervous system is facilitated by
repetition, and we can learn useful habits as easily as harmful. In the
special case of $s.r$ also we can train ourselves either way, though one may
have useful survival value; the other being harmful, with no survival
value. The problem is again one of *order,* and, among others, a problem
of extension and intension, as has already been mentioned several times.

Section D. Order and the problems of extension and intension.

The problems of extension and intension are not new, but have been treated as yet only casually by 'philosophers', 'logicians', and mathematicians, and it has not been suspected what profound, far-reaching, and important structural psycho-logical, semantic components they represent.

At this point, to avoid confusion, a warning is necessary. The problems with which we are dealing have never been analysed from the point of view of this work; namely, from that of structure. So, naturally, all that has been accomplished in these fields is *over-simplified,* and leaves out vital characteristics. Discrepancies have arisen between the structure of the old verbiage and that of the new. Before it is possible to formulate the general theory of this work, it will be necessary to go ahead in spite of discrepancies, and then to formulate the general theory and show how these discrepancies had a perfectly *general* origin in the stratified—and, therefore, ordered—structure of human knowledge. The discrepancies were inherent and *unavoidable* in the old way, but are avoidable in the new. It is the main aim of the present theory to elucidate structural issues in connection with s.r and many problems of human and scientific conduct, mathematics and insanity included, and, in general, with all known problems of scientific method and theory of knowledge. But we ought not to be surprised if such a pioneering enquiry proves to need many corrections and elaborations in the future. Psychiatrists are the least likely to disregard the problems of structure and s.r, since their science is young and still flexible. Besides, the psychiatrists know a great deal about 'human nature' and behaviour, though they are handicapped by insufficient knowledge of the exact sciences and the absence of $\bar{A}$, *non-el* semantics. The opposite, perhaps, would be true about mathematicians. They know a great deal about how to play with symbols. Their work is engrossing and exacting. But very few are capable of separating themselves enough from this play to contemplate the broader, more 'human', aspects of their own science, the interplay of symbols in language, their structure, and the bearing of structure on s.r and adaptation.

Some of these specialists might say that the author uses their terms in a sense different from that in which they use them, and that, consequently, from their point of view, this work is not strictly legitimate. However, when a mathematician lays down a definition, such as $1+1=2$, this has nothing to do with the *application* we make of it when we say that one penny and one penny make two pennies. Neither can he object when we add one gallon (of water) to one gallon (of alcohol) and do not obtain two gallons (of the alcohol-water mixture), but slightly less. This last is a profound experimental fact, intimately connected with the structure of 'matter' and, therefore, of the world around us. The mathematician has nothing to do with the fact that his *additive* definitions, important as they are, do not cover the facts of the world around us, which happens not to be additive in its more fundamental aspects.

Also, the mathematical definition, one and one make two, is *not* invalidated by such non-additive facts. The mathematician does not claim, but rather disclaims, content in its formulae. There is no mention of pennies or apples or gallons of alcohol,. It is simply a definite language of definitive structure for talking about anything which *can* be covered. If facts cannot be covered by given linguistic forms and methods, new forms, new structures, new methods are invented or created to cover the structure of facts in nature.

The mathematician created such a different language long ago. He now calls his additive language 'linear', and the corresponding equations are of 'first degree'. He calls his non-additive language 'non-linear' and the equations are of 'higher degrees'. These latter equations happen to be much more difficult than the former and of complex structure, so that very often they can be solved only by approximation to linear equations. Now, without anybody's fault, the world around us does not happen to be an additive affair in its more fundamental structural aspects. Perhaps the most important and beneficial results of the new physical theories is that they point out this structural fact, and take it into consideration. The reader should recall the example about the man-made green leaf and the non-man-made green leaf, which differ in structure, and he will understand how our additive tendencies are the result of our primitive state of development and of this projection of our anthropomorphic point of view on the world. We reversed the natural order and imposed on the world the structure of our verbal forms, instead of the *natural order* of patterning the structure of language after the structure of the world.

This digression was especially necessary before approaching the problem of 'extension' and 'intension'. These have never been analysed from the point of view of structure and order, and whatever is known about them is taken for granted. It is true that we hear now and then casual remarks that mathematicians had a predilection for extension and 'philosophers' for intension, but these true remarks are not further analysed.

We usually forget that whenever a mathematician or a 'philosopher' produces a work, this involves his 'attitude', which represents an extremely complex psycho-logical *s.r* of the organism-as-a-whole. In most cases, these attitudes determine not only the character of our work, but also other reactions which make up our individual and social life. Historically, the mathematicians have a steady record of achievement, and 'philosophers' (excluding epistemologists) one of uselessness or failure. Has this record something to do with the extensional and intensional *attitudes?* In fact, it has. It is easy to show that the extensional attitude is the *only one* which is in accordance with the *survival order and nervous structure,* and that the intensional attitude is the reversal of the natural order, and, therefore, must involve non-survival or pathological *s.r.*

One of the simplest ways of approaching the problems of 'extension' and 'intension' is perhaps to point its connection with definitions. A collection may be defined, so we are told, by enumeration of its members, as, for instance, when we say that the collection contains Smith, Brown, Jones,. Or we may define our collection by giving a defining 'property'. We are told that the first type of definitions which enumerates individual members is to be called a definition by *extension,* the second, which gives a defining 'property', is to be called a definition by *intension.*

We can easily see that a 'definition by extension' uniquely characterizes the collection, $Smith_1$, $Brown_1$, $Jones_1$,. Any other collection, $Smith_2$, $Brown_2$, $Jones_2$., would obviously be different from the first one, since the individuals differ. If we 'define' our collection by intension; that is, by ascribing some characteristic to each of the individuals, for instance, that they have no tails, many collections of individuals without tails might be selected. Since these collections would be composed of entirely different individuals, they would be entirely different, yet by 'intension', or defining characteristic, they all would be supposed to be one collection.

Similar contrast exists between relations in extension, and relations in intension. These relations have been defined more or less as follows: 'Intensional relations are relations of "concepts"; *extensional relations are relations of denoted facts'*. Or, 'relations of intension are those which are ascertainable *a priori;* a relation of extension is discoverable *only by inspection of the existent'*. Or, 'intension covers the relations which hold for all the possible individuals, while *extension holds only for the existent'*. 'A relation of intension is one which is only discoverable by logical analysis; a relation of extension is one which is only discoverable by the *enumeration of particulars',* .[4]

All that has just been said are perhaps standard definitions; but, for my purpose, they are profoundly unsatisfactory. Because we have had no better understanding of this most important question of order, a great deal of confusion has occurred in human 'thought' and many of our disciplines have become twisted in undesirable directions. This, to a large degree, accounts for the obscurity which characterizes the problems we are dealing with in this book. But it should be emphasized that even in this very unsatisfactory form, 'intension' and 'extension', as they were *felt and applied* (*s.r*, largely disregarding the verbal formulations), have played an enormous role in the development of our forms of representation, and our 'civilization'. Unfortunately, without an ordinal analysis, it was impossible to evaluate properly the relative importance of these semantic attitudes, and to realize the serious importance of these problems for a theory of sanity and resulting consequences.

Here, again, the knowledge which psychiatry gives will help a great deal. We know, in the rough, quite a little about two semantic mechanisms which are called extroversion and introversion. In the rough, again, the extrovert projects all that is going on within himself upon the outside world, and believes that his personal projections have some kind of non-personal objective existence, and so have 'the same' validity and value for other observers. As a result, quite naturally, the extrovert is due to receive a great many unpleasant shocks, for the other observer does not necessarily observe or 'perceive' in the external events the characteristics which the first observer 'finds'. He has often projected them from his 'inside', but they were entirely personal. The first observer, in his semantic conviction that his observations are the only, uniquely correct, observations, feels that the second observer is either blind in some way, or unfair to him. In acute cases, he develops a mania of persecution. He feels that everybody misunderstands him; nobody is fair to him; everybody is hostile to him; he will get even with them; in the name of 'justice', he will punish them,. A dangerous and quite often incalculable bitterness and hostility follow. Such types are usually troublesome, and, if the affective components are strong, then such types are dangerously ill and liable to produce bloodshed or make other attacks. The most pronounced types in this extreme direction are called paranoids and paranoiacs.

The introvert type is different. He is mostly concentrated on what is going on inside of his own skin. Almost all of what is going on outside of his skin, he interprets in personal terms and feelings. Whatever unpleasant happens, he is always guilty; always willing to take the blame, which is quite often just, for many psycho-logical reasons upon which we cannot dwell here. This type, in its extreme development, quite often finds a solution in suicide. The most extreme cases are called 'dementia praecox' or 'schizophrenic' types.

In everyday experience, it is seldom that such clear-cut types as just described can be observed. For the purpose of studying such extreme types, one has to do researches in asylums. Even there we find a great number of mixed cases. In daily life we find in practically everybody a *predominance* of one or the other of these types of *s.r*, but in some the two types appear to be inextricably mixed. Observations upon this problem among so-called 'normal' men is difficult, as they represent great complexities.

It has been already mentioned that the well-balanced man, a man who has survival value, should be a well-balanced mixture of both tendencies; namely, an extroverted-introvert, or, if we wish, an introverted-extrovert. As yet, these problems, no matter how important they may be, are beyond our educational methods, and only in acute cases are they taken care of by physicians, and then mostly in asylums. It is important to have simple means to deal with these semantic problems in elementary education as a *preventive* method, or as a branch of semantic hygiene.

Even this brief analysis shows how tremendously powerful the affective factors are which may be behind the unbalanced semantic attitudes. The reader should not miss the fact that in both types, when well developed, there is material for an extreme amount of self-imposed suffering. Then the nervous energy produced by the organism is absorbed in fighting phantoms, instead of being directed toward useful ends, such as regulating the normal activities of the organism, or fighting internal enemies, whereas, there should also be some energy left for activities and interests useful socially or for the survival of the race.

While the majority of individuals present different degrees of prevalence of one mechanism over the other, yet fairly clear-cut cases are to be found. The extreme complexity of the structure of the nervous system of man justifies the enormous number of degrees recognized. So large, indeed, is this number of possibilities, that we have little difficulty in understanding that the individuality of every one is unique.

Extroverted and introverted individuals are usually born such; at least, they usually have a predisposition to be the one or the other. To what extent these tendencies can be aggravated or improved by education is not yet solved, and, indeed, has never been much bothered about. To consider our activities merely as results of inborn tendencies is too narrow a view. The human nervous system is not finished at birth, and it continues its development for some time after the birth of the child. So it is much more influenced by environmental conditions, the verbal included, than is the nervous system of an animal. The make-up of the individual is thus some function of different variables, among which the hereditary inclinations and the environmental conditions appear in a relation, at present, not fully known. The individual *feels and acts* according to his complex make-up, including *acquired s.r*, no matter what factors have played a role in its moulding, and, as a rule, he is little influenced by the way he rationalizes his activities. From this point of view, we may consider that the extensional make-ups and intensional characters are bound to show themselves later on in life, no matter how the subject may have rationalized them, if his *s.r* have not been modified.

It seems evident that the extroverted and introverted tendencies have some connection with extensional and intensional types of reaction; but, of course, they are not identical. They influence the individual in the selection of a profession, and in the preference for some special trend of activity. Thus, mathematicians, generally, have an inclination toward extension, 'philosophers' toward intension. Now, it is interesting to note that mathematicians have a record of continuous constructive progress, and at each epoch have produced the highest kind of language known. Also, the most important achievements in the fields which traditionally belonged to 'philosophers' have actually been produced by mathematicians. The 'philosophers', in the main, have a record of failure.

The reason for this difference, which is too remarkable to be a mere coincidence, may be found by application of the term 'order' in our analysis. *The extensional method is the only method which is in accordance with the structure of our nervous system as established by survival.* Reversed intensional methods disorganize this normal mode of activity of the nervous system, and so lead toward nervous and 'mental' illnesses.

As explained before, the structure of our nervous system was established with 'senses' first, and 'mind' next. In neurological terms, the nervous impulses should be received first in the lower centres and pass on through the sub-cortical layers to the cortex, be influenced there and be transformed in the cortex by the effect of past experiences. In this transformed state they should then proceed to different destinations, as predetermined by the structure established by survival values. We know, and let us remember this, that the reversed order in semantic manifestation—namely, the *projection into 'senses' of memory traces* or doctrinal impulses—is against the survival structure, and hallucinations, delusions, illusions, and confusion of orders of abstractions are to be considered pathological. In a 'normal' human nervous system with *survival* value, the nervous impulses should not be lost in the sub-cortical layers. In such a case, the activity of our human nervous system would correspond to the activity of the less-developed nervous systems of animals which have no cortex at all. It must be remembered, also, that the sub-cortical layers which have a cortex, as in man, are quite different from corresponding layers of those animals which have never developed a cortex. It is impossible to avoid the conclusion that survival values are *sharply* characterized by *adequacy,* and that animals without cortex have nervous systems adequate for their needs under their special conditions; otherwise, they would not have survived. This applies, also, to those animals who have a cortex. Their activities for survival depend on this cortex; and when the cortex is removed, their activities become inadequate. Their sub-cortical layers alone are not adequate to insure survival. For survival, such animals must use not only their lower centres and their sub-cortical layers, but also their cortices.

Among animals, as all evidence shows, the enormous majority have, without human interference, nervous systems working usually in the 'normal' way; that is, according to the survival structure. 'Insanity' and kindred nervous disturbances are known only among ourselves (however, see Part VI). Apparently, the cortex, through its enormous internal complexity, which provides many more pathways, and through its complex interconnections, which offer many more possibilities, with a greater number of degrees of 'inhibition', of excitability, of delayed action, of activation., introduces not only a much greater flexibility of reaction, but, through this flexibility, a possibility for abuse, for reversal of manifestations, and so for a deterioration of the survival activity of the nervous system as-a-whole. The sub-cortical layers and other parts of the brain of man are different from the corresponding parts of the animal brain, which has a less-developed cortex. The nervous system works as-a-whole, and the anatomical homology of the parts of different nervous systems is a very inadequate, perhaps even a misguiding, foundation for inferring *a priori* the *functioning* of these systems, which ultimately depend not only on the macroscopic but also on the microscopic and sub-microscopic structures. For instance, we can cut off the head of some insects, and they go along quite happily and do not seem to mind the operation much. But we could not repeat this with higher animals. The behaviour is changed. A decorticated pigeon behaves differently from a decorticated rat, though neither of them seems affected greatly by the operation. A decorticated dog or ape is affected much more. Man is entirely changed. None of the higher types is able to survive long if decorticated.

There is on record the medical history, reported by Edinger and Fischer, of a boy who was born entirely without cerebral cortex. There

12

were apparently no other important defects. This child never showed any development of sensory or motor power, or of 'intelligence', or signs of hunger or thirst. During the first year of his life, he was continually in a state of profound stupor, without any movements, and from the second year on, until his death (at three years and nine months), was continually crying.[5]

Although many animals, for instance fishes, have no differentiated cortex, yet their nervous system is perfectly *adequate* for their lives and conditions. But in a more complex nervous system, the relative functions of different parts of the brain undergo fundamental transformation. In the most complex nervous system, as found in man, the older parts of the brain are much more under the control of the cerebral cortex than in any of the animals, as is shown in the example above. The absence of the human cortex involves a much more profound disturbance of the activities of the other parts of the brain. Since the cortex has a profound influence upon the other parts of the brain, the insufficient use of the cortex must reflect detrimentally upon the functioning of the other parts of the brain. The enormous complexity of the structure of the human brain and the corresponding complexity of its functioning accounts not only for all human achievements, but also for all human difficulties. It also explains why, in spite of the fact that our anatomy differs but little from that of some higher animals, veterinary science is more simple than human medicine.

Because of the structure of the nervous system, we see how the completion of one stage of the process which originated by an external stimulus (A) and has itself become a nervously elaborated *end-product* (B), may, in its turn, become the stimulus for a still further nervously elaborated *new end-product* (C), and so on. When association or relation neurons enter, the number of possibilities is enormously increased.

It must be emphasized that A, B, C., are, fundamentally, entirely *different*. For instance, the external event A' may be a falling stone, which is an entirely different affair from the pain we have when this stone falls on our foot. It thus becomes clearer what is meant by a statement that the 'senses' abstract in their own appropriate way, determined by survival value, the external events; give these abstractions their special colouring (a blow on the eye gives us the feeling of *light*) ; discharge these transformed stimuli to further centres, in which they become again abstracted, coloured, transformed,. The end-product of this second abstracting is again an entirely different affair from the first abstraction.

Obviously, for survival value, this extremely complex nervous system should work in complete co-ordination. Processes should pass the *entire cycle*. If not, there must be something wrong with the system. The activities of the organism are then regressive, of lower order, a condition known as 'mental' illness. The gross anatomical divisions of the nervous system should not be relied upon too much as an index of function. Perhaps these anatomical speculations are even harmful for understanding, because they falsify the facts, emphasize the macroscopic similarities unduly and disregard subtle yet fundamentally important microscopic and sub-microscopic structures and differences, which are perfectly manifest in the functioning, but which are difficult to observe directly on their level.

The term 'abstracting' is a multiordinal term, and hence has different meanings, depending on the order of abstractions. It is a functional term and, to indicate the differences in meanings, it is necessary to indicate the different orders. It is structurally a *non-el* term, built upon the extensional mathematical pattern x', x'', x'''., or x_1, x_2, x_3, ... x_n, or x_k ($k = 1, 2, 3, \ldots n$). This allows us to give the term 'abstracting of different orders' a perfectly *unique* meaning in a given problem and yet to keep in a fluid state its most important *functional implications*. Something similar happens when the mathematician discusses his x_n. No one can miss the fact that he deals with a variable which can take n values; so this symbol has a quite definite descriptive structural and semantic value. So has the 'abstracting of different orders'.

It is desirable to introduce consciously and deliberately *terms* of a *structure* similar to the *structure* of human knowledge, of our nervous system, and of the world, involving appropriate *s.r.* Multiordinal terms are uniquely appropriate, since they take their ∞-valued structure from the structure of events (1933) and do not reflect their older, one-valued, false to fact character on the events. (Note the order.)

Now we are ready to reformulate the problem of extension and intension in terms of *order*.

If the natural survival order is lower abstractions first and higher next, then extension starts with absolute individuals, and conforms to the proper survival order. Extension recognizes the uniqueness, with corresponding one-value, of the individual by giving each individual a unique name, and so makes confusion impossible. Training in *s.r* of sanity becomes a possibility, and order becomes paramount. Extension and order cannot be divided. When we speak about 'order', we imply extension, and, when we speak about extension, order is implied. That modern mathematics and mathematical 'logic' has so much to do with order, as to make this term fundamental, is a necessary consequence of the extensional method which starts with unique individuals, labels them by unique names and only then generalizes or passes to ∞-valued higher order abstractions like 'numbers',. The direction of the process of abstraction is here in the survival order, from lower abstractions to higher. It hardly needs to be emphasized that, to the best of our knowledge in 1933, it is the only possible way to follow the natural order and to evade reading into a fundamentally one-valued external event, our older *undifferentiated* ∞-valued fancies (which happens if the process is reversed in order) involving powerful factors in our *s.r.*

Intension means structurally the reversal of the survival order, since it starts with undifferentiated ∞-valued higher abstractions and distorts or disregards the essential one-values of the individuals and reads into them as *uniquely* important undifferentiated ∞-valued characteristics.

Historically, mathematicians have had a predilection and, because of the character of their 'element' (numbers) and their technique, a structural necessity, for the use of extensional methods. It does not need much imagination to see why they have produced results of utmost (although relative) importance and validity at each date.

'Philosophers' and reasoners of that class have had a predilection for intension, and this also explains why, in spite of tremendously acute verbal exercises, they have not produced anything of lasting value, for they were carried away by the structure of the language they used. This predilection being already based on the reversal of the survival order, it was bound to lead in the less-resistant individuals to nervous and 'mental' defects.

The issues, as presented here, are very clear-cut, and, in fact, too clear-cut, as we have disregarded for the present the *cyclic* order of the nervous process. This last fact first abolishes the sharp distinction between 'pure' extension and 'pure' intension, each process never being 'pure', but always 'impure', one influencing the other. 'Pure' intension and 'pure' extension are delusional, to be found only in 'mentally' ill, with no survival value. This explains why we have to use the terms of *preference* and *order*. Without these terms I would not have been able to carry through this analysis at all. This reversal of order in *s.r* implies different distribution, diffusion, intensity of nerve currents in the sub-microscopic field, and so involves important, different semantic components of non-survival value. It is most desirable to learn to control the activities on the sub-microscopic level by means of training on the macroscopic level, if means to do this can be devised.

The writer is not at all convinced that, acting as we do under the spell of intensional and ignorant 'philosophers', the existing systems and educational methods are not largely following the reversal of the survival order of our nervous processes. It seems unnecessary to point out that a structural and semantic enquiry on this particular question might be important and beneficial. It seems, without much doubt, that human institutions and activities should be in accord with 'human nature', if we are to expect them to survive without crushing us, and a scientific enquiry in this 'human nature' would be not only desirable but exceedingly important.

The reader, with the help of another person, should perform a very simple experiment. Let the assistant select secretly a dozen newspaper headlines of letters of equal size. Let the reader then sit in a chair without altering his distance from the assistant and let the assistant show him one of these headlines. If he is able to read the headline, it should be rejected, and a new one selected by the assistant and put a foot or more farther away. If this one is read correctly, it should be rejected, and a third one placed still farther away. By such trials we can finally find a distance which is slightly greater than the maximum range of clear visibility for the reader, so that, although the headline is only slightly beyond the distance at which one could read it, yet it would be illegible. Let the reader then try as hard as possible to read headlines which are just beyond his visual range. When he is convinced that he cannot read the headline, let the assistant *tell* him the content of it. Then the sitter can usually *see* with his *eyes* the letters, when he *knows* what is supposed to be there. The question arises, what part in the 'seeing' is due to 'senses', and what to 'mind'? The answer is, that, structurally, the 'seeing' is the result of a cyclic *interdependent* process, which can be *split only verbally.* The independent elements are fictitious and, structurally, have little or nothing to do with actual facts. The human nervous system represents, structurally, a mutually interdependent cyclic chain, where each partial function is in the functional chain, together with enforcing and 'inhibiting', and other mechanisms.

Up to this stage, we have used the term 'cyclic order', but, in reality, the order is *recurrent,* though of a character better described by the 'spiral theory', as explained in my *Manhood of Humanity* on p. 233. In the 'spiral theory', we find the foundation for this peculiar stratification in levels and orders, which is necessitated by the structure and function of the human nervous system. It should be noticed that the equations of the circle and spiral are non-linear, non-plus equations.

The above relation underlies a fundamental mechanism, known in psychiatry as 'sublimation', in which, and by which, quite primitive impulses, without losing their intensity and fundamental character, quite often are transformed from very primitive levels, which frequently represent vicious and anti-social effects, into desirable characteristics, socially useful. Thus, a sadistic impulse may be sublimated into the socially useful vocation of the butcher, or, still further, into the skill and devotion to the service of their fellowmen, shown by many surgeons. We see that this mechanism is of tremendous importance, and responsible for what we call 'culture' and many other values. Our educational methods should understand this mechanism and apply this knowledge in the semantic training of youth. It is important to realize that this mechanism appears as the only semantic mechanism of correction which is in accord with the structure of the human nervous system, and so it seems workable. Various metaphysical preachings usually start by disorganizing the proper survival working of the human nervous system, and then we wonder that they fail, and that we cannot change 'human nature'. To deal with 'human nature', which is not something static and absolute, we need to approach it with more structural understanding and less prejudice. Then, and then only, can we eventually look for better semantic results.

The writer does not want the reader to conclude that, because in mathematics we have followed the survival order through extension, the mathematicians must, by necessity, be the sanest of the sane. Quite often this is not true, since many complexities exist which will be taken under analysis later.

Section E. Concluding remarks on order.

One thing remains fundamental; namely, that the problems of order and extension are of paramount structural importance for sanity and our lives. They should be worked out and applied to the semantic training of the young in elementary education,. This would certainly produce a new generation saner than we are, and one which would, perhaps, lead lives less troubled than our own, and so, perhaps, of better survival value.

To appreciate fully the immensity of the task of a more detailed analysis of the problems of 'extension' and 'intension', the reader is advised to read the *Survey of Symbolic Logic,* by Professor C. I. Lewis, University of California Press, 1918, in which Chapter V is devoted to an important attempt at a formulation of strict implication of *both* extensional and intensional character, which is the *only* organism-as-a-whole, *non-el* possibility. Lewis's theory of 'strict implication' introduces the notions of *impossible* propositions and so throws considerable light on the problem of non-sense, a light which is very seriously needed.

In concluding, it must be mentioned that a theory of *sanity*, because of the survival value of *order*, cannot *start* with the older, undifferentiated similarities, which are a product of *higher* abstractions, and thus of later origin, but *must* start with *differences* as fundamental, and so preserve the structure and order of the survival trend as applied in this work.

Animals do not possess such a highly differentiated nervous system as human beings. The difference between their higher and lower abstractions is thus not so fundamental, as we shall see later on. With them the question of *order* is less important, as they cannot alter it. Animals have the benefit of better co-ordination, since in them the above-described structural difficulties do not arise. They have normally no 'insane'. But, also, for the same reason, animals are not able to start every generation where the older left off. In other words, animals are not time-binders.

The structural complexity and differentiation of the nervous system in man is responsible, as is well known, not only for all our achievements and control over the world around us, but also for practically all our human, mostly semantic difficulties, many 'mental' ills included. The analysis in terms of *order* on the macroscopic level (semantic manifestations) reveals a profound connection with sub-microscopic processes of distribution., of nervous energy. When the mechanism which controls these processes is properly understood, then they can be controlled and educated by special semantic training. In other words, theoretical, doctrinal, higher abstractions may have a stabilizing and regulating *physiological* effect on the function of our nervous system.

The reader may be interested to know that 'order' is very important in animal life. An analysis of nest-building and the rearing of young among birds shows that each step of the cycle is necessary before the next step is taken. If the cycle is broken, they usually cannot adjust themselves to the new state of affairs, but must start from the beginning.[6] This is a situation similar to our own when we cannot recall a line in a poem, but have to start from the beginning of the piece in order to recapture it. Pavlov was able, by the change of four-dimensional order of stimuli, to induce profound nervous disturbances in the nervous systems of his dogs,.

It appears, also, that in mild cases of aphasia, which is a neurological disturbance of linguistic processes, with word-blindness, word-deafness., the notion of 'order' and 'relations' is often the first to be disturbed. In some cases, lower order abstractions are carried out successfully, but calculation, algebra, and other higher order abstractions, which require *ordered* chains, become impossible. The aphasic seems to have a general incapacity for grasping *relations,* realizing *ordered series,* or grasping their succession.[7]

We see that the problems of *order* are somehow uniquely important; and so the investigation of the psycho-logics of mathematics, which is based on *order,* might give us means of at least partial control of different undesirable human semantic afflictions.

But, after all, we should not be surprised that it is so. The structure of nervous systems consists of *ordered* chains produced by the impact of external and internal stimuli in a four-dimensional space-time manifold, which have a spatial and also a temporal *order.* The introduction of the finite velocity of nerve currents, which, although known, was, as a rule, disregarded by all of us, introduces automatically our *ordering* in 'space' and 'time' and, therefore, in space-time. That is why the old anatomical three-dimensional analogies are vicious and false to facts when generalized. For better or worse, we happen to live in a four-dimensional world, where 'space' and 'time' cannot be divided. Whoever does this splitting must introduce fictitious, non-survival entities and influences into his system, which is moulded by this actual world and unable to adjust itself to fictions.

It seems obvious that all these problems of 'adjustment' and 'non-adjustment', 'fictitious' or 'actual' worlds., are strictly connected with our *s.r* toward these problems, and so ultimately with some structural knowledge about them. But *attitudes* involve lower order abstractions, 'emotions', affective components, and other potent semantic factors which we have usually disregarded when dealing with science and with scientific problems and method. For adjustment, and, therefore, for *sanity,* we must take into account the neglected aspects of science, of mathematics, and of scientific method; namely, their semantic aspects. In this way we shall abandon that other prevalent structural fiction referred to at the beginning of the present chapter; namely, that science and mathematics have an *isolated* existence.

The above considerations of order lead to a formulation of a fundamental principle (a principle underlying the whole of the *non-aristotelian system*) ; namely, that organisms which represent *processes* must develop in a certain *natural survival four-dimensional order,* and that the *reversal* of that order must lead to pathological (non-survival) developments. Observations disclose that, in all human difficulties, 'mental' ills included, a *reversal of the natural order* can be found as a matter of fact, once we decide to consider order as fundamental. Any identification of inherently different levels, or confusion of orders of abstractions, leads automatically to the reversal of natural order. As a method of preventive education and psychotherapy, whenever we succeed in reversing the reversed order or restoring the natural survival order, serious beneficial results are to be expected. These theoretical conclusions have been fully justified by experience and the work of Doctor Philip S. Graven in psychotherapy. It should be noticed that different primitive 'magic of words', or modern 'hypostatizations', 'reifications', 'misplaced concreteness', 'objectifications'., and all semantic disturbances represent nothing else but a confusion of orders of abstractions, or identifications in value of essentially different orders of abstractions.

The above considerations are entirely general, but, because of their novelty, they have not, as yet, been applied in the *non-aristotelian* simple and workable form to psychiatry or education. In a very instructive paper on *The Language of Schizophrenia*,[8] Doctor William A. White applies some of these notions. Because of the method of approach, I will quote from this paper. It should be understood that this paper deals, also, with other issues, and the quotations do not do justice to the author, because I quote only those passages which are of particular interest here, omitting the literature given by Doctor White. The italics and one footnote are mine.

'It requires but a moment's serious consideration to realize that the subject of schizophrenic language must be immense if for no other reason than that it involves an understanding of the whole subject of language of which it is but a part. The extent and depth of the subject of language may be further appreciated from the fact that the single feature of its neuronic background as it is brought to attention in aphasia constitutes one of the most complex problems in the whole field of neurology and one with respect to which we are still hopelessly ignorant, especially when the enormous amount of work that has been done in this field is considered. . . .

. . . .

'There have been a few other recent contributions to the subject of schizophrenic thought and speech which, as they run more in line with my own thinking on the subject, I will refer to more fully. These studies *equate the processes of thinking of schizophrenic persons with those of primitive peoples and of children.*

. . . .

'. . . In the archaic thinking of a prelogical kind, found among primitive savage races, the vividness of the images is greater than among more highly developed races, and the effect produced in the observer is projected and believed to be an inherent attitude of the object, which thus acquires a "demonic" character. All things which arouse a similar emotion are thought of as being actually the same. In dementia praecox there is a similar loss of objectivity; hallucinations and reality are imperfectly distinguished, and every happening has a meaning and effect on the observer; the idea of an action produces the action directly, instead of offering a possibility of action, and this is interpreted as a compulsion from without. Paralogical thinking is a stage beyond this; identification of objects is based on similarities, differences being neglected. . . . This form of thought is common in dementia praecox.

. . . .

'. . . While for the normal person the chief criterion of the world of real objects is their independence of him, whereas imaginary things depend for their existence on him, the general characteristic of the schizophrenic patient's experience is that his mental and imaginary experiences have a substantial and concrete nature when the normal person would see only symbols and analogies. His thoughts have magic power and can produce real results; they have for him a substance and he can manipulate them physically. . . .

. . . .

'Many other childish manifestations resemble those of schizophrenic persons: children's jokes, tricks, and plays in words have a similar autistic character, with no apparent meaning in relation to actualities, and this changes at puberty. Children, like the patients, love to make up a sort of neoplastic language of their own, having meanings known only to themselves or their immediate circle. Perseveration and stereotypy in speech and actions are often observed in children. Their musical performances show the same mechanical rendering, and the same preference for simple melodies and rhythms as are found in schizophrenic patients. . . .

. . . .

'Just in the same way we must be careful *not to equate* the regressive psychotic and the primitive *too literally*. That there are close analogies in their respective ways of thinking there is no doubt and that the recognition of these analogies has been of the utmost importance in enabling us to understand schizophrenic thinking must be acknowledged but perhaps, to be on the safe side, the matter should rest there for the present at least.

. . . .

'Another point should be made at this time, after what has been said of the loss of the boundaries of the ego, its indefiniteness, etc. These expressions are apt to be equated, if we do not think carefully, with such concepts as disintegration and dementia. We must not lose sight of the magic of words and not be led astray by the old meanings when we are striving for new ones but are forced to use the words in current use. . . . Now regression to a childish or primitive level of this sort, which is what occurs in those conditions in which the ego is said to lose its clearness of definition, does *not imply disintegration* in the sense of disorder *but regression* to a different kind of order or, as my friend Korzybski would say, *to a lower order of abstraction.*

'This is important for a *principle* is embodied in the nature of this change. . . .

. . . .

'This *principle,* namely, that the schizophrenic thought processes and language are of a *lower order of abstraction,* accounts, in part at least, for another phenomenon. If by a process of regression the mechanisms of thinking tend to ever more primitive levels then we should expect them ultimately to arrive at a concrete perceptional level, and when this occurs hallucinations, which have long been regarded as evidence of the schizophrenic splitting, come into the picture. While I believe that there must be other factors to account for the hallucinations they are at least to be expected as the natural outcome of regression—as are the forms of thought already referred to. . . .

'It comes about, therefore, that we cannot understand the language of the schizophrenic patient without the aid of these principles, because the language of a lower level of psychologic development, or a lower order of abstraction, must remain unintelligible to those who think in the terms of higher levels. The whole problem of the understanding of the psychoses, from this point of view, might be well considered as the problem of the translation of the language of the psychoses.

. . . .

'Summary and Conclusions

'1. A complete understanding of the language of schizophrenia would imply an understanding of language in general of which schizophrenic language is only a part. This would further imply an understanding of thought in general of which language is largely an expression. Because of its extent this program is quite impossible, but certain principles need to be clearly in mind in order to avoid taking over, in any attempt to understand the language of schizophrenia, certain misconceptions in both of these territories which are still rife, not having been as yet fully replaced by the newer ways of thinking about the matters involved. . . .

• •

'2. There is one psychiatric assumption I have made and which is fundamental to my approach to the problem of the language of schizophrenia. It is that schizophrenia is a regression psychosis. This is of the greatest importance for, if it is true, we should expect to find in the thinking and in the language of schizophrenic patients characteristics of earlier stages of development, earlier genetic levels.

'3. The development or evolution of thought and speech, the assumption of genetic levels, implies that there must be a *law in accordance with which this development proceeds.* This law is that thought and language in their development change from feeling, concreteness and perception in the direction of reasoning, differentiation and abstraction.

'4. The law of schizophrenic thought and language must be the *reverse of the law for their development,* on the assumption that schizophrenia is a regression psychosis.

'5. *This reversal of the law of development implies results that are very different from those implied in the old terms "disintegration" and "dementia."*

'6. This *reversal* can be briefly and simply indicated. The language of schizophrenia is of a lower order of abstraction than normal adult language.

• •

'7. The thinking and the speech of schizophrenia while of a *lower order of abstraction* nevertheless make use of words which we are accustomed to use to express a higher order. This *discrepancy* is one reason why such language is so hard for us to understand. Another reason for our difficulty in understanding the schizophrenic patient is that while some of his symbols are of a lower order of abstraction by no means all of them are so that there is a *strange mixture* which further confuses our understanding. Still another difficulty is due to the magic of words.* We are still far from free of this influence and are therefore forced to think that when there is a word there must be a thing corresponding to it and also forced to think of the wording as necessarily meaning what it usually has meant in our experience.

'8. *The reversal of the law of development* in schizophrenia also accounts, in part at least, for hallucinations which have long been regarded as signs of the schizophrenic splitting. Regression must lead ultimately to concrete, perceptional configurations and all that that implies.

• •

'12. For the understanding of the language of schizophrenia, therefore, the whole dynamic situation needs to be comprehended. The main obstacle to this understanding has been, in the past, the magic of words.* ...'

Identification, or the confusion of orders of abstractions, in an *aristotelian* or *infantile* system, plays a much more pernicious role than the present official psychiatry recognizes. *Any* identification, at *any* level, or of *any* orders, represents a non-survival *s.r* which leads invariably to the reversal of the natural survival order, and becomes the foundation for *general* improper evaluation, and, therefore, *general* lack of adjustment, no matter whether the maladjustment is subtle as in daily life, or whether it is aggravated as in cases of schizophrenia. A *non-aristotelian system,* by a complete elimination of 'identity' and identification, supplies simple yet effective means for the elimination by preventive education of this general source of maladjustment. Book II is entirely devoted to this subject.

*['Magic of words' represents only a minor yet very complex manifestation of *aristotelian s.r* of identification and, naturally, exhibits, also, the reversed natural order in evaluation.—A. K.]

CHAPTER XIII

ON RELATIONS

To be is to be related. (266)   CASSIUS J. KEYSER

Science, in other words, is a system of relations. (417)   H. POINCARÉ

Asymmetrical relations are involved in all series—in space and time, greater and less, whole and part, and many others of the most important characteristics of the actual world. All these aspects, therefore, the logic which reduces everything to subjects and predicates is compelled to condemn as error and mere appearance. (453)   BERTRAND RUSSELL

My own investigations in this field, extending over some fifteen years, together with the facts already at hand, as I see them, have forced me to the conclusion that the organic individual is fundamentally . . . a system of relations between a physical substratum or structure and chemical reactions. (90)   CHARLES M. CHILD

The thalamus, which in the lower vertebrates deprived of the cortex ensures the general reactions of the organism and the elementary mental functions, possesses an affective excitability in relation with the profound biological tendencies of the organism; among the higher mammals, indeed, it seems to preserve this rôle of affective regulation, whose importance in the behaviour of the organism and mental life is so often misunderstood. (411)   HENRI PIÉRON
. . . organic impressions ('interoceptive' sensibility) appear in all cases to arrive at the cortex only when translated by the thalamus, with its own affective elaboration. (411)   HENRI PIÉRON

Nevertheless, the consuming hunger of the uncritical mind for what it imagines to be certainty or finality impels it to feast upon shadows in the prevailing famine of substance. (22)   E. T. BELL

In the foregoing chapters I made use of an expression, 'the organism-as-a-whole', which is employed continually in biology, psychiatry, and other branches of science. This expression is a restricted form of the general structural principle of non-elementalism. This expression implies that an organism is *not* a mere algebraic *sum* of its parts, but is *more* than that, and must be treated as an integrated whole. It was mentioned that the non-additivity and the 'more' than a mere 'sum' are complex problems which call for a new method of analysis. We have already seen that a simple analysis of the expression, 'Smith kicks Brown', involves a full-fledged structural metaphysics, or set of assumptions and terms which are taken on faith, since they cannot be defined, except circularly. In the present chapter, these subjects of great semantic importance will be developed further.

One of the fundamental structural defects and insufficiencies of the traditional *A*-system was that it had no place for 'relations', since it assumed that everything could be expressed in a subject-predicate form. As we shall see, this is not true. Restriction to the subject-predicate form leaves out some of the most important structural means we have for representing this world and ourselves and has resulted in a general state of un-sanity. The explicit introduction of 'relations' is rather a recent innovation. A few words may be said about them, although the term 'relation' is one of the terms that we may accept as undefined, or that we may define in terms of multi-dimensional order.

Some relations, when they hold between A and B, hold also between B and A. Such relations are called *symmetrical*. For instance, the relation 'spouse'. If it holds between A and B, it holds also between B and A. If A is the spouse of B, B is the spouse of A. Terms like 'similarity' and 'dissimilarity' also designate relations of this kind. If A is similar or dissimilar to B, so is B similar or dissimilar to A. In general, a symmetrical relation is such that, if it holds between A and B, it also holds between B and A. In other words, the *order* in which we consider the relation of our entities is immaterial.

It is easy to see that not all relations are of such a character. For instance, in the relation 'A is the brother of B', B is not necessarily a brother of A, because B might be the sister of A. In general, relations which hold between A and B, but not necessarily between B and A, are called *non-symmetrical*. In these relations *order* becomes important. It is not a matter of indifference in what *order* we consider our entities.

If a relation is such that, if it holds between A and B, it *never* holds between B and A, it is called *asymmetrical*. Let us take, for instance, the relations 'father', 'mother', 'husband',. We readily see that if A is a father, or mother, or husband of B, B is *never* a father, or mother, or husband of A. The reversal of *order* is impossible in asymmetrical relations, and so any asymmetrical relation establishes a definite order.

Relations such as *before, after, greater, more, less, above, to the right, to the left, part,* and *whole,* and a great many others of the most important terms we have, are asymmetrical. The reader may easily verify this for himself. For instance, if A is *more* than B, B is *never* more than A,. We see at once that the troublesome little words, which are necessary to express *order* as 'before' and 'after'; terms of *evaluation,* such as 'more' and 'less'; and terms on which elementalism or non-elementalism depends, such as 'part' and 'whole', are in the list of asymmetrical relations.

Relations can be classified in another way, when three or more terms are considered. Some relations, called *transitive,* are such that, whenever they hold between A and B and also between B and C, they hold between A and C. For example, if A is before, or after, or above, or more., than B, and B is before, or after, or above, or more., than C, then A is before, or after, or above, or more., than C.

It should be noted that all relations which give origin to series are transitive. But so are many others. In the above examples, the relations were transitive and asymmetrical, but there are numerous relations which are transitive and symmetrical. Among these are relations of equality, of being equally numerous,.

Relations which are not transitive are called non-transitive. For instance, dissimilarity is not transitive. If A is dissimilar to B, and B dissimilar to C, it does not follow that A is dissimilar to C.

Relations which, whenever they hold between A and B, *never* hold between A and C are called *intransitive*. 'Father', 'one inch longer', 'one year later'., are intransitive relations.

Relations are classified in several other ways; but, for our purpose, the above will be sufficient.

It is necessary now to compare the relational forms with the subject-predicate form of representation, which structurally underlies the traditional A-system and two-valued 'logic'. The structural question arises whether all relations can be reduced to the subject-predicate forms of language.

Symmetrical relations, which hold between B and A whenever they hold between A and B, seem plausibly expressed in the subject-predicate language. A symmetrical and transitive relation, such as that of 'equality', could be expressed as the possession of a common 'property'. A non-transitive relation, such as that of 'inequality', could also be considered as representing 'different properties'. But when we analyse *asymmetrical* relations, the situation becomes obviously different, and we find it a structural impossibility to give an adequate representation in terms of 'properties' and subject-predicates.

This fact has very serious semantic consequences, for we have already seen that some of the most important relations we know at present belong to the asymmetrical class. For example, the term 'greater' obviously differs from the term 'unequal', and 'father' from the term 'relative'. If two things are said to be unequal, this statement conveys that they differ in the magnitude of some 'property' without designating the greater. We could also say that they have different magnitudes, because inequality is a symmetrical relation; but if we were to say that a thing is unequal to another, or that the two have different magnitudes, when one of them was greater than the other, we simply should *not give an adequate account of the structural facts at hand*. If A is greater than

B, and we merely state that they are unequal or of different magnitudes, we *imply the possibility* that B is greater than A, *which is false to facts*. To give an adequate account, and to prevent *false implications,* there is no other way than to say which one is greater than the other. We see that it is impossible to give an *A adequate* account when asymmetrical relations are present. The possession of the 'same' 'property', or of different 'properties', are both *symmetrical relations* and seem covered by the subject-predicate form. But it is impossible to account adequately for asymmetrical relations in terms of 'properties'. In other words, we see that a language and 'logic' based upon subject-predicate structure may perhaps express symmetrical relations, but fail to express adequately asymmetrical relations, because both 'sameness' and difference of predicates are symmetrical.[1] Asymmetrical relations introduce a language of *new structure,* involving new *s.r.* Yet asymmetrical relations include many of the most important ones. They are involved in all *order,* all *series,* all *function,* in 'space', in 'time', in 'greater' and 'less', 'more' and 'less', 'whole' and 'part', 'infinity', 'space-time',. If we are restricted to the use of forms of representation unfitted for the expression of asymmetrical relations, ordinal, serial, functional, and structural problems could not be dealt with adequately. We should also have many insoluble semantic puzzles in connection with 'space', 'time', 'cause and effect', and many other relations in the world around us, and ourselves.

A very interesting structural and semantic fact should be noticed: that in symmetrical relations *order* is immaterial, in non-symmetrical relations it is important, and in asymmetrical relations *order* plays an all-important role and cannot be reversed. Order itself is expressed in terms of asymmetrical relations; as, for instance, 'before' or 'after', which apply to 'space', to 'time', 'space-time', 'structure'., and also to *all processes* and activities, the activities of the nervous system included. The asymmetrical relations 'greater', 'father'., imply ordering, while the 'unequal' (having different 'properties') or a 'relative'., do *not* imply ordering. If we consider subject-predicate forms as expressing a relation between the 'observer' and the 'observed', excluding humans, this last relation is also asymmetrical. Applying correct symbolism: if a leaf appears green to me, I certainly do not 'appear green' to the leaf! The last remark suggests that any *A* revision of the *A*-system is structurally impossible. To attempt a revision, we must begin with the formulation of a $\bar{A}$-system of different structure.

The above simple considerations have very far-reaching consequences, as without relations, and particularly without asymmetrical relations, we cannot have *order,* and without order, in the analysis of processes, we are bound to introduce explicity or implicitly some objectively meaningless 'infinite velocities' of the propagation of the process. Thus, the 'infinite velocity' of light, which is known to be false to facts, is at the very foundation of the N-system. The equally false to facts silent assumption of the 'infinite velocity' of nerve currents underlies A animalistic 'psychology' and results in elementalism. This el 'psychology', until this day, vitiates all human concerns and even all science, the newer quantum theories not excluded.

General non-elementalism and, in particular, its restricted aspect, the 'organism-as-a-whole', implies the relation of the 'parts' to the 'whole', for which we need asymmetrical relations. In the statement '$more$ than an algebraic sum', 'more' is also an asymmetrical relation. When we analysed the statement, 'Smith kicks Brown', we saw that the problems of 'space', 'time', 'infinity'., entered, the solution of which requires $serial$ notions, which evade analysis without asymmetrical relations.

The solution of the problems of 'space' and 'time' are fundamental for a theory of sanity, as they are potent structural factors in all $s.r.$ In the majority of 'mentally' ill, we find a disorientation as to 'space' and 'time'. Similar milder forms of disorientation appear in all forms of semantic disturbances, as they are disturbances of evaluation and meanings in the form of delusional 'absolute space' and 'absolute time'. These semantic disturbances can be eliminated only by considerations of multi-dimensional order, which are impossible without asymmetrical relations, and so could not have been accomplished in an A-system.

The problems of multi-dimensional order and asymmetrical relations are strictly interdependent and are the foundation of structure and so of human 'knowledge'; and they underlie the problems of human adjustment and sanity. Without going into details, I shall suggest some relational and ordinal aspects as found in the structure and function of the human nervous system and their bearing on semantic reactions and sanity. I shall also apply these considerations to the analysis of a historically very important delusional factor which has influenced, until now, the $s.r$ of mankind away from sanity. I am dealing only with selected topics, important for my purpose, which, to the reader, may appear one-sided and unduly isolated. In fact, all issues involved are strictly interconnected in a circular way, and no verbal analysis of objective levels can ever be 'complete' or 'exhaustive', and this should be remembered. On the A silent assumption of the infinite velocity of nervous impulses, that the nervous impulses spread 'instantaneously', 'in no time' (to use an Alice-in-Wonderland expression), order was of no importance. But when we take into account the $finite$ and known velocity of nervous impulses, and the *serial,* chain structure of the nervous system, order becomes paramount. In such a serial structure, the problems of resistance, 'inhibitions', blockage, activation., become intelligible, so that some sane orientation is possible in this maze. It may be added that the intensity and the transformation of nervous impulses must somehow be connected with the paths they travel and are, therefore, problems to be spoken about in terms of order.

What has just been said may be illustrated by a rough and over-simplified hypothetical diagram. Fig. 1 shows how the normal (survival in man) impulse should travel. It should pass the thalamus, pass the sub-cortical layers, reach the cortex, and return. That the impulse is *altered* in passing this complicated chain is indicated in the diagram by the arbitrarily diminishing thickness of the line of the impulse.

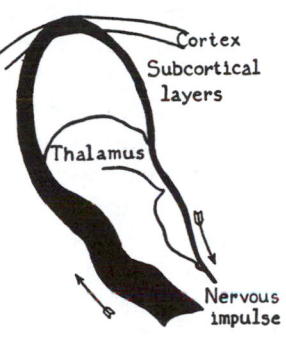

FIG. 1

Fig. 2 illustrates an hypothetical abnormal (non-survival in man) impulse. It emerges from the lower centres. For some reason or other, the main impulse is blocked semantically, or otherwise, and does not reach the cortex; only a weak impulse does. What should be expected in such a case? We should expect regression to the level of activities of organisms which have no cortex, or a cortex very little developed. But this could not be entirely true, as organisms without a cortex have a nervous system adequate for their lives, activities., in their environment, with survival values. But a higher organism with a cortex, no matter how rudimentary, has the other parts of the nervous structure quite different in function, and without the cortex they are *inadequate* for survival, as experience shows. We see that the *order* in which the impulses pass, or are deviated from their survival path, is paramount. A great many different reasons may produce such deviation, too many to

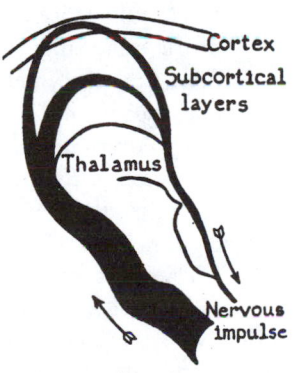

FIG. 2

list conveniently. A great many of them are known, in spite of the fact that, in general, we know very little about nerve mechanisms. Suffice to say, that we know, on colloidal grounds and from experience, that macroscopic or microscopic lesions, drugs, and *false doctrines* affecting

13

the sub-microscopic levels, may often produce similar end-results. Here I use the term 'false doctrines' in the *non-el* sense, and, therefore, take into account affective and *evaluation*-components, which are usually disregarded when we speak about 'false doctrines'.

Here we must consider a problem of crucial, general human significance. It seems evident that *evaluation* in life, and particularly in human life, represents a most fundamental psycho-logical process underlying motivation and, in general, *s.r*, which shape our behaviour and result in collective structures which we may call 'stages of civilization'.

We may distinguish three periods of human development as characterized by their standards of evaluation:

1) The pre-human and primitive period of literal, general, and unrestricted identification. The semantics of this period could be formulated roughly as 'everything is everything else', which might be called one-valued semantics.

2) The infantile, or A period of partial or restricted identification, allowing symmetrical relations, to the exclusion of asymmetrical relations. Its semantics involve, among others, the 'law of identity'—'everything is identical with itself', its two-valued character being expressed by the postulate 'A is B or not B'.

3) The adult, or $\bar{A}$, or scientific period based on the complete elimination of identification, by means of asymmetrical and other relations, which establishes *structure* as the foundation of all 'knowledge'. Its semantics follow the ∞-valued semantics of probability and recognize 'equality', 'equivalence'., but no 'identity'.

Before analysing the above three periods separately, it must be stated that 'identity', defined as 'absolute sameness', necessitates 'absolute sameness' in '*all*' aspects, never to be found in this world, nor in our heads. Anything with which we deal on the objective levels represents a process, different all the 'time', no matter how slow or fast the process might be; therefore, a principle or a premise that 'everything is identical with itself' is *invariably false to facts*. From a structural point of view, it represents a foundation for a linguistic system nonsimilar in structure to the world or ourselves. All world pictures, speculations and *s.r* based on such premises must build for us delusional worlds, and an optimum adjustment to an *actual* world, so fundamentally different from our fancies, must, in principle, be impossible.

If we take even a symbolic expression $1 = 1$, 'absolute sameness' in '*all*' aspects is equally impossible, although we may use in this connection terms such as 'equal', 'equivalent',. 'Absolute sameness in all aspects' would necessitate an *identity* of different nervous systems which produce and use these symbols, an *identity* of the different states of the nervous system of the person who wrote the above two symbols, an identity of the surfaces ., of different parts of the paper, in the distribution of ink, and what not. To demand such impossible conditions is, of course, absurd, but it is equally absurd and very harmful for sanity and civilization to preserve until this day such delusional formulations as *standards of evaluation,* and then spend a lifetime of suffering and toil to evade the consequences. This may be comparable to the spending of many years in teaching and training our children that one and one *never* equal two, that twice two *never* equal four ., and then they would have to spend a lifetime full of surprises and disappointments, if not tragedies, to learn, when they are about to die, that the above statements are always correct in mathematics and very often true in daily life, and finally acquire the sadly belated wisdom that they were taught false doctrines and trained in delusional *s.r* from the beginning.

If we revised these false doctrines, we would not twist the lives of younger generations to begin with. It seems that, for the sake of sanity, the term 'identity', symbolizing such a fundamental false structural doctrine, should be entirely eliminated from the vocabulary, but the term 'identification' should be retained in psychiatry as a label for extremely wide-spread delusional states which, at present, in a mild form, affect the majority of us.

If we investigate the standards of evaluation of animals, the experiments of Pavlov and his followers show that, after establishing a 'conditional reflex' (which means a physiological relating of a signal with food, for instance), the *physiological* effect of the signal on the nervous system of the animal is to produce secretions similar in quantity and quality to those the food produces. We can thus say that, from a physiological point of view, the animal organism *identifies* the signal with food. That represents the *animal standard of evaluation* at that given period. But even the animal nervous system is flexible enough to learn by experience that identification has no survival value, for, if, after the signal, food is repeatedly not forthcoming, he identifies again the signal with the absence of food. In more complex experiments, when both these identifications are interplayed, the result is a real physiological dilemma, culminating, usually, in a more or less profound nervous disturbance, corresponding to 'mental' ills in humans.

Identification represents a comparatively unflexible, rigid form of adaptation, of low degree conditionality, so to say, and, by neurological necessity, represents the processes of *animal* adaptation, inadequate for modern man. On human levels, it is found best exemplified in primitive peoples and in cases of 'mentally' ill. In less severe cases of semantic disturbances, we also find identification of different degrees of intensity. The milder cases are usually considered as 'normal', which, in principle, is very harmful, because it establishes an animalistic, or primitive, standard of evaluation for 'normal'. 'Identity', as we have seen, is invariably false to facts; and so identification produces, and must produce, non-survival $s.r$, and, therefore, must be considered *pathological* for modern man.

That identification afflicts the majority of us today is also shown by experiments with conditional reflexes and the psychogalvanic experiments which show clearly that the majority of humans *identify* the symbol with actualities, and *secretions very often follow*. In other words, the reactions are of such a low order of conditionality as we find in animals and in primitive men. In principle, it makes no difference whether a sound (or word), or other signal (symbol) is identified with food or other actualities which are not symbols, and the secretions are produced by the adrenal glands, for instance, resulting in fear or anger, instead of by the salivary or sweat glands. In all such cases, *in experiments* with humans, the evaluation is false to facts, and the *physiological secretion* is uncalled for if the *evaluation would be appropriate* to the situation. In very few instances, the human experiments with conditional and psychogalvanic reflexes break down, in the sense that the signal-symbol is *not identified* with first order actualities, and so such an organism has no uncontrolled glandular secretions for signal-symbols *alone*. In a $\bar{A}$-system of evaluation, which involves on semantic levels the consciousness of abstracting, these exceptional persons (1933), with proper evaluation and controlled reactions, *prove the rule* for modern man. In other words, modern man, when he stops the pre-human and primitive identification, will have a much-increased and *conscious control* of his secretions, colloidal states of his nervous system., and so of his reactions and behaviour. The above applies to all $s.r$, 'logical' processes included.

Identification is found in all known forms of 'mental' ills. A symbol, in any form, or any $s.r$ may be identified in value with some fictitious 'reality' at a given date, resulting in macro-physiological (glandular, for instance) or micro-physiological (colloidal.,) activities or disturbances which result in particular semantic states and behaviour. It is impossible to deny that 'mentally' ill have inappropriate standards of evaluation, and that identification appears always as an important factor in pathological evaluations. Experiments with 'mentally' ill show clearly that this evaluation can be altered or improved by different chemical agencies which affect the colloids of the nervous system, by environmental changes . , and by *changing the standards of evaluation* which, at present, is usually called 'psychotherapy'. The analysis of the mechanism of evaluation leads, naturally, to a generalized and *simplified* method, which may have not only a therapeutic but also an important new *preventive value*.

Literal identification is found in all primitive peoples and accounts for their semantic states, reactions, their metaphysics, low development . , but it is impossible, for lack of space, to go into details here.

The A standard of evaluation departed from literal identification to some extent. We still preserve in our school books as the most fundamental 'law of thought'—the 'law of identity'—often expressed in the form 'everything is identical with itself', which, as we have seen, is invariably false to facts. We do not realize that, in a human world, we are dealing at most only with 'equality', 'equivalence'. , at a given place and date, or by definition, but never with 'identity', or 'absolute sameness', disregarding entirely space-time relations, involving 'all' the indefinitely many aspects which, through human ingenuity, we often manufacture at will. In an actual world of four-dimensional processes and the indefinitely many 'aspects' manufactured by ourselves, adjustment in principle is impossible, or, at best, only accidental, if we retain 'identity'. The A evaluation was based on symmetrical relations of 'identity' and also partial 'identity', expressed even in our political, economic. , doctrines and corresponding behaviour, the analysis of which would require a special volume to be written, I hope soon, by some one.

Under the pre-human and primitive standards of evaluation, science was not possible. Under the A standards the beginnings of science became possible, but if science had not departed from those standards, we would have had no modern science. Lately, when the persecution of science has increasingly relaxed (not in all countries in a similar degree) and scientists were allowed to develop their disciplines with much less fear of persecution, sometimes even encouraged and helped by public interest, scientists found that they invariably had to build their own vocabularies of a distinctly, although unrealized, $\bar{A}$ character. The chasm between human affairs and science became wider and wider. The reason for it was that, in life, even at present, we preserve A standards of evaluation, and science mainly depends on subtler $\bar{A}$ means involving asymmetrical relations which alone can give us *structure*. I will return repeatedly, later on, to the $\bar{A}$ re-evaluation of the A standards of values.

The $\bar{A}$ evaluation is based on asymmetrical and other relations. I shall not attempt to summarize it here because the problems are very large and this whole volume is devoted to that subject. Here I shall mention, once more, that only with $\bar{A}$ standards of evaluation does a scientific treatment of man and his affairs become possible. A $\bar{A}$-system depends on a complete elimination of identification which affects beneficially all our *s.r*, as experience and experiments show.

It has been already emphasized that in the human child the nervous system is not physically finished at birth, and that for some years thereafter it is plastic. Hence, the 'environment'—which includes languages, doctrines, with their structure, all connected with evaluation-components —conditions the future functioning of the system. The way in which the nervous system works, the 'sanity', 'un-sanity' and 'insanity' of the individual depends to a large extent on how this plastic and sensitive apparatus is treated, particularly in childhood. Because of the *serial structure* of the nervous system, the language and doctrines supplied should be of the structure necessary for the adequate representation of serial structures and functions. With the old A means this could not be accomplished.

At this point, it will be well to introduce an important semantic subject, to which we shall return later; namely, the connection between the primitive subject-predicate language and identification. For example, the statement, 'the leaf is green', is taken to imply 'greenness', which, by its verbal structure, has the character of a 'substantive' and implies some sort of objective independence. It is *not* considered as an *asymmetrical relation* between the observer and the observed and, accordingly, tends toward an *additive* implication. 'Greenness' is thus objectified and *added* to the leaf in describing a 'green leaf'. The objectified 'greenness' leads to an anthropomorphic mythology, which, in turn, involves and develops the undifferentiated projecting mechanism so fundamental in semantic disturbances. The objectification is evaluated structurally as a 'real' situation, and this introduces the non-survival *reversed* order evaluation in which the use of the 'is' of identity, resulting in identification, is the main factor. The stronger the structural 'belief' in the 'truth' of the representation, or, in other words, the more we identify the higher order abstractions with the lower, which, in fact, are different, the more dangerous becomes the 'emotional' tension in the form of unjustified *evaluation,* which, ultimately, must involve *delusional* factors, no matter how slight, and result in semantic disturbances. *Ignorance,* involving strong faith in the erroneous structural belief, is dangerously akin to more developed symptoms of 'mental' illness called illusions, delusions, and hallucinations. We are mostly semantic victims of the primitive doctrines which underlie the A structure of our language, and so we populate the world around us with semantic phantoms which add to our fears and worries, or which lead to abnormal cheerfulness, well known among some 'mentally' ill.

It should be realized that in the A system of evaluation many individuals profit in various ways by what amounts to distracting the attention of mankind from actual life problems, which make us forget or disregard actualities. They often supply us with phantom semantic structures, while they devote their attention to the control of actualities not seldom for their personal benefit. If one surveys the A situation impartially, one occasionally feels hopeless. But, no matter how we now conspire one against another, and thus, in the long run, against ourselves, the plain realization that the difficulty is found in the standards of evaluation, establishes the necessary preliminary step to the escape.

It is a well-known fact that, in a large proportion of 'mental' ills, we find a semantic flight from 'reality' ($m.o$) when their 'reality' becomes too hard to endure. It is not difficult to see that different mythologies, cults ., often supply such structural semantic 'flights from reality'; and that those who actually help, or who are professionally or otherwise engaged in producing and promulgating such semantic flights, help mankind to be un-sane, to deal with phantoms, to create dream states ,. There is no longer any excuse in the old animalistic law of supply and demand —that, because there is a demand for such flights, they should be supplied. That argument is not held to apply to those who peddle drugs or wood alcohol. The flights from reality always have the earmarks of 'mental' illness. Very often such actively engaged individuals are themselves ill to the point of hallucinations; they often 'hear voices', 'see visions', 'speak tongues',. Very often other morbid symptoms occur which are similar to those shown by the 'mentally' ill of the usual hospital types. It is not generally realized that, although the patient suffers intensely, he usually shows marked resistance to any attempt to relieve him of his semantic affliction. Only *after* he is relieved by semantic re-education does the patient realize how very *unhappy he was*.

The situation is very serious. There is a powerful well-organized system, with enormous wealth behind it, based on A and pre-aristotelian standards of evaluation which keeps mankind in delusional semantic states. Its members do their best, better than they know, to keep mankind *un-sane* in flights from 'reality', instead of helping to revise the A standards of evaluation and to reorganize the horrible 'realities', *all of our own making,* into realities less painful. The comparatively few psychiatrists are naturally not a match for such vast numbers of well-organized men and women who, in their blissful ignorance, work in the opposite direction; and all of us pay the price.

The activities of these individuals often promulgate something similar to the well-known 'induced insanity'. Quite often paranoid or paranoiac and, more rarely, hypomanic patients can influence their immediate companions to such an extent that they join in believing in their delusions and copy their s.r. Susceptible associates begin to develop similar delusions and hallucinations and to pass through episodes themselves, perfectly oblivious to contradictions with external m.o reality. There are many paranoiac-like semantic epidemics of this kind on record. It is instructive to visit some 'meetings' and watch the performer and the audience. The pathetic side of it is that these performers, themselves not realizing the harmfulness of the situation, often pretend, or genuinely believe, that they are helping mankind by preaching some metaphysical 'morals'. What they *actually* produce is a disorganization of the survival-working of the human nervous system, particularly if they train the structurally undeveloped nervous system of children to delusional evaluations and s.r, and, in general, make sanity and higher and effective ethical standards very difficult or impossible. It is positively known that s.r are inextricably connected with electrical currents, secretions of different glands., which, in turn, exert a powerful influence on colloidal structure and behaviour, and so condition our neurological and physiological development. There can be no doubt that imposing delusional s.r on the undeveloped child must result in at least colloidal injury, which later on facilitates arrested development or regression, and, in general, leads away from adjustment and sanity.

Lack of space and the essentially constructive aims of the present system do not allow me to analyse many fundamental interrelations in the development of man, but a brief list, worthy of analysis, may be suggested:

1) The relation between the pre-human reactions and the reactions of the primitive man, involving always some *copying* by mutants of the responses of the prevailing simpler organisms.

2) The interrelation between the reactions of the primitive man, his animism, anthropomorphism, his other s.r and the *structure* of his language and semantics.

3) The relation between the structure of primitive languages and the structure of the 'philosophical grammar' formulated by Aristotle, generally called 'logic'.

4) The relation between this grammar, the structure of language, and the further development of our structural metaphysics and s.r.

5) The influence the last conditions exerted on the structure of our institutions, doctrines, and the *s.r* related to them.

6) The relation between the 'copying animals in our nervous processes' and semantic blockages., preventing an adult civilization, agreement, sanity, and other desirable human reactions.

This brief list suggests an enormous field for further research, but, even now, the formulation of a $\bar{A}$-system of evaluations makes a few points clearer.

An infant, be it primitive or modern, begins life with *s.r* of identity and confusion of orders of abstractions, natural to his age, yet false in principle, and structurally false to fact. At present, parents and teachers seldom check or counteract this tendency, mostly not realizing the importance of this semantic factor and its role in the future adjustment of the individual. In the rough, to a baby, his cry 'is' food. Words 'are' magic. This identification is structurally false to facts, but in *babyhood it mostly works*. To the infant, experience proves that the noises he makes, a cry or a word, have the objective value,—food. The semantic identity of the symbol and the un-speakable object level,—food,—has been established. This infantile attitude or *s.r* is carried on into grown-up life.

Under very simple conditions of primitive peoples, in spite of many difficulties, this attitude of identification is not always checked by experience, and experimenting is non-existent at this stage. If it is, then such checking of identification is 'explained' by some sort of demonology and 'good' or 'evil' 'spirits',. Delusional, from the modern point of view, *s.r* are compensated by mythologies, making the two sides of the semantic equation equivalent. This equating tendency is *inherent* in all human *s.r*. It expresses the instinctive 'feel' for the similarity of structure as the base of 'knowledge', and it ultimately finds its expression in mathematical equations. In all psycho-logical processes of 'understanding', we must have some standards of evaluation and 'equivalence'. On primitive levels, this is accomplished by literal identification and delusional mythology of the type, that a storm at sea is 'caused' by a violent quarrel between a 'god' and his 'wife'; or, in contemporaneous mythology, a draught, or fire, or death by lightning, is explained as 'punishment' for 'sins',. Semantic compensation is needed and produced. A similar semantic process produces scientific theories, but with different standards of evaluation. At present, scientific theories do not cover all semantic needs and urges of mankind, owing to the prevailing false to fact identification of different orders of abstractions. With the full consciousness of abstracting, which means proper evaluation or differentiation between orders of abstractions, science will then cover all our non-pathological semantic needs, and different primitive mythologies will become unnecessary. A very harmful, primitive, delusional semantic factor of blockages would be eliminated.

The 'is' of identity plays a great havoc with our $s.r$, as any 'identity' is structurally false to fact. An infant does not know and cannot know that. In his life, the 'is' of identity plays an important semantic role, which, if not checked intelligently, becomes a pernicious semantic factor in his grown-up reactions, which preserve the infantile character and with which *adult* adjustment and semantic health is impossible. The infant begins to speak and again he is trained in the 'is' of identity. Symbols are identified with the un-speakable actions, events and objects under penalty of pain or even death. The magic of words begins its full sway. As a rule, parental, crude disciplining of the infant, particularly in former days, trained the $s.r$ of the infant again in the delusional 'is' of identity. The results are semantically and structurally very far-reaching and are found to underlie modern mythologies, militarism, the prevailing economic and social systems, the control by fear (be it 'hell' or machine guns), illusory gold standards, hunger ,.

Experience shows that such identification of symbols with the un-speakable levels works very well with animals. With man, it leads only to the misuse of the human nervous system, semantic disturbances of evaluation, and the prevailing unstable animalistic systems in practically all fields, resulting in the general chaos in human affairs.

It should be noticed that the 'is' of predication also expresses a sort of *partial identity,* leading to primitive anthropomorphism and general confusion of orders of abstractions. By an inherent necessity, our lives are lived on the un-speakable objective levels, which include not only ordinary objects but also actions and immediate feelings, symbols being only auxiliary means. The natural ordinal evaluation, which should be the foundation for healthy $s.r,$ appears as the event-process level first, the object next in importance; the objective level first, the symbolic next in importance; the descriptive level first, the inferential level next in importance ,. The semantic *identification* of these different levels not only abolishes the natural evaluation, but, in fact, reverses the natural order. Once this is realized, we see clearly that all statements about the objective level, which is made up of absolute individuals, are only *probable* in different degrees and can never be certain. The 'is' of identity underlies, also, the two-valued, too primitive, too restricted, and structurally fallacious A 'logic'.

The crucial semantic importance of asymmetrical relations becomes obvious when we consider that all *evaluation* and *non-el* meanings depend ultimately on asymmetrical relations. In the technical fields, mathematics and the exact sciences; in the semi-scientific fields, economics, politics, sociology.; in the as yet non-scientific fields, 'ethics', 'happiness', 'adjustment'., represent ultimately different forms of *evaluation,* impossible to formulate adequately under aristotelianism.

Obviously, a $\bar{A}$-system based on proper semantic evaluation leading to non-pathological reactions, adjustments., must make relations and multi-dimensional order fundamental for sanity. The semantic connection between mathematical methods and all the other concerns of man becomes also necessary and obvious.

In mathematics, recently, the notion of equality needed a refinement and the notion of 'identity' has been introduced. The present analysis discloses that, although the refinement and the symbol may be retained, yet the name should be entirely abandoned, because it conceals a very semantically vicious confusion of orders of abstractions. If, by definition, we produce new terms, these new terms are of a higher order abstraction than the terms used in the definition, and so the *identification* of them as to the orders of abstractions is *physiologically* and structurally false to facts.

The problems discussed in the present chapter have been felt vaguely for more than two thousand years and found their first historical expression in the rift between Aristotle, the biologist, and Plato, the founder of mathematical philosophy. Mathematics is, in principle, $\bar{A}$, and so, in the study of mathematics, we can learn most about the principles of non-aristotelianism. In physics, only very recently, do we begin to eliminate the 'is' of identity and elementalism which resulted in the $\bar{N}$ systems. All sciences strive to become more mathematical and exact and so $\bar{A}$. In fact, all advances in science are due to the building of new $\bar{A}$ languages, usually called 'terminology'. We can go further and say, definitely, that, to have any science, we must make a $\bar{A}$ revision of the languages used. Similarly with 'man', either we decide to introduce into human affairs scientific evaluation, and so part company with the A and pre-aristotelian system of evaluation, or preserve A structure, and have no science of man, or science of sanity, but continue in the prevailing chaos.

CHAPTER XIV

ON THE NOTION OF INFINITY

The questions on which there is disagreement are not trivialities; they are the very roots of the whole vast tree of modern mathematics. (22)

E. T. BELL

The task of cleaning up mathematics and salvaging whatever can be saved from the wreckage of the past twenty years will probably be enough to occupy one generation. (22)

E. T. BELL

The intention of the Hilbert proof theory is to atone by an act performed once for all for the continual titanic offences which mathematics and all mathematicians have committed and will still commit against mind, against the principle of evidence; and this act consists of gaining the insight that mathematics, if it is not true, is at least consistent. Mathematics, as we saw, abounds in propositions that are not really significant judgments. (549)

HERMANN WEYL

An objectivated property is usually called a *set* in mathematics. (549)

HERMANN WEYL

If the objects are *indefinite* in number, that is to say if one is constantly exposed to seeing new and unforseen objects arise, it may happen that the appearance of a new object may require the classification to be modified, and thus it is we are exposed to antinomies. *There is no actual (given complete) infinity.* (417)

H. POINCARÉ

The structural notion of 'infinite', 'infinity', is of great semantic importance and lately has again become a subject of heated mathematical debates. My examination of this subject is from the point of view of a A-system, general semantic, and a theory of sanity which completely eliminates identification. In Supplement III, I give a more detailed A analysis of the problem already anticipated by Brouwer, Weyl, Chwistek, and others. These problems are not yet solved, because mathematicians, in their orientations and arguments, still use el, A 'logic', 'psychology', and epistemology, which involve and depend on the 'is' of identity, making agreement impossible.

Mathematical infinity was first put on record by the Roman poet, Titus Lucretius, who, as far back as the first century B.C., wrote very beautifully about it in his *De Rerum Natura*.[1] As the author was a poet, and his work poetry, a few privileged literati had great pleasure in reading it; but this discovery, not being rigorously formulated, remained inoperative, and so practically worthless for mankind at large, for 2000 years. Only about fifty years ago, mathematical infinity was rediscovered by mathematicians, who formulated it rigorously, without poetry. Since then, mathematics has progressed with all other sciences in an unprecedented way. That this structural linguistic discovery was made so late is probably due to the usual blockage, the old *s.r*, old habits of 'thought', and prejudices.

In all arguments about infinity, from remote antiquity until Bolzano (1781-1848), Dedekind (1831-1916), and Cantor (1845-1918), there was a peculiar maxim involved. All arguments against infinity involved a certain structural assumption, which, at first inspection, seemed to be true and 'self-evident', and yet, if carried through, would be quite destructive to all mathematics existing at that date. Arguments favorable to infinity did not involve these tragic consequences. Quite naturally, mathematicians, and particularly Cantor, began to investigate this peculiar maxim and the *s.r* which were playing havoc. The structural assumption in question is that 'if a collection is part of another, the one which is a part must have fewer terms than the one of which it is a part'. This *s.r* was deeply rooted, and even found a scholarly formulation in Euclid's wording in one of his axioms: 'The whole is greater than any of its parts'. This axiom, although it is not an exact equivalent of the maxim stated above, by loose reasoning, which was usual in the older days, could be said to imply the troublesome maxim. It is not difficult to see that the E axiom, as well as our troublesome maxim, expresses a structural generalization taken from experience which applies only to *finite* processes, arrays,. Indeed, both can be taken as a definition of finite processes, arrays,. It does *not* follow, however, that the one definition and structure must be true of infinite processes, arrays,. As a matter of fact, the break-down of this maxim gives us the precise definition of mathematical infinity. A process of generating arrays., is called infinite when it contains, as parts, other processes, arrays., which have 'as many' terms as the first process, array,.

The term 'infinite' means a process which does not end or stop, and it is usually symbolized by ∞. The term may be applied, also, to an array of terms or other entities, the production of which does not end or stop. Thus we may speak of the infinite process of generating numbers because every positive integer, no matter how great, has a successor; we can also speak of infinite divisibility because the numerical technique gives us means to accomplish that. The term 'infinite' is used here as an adjective describing the characteristics of a *process,* but should never be used as a noun, as this leads to self-contradictions. The term 'infinity', as a noun, is used here only as an abbreviation for the phrase 'infinite process of generating numbers',. If used in any other way than as an abbreviation for the full phrase, the term is meaningless in science (not in psychopathology) and should never be used. The above semantic restrictions are not arbitrary or purely etymological, but they follow the rejection of the 'is' of identity of a $\bar{A}$-system.

Before we can apply the term 'infinite' to physical processes, we must first theoretically elucidate this term to the utmost, and only then find out by experiment whether or not we can discover physical processes to which such a term can be applied. Fortunately, we have at our disposal a *semantic process* of generating numbers which, by common experience, by definition, and by the numerical technique, is such that every number has a successor. Similarly, our semantic processes are capable by common experience, by definition, and by the numerical technique to divide a finite whole indefinitely. Thus, if we do not identify external physical objective processes with internal semantic processes, but differentiate between them and apply correct symbolism, we can see our way clear. If we *stop* this semantic process of generating numbers at any stage, then we deal with a finite number, no matter how great; yet the process remains, by common experience, by definition, and by the numerical technique, such that it can proceed indefinitely. In the $\bar{A}$ sense, 'infinite', as applied to processes, means as much as 'indefinite'. We should notice that the semantic *process* of *generating* numbers should not be identified with a *selection* of *a* definite number, which, by necessity, is finite, no matter how great. The identification of the semantic process of generating numbers with a definite number; the identification of the semantic process of infinite divisibility of finites in the direction of the small with the generating of numbers in the direction of the great; and the identification of semantic internal processes with external physical processes., are found at the foundation of the whole present mathematical scandal, which divides the mathematical world into two hostile camps.

The process of infinite divisibility is closely connected with the process of the infinite generation of numbers. Thus we may have an *array* of *numbers* 1, 2, 3, . . . *n*, all of which are finite. The *semantic process* of *passing* from n to $n + 1$ *is not* a number, but constitutes a characteristic of the semantic process. The *result* of the semantic process; namely, $n + 1$, again becomes a finite number. If we take a fraction, a/n, the greater an n is selected, the smaller the fraction becomes, but with each selection the fraction again is finite, no matter how small.

Although the two processes are closely connected on the formal side, they are very different from the semantic point of view. The process of generating numbers may be carried on indefinitely or 'infinitely' and has no upper limit, and we cannot assign such a limit without becoming tangled up in self-contradiction in terms. Not so with the process of indefinite or infinite divisibility. In this case, we start with a *finite*.

Existing mathematical symbolism and formalism lead to identification of both fundamentally different semantic processes and introduce a great deal of avoidable confusion. A $\bar{A}$ orientation will allow us to retain mathematical symbolism and formalism, but will not allow the identification of the semantic process of *passing* from number to number, which passing *is not* a number, with the *result* of this process which, in each case, becomes a *definite and finite number*.

It becomes obvious that the A terminology and present standard notions of 'number' identify the semantic *process* with its *result,* an identification which must ultimately be disastrous. The semantic process is thus potentially infinite, but the *passing* from n to $n + 1$ characterizes the semantic process, not number; number*s* representing only finite results of the indefinitely extended semantic process.

A $\bar{A}$ analysis without identification discloses, then, that only the semantic process can be indefinitely extended, but that the *results* of this process, or a number in *each case,* must be *finite.* To speak about an 'infinite' or, as it is called, 'transfinite' 'number', is to identify entirely different issues, and involves very definite self-contradictions in *m.o* terms. The existing mathematical terminology has been developed without the realization of $\bar{A}$ issues and the multiordinality of terms and leads automatically to such identifications. As long as mathematicians do not consider $\bar{A}$ issues, the problems of mathematical infinity will remain unsolved and hopeless; and yet, without a scientific theory of infinity, all of mathematics and most of science would be entirely impossible. A $\bar{A}$ clarification of these problems involves a new semantic definition of number*s* and mathematics, given in Chapter XVIII, which eliminates a great many mysteries in connection with mathematics and does not allow these dangerous and befogging identifications.

From a $\bar{A}$ point of view, we must treat infinity in the first cantorian sense; namely, as a *variable finite,* the term *variable* pertaining to the semantic process but *not* to number, the term *finite* pertaining to both the semantic *arrest* of the infinite semantic process, and so characterizing also its result; namely—a number.

In the meantime, the numerical technique is *indefinitely flexible* in the sense that no matter how great a number we take, we always can, by a semantic process, produce a greater number, and no matter how small the difference between two numbers might be, we always can find a third number which will be greater than the smaller, and smaller than the given greater number. Thus, we see that the *numerical technique* is such as to correspond in flexibility exactly to the *semantic processes,* but there is nothing flexible about a definite number once it is selected. What has been already said about a variable applies, also, to a number; namely, that a 'variable' does not 'vary' in the ordinary sense; but this term applies only to the semantic processes of the mathematician. The older intensional A definition of 'number' must have led to the older identifications. The $\bar{A}$, extensional, and *non-el* semantic definition of number*s* does not allow such identifications. The A term 'number' applied to a definite number, but also to an intensional definition of number*s*. The $\bar{A}$, or semantic definition of number*s*, is different in the sense that it finds extensional characteristics of each number, applicable to all number*s*, and so helps not to identify a definite number with the process of generating number*s*, which the use of one term for two entirely different entities must involve.

Cantorian *alephs,* then, are the result of identification or confusion of entirely different issues and must be completely eliminated. The rejection of *alephs* will require a fundamental revision of those branches of mathematics and physics which utilize them; yet, as far as I know, with a very few exceptions, the *alephs* are not utilized or needed, although the *'name'* is used, which spell-mark has become fashionable in many mathematical and physical circles. In the case of *alephs,* history may repeat itself and the *alephs,* like the 'infinitesimal', when their self-contradictory character becomes understood, will be eliminated without affecting the great body of mathematics, but only the small portions which are built on the *alephs.*

As to the existence of infinite processes, we know positively *only* about the *semantic process* of generating number*s* and the *semantic process* of infinite divisibility. These processes are evident in our common experience. We cannot *a priori* know if such infinite processes can be found in the world which must be discovered by investigation and experimentation.

The existing terminology is still A and is based on, and leads to, identification, and so in my $\bar{A}$ presentation I cannot use it and expect to clear up some of the issues involved. The terms such as 'class', 'aggregate', 'set'., imply a definite static collection. The term 'infinite', in the meantime, can only be correctly and significantly used as applied to a dynamic semantic process. We cannot speak of 'infinite' classes, aggregates, sets., and evade the issues of identification of entirely different entities. The term 'series' has a technical meaning in connection with numbers and so, for a general discussion of processes, is a little too specific. The term 'array' is more general, yet extensional, of which 'series' would be a special case. The general term 'number' is multiordinal and intensional and so, in the $\bar{A}$ extensional system, ∞-valued, and must be used in the plural; namely, 'numbers'. The term 'number' in the singular will be used to indicate a definite number. The term 'denumerable' has been introduced by Cantor and means any extensional array of terms, facts, states, observables., which can be put in one-to-one correspondence with the infinite array of positive integers.

Let me repeat once more: the semantic process may be carried on without limits, and the infinite series of positive integers is an extensional, technical, and verbal expression of this semantic process and the only infinite array of which existence we are certain.

We shall be able to explain, and to give a better definition of, mathematical infinity if we introduce an extremely useful structural term, 'equivalence'. Two processes, arrays., between which it is possible to set up, by some law of transformation, a *one-to-one* correspondence are said to be *equivalent*. A process, array., which is equivalent to a part of itself, is said to be infinite. In other words, a process, array., which can be put into a *one-to-one* correspondence with a part of itself is said to be infinite. We can define a *finite* process, array., (class, aggregate.,) as one which is not infinite. The following is valid *exclusively* because of the use of the '**e t c .**'

A few examples will make this definition clearer. If we take the series of positive integers, 1, 2, 3, 4, . . . **e t c .**, we can always double every number of this row *provided* we retain the *process-character, but not otherwise*. Let us write the corresponding row of their doubles under the row of positive integers, thus:

$$1, 2, 3, 4, 5, ... \textbf{e t c .}$$
$$2, 4, 6, 8, 10, . . . \textbf{e t c .}$$

or we can treble them, or *n*-ble them, thus:

$$1, 2, 3, 4, 5, ... \textbf{e t c .}$$
$$3, 6, 9, 12, 15, . . \textbf{e t c .}$$

there are obviously as many numbers in each row below as in the row above, *provided we retain the* '**e t c .**', so the numbers of numbers in the two rows compared must be equal. All numbers which appear in each bottom row also occur in the corresponding upper row, although they only represent a *part* of the top row, *again provided that we retain the* '**e t c .**'

The above examples show another characteristic of infinite processes, arrays,. In the first example, we have a *one-to-one* correspondence between the natural numbers and the *even* numbers, which are equal in number at each stage. Yet, the second row results from the first row by

14

taking away all odd number*s*, which, itself, represents infinite number*s* of number*s*.

This example was used by Leibnitz to prove that infinite arrays cannot exist, a conclusion which is not correct, since he did not realize that both finite and infinite arrays depend on definitions. We should be careful not to approach *infinite* processes, arrays., with prejudices, or silent doctrines and assumptions, or, in general *s.r*, taken over from *finite* processes, arrays , .

Thus we see that the process of generating natural number*s* is structurally an infinite process because its *results* can be put in a *one-to-one* correspondence with the results of the process of generating even numbers., which is only a part of itself. Similarly, a line AB has infinitely many points, since its points can be put into a *one-to-one* correspondence with the points on a segment CD of AB. Another
see page xii example can be given in the Tristram Shandy paradox of Russell. Tristram Shandy was writing his autobiography, and was using one year to write the history of one day. The question is, would Shandy ever complete his biography? He would, provided he never died, or he lived infinite number*s* of years. The hundredth day would be written in the hundredth year, the thousandth in the thousandth year, **e t c .** No day of his life would remain unwritten, *again provided his process of living and writing would never stop*.

Such examples could be given endlessly. It is desirable to give one more example which throws some light on the problems of 'probability', 'chance',. The theory of probability originated through consideration of games of chance. Lately it has become an extremely important branch of mathematical knowledge, with fundamental structural application in physics, general semantics, and other branches of science. For instance, Boltzman based the second law of thermodynamics on considerations of probability. Boole's 'laws of thought', and the many-valued 'logic' of Łukasiewicz and Tarski are also closely related to probability; and the new quantum mechanics uses it constantly , .

The term 'probability' may be defined in the rough as follows: If an event can happen in a different ways, and fails to happen in b different ways, and all these ways are equally likely to occur, the probability of the happening of the event is $\dfrac{a}{a+b}$, and the probability of its failing is $\dfrac{b}{a+b}$.

Let us assume that in a certain city a lecture is held each day, and that, though the listeners may change each day, the numbers of listeners are always equal. Suppose that one in each twenty inhabitants of this town has M as the first letter of his name. What is the probability that, *'by chance'*, all the names of the audience would begin with M? Let us call such a happening the M-event. In the simplest case, when the daily ∗ _{see page xii} number of listeners is only one, the probability of an M-event is 1 in 20, or 1/20. The probability of an M-event for an audience of 2 is 1 in $20 \times 20 = 400$, or 1/400. The probability that an audience of three members should have all three names begin with M would decrease twenty times further. Only once in 8000 lectures, on an average, would an M-event happen. For five people it would amount to 1 in $20 \times 20 \times 20 \times 20 \times 20 = 3,200,000$ days, or 1/3,200,000, or once in approximately 9000 *years;* for ten people, about once in thirty billion years; for twenty people, about once in a third of a quadrillion years. For one hundred people, the recurrence period of the M-event would be given as once in a number of years represented by more than a hundred figures. If the town, in this last example, should be as old as the solar system, and if the lectures had been delivered daily to an audience of one hundred people through this inconceivably long period, the probability is extremely small that the M-event would happen at all.[2]

From the human, *anthropomorphic,* point of view, we would say that such an event is impossible. But it must be remembered that this is only an anthropomorphic point of view, and our judgements are coloured by the temporal scale of our own lives. Of course, to carry such an anthropomorphic viewpoint into cosmic speculations is simply silly, a survival of the primitive structure of language and its progeny—metaphysics and mythologies.

The theory of infinity throws considerable structural light on such primitive speculations. In this external world, we deal with processes, and, as we measure 'length' by comparison with freely selected convenient units of 'length', let us say, an inch; or we measure 'volume' by freely selected convenient units of 'volume'; so, also, we compare *processes* with some freely selected and convenient *unit-process.* The diurnal rotation of our earth is such a process, and, if we choose, we can use it as a measuring unit or as a comparison standard. Of late, we have become aware that the rotation of the earth is not quite regular, and so, for accurate measurements, the old accepted unit-process of a day, or its subdivision, a second, is not entirely satisfactory. For scientific purposes, we are trying to find some better unit-process, but we have difficulty, as the problem is naturally circular. When we speak in terms of a 'number of years', or of seconds, we speak about perfectly good observational experimental facts, about quite definite relations, the best we know in

1933. We do not make any metaphysical assertions about 'time' and we should not be surprised to find that statements involving 'years' are generally propositions, but that statements involving 'time' often are not. It is necessary not to forget this to appreciate fully what follows.

The theory of infinity will clear away a troublesome stumbling-block. We will use the expression 'infinite numbers of years', remembering the definition of 'infinite numbers' and what was said about the *unit-process* which we call a year. We have seen in an example above that if only a hundred individuals attend a lecture, and all 'by chance' have their names begin with M, such an event happens, on an average, only once in an inconceivably large number of years, represented by a number with a hundred figures. If we would ask *how many* times an occurrence would happen, we would have to state the period in years for which we ask the *how many*. It is easy to see that in infinite numbers of years, this humanly extremely rare occurrence would happen precisely *infinite numbers of times,* or, in other words, 'just as often', this last statement being from a non-anthropomorphic point of view. An event that appears, from our human, limited, anthropomorphic point of view, as 'rare', or as 'chance', when transposed from the level of finite process, arrays., to that of infinite processes, arrays., is as 'regular', as much a 'law', involving 'order', as anything else. It is the old primitive *s.r* to suppose that man is the only measure of things.

Here the reader might say that infinite numbers of years is a rather large assumption to be accepted so easily. This objection is indeed serious, but a method which can dispose of it is given later on. At this stage, it is sufficient to say that, on the one hand, this problem is connected with the semantic disturbance, called identification (objectification of 'time'), which afflicts the majority of us, excepting a few younger einsteinists; and that, on the other hand, it involves the structurally reformulated law of the 'conservation of energy', 'entropy',.

Before parting with the problem of infinity, let me say a word about the notion of 'continuity', which is fundamental in mathematics. Mathematical continuity is a structural characteristic connected with ordered series. The difficulties originated in the fact that a 'continuous' series must have infinite numbers of terms between any two terms. Accordingly, these difficulties are concerned with infinity. That mathematicians need some kind of continuity is evident from the example of two intersecting lines. If the lines have gaps, as, for instance, — — — — , there would be the possibility that two gaps would coincide, and the two lines not intersect; although in a plane the first line would pass to the other side of the second line. At present, we have two kinds of 'continuity' used in mathematics. One is a supposedly 'high-grade' continuity; the other, supposedly, is a 'low-grade' continuity, which is called 'compactness' or 'density', with the eventual possibility of gaps. I am purposely using rather vague language, since these fundamental notions are now being revised, with the probability that we shall have to be satisfied with 'dense' or 'compact' series and abandon the older, perhaps delusional, 'high-grade' continuity. It is interesting to note that the differential and integral calculus is supposedly based on the 'high-grade' continuity, but the calculus will not be altered·if we accept the 'low-grade' compactness, all of which is a question of an A or $\bar{A}$ orientation.

Vague feelings of 'infinity' have pervaded human $s.r$ as far back as records go. Structurally, this is quite natural because the term infinity expresses primarily a most important semantic process. The majority of our statements can also be reformulated in a language which explicitly involves the term 'infinity'. An example has been already given when we were speaking about the universal propositions which were supposed to be of *permanent* validity, in other *language,* valid for 'infinite numbers of years'. We see how the trick is done—a vague quasi-qualitative expression like 'permanent' or 'universal' is translated into a quantitative language in terms of 'numbers of years'. Such translation of qualitative language into quantitative language is very useful, since it allows us to make more precise and definite the vague, primitive structural assumptions, which present enormous semantic difficulties. This brings to our attention more clearly the structural facts they supposedly state, and aids analysis and revision. In many instances, such translations make obvious the illegitimacy of the assumptions of 'infinite velocities' and so clear away befogging misunderstandings, and beneficially affect our $s.r$.

CHAPTER XV

THE 'INFINITESIMAL' AND 'CAUSE AND EFFECT'

But we are not likely to find science returning to the crude form of causality believed in by Fijians and philosophers, of which the type is "lightning causes thunder." (457)   BERTRAND RUSSELL
The notion of causality has been greatly modified by the substitution of space-time for space and time. . . . Thus geometry and causation becomes inextricably intertwined. (457)   BERTRAND RUSSELL

In classical mechanics, and no less in the special theory of relativity, there is an inherent epistemological defect which was, perhaps for the first time, clearly pointed out by Ernst Mach. . . . No answer can be admitted as epistemologically satisfactory, unless the reason given is an *observable fact of experience*. The law of causality has not the significance of a statement as to the world of experience, except when *observable facts* ultimately appear as causes and effects. (155)   A. EINSTEIN

The chain of cause and effect could be quantitatively verified only if the whole universe were considered as a single system—but then physics has vanished, and only a mathematical scheme remains. The partition of the world into observing and observed system prevents a sharp formulation of the law of cause and effect. (215)   W. HEISENBERG

Of late, another perplexing semantic problem concerning 'causality' or 'non-causality' has arisen in connection with the newer quantum mechanics. It is possible to examine this question by different methods. The simpler one is connected with vague feelings of 'infinity' and its supposed opposite, the 'infinitesimal'; the more fundamental method is based on the orders of abstractions leading toward the ∞-valued semantics of probability.

Because of man's natural tendency to speak in terms of 'infinity', and his further marked tendency of having opposites, such as 'yes', 'no', 'right', 'left', 'positive', 'negative', 'love', hate', 'honesty', 'dishonesty'., quite naturally the notion of 'infinity' carried with it the tendency to invent the 'infinitesimal'. Even mathematicians have had great semantic difficulties in breaking away from this habit. Analysis persistently reveals that structurally no matter how far we go in dividing something, let us say an inch, whatever is left may be extremely small, but yet it is a perfectly good *finite* quantity. Thus, structural difficulties were encountered with the postulated 'infinitesimal'. The *name* implies that they are not finite, yet analysis shows only finites. Mathematicians supposed that an 'infinitesimal' was necessary for mathematics, and so they were reluctant to abandon it.

The 'infinitesimal', like so many other baffling suppositions, was invented by the Greeks, who regarded a circle as differing 'infinitesimally'

214

from a polygon with a very large number of very small equal sides. With the invention of the differential and integral calculus, 'infinitesimal calculus', as it was called, the importance of the 'infinitesimal' increased, and even mathematicians used it as a fundamental notion. Finally, Weierstrass succeeded in showing the meaningless character of the 'infinitesimal', and also that the 'infinitesimal' was not structurally necessary for the calculus. Up to that date, the problem was baffling; we knew that the calculus required 'continuity', which, in turn, seemed to require 'the infinitely little', and yet no one could tell what this 'infinitely little' might represent. It was quite obviously not zero, because a sufficient number of them was able to make up a finite whole; and we knew no fraction which was not zero and yet not finite. The discovery by Weierstrass that the calculus does not require the 'infinitesimal', and that all deductions could be made without it, abolished a very serious structural, verbal, metaphysical, and semantic bugaboo. Common sense, of course, is much simpler, although unreliable in such matters, and was satisfied also.

The elimination of the 'infinitesimal' is a great semantic step forward, and helps to clarify structurally some deeply rooted, vague, fallacious notions, which are overloaded with affective components and are extremely vicious in their effects.

If there is no 'infinitesimal', there is no 'next moment'; for the interval between any two moments must be finite, and so there are always other moments in the interval between them. Also, two moments cannot be consecutive, for between any two there are always other moments, no matter how far we go; similarly, the 'present' becomes a very vague notion.

For our purpose, the most fundamental semantic application of what has been said above is in the vast field embraced by the old structural notions of 'cause' and 'effect'. These terms are of great antiquity, of a distinctly pre-scientific one-, two-valued semantic epoch. They originated in the rough experience of our race, and are firmly rooted in the habits of 'thought' and the structure of our old two-valued 'logic' and language, and because of that are even now unduly baffling. These terms, in the *two-valued sense,* were and are the structural assumptions of our 'private' and 'official' 'philosophies'. The unenlightened use of these terms has done much to prevent the formulation of a science of man and to build up vicious anti-scientific metaphysics of various sorts involving pathological *s.r.* With the new quantum mechanics, a better understanding of these notions, based on the ∞-valued semantics of probability, becomes a paramount issue for all science. In daily life, the indiscriminate use of two-valued 'cause' and 'effect' leads structurally to a great deal of absolutism, dogmatism, and other harmful semantic disturbances, which I call confusion of orders of abstraction.

We usually follow the 'philosophers' and ascribe—or, rather feel, as conscious ascribing would not stand criticism—some mysterious structural continuity, some mysterious overlapping of 'cause' and 'effect'. We 'feel', and try to 'think', about 'cause and effect' as *contiguous* in 'time'. But 'contiguous in time' involves the impossible 'infinitesimal' of some unit of 'time'. But, since we have seen that there is no such thing, we must accept that the interval between 'cause' and 'effect' is finite. This structural fact changes the whole situation. If the interval between 'cause' and 'effect' is finite, then always something might happen between, no matter how small the interval may be. The 'same cause' would not produce the 'same effect'. The expected result would not follow. This means only that in this world, to be sure of some expected effect, requires that there must be nothing in the environment which can interfere with the process of passing from the conditions labelled 'cause' to the conditions labelled 'effect'. In this world, with the structure which it has, we can never suppose that a 'cause', as we know it, is *alone* sufficient to produce the supposed 'effect'. When we consider the ever-changing environment, the number of possibilities increases enormously. If it were possible to take into account the *whole* of the environment, the *probability* that some event would be repeated, in all details, thus exhibiting the assumed two-valued relation of 'cause' and 'effect', which we took for granted in the old days, would practically be nil. The principle of non-elementalism, as we see, requires an ∞-valued semantics of probability.

The reader should not take what is said here as a denial that in this external world some regularities of sequence occur; but the above analysis, which is mainly due to Russell,[1] shows clearly that the verbal principle of 'same cause, same effect' is structurally untenable. We can never manage to observe the 'same cause' in detail. As soon as the antecedents have been sufficiently ascertained, so as to calculate the consequences with some plausible accuracy as to details, the relations of these antecedents have become so complex that there is very little *probability* that they will ever occur again.

The clearing up of the problems of 'cause' and 'effect' is of serious importance, because powerful semantic reactions are connected with it. To begin with, we must differentiate between the terms 'cause' and 'effect', which, linked together, imply a two-term relation nowhere to be found in this world, and thus represent a language and a two-valued 'logic' of a structure not similar to the structure of the world, and the *general ∞-valued notion of causality*. This last notion is the psychological foundation of all explanations leading toward ∞-valued determinism, and is an exclusive test for structure; and so of extreme semantic importance.

Besides the analysis from the point of view of the impossible 'infinitesimal', the term 'cause-effect' represents a two-term relation, and, as such, is a primitive generalization *never* to be found in this world, as all events are *serially* related in a most complex way, independent of our way of speaking about them. If we expand our two-term relation 'cause-effect' into a *series,* we pass from the inferential level to the *descriptive* level, and so can apply a behaviouristic, functional, actional language of order. In such series, we could only use the language of 'cause' and 'effect' if we could select neighbouring factors, a selection which is often impossible. Also, if we pass from macroscopic to microscopic or sub-microscopic levels, we could use such language, but then the terms would have different meanings, supplied by the theory of probability.

The semantic side of this problem is of importance, because, in the old *el* way, it was neglected. General speculations about such *m.o* terms as 'cause' and 'effect' are useless. Such statements are not propositions, but involve variable meanings and, therefore, generate propositional functions which are neither true nor false. Our expanding of the too simple, two-term relation 'cause-effect' into a complex series is closer to the structure of this world, as far as we know it.

The understanding and habitual application of what has just been said would not only save us from silly dogmatizing and inappropriate *s.r,* but would teach us not to disregard any regularity, and to investigate any relation which might appear. Then, in a *specific case,* we could again use the restricted principle of causality, based on probability and averages. The old absolute and objectified semantic attitude toward 'cause-effect' was and often is a serious hindrance in observing impartially the sequence of events (order) and relations. Preconceived notions and old *s.r* played havoc, for it is well known that we usually find what we want to find. If we approach a problem with definite unconscious '*emotional*' wants, and cannot satisfy these *s.r,* we become bewildered, down-hearted, and perhaps utter some such non-sense as the 'finite mind', or the like. Under such semantic pressure, our power of observation and analysis is reduced by a kind of 'emotional stupor'. Such an occurrence is harmful in science and in life. 'Human knowledge' depends on human ingenuity, power of observation, power of abstraction,. It is an activity

218 IV. STRUCTURAL FACTORS IN $\bar{A}$ LANGUAGES

of the human nervous system inside of our skin and can never be the events themselves.

We see that the old two-valued verbal structure of 'cause' and 'effect' is not similar to the structure of the world, but a rash limiting generalization from probability. Since these expressions belong to the class of statistical averages and depend on the scale of the events and intervals dealt with, we must not expect that such terms as two-valued 'causality', which is a term of statistical macroscopic averages, will apply in that sense to small-scale events when the intervals are much smaller and when entirely different conditions and 'causes' prevail. Today we have structural evidence that even 'space' and 'time' represent statistical averages and do not apply to the smallest scale events. It is natural that 'cause' and 'effect' should join their company. The above involves epistemologically the passing from the A two-valued system to a $\bar{A}$ ∞-valued system. Psychophysiologically, it involves new *s.r.*

In mathematics, the old religious attitude toward the 'infinitesimal' is rapidly vanishing. Many mathematicians deliberately, and justly, avoid the use of the word. A term like 'indefinitely small' or 'indefinitesimal' is a better descriptive term, truer in its implications. We even see scientists like Eddington, who had the pluck—it is still pluck, unfortunately— to treat enormous stellar distances as 'infinitesimals of second order'. ('Infinitesimal' is used here in a mathematical sense of indefinitesimal.)

It has been already mentioned that most of the important discoveries of mathematics were due to a special semantic attitude on the part of those who made them. This attitude was an unconscious or conscious treatment of mathematics as a form of human behaviour. We see an example in the work of Weierstrass and his analysis of the 'infinitesimal'. He did not take the 'infinitesimal' as some objectified metaphysical structure and remain content; he analysed the *genetic process* by which the 'infinitesimal' was *made* by Smith and Brown, and so treated mathematics structurally as a form of human behaviour. Any deepening of the foundations, or clarification of fundamental notions, or investigation of underlying assumptions . , *must,* by necessity, have this characteristic. The man who does it must take into account how the given process was produced—analyse its structure, and so start with the ways and methods of production. In other words, he must treat the given problem as a form of human behaviour. The fact that this simple and quite obvious method has been formulated and structurally explained as *desirable* is helpful. It shows the *method* and structure of the path by which advances can be reached. We can *train* the semantic reactions of students to it and make progress inevitable; but now, instead, it takes a genius to break, by himself, through the old semantic habits which have been produced by the lack of scientific psycho-logics and training.

The term 'correct symbolism' has already been used. In this world of structurally absolute individuals, the minimum of structurally desirable correct symbolism must provide for the possibility of labelling these absolute individuals by separate names. For scientific purposes, we must use terms built on the pattern of mathematical symbolism; i.e., according to the *extensional* methods. We must adopt a behaviouristic attitude and habits in our term-making. As we proceed, we must emphasize *order*, considering what comes first and what next. This is semantically important, for the usual procedure is entirely different: first, we have our structurally 'preconceived' doctrines and languages; next, we observe the structure of the world; and *then* we try to force the observed facts into the linguistic structural patterns. But, in the new way, we *start* with silent observations, and search empirically for structure; next, we invent verbal structures similar to them; and, finally, we see what can be said about the situation, and so test the language. Experience shows that the old habits of labels first, objects next, instead of the structurally natural order of *objects first, labels next*, is semantically pernicious and harmful. In Part VII, it is shown that the semantic structural reversal of the unnatural reversed order is crucial for sanity.

From the days of the Greeks an acute difficulty has made itself felt; namely, how to reconcile the world of physics with the world of mathematics. For mathematics, we need 'extensionless' points; for physics, we need finite-sized elements. Whitehead and Russell have suggested different structures by which this may be accomplished. It seems possible to demand that none of the material dealt with shall be smaller than an assigned finite size. That this condition can be reconciled with mathematical continuity seems to be novel. Whether this device is valid or not, it is yet too early to decide. This problem of reconciliation will become important further on when we come to speak of events as made up from point-events.[2]

CHAPTER XVI

ON THE EXISTENCE OF RELATIONS

We cannot choose to do without them, without seeking to choose, since choice is action, and involves, for instance, the aforesaid difference between affirming and denying that we mean to do thus and thus. (449)

JOSIAH ROYCE

In concluding the foregoing remarks, I must explain one more general consideration. This concerns an extremely profound structural psycho-logical discovery, made by Prof. Royce,[1] which underlies any and all semantic problems of human 'mentality'. Royce, although a 'philosopher', was a lover of mathematics and was much interested in the problems of *order*. He was trying to reformulate 'logic' in terms of order. We had already encountered the inherent circularity in the structure of human knowledge, which admittedly is semantically disconcerting if not faced boldly. But, when recognized, this circularity is not only not vicious, but even adds to the interest and beauty of life and makes science more interesting. Besides, the structure of human knowledge is such that there are activities of man which are not only circular but also 'absolute', or 'necessary'. Whatever we do, we cannot get away from them—a fact of serious semantic importance. Except from Royce and a few of his students, these problems have as yet received little attention.

Royce shows that there are certain activities which we reinstate and verify through the very fact of attempting to assume that these forms of activity do not exist, or that these laws are not valid. If any one attempts to say that there are no classes whatsoever in his world, he thereby inevitably classifies. If any one denies the existence of relations, and, in particular, a semantic relation between affirmation and denial, or affirms that 'yes' and 'no' have *one* meaning, in that breath he affirms and denies. He makes a difference between 'yes' and 'no', and emphatically asserts relational equivalence even in denying the difference between 'yes' and 'no'. To use Royce's own remarkable words: 'In brief, whatever actions are such, whatever types of action are such, whatever results of activity, whatever conceptual constructions are such, that the very act of getting rid of them, or of thinking them away, logically implies their presence, are known to us indeed both empirically and pragmatically (since we note their presence and learn of them through action); but they are also absolute. And any account which succeeds in telling what they are has absolute truth. Such truth is a "construction"

220

or "creation", for activity determines its nature. It is "found", for we observe it when we act.'

We see that we have definite semantic guides in this enquiry. One guide to follow is these unescapable characteristics of the structure of human knowledge, which Royce called 'absolute', but which I prefer to call 'necessary'. The other guide leads us to avoid 'impossible' or absurd statements, or statements which have no 'logical existence'; which, in the rough, means statements which abuse symbolism and produce noises., instead of symbols. As we have already seen, both guides have sound *neurological* justification, to be expressed in terms of *order* and *circularity,* terms uniquely fit structurally to speåk about processes, stages of processes, orders of abstractions,. Obviously, our task of formulating a theory of sanity can proceed along these structural and semantic lines. It should be noticed that mathematics, considered as a form of human behaviour, and 'mental' illnesses, also considered as definitely human behaviour, have yielded their share for our structural guidance.

Although many a scientist has instinctively proceeded in the way indicated, yet the instinctive successful procedure of an isolated scientist is usually not capable of being transmitted to others. It is his personal benefit. Only a *methodological structural* formulation of such private routes to semantic success can become a *public* fact, to be analysed, criticized, improved, and transmitted or rejected.

It must be noticed that terms like 'chance' or 'law' are fundamentally connected with discussions of determinism versus indeterminism, and so involve problems in connection with 'necessary' semantic processes. In the example about the probability of the M-event, it was shown how a 'chance' event on one level may become a 'law' on another. The structural *possibility* of such transformations is very interesting and of basic semantic importance. For scientific purposes, we must accept ∞-valued determinism on the scientific level as *it is the test of structure;* but this has nothing to do with the *apparent,* mostly two-valued indeterminism in our daily lives. To solve a number of equations, we must have as many equations as we have unknowns. If we have fewer equations than unknowns, we do not get definite values; our unknowns are still undetermined. The origin of 'indeterminism' is similar; we lack knowledge; the number of equations is less than the number of unknowns. Hence, it is impossible to discover determined values in all cases. This gives an appearance of two-valued indeterminism, but with the increase of our knowledge, or with additional equations, the unknown may be determined. Determinism is a more fundamental point of view than indeterminism; in it we find a *test for structure*. It is also a more general point of view, in which indeterminism is only a particular case and does not allow of the structural test. In a science of man, in a $\bar{A}$-system, we must start with the more fundamental and general. Accordingly, we have to accept ∞-valued determinism, which, in 1933, becomes the broad scientific point of view. The unnecessary semantic war between the advocates of the different points of view has been unduly bitter and necessarily futile.

As words *are not* the things we speak about, and the only link is structural, the 'human mind' must require linguistic structural ∞-valued determinism as a condition of rationality. As soon as we find that any linguistic issues are not deterministic, it is an unmistakable sign that the language or the 'logic' we are using is *not* similar in structure to the empirical world and so should be changed.

This statement seems to be general. In application to the new quantum mechanics' special problem, it would appear that the old macroscopic language of 'space', 'time'., is not similar to the sub-microscopic structure and should, therefore, be changed. Perhaps the electrodynamic language, instead of the macro-mechanistic, would fare better.

CHAPTER XVII

ON THE NOTIONS OF 'MATTER', 'SPACE', 'TIME'

Common sense starts with the notion that there is matter where we can get sensations of touch, but not elsewhere. Then it gets puzzled by wind, breath, clouds, etc., whence it is led to the conception of "spirit"—I speak etymologically. After "spirit" has been replaced by "gas," there is a further stage, that of the aether. (457)   BERTRAND RUSSELL

The supposition of common sense and naive realism, that we see the actual physical object, is very hard to reconcile with the scientific view that our perception occurs somewhat later than the emission of light by the object; and this difficulty is not overcome by the fact that the time involved, like the notorious baby, is a very little one. (457)

BERTRAND RUSSELL

We have certain preconceived ideas about location in space which have come down to us from ape-like ancestors. (149)   A. S. EDDINGTON

But it does not seem a profitable procedure to make odd noises on the off-chance that posterity will find a significance to attribute to them. (149)   A. S. EDDINGTON

There is a blessed phrase "hidden reserves"; and generally speaking the more respectable the company the more widely does its balance-sheet deviate from reality. This is called sound finance. . . .

Thanks to Minkowski a way of keeping accounts has been found which exhibits realities (absolute things) *and balances*. (149)   A. S. EDDINGTON

The quest of the absolute leads into the four-dimensional world. (149)

A. S. EDDINGTON

The views of space and time which I wish to lay before you have sprung from the soil of experimental physics, and therein lies their strength. They are radical. Henceforth space by itself, and time by itself, are doomed to fade away into mere shadows, and only a kind of union of the two will preserve an independent reality. (352)   H. MINKOWSKI

It is a *thing;* not like space, which is a mere negation; nor like time, which is—Heaven knows what! (149)   A. S. EDDINGTON

Newton objectivises space. Since he classes his absolute space together with real things, for him rotation relative to an absolute space is also something real. (151)   A. EINSTEIN

Space is only a word that we have believed a thing. (417)   H. POINCARÉ

In fact, our ordinary description of nature, and the idea of exact laws, rests on the assumption that it is possible to observe the phenomena without appreciably influencing them. (215)   W. HEISENBERG

Even when this arbitrariness is taken into account the concept "observation" belongs, strictly speaking, to the class of ideas borrowed from the experiences of everyday life. It can only be carried over to atomic phenomena when due regard is paid to the limitations placed on all space-time descriptions by the uncertainty principle. (215)   W. HEISENBERG

Section A. Structural considerations.

The facts at hand in 1933 show that the language we use for the purpose of describing events *is not* the events; the representation symbolizes what is going on inside our skins; the events are outside our skins and *structural similarity* is the only link between them. Historically, as a race, we learned sooner and more about the events outside our skins than about the events inside our skins; just as a fish or a dog 'knows' a lot about his world, lives sometimes happily and abundantly, and yet 'knows' nothing about biology, or physiology, or psycho-logics. Only recently did we begin to study ourselves scientifically. At some stage of our development, we introduced structurally simple forms of representation, such as a *language* of subject-predicate, of additivity,. We are still perplexed when we find that the events outside our skins cannot be pressed into schemes which are manufactured inside our skins. Our nervous system, with its ordered and cyclic structure and function, manufactures abstractions of different orders, which have quite distinct structure and different characteristics. On different levels, we manufacture different abstractions, dynamic and static, continuous and discontinuous., which have to take care of our needs. If the verbal schemes we invent do not fit structurally the world around us, we can always invent new schemes of new structure which will be more satisfactory. It is not a problem of the world around us, for our words cannot change that, but of *our ingenuity*. In the meantime, we learn something very important; namely, about the world's *structure*, which is the only content of knowledge.

There are good structural reasons why the world should, or should not, be accounted for in terms of differential equations, or in *terms and language* of 'causality',. The term *order* is structurally fundamental and will help us in a radical and constructive way, in our quest.

First, however, we will investigate some further semantic problems, remembering that a theory of sanity, which means a theory of adjustment, should emphasize the methodological and structural means for such semantic adjustment. The dynamic-static translations are fundamentally connected with different orders of abstractions and involve psycho-logical issues connected with 'emotions' and 'intellect', linearity versus non-linearity, 'straight' versus 'curved'., explained in Parts VII and VIII.

In life, as well as in science, we deal with different happenings, objects, and larger or smaller bits of materials. We have a habit of speaking about them in terms of 'matter'. Through a *semantic disturb-*

ance, called identification, we fancy that such a thing as 'matter' has separate physical existence. It would probably be a shock to be invited seriously to *give* a piece of 'matter' (give and *not* burst into speech). I have had the most amusing experiences in this field. Most people, scientists included, hand over a pencil or something of this sort. But did they actually give 'matter'? What they gave *is not* to be *symbolized* simply 'matter'. The object, 'pencil', which they *handed,* requires linguistically 'space'; otherwise, there would be no pencil but a mathematical point, a fiction. It also requires verbally 'time'; otherwise, there would be no pencil but a 'flash'.

Similarly, if any one is invited to *give* a piece of 'space' (again *give* it, and *not* burst into speech), the best he could do would be to wave his hand and try to show 'space'. But the waving of the hand referred to what we call air, dust, microbes, gravitational and electromagnetic fields,. In other words, structurally, the supposed 'space' was *fulness* of some materials already 'in space' and 'in time'.

In the case of *giving* 'time', one could *show* his watch. A similar objection holds, also; namely, that he has shown us so-called 'matter' which is 'moving' in 'space'. It is very important to acquire the *s.r* that when we use the term 'matter' we refer to something, let us say, the pencil, which, according to the accepted *el language,* also involves 'space' and 'time', which we disregard. When we use the term 'space', we refer to a fulness of some materials, which exists in 'time'. But because these materials are usually invisible to the 'senses', we again disregard them. In using the term 'time', we refer to 'matter' moving in 'space', which again we disregard.

What is said here and what will follow is structurally unconditionally fundamental for a theory of sanity, because in most cases of 'insanity' and un-sanity, there is a disorientation as to 'space' and 'time'. In identification, the semantic disturbance which affects nearly all of us, and is at the foundation of the majority of human difficulties, private or public, there invariably appears a special semantic disorientation in our feelings toward 'matter', 'space', and 'time'. This is only natural, for the 'insane' and un-sane are the unadjusted; the 'sane' are the supposedly adjusted.

Adjusted to what? To the world around us and ourselves. Our *human* world differs from the world of animals. It is more complex and the problems of human adjustment become also more subtle. In animal life, attitudes toward the world do not matter in a similar sense; with us, they become important; hence the need of analysis of the new human 'semantic universe', which involves the 'universe of discourse'. This

15

'universe of discourse' is strictly connected with the *terms* of 'matter', 'space', and 'time', structure, and our semantic *attitude* toward these terms.

Let us return to the analysis of our object which we call the 'pencil'. We have seen that the *object* pencil *is not* 'matter', nor 'space', nor 'time'. A question arises, which has been asked very often and has *never*, to my knowledge, been answered satisfactorily: what 'is' the *object* pencil, and what 'are' the *terms* 'matter', 'space', and 'time'? Here and there some one has given fragments of answers or some satisfactory detached statements. But in every case I know, the semantic disturbance called identification appears, and so even the casual correct answer is not applied but remains enmeshed in some other identifications. I have spent much 'time' and labour in overcoming my own identifications, and now confront the situation that nearly every work I read from this point of view cannot be criticized, but requires rewriting. This task is impossible for me, technically and otherwise. So, finally, I decided to formulate the present $\bar{A}$-system, and then see what kind of reconstruction can be accomplished with the new evaluation.

The answer to the questions set above is childishly simple, yet I will carry it all through and let the semantic consequences speak for themselves. The chunk of nature, the specially shaped accumulation of materials., which we call a pencil, 'is' fundamentally and *absolutely un-speakable,* simply because whatever we may *say* about it, *is not it.* We may write with this something, but we cannot write with its name or the *descriptions* of this something. So the object *is not words.* It is important that the reader be entirely convinced at this point, and it requires some training, performed repeatedly, before we get our *s.r* adjusted to this simple fact. Our statement had two parts. One was rather unpromising; namely, that the object was absolutely un-speakable, because no amount of words will make the object. The other was more promising, for we learned an extremely important, perhaps crucial, semantic fact; namely, what the *object* pencil *is not;* namely, that the *object is not words.* We must face here an important semantic fact. If we are told that we cannot get the moon, we stop worrying about it, and we regard any dream about getting the moon as an infantile phantasy. In this example, we could not even say that such news as the impossibility of getting the moon was sad, or unpleasant, news. We might say so, jokingly, to an infant, but the majority of the grown-ups would not have their *s.r* perturbed by it. A similar situation arises with the object called pencil. The *object is not words.* There is nothing sad or depressing about this fact. We accept it as a fact and stop worrying about it, as an infant would. The majority of the old 'philosophical' speculations about this subject belong to the semantic period of our infancy, when we live in phantasies and structurally gamble on words to which we affectively ascribe objective existence,. This represents full-fledged un-sanity due to identification. The answer to the question, what 'are' the *terms* 'matter', 'space', and 'time', is, as usual, given in the properly formulated question. They 'are' *terms,—'Modi considerandi'*, as Leibnitz called them without fully realizing the semantic importance of his own statement. Incidentally, it must be noticed that it was the psycho-logical characteristic of Leibnitz who was capable of such a statement, that was probably responsible for his whole work, as will become more apparent later on. When we abandon primitive standards of evaluation, geniuses will be *made* by a semantic education which relieves the race from the older blockages.

So we see clearly that outside of our skins there is something going on, which *we call* the world, or a pencil, or anything, which is *independent* of our words and which is *not* words. Here we come across a very fundamental irreversible process. We can say that in this world a man and his words have happened. There is a 'causal' eventual complex series between the world, us, and our words, but in unaided nature this process is, in the main, irreversible, a fact unknown to primitives who believe in the magic of words. Through our ingenuity we can make this process partially reversible; namely, we can produce gramophones, telephones in all their developments, electromechanical men who obey orders,.

We know in 1933 that in the semantic world this process is dramatically effective. Words are the result of the activity of one organism, and they, in turn, activate other organisms. On the macroscopic level of ordinary behaviour, this last was known long ago, but only in the last few years has psychiatry discovered what kind of semantic and psycho-physiological disasters words and their consequences may produce in the human organism. These last are already on sub-microscopic levels, not obvious and, therefore, only recently discovered.

The structure of the language of 'matter', 'space', and 'time' is ancient. The primitive saw something, ate something, was hurt by something,. Here was an occasion for a grunt of satisfaction, or of pain. The equivalents of words like 'matter', 'substance'., originated. Neither he, nor the majority of us, realized that the small or large bits of materials we deal with appear as extremely complex *processes* (explained in Part X). For him, as for most of us, these bits of materials 'are' 'concrete', whatever that means, and he might know 'all about them', which must have led to identification., and other delusional evaluations. Of course, these were phantasies of human infancy, and, in lives lived in a world of phantasies, adjustment, and, therefore, sanity, is impossible. Since he did not see or feel, or know about the material he was immersed in, the *fulness* he was living in, he invented the term 'space', or its equivalent, for the *invisible* materials which were present. Knowing nothing of fulness, he objectified what appeared to him to be empty 'space' into 'absolute emptiness', which later became 'absolute space', 'absolute nothingness', by 'definition'.

There are several important remarks which can be made about this 'absolute emptiness' and 'absolute nothingness'. First of all, we now know, theoretically and empirically, that such a thing does not exist. There may be more or less of something, but never an *unlimited* 'perfect vacuum'. In the second place, our nervous make-up, being in accord with experience, is such that 'absolute emptiness' requires 'outside walls'. The question at once arises, is the world 'finite' or 'infinite'? If we *say* 'finite', it *has* to have outside walls, and then the question arises: What is 'behind the walls'? If we say it is 'infinite', the problem of the psychological 'walls' is not eliminated, and we still have the semantic need for walls, and then ask what is beyond the walls. So we see that such a world suspended in some sort of an 'absolute void' represents a *nature against human nature,* and so we had to invent something *supernatural* to account for such assumed nature against human nature. In the third place, and this remark is the most fundamental of all, because a symbol must stand for something to be a symbol at all, *'absolute nothingness' cannot be objective and cannot be symbolized at all.* This ends the argument, as all we may say about it is neither true nor false, but *non-sense.* We can make noises, but say nothing about the external world. It is easy to see that 'absolute nothingness' *is a label for a semantic disturbance,* for verbal objectification, for a pathological state inside our skin, for a fancy, but not a symbol, for a something which has *objective* existence outside our skin.

Some other imaginary consequences of this semantic disturbance are far-fetched and very gloomy. If our world and all other worlds (island universes) were somehow suspended in such an 'absolute void', these universes would radiate their energy into this 'infinite void', whatever that means, and so sooner or later would come to an end, their energy being exhausted. But, fortunately, when we eliminate this pathological semantic state by proper education all these gloomy symptoms vanish as mere fancies. It must be noticed that this 'absolute space', 'absolute void', 'absolute nothingness' with its difficulties, which are due to very primitive structural speculations on *words,* and to some un-sane ascribing of objectivity to words, can be abolished quite simply if we decide to investigate and re-educate our *s.r.*

What we know positively about 'space' is that it is not 'emptiness', but 'fulness', or a 'plenum'. Now 'fulness' or a 'plenum', first of all, is a term of entirely different *non-el* structure. When we have a plenum or fulness, it must be a plenum of 'something', 'somewhere', at 'some time', and so the *term implies,* at least, *all three of our former elementalistic terms.* Furthermore, fulness by some psycho-logical process does not require 'outside walls'. If we ask if such a universe of fulness is 'finite' or 'infinite', without any psycho-logical difficulties, we may reply that we do not know, but if we study enough of the materials of this universe we *may know.* A universe of fulness may be assumed to have boundaries, and then we may ask again the annoying question: What is beyond? With the proper use of *language,* this difficulty is again eliminated.

Without going into unnecessary details, we see that a boundary, or a limit, or a wall, is something by *definition,* beyond which we cannot go. If there is nothing to restrict our progress, there are *no boundaries.* Let us fancy some cosmic traveller with some extraordinary flying machine, and let us assume that he flies without stopping in a 'definite direction'. If he never encounters any boundary, he is surely entitled to say that his universe is unbounded. The question may arise: Is such an unbounded universe finite or infinite in size? Again let us apply correct language and a little analogy. A traveller on a sphere, like our own earth, could travel *endlessly* without ever coming to a boundary, and yet we know that the sphere, our earth, is of finite size. Mathematicians have worked out this point, and it is embodied in the Einstein theory. The universe is unbounded, an answer which satisfies our feelings; yet it is finite in size, although very large, an answer which satisfies our rationality.* The visualization of such a universe is quite difficult. It should not be visualized as a sphere, but, at a later stage, we will see that it *can* be visualized satisfactorily. The condition for visualization is to eliminate identification, that *semantic disturbance* which is strictly connected with primitive ways of 'thinking'.

The problems of 'time' are similar, although they have a different neurological background. The rough materials we deal with mostly affect our sight, touch,. Invisible materials, like air., affect these 'senses'

---

*I do not introduce here the latest speculations in this field, because, from a non-aristotelian point of view, they appear meaningless.

less, but more the kinesthetic 'senses' by which muscular movements are appreciated, and so 'space' and 'time' have different neurological backgrounds. 'Time' seemingly represents a general characteristic of *all* nervous tissue (and, perhaps, living tissue in general) connected with summarizing or integrating. What we have to deal with in this world and in ourselves appears as periods and periodicity, pulsations,. We are made up of very long chains of atomic pulsating clocks, on the submicroscopic level. On the macroscopic level, we have also to deal with periodic occurrences, of hunger, sleep, breathing, heart-beats,. We know already that, beyond some limits, discontinuous time*s*, when rapid enough, are blended into continuous feelings of pressure, or warmth, or light,. On objective levels we deal with time*s*, and we feel 'time', when the time*s* are rapid enough.

Again, the moving pictures are a good illustration. The normal moving-picture film shows sixteen pictures a second. The film gives us static pictures with finite differences. When we put it on the projector, the differences vanish. Our nervous system has summarized and integrated them, and we see 'continuous motion'. If pictures are taken at the rate of eight a second and then run on the normal projector for the speed of sixteen a second, we summarize and integrate again, but we see a fast moving picture. If the pictures are taken at the rate of 128 exposures a second and run on the normal projector of sixteen pictures to a second, we have what is called a slow moving picture. It should be noticed that the order of the semantic rhythmic processes is fourfold; it involves order not only in 'space' (three dimensions) but in 'time' also. Periods of contraction alternate with periods of rest, and this occurs at nearly regular intervals.

This rhythmic tendency is, indeed, so fundamental and so inherent in living tissue that we can, at pleasure, make voluntary muscles; for example, exhibit artificially induced rhythmic contractions by immersing them in special saline solutions, as, for instance, a solution of sodium chloride. We should also not wonder why modern science assumes that life may have originated in the sea. The physico-chemical conditions of saline solutions are such that they favour rhythmic processes; they not only may originate them, but may also keep them up, and life seemingly is very closely connected with autonomous rhythmic processes.

Such rhythmic processes are *felt* on lower orders of abstraction as 'continuous time', probably because of the rapidity and overlapping of periods. On higher order abstractions, when structurally proper linguistic and extra-neural means are developed, they appear as time*s*.

Perhaps, neurologically, animals *feel* similarly as we do about 'time', but they have no neurological means to elaborate linguistic and extra-neural means which alone allow us to extend and summarize the manifold experience of many generations (time-binding). They cannot pass from 'time' to 'times'. Obviously, if we do not, we then renounce our human characteristics, and copy animals in our evaluating processes, a practice which must be harmful.

In nature the visible and invisible materials seemingly consist of recurring pulsations of extremely minute and rapid periods, which, in some instances, become macroscopic periods. In the first case, we cannot see them or feel them, so we talk about 'concreteness',. In the second case, we see the periodic movements, as of the earth around the sun., or we feel our heart-beats,. We see that the visible or invisible materials in nature are compounded of periodic pulsations and are simply two aspects of one process. The splitting of these processes into 'matter', 'space', and 'time' is a characteristic function of our nervous system. These abstractions are *inside* our skins, and are methods of representation for ourselves to ourselves, and *are not* the objective world around us.

It must be realized that under such circumstances we cannot speak about 'finiteness' or 'infiniteness' of 'matter', 'space', and 'time', as all the old 'philosophers' have done, Leibnitz included, because these terms 'finite' and 'infinite', though they may be conceivably applied to *numbers* of aspects of objective entities, have *no meaning* if applied to linguistic issues, that is, to *forms of representation* outside of numbers. Of course, if, through a semantic pathological disturbance (objectification), we do ascribe some delusional objective existence to verbal terms, we can then talk about anything, but such conversations have no more value than the deliria of the 'mentally' ill. The terms 'finite' or 'infinite' are only legitimately applied to *numerical* problems, and so we can speak legitimately of a finite or infinite numbers of inches, or pounds, or hours, or similar entities, but statements about the 'finite mind' or the 'understanding of the infinite'., have *no* meanings and only reveal the pathological semantic disturbance of the patient.

The objectification of *our feeling* of 'time' has had, and has at present, very tragic consequences strictly connected with our un-sanity. It must be remembered that particularly in 'mental' and nervous difficulties the patient seldom realizes the character of his illness. He may feel pains, he may feel very unhappy, and what not, but he usually does not understand their origin. This is particularly true with semantic disturbances. One may explain endlessly, but, in most cases, it is perfectly hopeless to try to help. Only a very few benefit. Here lies, also, the main difficulty in writing this book. Readers who identify, that is, who believe unconsciously with all their affective impulses in the objectivity of 'matter', 'space', and 'time', will have difficulty in modifying their *s.r* in this field.

Let us see now what consequences the objectification of 'time' will have for us. If we do *not* objectify, and *feel* instinctively and permanently that words *are not* the things spoken about, then we could not speak about such *meaningless* subjects as the 'beginning' or the 'end' of 'time'. But, if we are semantically disturbed and objectify, then, of course, since objects have a beginning and an end, so also would 'time' have a 'beginning' and an 'end'. In such pathological fancies the universe must have a 'beginning in time' and so must have been made., and all of our old anthropomorphic and objectified mythologies follow, including the older theories of entropy in physics. But, if 'time' is only a *human form of representation* and *not an object,* the universe has no 'beginning in time' and no 'end in time'; in other words, the universe is 'time'-less. It was not made, it just 'was, is, and will be'. The moment we realize, feel permanently, and utilize these realizations and feelings that words *are not* things, then only do we acquire the semantic freedom to use different forms of representation. We can fit better their structure to the facts at hand, become better adjusted to these facts which *are not* words, and so evaluate properly *m.o* realities, which evaluation is important for sanity.

According to what we know in 1933, the universe is 'time'-less; in other words, there is no such *object* as 'time'. In terms of periods, or years, or minutes, or seconds, which is a *different language,* we may have infinite numbers of such time*s*. This statement is another form of stating the principle of conservation of energy, or whatever other fundamental higher abstraction physicists will discover.

Because 'time' is a *feeling,* produced by conditions of this world outside and inside our skins, which can be said to represent time*s*, the problem of 'time' becomes a neuro-mathematical issue. It must also be noticed that time*s*, as a term, implies time*s* of something, somewhere, and so, as with plenum or fulness, it is structurally a *non-el*, $\bar{A}$ term.

Time*s* has also many other most important implications. It implies *numbers* of time*s*, it implies periods, waves, vibrations, frequencies, units, quanta, discontinuities, and, indeed, the whole structural apparatus of modern science.

Euclidean 'space' had the semantic background of 'emptiness'. In it we moved our figures from place to place and always assumed that this could be done quite safely and accurately. Newtonian mechanics also followed this path and even postulated an 'absolute space' (emptiness). All of which harks back to the old aristotelianism.

$\bar{E}$, $\bar{N}$, and $\bar{A}$ systems have the semantic background of fulness or plenum, although, unfortunately, this background is, as yet, mainly unrealized, not fully utilized; it has not, as yet, generally affected our *s.r.*

A simple illustration will make the difference clear. Imagine that in one part of a large room we have an open umbrella which we would like to compare with another 'unit' open umbrella. Let us imagine that the room has the air pumped out and also that all other eventual disturbing factors are eliminated. We can move our open 'unit' umbrella from one part of the room to another, and this movement will not considerably distort our 'unit' umbrella. Now let us perform a similar experiment in two houses, separated some distance, during a storm, a storm implying, of course, *fulness*. Can we transport our 'unit' umbrella through the storm and preserve its shape., in a fulness, without taking the fulness into account? Of course not. We see what serious difference it makes if our theories presuppose 'emptiness' or 'fulness'.

This shows also why the non-euclidean geometries which deal with a plenum are structurally preferable and semantically sounder and more in accord with the structure of the world, than the language of euclidean 'emptiness', to which there is nothing in nature to correspond. Should we wonder that modern linguists (mathematicians) work in the direction of fulness and of fusing geometry with physics. It is obviously the only thing to do. Differential geometry is the foundation of this new outlook, but, even in this geometry, lines could legitimately be transported over great distances. Weyl introduced a semantic improvement of this point of view by assuming that for a differential geometry it is illegitimate to use comparisons at large distances, but that all operations should be between indefinitely near points.[1]

It should be noticed that scientists, in general, disregard almost completely the verbal and semantic problems explained here, a fact which leads to great and unnecessary confusion, and makes modern works inaccessible to the layman. Take, for instance, the case of the 'curvature of space-time'. Mathematicians use this expression very often and, inside their skins, they know mostly what they are talking about. Millions upon millions of even intelligent readers hear such an expression as the 'curvature of space-time'. Owing to nursery mythology and primitive *s.r,* 'space' for them is 'emptiness', and so they try to understand the 'curvature of emptiness'. After severe pains, they come to a very *true,* yet, for them, hopeless, conclusion; namely, that 'curvature of emptiness' is either *non-sense* or 'beyond them', with the semantic result that either they have contempt for the mathematicians who deal with non-sense or feel hopeless about their own capacities—both undesirable semantic results.

The truth is that 'curvature of emptiness' has no meanings, no matter *who* might say it, but curvature of fulness is entirely different. Let the reader look at the cloud of smoke from his cigarette or cigar, and he will at once understand what 'curvature of fulness' means. Of course, he will realize, as well as the mathematicians do, that the problem may be difficult, but, at least, it *has sense* and represents a problem. It is not non-sense.

Similar remarks apply to higher dimensions in 'space'. Higher dimensions in 'emptiness' is also non-sense; and the layman is right in refusing to accept it. But higher dimensions in fulness is entirely a different problem. A look at the cloud of smoke from our cigarette will again make it completely plain to everybody that to give an account of fulness, we may need an enormous number of data or, as we say roughly, of dimensions. This applies, also, to the new four-dimensional world of Minkowski. It is a fulness made up of world lines, a network of events or intervals., and it is not non-sense.

Lately, there has appeared an excellent book by Bertrand Russell, published by the International Library of Psychology, Philosophy and Scientific Method; and yet the title is *The Analysis of Matter,* without any quotation marks.

This book is really an unusually fine and fundamental work which has no defects which could be implied by the title. This title simply disregards the issues explained here; it should be *The Analysis of 'Matter'*.

It is with some pleasure that one sees such an authority as Eddington, in his *The Mathematical Theory of Relativity,* on p. 158, making the statement: 'In using the word "space" it is difficult to repress irrelevant ideas; therefore let us abandon the word and state explicitly that we are considering a *network of intervals'*.

For the reasons already given, I do not use the terms 'matter', 'space', or 'time' without quotation marks; and, wherever possible, shall use, instead, the terms 'materials', 'plenum', 'fulness', 'spread', and 'times', (say seconds). Indeed, these semantic problems are so serious that they should be brought to the attention of International Mathematical and Physical Congresses, so that a new and *structurally correct* terminology could be established. It is *not* desirable that science should *structurally mislead* the layman and disturb his *s.r.* It is easier for trained specialists to change their terminology than to re-educate semantically the rest of the race. I would suggest that terms 'matter', 'substance', 'space', and 'time' should be completely eliminated from science, because of their extremely wide-spread and vicious structural and so semantic implications, and that the terms 'events', 'space-time', 'material', 'plenum', 'fulness', 'spreads', 'times'. , be used instead. These terms not only do not have the old structural and semantic implications, but, on the contrary, they convey the *modern* structural notions and involve new *s.r*. The use of the old terms drags in, unconsciously and automatically, the old primitive metaphysical structure and *s.r* which are entirely contradicted by experience and modern science. I venture to suggest that such a change in terminology would do more to render the newer works intelligible than scores of volumes of explanations using the old terminology.

Before summarizing in Parts IX and X what modern science has to tell about the structure of the world around us, it will be profitable to enquire what are the means by which we can recognize this structure.

Section B. The neurological function of abstracting.

Protoplasm, even in its simplest form, is sensitive to different mechanical and chemical stimulations; and, indeed, undifferentiated protoplasm has already all the potentialities of the future nervous system. If we take an undifferentiated bit of protoplasm, and some stimulation is applied to some point, the stimulus does not spread somehow 'all over at once', with some mysterious 'infinite' velocity, but propagates itself with finite velocity and a diminishing gradient from one end of our bit of protoplasm to the opposite end.

Because of the *finite velocity* of propagation and the fact that the *action is by contact in a plenum,* the impulse has a definite direction and diminishing intensity, or, as we say, the bit of protoplasm acquires a temporal polarity (head-end). Such polarity conditions produce a directed wave of excitement of diminishing intensity, which Child calls a dynamic gradient. If such a stimulation were applied to one spot for a considerable length of 'time', some kind of polarization may become lasting. In some such way those dynamic gradients have become *structuralized* in the forms of our nervous system, which represent the preferred paths by which the nervous impulses travel.

The bodies of most organisms are protected from outside stimulation by some kind of membrane or cuticle and the parts of the surface have developed so as to be sensitive to one form of stimulation and not to others. For instance, the eye registers the stimulations of light waves, while it is insensitive to sound., and, even if hit, it gives only the feeling of light. Each 'sense-organ' has also the nervous means of concentrating stimuli, intensifying them., and so of effecting the most efficient response of the corresponding end-organ.

In our school days we were taught that we have five 'senses'. Modern researches show that there are more than twenty different 'senses'. Besides, as far as 'Smith' is concerned, we know that 'senses' and 'mind' cannot be divided.

The main stimulations which we find in the outside world may be divided into three groups. The first are connected with the roughest macroscopic manifestations of the outside world; they are mechanical impacts which we abstract as 'tactile sensations', which range from a single mechanical contact to rhythmically repeated contacts with our skin as frequently as 1552 vibrations per second. Above this limit 'times' begin to be registered as a 'duration'; that is, the individuality of 'times' is lost, and we feel pressure. At this level we deal with gross macro scopic manifestations, which are not only felt but can also be seen.

The second group of manifestations is, in the main, no more on the gross level. Here belong the vibratory manifestations which are no more visible to the unaided eye. We may speak of them as on the microscopic level. They are mechanical vibrations of the air., and we become acquainted with them in the form of sound. The vibrations which the average ear is able to register range from about 30 (sometimes even 12) to about 30,000 or even 50,000 vibrations a second. The ear does not register any other vibrations.*

The third group of vibratory manifestations belongs to a still subtler level. They are electromagnetic waves of an enormous variety of wavelengths and number of vibrations per second. The lower members of this series are the Hertzian electric waves, the higher members are the X-, or Röntgen-rays. Our nervous systems are capable of registering only a very limited range of these vibrations; namely, the waves called radiant heat, the light waves, and the ultra-violet rays, these last only on a chemical level. It seems that we have no organ which responds directly to electric waves, ultra-violet rays, X-rays, and the many other rays which we know from laboratory work.

Similarly, the chemical 'senses' of taste and smell register only a very small number of actual excitations to which they are exposed.

Animals have different limits of nervous susceptibility, but we can have no idea how the world looks to them unless their nervous system is quite similar to our own. The above statements will become clearer if we tabulate some of them. The following table is taken from *An Introduction to Neurology* by Professor C. Judson Herrick, p. 85 (Fifth Edition):

---

*Latest researches seem to modify these data.

TABLE OF PHYSICAL VIBRATIONS*

Physical process.	Wave length.	Number of vibrations per second.	Recep- tor.	Sensa- tion.
Mechanical contact.	. . . . .	From very slow to 1552 per second.	Skin.	Touch and pressure.
Waves in material media.	Above 12,280 mm.	Below 30 per second.	None.	None.
	12,280 mm. to 13 mm.	30 per second to 30,000 per second.	Internal ear.	Tone.
	Below 13 mm.	Above 30,000 per second.	None.	None.
Ether waves.	∞ to .2 mm. (electric waves).	0 to 1500 billion (1.5×10^{12}).	None.	None.
	.1 mm. to .0004 mm.	3000 billion (3×10^{12}) to 800,000 billion (8×10^{14}).	Skin.	Radiant heat.
	.0008 mm. to .0004 mm.	400,000 billion (4×10^{14}) to 800,000 billion (8×10^{14}).	Retina.	Light and color.
	.0004 mm. to .000008 mm. (ultra-violet rays).	800,000 billion (8×10^{14}) to 40,000,000 billion (4×10^{16}).	None.	None.
	.00002 mm. to .00000001 mm. (x-rays).	15,000,000 billion (1.5×10^{16}) to 30,000,000,000 billion (30×10^{18}).	None.	None.
	.00000014 mm. to .0000000005 mm. (γ-rays).	2 billion billion (2×10^{18}) to 600 billion billion (6×10^{20}).	None.	None.
	.00000000005 mm. to .000000000008 mm. (cosmic rays).	6,000 billion billion (6×10^{21}) to 40,000 billion billion (4×10^{22}).	None.	None.

*The use of names for large numbers is not uniform in different countries, and so I give, in brackets, the United States and French equivalents to the English names.

Million $1,000,000 = 10^6$; (million).
Milliard $1,000,000,000 = 10^9$; (billion or milliard).
Billion $10^6 \times 10^6 = 10^{12}$; (trillion).
In this table 1 billion $= 10^9$.

As a further illustration of the mechanism of abstracting, we may suggest the observation of Weber that if, for instance, a room is lighted with 100 candles, and if one more candle is brought in, the increased illumination will be appreciated very slightly. But not so if we had a room illuminated with 1000 candles. In this case, we should not appreciate the addition of one candle at all. Ten candles should be introduced to make an appreciable difference in our perceptions. The Weber law, as it is called, stated that in the above case 1/100 of the original strength of the stimulus is needed to make a change appreciable. For light, the fraction is about 1/100; for noise, about 1/3; for pressure, it varies between 1/30 and 1/10; for weight, between 1/70 and 1/40 in various parts of the body.

If we use compasses and experiment with pricks, we find that in different parts of the body the limit of the distance apart of the points when we *feel one prick* and yet have two, is different.

On the tip of the tongue this limit is........................... 1 mm.
On the palmar surface of third phalanx of forefinger.......... 2 mm.
On the palmar surface of second phalanges of fingers.......... 4 mm.
On the palm of the hand...................................... 10 mm.
On the dorsal surface of first phalanges of fingers............. 14 mm.
On the back of hand... 25 mm.
On the upper and lower parts of forearm..................... 37 mm.
On the middle thigh and back............................... 62 mm.[2]

A 'sensation' requires appreciable 'time' (times by a clock) for its development. Part of the 'time' is spent at the end-organ, part in conveying the nervous impulse along the nerves to the brain and part in the brain. A 'sensation' usually outlasts the stimulus, and often a single stimulus produces a whole series of 'after-sensations'.

As compared with the 'sensations' obtained from pain spots, touch is quicker in its development and persistence. With a vibrating string, 1500 vibrations a second are recognizable by touch as vibrations. At over 1552 vibrations a second, the vibratory character is lost, and we feel only continuous pressure. A toothed revolving wheel gives the feeling of smoothness (and 'continuity') when the teeth meet the skin at the rate of from 480 to 680 per second.[3]

The above given tables and facts are deeply significant. We see, first of all, that structurally we are immersed in a world full of energy manifestations, out of which we abstract directly only a very small portion, these abstractions being already coloured by the specific functioning and structure of the nervous system—the abstractors. Very probably, there are many more energy manifestations which, as yet, we have not discovered. Every few years we discover some new form of energy manifestation, and, at present, our knowledge is already so advanced that it is highly probable that the list is much longer. Finally, and here the whole 'structure of human knowledge' begins to play its role; for sanity *we have to know and evaluate this world* around us, if we want to adjust ourselves satisfactorily to it.

Section C. Problems of adjustment.

Is the problem of adjustment in the animal world similar to that in the human world? No, it is entirely different. Animals do not alter their environment so rapidly, nor to such an extent as humans do. Animals are not time-binders; they have not the capacity by which each generation can start where the former left off. Neurologically, animals have no means for extra-neural extensions, which extensions involve the complex mechanism with which we are dealing throughout this work.

The example of the caterpillar, already cited, shows clearly how organisms not adapted to their environment perish and do not propagate their special, non-survival characteristics. Similar remarks apply to hens, their eggs, and chicks which are kept in buildings without sunlight or with ordinary glass windows; these, also, do not survive, and so pass out of the picture.

With humans, the situation is entirely different. We are able to produce conditions which do not exist in unaided nature. We produce artificial conditions and so *our numbers and distribution* are not regulated by unaided nature alone. Animals cannot over-populate the globe, as they do not produce artificially. We do over-populate this globe because we produce artificially. With *animals, selfishness comes before altruism, and the non-selfish perish. An animal has to live first, then act.* With man, the reverse is true. The selfish may produce such conditions that they are destroyed by them. We can over-populate the globe because of artificial production, and so we are actually born nowadays into a world where we must *act first before we can live.* As I have already shown in my *Manhood of Humanity* (p. 72), the old animalistic, fallacious generalizations have been, and are, the foundation of our 'philosophies', 'ethics', systems., and naturally such animalistic doctrines must be disastrous to us. Neurologically, we build up conditions which our nervous systems cannot stand; and so we break down, and, perhaps, shall not even survive.

Animals have no 'doctrines' in *our meaning* of the term; thus, doctrines are no part of their environment, and, accordingly, animals cannot perish through false doctrines. We do have them, however, and, since they are the most vital environmental semantic conditions regulating our lives, if they are fallacious, they make our lives unadjusted and so, ultimately, lead to non-survival.

So we see that 'human adjustment' is quite a different and much more complex affair than 'animal adjustment'. The 'world' of 'man' is also a different and much more complex 'world' than that of the animal. There seems to be no escape from this conclusion. We see, also, that what we used to call 'senses' supply us with information about the world that is very limited in quantity, *specific* in quality, an abstraction of low order, never being 'it'. Being often unaffected, our 'senses' are not able to abstract, obviously, some of the most fundamental manifestations of energy to be found in the external world. If we speak of the older so-called 'sense perceptions' as lower order abstractions, then we find that we learn about the other subtler manifestations of energy through science, higher neural and extra-neural means, which we may call higher order abstractions. In the older days, we called this kind of knowledge 'inferential knowledge'. The animals do not have these higher order abstractions in that sense, and so their world is for them devoid of these extra-sensible manifestations of energy.

It must be remembered that these higher order abstractions and the 'inferential knowledge' of the old theories (they are not equivalent by definition) have a very similar status. Organisms work as-a-whole, and to separate completely higher and lower order abstractions is impossible. All that is said here justifies the new terminology. Our nervous system does abstract, does summarize, does integrate on different levels and in different orders, and the *result* of a stimulus *is not* the stimulus itself. The stone *is not* the pain produced by the stone dropping on our foot; neither is the flame we see, nor the burn we feel. The actual process goes on outside of our skin, as represented by the 'realities' of modern science.

We have already spoken frequently of the different order abstractions, their special characteristics, dynamic versus static., and the means of translation of lower orders into higher, and vice versa. Events which are going on and for which we have no direct 'senses' of abstraction, as, for instance, electric waves, Röntgen-rays, wireless waves., we know only through extra-neural extensions of our nervous system given by science and scientific instruments. Naturally, we should expect that the structure of our abstracting mechanism would be also reflected in these higher order abstractions. Facts show this to be true; and practically all modern science proves it directly or indirectly. This is why, for instance, we have the mathematical methods for passing from dynamic to static, and vice versa; why we have quantum theories and conditions; and why we have problems of continuity versus discontinuity, atomic theories,.

The above is not a plea for certain old-fashioned 'idealistic philosophies' and still less for 'solipsism'. Far from it. The object of this present work is to face hard structural experimental $m.o$ facts, analyse these facts in a language of a similar *structure* $(\bar{A})$, and so to reach tentatively new conclusions which again can be verified by experiments. Once more the reader must be warned against carelessly translating the structurally *new* terms into the *old* terms. The complete structural, psycho-logical, semantic, and neurological analysis of one single such new term would afford material for several volumes and so is impossible in this work. The usefulness of the old terms has been exhausted. The structural consequences of the old terms have been practically all worked out, and, as a rule, we cannot have much quarrel with the older conclusions in the *old language*. If we reach *different* conclusions, or get some new emphasis, it will be due to the use of the structurally *new* language. If we translate the new into the old, the old conclusions are usually *truer* than the new ones. The reverse is also true; the old conclusions become false or, at best, only gain emphasis because of the structure of the new language. The problem of all theories, old or new, is to give a structural account of the facts known, to account for exceptions, and to predict new experimental structural facts which again may be verified empirically.

Section D. Semantic considerations.

We speak much and vaguely about the 'structure' of language, but extremely little work has been done in this field. In the present work, we not only tackle this problem as best we can, theoretically, but we also use a language of a new *non-el,* functional structure, and the results, whatever their value, are actually the results of such procedure.

A short while ago we did not even know that such problems existed. Dreams, alone, about such problems did not help, for, before the structures of two different entities can be compared, these entities must first be produced. Then, and only then, can we compare and evaluate them. Before we could compare the A, the E, and the N systems with $\bar{A}$, $\bar{E}$, and $\bar{N}$ systems the last had to be produced, no matter how imperfect they might be at the beginning.

Something similar can be said about languages. Before we can speak of them in the plural and compare them, we must have more than one for comparison. Mathematics has pointed out this problem for us in

16

geometry. For instance, we have to deal with different frames or refer- ences, or different systems of co-ordinates. We find that they represent different languages and that they may introduce purely verbal statements which have nothing to do with the subject of our analysis (extrinsic characteristics). We have also found that some characteristics may appear in one form of representation and not appear so readily in an- other. For instance, we know that every line, except the X axis through a point 0 which is the intersection of a parabola with its X axis, cuts the curve a second time. This fact, important for us to know, appears clearly in the polar co-ordinate form of representation, but does not appear in the rectangular form of equations, although, when once a characteristic of a curve is discovered, it can be usually translated into the other co-ordinate languages. In such cases as this, a language of new structure has a kind of creative character, in that it makes some structural discoveries easier.

But the co-ordinate methods were not quite satisfactory. They introduced, too easily, too many extraneous, extrinsic characteristics, belonging to the language and not to the subject. Mathematicians de- cided to get away from these metaphysical 'outside' references by refer- ring the entity to itself to become more experimental. They invented the internal theory of surfaces, a vector language where they *refer the entity to itself,* its curvature, its length and direction. Finally, in the extension of the vector language which is called the tensor calculus, they achieved a still larger kind of independence. Having invented *three* languages in which we can speak about *one* issue, we are now able to meet the problem of *comparison* of these languages. At once, most im- portant structural and methodological problems arise.

The newer quantum mechanics are also an epoch-making linguistic structural innovation, not only in physics, but also in *methodology*. We have, at present, three, or, perhaps, more, quantum mechanics which speak about one subject, but in entirely different languages. I say 'three or more', because, from a methodological point of view, it is very hard, at present, to be precise, as the problems are too new and, as yet, too little analysed. Similar remarks apply to systems. Before two systems can be compared, a second system must be produced. Then we can compare them.

In our brief verbal analysis of 'space', and 'time', and 'matter', we have seen that these represent *terms,* or *linguistic means, not objects.* We have seen, also, that these antique forms of representation have very unsatisfactory structural implications. They introduce a verbal *elementalism* structurally *absent* in nature, and by a process of objectification lead to many kinds of fanciful semantically harmful metaphysics. Since Einstein and Minkowski, the excellent term 'event' has been introduced into scientific literature. It is a term of such epoch-making semantic importance that it should become a term of daily use and should be introduced into elementary schools. Teachers do not perform their duties honestly or intelligently if they disregard such structural, linguistic, and semantic issues, which, as we have seen, are the central problem of all possible education.

Likewise, we have already seen that the chunk of nature which we call a 'pencil' *is not* 'matter' nor 'space' nor 'time', the terms being only *terms*. Is such *el language* structurally appropriate for the purpose of speaking about the world around us? It seems undeniable that such language is quite out-dated and very unsatisfactory. It introduces structurally an artificial elementalism of a verbal character, in spite of the fact that even the most elementary consideration shows that structurally the opposite is true; namely, that 'matter', 'space', and 'time' can *never* be experimentally divided. The term 'event' is precisely the term which does away with this old and vicious elementalism.

All that we deal with in the outside world involves indivisibly 'matter', 'space', and 'time'. Using the old language, there cannot be something somewhere at 'no time', or something at some 'time' and 'no where' or 'nothing' 'somewhere' at 'some time'. Everything which happens must be structurally represented as something, somewhere, at some 'time'. If the structure of the world happened to be such that 'nothing' would happen 'nowhere' at 'no time', then we should have nothing to talk about, and all we would or could say would deal with our fancies. The four-dimensional language, which describes happenings structurally more nearly as we experience them, is precisely the language of 'events'. It should be remembered that in daily life we live by four-dimensional event-conditions. That is, the events which interest us are something, somewhere, and some 'time'. If we want two of our friends to become acquainted with each other, we invite them to our home. The appointment is in three dimensions in 'space' (to the left or to the right, forward or backward, up or down), and at a given hour. So we see that our daily life is lived in a four-dimensional space-time manifold, and we begin to appreciate the fact that science has lately caught up with such fundamental structural 'realities'. It must be noticed that the new four-dimensional space-time *language* does not, or should not, use the *term* 'matter' as we used it in the old way. In the new language, the bits of materials we deal with are connected analytically with the 'curvature' of this space-time manifold.

The reader should realize that the structurally new language is similar to the structure of our experience, and involves profound methodological and, therefore, psycho-logical, semantic factors. It has entirely different semantic values; and, perhaps, because of this fact, it is an irreversible advance, no matter how details may be altered.

The newtonians, for the most part, overlook the fact that all theories, their own included, are a semantic product of the functioning of the nervous system, and so involve some 'logic' and 'psychology'. In the new theories, a kind of *physical* subjectivity always appears which ought to be taken care of. We know, for instance, that if we immerse a part of a straight stick in water, the stick appears to be broken, although actually it has not been broken. A photographic camera gives a similar record. So we see that, besides the psycho-logical subjectivity, there is a most important *physical subjectivity*, which is introduced by the use of instruments. The main difficulties in modern science are precisely in the elimination of this physical subjectivity, particularly when we deal with such minute entities that the light waves miss them. In the case of an hypothetical gamma-ray microscope, for instance, the rays would produce what is called a Compton effect,* and so the results of the experiment would be altered by the instrument and procedure.

We ought not to be surprised that the old systems of 'motion' and 'emotion' in science (Newton) and 'philosophy' (Bergson) should result from speculations on the old $\bar{A}$ el language and the introduction of fanciful and fallacious assumptions of an 'infinity' somewhere, and other fancies. The realization of this marks a new semantic epoch in our lives. It is to the credit of these two men that they have summarized these old tendencies in such a masterly way that we are enabled to go beyond them. We shall return to this subject when we analyse the four-dimensional 'world' of Minkowski, and, then, we shall summarize briefly what we now know about 'space', 'time', and 'matter' (see Parts IX and X).

---

*Compton discovered, in 1923, that the generation of secondary continuous Röntgen radiation by a primary radiation is accomplished by an increase in the wave-length.

PART V

ON THE NON-ARISTOTELIAN LANGUAGE
CALLED MATHEMATICS

Once a statement is cast into mathematical form it may be manipulated in accordance with these rules and every configuration of the symbols will represent facts in harmony with and dependent on those contained in the original statement. Now this comes very close to what we conceive the action of the brain structures to be in performing intellectual acts with the symbols of ordinary language. In a sense, therefore, the mathematician has been able to perfect a device through which a part of the labor of logical thought is carried on outside the central nervous system with only that supervision which is requisite to manipulate the symbols in accordance with the rules. (583)   HORATIO B. WILLIAMS

The toughminded suggest that the theory of the infinite elaborated by the great mathematicians of the Nineteenth and Twentieth Centuries, without which mathematical analysis as it is actually used today is impossible, has been committing suicide in an unnecessarily prolonged and complicated manner for the past half century. (22)   E. T. BELL

The solution goes on famously; but just as we have got rid of the other unknowns, behold! V disappears as well, and we are left with the indisputable but irritating conclusion—
$$0 = 0$$
This is a favourite device that mathematical equations resort to, when we propound stupid questions. (149)   A. S. EDDINGTON

Who shall criticize the builders? Certainly not those who have stood idly by without lifting a stone. (23)   E. T. BELL

. . . let me remind any non-mathematicians . . . that when a mathematician lays down the elaborate tools by which he achieves precision in his own domain, he is unprepared and awkward in handling the ordinary tools of language. This is why mathematicians always disappoint the expectation that they will be precise and reasonable and clear-cut in their statements about everyday affairs, and why they are, in fact, more fallible than ordinary mortals. (529)   OSWALD VEBLEN

CHAPTER XVIII

MATHEMATICS AS A LANGUAGE OF A STRUCTURE SIMILAR TO THE STRUCTURE OF THE WORLD

To-day there are not a few physicists who, like Kirchhoff and Mach, regard the task of physical theory as being merely a mathematical description (*as economical as possible*) of the empirical connections between observable quantities, *i. e.* a description which reproduces the connection, as far as possible, without the intervention of unobservable elements. (466)   E. SCHRÖDINGER

But in the prevalent discussion of classes, there are illegitimate transitions to the notions of a 'nexus' and of a 'proposition.' The appeal to a class to perform the services of a proper entity is exactly analogous to an appeal to an imaginary terrier to kill a real rat. (578)   A. N. WHITEHEAD

Roughly it amounts to this: mathematical analysis as it works today must make use of irrational numbers (such as the square root of two); the sense if any in which such numbers exist is hazy. Their reputed mathematical existence implies the disputed theories of the infinite. The paradoxes remain. Without a satisfactory theory of irrational numbers, among other things, Achilles does not catch up with the tortoise, and the earth cannot turn on its axis. But, as Galileo remarked, it does. It would seem to follow that something is wrong with our attempts to compass the infinite. (22)   E. T. BELL

The map is not the thing mapped. When the map is identified with the thing mapped we have one of the vast melting pots of numerology. (604) E. T. BELL

The theory of numbers is the last great uncivilized continent of mathematics. It is split up into innumerable countries, fertile enough in themselves, but all more or less indifferent to one another's welfare and without a vestige of a central, intelligent government. If any young Alexander is weeping for a new world to conquer, it lies before him. (23)   E. T. BELL

The present work—namely, the building of a *non-aristotelian system,* and an introduction to a theory of sanity and general semantics—depends, fundamentally, for its success on the recognition of mathematics as a language similar in structure to the world in which we live.

The maze of often unconnected knowledge we have gathered in the fields with which this part is dealing is so tremendous that it would require several volumes to cover the field even partially. Under such conditions, it is impossible to deal with the subject in any other way than by very careful selection, and so I shall, therefore, say only as much as is necessary for my present semantic purpose.

It is a common experience of our race that with a happy generalization many unconnected parts of our knowledge become connected; many 'mysteries' of science become simply a linguistic issue, and then the mysteries vanish. New generalizations introduce new *attitudes* (evaluation) which, as usual, seriously simplify the problems for a new generation. In the present work, we are treating problems from the point of view of such a generalization, of wide application; namely, *structure,* which is forced upon us by the denial of the 'is' of identity.; so that structure becomes the only link between the objective and verbal levels. The next consequence is that structure alone is the only possible content of knowledge.

247

Investigating structure, we have found that structure can be defined in terms of relations; and the latter, for special purposes, in terms of multi-dimensional order. Obviously, to investigate structure, we must look for relations, and so for multi-dimensional order. The full application of the above principles becomes our guide for future enquiry.

In the recent past, we have become accustomed to such arguments as, for instance, that the theory of Einstein has to be accepted on 'epistemological' grounds. Naturally, the scientist or the layman who has heard the last term, but never bothered to ascertain that it means 'according to the structure of human knowledge', would recognize no necessity to accept something which violates all his habitual $s.r$, for reasons about which he does not know or care. But if we say that the Einstein theory has to be accepted, for the 'time' being, at least, as an irreversible *structural linguistic progress,* this statement carries for many quite a different verbal and semantic implication, and one worth considering.

Mathematics has, of late, become so extremely elaborate and complex that it takes practically a lifetime to specialize in even one of its many fields. Here and there notions of extreme creative generality appear, which help us to see relations and dependence between formerly non-connected fields. For instance, the arithmetization of mathematics, or the theory of groups, or the theory of aggregates, has each become such a supreme generalization. At present, there is a general tendency among all of us, scientists included, to confuse orders of abstractions. This results in a psycho-logical semantic blockage and in the impossibility of seeing broader issues clearly.

Some of the structural issues are still but little understood, and, in writing this chapter, I lay myself open to a reproach from the layman that I have given too much attention to mathematics, and from the professional mathematician that I have given too little. My reply is that what is said here is necessary for rounding up the semantic foundations of the system, and that I explain only enough to carry the main points of structure and as semantic suggestions for further semantic researches.

I have found that among some physicists and some mathematicians the thesis that mathematics is the only language which, in 1933, is similar in structure to the world, is not always acceptable. As to the second thesis, the similarity of its structure to our nervous system, some even seem to feel that this statement borders on the sacrilegious! These objectors, apparently, believe that I ascribe more to mathematics than is just. Some physicists point out to me the non-satisfactory development of mathematics, and they seem to confuse the inadequacy of a given mathematical theory with the general $m.o$ structure of mathematics. Thus, if some physical experimental investigation is conducted—for instance, on high pressure—and the older theories predict a behaviour exemplified by the curve (A), while the experimental new data show that the actual curve is (B), such a result would show unmistakably that the first theory is not structurally correct. But, in itself, this result does not affect the

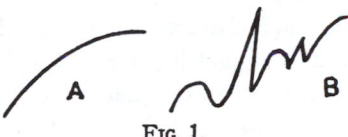

FIG. 1.

correctness of a statement about the general structure of mathematics which can account for *both curves*.

Until very lately, we had a very genuine problem in physics with the quantum phenomena which seem to proceed by discrete steps, while our mathematics is fundamentally based on assumptions of continuity. Here we had seemingly a serious structural discrepancy, which, however, has been satisfactorily overcome by the wave theory of the newer quantum mechanics, explained in Part X, where the discontinuities are accounted for, in spite of the use of differential equations and, therefore, of continuous mathematics.

But, if we start with fundamental assumptions of continuity, we always can account for discontinuities by introducing wave theories or some similar devices. Therefore, it is impossible, in our case, to argue from the wave theory (for instance) to the structure of mathematics, or vice versa, without a fundamental and *independent* general structural analysis, which alone can elucidate the problem at hand.

Mathematicians may object on the ground that the new revision of the foundation of mathematics, originated by Brouwer and Weyl, challenges the 'existence' of irrational numbers., and, therefore, destroys the very foundations of continuity and the legitimacy of existing mathematics.

In answer to such a criticism, we should notice, first, that the current 'continuity' is of two kinds. One is of a higher grade, and is usually called by this name; the other continuity is of a lower grade and is usually called 'compactness'. The new revision challenges the higher continuity, but does not affect compactness, which. as a result, will, perhaps, have to suffice in the future for all mathematics, since compactness is sufficient to meet all psycho-logical requirements, once the problems of 'infinity' are properly understood.

A structural independent analysis of mathematics, treated as a language and a form of human behaviour, establishes the similarity of this language to the undeniable structural characteristics of this world and of the human nervous system. These few and simple structural foundations are arrived at by inspection of known data and may be considered as well established.

The existing definitions of mathematics are not entirely satisfactory. They are either too broad, or too narrow, or do not emphasize enough the main characteristics of mathematics. A semantic definition of mathematics should be broad enough to cover all existing branches of mathematics; should be narrow enough to exclude linguistic disciplines which are not considered mathematical by the best judgement of specialists, and should also be *flexible* enough to remain valid, no matter what the future developments of mathematics may be.

I have said that mathematics is the only language, at present, which in structure, is similar to the structure of the world *and* the nervous system. For purposes of exposition, we shall have to divide our analysis accordingly, remembering, in the meantime, that this division is, in a way, artificial and optional, as the issues overlap. In some instances, it is really difficult to decide under which division a given aspect should be analysed. The problems are very large, and for full discussion would require volumes; so we have to limit ourselves to a suggestive sketch of the most important aspects necessary for the present investigation.

From the point of view of general semantics, mathematics, having symbols and propositions, must be considered as a language. From the psychophysiological point of view, it must be treated as an activity of the human nervous system and as a form of the behaviour of the organisms called humans.

All languages are composed of two kinds of words: (1) Of *names* for the somethings on the un-speakable level, be they external objects., or *internal feelings,* which admittedly are *not* words, and (2) of *relational terms,* which express the actual, or desired, or any other relations between the un-speakable entities of the objective level.

When a 'quality' is treated physiologically as a reaction of an organism to a stimulus, it also becomes a relation. It should be noticed that often some words can be, and actually are, used in both senses; but, in a given context, we can always, by further analysis, separate the words used into these two categories. Numbers are not exceptions; we can use the labels 'one', 'two'., as numbers (of which the character will be explained presently) but also as names for anything we want, as, for instance, Second or Third Avenue, or John Smith I or John Smith II. When we use numbers as names, or labels for anything, we call them numerals; and this is *not* a mathematical use of 'one', 'two'., as these names do not follow mathematical rules. Thus, Second Avenue and Third Avenue cannot be added together, and do not give us Fifth Avenue in any sense whatever.

Names alone do not produce propositions and so, by themselves, say nothing. Before we can have a proposition and, therefore, meanings, the names must be related by some relation-word, which, however, may be explicit or implied by the context, the situation, by established habits of speech,. The division of words into the above two classes may seem arbitrary, or to introduce an unnecessary complication through its simplicity; yet, if we take modern knowledge into account, we cannot follow the grammatical divisions of a primitive-made language, and such a division as I have suggested above seems structurally correct in 1933.

Traditionally, mathematics was divided into two branches: one was called arithmetic, dealing with numbers; the other was called geometry, and dealt with such entities as 'line', 'surface', 'volume',. Once Descartes, lying in bed ill, watched the branches of a tree swaying under the influence of a breeze. It occurred to him that the varying distances of the branches from the horizontal and vertical window frames could be expressed by numbers representing measurements of the distances. An epoch-making step was taken: geometrical relations were expressed by numerical relations; it meant the beginning of analytical geometry and the unification and arithmetization of mathematics.

Further investigation by the pioneers Frege, Peano, Whitehead, Russell, Keyser, and others has revealed that 'number' can be expressed in 'logical terms'—a quite important discovery, provided we have a valid 'logic' and structurally correct *non-el* terms.

Traditionally, too, since Aristotle, and, in the opinion of the majority, even today, mathematics is considered as uniquely connected with quantity and measurement. Such a view is only partial, because there are many most important and fundamental branches of mathematics which have nothing to do with quantity or measurement—as, for instance, the theory of groups, analysis situs, projective geometry, the theory of numbers, the algebra of 'logic',.

Sometimes mathematics is spoken about as the science of relations, but obviously such a definition is too broad. If the only content of knowledge is structural, then relations, obvious, or to be discovered, are the foundation of all knowledge and of *all* language, as stated in the division of words given above. Such a definition as suggested would make mathematics co-extensive with *all* language, and this, obviously, is not the case.

Before offering a semantic definition of mathematics, I introduce a synoptical table taken from Professor Shaw's *The Philosophy of Mathematics,* which he calls only suggestive and 'doubtless incomplete in many ways'. I use this table because it gives a modern list of the most important mathematical terms and disciplines necessary for the purpose of this work, indicating, also, in a way, their evolution and structural interrelations.

CENTRAL PRINCIPLES OF MATHEMATICS*

					Objects	WORLDS	MORPHOLOGY	INVARIANCY	FUNCTION-ALITY	IDEALITY
MATHEMATICS ≡ QUALITATIVE STUDY OF STRUCTURE	PATTERNS	Static Mathematics	Chaotic	Rhythm Patterns—Arithmetic	Numbers	Integers Rationals Irrationals Ensembles — ATOMICITY	Integers Ensemble theory Literal arithmetic Complexes Point geometries Functional spaces	Congruences Arithmetic invariants Modular geometry	Arithmetic functions Algebraic functions Functions of real variable Infinitesimal calculus General analysis	Arithmetic ideals Galois ideals Higher number theory Picard-Vessiot theory
				Space Patterns—Geometry	Figures	Points Lines Surfaces Varieties Higher elements — CONTINUITY	Point space of two or more dimensions Line geometry Surface geometry Absolute geometry Higher elements Expansions	Geometric invariants Algebraic invariants Symmetric and alternating forms Modular systems	Real functions of N variables Vector fields Functions of lines Partial derivatives Differential geometry	Systems of differential equations Mathematical physics Relativity Theory
			Ordered	Design Patterns—Tactic	Forms	Arrangements Configurations Constellations — DESIGN	Combinatory analysis Stereochemistry Theory of symbols Poincaré's theory of differential equations	Stable systems Irreducibles	Functions of arrangements, configurations, constellations, and structures	Ideal elements of construction Mathematical chemistry
				Idea Patterns—Logistic	Propositions	Concepts Relatives — CONSISTENCY	Foundations Postulational systems Calculus of classes, or of relatives	Equivalent systems Logical invariants	Classes functions of classes Functions of relatives Implications	Ideal classes or relatives Classificatory schemes
	MOTRICITIES	Kinematic Mathematics	Chaotic	Mutation Motricities—Operators	Operators	Substitutions Transformations Groups — MUTATION	Theory of finite or infinite groups Calculus of operations	Projective geometry Inversion geometry Differential and integral invariants Analysis situs	Geometric transformations Homomorphisms Transmutations	Automorphic functions Functional equations Calculus of variations Functional analysis Integral equations Difference equations Differential equations
				Quality Motricities—Algebras	Vids	Negatives Imaginaries Hyper-numbers — QUALITY	Linear algebras Functional transformations	Invariant equations Invariants of expressions	Functions of complex and hyper-complex variables General function theory	Hyperideals
			Ordered	Action Motricities—Processes	Processes	Routes Displacements Combinations — ACTION	Composite actions Actional structures	Equivalents of actions Invariants of processes	Processes dependent on processes Functions of action	Ideal processes
				Inferences	Thoughts	Transpositions Syllogisms Implications — DEDUCTION	Syllogisms Calculus of deduction	Laws of inference Equivalent deductions Invariants	Functions of inferences	Ideal entities that satisfy inferences Scientific theories

*This table differs slightly from the one printed in *The Philosophy of Mathematics*. The corrections have been made by Professor Shaw and kindly communicated to me by letter.

A semantic definition of mathematics may run somehow as follows: Mathematics consists of limited linguistic schemes of multiordinal relations capable of exact treatment at a given date.

After I have given a semantic definition of number, it will be obvious that the above definition covers all existing disciplines considered mathematical. However, these developments are not fixed affairs. Does that definition provide for their future growth? By inserting as a fundamental part of the definition 'exact treatment at a given date', it obviously does. Whenever we discover any relations in any fields which will allow exact 'logical' treatment, such a discipline will be included in the body of linguistic schemes called mathematics, and, at present, there are no indications that these developments can ever come to an end. When 'logic' becomes an ∞-valued 'structural calculus', then mathematics and 'logic' will merge completely and become a general science of *m.o* relations and multi-dimensional order, and all sciences may become exact.

It is necessary to show that this definition is not too broad, and that it eliminates notions which are admittedly non-mathematical, without invalidating the statement that the content of all knowledge is structural, and so ultimately relational. The word 'exact' eliminates non-mathematical relations. If we enquire into the meaning of the word 'exact', we find from experience that this meaning is not constant, but that it varies with the date, and so only a statement 'exact at a given date' can have a definite meaning.

We can analyse a simple statement, 'grass is green' (the 'is' here is the 'is' of predication, not of identity), which, perhaps, represents an extreme example of a non-mathematical statement; but a similar reasoning can be applied to other examples. Sometimes we have a feeling which we express by saying, 'grass is green'. Usually, such a feeling is called a 'perception'. But is such a *process* to be dismissed so simply, by just calling it a name, 'perception'? It is easy to 'call names under provocation', as Santayana says somewhere; but does that exhaust the question?

If we analyse such a statement further, we find that it involves comparison, evaluation in certain respects with other characters of experience., and the statement thus assumes relational characteristics. These, in the meantime, are non-exact and, therefore, non-mathematical. If we carry this analysis still further, involving data taken from chemistry, physics, physiology, neurology., we involve relations which become more and more exact, and, finally, in such terms as 'wave-length', 'frequency'., we reach structural terms which allow of exact 1933 treatment. It is true that a language of 'quality' conceals relations, sometimes very effectively; but once 'quality' is taken as the reaction of a given organism to a stimulus, the term used for that 'quality' becomes a name for a very complex relation. This procedure can be always employed, thus establishing once more the fundamental character of relations.

These last statements are of serious structural and semantic importance, being closely connected with the $\bar{A}$, fundamental, and undeniable negative premises. These results can be taught to children very simply; yet this automatically involves an entirely new and modern method of evaluation and attitude toward language, which will affect beneficially the, as yet entirely disregarded, s.r.

We must consider, briefly, the terms 'kind' and 'degree', as we shall need them later. Words, symbols., serve as forms of representation and belong to a different universe—the 'universe of discourse'—since they are not the un-speakable levels we are speaking about. They belong to a world of higher abstractions and not to the world of lower abstractions given to us by the lower nerve centres.

Common experience and scientific investigations (more refined experience) show us that the world around us is made up of absolute individuals, each different and unique, although interconnected. Under such conditions it is obviously optional what language we use. The more we use the language of diverse 'kind', the sharper our definitions must be. Psycho-logically, the emphasis is on difference. Such procedure may be a tax on our ingenuity, but by it we are closer to the structural facts of life, where, in the limit, we should have to establish a 'kind' for every individual.

In using the term 'degree', we may be more vague. We proceed by similarities, but such a treatment implies a fundamental interconnection between different individuals of a special kind. It implies a definite kind of metaphysics or structural assumptions—as, for instance, a theory of evolution. As our 'knowledge' is the result of nervous abstracting, it seems, in accordance with the structure of our nervous system, to give preference to the term 'degree' first, and only when we have attained a certain order of verbal sharpness to pass to a language of 'kind', if need arises.

The study of primitive languages shows that, historically, we had a tendency for the 'kind' language, resulting in over-abundance of names and few relation-words, which makes higher analysis impossible. Science, on the other hand, has a preference for the 'degree' language, which, ultimately, leads to mathematical languages, enormous simplicity and economy of words, and so to better efficiency, more intelligence, and to the unification of science. Thus, chemistry became a branch of physics; physics, a branch of geometry; geometry merges with analysis, and analysis merges with general semantics; and life itself becomes a physico-chemical colloidal occurrence. The language of 'degree' has very important *relational, quantitative,* and *order* implications, while that of 'kind' has, in the main, qualitative implications, often, if not always, concealing relations, instead of expressing them.

The current definition of 'number', as formulated by Frege and Russell, reads: 'The number of a class is the class of all those classes that are similar to it'.[1] This definition is not entirely satisfactory: first, because the multiordinality of the term 'class' is not stated; second, it is *A*, as it involves the ambiguous (as to the order of abstractions) term 'class'. What do we mean by the term 'class'? Do we mean an extensive array of absolute individuals, un-speakable by its very character, such as some *seen aggregate,* or do we mean the *spoken definition* or *description* of such un-speakable objective entities? The term implies, then, a fundamental confusion of orders of abstractions, to start with—the very issue which we must avoid most carefully, as positively demanded by the non-identity principle. Besides, if we explore the world with a 'class of classes'., and obtain results also of 'class of classes', such procedure throws no light on mathematics, their applications and their importance as a tool of research. Perhaps, it even increases the mysteries surrounding mathematics and conceals the relations between mathematics and human knowledge in general.

We should expect of a satisfactory definition of 'number' that it would make the semantic character of numbers clear. Somehow, through long experience, we have learned that numbers and measurement have some mysterious, sometimes an uncanny, importance. This is exemplified by mathematical predictions, which are verified later empirically. Let me recall only the discovery of the planet Neptune through mathematical investigations, based on its action upon Uranus, long before the astronomer actually verified this prediction with his telescope. Many, a great many, such examples could be given, scientific literature being full of them. Why should mathematics and measurement be so extremely important? Why should mathematical operations of a given Smith, which often seem innocent (and sometimes silly enough) give such an unusual security and such undeniably practical results?

Is it true that the majority of us are born mathematical imbeciles? Why is there this general fear of, and dislike for, mathematics? Is mathematics really so difficult and repelling, or is it the way mathematics is treated and taught by mathematicians that is at fault? If some light can be thrown on these perplexing semantic problems, perhaps we shall face a scientific revolution which might deeply affect our educational system and may even mark the beginning of a new period in standards of evaluation, in which mathematics will take the place which it ought to have. Certainly, there must be something the matter with our epistemologies and 'psychologies' if they cannot cope with these problems.

A simple explanation is given by a new $\bar{A}$ analysis and a *semantic* definition of numbers. What follows is written, in the main, for non-mathematicians, as the word 'semantic' indicates, but it is hoped that professional mathematicians (or some, at least) may be interested in the *meanings* of the term 'number', and that they will not entirely disregard it. As semantic, the definition seems satisfactory; but, perhaps, it is not entirely satisfactory for technical purposes, and the definition would have to be slightly re-worded to satisfy the technical needs of the mathematicians. In the meantime, the gains are so important that we should not begrudge any amount of labour in order to produce finally a mathematical and, this time, $\bar{A}$ semantic definition of numbers.

As has already been mentioned, the importance of notation is paramount. Thus the Roman notation for number—I, II, III, IV, V, VI.,—was not satisfactory and could not have led to modern developments in mathematics, because it did not possess enough positional and structural characteristics. Modern mathematics began when it was made possible by the invention or discovery of positional notation. We use the symbol '1' in 1, 10, 100, 1000., in which, because of its place, it had different values. In the expression '1', the symbol means 'one unit'; in 10, the symbol '1' means ten units; in 100, the symbol '1' means one hundred units,.

To have a positional notation, we need a symbol '0', called zero, to indicate an empty column and, at least, one symbol '1'. The number of special symbols for 'number' depends on what base we accept. Thus, in a binary system, with the base 2, our 1 is represented by 1; our 2, by a unity in the second place and a zero in the first place, thus by 10; our 3, by 11; our 4, by a unity in the third place and two zeros in the first and second places; namely, 100; our 5, by 101; our 6, by 110; 7, by 111; 8, by 1000; 9, by 1001; ... 15, by 1111; 16, by 10000,. In a binary system, we needed only the two symbols, 1 and 0. For a system with the base 3, we would need three symbols, 1, 2, 0: our 1 would be represented by 1; our 2, by 2; 3, by 10; 4, by 11; 5, by 12; 6, by 20,. In our decimal system, obviously, we need 10 symbols, 1, 2, 3, 4, 5, 6, 7, 8, 9, 0.

For more details on notation, the interested reader is referred to the fascinating and elementary book of Professor Danzig, *Number the Language of Science*. Here we only emphasize what is necessary for our purpose. Every system has its advantages and drawbacks. Thus, in the binary system, still used by some savage tribes, of which we retain traces when we speak of couples, or pairs, or braces, we get an enormous simplicity in operations by using only two symbols, 1 and 0. It should be remembered that, in every system, the tables of addition and multiplication must be memorized. In the binary system, these tables are reduced to $1 + 1 = 10$ and $1 \times 1 = 1$, while in our decimal system, each table has 100 entries. But what we gain in simplicity by a low base-number is offset very seriously by the cumbersomeness of the notation. As Danzig tells us, our number 4096 is represented in a binary system by 1,000,000,000. That we adopted the decimal system is probably a physiological accident, because we have ten fingers. The savage, with his binary system, did not reach even the finger stage; he is still in the fist stage.

For practical purposes, it is simpler to have a base which has many divisors, as, for instance, 12. We still use this duodecimal system when we divide a foot into twelve inches, or a shilling into twelve pennies, or count by dozens or gross. It seems that mathematicians would probably select a prime number for a base, but the gain would be so slight and the difficulty of offsetting a physiological habit so tremendous, that this will probably never happen.[2]

From what was said already, it is, perhaps, clear that mathematics requires a positional notation in which we must have a symbol for '1' and zero, at least. For these and other reasons, the two numbers 1 and 0 are somehow unusually important. Even in our decimal system we generate numbers by adding 1 to its predecessor. Thus $1 + 1 = 2$, $2 + 1 = 3.$, and we must enquire into the semantic character of these numbers.

The notions of matching, comparing, measuring, quantity, equality., are all interwoven and, by necessity, involve a circularity in definitions and implications if the analysis is carried far enough. The interested reader may be referred to the chapter on equality in Whitehead's *The Principle of Relativity* to learn more on this subject.

In the evolution of mathematics, we find that the notions of 'greater', 'equal', and 'less' precede the notion of numbers. Comparison is the simplest form of evaluation; the first being a search for relations; the second, a discovery of exact relations. This process of search for relations and structure is inherent and natural in man, and has led not only to the discovery of numbers, but also has shaped their two aspects; namely, the cardinal and the ordinal aspects. For instance, to ascertain whether the number of persons in a hall is equal to, greater than, or less than the number of seats, it is enough to ascertain if all seats are

17

occupied and there are no empty seats and no persons standing; then we would say that the number of persons is equal to the number of seats, and a *symmetrical relation* of equality would be established. If all seats were occupied and there were some persons standing in the hall, or if we found that no one was standing, yet not all were occupied, we would establish the *asymmetrical relation* of greater or less.

In the above processes, we were using an important principle; namely, that of *one-to-one* correspondence. In our search for relations, we assigned to each seat one person, and reached our conclusions without any counting. This process, based on the *one-to-one* correspondence, establishes what is called the cardinal number. It gives us specific relational data about this world; yet it is not enough for counting and for mathematics. To produce the latter, we must, first of all, establish a definite system of symbolism, based on a definite relation for generating numbers; for instance, $1 + 1 = 2$, $2 + 1 = 3$., which establishes a definite *order*. Without this ordinal notion, neither counting nor mathematics would be possible; and, as we have already seen, order can be used for defining relations, as the notions of relation and order are interdependent. Order, also, involves asymmetrical relations.

If we consider the two most important numbers, 0 and 1, we find that in the accepted symbolism, if $a = b$, $a - b = 0$; and if $a = b$, $a/b = b/b = 1$; so that both fundamental numbers express, or can be interpreted as expressing, a *symmetrical relation* of equality.

If we consider any other number—and this applies to all kinds of numbers, not only to natural numbers—we find that any number is not altered by dividing it by one, thus, $2/1 = 2$, $3/1 = 3$., in general, $N/1 = N$; establishing the *asymmetrical* relation, *unique* and *specific* in a given case that N is N times more than one.

If we consider, further, that $2/1 = 2$, $3/1 = 3$, *and so on,* are all *different, specific,* and *unique,* we come to an obvious and $\bar{A}$ semantic definition of number in terms of relations, in which 0 and 1 represent *unique* and *specific symmetrical* relations and all other numbers also *unique* and *specific asymmetrical* relations. Thus, if we have a result '5', we *can always say* that the number '5 is five times as many as one. Similarly, if we introduce apples. Five apples are five·times as many as one apple. Thus, a number in any form, 'pure' or 'applied', can always be represented as a relation, *unique* and *specific* in a given case; and this is the foundation of the exactness of dealing with numbers. For instance, to say that a is greater than b also establishes an asymmetrical relation, but it is *not unique* and *specific;* but when we say that a is five times greater than *b*, this relation is *asymmetrical, exact, unique,* and *specific.*

The above simple remarks are not entirely orthodox. That $5/1 = 5$ is very orthodox, indeed; but that numbers, in general, represent indefinitely many *exact, specific,* and *unique,* and, in the main, *asymmetrical* relations is a structural notion which necessitates the revision of the foundations of mathematics and their rebuilding on the basis of new semantics and a future structural calculus. When we say 'indefinitely many', this means, from the reflex point of view, 'indefinitely flexible', or 'fully conditional' in the semantic field, and, therefore, a prototype of human semantic reactions (see Part VI). The scope of the present work precludes the analysis of the notion of the lately disputed 'irrational'; but we must state that this revision requires new psycho-logical and structural considerations of fundamental 'logical' postulates and of the problems of 'infinity'. If, by an arbitrary process, we *postulate* the existence of a 'number' which alters all the while, then, according to the definition given here, such expressions should be considered as functions, perhaps, but not as a number, because they do not give us *unique* and *specific relations.*

These few remarks, although suggestive to the mathematician, do not, in any way, exhaust the question, which can only be properly presented in technical literature in a postulational form.

It seems that mathematicians, no matter how important the work which they have produced, have never gone so far as to appreciate fully that they are willy-nilly producing an ideal human relational language of structure similar to that of the world *and* to that of the human nervous system. This they cannot help, in spite of some vehement denials, and their work should also be treated from the semantic point of view.

Similarly with measurement. From a functional or actional and semantic point of view, measurement represents nothing else but a search for *empirical structure* by means of extensional, ordered, symmetrical, and asymmetrical relations. Thus, when we say that a given length measures five feet, we have reached this conclusion by *selecting* a unit called 'foot', an *arbitrary and un-speakable* affair, then laying it end to end five times in a definite *extensional order* and so have established the asymmetrical, and, in each case, *unique* and *specific* relation, that the given entity represents, in this case, five times as many as the arbitrarily selected unit.

Objection may be raised that the formal working out of a definition of numbers in terms of relations, instead of classes, would be very laborious, and would also involve a revision of the foundations of mathematics. This can hardly be denied; but, in the discussions of the foundations, the confusion of orders of abstractions is still very marked, thereby resulting in the manufacture of *artificial semantic difficulties*. Moreover, the benefits of such a definition, in eliminating the mysteries about mathematics, are so important that they by far outweigh the difficulties.

As the only possible content of knowledge is structural, as given in terms of relations and multiordinal and multi-dimensional order; numbers, which establish an endless array of exact, specific, and, in each case, unique, relations are obviously the most important tools for exploring the *structure* of the world, since structure can always be analysed in terms of *relations*. In this way, all mysteries about the importance of mathematics and measurement vanish. The above understanding will give the student of mathematics an entirely different and a very natural feeling for his subject. As his only possible aim is the study of structure of the world, or of whatever else, he must naturally use a *relational tool* to explore this complex of relations called 'structure'. A most spectacular illustration of this is given in the internal theory of surfaces, the tensor calculus., described in Part VIII.

In all measurements, we select a unit of a necessary kind, for a given case, and then we find a *unique* and *specific* relation as expressed by a number, between the given something and the selected unit. By relating different happenings and processes to the same unit-process, we find, again, *unique* and *specific* interrelations, in a given case, between these events, and so gather structural (and most important, because uniquely possible) wisdom, called 'knowledge', 'science',.

If we treat numbers as relations, then fractions and all operations become relations of relations, and so relations of higher order, into the analysis of which we cannot enter here, as these are, of necessity, technical.

It should be firmly grasped, however, that some fundamental human relations to this world have not been changed. The primitive may have believed that words *were* things (identification) and so have established what is called the 'magic of words' (and, in fact, the majority of us still have our *s.r* regulated by some such unconscious identifications); but, in spite of this, the primitive or 'civilized' man's words *are not,* and never could be, the things spoken about, no matter what semantic disturbances we might have accompanying their use, or what delusions or illusions we may cherish in respect to them.

At present, of all branches of mathematics the theory of numbers is probably the most difficult, obscure., and seemingly with the fewest applications. With a new $\bar{A}$ definition of numbers in terms of relations, this theory may become a relational study of very high order, which, perhaps, will some day become the foundation for epistemology and the key for the solution of all the problems of science and life. In the fields of cosmology many, if not the majority, of the problems, by necessity, cannot be considered as directly experimental, and so the solution must be epistemological.

At present, in our speculations, we are carried away by words, disregarding the simple fact that speaking about the 'radius of the universe', for instance, has *no meaning,* as it cannot possibly be observed. Perhaps, some day, we shall discover that such conversations are the result of our old stumbling block, identification, which leads to our being carried away by the sounds of words applicable to terrestrial conditions but meaningless in the very small, as discovered lately in the newer quantum mechanics, and, in the very large, as applied to the cosmos. An important illustration of the retardation of scientific progress, blocked by the confusion of orders of abstractions, is shown in the fact that the newer quantum mechanics were slow in coming, and though astronomers probably know about it, yet they still fail to grasp that expressions such as the 'radius of the universe', the 'running down of the universe'., are meaningless outside of psychopathology.

In this connection, we should notice an extremely interesting and important semantic characteristic; namely, that the term 'relation' is not only multiordinal but also *non-el,* as it applies to 'senses' and 'mind'. Relations are usually found empirically; so in a language of relations we have a language of similar structure to the world and a unique means for predictability and rationality.

Let me again emphasize that, from time immemorial, things have *not* been words; the only content of knowledge has been structural; mathematics has dealt, in the main, with numbers; no matter whether we have understood the character of numbers or not, numbers have expressed relations and so have given us structural data willy-nilly,. This explains why mathematics and numbers have, since time immemorial, been a favorite field, not only for speculations, but also why, in history, we find so many religious semantic disturbances connected with numbers. Mankind has somehow felt instinctively that in numbers we have a potentially endless array of *unique* and *specific exact relations,* which ultimately give us structure, the last being the only possible content of knowledge, because words are *not* things.

As relations, generally, are empirically present, and as man and his 'knowledge' is as 'natural' as rocks, flowers, and donkeys, we should not be surprised to find that the unique language of exact., relations called mathematics is, by necessity, the natural language of man and *similar in structure* to the world *and* our nervous system.

As has already been stated, it is incorrect to argue from the structure of mathematical theories to the structure of the world, and so try to establish the similarity of structure; but that such enquiry must be *independent* and start with quite ordinary structural experiences, and only at a later stage proceed to more advanced knowledge as given by science. Because this analysis must be independent, it can also be made very simple and elementary. All exact sciences give us a wealth of experimental data to establish the first thesis on similarity of structure; and it is unnecessary to repeat it here. I will restrict myself only to a minimum of quite obvious facts, reserving the second thesis—about the similarity of structure with our nervous system—for the next chapter.

If we analyse the silent objective level by objective means available in 1933, say a microscope, we shall find that whatever we can see, handle., represents an *absolute individual,* and *different from anything else in this world.* We discover, thus, an important *structural* fact of the external world; namely, that in it, everything we can see, touch., that is to say, all lower order abstractions represent absolute individuals, different from everything else.

On the verbal level, under such empirical conditions, we should then have a language of *similar structure;* namely, one giving us an indefinite number of *proper names, each different.* We find such a language *uniquely* in numbers, each number 1, 2, 3., being a unique, sharply distinguishable, *proper name* for a relation, and, if we wish, for anything else also.

Without some higher abstractions we cannot be human at all. No science could exist with absolute individuals and no relations; so we pass to higher abstractions and build a language of say x_i, $(i = 1, 2, 3, \ldots n)$, where the x shows, let us say, that we deal with a variable x with many values, and the number we assign to i indicates the individuality under consideration. From the structural point of view, such a vocabulary is similar to the world around us; it accounts for the individuality of the external objects, it also is similar to the structure of our nervous system, because it allows generalizations or higher order abstractions, emphasizes the abstracting nervous characteristics,. The subscript emphasizes the differences; the letter x implies the similarities.

In daily language a similar device is extremely useful and has very far-reaching psycho-logical semantic effects. Thus, if we say 'pencil$_1$', 'pencil$_2$', . . . 'pencil$_n$', we have indicated structurally two main characteristics: (1) the absolute individuality of the object, by adding the indefinitely individualizing subscript 1, 2, . . . n; and (2) we have also complied with the nervous higher order abstracting characteristics, which establish similarity in diversity of different 'pencils'. From the point of view of *relations,* these are usually found empirically; besides, they may be invariant, no matter how changing the world may be.

In general terms, the structure of the external world is such that we deal always on the objective levels with absolute individuals, with absolute differences. The structure of the human nervous system is such that it abstracts, or generalizes, or integrates., in higher orders, and so finds similarities, discovering often invariant (sometimes relatively invariant) relations. To have 'similar structure', a language should comply with both structural exigencies, and this characteristic is found in the mathematical notation of x_i, which can be enlarged to the daily language as 'Smith$_i$', 'Fido$_i$'., where $i = 1, 2, 3, . . . n$.

Further objective enquiry shows that the world and ourselves are made up of *processes,* thus, 'Smith$_{1900}$' is quite a different person from 'Smith$_{1933}$'. To be convinced, it is enough to look over old photographs of ourselves, the above remark being structurally entirely general. A language of 'similar structure' should cover these facts. We find such a language in the vocabulary of 'function', 'propositional function', as already explained, involving also four-dimensional considerations.

As words are not objects—and this expresses a structural fact—we see that the 'is' of identity is unconditionally false, and should be entirely abolished as such. Let us be simple about it. This last semantic requirement is genuinely difficult to carry out, because the general *el* structure of our language is such as to facilitate identification. It is admitted that in some fields some persons identify only a little; but even they usually identify a great deal when they pass to other fields. Even science is not free from identification, and this fact introduces great and artificial semantic difficulties, which simply vanish when we stop identification or the confusion of orders of abstractions. Thus, for instance, the semantic difficulties in the foundations of mathematics, the problems of 'infinity', the 'irrational'., the difficulties of Einstein's theory, the difficulties of the newer quantum theory, the arguments about the 'radius of the universe', 'infinite velocities', the difficulties in the present theory., ., are due, in the main, to semantic blockages or commitment to the structure of the old language—we may call it 'habit'—*which says structurally very little,* and which I disclose as a semantic disturbance of *evaluation* by showing the *physiological mechanism in terms of order*.

If we abolish the 'is' of identity, then we are left only with a functional, actional., language elaborated in the mathematical language of function. Under such conditions, a *descriptive* language of ordered happenings on the objective level takes the form of 'if so and so happens, then so and so happens', or, briefly, 'if so, then so'; which is the prototype of 'logical' and mathematical processes and languages. We see that such a language is again similar in structure to the external world descriptively; yet it is similar to the 'logical' nervous processes, and so allows us, because of this similarity of structure, predictability and so rationality.

In the traditional systems, we did not recognize the complete semantic interdependence of differences and similarities, the empirical world exhibiting differences, the nervous system manufacturing primarily similarities, and our 'knowledge', if worth anything at all, being the *joint product* of both. Was it not Sylvester who said that 'in mathematics we look for similarities in differences and differences in similarities'? This statement applies to our whole abstracting process.

The empirical world is such in structure (by inspection) that in it we can add, subtract, multiply, and divide. In mathematics, we find a language of similar structure. Obviously, in the physical world these actions or operations alter the relations, which are expressed as altered unique and specific relations, by the language of mathematics. Further, as the world is full of different shapes, forms, curves., we do not only find in mathematics special languages dealing with these subjects, but we find in analytical geometry unifying linguistic means for translation of one language into another. Thus any 'quality' can be formulated in terms of relations which may take the 'quantitative' character which, at present, in all cases, can be also translated into geometrical terms and methods, giving structures to be *visualized*.

It is interesting, yet not entirely unexpected, that the activities of the higher nervous centres, the conditional reflexes of higher order, the semantic reactions, time-binding included, should follow the exponential rules, as shown in my *Manhood of Humanity*.

In our experience, we find that some issues are additive—as, for instance, if one guest is added to a dinner party, we will have to add plates and a chair. Such facts are covered by additive methods and the language called 'linear' (see Part VIII). In many instances—and these are, perhaps, the most important and are strictly connected with submicroscopic processes—the issues are not additive, one atom of oxygen 'plus' two atoms of hydrogen, under proper conditions, will produce water, of which the characteristics are *not* the sum of the characteristics of oxygen and hydrogen 'added' together, but entirely *new* characteristics *emerge*. These may some day be taken care of by non-linear equations, when our knowledge has advanced considerably. These problems are unusually important and vital, because with our present low development and the lack of structural researches, we still keep an additive A language, which is, perhaps, able to deal with additive, simple, immediate, and comparatively unimportant issues, but is entirely unfit structurally to deal with principles which underlie the most fundamental problems of life. Similarly, in physics, only since Einstein have we begun to see that the primitive, simplest, and easiest to solve linear equations are not structurally adequate.

One of the most marked structural characteristics of the empirical world is 'change', 'motion', 'waves', and similar dynamic manifestations. Obviously, a language of similar structure must have means to deal with such relations. In this respect, mathematics is unique, because, in the differential and integral calculus, the four-dimensional geometries and similar disciplines, with all their developments, we find such a perfect language to be explained more in detail in the chapters which follow.

It will be profitable for our purpose to discuss, in the next chapter, some of the mathematical structural characteristics in connection with their similarity to the human nervous system; but here I will add only that, for our purposes, at this particular point, we must specially emphasize *arithmetical* language, which means numbers and arithmetical operations, the theory of function, the differential and integral calculus (language) and different geometries in their two aspects, 'pure' and applied. Indeed, Riemann tells us bluntly that the *science* of physics originated only with the introduction of differential equations, a statement which is quite justified, but to which I would add, that physics is becoming scientific since we began to eliminate from physics semantic disturbances; namely, identification and elementalism. This movement was originated, in fact, although not stated in an explicit form, by the Einstein theory and the new quantum theories, the psycho-logical trend of which is formulated in a *general semantic theory* in the present work.

It is reasonable to consider that metric geometry, and, in particular, the E-system, was derived from touch, and, perhaps, the 'kinesthetic sense', and projective geometry from sight.

Although the issues presented here appear extremely simple, and sometimes even commonplace, yet the actual working out of the verbal schemes is quite elaborate and ingenious, and impossible to analyse here more fully; so that only one example can be given.

The solution of mathematical equations is perhaps to be considered as the central problem of mathematics. The word 'equation' is derived from Latin *aequare*, to equalize, and is a statement of the symmetrical relation of equality expressed as $a = b$ or $a - b = 0$ or $a/b = 1$. An equation expresses the relation between quantities, some of which are known, some unknown and to be found. By the solution of an equation, we mean the finding of values for the unknowns which will satisfy the equation.

Linear equations of the type $ax = b$ necessitated the introduction of fractions. Linear equations with several variables led to the theory of determinants and matrices ., which underwent, later, a tremendous independent development; yet they originated in the attempt to simplify the solution of these equations.

Quadratic equations of the type $x^2 + ax + b = 0$ can be reduced to the form $z^2 = A$, the solution of which depends on the extraction of a square root. Here, serious difficulties arose, and seemingly necessitated the introduction of 'irrational' numbers and ordinary complex numbers, involving the square root of minus one $(\sqrt{-1} = i)$, a notion which revolutionized mathematics.

Cubic equations of the form $x^3 + ax^2 + bx + c = 0$ necessitated the extraction of cube roots, in addition to the problems connected with the solution of quadratic equations, and involved more difficulties, which have been analysed in an extensive literature.

Biquadratic equations of the type $x^4 + ax^3 + bx^2 + cx + d = 0$ involve the problems of quadratic and cubic equations. When we consider equations of a degree higher than the fourth, we find that we cannot solve them by former methods; and mathematicians have had to invent theories of substitutions, groups, different special functions and similar devices. The solution of differential equations introduced further difficulties, allied with the theory of function.

The linear transformations of algebraic polynomials with two or more variables in connection with the theory of determinants, symmetrical functions, differential operations ., necessitated the development of an extensive theory of algebraic forms which, at present, is far from being complete.

In the above analysis, I have refrained from giving details, most of which would be of no value to the layman, and unnecessary for the mathematician; but it must be emphasized that the theory of function and the theory of groups, with their very extensive developments, involving the theory of invariance, and, in a way, the theory of numbers, rapidly became a unifying 'foundation upon which practically the whole of mathematics is being rebuilt. Many branches of mathematics have become, of late, nothing more than a theory of invariance of special groups.

As to practical applications, there is no possibility to list them, and the number increases steadily. But, without the theory of analytic function, for instance, we could not study the flow of electricity, or heat, or deal with two-dimensional gravitational, electrostatic, or magnetic attractions. The complex number involving the square root of minus one was necessary for the development of wireless and telegraphy; the kinetic theory of gases and the building of automobile engines require geometries of n dimensions; rectangular and triangular membranes are connected with questions discussed in the theory of numbers; the theory of groups has direct application in crystallography; the theory of invariants underlies the theory of Einstein, the theory of matrices and operators has revolutionized the quantum theory; and there are other applications in an endless array.[3]

In Part VIII, different aspects of mathematics are analysed, but the interested reader can be referred also to the above-mentioned book of Professor Shaw for an excellent elementary, yet structural, view of the progress of mathematics.

CHAPTER XIX

MATHEMATICS AS A LANGUAGE OF A STRUCTURE SIMILAR TO THE STRUCTURE OF THE HUMAN NERVOUS SYSTEM

In recent times the view becomes more and more prevalent that many branches of mathematics are nothing but the theory of invariants of special groups. s. LIE

A natural law,—if, strictly speaking, there be such a thing outside the conception thereof,—is fundamentally nothing more nor less than a constant connection among inconstant phenomena: it is, in other words, an invariant relation among variant terms. (264) CASSIUS J. KEYSER

Whatsoever things are invariant under all and only the transformations of some group constitute the peculiar subject-matter of some (actual or potential) branch of knowledge. (264) CASSIUS J. KEYSER

The general laws of nature are to be expressed by equations which hold good for all systems of co-ordinates, that is, are co-variant with respect to any substitutions whatever (generally co-variant). (155) A. EINSTEIN

The things hereafter called tensors are further characterized by the fact that the equations of transformation for their components are linear and homogeneous. Accordingly, all the components in the new system vanish, if they all vanish in the original system. If, therefore, a law of nature is expressed by equating all the components of a tensor to zero, it is generally covariant. By examining the laws of the formation of tensors, we acquire the means of formulating generally covariant laws. (155) A. EINSTEIN

The thalamus is a centre of affective reactivity to sensory stimuli, while the cortex is an apparatus for discrimination. (411) HENRI PIÉRON

Section A. Introductory.

It becomes increasingly evident that we have come to a linguistic impasse, reflected in our historical, cultural, economic, social, doctrinal., impasses, all these issues being interconnected. The structural linguistic aspect is the most fundamental of them all, as it underlies the others and involves the *s.r*, or psycho-logical responses to words and other events in connection with meanings.

One of the benefits of building a system on undeniable negative premises is that many older and controversial problems become relatively simple and often uncontroversial, disclosing an important psycho-logical mechanism. Such formulations have often the appearance of the 'discovery of the obvious'; but it is known, in some quarters, that the discovery of the obvious is sometimes useful, not always easy, and often much delayed; as, for instance, the discovery of the equality of gravitational and inertial mass, which has lately revolutionized physics.

As words are *not* the things we are talking about, the only possible link between the objective world and the verbal world is structural. If the two structures are similar, then the empirical world becomes intelligible to us—we 'understand', we can adjust ourselves,. If we carry out verbal experiments and predict, these predictions are verified empirically. If the two structures are not similar, then our predictions are not verified —we do not 'know', we do not 'understand', the given problems are 'unintelligible' to us., we do not know what to do to adjust ourselves,.

Psycho-logically, in the first case we feel security, we are satisfied, hopeful.; in the second case, we feel insecure, a floating anxiety, fear, worry, disappointment, depression, hopelessness, and other harmful *s.r* appear. The considerations of structure thus disclose an unexpected and powerful semantic mechanism of individual and collective happiness, adjustment., but also of tragedies, supplying us with *physiological* means for a certain amount of desirable control, because relations and structure represent fundamental factors of all meanings and evaluations, and, therefore, of all *s.r*.

The present increasing world unrest is an excellent example of this. The structure of our old languages has shaped our *s.r* and suggested our doctrines, creeds., which build our institutions, customs, habits, and, finally, lead fatalistically to catastrophes like the World War. We have learned long ago, by repeated sad experience, that predictions concerning human affairs are not verified empirically. Our doctrines, institutions, and other disciplines are unable somehow to deal with this semantic situation, and hence the prevailing depression and pessimism.

We hear everywhere complaints of the stupidity or dishonesty of our rulers, as already defined, without realizing that although our rulers are admittedly very ignorant, and often dishonest, yet the most informed, gifted, and honest among them cannot predict or foresee happenings, if their arguments are performed in a language of a structure dissimilar to the world *and* to our nervous system. Under such conditions, calling names, even under provocations, is not constructive or helpful enough. Arguments in the languages of the old structure have led fatalistically to systems which are structurally 'un-natural' and so must collapse and impose unnecessary and artificial stress on our nervous system. The self-imposed conditions of life become more and more unbearable, resulting in the increase of 'mental' illness, prostitution, criminality, brutality, violence, suicides, and similar signs of maladjustment. It should never be forgotten that human endurance has limits. Human 'knowledge' shapes the human world, alters conditions, and other features of the environment—a factor which does not exist to any such extent in the animal world.

We often speak about the influence of heredity, but much less do we analyse what influence environment, and particularly the *verbal environment*, has upon us. Not only are all doctrines verbal, but the structure of an old language reflects the structural metaphysics of bygone generations, which affect the *s.r*. The vicious circle is complete. Primitive mythology shaped the structure of language. In it we have discussed and argued our institutions, systems., and so again the primitive structural assumptions or mythologies influenced them. It should not be forgotten that the affective interplay, interaction, interchange is ever present in human life, excepting, perhaps, in severe and comparatively rare (not in all countries) 'mental' ills. We can stop talking, we can stop reading or writing, and stop any 'intellectual' interplay and interaction between individuals, but we cannot stop or entirely abolish some *s.r*.

A structural linguistic readjustment will, it is true, result in making the majority of our old doctrines untenable, leading also to a fundamental scientific revision of new doctrines and systems, affecting all of them and our *s.r* in a constructive way. It is incorrect, for instance, to use the terms 'capitalism' as opposed to 'socialism', as these terms apply to different non-directly comparable aspects of the human problem. If we wish to use a term emphasizing the *symbolic character* of human relations, we can use the term 'capitalism', and then we can contrast directly individual, group, national, international., capitalisms. If we want to emphasize the psycho-logical aspects, we can speak of individualism versus socialism,. Obviously, in life the issues overlap, but the verbal implications remain, preventing clarity and inducing inappropriate *s.r* in any discussion.

In vernacular terms, there is at present a 'struggle' and 'competition' between two entirely different 'industrialisms' and two different 'commercialisms', based ultimately on two different forms of 'capitalism'. One is the 'individual capitalism', rapidly being transformed into 'group capitalism', in the main advanced theoretically to its limits in the United States of America and to a lesser extent in the rest of the civilized world, and 'social capitalism', proclaimed in the United Socialistic Soviet Republics. Both these extreme tendencies, connected also with semantic disturbances, are due to a verbal or doctrinal 'declaration of independence' of two, until lately, much isolated countries. The United States of America proclaimed the doctrine that man is 'free and independent', while, in fact, he is *not* free, but is inherently *interdependent*. The Soviets accepted uncritically an unrevised antiquated doctrine of the 'dictatorship of the proletarians'. In *practice,* this would mean the dictatorship of unenlightened masses, which, if left *actually to their creeds,* and deprived of the *brain-work* of scientists and leaders, would revert to primitive forms of animal life. Obviously, these two extreme creeds violate every typically *human* characteristic. We are interdependent, time-binders, and we are interdependent because we possess the higher nervous centres, which complexity animals do not possess. Without these higher centres, we could not be human at all; both countries seem to disregard this fact, as in both the brain-work is exploited, yet the brain-workers are not properly evaluated. The ignorant mob, with its historically and psycho-logically cultivated animalistic *s.r,* retards human progress and agreement. Leaders do not lead, but the majority play down to the mob psycho-logics, in fear of their heads or stomachs.

In both countries, the *s.r* are such that brain-work, although commercially exploited, is not properly evaluated, and is still persecuted here and there. For instance, in the United States of America, we witness court trials and resolutions against the work of Darwin, in spite of the fact that without some theory of evolution most of the natural sciences, medicine included, would be impossible. In Russia, we find decrees against researches in pure science, without which *modern* science is impossible. Both countries seemingly forget that all 'material' progress among humans is due uniquely to the *brain-work* of a few mostly under-paid and overworked workers, who exercise properly their higher nervous centres. With science getting hold of problems of *s.r* and sanity, our human relations and individual happiness will also become the subject matter of scientific enquiry. If international and *inter*dependent brain-workers produce discoveries and inventions, any one, even of the lowest development, can use or misuse their achievements, no matter what 'plan', or 'no-plan', is adopted. Both countries seem at present not to understand that a great development of mechanical means and the application of scientific achievements exclusively for animal comfort fail to lead to greater happiness or higher culture, and that, perhaps, indeed, they lead in just the opposite direction. Personally, I have no doubt that some day they will understand it; but an earlier understanding of this simple semantic fact would have saved, in the meantime, a great deal of suffering, bewilderment, and other semantic difficulties to a great number of people, if the rulers in both countries would be enlightened enough and could have foreseen it soon enough.

The future will witness a struggle between the individual and group capitalism, as exemplified in the United States of America, and the collective or social capitalism, as exemplified in the Soviet Republics. It does not require prophetic vision to foresee that some trends of history are foregone conclusions because of the structure of the human nervous system. As trusts or groups have replaced the theoretically 'individual' capitalism in the United States of America, so will the state capitalism replace the trusts, to be replaced in its turn by international capitalism.

We are not shocked by the international character of science. We are not '100 per cent patriotic' when it comes to the use in daily life of discoveries and inventions of other nations. Science is a semantic product of a *general human symbolic characteristic;* so, naturally, it must be general and, therefore, 'international'. But 'capitalism' is also a unique and *general* semantic product of symbolism; it is also a unique product of the human nervous system, dependent on mathematics, and, as such, by its inherent character, must become some day international. There is no reason why our *s.r* should be disturbed in one case more than in the other. The ultimate problem is not whether to 'abolish capitalism' or not, which will never happen in a symbolic class of life, but to transfer the control from private, socially irresponsible, uncontrolled, and mostly ignorant, leaders to more responsible, *professionally trained,* and socially controlled *public servants,* not bosses. If a country cannot produce honest, intelligent, and scientifically trained public men and leaders, that is, of course, very disastrous for its citizens; but this is not to be proclaimed as a rule, because it is an exception. Thus, in the Soviet Republics, graft is practically non-existent in the sense that it exists in the United States; but the mentality of the public men is practically at a similar standstill because of a deliberate minimizing of the value of brainwork. I wonder if it is realized at all, in either country, that *any* 'manual worker', no matter how lowly, is hired *exclusively* for his *human brain,* his *s.r,* and *not* primarily for his hands!

The only problem which the rest of mankind has to face is how this struggle will be managed and how long it will last, the outcome admitting of no doubt, as the ruthless elimination of individual capitalism by group capitalism (trusts) in the United States is an excellent example. In the Soviet Republics, they simply have gone further, but in a similar direction. Struggles mean suffering; and we should reconcile ourselves to that fact. If we want the minimum of suffering, we should stop the animalistic methods of contest. Human methods of solving problems depend on higher order abstractions, scientific investigations of structure and language, revision of our doctrines., resulting in peaceful adjustment of living facts, which are actualities whether we like it or not. If we want the maximum of suffering, let us proceed in the stupid, blind, animalistic and unscientific way of trial and error, as we are doing at present.

My aim is not to be a prophet, but to analyse different structural and linguistic semantic issues underlying all human activities, and so to produce material which may help mankind to *select* their lot *consciously*. What they will do is not my official concern, but it seems that both countries, which have so much in common, and which are bound to play an important role in the future of mankind, owing to their numbers, their areas, and their natural resources., will have to pay more attention to the so-called 'intellectual' issues, or, more simply, not disregard the difference between the reactions of infants and adults. Otherwise, very serious and disastrous cultural results for all of us will follow.

The problems of the world 1933 are acute and immediate, **over-**loaded with confusion, bitterness, hopelessness, and other forms of semantic disturbances. Without some means—and, in this case, scientific and physiological means—to regulate our *s.r*, we shall not be able to solve our problems soon enough to avoid disasters. The similarity in structure of mathematics, and our nervous system, once pointed out and *applied*, gives us a unique means to regulate the *s.r*, without which it is practically impossible to analyse dispassionately and wisely the most pressing problems of immediate importance.

The present investigation shows that the old languages which, in structure, are *not* similar to the world and our nervous system, have automatically reflected their structure on our doctrines, creeds, and habits, *s.r*, and also on those man-made institutions which result from verbal arguments. These, in turn, shape further *s.r* and, as long as they last, control our destinies.

Four important issues could be shown in detail, but, for lack of space, I give only a suggestive sketch of them here.

1) In the *A*-system, all our existing older sub-systems, with all their benefits as well as shortcomings, follow as an *A* psycho-logical structural semantic necessity.

2) The tremendous handicap for any new and less deficient systems consists in the fact that such systems lack new constructive ∞-valued semantics, and are carried on the one side by linguistic two-valued arguments in the language of old *el* structure; yet they aspire 'emotionally' to something new and better, while the two cannot be reconciled.

3) An argument carried on in the old *el* and two-valued way, no matter how fundamentally true and eventually beneficial, can be easily defeated on verbal grounds if it follows the old structure of language. Our decisions are never well-grounded psycho-logically, and so can never command the respect or achieve the reliability of scientific reason-

ing. That is why we are groping—the only method possible under such conditions being the animalistic trial-and-error methods, swaying masses by inflammatory speeches because reason has nothing to offer, being tied up by the old verbal structure to the older consequences based on animalistic and fundamentally false-for-man assumptions.

4) In the old A, el, two-valued system, agreement is theoretically impossible; so one of the main, and perhaps revolutionary, semantic departures from the old system is the fact that a *non-el* ∞-valued $\bar{A}$-system, based on fundamental *negative* premises, leads to a theory of *universal agreement,* which is based on a structural revision of our languages, producing new and undisturbed *s.r*, which eliminate the copying of animals in our nervous reactions.

The subject matter of this chapter divides, naturally, into three interconnected semantic parts. In the first, we shall recall a few general notions, known in the main but seldom taken into consideration, reformulated in a language of different structure. In the second, I shall indicate how the most important mathematical disciplines, which traditionally and, in the opinion of the majority, could hardly be called mathematical, represent a scientific and exact formulation of the *general* 'thinking' process. In this connection, a few words will be said about the theory of aggregates, and a little more about the theory of groups. This latter theory, with its implications and applications, leads to a reformulation of mathematics on quite obvious psycho-*logical* grounds, bringing mathematics into the closest relationship to the general processes of mentation. Finally, in the third part, I shall indicate the astonishing and quite unexpected *physiological* fact of the similarity of the structure of mathematics with the structure and function of our nervous system.

The intelligent layman should be reminded that, although he needs to know *about* mathematics, the minimum given here, supplemented, perhaps, by a few most elementary and fascinating books on mathematical philosophy, given in the bibliography to this volume, yet he does not need, and probably never will need, more technical mathematics than is given in the high schools and supplemented by the fundamental notions of the differential calculus. For directly we treat all languages, mathematics included, from a more general (and, at present, perhaps, the most general) aspect; namely, *structure;* the reader will obtain all the essential psycho-logical benefits of modern science by absorbing the $\bar{A}$-system and habits, which will result in completely novel standards of evaluation and distinctly modern and adult *s.r*.

The last is of extreme and unrealized importance. In fact, its importance cannot be fully appreciated until we actually acquire such reactions, because only then shall we have semantic disturbances eliminated, so that all problems can be analysed properly, and, therefore, agreement *must* be reached.

The future generations, of course, will have no difficulties whatsoever in establishing the healthy *s.r*; neither at present have very young children. These do not need such treatises as the present work. But, before the grown-up parents or teachers can train their children, they must first unlearn a great deal and train themselves to new habits involving the $\bar{A}$ *standards* of evaluation. So, for them, such a book, in order to be convincing, must deal with the foundations of their difficulties. The last task is difficult for the writer as well for the reader.

What has been said here does not apply, I am sorry to say, to *professional* 'philosophers', 'logicians', 'psychologists', psychiatrists, and teachers. These, to be adequate at all for their responsible and difficult professional duties, *must become thoroughly acquainted* with *structure* in general, and with the *structure of mathematics* in particular, as factors in *s.r*, and must work out the present outline much further.

Section B. General.

Mathematics in the twentieth century is characterized by an enormous productiveness, by the revision of its foundations, and the quest for rigour, all of which implies material of great and unexplored psychological value, a result of the activity of the human nervous system. Branches of mathematics, as, for instance, mathematical 'logic', or the analytical theory of numbers, have been created in this period; others, like the theory of function, have been revised and reshaped. The theory of Einstein and the newer quantum mechanics have also suggested further needs and developments.

Any branch of mathematics consists of propositional functions which state certain structural relations. The mathematician tries to discover new characteristics and to reduce the known characteristics to a dependence on the smallest possible set of constantly revised and simplest structural assumptions. Of late, we have found that no assumption is ever 'self-evident' or ultimate.

To those structural assumptions, we give at present the more polite name of postulates. These involve undefined terms, not always stated explicitly, but always present implicitly. A *postulate system gives us the structure of the linguistic scheme.* The older mathematicians were less particular in their methods. Their primitive propositional functions or postulates were less well investigated. They did not start explicitly with undefined terms. The twentieth century has witnessed in this field a marked progress in mathematics, though much less in other verbal enterprises; which accounts for the long neglect of the structure of languages. Without tracing down a linguistic scheme to a postulate system, it is extremely difficult or impossible to find its structural assumptions.

A peculiarity of modern mathematics is the insistence upon the formal character of all mathematical reasoning, which, with the new *non-el* theory of meanings, ultimately should apply to all linguistic procedures.

The problems of 'formalism' are of serious and neglected psychological importance, and are connected with great semantic dangers in daily life if associated with the lack of consciousness of abstracting; or, in other words, when we confuse the orders of abstractions. Indeed, the majority of 'mentally' ill are *too formal* in their psycho-logical, one-, two-, or few-valued processes and so cannot adjust themselves to the ∞-valued experiences of life. Formalism is only useful in the search for, and test of, structure; but, in that case, the consciousness of abstracting makes the attitude behind formal reasoning *∞-valued and probable,* so that semantic disturbances and shocks in life are avoided. Let us be simple about it: the mechanism of the semantic disturbance, called 'identification', or 'the confusion of orders of abstractions' in general, and 'objectification' in particular, is, to a large extent, dependent on two-valued formalism without the consciousness of abstracting.

In mathematics, formalism is uniquely useful and necessary. In mathematics, the formal point of view is pressed so far as to disclaim that any meanings, in the ordinary sense, have been ascribed to the undefined terms, the emphasis being on the postulated relations between the undefined terms. The last makes the majority of mathematicians able to adjust themselves, and mathematics extremely general, as it allows us to ascribe to the mathematical postulates an indefinite number of meanings which satisfy the postulates.

This fact is not a defect of mathematics; quite the opposite. It is the basis of its tremendous practical value. It makes mathematics a linguistic scheme which embodies the possibility of perfection, and which, no doubt, satisfies semantically, at each epoch, the great majority of properly informed individual Smiths and Browns. There is nothing absolute about it, as all mathematics is ultimately a product of the human nervous system, the best product produced at each stage of our development. The fact that mathematics establishes such linguistic relational patterns without specific content, accounts for the generality of mathematics in applications.

If mathematics had physical content or a definite meaning ascribed to its undefined terms, such mathematics could be applied only in the given case and not otherwise. If, instead of making the mathematical statement that one and one make two, without mentioning what the one or the two stands for, we should establish that one apple and one apple make two apples, this statement would not be applied safely to anything else but apples. The generality would be lost, the validity of the statement endangered, and we should be deprived of the greatest value of mathematics. Such a statement concerning apples is not a mathematical statement, but belongs to what is called 'applied mathematics', which has content. Such experimental facts as that one gallon of water added to one gallon of alcohol gives *less* than two gallons of the mixture, do *not* invalidate the mathematical statement that one and one make two, which remains valid by definition. The last mentioned experiment with the 'addition' of water to alcohol is a deep sub-microscopic structural characteristic of the empirical world, which must be discovered at present by experiment. The most we can say is that we find the above mathematical statement applicable in some instances, and non-applicable in others.

Not assigning definite meanings to the undefined terms, mathematical postulates have variable meanings and so consist of propositional functions. Mathematics must be viewed as a manifold of patterns of exact relational languages, representing, at each stage, samples of the best working of the human 'mind'. The application to practical problems depends on the ingenuity of those desiring to use such languages.

Because of these characteristics, mathematics, when studied as a form of human behaviour, gives us a wealth of psycho-logical and semantic data, usually entirely neglected.

As postulates consist of propositional functions with undefined terms, all mathematical proof is formal and depends exclusively on the form of the premises and not on special meanings which we may assign to our undefined terms. This applies to all 'proof'. 'Theories' represent linguistic structures, and must be proved on semantic grounds and never by empirical 'facts'. Experimental facts only make a theory more plausible, but no number of experiments can 'prove' a theory. A proof belongs to the *verbal* level, the experimental facts do not; they belong to a different order of abstractions, not to be reached by language, the connecting link being *structure*, which, in languages, is given by the systems of postulates.

Theories or doctrines are always linguistic. They formulate something which is going on inside our skin in relation to what is going on on the un-speakable levels, and which is not a theory. Theories are the rational means for a rational being to be as rational as he possibly can. As a fact of experience, the working of the human nervous system is such that we have theories. Such was the survival trend; and we must not only reconcile ourselves with this fact, but must also investigate the structure of theories.

Theories are the result of extremely complex cyclic chains of nerve currents of the human nervous system. Any semantic disturbance, be it a confusion of orders of abstractions, or identification, or any of their progeny, called 'elementalism', 'absolutism', 'dogmatism', 'finalism'., introduces some deviations or resistances, or semantic blockages of the normal survival cycles, and the organism is at once on the abnormal non-adjustment path.

The structure of protoplasm of the simplest kind, or of the most elaborate nervous system, is such that it abstracts and reacts in its own specific way to different external and internal stimuli.

Our 'experience' is based normally on abstractions and integrations of different stimuli by different receptors, with different and specific reactions. The eye produces its share, and we may see a stone; but the eye does not convey to us the *feel* of weight of the stone, or its temperature, or its hardness,. To get this new wisdom, we need other receptors of an entirely different kind from those the eye can supply. If the eye plays some role in establishing the weight, for instance, without ever giving the *actual feel* of weight, it is usually misleading. If we would try to lift a pound of lead and a pound of feathers, which the balance would register as of equal weight, the pound of lead would feel heavier to us than the pound of feathers. The eye saw that the pound of lead is smaller in bulk, and so the doctrinal, semantic, and muscular expectation was for a smaller weight, and so, by contrast, the pound of lead would appear unexpectedly heavy.

As the eye is one of the most subtle organs, in fact, a part of the brain, science is devising methods to bring all other characteristics of the external world to direct or indirect inspection of the eye. We build balances, thermometers, microscopes, telescopes, and other instruments, but the character and *feel* of weight, or warmth., must be supplied directly by the special receptors, which uniquely can produce the special 'sensations'. The swinging of the balance, or the rise of the column of the thermometer, establishes most important *relations,* but does not give the immediate specific and un-speakable feel of 'weight' or of 'warmth'. Our first and most primitive contact with a stone, its feel., is a personal abstraction from the object, full of characteristics supplied by the peculiarities of the special receptors. Our primitive picture 'stone' is a summary, an integration, of all these separate 'sense' abstractions. It is an abstraction from many abstractions, or an abstraction of a higher order.

Theories are relational or structural verbal schemes, built by a process of high abstractions from many lower abstractions, which are produced not only by ourselves but by others (time-binding). Theories, therefore, represent the shortest, simplest structural summaries and generalizations, or the highest abstractions from individual experience and through symbolism of racial past experiences. Theories are mostly not an individual, but a collective, product. They follow a more subtle but inevitable semantic survival trend, like all life. Human races and epochs which have not revised or advanced their theories have either perished, or are perishing.

The process of abstracting in different orders being inherent in the human nervous system, it can neither be stopped nor abolished; but it can be deviated, vitiated, and forced into harmful channels contrary to the survival trend, particularly in connection with pathological $s.r.$ No one of us, even when profoundly 'mentally' ill, is free from theories. The only selection we can make is between antiquated, often primitive-made, theories, and modern theories, which always involve important semantic factors.

The understanding of the above is of serious importance, as, by proper selection of theories, all wasteful semantic disturbances, which lead even to crimes, and such historical examples of human un-sanity as the 'holy inquisition', burning at the stake, religious wars, persecution of science, the Tennessee trial., could have been avoided.

Whenever *any one* says *anything,* he is indulging in theories. A similar statement is true of writing or 'thinking'. We *must* use terms, and the very selection of our terms and the structure of the language selected reflect their structure on the subject under discussion. Besides, words are *not* the events. Even simple 'descriptions', since they involve terms, and ultimately undefined terms, involve structural assumptions, postulates, and theories, conscious or unconscious—at present, mostly the latter.

It is very harmful to sanity to teach a disregard for theories or doctrines and theoretical work, as we can never get away from them as long as we are humans. If we disregard them, we only build for ourselves semantic disturbances. The difference between morbid and not so obviously morbid confusions of orders of abstractions is not very clear. The strong affective components of such semantic disturbances must lead to absolutism, dogmatism, finalism, and similar states, which are semantic factors out of which states of un-sanity are built.

We know that we must start with undefined terms, which may be defined at some other date in other undefined terms. At a given date, our undefined terms must be treated as postulates. If we prefer, we may call them structural assumptions or hypotheses. From a theoretical point of view, these undefined terms represent not only postulates but also variables, and so generate propositional functions and not propositions. In mathematics, these issues are clear and simple. Every theory is ultimately based on postulates which consist of propositional functions containing variables, and which express relations, indicating the structure of the scheme.

It appears that the main importance of the linguistic higher order abstractions is in their *public* character, for they are capable of being transmitted in neural and extra-neural forms. But our private lives are influenced also very much by the lower order abstractions, 'feelings', 'intuitions',. These can be, should be, but seldom are, properly influenced by the higher order abstractions. These 'feelings'., are personal, unspeakable, and so are non-transmittable. For instance, we cannot transmit the actual feeling of pain when we burn ourselves; but we can transmit the invariant relation of the extremely complex fire-flesh-nerve-pain manifold. A relation is present empirically, but also can be expressed by words. It seems important to have means to translate these higher order abstractions into lower, and this will be the subject of Part VII.

Section C. The psycho-logical importance of the theory of aggregates and the theory of groups.

Starting with the $\bar{A}$ denial of identity, we were compelled to consider structure as the only possible link between the empirical and the verbal worlds. The analysis of structure involved relations and *m.o* and multi-dimensional order, and, ultimately, has led us to a semantic definition of mathematics and numbers. These definitions make it obvious that all mathematics expresses general processes of mentation *par excellence*. We could thus review all mathematics from this psycho-logical point of view, but this would not be profitable for our purpose; so we will limit ourselves to a brief sketch connected with the theory of aggregates and the theory of groups, because these two fundamental and most general theories formulate in a crisp form the general psycho-logical process, and also show the mechanism by which all languages (not only mathematics) have been built. Besides, with the exception of a few specialists, the general public is not even aware of the existence of such disciplines which depart widely from traditional notions about mathematics. They represent most successful and powerful attempts at building exact relational languages in subjects which are on the border-line between psycho-logics and the traditional mathematics. Because they are exact, they have been embodied in mathematics, although they belong just as well to a general science of relations, or general semantics, or 'psychology', or 'logic', or scientific linguistics and psychophysiology. There are other mathematical disciplines, as, for instance, analysis situs, or the 'algebra of logic'., to which the above statements apply; but, for our present purposes, we shall limit ourselves to the former two.

Dealing with the theory of aggregates, I will give only a few definitions taken from the *Encyclopaedia Britannica,* with the purpose of drawing the attention of the 'psychologists', and others, to those psychological data.

The theory of aggregates underlies the theory of function. An aggregate, or manifold, or set, is a system such that: (1) It includes all entities to which a certain characteristic belongs; and (2) no entity without this characteristic belongs to the system; (3) any entity of the system is permanently recognizable as distinct from other entities.

The separate entities which belong to such a collection, system, aggregate, manifold, or set are called elements. We assume the possibility of selecting at pleasure, by a definite process or law, one or more elements of any aggregate A, which would form another aggregate B,.

The above few lines express how the human 'thought' processes work and how languages were built up. It is true that the exactness imposes limitations, and so the mathematical theories are not expressed in the usual antiquated 'psychological' terms, although they describe one of the most important psycho-logical processes.

Lately, the theory of aggregates has led to a weighty question: Does one of the fundamental laws of old 'logic'; namely, the two-valued law of the 'excluded third' (A is either B or not B), apply in all instances? Or is it valid in some instances and invalid in others?

This problem is the psycho-logical kernel of the new revision of the foundation of mathematics, which has lately been considerably advanced by Professor Łukasiewicz and Tarski with their many-valued 'logic', which merges ultimately with the mathematical theory of probability; and on different grounds has perhaps been solved in the present *non-el, $\bar{A}$*-system.

The notion of a group is psycho-logically still more important. It is connected with the notions of transformation and invariance. Without giving formal definitions unnecessary for our purpose, we may say that if we consider a set of elements a, b, c., and we have a rule for combining them, say O, and if the result of combining any two members of the set is itself a member of the set, such aggregate is said to have the 'group property'.

Thus, if we take numbers or colours, for instance, and the rule which we accept is '$+$', we say that a number or a colour is transformed by this rule into a number or a colour, and so both possess the 'group property'. Obviously, by performing the given operation, we have transformed one element into another; yet some characteristics of our elements have remained invariant under transformation. Thus, if 1 is a number and 2 is a number, the operation '$+$' transforms 1 into 3, since $1 + 2 = 3$; but 3 has the character of being a number; so this characteristic is preserved or remains invariant. Similarly, with colours, if we add colours, these are transformed, but remain colours, and so both sets have the 'group property'. Keyser suggests that the 'mental' processes have the group property, which is undoubtedly true.[1]

The role this theory plays in our language is of great importance, because in it we find a method of search for structure, and a method by which we can establish a similarity of structure between the un-speakable objective level and the verbal level, based on invariance of relations which are found or discovered in both.

The role of groups in physical theory is best described by quoting Professor Rainich. (Remarks in brackets are mine.) 'A physicist, we may take it, is a person who measures according to certain rules. Let us denote by a the number he obtained in a given situation by applying the rule number one, by b the number obtained in the same situation by measuring according to rule number two and so on (a may be e.g. the volume, b the pressure, c the temperature of gas in a given container). The physicist finds further that the results of measurements of the same kind undertaken in different situations satisfy certain relations, we may write, for instance:

$$r(a,b) = c.$$

'A mathematician is busy deducing from some given propositions other propositions; this usually leads to numbers which we may call A, B, C, ... These numbers also satisfy certain relations, say

$$R\ (A,B) = C.$$

'Then comes, as Professor Weyl says, a messenger, a go-between who may be a mathematician or a physicist, or both, and says: "If you establish a correspondence between the physical quantities and the mathematical quantities, if you assign A to a, B to b, etc., the *same* relations hold for the physical quantities as for the corresponding mathematical quantities so that $R \equiv r$." [*Similarity of structure.*]

'In the course of time new procedures of measurement are invented, some physical relations do not find their counterpart in the mathematical theory, the mathematical theory has to be patched up by introducing new quantities till too many quantities appear in it which do not correspond to physical quantities; then comes the phenomenological point of view and sweeps the theory out of applied mathematics—the theory becomes pure mathematics once more, and physicists begin to look around for a new theory. Everybody can find examples for this situation; it is enough to mention the Bohr atom which was not even mentioned today only fifteen years after its introduction.

'However the theory of groups which is being applied to physics is not just one of many mathematical theories of the character described above; its application is of a far more fundamental nature and we shall be able to indicate what it is by analysing further the scheme outlined above.

'It may happen, and in fact it happens often, that the same mathematical theory can be applied to the same physical facts in more than one way; for instance, instead of assigning to the physical quantities $a,b, \ldots$ the mathematical quantities $A,B, \ldots$ we might have assigned to them $A',B', \ldots$ with the same results, that is, the relations for physical quantities are the same as for the mathematical quantities corresponding to them now (think of space considered from the experimental point of view—and of coördinate geometry; different ways of establishing a correspondence result from different choices of coördinate axes). If this happens it means that the mathematical theory possesses a peculiar property, namely, that if A' is substituted for A, B' for B and so on, no relation of the type $R(A,B) = C$ which was correct before the substitution is destroyed; in other words, there are substitutions or transformations for which all relations are invariant. All such transformations constitute what we call a group; the existence and the properties of such a group present a very important characteristic of the mathematical theory. Moreover it is clear that if two different mathematical theories can be applied—in the sense described above—to the same physical theory, the groups of these two theories will be essentially the same, so that the groups reflect some of the most fundamental properties of physical systems.'[2]

The connection between groups and structure is described by Professor Shaw as follows: 'The first branch of dynamic mathematics is the theory of operations. It includes the general theory of operators of any type and in particular the theory of groups of operators. The structure of such groups is evidently a study of form. It may often be exemplified in some concrete manner. Thus the groups of geometric crystals exemplify the structure of thirty-two groups of a discontinuous character, and the 230 space-groups of the composition of crystals exemplify the corresponding infinite discontinuous groups. The study of the composition series of groups, the subgroups and their relations, whether in the case of substitution groups, linear groups, geometric groups, or continuous groups, is a study of form. Also, the study of the construction of groups, whether by generators, or by the combination of groups, or in other ways, is also a study of structure or form. The calculus of operations in general, with such particular forms as differential operators, integral operators, difference operators, distributive operations in general, is for the most part a study of structure. In so far as any of these is concerned with the synthesis of compound forms from simple elements, it is to be classed as a study of form, as the term is here used.'[3]

In the notion of a group, we have become acquainted with two terms ; namely, transformation and invariance. The first implies 'change' ; the other, a lack of 'change' or 'permanence'. Both of these characteristics are semantically fundamental, but involve serious complexities.

The world, ourselves included, can be considered as processes which can be analysed in terms of transformed stages with all their derivative notions. In the objective world, 'change' is ever present and is, perhaps, the most important structural characteristic of our experience. But when a highly developed nervous system, a process itself, is acted upon by other processes, such nervous system discovers, at some stage of its development, a certain relative permanence, which, at a still later stage, is formulated as invariance of function and relations. The latter formulation is *non-el* because it can be discovered empirically, which means by the lower nerve centres, but also is the main necessity and means of operating of the higher nerve centres, so-called 'thought'. All that we usually call a process of 'association' is nothing else than a *process of relating,* a direct consequence of the structure of the nervous system, where stimuli are registered in a certain four-dimensional order, which, on the psycho-logical level, take the form of relations. From this point of view, it is natural that the higher nerve centres, as a limit of integrating processes, should produce *and* discover invariance of relations, which appears then as the supreme product and so, ultimately, a necessity of the activity of the higher centres. Obviously, if the invariance of relations has any objective counterpart whatsoever in the external world, this invariance is impressed on the nervous system more than other characteristics; and so, at a certain stage, a nervous system which is capable of producing and using a highly developed symbolism, must discover and formulate this invariance.

It seems that *relations*, because of the possibility of discovering them and their invariance in *both* worlds, are, in a way, more 'objective' than so-called objects. We may have a science of 'invariance of relations', but we could not have a science of permanence of things; and the older doctrines of the permanence of our institutions must also be revised. Under modern conditions, which change rather rapidly nowadays, obviously, some relations between humans alter, and so the institutions must be revised. If we want *their invariance,* we must build them on such *invariant relations between humans* as are not altered by the transformations. This present work, indeed, is concerned with investigating such relations, and they are found in the *mechanism of time-binding,* which, once stated, becomes quite obvious after reflection.

As Professor Shaw says: 'We find in the invariants of mathematics a source of objective truth. So far as the creations of the mathematician fit the objects of nature, just so far must the inherent invariants point to objective reality. Indeed, much of the value of mathematics in its applications lies in the fact that its invariants have an objective meaning. When a geometric invariant vanishes, it points to a very definite character in the corresponding class of figures. When a physical invariant vanishes or has particular values, there must correspond to it physical facts. When a set of equations that represent physical phenomena have a set of invariants or covariants which they admit, then the physical phenomena have a corresponding character, and the physicist is forced to explain the law resulting. The unnoticed invariants of the electromagnetic equations have overturned physical theories, and have threatened philosophy. Consequently the importance of invariants cannot be too much magnified, from a practical point of view'.[4]

It should be noticed that the *non-el* character of the terms relation, invariance . , which apply both to 'senses' and 'mind', is particularly important, as it allows us to apply them to all processes; and that such a language is similar in structure not only to the world around us, but also to our nervous processes. Thus, a process of being iron, or a rock, or a table, or you, or me, may be considered, for practical purposes, as a temporal and average invariance of function on the sub-microscopic level. Under the action of other processes, the process becomes structurally transformed into different relational complexes, and we die, and a table or rock turns into dust, and so the invariance of this function vanishes.

The notion of a function involves the notion of a variable. The functional notion has been extended to the propositional function and, finally, to the doctrinal function and system-function. The term transformation is closely related to that of function and relation. This notion is based on our capacity to associate, or relate, any two or more 'mental' entities. We can, for instance, associate a with b or b with a. We say that we have transformed a into b, or vice versa.

An excellent example of transformation, given by Keyser, is an ordinary dictionary, which would be genuinely mathematical if it were more precise. In a dictionary, every word is transformed into its verbal meaning, and vice versa. A telephone directory is another example. Quite obviously, the term 'transformation' has far-reaching implications. If a is transformed into b, this implies that there is a relation between a and b which is being established, by the fact of transformation. Once a relation is established, we have a propositional function of two or more variables which define an extensional set of all elements connected by this relation.[5]

We see that these three terms are inseparably united and are three aspects of one psycho-logical process. If we have a transformation, we have a function and a relation; if we have a function, we have a relation and a transformation; if we have a relation, we have a transformation and a function. Transformation, as we see, is a psycho-logical term of action. A relation has a psycho-logically mixed character. A propositional function is a static statement, on record, with blanks for the values of the variables. In it the form is invariant, but it may take an indefinite number of values. The *extensional manifold* of the values for the variable is static, given once for all in a given context. It is extensional and, therefore, may be empirical and experimental.

Let us take as an example, for instance, the transformation of a set of integers 1, 2, 3,. Let us suppose that the given law of transformation is given by the function $y = 2x$. The result would be the manifold of even integers 2, 4, 6,. We see that integers are transformed into integers; therefore, the characteristic of being an integer is preserved; in other words, this characteristic is an invariant under the given transformation $y = 2x$, but the values of the integers are not preserved.

The theory of invariance is an important branch of mathematics, made famous of late through the work of Einstein. Einstein fulfilled the dearest dream of Riemann and attained the methodological and scientific ideal, that a 'law of nature' should be formulated in such a manner as to be invariant under groups of transformations. Such a semantic ideal, once stated, cannot be denied; it expresses exactly a necessity of the proper working of the human nervous system.   In fact, a 'law' of nature represents nothing else than a statement of the invariance of some relations.   When the Einstein criterion is applied, it renders most of the old 'natural laws' invalid, as they cannot stand the test of invariance.   The older 'universal laws' then appear as local private gossips, true for one observer and false for another.

The method of the theory of invariance gives us the trend of relations that abide, and so expresses important psycho-logical characteristics of the human 'mind'.   Its further significance is revealed by Keyser in the suggestion that when a group of transformations leaves some specified psycho-logical activity invariant, it defines perfectly some actual or potential branch of science, some actual or potential doctrine.[6]

We all know how deeply rooted in us is the feeling, the longing for stability, how worried we are when things become unstable.   Worries and fear are destructive to semantic health and should be taken into account in a theory of sanity.   A similar semantic urge apparently moved mathematicians when they worked out the theory of invariance; it was a formulation of a necessity of the activities of the human nervous system.   That similar semantic methods, if applied, would give similar results in our daily lives, scarcely needs to be emphasized.

We have already spoken of the mathematical theory of invariance as a mathematical species of a semantic theory of universal agreement. Similarly, in a $\bar{A}$-system based on relations and structure, it is possible to formulate a theory of universal agreement which would be structurally impossible in the A-system, and so the dreams of Leibnitz become a sober reality; but we must first re-educate our *s.r.*

Section D.   Similarity in structure of mathematics and of our nervous system.

In the chapter on the Semantics of the Differential Calculus, the fundamental notions and method of this calculus are explained. Here we may say, briefly, that it consists in stratifying, or expanding into a series, of an interval of any sort which proceeded by large steps. The large steps are divided into a great number of smaller and smaller steps, which, in the limit, when the numbers of steps become infinite, take on the aspect of 'continuity' so that we can study the 'rate of change'. When 'time' is taken into consideration, the dynamic may be translated into static, and vice versa; processes can be analysed at any stage,. This short description is far from exact or exhaustive; I emphasize only in an intuitive way what is of main semantic importance for our purpose.

The main object of the present chapter is to explain that the structure of the human nervous system is such that, on some levels, we produce dynamic abstractions; on others, static. As the organism works as-a-whole, for its optimum working, and, therefore, for sanity, we need a language, a method, which may be translated into a $s.r$ by which to translate the dynamic into the static, and vice versa; and such a language, such a method, is produced and supplied by mathematicians. To some readers, these remarks may appear so obvious as to make it unnecessary to write them, but I have found, through personal observation of reactions of different individuals, and by a careful survey of the literature of the subject, that even many mathematicians and physicists do not have this $s.r$ in all problems—or, at least, they do not know how to apply it.

In Part VII, elementary $\bar{A}$ methods are worked out, which supply the neurological semantic benefits of the calculus, very easily imparted to even small children *without any mathematical technique,* and establishing in them a mathematical attitude toward all language in general, training them in the only structural psycho-logics of sanity; namely, that of the calculus, which thus becomes the foundation of healthy and normal human $s.r$. And this, let us repeat again, without any mathematical technique. We find, also, that there are simple and *physiological* means, based on structure, of training our $s.r$ and imparting the feel for the structural stratification inherent in the consciousness of abstracting.

To start with, let me mention briefly a quite unexpected, unconscious, structural *biological* characteristic of mathematics; namely, its (in the main) *non-el,* organism-as-a-whole character.

From the time of Aristotle. biologists, physiologists, neurologists, 'psychologists', psychiatrists and others have spoken a great deal about the organism-as-a-whole; yet, they have not seemed to realize that if they produce *el* terms. they cannot apply the *non-el* principle.

It will probably not be an exaggeration to say that the majority of mathematicians have never heard of this principle, and that, if they have, they paid no attention to it; *yet,* in practice, they have applied it very thoroughly. The main mathematical terms are *non-el,* organism-as-a-whole terms which apply to 'senses' as well as to 'mind'. For instance, relation, order. difference, variable, function, transformation, invariance ., can mostly be seen as well as 'thought' of. The use of such terms prevents our speculation from degenerating into purely *el* speculations on words. a process always closely related to the morbid semantic manifestations of the 'mentally' ill. and obviously based on the pathological confusion of orders of abstractions, involving inappropriate evaluation.

This fact alone is of serious importance, as it indicates that mathematics is a language of similar structure to the structure of organisms and is a correct language, not only neurologically, but also *biologically*. This characteristic of mathematics, quite unexpectedly discovered, made the fusion of geometry and physics possible. It underlies, also, the theory of space-time and the Einstein theory. It will be seen later that it has also serious psycho-neurological importance.

It was already emphasized that the existing 'psychologies' are animalistic or metaphysical, because either they disregard one of the most unique human characteristics, such as the behaviour called mathematizing, or they indulge in speculations on, and in, *el* terms. It was suggested that no *human* 'psychologist' can actually perform his official task unless he is an equipped student of mathematics. Unless we actually apply the *non-el* principle, and take into account that the structure of languages introduces implications, unconscious in the main, and that no man is ever free from some doctrines and some so-called 'logical' processes involving physiological and semantic concomitants, no general theory of *human* 'psychology' can be produced.

The above solves a very knotty semantic problem, for we see that if we apply the *non-el* principle, any 'psychology' on the human level must become *psycho-logics,* though the old term 'psychology' could be retained as applying to animal researches only. The very name 'psychology', or the 'theory or science of mind', is obviously *el*, and treats 'mind' as an objective separate entity. As these results were originally reached independently, it is interesting to notice that the modern methods and the application of the structural positive knowledge 1933 lead to very many analogies and similarities, though this, after all, might be expected.

Notice the hyphen which, out of the *el* and delusional objectified 'space' and 'time', made the einsteinian space-time a language of *non-el* structure similar to the world around us ; and the hyphen which out of *el* 'psychology' makes a *non-el* human discipline of psycho-logics. It seems that a little dash here and there may be of serious semantic importance when we deal with symbolism.

To facilitate exposition, it is useful to stress, in the present section, the neurological and psychiatrical side, as an outline of the methods of the calculus, and related subjects will, of necessity, require separate treatment.

When rats are trained to perform a simple experiment requiring some 'mentality' and afterwards a large part of the cerebral cortex is removed, their training may be wholly lost. If such decorticated rats are trained again, they re-acquire the habit as readily as before. It appears

19

that, with rats, the cortex is not essential for these learning processes. They 'learn' as well, or nearly as well, with their sub-cortical and thalamic regions.[7]   In what follows, to avoid misstatements, I will use the rather vague term, yet sufficient for my purpose, 'thalamic region' or 'lower centres' instead of more specific terms, the use of which would complicate the exposition unnecessarily. With dogs, apes, and men, the situation is increasingly different. Their nervous systems are more differentiated. Their functional interchangeability is impaired. In the most complex human brain there still exists some interchangeability of function. When an arm, for instance, is paralysed through a brain-lesion, the arm may re-acquire a nearly normal function, though there is no regeneration of the destroyed brain tissue.   However, the interchangeability is less pronounced than in the lower brains. There seems to be no doubt that the thalamic regions are not only a vestibule through which all impulses from the receptors have to pass in order to reach the cortex, but also that the affective characteristics are strictly connected with processes in these regions.   It seems that some very primitive and simple associations can be carried on by the thalamic regions.

The cortex receives its material as elaborated by the thalamus. The abstractions of the cortex are abstractions from abstractions and so ought to be called abstractions of higher order. In neurology, similarly, the neurons first excited are called of 'first order'; and the succeeding members of the series are called neurons of the 'second order',. Such terminology is structurally similar to the inherent structure and function of the nervous system. The receptors are in direct contact with the outside world and convey their excitation and nerve currents to the lower nerve centres, where these impulses are further elaborated and then abstracted by the higher centres.

According to our daily experience and scientific knowledge, the outside world is an ever-changing chain of events, a kind of flux; and, naturally, those nerve centres in closest contact with the outside world must react in a shifting way. These reactions are easily moved one way or another, as in our 'emotions', 'affective moods', 'attention', 'concentration', 'evaluation', and other such semantic responses. In these processes, some associative or relational circuits exist, and there may be some very low kind of 'thinking' on this level. Birds have a well-developed, or, perhaps, over-developed, thalamus but under-developed and poor cortex, which may be connected with their stupidity and excitability.

Something similar could be said about the 'thalamic thinking' in humans; those individuals who overwork their thalamus and use their cortex too little are 'emotional' and stupid. This statement is not exaggerated, because there are experimental data to show how through a psycho-neural training the *s.r*, in some cases, can be re-educated, and that with the elimination of the semantic disturbances there is a marked development of poise, balance, and a proportional increase of critical judgement, and so 'intelligence'. Idiots, imbeciles, and morons are usually 'emotional' and excitable, as well as deficient in their 'mental' processes. A similar characteristic can be found in other unclassified 'mentally' deficient, and their name is legion—a characteristic strictly connected with, and often produced by, disturbances of the *s.r*. When these shifting, dynamic, affective, thalamic-region, lower order abstractions are abstracted again by the higher centres, these new abstractions are further removed from the outside world and must be somehow different.

In fact, they *are* different; and one of the most characteristic differences is that they have *lost* their *shifting* character. These new abstractions are relatively static. It is true that one may be supplanted by another, but they do not change. In this fact lies the tremendous value and danger of this mechanism, as disclosed clearly by the disturbances of the *s.r*. The value is chiefly in the fact that such higher order abstractions represent a perfected kind of memory, which can be recalled exactly in the form as it was originally produced. For instance, the circle, *defined* as the locus of points in a plane at equal distance from a given point called the centre, remains permanent as long as we wish to use this definition. We can, therefore, recall it perfectly, analyse it., without losing the definiteness and the stability of this memory. Thus, critical analysis, and, therefore, progress, becomes possible. Compare this perfected memory, which may last indefinitely unchanged, with memories of 'emotions' which, whether dim or clear, are always distorted. We see that the first are reliable, that the others are not.

Another most important characteristic of the higher order abstractions is that, although of neural origin, they may be preserved and used over and over again in extra-neural forms, as recorded in books and otherwise. This fact is never fully appreciated from a neurological point of view. Neural products are stored up or preserved in extra-neural form, and they can be put back in the nervous system *as active neural processes*. The above represents a fundamental mechanism of time-binding which becomes overwhelmingly important, provided we discover the physiological mechanism of regulating the *s.r*, on the one hand, and discover the mechanism by which these extra-neural factors can be made physiologically effective, on the other.

If humans are characterized by the fact that they build up this cumulative affair called 'civilization', this is possible through those higher order abstractions and the time-binding ability to extend our nervous system by extra-neural means, which, in the meantime, may play a most important neural role and become active nervous impulses. The last is only possible if some abstractions are static, and so can be recorded, leading ultimately to further extensions of the human nervous system by extra-neural means, such as microscopes, telescopes, and practically all modern scientific instruments, books, and other records.

To illustrate what has been said here, I know of no better example than is found in moving pictures. When we watch a moving picture representing some life occurrence, our 'emotions' are aroused, we 'live through' the drama; but the details, in the main, are blurred, and a short time after seeing it either we forget it all or in parts, or our memory falsifies most effectively what was seen. It is easy to verify the above experimentally by seeing one picture twice or three times, with an interval of a few days between each seeing. The picture was 'moving', all was changing, shifting, dynamic, similar to the world *and* our feelings on the un-speakable levels. The impressions were vague, shifting, non-lasting, and what was left of it was mostly coloured by the individual mood., while seeing the moving picture. Naturally, under such conditions, there is little possibility of a rational scientific analysis of a situation.

But if we *stop* the moving film which ran, say, thirty minutes, and analyse the static and extensional series of small pictures on the reel, we find that the drama which so stirred our 'emotions' in its moving aspect becomes a series of slightly different static pictures, each difference between the given jerk or grimace being a *measurable* entity, establishing relations which last indefinitely.

The *moving* picture represents the usually brief processes going on in the lower nerve centres, 'close to life', but unreliable and evading scrutiny. The *arrested* static film which lasts indefinitely, giving *measurable* differences between the recorded jerks and grimaces, obviously allows analysis and gives a good analogy of the working of higher nerve centres, disclosing also that all life occurrences have many aspects, the selection of which is mostly a problem of our pleasure and of the selection of language. The moving picture gives us the process; each static film of the reel gives us stages of the process in chosen intervals. In case we want a moving picture of a growing plant, for instance, we photograph it at given intervals and then run it in a moving-picture projector, and then we see the process of growth. These are empirical facts, and the calculus supplies us with a language of similar structure with many other important consequences.

It is characteristic that those who claim to be most interested in human affairs and human processes, whom we call, among others, 'philosophers', 'psychologists'., should not have discovered much of value in these fields. But mathematicians, who disclaim meaning in their undefined terms, or 'truth' in their postulates, or interest in human affairs, have had a most astonishing and unique success by elaborating methods for the translation of the dynamic into static and the static into dynamic. Claims and disclaims matter little, but working in accordance with the survival order of the nerve structure and currents has produced most valuable results.

The different methods of mathematics and the four-dimensional 'world' of Minkowski form the means for translating the dynamic into static and vice versa. Minkowski established a language of a new structure, closer to actual facts of the world around us and ourselves, making the general theory of Einstein possible. Further analysis of these issues is carried out in Part IX, and it is one of the semantic foundations upon which a positive theory of sanity can be built.

Disclaiming definite meanings, mathematicians have an intuitive predilection for selecting their terms and pursuing their line of enquiry among *possible meanings*, although formally these meanings are disregarded. The feeling which directs the selection of material which is formally interesting and important is akin to the artistic sense, but, unfortunately, in spite of its importance, it has been neglected by 'psychologists'. Quite often it is the 'feel' which directs the mathematicians in their researches and suggests or modifies lines of development or the selection of one set of postulates in preference to other sets. This is why the ordinary sense of the terms used in mathematics is so important, although it represents only some of the possible meanings. These, with their implications, usually represent most important structural characteristics of the human nervous system and the world.

This is to be expected because of the reasons given above; the more so that invariance in this shifting world is a characteristic of relations, and mathematics is a language of exact relations which, in the meantime, have mostly objective counterparts. The highest abstractions at every date are detached from the outside world neurologically, and *should remain detached*, to represent 'pure mind' in action. These higher abstractions are on the public level, as they are transmittable verbally with all characteristics included. They are static, unhampered directly by the outside events, although they normally originate in them. These higher order abstractions are 'digested' and translated into lower order abstractions and returned to the lower centres; and they receive their meanings close to life. Such meanings are enlightened meanings, a survival process, and each nervous level did its work properly.

We know that a number of human races have perished without leaving many traces of their existence. This process is going on continually, even now. Some races are progressing; some are regressing; some are at a seeming standstill. It would appear that the mechanism of higher order abstractions had and has survival value, and, therefore, should not be neglected but cultivated. In this special case, cultivation is a condition inherent in the process and a necessity for time-binders.'

Serious semantic dangers are also revealed by analysis and verified by observation. These higher order abstractions, let us repeat, are static and may last indefinitely, as long as for structural reasons we do not replace the old by new ones. Even then, though rejected, they remain as a permanent fact on record. Obviously, these higher abstractions have only a 'second-hand' connection with the outside world. Even their character is changed, they are static while the world is dynamic. The lower 'sense' world has 'characteristics left out', owing to the mechanism of abstracting of the lower centres; and the abstractions of higher orders have 'all characteristics included', because these are abstractions from abstractions, an *intra-organismal* process in its entirety, their starting material being already an end-product of the activities of the lower centres. This mechanism is only under full control if we are conscious of abstracting, because the higher order abstractions in the nervous chain affect, in their turn, the lower centres, and, in pathological cases, impress on them a semantic *delusional* or *illusional* evaluation as if a character of experience. In severe cases, even the lower nerve centres are stimulated to such an extent that hallucinations appear.

If we do not know how to handle different order abstractions, this results in serious semantic dangers. If the distribution of the returning nerve currents is a non-survival one, we exhibit semantic disturbances, such as identification or confusion of orders of abstractions, delusions, illusions, and hallucinations. Thus, we ascribe to the products of the lower nerve centres, the lower order abstractions, characteristics fictitious and impossible for them, such as 'immutability', 'permanence', involving disorientation about 'time'., ., which are characteristics of the higher order abstractions, but do not belong to the world as given by the lower abstractions, and result in an improper evaluation disturbing to the *s.r.* Such disturbances make us, naturally, absolutists and dogmatists, involve serious affective disturbances, and lead to non-adaptive behaviour and reactions, and other semantic manifestations of un-sanity. These, in their turn, make adjustment more difficult, often affecting the structure of man-made institutions, which again make adjustments more complex and often impossible. We become un-sane, 'insane', and life, whether public or private, becomes a mess. In such a vicious semantic circle, we distort our education, our systems, and institutions. Often the morbid reactions of powerful individuals are forced upon masses, who are then ruled by these morbid products, with injury to their nervous systems. Different mass hysterias, 'revivals', wars, political and religious propaganda, very often commercial advertisements, offer notable examples.

The morbid semantic influence of commercialism has not been investigated, but it does not take much imagination to see that commercial psycho-logics, as exemplified by the theories of commercial evaluation, 'wisdom', appeal to selfishness, animal cunning, concealing of true facts, appeal to 'sense' gratification., produce a *verbal and semantic environment* and slogans for the children which, if preserved in the grown-ups, must produce some pathological results. It is hoped that some day a psychiatrist will investigate this large, neglected, and very important semantic problem.

The lack of structural linguistic researches and investigation of our *s.r*, and the ignorance of those who rule, make us nearly helpless. Malaria or other germ diseases would never be eliminated were we to preserve religiously the sources of infection. The semantic sources of un-sanity are not only defended but are actively sponsored by organized ignorance and the power of merchants, state, and church.

The situation is acute. If we could entirely eliminate our cortex, it would, perhaps, not be so serious. We could, perhaps, live as complex a life as a fish and have a nervous system perfectly adjusted to such a life. But, unfortunately, with a structural change, or. according to Lashley, with the change even in the total mass of the brain, the activities and the role of the whole, including other parts, are profoundly altered.[8] These become inadequate, as shown by the boy born without the cortex, already described. His nervous system was much more complex than that of fishes or of some lower animals which lead *adequately* a rather complex life. But the boy was less equipped for life than they. Even his 'senses', though apparently 'normal' on macroscopic levels, must have been pathological on colloidal and sub-microscopic levels and did not function properly. We know, also, that in many cases of 'mental' ills the 'sense reactions' are abnormal; sometimes the patients seem to be entirely insensitive to stimuli which would produce most acute pain to other less pathological individuals.

It is impossible to eliminate completely from our lives or nerve currents the higher abstractions and their psycho-neural effect. Curiously enough, this elementary fact has never been emphasized or taken into account seriously; yet it is a crucial semantic factor in our attitude toward science and our future. Those who attempt such elimination, whether by actively persecuting science, or by emitting propaganda against science, or by the cynical or ignoring attitude toward 'mental' achievements, whether personally, or in education, or in public prints, or other public activities, do not succeed in eliminating the higher order abstractions, but simply introduce *pathological semantic reactions* and succeed in disorganizing their own nervous systems and those of others. I intended this implication when I said that our existing educational., systems *produce* morons, but 'geniuses' are born. Such very general semantic directives are, perhaps, responsible for the extremely low level of our non-technical development. Humans are not to be judged simply by the ability to drive an automobile or by the knowledge of how to use a bathtub; nor yet by their capacity for buying and selling things produced by others.

The tendency of some public prints to appeal to the morbidity of mob psycho-logics and to its ignorance, insisting that all that is said should be said in 'one-syllable' words, so that the mob can understand, in a *human* class of life, is an *arresting* or *regressive tendency*. What should be urged for sanity, and for humans, is that the mob should also learn the use of at least two-syllable words! Then, perhaps, the day would come when they could follow easily and habitually the use of *non-el* terms and, perhaps, even of words connected by a hyphen.

This appeal to mob psycho-logics and ignorance affects profoundly our *s.r* and should be investigated. It definitely appears that in countries where the majority reads only the sort of publications referred to above and commercial advertisements, their psycho-logical equipment and standards are lower than those of perfectly illiterate peasants of other countries. It is not fully realized that in a symbolic class of life, symbolism of any sort—e.g., public prints—plays an environmental role and creates *s.r* which may be distinctly morbid. The problems of public prints, commercialism., and their psycho-logical effect on the *s.r* should undergo a searching analysis by psychiatrists, and definite suggestions should be formulated by psychiatric scientific organizations or congresses.

Under the conditions prevailing at present, it is futile to preach 'morals' of any metaphysical kind. They have never worked satisfactorily, and increasingly they cannot work, particularly under the present much more complex conditions of life. They disorganize the survival activities and processes of the human nervous system. The imposed and delusional dogmas are themselves the result of pathological evaluation in their originators; a necessity, perhaps, on a primitive level, but profoundly semantically harmful under the complexities of life-conditions 1933.

As it is impossible to eliminate the influence of the higher order abstractions, we should investigate whether or not we can control these processes and the related s.r. We can learn to regulate these processes, which otherwise may become pathological, and to redirect the currents into constructive survival channels. I can state definitely that this is possible. We can control physiologically the s.r through the elimination of identification, by training in order, in consciousness of abstracting, and similar disciplines, and thus eliminate the pathological semantic disturbances of confusion of orders of abstractions. Such training, whenever possible, has seemingly a beneficial influence even on the more extreme pathological states listed above, and suggests general preventive value.

Let me briefly restate the fundamental differences between lower order abstractions and higher. The lower order abstractions are manufactured by the lower nerve centres, which are closer to, and in direct contact with, actual life experiences. These are non-permanent, shifting, vague and un-speakable, but often very intense. They play a most important role in our daily lives. They cannot be transmitted, as they are essentially of a non-transmittable character, and have a private, non-public character. All 'sense' impressions, 'feelings', 'moods'., are representative of them. We should remember that, detached, they are fictions, manufactured verbally, because our language happens to be el. Actually, these lower centres are in the cyclic chain and so influence, and are influenced by, the full cycle, including the higher order abstractions, whatever the latter may be in a given individual. The main point is that they are shifting, changing, non-permanent, non-stable—'moving', so to say—and remain un-speakable.

The higher order abstractions are abstractions from the lower order abstractions, being further removed from the outside world, and are of a distinctly different character. These are static, 'permanent', and cannot be entirely eliminated from any one.

From the point of view of sanity, the problem of how we can handle these functions becomes paramount. In the cyclic nervous chain, we always must translate one level into the other. Obviously, if, in the *higher* centres, we elaborate shifting, changing, non-permanent material, this material is not appropriate for them; they cannot work properly, and some pathological processes may set in.

If we elaborate in the *lower* nerve centres abstractions that are static, permanent., in character, and hence inappropriate for the lower centres, we build up morbid non-survival identifications, delusions, illusions, hallucinations, and other disturbances of evaluation, resulting in milder cases in absolutism, dogmatism, fanaticism., and, in heavier cases, in a neurosis or even a psychosis.

It seems quite obvious that each nervous level has its own specific kind of material to deal with. As they are in a cyclic nervous chain and are interconnected in a bewilderingly complex way, the problem of appropriate translation of one level of abstractions into the other becomes a semantic foundation for a well-balanced functioning of the nervous system. In this respect, we differ fundamentally from animals. The above difficulties do not arise in animals to that extent, because their nervous systems are not differentiated enough for such sharp differentiation in the functioning. For this reason, without human interference, there could be no 'insane' animals which could survive (see Part VI). But, having no static higher order abstractions in the human sense, they cannot pass on their 'experiences', which are transmittable *only* in the higher order formulations in neural and extra-neural forms to the next generations. Animals are not time-binders.

For humans, the proper translation of dynamic into static and static into dynamic becomes paramount for sanity, on psycho-logical levels, affecting, probably by colloidal processes, the psycho-neural foundation of semantic responses.

Psychiatry informs us that most of the 'mentally' ill have their main disturbances in the dynamic affective field. It is a very difficult field to reach by the older methods, the more so that the older *el* sharp distinction between 'intellect' and 'emotions' prevented the discovery of workable means. 'Thinking' and 'feeling' are not to be divided so simply. We know how 'thinking' is influenced by 'feeling'; but we know very little how 'feeling' is influenced by 'thinking'—perhaps, because we have not analysed the semantic issues in *non-el* terms.

All psychotherapy, with its manifold theories, each contributing its share, is a semantic attempt to influence 'feeling' by 'thinking'. A large number of successful cases seems to show clearly that some such means are possible. Large numbers of failures show equally that the methods used are not structurally satisfactory. The need of more scientific investigations of a more general and fundamental, *non-el* character becomes emphatic. The present enquiry shows that such structural investigations suggest that the method can be found in the psycho-logics of the 'mind' at .its best; namely, in mathematics, which unexpectedly leads to a physiological control of the *s.r*, effective not only as a therapeutic, but also as a preventive, educational means.

Identification as a factor of un-sanity seems to be a natural consequence of the evolution from 'animal' to 'man', particularly at our present stage, while the human race is so recent a product. The human cortex appeared only comparatively lately and is a young structure; the thalamic regions have a much longer history of functioning. It seems natural that the nervous impulses should pass the shorter, more phylogenetically travelled, paths in preference to comparatively newer and longer paths, a principle well known in neurology in connection with so-called 'Bahnung'. If education, and on human levels any kind of adjustment involving *s.r* involves some education, fails to force the nerve currents into their proper channels, or actively establishes in them semantic psycho-neural blockages through pathological evaluation acquired because of faulty training, we should expect either infantilism or regression to still lower levels. Whatever the correct explanation of the distribution of nerve currents, semantic blockages ., may be, observation shows unmistakably that some such assumptions are necessitated by observed manifestations in behaviour. Experiments show, also, that such defects can be helped greatly by the proper re-training and re-education of the *s.r*.

To understand the structure of these semantic disturbances, we must become acquainted with the affective components which underlie mathematics and mathematical methods, hitherto disregarded, because of the *el* character of our old terminology. There is another striking connection. In severe 'mental' illnesses, we usually find a disorientation in 'space' and 'time', which are, by necessity, *relational data* of experience. In the semantic disturbances called identification, we also find, as a rule, relational disorientation *about* 'space' and 'time', more subtle but very vicious in effect, bordering on what are called 'philosophical' problems, which, as a matter of fact, represent psycho-neural disturbances. Since Einstein, the disturbances can be easily eliminated, provided we take into account structural *non-el* issues in connection with *s.r* and a *Ā*-system.

It is instructive to make a short survey of the methods by which the mechanism of the nervous cycle—'senses', 'feelings'., first; 'mind', which again influences the 'feelings', next—works in mathematics. Weierstrass, the famous mathematician, says, in one of his writings, that a mathematician is a kind of poet. This is largely true. Mathematics is not only a rigorous linguistic relational pattern, but it uses the highest abstractions which we have reached at a given period from the data given by the lower nerve centres, which are closer to experience, or rather which constitute experience. The older arguments about the connection or lack of connection between the lower order abstractions ('sense' data.,), and mathematics are due solely to a confusion of orders of abstractions and are a useless gambling in *el* terms. Only in severe 'mental' ills is the speech of the patients entirely unconnected with first order external 'realities', and so the study of relations of many kinds and orders, called 'mathematics', cannot, as long as it is sane, be entirely detached from 'reality'. In fact, it is useless for mathematicians to try to produce disciplines which have no practical applications. As long as it is professionally accepted as mathematics, and, therefore, a science and sane, whatever mathematicians produce will always be connected with lower order abstractions, and must have an application sooner or later. When these higher order abstractions, produced very often by many individuals, are absorbed and returned in a modified form to the lower centres as 'visualization', 'intuition', 'feelings'., the given individual is closer to the external world than he was before, because he has absorbed, digested, and appropriated the nervous results of many more experiences than he himself could have gathered alone. He is able to compare, evaluate, and relate, revise and adjust his private experiences and observations with the *translated* experiences from higher abstractions of many more individuals. The *translation* is indispensable, because the reactions of both levels are entirely different, and comparable only when they are on one level. *Creative work has begun.*

Experiences given by the lower centres and lower abstractions are full of meanings, colouring, affective and semantic components, and these are not directly comparable with the higher abstractions produced by the higher nerve centres. They must be first transformed, 'digested', and translated into terms of the lower centres, which are the only ones which are effective on the lower levels. We call them 'visualization', 'intuition', 'feeling', 'culture',. The exact mechanism is not well known, but we have a number of data which show that the lower nerve centres are somehow engaged in these processes.

When this is accomplished, the mathematician has at his disposal an enormous amount of data; first, his personal experiences and observation of actual life (lower centres and lower order abstractions), and also all the personal experiences and observations of past generations. Although the latter were stored in the form of higher order abstractions only as an *account* of past experiences in neural or extra-neural forms, his nervous cycle was affected by them, and they were translated back into experiences of the lower levels.

With such an enormous amount of data of experience, he can *re-evaluate* the data, 'see' them anew, and so produce new and more useful and structurally more correct higher order abstractions. In their turn, these will produce similar semantic effects with other individuals,. The mechanism is, after all, well known and general, obvious even in the relations between some feeble-minded parents and their eventually feeble-minded children. It is entirely obvious on racial grounds; but, at present, it is not so obvious, and often but slightly effective, on personal and individual grounds, because we have had no means of training structurally and effectively the *s.r* in proper evaluation. The mechanism is entirely general, but it is obvious and seen at work in the majority of *creative* scientists and so-called 'geniuses'. These processes have not been analysed in terms of order, and so, although we use them often, we are not conscious of their mechanism and have no means of training our *s.r*. The *s.r* are a product of training, education., and are not inborn in *a given form*. Even birds bred in a laboratory which have never heard their parents or other birds sing will sing, as this is an inborn reflex, but the melody produced is different from that of their parents. Under normal conditions, the form of the song is standardized and is a result of *copying* parents. In other words, the melody-environment has affected them. With humans, it is not only a question of the given noises, the 'melody-environment' which we relate with some experiences, but the *s.r* involve affective responses to meanings, and this depends on the structure of language, involving unconscious, yet vital, evaluation factors and our *attitude* toward language, which ultimately depends on our knowledge of the mechanism and use of language.

These problems are extremely complex and subtle, and, at this stage, we are not ready to go into further details, the more so that there is a very simple and effective physiological structural method given in Part VII, which in practice eliminates enormous theoretical difficulties. There is little doubt that this mechanism of recasting, or translation of abstractions, is present in all of us, but this mechanism requires knowledge of the proper way to handle it, and that knowledge is not inborn, but has to be acquired by education. Up to the present date, these problems have been disregarded, and the *s.r* treated in a haphazard way; once the physiological mechanism of these reactions is discovered, however, we shall be able to use its benefits without the inherent dangers of disturbances.

Here we must face a rather unexpected fact.

Mathematics is alone and unique in that it has no content or definite meanings ascribed to the undefined terms; and, therefore, only in mathematics can we avoid the vicious influencing of lower centres through the feeling of false analogies which distort and disorganize the process. It is important to notice that the main and only lasting advances in 'philosophy' have been made by mathematicians; and, as a rule, whenever a trained mathematician attempts to work at any other profession not requiring mathematics, he shortly becomes an outstanding worker in the new field. It must be obvious that the returning nerve currents, when they produce the 'feel' (language of the lower centres) of physics, or chemistry, or biology, or other sciences with a definite content, must have a most pronounced semantic effect. Because of this physical content, identification and other semantic disturbances are usually present, instead of the highly beneficial visualization.

Empirically, this is quite obviously true. Let us survey the character of this process in physicists and chemists. Their problems, the content of their abstractions, are obviously not so closely related to human lives as the problems of biology. History shows that the attitude (affective) of those scientists toward human affairs is often shallow, but very seldom vicious or harmful. But let us take the attitudes of biologists, whose subject is seemingly much closer, or, at least, more affectively related to our problems, and we see, from Aristotle on, the brutalizing and *unscientific* (1933) effect of the false biological analogies. Practically all the vicious, unjustified, and unscientific generalizations which have made the white race the most animalistic, selfish, cruel, hypocritical, and un-sane race on earth are mainly due to the biological, A, distorted reasonings and *s.r* produced by false analogy.

In all this 'philosophy', they always reasoned from pigs, cats, and dogs to man. Since they were 'scientific', we blindly assumed that they must know what they were talking about. Even today, the majority of the older biologists refuse to investigate the structure of their language. They do not seem to be able to realize that most biological 'philosophies' are structurally fallacious and unscientific in 1933. They still unconsciously follow Aristotle. They refuse to understand that life is made up of absolute and *unique* individuals, and that 'man' or 'animal' *is not* an object, but labels verbal fictions.

In actual life, the differences between individuals are absolute, and father and son are different. These are the empirical facts of their sciences, the rest being verbal fictions. The notorious Tennessee trial demonstrated that in a large country like the United States of America, with a few good universities, there was no biologist to voice these points about 'evolution'. It is true that, through the work of neurologists and some others, biologists, of late, are beginning to see that they cannot generalize in the way they have done for more than two thousand years. Naturally, there are notable exceptions; yet even these do not realize the structural linguistic and semantic issues involved.

I most emphatically do not deny that animal researches are extremely useful and necessary; but I question the right of biologists to remain innocent of the importance of linguistic and semantic issues, and to indulge in vicious, unwarranted generalizations which, although they may express their own metaphysics and *s.r*, should not be advanced as 'scientific' results. Biologists ought to be informed enough to understand that 'man' and 'animal' are verbal fictions, and labels for something going on inside our skins—not labels for the unique individuals with which they have to deal outside their skins.

An example may, perhaps, be useful. We know that rats, prairie dogs, and some other animals are mostly immune to scurvy, but that man, monkeys, and guinea pigs are mostly not immune. How can we generalize from a rat to a man *or* a guinea pig? Or how much can we learn about the behaviour of a bee from the behaviour of an oyster, to use the example of Professor Jennings? Even in 'man', what helps one 'man', kills another.

Similar false analogies occur in the A classification of 'man' as an 'animal'. This classification disregards completely the *s.r* and twists the generally accepted folk-meaning of the term 'animal' into a special meaning which introduces very vicious semantic implications. If we classify 'man' as an 'animal', the structural A 'plus' elementalism is automatically introduced, since 'man', obviously, has many characteristics of behaviour not shown by the 'animal', taken in its folk-meaning. The disregard of the folk-meaning in our terminology shows clearly the complete disregard for *s.r* which are very strongly related to those folk-meanings. If we are to call 'man' an 'animal', then 'man' must be an 'animal' 'plus' something. If we were to call him some sort of a junior 'god', he would be a 'god' 'minus' something. The latter structural fallacy would be just as vicious in its implications, and would again deliver our speculations into the semantic clutches of the structure of a primitive-made *el* language.

Similar objections could be raised to that class of 'biological psychologies' exemplified by the 'behaviourists'—(not to be confused with the illuminating and highly constructive biological psychiatry or psychobiology introduced by Professor Adolf Meyer).[9] The 'behaviourists' try to be ultra-'scientific', not realizing that their knowledge of scientific method and structure belongs somewhere to the sixteenth century.

Creative mathematicians, after becoming acquainted with the work of their predecessors and contemporaries, achieve their own results, at first, through 'intuition', 'feeling',. They 'visualize' the most abstract theories, though sometimes it takes the invention of new means to achieve this result. Their lower nervous centres are affected by the higher abstractions made by themselves and others. This process accounts for the fact that no mathematical achievement is ever detached, or possibly can be detached, from life. The source of all creative work is always in the lower centres, which are in more direct contact with the world around us, through 'feelings', 'intuitions', 'visualization', and other first order reactions.

Mathematics and what is called 'sublimation' in psychiatry have a similar neural mechanism, which is expressed structurally in the spiral theory, or in the cyclic chain of nerve currents, where the end-product of one process becomes the starting point of the next. As was said before, this is quite obvious on racial grounds, but more difficult to discover or apply in individual experience, if we disregard structure and *non-el s.r.*

If we can, let us discover means by which the 'feel' of modern science can be imparted without falsification and technicalities, which, perhaps, may be only auxiliary means to get the more fundamental life results. We may at once anticipate the means which we shall discover. The key problem is to eliminate, first, the semantic disturbance called identification or the confusion of orders of abstractions, and similar disturbances of evaluation. This elimination is attained physiologically through the development of the consciousness of abstracting, which leads to proper evaluation, visualization *without* semantic disturbances. In other words, we must find means by which higher abstractions can be translated physiologically into lower abstractions, uniquely connected with the translation of the dynamic into the static and vice versa.

The present status of the white race—I do not know enough about the structures of languages of other races and their *s.r* to speak about them—is such that a majority of our self-imposed difficulties is due to the lack of scientific structural analysis, which lack makes it impossible to control or regulate physiologically and adequately the semantic evaluation through education. Under such conditions, everything based on arguments involving the 'is' of identity and the older *el* 'logic' and 'psychology', such as the prevailing doctrines, laws, institutions, systems., cannot possibly be in full accordance with the structure of our nervous system. This, in turn, affects the latter and results in the prevailing private and public un-sanity. Hence, the unrest, unhappiness, nervous strain, irritability, lack of wisdom and absence of balance, the instability of our institutions, the wars and revolutions, the increase of 'mental' ills, prostitution, criminality, commercialism as a creed, the inadequate standards of education, the low professional standards of lawyers, priests, politicians, physicians, teachers, parents, and even of scientists —which in the last-named field often lead to dogmatic and antisocial attitudes and lack of creativeness.

This is, naturally, an unsatisfactory semantic state of affairs, and, in consequence, our nervous systems do not function properly, according to the potentialities of proper evaluation inherent in their structure. False creeds or doctrines underlying the $s.r$, particularly when connected with strong affective tension, play as great a havoc with our responses and capacities on sub-microscopic colloidal levels as any macroscopic organic lesion of our nervous system. If our $s.r$ are pathological, invariably some affective disturbance, and psycho-neural blockages on the colloidal level, must be present. The nervous currents are then deviated and forced into lower, non-survival-for-man channels, resulting in various forms of arrested development or regressive symptoms. Through this we are deprived of the higher (*human*) 'intelligence', which is the result of the optimum working of the nervous system on all levels; we become 'mentally' deficient in various aspects and degrees, and we have to copy animals, primitives, and infants, and so present, in milder disturbances, the pathetic picture—so often seen—of adult infantilism, or display other regressive manifestations. Thousands of such cases have been analysed and recorded in psychiatrical literature. The mechanism of these disturbances is quite clear, because, after the re-education of the $s.r$, if this is at all successful, the psycho-neurological colloidal blockage is eliminated, and the patient is relieved from his semantic afflictions.

Instances of infantilism and animalistic reactions are abundant everywhere; but as this problem is analysed further in Part VII, here we shall not pursue the matter further.

It should be noticed, however, in this connection, that sex abnormalities of every description and most sex disturbances are also interconnected with infantilism in adults. In public life and activities, the results are equally pathetic. Instead of analysing and foreseeing, we proceed by trial and error, as animals do, a wasteful and painful method. The possession of an adequate physiological method for the translation from the appropriate reactions of one level to that of another, therefore, becomes paramount. The *non-el* language and the methods of mathematics appear, then, to be of neurological value. The terms are easily

20

and correctly applied to both levels, and thus facilitate passing from the language appropriate to one level to the language appropriate to the other. But, in this case, to avoid confusion, we should have to make clear the multiordinality of terms and to embody recognition of this multiordinality in every, even the most elementary, education, as *any* education shapes and moulds some *s.r*. This will aid the working of the human nervous system, which, at present, is blocked, sometimes very effectively, by disturbances of evaluation. The old *el*, subject-predicate language has a structure dissimilar to the structure of this world as we know it in 1933, and also dissimilar to the structure and function of the human nervous system, and so, by necessity, hampers the *s.r* and deviates them from their natural course.

That the problems before us are subtle, and that the demarcation line between 'sanity', 'un-sanity', and 'insanity' is extremely thin, is no reason for neglecting this neurological benefit of psychophysiological investigation. It seems obvious that the attitude toward our forms of representation, and toward our *s.r*, are fundamentally affected by the disturbances of evaluation called identification or confusion of orders of abstractions, and, in particular, by objectification, which ascribe unjustified and delusional values and meanings to these forms.

Up to this point, we have been emphasizing the beneficial structural aspect of mathematics, and it is now necessary to explain why mathematizing, when considered as a formal interplay of contentless symbols, should not be considered a high-class 'mental' activity, no matter how useful and important it may be, and why the majority of mathematicians do not get the *full* psycho-logical semantic benefit of their training and activities. The nervous systems of many such mathematicians do not act fully and successfully, nor pass normally through the cycle of their natural activities. Such a technician is seldom, if ever, what we call a great man. He seldom has a direct creative influence on our lives. But, in the case of a man with a more efficient nervous system, the cycle is completed successfully, the higher abstractions are translated back into new lower abstractions, which are closer to life. Such an individual 'sees', 'visualizes', has 'intuitions'., in his symbolic interplays. He then has a new structural vision through a new survey of his own experiences and all the experiences of others when translated in terms of lower centres. He gains a deeper insight, which he ultimately makes useful to all of us.

Immediate experience, always un-speakable, is strictly connected with the lower centres. In the translation of experience into higher order abstractions and language, the un-speakable character of experience is lost, and a *new neurological process* is needed to re-translate these higher order abstractions into new lower abstractions, and thus fully and successfully to complete the nervous cycle. One can learn to play with symbols according to rules,. but such play has little creative value. If the translation is made into the language of lower centres—namely, into 'intuitions', 'feelings', 'visualizations'.,—the higher abstractions gain the character of experience, and so creative activity begins. Individuals with thoroughly efficient nervous systems become   hat we call 'geniuses'. They create new values by inventions of new methods and in other ways, which give us a new structural means of exploring, and thus of dealing with, the world around us and ourselves, and so, ultimately, human adjustment is helped.

It is important for the reader to become thoroughly familiar with the simple division of our nervous processes into terms of order in a cyclic chain. Even neurology calls the neurons excited first of 'first order', and the succeeding members of the series, of 'second order',. The above considerations have an important practical semantic bearing for all of us, since many of the processes which we are describing can be influenced educationally by simple methods, because the term 'order', when applied, acquires a *physiological* character for *evaluation*. The description and verbal analysis of the process is, naturally, complex, but once the physiological base of evaluation is discovered, the training becomes very simple, although not easy.

The principal aim of this present work is to make available a simple and practical physiological means for accomplishing what is highly desirable, and, at the same time, for eliminating what is semantically undesirable. We deal with mathematics, because mathematics is *unique,* and, being unique, has no substitute. When discussing the theory of meanings, we have shown that all verbalism is, ultimately, similar to mathematics in structure. This conclusion contradicts many current theories of language and meanings, and so, at this stage of our argument, we lay special emphasis on the only discipline in which these issues are clear and obvious; namely, mathematics. The older theories, based on ignorance of mathematics, have led to serious abuses of our linguistic capacities and to *s.r* which are mostly pathological, with the result that practically 99 per cent of us are semantically disturbed and un-sane. Many of us, even, are on the verge of more serious 'mental' illnesses.

It will be well to give a rough picture of the similarities of, and differences between, the working of the human 'mind' at its worst ('insanity'), and its working at its best (mathematics). We shall find that the average man is between the two, often dangerously close to the first. The following picture is rough and one-sided, but suggestive, and should be worked out more fully.

The 'insane' have structural, conscious or unconscious, 'premises', which are 'false', or, in general, semantically inappropriate. Their *s.r* are shifting when they should be static, or static when they should be flexible. In the main, the difficulty of evaluation lies in the lower abstractions and the affective field. These abstractions are not properly transmitted or translated or regulated by the higher centres; or else, the higher order static abstractions are projected with too strong affective components on the lower centres. Hence, different identifications, delusions, illusions, and hallucinations result. Their 'ideas' are evaluated as things or experience, and affectively objectified in different degrees, which results in the above mis-evaluating manifestations. These semantic disturbances and tensions make the 'mentally' ill believe irresistibly in the 'truth' of their 'premises' and their inductions and deductions, which they follow blindly. In them, as in the rest of us, some internal affective pressure comes first, but because in humans the effect of higher nerve centres cannot be entirely abolished, this affective pressure is rationalized somehow into some sort of 'premises'. This organism-as-a-whole process is entirely general and applies to all of us in all our activities, but is most clearly seen in the ordered details in the work of creative scientists and 'geniuses', and in the more severe cases of 'mental' illness. To the 'mentally' ill these 'premises' have the value of 'the' and not 'a' premise. *They act upon them,* and so cannot adjust themselves to a world different from their fancies. They would seldom survive at all if left alone by themselves, particularly in a complex 'civilization'.

Mathematicians, also, have structural premises, often called postulates, but they *never* evaluate them to be 'true'; wherefore their premises *cannot* be 'false'. They have no claims, and claims are always affective. Like the 'insane', they follow up these premises blindly, but, being generally conscious of abstracting in the field of their profession, they are not usually subject to semantic disturbances *in this field* and do not live out their theories in life, the theories thus remaining affectively hypothetical. If a mathematician were to believe, with strong affective evaluation, that his premises are 'true', these premises then would become mostly false, or meaningless, or, in general, inappropriate. If he lived through them, the given individual would then be 'mentally' ill, *not* because of his premises, but because of the semantic disturbance, which would involve erroneous evaluation, identifications, confusion of orders of abstractions in his affective *attitude toward his premises*. This subtle organism-as-a-whole mechanism, in which all affective pressure can be rationalized, and all rationalization can produce affective manifestations, not only makes the present *non-el* analysis possible and legitimate, but also offers some explanation of those remarkable cases of 'mental' illness in a number of mathematical geniuses. Under such organism-as-a-whole structural conditions, a *general* consciousness of abstracting not restricted to a special field is the only possible safeguard against the semantic disturbances which lead to an unbalanced 'mental' condition.

As we have seen, the difference between 'sanity' and 'insanity' is subtle. The reader must be reminded that it takes a good 'mind' to be 'insane'. Morons, imbeciles, and idiots are 'mentally' deficient, but could not be 'insane'.

The so-called 'sane' also have structural premises; we all have some standards of evaluation. These are also usually false, or, in general, inappropriate, being mostly due to our savage inheritance. But the saner we are, the less we abide by them. Therefore, in a world quite different structurally from our fancies, we are often able to adjust ourselves for all practical purposes, often avoiding major disasters for a number of years.

For instance, the believers in extraordinary blisses in the 'other life' or the 'other world' should welcome death. Why be so unhappy here, when, according to their doctrines, there is such an ideally happy future after death? Why make use of medicine and doctors, when a deadly illness should open the door to everlasting bliss! In conflict with such a creed, he lives as long as he can, often most unhappily, and is generally willing to spend fortunes on doctors and medicines to delay the bliss! The genuine and very serious danger to all of us of such creeds is that when the *s.r* of an individual are trained in this way he finally does become indifferent, or apathetic toward actualities in *this world*, so that cunning, and often pathological, individuals are thus given an opportunity of directing human affairs toward their personal ends.

Naturally, with the increase of the complexities of conditions, the dangers also increase in a geometrical ratio, because when *m.o* realities become too unbearable, the masses cease to be influenced by these semantic illusions, and they break all barriers, only to fall again under the influence of new leaders very often equally irresponsible and ignorant.

Unfortunately, the failure to understand these semantic issues, based on animalistic lack of foresight, results invariably in a great deal of unnecessary suffering. There is little doubt that without these delusions and illusions we should look after the conditions of our actual lives more closely, and many of our pressing needs would be adjusted.

The difficulties which we have are mostly man-made, and so only mankind can remedy them, and any attempts to escape from *m.o* reality only aggravate the situation.

Lack of space does not allow me to dwell here on many other aspects of mathematics which are of neurological structural importance, except to mention the theory of statistics and probability. All human knowledge is neurologically due to a process of abstracting in different orders, giving us the only structural knowledge of processes, which, in 1933, must always be considered on three levels, the macroscopic, the microscopic, and the sub-microscopic.

Because the nervous system is an abstracting, integrating mechanism, all human psycho-neurological reactions and, particularly, psychological, to be similar in structure, *must* be based on the mathematical theories of statistics and *probability*. On the objective level, we deal with absolute individuals, and so all statements, or higher order abstractions, can only be probable. Historically, mathematicians have elaborated not only both theories, but Boole, in his *Laws of Thought*, extended the mathematical approach to 'logic' in connection with the theory of probability. Finally, the difficulties of the law of excluded third have been solved by Łukasiewicz and Tarski[10] in their 'many-valued logic', which, when N increases indefinitely, merges with the mathematical theory of probability, a result reached independently by a different type of analysis in the present system. Any possible future scientific $\bar{A}$, *non-el* 'logic', which I call general semantics, must be built on this structurally more correct foundation. It should be noticed that the notions of probability are very flexible, and entirely cover our structural needs, the field of degrees of probability ranging from impossibility to certainty. This new semantics involves entirely new affective attitudes, and underlies new and better balanced *s.r*.

Under such conditions, the restricted 'uncertainty principle' of Heisenberg becomes a structural, most revolutionary, and creative *general principle*, transferring the laws of two-valued 'cause and effect' from the realm of gambling on words by 'philosophers' to the scrutiny of scientists, and establishing ∞-valued 'determinism' on a neuro-mathematical base of 'the greatest probability'. Methodologically and psychologically, this requires *full consciousness of abstracting*, achieved, as yet, by extremely few of us, even among physicists and mathematicians. Then the 'law' of two-valued 'cause and effect', instead of depending on the *el* and objectified older interpretations, will be based on the mathematical, and much more reliable, ∞-valued principle of greatest probability. This will eliminate to a large extent semantic disturbances, and so the problems of sanity will be greatly helped towards solution.

To those who are accustomed to the disclaimers made by many mathematicians of human values in their work, such an analysis as I have given in the present chapter must seem unexpected. But, upon reflection, we may see that, after all, it is only a natural evaluation. Language is a unique, and, therefore, most important, human characteristic. Ought we to wonder that these linguists of exact sciences, whom we call mathematicians, should have produced unknowingly and unwittingly great human values, fundamentally affecting the $s.r$? They could not help it. Once they worked out their own problems properly—and no one doubts that they did it well—the results were bound to have broad human significance. Their activities were kept on the proper levels and so were naturally a help toward sanity. In Part VII, I shall discuss another mathematical discovery, known as the 'theory of mathematical types', of Russell, which, when generalized, becomes a *physiological* theory of enormous semantic importance and of fundamental and constant human application.

In spite of popular belief, mathematics is the simplest language in existence. Our daily language is so very complex in its structure that for many thousand years it evaded analysis. Probably, the writer, without the study of mathematics, would not have been able to discover the ultimately extremely simple yet workable principles outlined in the present work.

PART VI

ON THE FOUNDATION OF PSYCHOPHYSIOLOGY

It is an important principle of physiological epistemology that a phenomenon which occurs generally, cannot possibly be the *specific* function of an organ which is peculiar to a few forms only. (306)   JACQUES LOEB

I have observed, as a consequence of serious and repeated emotional shocks, very curious cases of infantilism in adults, where complete amnesia, accompanying sexual inhibition and disturbances of the affective area, produced the mentality and conduct of a little child. (411)   HENRI PIÉRON

The organism is reacting as a whole to its environment as a whole, and it is doing so in ways that cannot be formulated in terms of an algebraic sum or simple mechanical resultant of the interplay of the simple reflex responses to external stimulation. . . . the mechanisms of traditional reflexology seem hopelessly inadequate. (224)   C. JUDSON HERRICK

CHAPTER XX

GENERAL CONSIDERATIONS

We know also how different extra stimuli inhibit and discoordinate a well-established routine of activity, and how a change in a pre-established order dislocates and renders difficult our movements, activities and the whole routine of life. (394)   I. P. PAVLOV

The experiments show that a compound stimulus the component units of which remain themselves unaltered, and consequently most probably affect the same cells of the cerebral cortex, behaves in different modifications as a different stimulus, evoking in these cells now an excitatory process and now an inhibitory one. (394)   I. P. PAVLOV

We thus come to the following conclusion: when perfectly neutral stimuli fall upon the hemispheres at a time when there prevails a state of inhibition they acquire an inhibitory function of their own, so that when they act subsequently upon any region of the brain which is in a state of excitation they produce inhibition. (394)   I. P. PAVLOV

Some of the most important researches in the function of the higher nervous centres have been done lately by Professor Pavlov in his work on the so-called 'conditioned reflexes'. This work was developed in a series of papers covering a period of nearly thirty years of experimentation, but the average international scientist did not know this work as an entirety, because the papers were scattered and written mostly in Russian. Only in 1927 did the Oxford Press publish Pavlov's *Conditioned Reflexes, an Investigation of the Physiological Activity of the Cerebral Cortex* in the English translation of Doctor G. V. Anrep; and in 1928 The International Publishers (New York) published Pavlov's *Lectures on Conditioned Reflexes, Twenty-five Years of Objective Study of the Higher Nervous Activity (Behaviour) of Animals* in the translation of Doctor W. Horsley Gantt. Both translators were collaborators of Professor Pavlov in Leningrad for a number of years. In these two books, the latest experiments and interpretations are given.

Hitherto, most of the researches on the function of the higher nervous systems were formulated in 'psychological' languages, which, obviously, are not fit for physiological disciplines. Professor Pavlov, himself, suggests this fact as an explanation why, until his work, the physiology of the cerebral cortex was so little known. There is no doubt that the descriptive physiological language of happenings, functionings., used by him exclusively, is responsible for his results. This language suggests structurally new experimentations, which suggestions are lacking in other accounts of the kind where antiquated 'psychological' terms are used.

Although I knew as much as the average scientist about the work of Pavlov, this knowledge was not integrated enough to make some issues clear. But, after I had formulated my $\bar{A}$-system, I read the books of Pavlov and found, to my great satisfaction, that a neurological mechanism, the analysis of which underlies my own work, and the existence of which was independently discovered by me on *theoretical* grounds, had been discovered by Professor Pavlov and his co-workers on *experimental* grounds, thus supplying additional experimental verification for my system.

It seems that the so-called 'ethics'., in general, sanity, which underlie desirable human characteristics have a definite *physiological* mechanism, automatically involving on psycho-logical levels these desirable semantic attitudes. It appears that some of the psycho-logical problems enormously complex and difficult to reach, or even inaccessible, are solved, not by preaching, but by the most simple and elementary *physiological* training, a fact which has been verified empirically. Obviously, such simplification, if at all possible, must be of fundamental importance.

Physiology deals, in the main, with the functioning of organs in organisms, and results in various formulations. Thus, there might be an hypothetical 'physiological theory of most effective feeding', for instance, stating that food should be secured first in one's hand, or spoon, or fork, before putting it in the mouth,. A group of people who habitually disregarded the 'physiological theory' and abandoned attempts to act in accordance with it after the first unsuccessful one, would be badly underfed or would simply perish. Facts of experience show that some such 'physiological theory' must have been known and applied from time immemorial, and that, perhaps, because of it we survive at all!

How about the 'mental' field? As I demonstrate—and close observation will verify this very generally—the existing theories of 'mental' life, closely related with our linguistic habits, are A, grossly inadequate, and lead to a wholesale production of morons, imbeciles, 'emotionally' disturbed, and, in general, un-sane individuals. Investigation shows the possibility of a simple and obvious *physiological* theory of the use of our nervous system, which automatically leads to desirable psycho-logical, semantic states of general sanity.

In the frivolous example of a 'physiological theory of feeding' given above, the problems of *order* were important. In the physiological theory of sanity, order becomes paramount. Processes and function involve series of states, by necessity exhibiting order. Adjustment to life-conditions means adjustment of processes, and a physiological theory of sanity must be based structurally on four-dimensional order, where 'space' and 'time' are indivisibly interwoven.

Pavlov shows, in an unusually impressive variety and numbers of experiments, how 'order' and 'delay' (four-dimensional order, in the language used here in this connection) are intimately related with most fundamental processes in the higher nervous centres, and how, by the changes or interplays of them, we can produce or eliminate *pathological states* of the nervous system.

In the human field we find a quite similar situation, unanalysable by older methods, because all order involves asymmetrical relations, which, as we have already shown, cannot be dealt with by A means.

The issue is clear and definite: either we persist in our old A habits of speech, in which case asymmetrical relations and order evade our grasp, and proper evaluation and sanity are *physiologically* impossible, *or* we build a $\bar{A}$-system free, or at least more free, from these evaluational limitations, which allows us to deal with order, and sanity becomes *physiologically possible*.

'Stimuli are never "simple", and, by necessity, involve fourfold space-time structure and order. Survival values involve, also, this four-dimensional order. For instance, the natural survival order is "senses" first, "mind" next; object first, label next; description first, inference next ,. The reversal of the natural order appears pathological and pathogenic and is found as a symptom in practically all forms of "mental" ills, as well as in most human difficulties and disturbances which, at present, are still not considered abnormal. Thus, objectivity is ascribed to words, "mind" projected into "senses", inferences evaluated as descriptions . ,—quite common "symptoms" . . . Observations on human levels show that we still copy animals in our nervous responses, confuse orders of abstractions (non-existent for the animals), leading fatalistically to the reversal of the natural order and to pathological results, making the great majority of us un-sane.'*

A structural *non-el* enquiry into the objective world shows quite clearly that no event is ever 'simple'; it is, at least, a limited whole of interrelated factors. The eventual 'simplicity' is manufactured by a nervous process of higher and higher abstractions.

In our consideration of 'order' and 'delay', and the role they play in connection with the activities of the nervous system, we must first discriminate sharply between the objective level which is *un-speakable*, because anything that can be said *is not* the object, and the verbal level, on which we can, at will, concentrate attention on similarities, or differences, or both. Secondly, we must pay special attention to structure—

---

*From—Discussion by A. Korzybski. *Proceedings of the First International Congress of Mental Hygiene.* New York, 1932.

that is to say, search for structure in the empirical world, and, once this
has been found, adjust, accordingly, the structure of our language.

The structure of the daily, as well as of the 'philosophical', language, which we inherited, in the main, from our primitive ancestors, is
such that we have *separate terms* for factors which are not separable,
such as 'matter', 'space', 'time', or 'body', 'soul', 'mind',. Then, as it were,
we try to make out of the word, flesh, by reversing the natural order
and affectively ascribing a delusional *objectivity* to these terms.

If we deal with the silent, un-speakable, objective level and try to
divide according to the implications of the verbal division, we find a
brutal fact, which, until Einstein and Minkowski, has escaped scientific
verbal formulation, that this cannot be done at all. On the objective
level every dealing with 'matter' involves 'space' and 'time'; any dealing
with 'space' involves some fulness of something and 'time'; and every
dealing with 'time' involves 'something' and 'space'.

The structure of the world happens to be such that empirically 'matter', 'space', 'time', cannot be divided; wherefore, we should have a
non-el language of *similar structure*. This was accomplished by Einstein-
* see page xii Minkowksi, when they created a language of 'space-time', in which the
hard lumps against which we bump our noses are connected analytically
with the curvature of space-time.

In this new, *non-el*, four-dimensional language, every three-dimensional point of 'space' has a date, and so is different. For our purpose,
we do not need, at present, to bother much about its curvature or the
kinks in space-time, called in the old way 'matter', but we must emphasize that the fourfold order is of great importance, as it corresponds
structurally to *experience*, and is intimately connected with physiological
reactions, the semantic included.

There is a great deal of confusion about these problems among
laymen and also among scientists. From a structural point of view, the
issues are quite simple, and there is nothing sensational in the latest
announcement of Einstein that 'space' in its importance is displacing
'matter' (*Nottingham Lectures*). Naturally, the statement *in this form*
is rather baffling and attracted much—even newspaper—attention. Yet
it seems that even the einsteinists do not fully realize the *verbal, structural,* and semantic issues involved.

For the layman, as well as for the majority of the physicists in their
less sober, or metaphysical, moments, 'space' is 'emotionally' newtonian
and an 'absolute void', which, of course, being 'absolute nothingness',
cannot have *objective* existence, by definition. For Einstein, 'space-time'
is, semantically, 'fulness', not 'emptiness', and, in *his language,* he does

not need any term like 'ether', as his 'plenum', structurally, covers the ground, without his committing himself to a definite two-valued mechanistic ether. The confusion of orders of abstractions, from which we all suffer, is semantic, and is due to disregard of the structure and role of language. If we accept a *non-el* language of space-time, structurally we deal with fulness, and we should not use the term 'space', as its old semantic implications are 'emptiness' and so are very confusing. The 'sensation' of Einstein's declaration amounts to the fact that the submicroscopic fulness ('space') is more important than a few kinks or concentrations of that fulness ('matter'),—a fact which science has established, and which is quite obvious.

Experiments with 'conditioned reflexes' have established firmly the fact that stimuli can be compounded, and that, when established, the compound stimulus acts as a unit, and that a change in the four-dimensional order of factors (including delays) acts as a *different* stimulus, not necessarily resulting in the established reflex. This often introduces great complexities.

As an example of this, we will use the so-called 'delayed reflex'. When established, the 'conditioned reflex' does not appear at once after the stimulation, but after the stimulation combined with the usual 'time'-delay has occurred, thus showing that the 'time' factor plays a physiological role in a compound stimulus. Organisms live in, and consist of, periodic processes, such as the alternation of day and night, sleep, taking in food, heart-beats, breathing, electronic pulsations.; so that *any* stimulus, no matter how nominally 'simple', is, in reality, a compound stimulus of, let us say, x and y heart-beats and what not. An organism represents, invariably, a clock of some sort, and, when that clock stops, life ceases.

Under such actual structural conditions a four-dimensional analysis makes every 'simple' stimulus compound, and thus four-dimensional *order* becomes a potent *physiological* factor, exerting definite effects. The interplay of four-dimensional order of factors represents, in general, a new stimulus; we have an interplay of positive and negative excitations which may lead to clashes between the two that the nervous system finds difficult to resolve, and so pathological results follow.

If we pass to sub-microscopic levels and processes, we find that, although we may speak of them as 'chemical' or 'stimuli of greater physiological strength'., yet, by structural necessity, they represent different kinds of *multi-dimensional* order, because, as we say in 1933, the dynamic physical unit of that order is a quantum of action. The metaphysician should not get excited about this statement, because whatever *he might say* will also be a verbal statement of a given date, made mostly without any structural considerations, and based mostly upon the reversed survival order, confusion of orders of abstractions and other semantic disturbances. As the world, both outside and inside our skins, is invariably found to exhibit a fourfold space-time order, it is inevitable that this order should be structurally impressed on the nervous system, establishing a natural survival order. Therefore, changes in this order on the macroscopic level, the level of outward events, must have direct inward sub-microscopic effects, disturbing or restoring the nervous equilibrium. This statement may appear innocent; it is not; it has a vital human significance, as it involves standards of evaluation. In short, it means that, in the *actual application* of the consideration of order in education and training on the daily-life levels, we can affect the evasive (as yet) microscopic and sub-microscopic structural levels of the human nervous system, thus directly affecting our *s.r* and behaviour.

To make this clearer, let us recall some of the neurological researches of Bolton (as quoted by Herrick). The cortex has different layers, characterized by the difference in the number, size, shape, internal structure, and density of neural cells. Bolton's third layer of granules divides the cortex into two types of layers. Those closer to the base of the brain, or below the third, are called the infragranular; those above, the supragranular, layers.

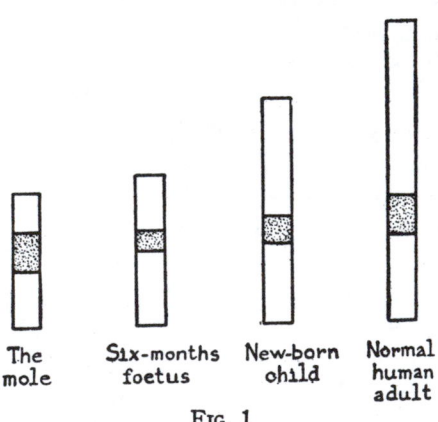

The   Six-months New-born Normal
mole   foetus   child   human
                                     adult

FIG. 1

Diagrams of the relative thickness of the supragranular, granular, and infragranular cerebral cortex in the six-months foetus, the new-born child, the normal human adult, and the adult mole. The granular layer is dotted. (Redrawn from G. A. Watson (1907), and adopted from Herrick.[1])

The lower mammals show a well-developed infragranular cortex, and a very poorly organized supragranular cortex, the latter increasing in relative size and complexity as we ascend the animal series. On the human level, we find a most important, and usually disregarded, fact—that the human nervous system is not completed at birth, but develops structurally years after birth.

The above explains why animalistic theories and methods, primitive-made languages of wrong structure, and similar relics, result in training the *s.r* of our children in the patho-

logical order, and bring about such great harm, individually and collectively. It enables us to understand, also, why all forms of 'mental' ills
invariably exhibit infantile characteristics of some sort.

If we train a child with a physically undeveloped nervous system in
animalistic doctrines strictly connected with a primitive-made language
of wrong structure, in the pathological reversed order of responses, such
semantic training must affect harmfully the still developing nervous
system. So, when we say, and demonstrate, that we still copy animals
in our nervous responses, we imply an undeveloped or thwarted nervous
system, of which the development has been arrested or made regressive.
Such a deficiency, of course, is superimposed functionally, and so structurally, upon whatever congenital deficiency there might have been in a
given case. We are nearly all in a situation of this kind. We continue
to be educated under animalistic conditions since we became time-binders,
which, from a biological point of view, is a very recent event, and it is
not rash to assume that our nervous system is still not fully developed,
the more so that we submit the cortex, which in childhood is still incomplete, to injurious semantic influences. Obviously, such a fundamental
human function as language, when used in a way not in correspondence
with the structure of the nervous system, must act detrimentally on its
development.

In congenital extreme imbecility, the cortex is poorly organized,
thin, and deficient in nerve cells, and the infragranular layers show less
impairment than the supragranular layers. It seems that Bolton's second
supragranular layer is the last to mature, and its relative development
corresponds to the relative development of an animal or human being,
and, in a way, it goes parallel with the so-called 'intelligence'.

In human defectives, its deficiency corresponds with the degree of
psycho-logical arrest, regression, or deterioration. Let us recall that
these 'mental' deficiencies, which, in behaviour as well as in nervous
structure, take us one step (or several) back toward the level of the
infant, or even to that of the animal, are always connected with infantile
behaviour in adults, and semantic disturbances.

Nervous as well as muscular tissues have differentiated from the
general protoplasm, and we know positively that, through training, we
can enlarge or otherwise improve muscular tissues, and there is no reason
to doubt that something similar can be done to nervous tissue. All education, and the establishing of any conditional or *s.r*, shows this, although
in a rather vague way.

If, by a *physiological* training based on order, we can alter a nervous
deficiency, as shown by behaviour, we may conclude that there are physio-

21

logical means by which we can effectively train and help the development of nervous cells and supragranular layers—or, at least, *not hinder* their natural development. On the macroscopic levels, this beneficial training consists in forming habits of proper evaluation through the natural survival order. The effect of this on the sub-microscopic level is neural, colloidal, and structural, a result which, by the older methods, could not be reached, either with ease or with effectiveness.

Although these conclusions are necessary, it is impossible directly to verify them empirically, because we should have to dissect the brain of a given person before, and during, the training. In this case, as in many others, we have to observe 'human nature' and semantic responses to stimuli of a given individual before and after the training; and, on the foundation of what we know of the development of the nervous system in animals, infants, 'mentally' defective and well-developed adults, build our eventual conclusion as preliminary hypotheses for further structural testing, improvement, and empirical verification.

To realize fully the importance and necessity of this conclusion, we must first understand that, in accordance with the modern discoveries of mathematics, physics, chemistry, colloidal chemistry, and other branches of science, all 'function' depends on *structure,* because the unit-brick of structure represents a dynamic unit of a quantum of action. In the remarks which follow, it is impossible to be as full and precise as I should like to be, because to be able to do so would be equivalent to being able to solve all scientific problems; yet the reader should realize that the considerations of structure will become extremely creative and helpful as long as we recognize a quantum of action, or any other *dynamic unit.*

We may recall that the characteristics of molecules are due to atomic structure, and that the characteristics of atoms are due to electronic structure. The latest quantum theories also seem to find that the characteristics of electrons are an outgrowth of structure; and, if the suggestions of Dirac are verified, even the difference between positive and negative electricity is structural. Even at present it appears that 'structure' is not only a term fit to handle and explain, but that it has an objective counterpart, allowing a similarity of structure, and, therefore, making the understanding of this world possible.

This last very important semantic point is based on the fact that relations of similar structure have similar 'logical' characteristics, and that, therefore, in structure we find means by which the events can always be made intelligible to us, and so properly evaluated.

Experience—my own, as well as that drawn from scientific literature—impresses me with the fact that we very seldom realize that our 'knowledge' (or, roughly, what we can *say*) *is not* '*it*', as the 'it' is always un-speakable.

Between two houses or two stones, there is some sort of sub-microscopic interaction; but on the macroscopic level, nothing definite happens. So we say that in the given context or configuration, the **units** under consideration are too heavy (implying gravitational structure), or the medium, the plenum in which they are immersed, is too light (again structural implications) and so, macroscopically, nothing obvious happens.

If that structure is changed, different conditions, different relations and results prevail. Thus, if the particles are very small, and the media not too heavy, the surface phenomena, electrical charges., begin to play an important role. We then have colloidal behaviour of enormous complexity and variability where we find, not necessarily life, but many inorganic forms, duplicating some forms found in life. Obviously, colloidal structure accounts for that.[2]

When little colloidal wholes, most probably of specialized internal structure, arise, we may have not only colloids, but also little wholes, separated by a membrane, or perhaps by surface phenomena, which represent a most generalized membrane. We may have a new structural fact, an interplay of the inside with the outside, and life begins.

The general irritability and conductivity of protoplasm is known to be strictly connected with permeability to the passage of ions and, therefore, is a structural phenomenon. On this structural foundation, physiological gradients result, forming a dynamic field of forces, again involving structure. The development of the differentiated tissue of muscles and nerves consists of higher order complex structures, based on more primitive structures; and, finally, function and behaviour of all life, man included, is due to sub-microscopic, microscopic, and macroscopic structure.

I may be reproached by specialists that, although what I have just said may possibly be true, yet, actually, to make these assertions is, perhaps, premature, in 1933, because we lack too many details.

My answer is sharp and definite, and may be considered a serious scientific suggestion, because it can be made legitimately in this form:

1) All science depends on human 'knowledge'.

2) All human 'knowledge' is structurally circular and self-reflexive, and so depends on some conscious or unconscious theory of knowledge and undefined terms.

3) Words are *not* the things we speak about; and, therefore,

4) The only possible connection between the objective and un-speakable levels and words is *structural;* introducing

5) Structural analysis of languages as fundamental; making

6) The *only possible content of 'knowledge' structural,* and

7) All science becomes a search for the unknown structure of the empirical world on all levels, and the matching of this unknown structure with the *potentially known* structure of languages; so that

8) All knowledge is hypothetical, in which

9) The most important facts must be *negative.* When the structures do not match, then we learn something quite definite about the empirical structures.

10) All predictability becomes possible because of similarity of structure; and so definitely making

11) All possible aims and quests of science uniquely structural; necessitating

12) Unique methods of translation of dynamic into static, and vice versa, in order to cover the structural exigency of both the dynamic world and the static languages.

13) Such unique methods of translation are given in the differential calculus and four-dimensional geometries, in which

14) What in a four-dimensional language is structure becomes in three-dimensional language 'importing time' function; showing once more that

15) Structural considerations are not only a modern necessity, but also the most creative and helpful for the future development of science and man, and justifying the above assertions, with the setback that

16) Full 1933 structural analysis, being one of the, or perhaps *the,* highest abstraction of this date, the mastering of that language may represent some difficulties.

The reader must be reminded (see, for details, Part VII) that the terms 'structure', 'function'., are multiordinal terms with many meanings, and so that they have no general meaning apart from context, but have definite meaning in each context. Without this realization of the multi-ordinality of terms, the statement above could not be made, for it is a structural statement about languages.

As an example of the immense and inherent importance of considerations of four-dimensional order, the following psychological experiment for which Doctor Harry Helson has suggested lately the name of Tau effect, is useful.[3]

If we stimulate three spots of the skin by touching them lightly with the end of a pencil in quick succession, and if the distance between the first and the second spots is, say, 20 mm., and that between the second and the third is 10 mm., but the 'time' interval between the second and the third stimulations is twice as long as that between the first and the second, the distance between the second and the third spots will be judged as nearly twice as great as that between the first and second. Similar results are obtained with other analysers, such as vision and hearing. If we change the conditions of the experiment, the results may be reversed. It is interesting to note that the effect does not depend on 'knowing', as similar results happen when the subject knows the conditions of the experiment. The last shows that the experiment deals with a physiological and neurological mechanism. In general terms, if we vary the time-interval in the opposite sense from the space-interval, the latter will be distorted, showing once more the structural fact that in actual life and experience we deal *exclusively* with the four-dimensional space-time order, which, as such, must have physiological and neurological significance, and an adapting mechanism.

CHAPTER XXI

ON CONDITIONAL REFLEXES

The conditioned reflex is conventionally regarded as differing essentially
from the unconditioned reflex, but this is contradicted by evidence drawn
from the development of behavior. (107)   G. E. COGHILL

The main experiments of Pavlov were made on dogs, animals with
a rather well-developed nervous system; and so most of what he has to
say is about dogs, although some general physiological facts apply to all
the higher nervous systems, man included. In some instances, because
of human complexities, some results must be re-interpreted, structural
linguistic re-adjustments made, and some obscuring, wrong-in-structure,
el 'psychological' terms analysed and rejected. My linguistic, structural,
non-el, theoretical revision leads to a new and important enlargement of
the application to man of the Pavlov *experimental* theory of 'conditioned'
reflexes. The fact that these independent discoveries reinforce and
support each other is a striking instance of the usefulness of theoretical
researches.

We must take care to notice and beware of the differences in languages. Any happening has as many aspects as there are sciences, or
even human interests. Thus, if we speak about an objective 'pencil', we
may speak about its chemistry, or methods of manufacturing, its uses,
prices, markets,. As the content of knowledge is structural, we must
search empirically for structure, understood nowadays always on three
levels (the term being multiordinal), the macroscopic gross structure,
the microscopic, and sub-microscopic structures.

When we deal with life phenomena, we have also different languages dealing with their different aspects. Thus, a biological language
would cover eventually the vital events in general; a physiological language would be narrower and cover the analysis of phenomena in an
organism, the function of its organs and the conditions and the mechanism which determine these functions; a neurological language would
be physiological as applied only to the nervous system. The day is not
distant when all these problems will be formulated in the language of
the quantum mechanics.

A psycho-logical language is legitimate only on human levels, as
we never know, or can know, what an animal 'thinks', 'feels'., and on
human levels it applies to so-called 'psychic' phenomena only.

Usually, one extremely fundamental semantic fact is disregarded;
namely, that what on the psycho-logical level is *objective* and in language
descriptive to one person (e.g., 'my toothache'), is *inferential* to the other

person, and vice versa. The lack of consciousness of abstracting introduces, by structural necessity, an identification of orders of abstractions; namely, the confusion of descriptions with inferences, and vice versa. This makes it imperative to avoid psycho-logical language as much as possible. It is also bad epistemology to use a language which applies to a few individuals (psycho-logics) for describing functions which are much more general, and which, fundamentally, apply to all organisms.

It is a striking fact that, although physiology is a fairly old and well-developed discipline, yet the purely physiological approach to the study of the brain-functions is very recent, and, in the main, has been carried on by Pavlov and his followers. Pavlov gives us a simple yet profoundly true explanation; namely, that the higher nervous centres have never been treated on equal footing with other organs, or other parts of the nervous system. The activities of the hemispheres have been treated from a 'psychological' point of view, and, by analogy, we have ascribed to animals similar 'psychological' states, a remnant of primitive animism. As such attitudes have become more and more obviously absurd, we have drifted into the opposite absurdity of animalism, ascribing animal characteristics to man, forgetting that the human nervous system is far more complex, matures later than in any animal, and is a non-additive affair. Naturally, reasoning by such analogies must be fallacious.

The prevalent complete disregard of the fact that these issues are linguistic and structural makes the advances in these fields very slow and halting, and only so-called 'geniuses' are capable of breaking through these semantic barriers. Once the linguistic character of the issues is fully realized, the psycho-logical, semantic blockage is removed, freedom of analysis is inwardly established, and even 'non-geniuses' will produce important creative work. Indeed, we may find that with this realization, particularly if embodied in early education, the 'normal' man would be, what we call at present, a 'genius'.

This conclusion naturally follows if we abandon animalistic analogies and face the fact that high-grade human intelligence happens to be not less 'natural' and inherent in the history of evolution than any other 'tropism'. By eliminating the psycho-logical semantic blockage due to copying animals in our nervous reactions, we may handle this important *human function* of language properly. Man will function as *man*, in accordance with the structure of his more complex nervous system. There is no doubt at present that some organisms called 'man' have an important function connected with *s.r* called 'speech', perhaps the most complex and involved and also *unique* function evolved by this class of life, and which it does not yet know how to use. Biologically and physiologically, this misuse of a function must be a *non-survival tendency* for *this* class of life.

Sanity must be based on methods for the most efficient use of the human nervous system, in accordance with its structure, and will thus bring about the full working of human capacities, which at present are still semantically blocked by faulty handling of the apparatus.

Before going further, I will analyse and suggest the complete elimination from the English language of the term 'conditioned' reflex, which is structurally false to facts, and suggest in its place the uniform use of the term 'conditional' reflex, introduced by Pavlov and used occasionally by some English writers. I will also suggest the elimination of a *psychological* term, 'inhibition', from physiology and neurology, in which it should have no place at all. Such a change in language leads to new results, and also suggests new experiments. It is little known and seldom taken into consideration that long ago Locke was quite clear on the point that the misuse of language has often been taken for deep mysteries of science; but Locke, unfortunately, did not take into consideration *structure*, and *s.r*; so his arguments were, in general, non-operative.

As everything in this actual world is structurally interrelated with everything else, we should consciously look for interrelations; in which case we have to build special languages for the eventual synthesis. As we must first ascertain empirical structure, and only then coin the languages, obviously to start with a descriptive, impersonal, non-'psychological' language of ordered events on a given level is most important.

In our case, we are investigating the structural and semantic problems in connection with language. We have to accept the structural facts as discovered by physicists, physiologists, neurologists, and other scientists, and then build a language similar in structure to the empirical world. The language in which the present theory is formulated is a physiological and neurological one, as it deals with observed impersonal functionings of the organisms called 'man'. When we reach results in a physiological language, these have, naturally on the human level, a psychological aspect, and perhaps the main importance, and even value, of the present work is that it reaches the very difficult psycho-logical, semantic level by purely functional and easily controlled physiological methods.

Thus the reader must translate for himself, as nobody else can do it for him, the physiological results into his psycho-logical feelings and attitudes and evoke the un-speakable *s.r*. These must be *evoked* by the reader, otherwise he will inevitably miss the point. For instance, if it is said that 'the objective level is un-speakable', the reader should try to become entirely 'emotionally' impassive, outwardly and *inwardly* silent about an object, or a feeling, as whatever we may say *is not 'it'*. This, obviously, involves a complete checking of affective responses, 'preconceived ideas'., making him an 'impartial observer'. In fact, to do this successfully is something very difficult to achieve, requiring long semantic training with the Structural Differential, and usually involving a complete reversal of our habitual modes of affective responses.

Similarly, when we speak of 'natural order' or reversal of this order, let the reader try actually to evoke these *s.r*, and he will find it is not so easy, as it involves a completely new process of re-evaluation. In both cases, we can gain physiological, easily operating means to re-educate the very stubborn semantic responses, by *functional and ordinal methods*. The difficulties are only serious with grown-ups; they present no difficulties in the early semantic training of infants, for whom this training becomes a powerful preventive method against future nervous disturbances (limited, of course, to this aspect of un-sanity).

Directly such a semantic re-education is accomplished, the formerly impossible is also performed, and 'human nature' has been changed. Obviously, the trouble has not been with 'human nature', but with the lack of physiological and educational means to affect the psycho-logical level and to change the *s.r*. The above applies to the so-called 'normal' man, as well as to the 'mentally' ill. It works with both types, provided the latter is in a condition to be at all accessible to approach.

The term 'nervous reflex' was originated by the mathematician Descartes. Structurally, it was a genuinely scientific notion. It implies necessity; namely, that a stimulus results in a response. Obviously, if such were not the case, an animal would not be in sufficient correspondence with its environment and could not survive. Thus an animal must be attracted and not repelled by food, it must avoid fire, and so forth.[1] The term 'reflex' is usually used in connection with A two-valued implications; this makes reflexology *el* and generally inadequate to account for organic *non-el* responses in the colloidal sense. In a $\bar{A}$-system based on the ∞-valued semantics of probability, I prefer to avoid the two-valued implication and use the term reaction instead.

The main function of the nervous system is the co-ordination of all activities of the organism for its preservation. Thus there must be no conflict between the opposing activities of different parts of the organism, and any action must ultimately benefit the whole. There may be a conflict of different excitations, but one must finally dominate the others, as otherwise co-ordination would be impossible.

If food or some noxious material is put into the mouth of a dog, a secretion of saliva is produced, either to alter the food chemically and help digestion, or to wash the mouth out and eliminate it. But observation shows further that other factors, not food or noxious materials alone, may produce similar secretions. Thus, for instance, the sight or smell of some such material, or of the person who usually administers it, or even the latter's footsteps may produce salivary secretions.

To make the experiments as exact as possible, the dogs observed by Pavlov were submitted to minor operations. Among others, the opening of the salivary duct was transplanted to the outside of the skin, so that all secretions could be carefully and exactly collected and measured.

The principle which underlies these experiments is the observation that if we combine some hitherto neutral stimulus, such as a definite tone, colour, or shape, with the presenting of the food or acid, after a few trials this neutral stimulus acquires the potentiality of producing similar secretory effects as the food or acid itself.

This fundamental and exact method of experimentation allows considerable freedom in the selection of neutral stimuli, affecting, as we prefer, the visual, or auditory, or tactile, or other nerve centres of the animal. We can also control their numbers, intensity, their combinations, the *order* and *delays* in their application,.

If food or noxious materials are placed in the mouth, the secretion of saliva is an almost automatic reaction, owing to the physico-chemical action of these materials. Such reaction is inborn and practically general for a given species, the nervous paths for such reactions being mostly completed at birth. Not so with the reactions produced by neutral stimuli, which acquire the secretory characteristics only after some experiences. These characteristics are reactions acquired during individual life, and the nervous paths and connections have to be completed during the lifetime of the individual.

Thus, when puppies are shown meat or bread, which they had never eaten before, usually no salivary secretion appears. Only after eating meat and bread on several occasions will the sight of them produce secretions.

Some of the effects of these acquired reactions are very strong and lasting. In some experiments, the dogs were given a hypodermic injection of morphine. The usual effect of the drug is to produce nausea with profuse secretion of saliva, followed by vomiting and deep sleep. In further experimenting, it was found that the preliminaries, or even seeing the experimenter, without injection, was often enough to produce the effects of the actual injection of the drug.

Pavlov studied the nervous mechanism of the functioning of the salivary glands, not because of any special physiological importance of these glands, but because such experimenting was the simplest, and the method used allowed him to conduct the most varied experiments under accurate control.

The experiments disclosed an amazingly subtle and complex nervous mechanism, probably typical for the functioning of other glands of internal secretions. These findings, when translated into a language applicable to the human level, disclose a great deal about the nervous mechanism underlying so-called 'associations' and other 'mental', relational, or psycho-logical semantic manifestations. Usually, the salivary glands are not supposed to be as closely connected with psycho-logical manifestations as the thyroid, the adrenals, and other glands are known to be. It is, therefore, a new and very important general discovery of Pavlov that the salivary glands have such intricate and far-reaching nervous interconnections.

The example of the dog reacting to the 'associations' (relations) of the experiment with morphine in a similar way as to the actual injection of the drug, is a close parallel to the example already given, of the patient who reproduced symptoms of hay fever at the *sight of paper roses*. In this case, the 'associations' were also uncritical, compulsory, almost automatic, of the type found in the animals. In fact, this statement is very nearly general, and we shall find later that most of 'mental' ills follow neurologically the patterns of animal responses, and so become pathological for man. This observation has very far-reaching consequences, to be explained later; but we want to emphasize it from the beginning, and to stress the fact that copying animals in our nervous reactions must be detrimental to humans.

The above narrows our problem considerably: we have to discover only the main differences between the nervous responses of animals and humans, and draw our conclusions.

The alimentary reactions to food and the mild defense-reaction to noxious materials may be roughly divided into two components, the secretory and the motor. It was found possible to link another neutral stimulus to an already acquired reaction. Thus, if a dog was trained to respond to a bell, which was a signal for food, he could be trained, further, to link a formerly neutral stimulus; let us say, the sound of a buzzer, with the bell, and the bell with food. Such a secondary acquired reaction may be called of the *second order*. Naturally, it is very instructive to find out if these responses could be extended to more orders. Experiments disclosed the important fact that, as far as dogs and *alimentary* reactions are concerned, it was impossible to go beyond the second order. However, when *defense* reactions were tested, it was found that it was possible to establish acquired reactions of the *third order*. But it was impossible to go beyond the third order, even in these cases.

In our field, where we have to formulate sharp differences between the nervous responses of 'man' and 'animal', we say that animals stop abstracting or linking of signals on some level, while humans do not. The latter abstract in indefinitely higher orders—at least potentially.

Here we encounter a fundamental and sharp far-reaching difference between the nervous functioning of 'animal' and 'man'. This abstracting in indefinitely higher orders no doubt conditions the mechanism of what we call human 'mentality'. If we stop this abstracting anywhere, and rest content with it, we copy animals in our nervous processes, involving animalistic *s.r.* As will be shown later, this is the actual case with practically all of us, owing to our *A* education and theories. This 'copying animals' in our nervous responses is, perhaps, a natural tendency at an extremely low level of development; but as soon as we understand the physiological mechanism, we can correct our education, with corresponding human semantic results. Naturally, such 'copying animals' by humans must be a process of arrested development or regression. It must be pathological for man, no matter how severe or how mild the affliction may be. Various absolutists, and the 'mentally' ill in general, show this semantic mechanism clearly.

The reactions can be divided into two groups, those which are *inborn*, almost automatic, almost unconditional, rather few and simple, belonging to the so-called 'species'; and those which are *acquired* during individual life, allow a great variety of complications, are *conditional in different degrees*, and are acquired by the individual. Pavlov suggests different terminologies; for instance, he calls the one 'inborn', the other 'acquired'; or as usually incorrectly translated into English as 'unconditioned' and 'conditioned' respectively. The two last terms have received a scientific general acceptance, yet I would suggest that in the English incorrect translation they are *structurally unsatisfactory*, and that particularly, when applied to humans, they carry harmful implications. Structurally, 'inborn' and 'acquired' are entirely satisfactory. Terms like 'conditional' and 'unconditional' (in the original language of Pavlov), although less satisfactory, are more appropriate, as they do not imply some sort of 'cause-lessness'. In fact, the 'unconditioned' salivary reactions *are conditioned* and produced by the physico-chemical effect of the food, and so to call them 'unconditioned' is structurally erroneous. The terms 'conditional' and 'unconditional' do not have similar implications, and carry others, as, for instance, the possibility of very important *degrees of conditionality,* establishing the ∞-valued character of the reactions; conditional meaning non-absolute, and non-one-valued.

For these structural reasons, I shall use the terms 'inborn' and 'acquired' or else 'unconditional' and 'conditional' reactions.

Under natural conditions, an animal, to survive, must respond not only to normal stimuli, which bring immediate harm or benefit, but also to different physical and chemical stimuli, in themselves neutral, such as waves of sound or light., which are *signals* for animals and *symbols* for man. The number of inborn reactions is comparatively small, and, alone, they are not sufficient for the survival of higher animals in their more complex environment. Experiments have made this point quite obvious. A completely decorticated animal may retain his inborn reactions and become a kind of automatic mechanism; but all his subtler means of adjustment, owing to acquired reactions, disappear, and if unaided he can not survive. Thus, a decorticated dog will only eat when food is introduced into his mouth, and would otherwise die of starvation though food be placed all around him.

Experimental evidence seems to show that all higher activities of the nervous system, the whole signalizing apparatus, which underlies the formation and maintenance of the acquired conditional reactions, depend on the integrity of the cortex. Stimuli which produce conditional reactions are acting as signals of benefit or danger. These signals are sometimes nominally 'simple', sometimes very complex, and the structure of the nervous system is such that it can abstract, analyse, and synthetize the factors of importance for the organism, and integrate them into excitatory complexes. The analysing and synthetizing functions, as usual, overlap, and cannot be sharply divided, both functions being only aspects of the manifestation of the activity of the nervous system as-a-whole. In general, one of the most important functions of the cerebral cortex is that of reacting to innumerable stimuli of variable significance, which act as signals in animals and symbols in humans, and give means of very subtle adjustment of the organism to the environment. In psycho-logical terms, we speak of 'associations', 'selection', 'intelligence'.; in mathematical terms, of relations, structure, order.; in psychophysiological terms, of semantic reactions.

The language of reactions is of special interest because its structure is similar to the structure of protoplasm in general and the nervous system in particular. This language can be expanded and supplemented by the following further structural observations:

1) That reactions in animals and humans exhibit *different degrees of conditionality;*

2) That the signals and symbols may have *different orders,* indicating superimposition of stimuli;

3) That animals cannot extend their responses to signals of higher order indefinitely;

4) That humans can extend their semantic responses to higher order symbols indefinitely, and, in fact, have done so through language which is always connected with *some* response, be it only repression or some other neurotic or psychotic manifestations.

The above extension is structurally fundamental, because we can extend the vocabulary of conditional reactions to humans in all their functions. Without it, we find ourselves saddled with a vocabulary which does not correspond in structure to the well-known elementary facts concerning *human* responses to stimuli, and we relapse into the old 'behaviourism', which is structurally insufficient.

The present system is based on such observations and extensions. It was reached independently from structural and physico-mathematical considerations. With this structural verbal extension, we can easily be convinced that everything that we call 'education', 'habits', 'learning'., on all levels is building up acquired or conditional and *s.r* of *different orders,* as one of the differences between 'man' and 'animal' consists in the fact that humans can extend their symbolism and responses to indefinitely high orders, while with animals this power of abstracting and response *stops somewhere.* We establish here a sharp distinction between the high abstractions 'man' and 'animal', and so build up a psychophysiological and structurally satisfactory language.

It is obvious that the fundamental means which man possesses of extending his orders of abstractions indefinitely is conditioned, and consists in general in symbolism and, in particular, in *speech.* Words, considered as symbols for humans, provide us with endlessly flexible conditional semantic stimuli, which are just as 'real' and effective for man as any other powerful stimulus.

Take, for instance, the example of the World War! Would the men in the trenches have endured at all the horrors they had to live through if it had not been for words, and, neurologically speaking, because of the conditional *s.r* connected with words?

'If any question why we died,

Tell them, because our fathers lied.'

said the poet truly, and experience shows it is not limited to the trenches.[2]

In interpreting the experiments on animals as applied to humans, it should be remembered that some of the experiments of Pavlov, *as they stand,* would be, at the least, *neurotic* for man. The reason for this is that the higher abstractions of man, which are due to the more developed complexities of his nervous system, would often make such simple experiments impossible. Once a conditional reaction is established with an animal, no amount of any sort of 'intellectual' persuasion, or the like, would disturb his glandular secretions, as the animal's range of 'meanings' is very limited. These secretions can be diminished or even abolished by other means, but not by 'intellectual' means alone. In the 'normal' man, his 'knowing' that the sound of the metronome or bell is part of an experiment and not a signal for actual food, would, *or should,* alter his nervous reactions and glandular secretions and make the experiments much more complex. The conditional reactions of the animals have still the *element of unconditionality.* In man, they may become *fully conditional* and depend on a much larger number of semantic factors called 'mental', 'psychic'., than we find in any animal.

On the human level, outside of the experiments with the salivary glands, we have in the psychogalvanic reaction a most subtle semantic means of experimenting with the effect of words as connected with some secretions, probably at least the sweat glands. Humans react to different events or words by minute electrical currents (among others) which can be registered by a very sensitive galvanometer and the curves photographed. It is interesting to notice that so-called 'self-consciousness' disturbs the success of the experiments, or makes them impossible, *at least with some individuals.* It should be remembered that general statements are invalidated if there are any exceptions.

In experiments, we are usually interested in their success. When analysing the ∞-valued *degrees of conditionality,* we are equally interested in their failures, which suggest a far-reaching revision of the *interpretation* of our experimental data in this field. Although some writers say that the reactions registered are 'beyond control' (unconditional), this statement, in general, is not correct, and should be amended to '*often* beyond control' (conditional of different degrees). It is impossible to go into details here, as the problems are extremely complex. In addition to this, the testing of *degrees of conditionality* presents an extremely wide *new semantic field* for experimentation which has not yet been attempted. It should be noted, however, in passing, that in these experiments different types of 'mentally' ill, as well as the 'healthy' persons, exhibit different types of curves.[3]

When psycho-logical events or *s.r* are interpreted, the difficulties become particularly acute. Thus, we seldom discriminate between the average and the 'normal' person. In the *animal world,* under natural conditions—by which is meant entirely without human interference—

the survival conditions are *two-valued* and very sharp. The animals survive or they die out. Because of this, it could be said, with regard to the animal world, with some sort of plausibility, that the average, with a long list of specifications, could be considered the 'normal' animal. We usually enlarge this notion to humans and land in fallacies, particularly in so-called 'psychological' problems, which admittedly are very difficult.

In general medical science, such mistakes are made more seldom. No physician, studying a colony of lepers or syphilitics, could conclude that a 'normal' man should be a leper or a syphilitic. He would say that probably, in a given colony, the average person is afflicted with such and such a disease, and he would keep as his medical standard for desirable health, a 'normal' man; that is to say, one free from this disease.

It is true that in the example given above, outside of such rare colonies, we have a majority which, in respect to the given disease, are healthy; so we are empirically forewarned against fallacies, although existing theories of knowledge do not forewarn us. But the main point remains true; namely, that in human life the average 1933 does *not* mean 'normal', and the standard for 'normal' will have to be established *exclusively* by scientific research. In our present work, we show that the average person copies animals in his psycho-logical and nervous processes, exhibits the unconditionality of nervous responses, confuses orders of abstractions, reverses the natural order., semantic symptoms of similar *structure* as found in obviously 'mentally' ill. Therefore, the average person 1933 *must be considered pathological.* If we take the animalistic average for 'normal', and apply it to man, we commit a similar fallacy as that of treating a colony of lepers as a 'normal' or 'healthy' group.

In conditional *s.r* of man, the average person cultivates, through inheritance and training in A doctrines, languages of inappropriate structure., animalistic, nervous, and so psycho-logical, *s.r.* But here, as in general medicine, the average pathological situation should not be considered 'normal'. Only a structural study can disclose what with man should be considered 'normal'. The present system performs this task to a limited degree and in various ways, among others, by the revision and the widening of the reaction vocabulary to a larger structural conditionality, as found in the, as yet, exceptional 'normal' man, and introduces the important notion of *non-elementalistic semantic reactions.*

Because of this 'average for normal' fallacy, the theories of 'conditional reflexes' in man should be thoroughly revised and enlarged to include *non-elementalistic semantic reactions;* and then we should find that often what is 'normal' with animals is quite pathological for man. The semantic difficulties are serious, because the accepted two-valued structure of language and semantic habits reflect the primitive mythologies; so there is always the danger of drifting either into animalism, or into some other sort of equally primitive mysticism.

The net psycho-logical result of such a revision appears to be that, on structural grounds, what on human level appears as desirable, and, at present, exceptional—as, for instance, the complete conditionality of conditional and *s.r*, based on the consciousness of abstracting—ought to be considered the rule for a 'normal' man. Then the older animalistic generalizations will become invalid and reactions transformed. But for this purpose, and to be able to apply these considerations in practice, we shall have to analyse 'consciousness of abstracting' and, therefore, 'consciousness' which must be defined in simpler terms, discussed in Part VII.

When we deal with 'mentally' ill persons, the reactions which would be conditional with 'normal' persons become, in a sense, unconditional, compulsory, and semi-automatic in effect, inwardly as well as outwardly. As with animals, no amount of 'intellectual' persuasion has any effect on them, and the reactions, secretions., follow automatically. From the physiological point of view, 'mental' ills in humans compare well with *conditional reactions in animals*. It seems that under such circumstances a physiological language of different orders of abstractions, different orders of conditional and *s.r* would be structurally satisfactory. In such a language, we should pass from the inborn reactions, which exhibit the maximum of persistence, unconditionality, and almost automatic character, to the acquired or conditional reactions *in animals*, which would be called *lower order conditional reactions*, still, to some extent, automatic in their working, and, finally, to the much more flexible, variable, ∞-valued and *potentially fully conditional reactions in man*, which we will call *conditional reactions of higher orders* which include the *semantic reactions*.

In such a vocabulary, the main term 'reaction' would be retained as a structural implication; yet the *degrees of conditionality* would be established by the terms of 'lower order' or 'higher order' conditional reactions. Such a language would have the enormous advantage of being physiological and ∞-valued. Structurally, it would be in accordance with what we know from psychiatry; namely, that the 'mentally' ill exhibit arrested development or regressive tendencies.

We would say that 'mental' illness exhibits not only arrested development or regression, but we could state definitely that the *fully conditional* (∞-valued) reactions of higher order have not developed enough, or have degenerated (regression) into *less conditional* (few-valued) reactions of lower orders as found in animals. All the 'phobias',

22

'panics', 'compulsory actions', identifications or confusions of orders of abstractions., show a similar semantic mechanism of mis-evaluation. Although they naturally belong to the so-called 'conditional reactions', yet, being impervious to reason, they have the one-valued character of *unconditionality*, as in animals.

Similarly with the difference between signals and symbols. The signal with the animal is *less* conditional, more one-valued, 'absolute', and involves the animal in the responses which we have named conditional reactions of lower order. Symbols with the *normally developed man* (see discussion of 'normal' above) are, or should be, ∞-valued, indefinitely conditional, not automatic; the *meanings,* and, therefore, the situation as-a-whole, or the context in a given case, become paramount, and the reactions should be fully conditional—that is to say, reactions of higher order. In human regression or undevelopment, human symbols have degenerated to the value of signals effective with animals, the main difference being in the *degree* of conditionality. Absolutism as a semantic tendency in humans involves, of necessity, one- or few-valued attitudes, the lack of conditionality, and thus represents a pre-human tendency.

To what extent the language of the *degrees of conditionality* is helpful in understanding the development of *human* 'intelligence', and why a fully developed *human* 'mind' should be related with *fully conditional* reactions of higher order, can be well illustrated by an example taken quite low in the scale of life.

This example is selected only because it is simple, and illustrates an important principle very clearly. We know that fishes have a well-developed nervous system, do not possess a differentiated cerebral cortex; but experiments show that they can learn by experience. If we take a pike (or a perch) and put it in a tank in which some minnows, its natural food, are separated from it by a glass partition, the pike will dash repeatedly against the glass partition to capture the minnows. After a number of such dashes it abandons the attempt. If we then remove the partition, the pike and the minnows will freely swim together and the pike will not attempt to capture the minnows.[4]

The dash for capturing the minnows was a positive and unconditional, inborn feeding reaction, unsuited for the environmental conditions as they happened to be at that moment. The (perhaps) painful striking of the glass was a negative stimulus, which abolished the positive reaction—speaking descriptively—and established a negative conditional reaction, the result of individual experience, which, as we observe by the actions of the fish, is not flexible, not adjustable, and quite rigid, one-valued, and semi-unconditional, or of low degree of conditionality, because, when the glass partition is removed, the pike swims freely among the minnows without adjusting itself to the new conditions and capturing the minnows.

A cat separated from a mouse by a glass partition also stops dashing against the glass, but this negative reaction is *more conditional*. In 'psychological' terms, the cat is 'more intelligent', *evaluates relations* better than the fish, and when the glass partition is removed, the cat captures the mouse almost immediately.

In this connection, an interesting experiment could be made, though I am not aware that it has been performed; namely, to separate the above fishes with a wire screen, which would be *visible* to the fishes, and repeat the experiments to test if the removal of a *visible* obstacle would alter the outcome of the experiment or the 'time' of the reactions. If the 'time' for capturing the minnows were reduced, this would mean that the conditionality of the reaction was increased, and so the seeing of the obstacle, or the increased power of abstraction, would play some role in it. Even humans are deceived by Houdinis. Are we so 'superior' to the 'poor fish'?

These problems of degrees of conditionality can also be studied in the life of insects, and the works of Professor Wm. M. Wheeler, for instance, furnish most instructive material, which we have not space to analyse here.[5]

In the process of human evolution from the lowest savage to the highest civilized man, it is natural that we should pass through a period in which the primitive doctrines and languages must be revised. The newest achievements in science indicate that the twentieth century may be such a period. Even in mathematics and physics, to say nothing of other disciplines, it is only the other day that the old elementalism and two-valued semantics were abandoned. Obviously, consciousness of abstracting produces *complete conditionality* in our conditional higher order reactions, and so must be the foundation on which a science of man, or a theory of sanity and human progress, must be built.

The suggested extension of the reaction vocabulary would allow us, at least, to apply a uniform physiological language to life, *man included*. We should have a general language for life and all activities, 'mind' included, of a structure similar to the known protoplasmic and nervous structure, not excepting the highest activities. 'Mental' ills would be considered as arrested development or regression to one-, or few-valued semantic levels; sanity would be in the other direction; namely, progression conditioned by larger and larger flexibility of conditional and semantic reactions of higher order, which, through ∞-valued semantics, would help adjustment under the most complex social and economic conditions for man. The maximum of conditionality would be reached, let us repeat, through the consciousness of abstracting, which is fundamental for sanity, and is the main object of the present work, explained in Part VII.

It seems that the aggregate of inborn, almost unconditional and acquired or conditional reactions of different orders and types constitute the foundation of the nervous activities of humans and animals. The mechanism is not an additive one. A little bit of cortex 'added' involves most far-reaching differences of behaviour in life; in fact, the number of possibilities probably follow the combinations of higher order.

Higher order combinations are constructed from groups which themselves are groups. Thus, out of twenty-six letters of the English alphabet, there are probably trillions of pronounceable combinations of letters. Sentences are groups of words which are groups of letters, and their number, therefore, exceeds enormously the original trillions. Books are combinations of sentences, and, finally, libraries are combinations of books. Thus, a library is a combination of fifth order, and the number of possible different libraries is inconceivably large. As a rule, we pay little attention to combinations of higher order, disregarding the fact that even materials and the possible variety of them have some such structure.

To give an intuitive feel how combinations of higher order increase, let me quote Jevons on the simplest case, starting with 2. 'At the first step we have 2; at the next 2^2, or 4; at the third, 2^{2^2}, or 16, numbers of very moderate amount. Let the reader calculate the next term, $2^{2^{2^2}}$, and he will be surprised to find it leap up to $65,536$. But at the next step he has to calculate the value of $65,536$ *two's* multiplied together, and it is so great that we could not possibly compute it, the mere expression of the result requiring $19,729$ places of figures. But go one step more and we pass the bounds of all reason. The sixth order of the powers of *two* becomes so great, that we could not even express the number of figures required in writing it down, without using about $19,729$ figures for the purpose.'[6]

In actual life, the number of possibilities of higher order combinations are limited by structural and environmental conditions; nevertheless, the numbers of possibilities which follow such a rule increase surprisingly fast.

CHAPTER XXII

ON 'INHIBITION'

... "destructive lesions never cause positive effects, but induce a negative condition which permits positive symptoms to appear," has become one of the hall-marks of English neurology. (212)   H. HEAD

Excitation rather than inhibition is important in correlation because from what has been said it appears that so far as known inhibition is not transmitted as such. The existence of inhibitory nervous correlation is of course a familiar fact, but in such cases the inhibitory effect is apparently produced, not by transmission of an inhibitory change, but by transmission of an excitation and the mechanism of the final inhibitory effect is still obscure. (92)   CHARLES M. CHILD

But since inhibition is not a static condition but a mode of action, the mechanism of the total pattern must be regarded as participating in every local reflex. (107)   G. E. COGHILL

It is highly probable that excitation and inhibition, the two functions of the nerve cell which are so intimately interwoven and which so constantly supersede each other, may, fundamentally, represent only different phases of one and the same physico-chemical process. (394)   I. P. PAVLOV

The term 'unconditional reflex' applies only under 'normal' or 'natural' conditions, as we know that different drugs, such as ether, which alter the conductivity of nervous tissue., can also alter its irritability. Similarly, with conditional reactions, the introduction of *degrees* of conditionality becomes an important ∞-valued structural refinement of language, depending on, and introducing explicitly or by implication, the number of factors, the degrees of freedom., which are observed empirically, and so should have a linguistic and semantic parallel.

If we disregard, for instance, the possibility of the use of a drug., then the '*un*conditional' reactions are largely *un*conditional. The 'conditional reflexes' in animals are a much subtler form of adjustment to many more factors, and if we call them 'conditional of lower orders' we cover structurally their limited conditionality, which with higher animals is considerable. For example, a fly in the laboratory might disturb the reactions, but merely 'intellectual' interference would be ineffective. And, finally, the 'conditional reactions of higher orders' in man involve still more factors, introduce more and new complexities, and necessitate that the human reactions should be *fully conditional,* requiring ∞-valued semantics. At present, this is an exceptional occurrence, although the potentiality for such *full* conditionality is present in the majority of us.

The mechanism of the unconditional reaction is, under ordinary circumstances, almost automatic. It is evolved on the background of general protoplasmic characteristics, combined with structural polarity, symmetry., and is not efficient enough for the survival of higher organisms.

Under more complex conditions, the adjustment for survival must be more flexible: a similar direct stimulation must, under different conditions, result in different reactions, or different stimuli, under other conditions, produce similar reactions, resulting, ultimately, not only in direct responses to stimuli, but also in the equally important holding in abeyance of the reaction, or even the abolishing of it. Let us assume that the direct response of a cat to a mouse would be clawing and chewing. If that given cat would just claw and chew when the mouse was some distance away, I am afraid such a cat would soon starve, for such an immediate response would not be a survival response, and this characteristic could not become hereditary. The cats which have survived and perpetuated their characteristics are, as a rule, different. When they see, hear, or smell the mouse at a distance, they flatten out, keep still, crouch., and get ready, until they are in such a position that a jump will procure the victim, and not merely frighten it away.

We see that, under more complex conditions, the nervous mechanism must produce not only direct responses to the stimuli but also equally important delays and temporal or permanent abolishments of these direct responses to stimuli.

Hitherto, we have analysed the simplest reactions of a positive character in which a stimulus produces a direct and obvious response; e.g., the showing of the food or the ringing of the bell results in an excitation in the nervous system and the secretion from the salivary glands. We are, however, acquainted with another type of fundamental nervous activity of equal importance. For instance, in experimenting with the positive reactions, we must be careful not to introduce any extra stimuli, as any new stimulus immediately excites an investigatory reaction, and the alimentary conditional reaction becomes temporarily abolished. From our personal experience, we know a large number of stimuli which have some such hindering effect on our respiration, circulation, locomotion., which we describe as 'paralysed with fear', 'speechless with rage', 'struck dumb', 'stupefied with pain',. The diminution, or deviation, or the lack of some function or response on the nervous level is usually called 'inhibition'.

The term 'inhibition' is structurally a profoundly unsatisfactory and a misleading psycho-logical term, and should be *completely* abandoned in physiology and neurology, although it could be retained in psycho-logics and psychiatry. This term is in general use, and the suggestion of abandoning a term in general use is always hard to accept. Therefore, it will be well to analyse it in some detail. In this case, it does not matter if the positive suggestion of a new term or terms is structurally acceptable; the analysis of the term 'inhibition' shows clearly that it has false to fact implications, and so should be rejected in neurology in any case.

This term is a favorite word in ecclesiastical and legal literature. and means, in the main, to forbid, to prohibit, to hinder, to restrain. It is a psycho-logical term; it implies anthropomorphic 'free will' and 'authority' notions perfectly unfit for *neurological* use. It is not an exaggeration to say that the structural implications of this term underlie the older animalistic prohibitive and punitive education, legal and ecclesiastical tendencies, which, in 1933, are known to be not only in a larger sense inoperative, but positively harmful. On the human level, this word is, perhaps, responsible for the fact that so much about our educational and social methods is uncertain and often harmful. Education is a process of building conditional and *s.r* of different orders. If the *neurological* terms dealing with conditional reactions are structurally unsatisfactory, our speculations which are carried on in these terms must involve these false implications. When the empirical results are unsatisfactory, as they must be, because of wrong structure of the arguments, and a scrutiny of our argumentation shows them to be correctly following the structural implications of the language used, then we usually blame 'human nature', which is a very unintelligent excuse, indeed.

The implications of the term 'inhibition' become a guidance for our conduct; we repress and, in consequence, breed un-sanity and maladjustment. On animal level, 'repression' is workable, but, on human levels, we need a subtler regulative mechanism, in accordance with the structure of the *human* nervous system, and this is found in the fuller conditionality of reactions, based upon consciousness of abstracting, and involving, of course, affective components, semantic factors of evaluation which regulate human impulses without the animalistic repression. In humans, the 'inhibited', repressed impulses *often remain as internal excitatory factors;* they are not eliminated by some 'supernatural' hocus-pocus, but remain active, sometimes very active, semantic sources of internal excitation, resulting in conflicts which generally have pathological results.

We are usually told that 'inhibition' plays an important role in conditional reactions. With the introduction of the *degrees* of conditionality, the importance of the possibility of altering, delaying, or abolishing some immediate response becomes much more accentuated. Indeed, it appears that this possibility of influencing responses is an important factor in the mechanism of conditionality of lower orders, but becomes the *main factor* in establishing the *degrees* of conditionality of higher orders. Obviously, the reactions become very labile, the adjustment to conditions very subtle, allowing the organism to survive under the most complex conditions, such as are found in highly 'civilized' life.

This mechanism is responsible not only for human intelligence, but also for all that is constructive in so-called 'civilization'. Vice versa, for survival in such complex civilization, one must possess these *fully* conditional reactions. At this point, it will suffice to mention that in organisms below humans, 'inhibition', which underlies the mechanism of conditionality of reactions, plays a most important biological and survival role, while on the human level it is the foundation on which all human *s.r*, 'intelligence', and desirable human characteristics are built. The present theory introduces methods to make the application of the above considerations possible in daily life.

All possible analysis depends not only on definitions of terms but also on *undefined terms,* which, outside of mathematics, have seldom, if ever, been investigated, thus making the structural assumptions which they introduce unconscious. In definitions, we also usually posit structure, though we seldom realize this fact. When we approach the experimental side of science, which is the search for empirical structure, the implications involved consciously or unconsciously in the defined and undefined terms play a very important role, and they direct, to a large extent, our efforts and ingenuity. This is why we still have so few genuinely creative scientists, although since the psycho-logically and semantically liberating work of Einstein, the number of creative physicists of the younger generation has increased surprisingly. Yet the majority of scientists do not realize to what extent their *s.r* are influenced by the terms they use and what enormous help and creative freedom they would have from being conscious of the role the structure of language plays.

With this realization, before we begin the constructive analysis of such an important term as 'inhibition', we must state clearly what the general biological presuppositions which underlie such an analysis are.

The present work is a $\bar{A}$-system, structurally very different from the older systems, which attempts to build a verbal system of similar structure to the empirical structures, as given by science 1933. The older systems had also a structure similar to the very limited knowledge of empirical structure which our primitive ancestors had. Hence, animism, anthropomorphism, 'psychologism', and the rest, and the persistence of such structural features in science as 'inhibition' in neurology, 'force' and 'heat' in physics..

According to scientific standards of 1933, there is, as far as I know, only one biological system in existence which can be called modern, and this is the $\bar{A}$ biology of Professor C. M. Child (see Chapter VIII). It is, therefore, necessary to accept this system, and also the $\bar{A}$ neurology of Professor Herrick, which is based on this biology.

Generally, the neurologists tell us that the structural aspects of 'inhibition' are unknown. To a large extent, this is true, although it is quite obvious that a *psychological* term cannot shed any light on its physiological structure. To get glimpses of this mechanism, we must start our analysis quite low in the scale of life and see what the most general characteristics of protoplasm are.

All protoplasm is irritable. In any undifferentiated bit of proto-plasm an excitation must (1933) spread in a diminishing gradient, establishing, by necessity, a region of highest excitation in contact with the stimulus, resulting in a polar orientation, with an eventual future head-end, and establishing a physiological gradient, long preceding the appearance of any differentiated tissue. The nervous system is a later outgrowth of such an oriented dynamic field, and its primary morpho-logical and physiological characteristics are, to that extent, predeter-mined, being, in the meantime, a joint phenomenon of the inherent characteristics of protoplasm, its irritability, conductivity, and what not, and of its reaction to the environment. The physiological gradient is, then, the simplest and the most general primary reaction arc in a given individual, and constitutes the physiological basis for the structural and functional development of all other arcs.[1]

Amoebas are primitive little aquatic animals of approximately spherical symmetry which have no differentiated organs at all; yet they show quite complex reactions and various adaptive activities to be found in higher animals. The amoebas can pursue their victim, show prefer-ence for stimuli, and move away from the prick of a pin, select their food,. This fact shows that protoplasm, so little differentiated, and, from the organic point of view, undifferentiated, exhibits both muscular as well as neural characteristics. This fact is fundamental. It shows that in colloids which happen to be sensitive and which possess a special type of conductivity, which, from a physico-chemical point of view, is only a special aspect of one mechanism, there is already present the potentiality for any further development. Professor Child's physiological gradients, the structural precursors of the nervous system, are a necessity, because of the dynamic potentialities of the plenum and the necessary relation to the environment, as there is no such thing as anything without environment. The stimulus, in the meantime, establishes structurally a functional polarity as a fundamental characteristic of all, even most primitive, protoplasm, and as the result of the contact of sensitive and conducting colloidal structures with the environment.[2]

In sponges, which have primitive muscular tissue but no nervous system, the muscular tissue exhibits also both characteristics, combining receptive and motor functions, showing that from the start the supposed muscles are, in reality, neuro-motor organs.[3] The actinians have no central nervous system. By the aid of an incision, we may produce in them special additional growths of tentacles sometimes with a mouth, sometimes without. If, in the last case, we place a piece of food in the tentacles, they will bend toward where the mouth should be. If we cut such a tentacle away from the body, we still find that in contact with food it will bend in the one direction. But here we are dealing not only with the sub-microscopic dynamic structure but with macroscopic structure, where the irritability and the structure of the peripheral organs determine the reaction.[4]

When we experiment with animals with a more developed nervous system, such as ascidians or worms, we come to new and very instructive facts. Loeb has removed the ganglion from a number of *Ciona intestinalis,* a large transparent ascidian, which normally, when touched at the oral or aboral opening, closes the openings, and the whole animal contracts into a small ball. It appears that a few hours after the operation mentioned they relax. If a drop of water falls on such an animal, the characteristic reaction appears again, showing that the reaction was not due to the ganglion but is determined by the structure and arrangement of the peripheral parts and the muscles. The nerves and the ganglion play only the main role as a quicker conductor for the stimulus.

Even in higher animals we find vestiges of such primitive generalized mechanisms. For instance, Loeb, in his experiments in removing the brain from sharks, found that, even after death and when signs of decomposition had already begun, light produced a contraction of the pupils.[5]

In a decapitated worm, practically all normal reactions are retained. If we cut the nervous system of a worm in two, the two parts of the worm move in a co-ordinated way as long as they are connected by a little bit of tissue. The experiments were carried further: a worm was cut in two completely, the two halves were connected by a string, and they still moved in a co-ordinated way, showing once more that originally the nervous system was a specialization of general protoplasmic characteristics of irritability and conductivity and structure, which, at present, are known to be strictly interconnected.[6] Multiordinal structure is the explanation of this behaviour. Similar examples could be given in great numbers, all of which would support the above well-established view.

Among the general protoplasmic characteristics we do not find 'inhibition', but only positive excitation and conductivity. This issue is fundamental and should be taken as a foundation for further analysis.

If a wandering amoeba comes to an illuminated spot, the animal will not remain in that region. Here is, seemingly, a new fact, and we must *select* the language we want to use in this connection. If we follow the old animism and anthropomorphism, we could say the animal 'knows'., or that some 'demon' has forewarned it, or, with equal justification, say that it is an example of 'internal inhibition' or 'prohibition'. The introduction of such terms, of course, explains nothing physiologically, but simply multiplies metaphysical identifications on the unconscious yet false to fact assumption that a word 'is' the thing we are talking about—a vestige of the primitive 'magic of words'.

Loeb pointed out long ago that to be forced to introduce animism and anthropomorphism is enough *to neglect the analysis of an external stimulus*. This is true not only in biology, physiology, neurology., but also in physics. The difference between the N and $\bar{N}$ systems depends on the fact that Newton did not take into consideration the character of the stimulus, the finite velocity of the ray of light, which is fundamental in any observation, but that Einstein did take this into consideration. The ∞-valued determinism (the restricted principle of uncertainty) in the newer quantum mechanics depends on taking into account the disturbing effects an 'observation' has on the 'observed',.

What are the known facts in the meantime? Let us start with the character of the stimulus, light. We know, positively, that light can be considered a very potent stimulus, and so the behaviour of the amoeba was a direct response to this stimulus. In fact, we know a little about this mechanism without introducing any 'demons' or 'internal inhibition'.

The starfish of a certain species has a symmetrical structure consisting of five arms. Its nervous system consists of a central ring around the mouth and peripheral nerves radiating from the ring into the arms. If such an animal is laid upon its back, it will right itself, but it is essential that not all arms should move simultaneously. In a normal animal, having five arms, usually three arms do the work and two of them remain quiet. If we destroy the nervous connection between the arms, this co-ordination is destroyed; all five arms begin to struggle, and the starfish cannot right itself, unless by accident. Should we again invoke 'demons' or 'inhibitions', or analyse the stimulus-complex and its effect?

Obviously, when the starfish is put on its back, a new stimulus-complex is operating upon it, resulting in a complex adjustment.[7]

As we already know, any stimulus applied to a bit of living protoplasm, because of the colloidal structure and of the inherent irritability and conductivity of the plenum, produces a physiological gradient, establishing, thus, some sort of polarity, symmetry, relations, order, and structure, and indicating what structure our language should have. Again, no trace of any 'inhibition' or 'prohibition' is found, and on the silent, un-speakable, objective level everything happens the usual way, without any regard to, or respect for, our *talking*. Talking only becomes a very genuine danger when on language of primitive structure we build our creeds, institutions, rules of conduct., and our methods of investigation. In the last case, our sciences are nearly as slow, halting, perplexing, difficult, non-co-ordinated, and, in a larger sense, ineffective, as our creeds and institutions have proven to be. Our sciences may have added to our comfort, but, outside of psychiatry, they have not contributed much to human happiness.

As structure seems so fundamental and can be discovered everywhere, we should not be surprised to find that in structure, or perhaps, still better, in the general structuro-sensitive-conductive dynamic complex with definite structure on different levels, we shall find the solution for obvious positive reactions of organisms, as well as for the lack of them.

It is not possible or necessary to go into further details here. The structural data, however, although they are not particularly emphasized, are given in handbooks of physics, colloidal chemistry, chemistry, biophysics, biochemistry, biology, physiology, neurology,. At present, it is realized in science that structure is of extreme importance; but, because of identification, it is not realized that structure is *the* only possible *content* of science and of all human 'knowledge'. This fact, of course, makes the quest of science uniquely structural. Because of it, we come to a very far-reaching general rule, that all 'understanding', to be such, must exhibit or assume structure, thus formulating the supreme aim, and, perhaps, uniquely indicating the only possible method, of science.

Two more simple examples may be helpful. *Mnemiopsis* or *Eucharis* have swimming plates which beat rhythmically, with considerable regularity. When the plates are stimulated mechanically, the movement ceases in the presence of sufficient calcium salts in the water. In similar media, but containing no calcium, a mechanical stimulus does not stop the movement of the plates, but just the opposite. It accelerates their motion, showing clearly that the effect of direct stimulation can be reversed when the structural relations are altered. Once more, no 'demons' and no 'inhibition'.[8]

In higher animals, we usually find a well-developed symmetry and muscles of which the activities oppose the results of the activity of other muscles. Such muscles are called antagonists. If two antagonists of equal strength are stimulated equally, no macroscopic effect of the stimulation of both muscles results. If one of the antagonists is stronger than the other, the macroscopic effect of the stimulation of both muscles results not in some general convulsion, but in a one-sided action of the stronger muscle. Obviously, these results are the necessary consequence of structure on different levels. We had, in the first case, a lack of obvious macroscopic reaction, although stimulation was present and did its work. This was due to structure.

It is known that some drugs, such as strychnine or the toxin produced by the tetanus bacillus, produce a state of *general* and high irritability of the nervous system. The slightest stimulus to the surface will produce a spasm which affects practically every muscle of the body. The pinching of the foot, instead of producing a withdrawal, results in the rigid extension of the legs, arms, and back. The extension is no longer a co-ordinated process, but is associated with strong contraction of the flexors, the final state of the limbs being determined by the surpassing strength of the stimulated extensors. The effect of the tetanus toxin is similar. In a monkey, under normal conditions, the electrical stimulus of a certain spot of the cortex will produce the opening of the mouth; similar stimulation of another spot will produce the closing of the jaws. But, under the effect of the toxin, the stimulation of *any* of these spots will produce the *closure* of the jaws, because any attempt to open the mouth will excite the stronger masseter muscles and effectively close the mouth.[9]

The above examples show again that no 'demon' or 'inhibition' has prohibited the withdrawal of the foot or opening of the mouth, but that the excitation of stronger antagonists is responsible for the result—or, if we wish, for the lack of results. All of which is obviously structural.

All the above discussions and examples—and they could be expanded and extended to fill volumes—show clearly: (1) That in the structurally more complex organisms the process of co-ordination and adjustment to more and more complex environmental conditions, leading to wider activities and fuller conditionality of reactions, is partially based —to the extent of one-half, or even more—on the lack of direct response to a stimulus, leading to delayed action and involving the four-dimensional order, all of this being a function of the entirely general characteristics of protoplasm; namely, its structure, excitability, and conductivity (the last two characteristics being also a result of sub-microscopic structure) without the intervention of 'demons' or of 'inhibition'; (2) that in every case there is an *excitation,* no matter whether the result is a positive or a negative reaction, or whether we can, at present, trace it in detail.

As Professor Herrick says: 'On this view of the situation the supposed inhibitory effect of the cerebral cortex resolves itself into a differential dynamogenic cortical influence. This is partly specific and phasic, acting upon particular subcortical functional systems while these are in process and tending to depress all conflicting activities either by withdrawing available nervous energy from their apparatus of control or by equal activation of agonist and antagonist systems with resulting stasis. It is partly a general and tonic activation or reinforcement of all lower reflex systems. Upon removal of the visual cortex the specific phasic activation of learned reactions is abolished. Upon removal of the entire cortex the general tonic cortical effect is abolished. The operation has not stimulated inhibitory fibers, as some have supposed; it has removed the sources of tonic activation which normally are always operating.'[10]

'The cerebral cortex from its inception exerts more or less inhibitory influence upon subcortical functions. In the simpler learning processes of rats there seems to be a differential activation of some key factor of a subcortical learning process . . . which in effect draws off all available cortical energy, leaving other and irrelevant sensori-motor processes relatively enfeebled so that they are subordinated. The effect is the same as if a specific inhibitory action were exerted by the cortex upon the inappropriate movements . . . It may be suggested, further, that all inhibition is in reality a differential activation, the mechanism being in some cases simply the "drainage" phenomenon . . . and in other cases this effect supplemented by positive activation of two antagonistic motor mechanisms so that their interference blocks all reactions of non-adaptive sorts.'[11]

In these statements of Professor Herrick, we find a language of similar structure to the known facts. The terms of *differential dynamogenic cortical influence* and *differential activation* cover all known facts, and may cover future facts, because the terms are structurally very flexible, and will always allow us to enlarge our knowledge of the mechanism of the so fundamental *differential activation.*

The difficulty in eliminating the term 'inhibition' and suggesting a new physiological term to take its place is considerable, because this term is used in many different forms and meanings. The term 'inhibit' is used in its various forms as a substantive, an adjective, a verb, an adverb, sometimes as a psycho-logical term, sometimes as a physiological one, yet *never* carrying physiological implications, but always psychological and anthropomorphic ones connected with its origin and standard use. It was introduced into science when physiology and neurology were in their infancy, and so were still under the influence of primitive animism and anthropomorphism.

The term, because of its character, is not scientifically descriptive. It does not suggest functional, actional, directional, or other structural implications, but suggests notions irrelevant to science connected with its origin and standard use, making it a far-fetched inferential term, the use of which must retard the advances of these sciences.

Once we introduce a physiological term with physiological and, therefore, structural implications, our expressions will have to be re-shaped to make the use of the term possible. Such re-wording will always carry quite definite structural implications, which, in turn, suggest further experiments in the search for structure and so have a *creative* character, not to be disregarded. Thus, as we have already seen, the term of 'degrees of conditionality' suggested further experiments and the revision of older data.

This statement is quite general and may be summed up as follows: The introduction of a new structural term may: (1) eliminate the improper implications of the older terms; (2) introduce new and creative implications which suggest the need of verification and so lead to new experiments.

At this point, I suggest a term which may be useful and will, perhaps, be acceptable for scientific use. As the fundamental character of 'inhibition' seems to be 'differential activation', the term to be coined should possess two main structural implications: (1) it should be directional, or indicate the sense of the reaction, and (2) it should imply activation.

We find such a term in 'negative excitation', 'negative stimulation', 'negative activation', 'negative phase'., and it is possible to extend the use of this term by making as many compound terms as we need.

If possible, we should have terms which help us to keep on one level of analysis, and so automatically prevent us from confusing levels, since modern science always deals, at least in principle, with not less than three levels, the macroscopic, the microscopic, and the sub-microscopic, thus making confusion quite easy. If we call the positive effect of a stimulation on the macroscopic level 'positive', any other stimulation which might fail to produce the positive effect on *this level,* or which might counteract it, would be negative. The implication would remain that there was some excitation, but that it did not produce the effect which we had called positive. Structurally, such a term would be satisfactory, especially as it would help us to keep on one level of analysis and not confuse the main levels through verbal structure.

Such a language would help us to study the mechanism of 'differential activation', and would carry helpful implications. If any cases appeared in which this term did not cover the field, either the term could be enlarged, keeping the implications, or the statements should be altered so as to be expressible by such terms. The last would always prove to carry interesting implications, suggesting experiments.

In the processes going on in the nervous system, there is no occasion for the application of terms like 'prohibition' or 'inhibition'. There is no standstill in these sub-microscopic processes, though the manifestation on the macroscopic levels can be either of a positive or of a negative character. On the sub-microscopic levels, there is a nervous excitation which often stimulates· antagonistic processes., with results which are not always obvious.

The implication of the term 'negative excitation', although limited, is structurally correct in 1933. Without going into full detail here, I merely suggest a few considerations. First of all, the term preserves its main implication; namely, that of excitation, 'negative' suggesting that this excitation takes an opposite course to the positive one. If, for instance, a positive excitation produces, let us say, the activities of the salivary glands, a negative excitation in this respect will not produce them but will produce other activities, such as, for instance, an investigatory reaction. With a negative excitation, there is an excitation, but it produces different results. There is no possibility of stopping or prohibiting or inhibiting nervous activities, short of death as-a-whole or destruction in parts; but only a possible deviation of activities, owing to enormous possibilities in establishing nervous connections, endlessly subtle dynamogenic effects,.

In some instances, 'inhibition' might be regarded as a form of nervous exhaustion; but such a notion cannot always be structurally correct, as there is much evidence at hand that 'inhibition' spreads to other cortical elements which were not functionally exhausted, or that it can be counteracted by some new excitation. 'Inhibition' thus preserves its *active* character. The origin of 'inhibition' is also very instructive, and a mass of experimental data shows that it can be produced experimentally. Among other ways, it can be produced by very weak, very strong, or unusual stimuli, *but stimuli, anyway*. As a rule, any extra *nervous excitation* in the central nervous system manifests itself at once, either in diminishing, or in completely abolishing (temporarily, at least) the conditional reflexes prevailing at the date.[12] If we find that exhaustion is, in some instances, the structurally correct term, there is no reason why we should not use it, instead of using a *psycho-logical* term of 'inhibition', on neurological levels.

That the terminology of positive and negative *excitation* is structurally appropriate finds its further support in the so-called 'disinhibition'. Thus, an 'inhibition' of an 'inhibition' reverses the neural process prevailing at a given 'time' and becomes a positive excitatory one. In our language, because of structural considerations, we should say that 'disinhibition' should be labelled as 'negative excitation of *second degree*', resulting in a positive excitation. If we were to 'inhibit' 'disinhibition', we should have, again, 'inhibition',. With the new terminology, it would be a negative excitation of the third degree, which would give negative results, and a general rule could be established, in complete accordance with the mathematical language in which the even degrees of a negative excitation would have positive characteristics and the uneven would remain negative ('inhibitory').

Such a language would not just borrow 'by analogy' some mathematical features. Once we take structure into consideration,—and linguistic issues represent an adjustment of structure—when a systematic analogy is found, it has always structural implications which should be used for testing structure. There can be no serious objection to the statement that mathematics is, at present, a limited language of which the structure in 1933 is similar, or the most similar we have, to the known structure of the world and our nervous system. The use of such language must be always desirable, as it is a test of structure and so leads to further discoveries of the unknown structure of this world. To the best of my knowledge, the above is a novel, very general, *structural* use of mathematics considered as a prototype of languages. Our emphasis is now on the *structure* of mathematics, and not on the numerical solutions of equations, the *possibility and usefulness* of which is precisely due to the fact that equations express relatedness, and so necessarily give us structural glimpses.

From a structural and linguistic point of view, the historical development of mathematics shows that it is a first successful attempt to develop a language with a structure similar to the empirical structures, and shows the ideal conditions of producing languages.

When we had only positive numbers, we could add two and three and make five, we could subtract two from three and have the remainder

23

one, but we could not subtract three from two. Yet the structure of this world is such that a further development in the structure of the language was imperative. Thus, if an object moves in a given direction with the velocity two feet per second, and some external factor imparts to it a velocity of three feet per second in the opposite direction, the original direction of motion will be reversed, and the object will move with the velocity of one foot per second in the opposite direction. Or, to give another example, some one has two units of money and he buys something which costs three units of money. He is then in debt one unit.

Such facts necessitated the introduction of negative numbers and so made subtraction always possible. If the motion in one direction or the amount of money in our pocket was called 'plus two' units, and we subtract from it three units, the results were 'minus one', meaning a conventional reversal of direction, or sense, for motion, or a debt, instead of a possession, for money.

Experimental facts of division again necessitated the expansion of this language. Thus, fractions were introduced so as always to allow of linguistic division. The 'imaginary' number, $i = \sqrt{-1}$, was introduced to allow, in all cases, the extraction of roots,. For a long 'time', the number $i = \sqrt{-1}$ was considered almost mystical, but, of late, when a physicist or an engineer finds it in his equations, it is almost an unmistakable indication for him to look for some wave-motion in the world. More extended observation of the empirical world and structure required further structural adjustment of our languages.

In the vector calculus we have the so-called scalar product which obeys the ordinary laws of multiplication and $a.b = b.a$ where the order of the factors is of no importance. The vector product does not follow these rules, as the order becomes important; thus, in a vector product, $a.b = -b.a$. In the newer quantum mechanics, to account structurally for the experiments, still newer numbers were introduced. Instead of the old arithmetical $qp = pq$ or $qp - pq = 0$, we introduce new numbers where $qp - pq = \frac{ih}{2\pi}.1$.

It is very significant that a similar linguistic evolution appears justifiable in the case of the function of the nervous system in general and in the structure and function of the conditional reactions in particular. As experience and theory show, the fundamental structures and functions we find in life are not 'plus' affairs, but represent some higher-degree functions of a non-additive character. The typical functioning of the human nervous system (time-binding) is represented by an exponential function of 'time'.[13] Now we see that the reversal of the sign of negative excitation also follows exponential rules, and experiments show that the change in order of abstractions which, by necessity, must be passing from even to uneven numbers of orders or vice versa, also reverses the sign of the reaction (see Part VII).

In the case of positive excitation, there is also a structural parallel with the newer languages of mathematics, but we do not need to analyse it here, because the foundation of the more flexible, adjustable responses begins with a negative effect; and, in this case, the language I suggest is fully justified without further explanations. The neurological importance of 'consciousness of abstracting' is based precisely on the fact that it automatically involves a fraction of a second of psycho-logical delay, and thus is fundamentally based on, and introduces in training, a wholesome 'inhibition'.

We come thus to a weighty structural conclusion that the fundamental processes of the nervous system are not only non-plus processes but that they follow the exponential rules of signs. As soon as we realize that from a structural point of view 'structure' and 'function' are only different types of language in which to speak about two aspects of what is going on, on the silent un-speakable level, and that on this level these two aspects *can never be divided,* we must also build a *non-el* language. Such a language is found in *dynamic structure,* out of which arises function, and even macroscopically relatively enduring structures as special aspects, and the exponential character of the fundamental activities of the nervous system becomes a necessity.

In modern mathematics numbers can be interpreted as operators., which, in our case, suggests great freedom of structural use, and widens the application of these notions.

To put the problems as simply as possible: all the more subtle forms of adjustment in organisms, 'intelligence', so-called 'civilization', our 'ethics', 'happiness'., and, finally, *sanity,* which is the evidence of semantic adjustment or proper evaluation on human levels, are based on the neurological interplay, the number, and multi-dimensional order of superimposed (not added) positive and negative excitations. The positive, or the direct and obvious, responses are the more primitive; the negative, resulting in not always obvious consequences, are the result of further structural complexities, which reach their culmination in the normally developed highly cultured man.

Such indefinitely superimposed negative excitations are found physiologically in the hierarchy of higher and higher orders of abstractions, which are able to reverse the sign of the $s.r$, and so, structurally, make these considerations extremely workable and neurologically sound, and justify their introduction and use. This accounts for the fact that what was evaluated as tragic or painful, or joyful, or shameful., to one generation or culture, does not seem so to another. Our personal difficulty usually is that, at present, we copy animals in the relative unconditionality of our responses, because we are not acquainted with this semantic mechanism. We are not prepared to change in one single generation the sign from a minus to a plus, or vice versa, without a great amount of struggle and semantic discomfort.

Now, such discomforts are usually harmful to the human nervous system, but the structural understanding of this mechanism helps us to eliminate these semantic pains, and so leads toward nervous balance and sanity.

It seems that the neurological mechanism operating in this connection is similar to the one formulated by Pavlov, thus: 'Two facts relating to the central nervous activities stand out clearly. The first is that the extraneous stimulus acting on the positive phase of the reflex inhibits, and acting on the negative phase dis-inhibits, in either case, therefore, reversing the nervous process prevailing at the time. The second is that the inhibitory process is more labile and more easily affected than the excitatory process, being influenced by stimuli of much weaker physiological strength.'[14]

Negative reactions or 'inhibition' must be interpreted as the neurological foundation of 'human mentality', and the result of external and internal stimulations. Because of structural interrelations, the main factor of building human 'mentality' and developing internal 'inhibition' must be more labile and must be influenced by stimuli of much weaker physiological strength.

This explains also why the solution of our problems in education, social life., must be not the animalistic external 'inhibition' alone, but must become, in the main, special internal 'inhibition', effective and yet harmless to the individual nervous system. All of us possess this most general nervous mechanism. The problem is to discover the means to operate it. We shall see later that in consciousness of abstracting we find a workable semantic solution, allowing an automatic change of sign of the reaction. It should be recalled here that all stimuli and all responses are complex, the word 'simple' being structurally false to facts. On the human, and particularly on the linguistic level, it is practically never possible to ascertain an 'absolute' order of abstraction, or the degree or order of an excitation. These are often the results of racial time-binding, and extremely complex, nervous processes, and every superimposition of a new neurological process (not addition) may fundamentally alter the whole character of the *s.r* and reverse the sign. In negative excitations, the passing from one degree to another changes the sign of the reaction. In practice, we are only interested in two neighbouring levels of abstractions or in two neighbouring degrees of negative excitation, simply because these involve, by necessity, a passing from an even to an uneven degree or vice versa—in both cases reversing the sign of the *s.r*.

The general organismal adjusting mechanism of the 'investigatory reaction' responds positively to a new stimulus, but with very important survival value acts negatively on established positive conditional reactions in animals. It is, at present, much weakened and often ineffective with man, resulting in non-survival, non-adjustment and 'mental' ills for man. It is a well-established fact that different stimuli either interfere with each other, resulting in modified behaviour, or reinforce each other and have cumulative effects. On the human level, different 'mental' factors play the role of internal positive or negative excitatory semantic complexes, which, because of verbal conditions (and all doctrines are *always* connected with an *affective* background), may reinforce a given stimulus, thus making its physiological effect variable and of different strength. Under such conditions a new stimulus does not produce the investigatory reaction with all its beneficial results. This mechanism is, perhaps, responsible for the well-known fact that primary instincts with humans are, by far, weaker and more variable than with animals; whence it comes that *humans seldom know by themselves, without science, what is best for them.*

We should not be surprised to find that under these more complex conditions human investigatory reactions may be of different types, culminating in the *typically human* investigatory reaction, which would introduce the natural, yet more important, *delay in an immediate reaction* to a former stimulus. We shall find, later, that consciousness of abstracting is such a distinctly human and very useful investigatory reaction that on the human complex semantic level brings relatively as much benefit to the human organism as it does on the animal level to animals.

It seems that the nervous mechanisms of both types are similar, except for the fact that on the human levels we have more factors which are external and internal stimuli than on the animal level. If we copy animals in our nervous processes, we are, in reality, worse off than the animals, because, with our more complicated nervous system, it means for us a pathological condition.

CHAPTER XXIII

ON CONDITIONAL REACTIONS OF HIGHER ORDERS AND PSYCHIATRY

> In the dog two conditions were found to produce pathological distur-bances by functional interference, namely, an unusually acute clashing of the excitatory and inhibitory processes, and the influence of strong and extraordinary stimuli. In man precisely similar conditions constitute the usual causes of nervous and psychic disturbances. (394)   I. P. PAVLOV
> The fact that the maximum disturbance in the central nervous activity does not appear immediately on administration of the causative stimulus, but after one or more days has been observed in many animals. (394)
>   I. P. PAVLOV

Psychiatrists will readily understand the structurally false to facts and harmful implications of the term 'inhibition' on the neurological level, when they consider that often 'pain', 'fright', and different 'prohi-bitions' and 'inhibitions' on the psycho-logical level result in nervous processes which are not *passive, eliminated factors,* but remain what they were originally—exciting semantic factors 'repressed' on human levels—and become very active and potent causative factors in many 'mental' and physical ailments.

If the *non-el* point of view and language are seriously applied, there seems to be no escape from the conclusion that the future physician, on perfectly scientific, structural, physico-chemical, and colloidal grounds, will never attempt to divide the 'physical' from the 'mental', and different nervous processes now called 'inhibition' will come prominently to the fore as active, to be taken care of and *never* to be disregarded.

That the mechanism of conditional reactions in animals bears an astonishing resemblance to the mechanism of 'mental' ills in humans, because of the relative unconditionality of both, is exemplified practically throughout the whole work of Pavlov, although he did not point out this particular connection. As soon as this is understood, we shall find that some of the experiments of Doctor Zavadzki, made in Pavlov's labora-tory *twenty-five years ago,* disclose a neurological mechanism which underlies practically all psychotherapy, and which, therefore, appears very important and to deserve special discussion.

I do not know the percentage of the successful application of psy-chotherapy of any scientific school, or of extra-medical cults, because the many cases of failure are very seldom recorded. We usually forget, or do not realize, that the successful cases teach us, structurally, *less* than the failures, because there is always an infinity of ways in which we can account for a positive result, which is structurally entirely invalidated as such by a single failure, if the possibility of it is not foreseen by the structural flexibility of the general method.

From what I gather (though I may be mistaken) of every hundred patients who seek relief in psychotherapy, fifty fail completely. The remaining fifty can, perhaps, be divided into two groups; the first one of, say, ten, who become entirely relieved; the other remaining forty, who improve in different degrees. The analysis in the present work may, perhaps, explain why the percentage of failures is so high. It seems that no school of psychotherapists has analysed 'mental' ills from the general *non-el* structural and semantic point of view; and, although the physicians struggle in every case to abolish the relative unconditionality of the reactions, their methods are neither neurological, nor physiological, nor fundamental enough.

The language used in these scientific theories includes such terms as 'conscious', 'unconscious', 'repression', 'inhibition', 'transfer', 'complex',. There seems no doubt that some such terms cover a few of the facts we know from experience and observation, and that they may be structurally correct on the psycho-logical level. The nervous mechanism involved, although discovered twenty-five years ago, has not generally attracted the attention of physicians, and the postulated theories, lacking neurological foundations, are often called 'far-fetched speculations', a fact which is ultimately harmful to the whole psychotherapeutic and semantic hygiene movement.

The 'psychologists' and the psychiatrists are very much divided as to the role 'introspection' plays. This is due to the confusion of the orders of abstractions. Animals may 'feel', may 'suffer', but they cannot *describe*. Humans differ in this respect; the given person may feel pain, the pain is very *objective* to the given individual, and it is *not words (objective level)*; but we can describe it, this description being valid on the *descriptive level,* a higher order abstraction than the objective level (which is un-speakable for the given individual). If we *ascribe* this process to others, this is no longer a description but an inference, or a still higher order abstraction, which statements have to be verified by averaging. Scientifically (1933), psycho-logics are impossible without the *description* of internal processes, and, therefore, some 'introspection', so that the United States Behaviourism becomes a very naïve discipline. The Behaviourists mean well, methodologically, without realizing fully what scientific methodology is. They completely condemn 'introspection', yet they continually use it. Consciousness of abstracting solves the riddles of pro- or anti-behaviouristic attitudes, because, when we are fully conscious of abstracting, we should never confuse description with inference, neurologically processes of different order.

Any discipline, to be a 'science', must start with the lowest abstractions available; which means descriptions of some objective, *un-speakable* level. In *human* psycho-logics, 'introspection' is the only *possible descriptive level,* all other methods being inferential.

The experiments of Doctor Zavadzki were conducted to investigate the mechanism of the so-called 'delayed reflexes'. Speaking roughly, in experiments in which the interval between the conditional stimulus and the reinforcement by food or acid is short, say, one to five seconds, the salivary secretion follows nearly immediately after the application of the conditional stimulus. If the delay between the two is longer, say, several minutes, the appearance of salivary secretions is also delayed, the length of this delay being proportional to the length of the interval between the two stimuli.

In these experiments there were two phases: the one in which the conditional stimulus has apparently no effect; the other in which the conditional stimulus becomes effective. The investigation was continued to discover what becomes of the excitation due to the conditional stimulus during its apparent inactivity.

New experiments finally disclosed an astonishing mechanism. A tactile stimulation was used for three minutes as a conditional stimulus for acid, and reinforced, as usual, by the application of acid, and a stable, delayed conditional reaction was obtained. But when a perfectly neutral stimulus, say, the sound of a metronome or a noiselessly rotating object, never connected with any alimentary stimulation, was superimposed upon the original conditional stimulus, immediately a copious secretion of saliva, together with the motor reactions peculiar to a given stimulus, were obtained.

We see that the excitatory process in the nervous system existed all the time in a concealed, non-manifest form and was released by an extra and neutral stimulus.[1]

Similar experiments show clearly that the structure and function of the central nervous system is such that some stimulations can be concealed and become macroscopically seemingly inactive, giving no obvious manifestation or response, yet preserving their active exciting characteristics which, by proper treatment, can be released at will. In physics, we have a similar phenomenon in the case of 'frozen' light, galvanic and storage batteries, pear-shaped drops of glass resulting from melting, which explode when the end is broken off, and many others, although probably the sub-microscopic mechanisms are different.

It does not take much explanation to see that the nervous mechanism revealed in the experiments of Doctor Zavadzki accounts on human levels for a great many 'mental' manifestations, including 'recall', 'unconscious', 'repression', 'complexes'., and allowing a further generalization, that a slight nervous disturbance of 'recall', in the sense of negative unconditionality, may be closely connected with a pathological semantic 'complex'.

Another experiment has close connection with the problems of the human 'unconscious', 'repressions', and 'complexes'. The positive conditional reactions were usually obtained by combining under certain conditions a formerly neutral stimulus with food or with a mild defense reaction to acid. If the neutral stimulus is not reinforced, it loses its significance for the organism, no secretion is obtained, and it becomes from this point of view a negative stimulus. If, with a given animal, a negative reaction is established, it can, under certain conditions, be transformed into a positive one by reinforcement. In the experiment we are describing, a dog was used, with a well-established negative alimentary reaction to the beats of the metronome at the rate of sixty beats per minute, while the rate of one hundred and twenty beats per minute was used as a positive stimulus. Both reactions were constant and precise. The process of transformation from negative to positive went slowly; after the seventeenth application with reinforcement, a small secretion of saliva was obtained; after the twenty-seventh reinforcement, the secretions of saliva were already considerable. No definite disturbances in other positive reactions were observed except for a tendency to the equalization of strong and weak conditional stimuli.

But the secretory reaction to the transformed stimulus of sixty beats did not remain constant, in spite of reinforcement; it diminished, and after the thirtieth application fell to zero. It was noticed, further, that immediately after the application of the metronome at the rate of sixty beats per minute, practically all the older positive reactions disappeared. After further experimenting, some of the positive effects of the metronome at sixty returned, but its negative or depressing effect on the positive reactions persisted. In all cases where the metronome at sixty was not used, all the positive conditional reactions maintained their strength, except that the weaker stimuli had an inclination to produce lesser effects toward the end of the experiment. Although the metronome at sixty or one hundred and twenty produced salivary secretions in varying quantities when used alone, whenever the metronome was used there followed a disturbance of all conditional reactions, varying from complete extinction to a diminution of secretions. The formerly positive stimulus of one hundred and twenty beats of the metronome produced greater disturbances than the formerly negative of sixty. Further experimentation disclosed that the disturbance of the cortex was profound, and that it could not withstand any kind of stronger stimuli without producing completely negative results. It became, also, obvious that the maximum disturbance in the central nervous activity in animals (and in man) does not appear immediately after the application of the injurious factor but after a delay.

Since other auditory stimuli acted during these experiments, Pavlov concludes that 'the disturbance must be regarded as a result of a strictly localized functional interference in the acoustic analyser, a chronic functional lesion of some circumscribed part, the stimulation of which produces an immediate effect upon the function of the whole cortex, and finally leads to a protracted pathological state', and, that 'it is obvious that the localized disturbance of the acoustic analyser is again the result of a clash between excitation and inhibition', which this particular nervous system finds difficult to adjust.[2]

These experiments were conducted upon a dog which had served in the laboratory for a long time and belonged to the type which has a very negatively excitable nervous system. Experiments on dogs with very positively excitable nervous systems, although different in details, led to similar general results; namely, that a clashing of the two antagonistic nervous processes led, usually, to a more or less protracted disturbance of the function of the cortex, in the form of a lasting predominance of one of the processes.[3]

Experimenting on the conditional reactions in animals, such as a dog, by inducing pathological states of the nervous system, gives us, in a *simplified form,* a means to understand the mechanism which underlies some of the human 'mental' illnesses, provided we realize the fundamental fact that these experiments on dogs correspond, in their simpler form, to 'mental' ills, and not to 'mental' health, in man. The above experiments would be impossible with a healthy person; yet they depict exactly what happens in the case of 'mentally' ill. The experiments started with a healthy animal, and they ended with a pathological case. If similar experiments were undertaken with a healthy person, no pathological results would follow, owing to the larger conditionality of reactions.; but similar pathological results are produced in humans by different means, the confusion of orders of abstractions being a standard semantic mechanism to bring about the 'clash' between the positive and negative excitations which the nervous system of man cannot resolve so easily.

The experiments of Pavlov disclose, also, a fact which, on human levels, introduces serious complications; namely, that some animals have highly excitable nervous systems, and that some have less excitable ones. Experiments conducted on some individuals produce one effect; similar experiments performed on individuals with different nervous systems produce different results. In some instances, the nervous systems are so sturdy that no disturbances appear at all.

To anticipate a little: it appears that under the present linguistic, educational, social, economic., conditions, nearly all of us suffer from nervous and semantic disturbances, produced by copying animals in our nervous responses. This last condition occurs because the larger conditionality of human responses has not been taken into consideration; its mechanism is unknown, and we do our best to teach and enforce the animalistic responses. As yet, we have had no physiological and simple methods by which to train in this larger conditionality. This is a simple explanation of our failure. Only a few of us have such sturdy nervous systems that they do not become semantically disturbed in any marked degree, and these are exceptions. Obviously, even an attempt to build a general theory dealing with these semantic problems may be useful, for the very mistakes made may serve others as an incentive for further enquiries in a field which is practically unexplored and extremely large.

In the formulation of the present general theory, theoretical considerations suggested necessary neurological mechanisms; yet the standard books on physiology and neurology did not give enough data. In the recent work of Pavlov, I found sufficient experiments and data to illustrate the neurological mechanisms which underlie the present theory. It seems likely that the work of Pavlov and the experiments described, together with the theoretical issues raised in the present system, will be of value to psycho-logicians and psychiatrists, provided that they pay attention to the semantic non-confusion of orders of abstractions, without which it is practically impossible to translate experiments dealing with nervous responses of animals to the human levels and escape verbal fallacies. The language of the structure, as introduced in the present work, is essential in this respect; in fact, the present writer could not have carried out his analysis without it.

Pavlov also suggests some applications to human pathological cases which are recognized as such, the average person being assumed 'normal'. The present work is an independent theoretical enquiry, and the results are much more general, as they show that the general neurological mechanism allows an almost universal misuse of our nervous systems, because of the disregard of structural, linguistic, and semantic issues.

In several chapters, Pavlov discusses a large number of experiments dealing with functionally induced pathological states of the nervous system, and suggests, also, some therapeutic measures. He concludes: 'This . . . and other observations suggest that a gradual development of internal inhibition in the cortex should be used for re-establishment of the balance of normal conditions in cases of an unbalanced nervous system . . . I do not know whether similar therapeutic measures . . . are applied in human neurotherapy.'[4]

The above remark is vitally important; it is not only a result of a lifetime of scientific work, but it expresses a principle which is used, *without being formulated explicitly,* all through psychotherapy. In the present volume, this principle is not only formulated in physiological terms, but is also made the foundation of a physiological method for its semantic application. This method is found in the training and development of the consciousness of abstracting (see Part VII), which, when applied, not only restores nervous balance as empirically shown, but also gives powerful *preventive semantic means* if used in early education.

Further consequences and conclusions are given in Part VII. At this point, we shall merely state that the above explanations also show why a theory of *universal agreement,* in the broadest sense; namely, agreement with one's self, eliminating internal 'conflict', and with others, eliminating family, social, and international conflicts., is neurologically not only possible, but also a necessary semantic consequence of using the human nervous system in its structurally appropriate way.

It is well known that the use of the terms 'positive' and 'negative' is optional; but the opposing character of the issues involved is not optional, because these are experimental and structural. In former days, we not only made our selection, and called some issues positive and some negative, but we naturally had and have some semantic responses connected with them. Thus something 'positive' implied certainty, 'reality', 'truth', 'absolute'.; something 'negative' implied the negation of these.

In 1933, it appears likely that we shall have to revise *in toto* these semantic orientations, which obviously we have been practicing since the days of savagery.

What are the facts? Curiously enough:

1) The electricity which lights our lamps or runs our dynamos, we call, in the old language, 'negative' electricity.

2) The numbers which are the foundation of the most important complex numbers in mathematics are formally based on negative numbers.

3) The foundation of so-called 'human mentality' is the 'negative' reaction.

4) Because words are never the things we speak about, the sole link between languages and the objective world being structural, the only 'positive' facts about this world are of the old 'negative' character.

5) Finally, the main difficulty of the A-system can be found in the positive 'is' of identity, involving us in false to fact evaluation and semantic disturbances; on the other hand, a $\bar{A}$-system is based on the complete elimination of identity formulated as a negative premise of the 'this *is not* this' type (see Part VII).

At present, only in *technical* mathematics can people behave semantically like 'gentlemen'. They analyse and agree; no quarrels are possible. Linguistic and semantic researches show that the structure of all languages can, and *must,* be made similar to empirical structures; and then, also, the rest of humans can, and probably will, behave in a less silly and futile way than they have done in the past and are doing in the present.

BOOK II

A GENERAL INTRODUCTION
TO NON-ARISTOTELIAN SYSTEMS
AND GENERAL SEMANTICS

Of all men, Aristotle is the one of whom his followers have worshipped his defects as well as his excellencies: which is what he himself never did to any man living or dead; indeed, he has been accused of the contrary fault. (354)   AUGUSTUS DE MORGAN

There is one very mportant fact on which we must be in no doubt, and that is that for any given deductive theory there is not *any one* system of fundamental notions nor *any one* system of fundamental propositions; there are generally several equally possible, *i. e.* from which it is equally possible to deduce correctly all the theorems. . . . This fact is very important, because it shows that there are in themselves no *undefinable* notions nor *indemonstrable* propositions; they are only so relatively to a certain adopted order, and they cease (at any rate partly) to be such if another order is adopted. This destroys the traditional conception of *fundamental ideas* an *fundamental truths*, fundamental, that is to say, absolutely and essentially. (120)   LOUIS COUTURAT

In this direction finality is not sought, for it is apparently unattainable. All that we can say is, in the words of a leading analyst, "sufficient unto the day is the rigor thereof." (23)   E. T. BELL
In mathematics it is new ways of looking at old things which seem to be the most prolific sources of far-reaching discoveries. (23)   E. T. BELL

The first will show us how to change the language suffices to reveal generalizations not before suspected. (417)   H. POINCARÉ
In sum, *all the scientist creates in a fact is the language in which he enunciates it.* (417)   H. POINCARÉ

This long discussion brings us to the final conclusion that the concrete facts of nature are events exhibiting a certain structure in their mutual relations and certain characters of their own. The aim of science is to express the relations between their characters in terms of the mutual structural relations between the events thus characterised. (573)

A. N. WHITEHEAD

We cease to seek resemblances; we devote ourselves above all to the differences, and among the differences are chosen first the most accentuated, not only because they are the most striking, but because they will be the most instructive. (417)   H. POINCARÉ

The materialistic theory has all the completeness of the thought of the middle ages, which had a complete answer to everything, be it in heaven or in hell or in nature. There is trimness about it, with its instantaneous present, its vanished past, its non-existent future, and its inert matter. This trimness is very medieval and ill accords with brute fact. (573)

A. N. WHITEHEAD

The existence of analogies between central features of various theories implies the existence of a general theory which underlies the particular theories and unifies them with respect to those central features.*

E. H. MOORE

Neither the authority of man alone nor the authority of fact alone is sufficient. The universe, as known to us, is a joint phenomenon of the observer and the observed; and every process of discovery in natural science or in other branches of human knowledge will acquire its best excellence when it is in accordance with this fundamental principle. (82)

R. D. CARMICHAEL

It is evident that if we adopt this point of view toward concepts, namely that the proper definition of a concept is not in terms of its properties but in terms of actual operations, we need run no danger of having to revise our attitude toward nature. (55)   P. W. BRIDGMAN

To say the facts are incomprehensible is a rationalization of individual ignorance.
Ignorance, however, may be no fault. It becomes so only when the individual permits himself to rationalize it, *i. e.*, give it a disguise, which effectually blocks him in the utilization of his intelligence, which might otherwise solve the problem in hand. (241)   SMITH ELY JELLIFFE

The symbol *A* is not the counterpart of anything in familiar life. To the child the letter *A* would seem horribly abstract; so we give him a familiar conception along with it. "*A* was an Archer who shot at a frog." This tides over his immediate difficulty; but he cannot make serious progress with word-building so long as Archers, Butchers, Captains, dance round the letters. The letters are abstract, and sooner or later he has to realise it. In physics we have outgrown archer and apple-pie definitions of the fundamental symbols. To a request to explain what an electron really is supposed to be we can only answer, "It is part of the A B C of physics". (149)   A. S. EDDINGTON

No previous existing system of thought had properly formed a working hypothesis to explain why for this or that individual it was necessary to "go up three steps or else be constipated," "or to take pills in multiples of three," or other analogous symptoms which will occur to the reader and which are found in bewildering profusion in all pathological cases, be they hysterias, or compulsion neuroses, phobias, schizophrenias, or what not. (241)   SMITH ELY JELLIFFE

The Dormouse . . . went on: "—that begins with an M, such as mousetraps, and the moon, and memory, and muchness—you know you say things are 'much of a muchness'—did you ever see such a thing as a drawing of a muchness!"
"Really, now you ask me," said Alice, very much confused, "I don't think—"
' "Then you shouldn't talk," said the Hatter.**   LEWIS CARROLL

4.1212   What *can* be shown *cannot* be said. (590)   L. WITTGENSTEIN

Introduction to a Form of General Analysis. Yale Univ. Press.
**Alice in Wonderland.*

PART VII

ON THE MECHANISM OF TIME-BINDING

There should be no theoretical objection to the hypothesis of the formation of new physiological paths and new connections within the cerebral hemispheres. (394)   I. P. PAVLOV

It seems desirable in this place to clearly emphasize the fact that in the use of psychoanalysis we are dealing solely with a method for gaining data. One occasionally hears the statement that psychoanalysis is nonsense. A method, or a tool, is not nonsense. (241)   SMITH ELY JELLIFFE

It is by means of internal inhibition that the signalizing activity of the hemispheres is constantly corrected and perfected. (394)   I. P. PAVLOV

We are dealing here with types of associative reaction peculiar to the cortical system, correctly opposed to the unqualified affective reactivity of the thalamus and usefully analysed by Head. (411)   HENRI PIÉRON

This example and other observations suggest that a gradual development of internal inhibition in the cortex should be used for re-establishment of the balance of normal conditions in cases of an unbalanced nervous system. (394)   I. P. PAVLOV

A self-satisfied rationalism is in effect a form of anti-rationalism. It means an arbitrary halt at a particular set of abstractions. (575)
                                           A. N. WHITEHEAD
. . . the 'fallacy of misplaced concreteness' . . . consists in neglecting the degree of abstraction involved when an actual entity is considered merely so far as it exemplifies certain categories of thought. (578)
                                           A. N. WHITEHEAD
In the Garden of Eden Adam saw the animals before he named them: in the traditional system, children named the animals before they saw them. (575)   A. N. WHITEHEAD
The negative judgment is the peak of mentality. (578)   A. N. WHITEHEAD

24

CHAPTER XXIV

ON ABSTRACTING

... to be an abstraction does not mean that an entity is nothing. It merely means that its existence is only one factor of a more concrete element of nature. (573)   A. N. WHITEHEAD

Aristotle, in building his theories, had at his disposal, besides his personal gifts, a good education according to his day and the science current in 400-300 B.C. Even in those days, the Greek language was a very elaborate affair. Aristotle and his followers simply took this language for granted. The problems of the structure of language and its effect on $s.r$ had not yet arisen. To them, the language they used was *the* (unique) language. When I use the expression '*the* language', I do not mean anything connected with the language, as *Greek;* I mean only the structure of it, which was much similar in the other national languages of this group. The language Aristotle inherited was of great antiquity, and originated in periods when knowledge was still more scanty. Being a keen observer, and scientifically and methodologically inclined, he took this language for granted and systematized the modes of speaking. This systematization was called 'logic'. The primitive structural metaphysics underlying this inherited language, and expressed in its structure, became also the 'philosophical' background of this system. The subject-predicate form, the 'is' of identity, and the elementalism of the A-system are perhaps the main semantic factors in need of revision, as they are found to be the foundation of the insufficiency of this system and represent the mechanism of semantic disturbances, making general adjustment and sanity impossible. These doctrines have come down to us, and through the mechanism of language the semantic disturbing factors are forced upon our children. A whole procedure of training in delusional values was thus started for future generations.

As the work of Aristotle was, at his date, the most advanced and 'scientific', quite naturally its influence was wide-spread. In those days, no one spoke of this influence as 'linguistic', involving $s.r$. Aristotle's work was, and still is, spoken of as 'philosophy', and we speak mostly of the influence of A 'philosophy' rather than of the A structure of language, and its semantic influence.

As we have already seen, when we make any proposition whatsoever we involve creeds, or metaphysics, which are embodied silently as structural assumptions and in our undefined terms. The use of terms not definable in simpler terms at a given date is inherent and seemingly unavoidable.

When our primitive ancestors were building their language, quite naturally they started with the lowest orders of abstractions, which are the most immediately connected with the outside world. They established a language of 'sensations'. Like infants, they identified their feelings with the outside world and personified most of the outside events.

This primitive semantic tendency resulted in the building of a language in which the 'is' of identity was fundamental. If we saw an animal and called it 'dog' and saw another animal roughly resembling the first, we said, quite happily, 'it *is* a dog', forgetting or not knowing that the objective level is un-speakable and that we deal only with absolute individuals, each one different from the other. Thus the mechanism of identification or confusion of orders of abstractions, natural at a very primitive stage of human development, became systematized and structurally embodied in this most important tool of daily use called 'language'. Having to deal with many *objects,* they had to have names for objects. These names were 'substantives'. They built 'substantives', grammatically speaking, for other feelings which were not 'substantives', ('colour', 'heat', 'soul',.). Judging by the lower order abstractions, they built adjectives and made a completely anthropomorphised world-picture. Speaking about speaking, let us be perfectly aware from the beginning that, when we make the simplest statement of any sort, this statement already presupposes some kind of structural metaphysics. The early vague feelings and savage speculations about the structure of this world, based on primitive insufficient scientific data, was influencing the building of the language. Once the language was built, and, particularly, systematized, these primitive structural metaphysics and $s.r$ had to be projected or reflected on the outside world—a procedure which became habitual and automatic.

Was such a language structurally reliable and safe? If we investigate, we can easily become convinced that it was not. Let us take three pails of water; the first at the temperature of 10° centigrade, the second at 30°, and the third at 50°. Let us put the left hand in the first pail and the right in the third. If we presently withdraw the left hand from the first pail and put it in the second, we feel how nicely *warm* the water in the second pail is. But, if we withdraw the right hand from the third pail and put it in the second, we notice how *cold* the water is. The temperature of the water in the second pail was practically not different in the two cases, yet our feelings registered a marked difference. The difference in the 'feel' depended on the former conditions to which our hands had been subjected. Thus, we see that a language of 'senses' is not a very reliable language, and that we cannot depend on it for general purposes of evaluation.

How about the term 'dog'? The number of individuals with which any one is directly acquainted is, by necessity, limited, and usually is small. Let us imagine that someone had dealt only with good-natured 'dogs', and had never been bitten by any of them. Next he sees some animal; he says, 'This *is* a dog'; his associations (relations) do not suggest a bite; he approaches the animal and begins to play with him, and is bitten. Was the statement 'this *is* a dog' a safe statement? Obviously not. He approached the animal with semantic expectations and *evaluation* of his verbal definition, but was bitten by the non-verbal, un-speakable objective level, which has different characteristics.

Judging by present standards, knowledge in the days of Aristotle was very meagre. It was comparatively easy 2300 years ago to summarize the few facts known, and so to build generalizations which would cover those few facts.

If we attempt to build a $\bar{A}$-system, 1933, can we escape the difficulties which beset Aristotle? The answer is that some difficulties are avoidable, but that some are inherent in the structure of human knowledge, and so cannot be entirely evaded. We can, however, invent new methods by which the harmful semantic effect of these limitations can be successfully eliminated.

There is no escape from the fact that we must start with undefined terms which express silent, structural creeds or metaphysics. If we state our undefined terms explicitly, we, at least, make our metaphysics conscious and public, and so we facilitate criticism, co-operation,. The modern undefined scientific terms, such as 'order', for instance, underlie the exact sciences and our wider world-outlook. We must start with these undefined terms as well as the modern structural world-outlook as given by science, 1933. That settles the important semantic point of our structural metaphysics. It need hardly be emphasized that in a human class of life, where creeds are characterized by having dates, they *should* always be labelled with this date. For sanity, the creeds utilized in 1933 should be of the issue of 1933.

Now as to the *structure* of our language. What structure shall we give to our language? Shall we keep the old structure, with all its primitive implications and corresponding *s.r*, or shall we deliberately build a language of new structure which will carry new modern implications and *s.r*? There seems to be only one reasonable choice. For a $\bar{A}$-system, we must build a new language. We must abandon the 'is' of identity, to say the least. We have already seen that we have an excellent substitute in an actional, behaviouristic, operational, functional language. This type of language involves modern asymmetrical implications of 'order', and eliminates the 'is' of identity, which always introduces false evaluation.

To these fundamental starting points, we must add the principle that our language should be of *non-el* structure. With these *minimum* semantic requirements, we are ready to proceed.

Let us take any object of ordinary experience, let us say the one we usually call a 'pencil', and let us briefly analyse our nervous relationship to it. We can see it, touch it, smell it, taste it., and use it in different ways. Is any of the relationships just mentioned an 'all-embracing' one, or is our acquaintance through any of them only *partial?* Obviously, each of these means provides an acquaintance with this object which is not only *partial,* but is also *specific* for the nerve centres which are engaged. Thus, when we look at the object, we do not get odor or taste stimuli, but only visual stimuli,.

If the object we call 'pencil' were lying on the surface of this paper and we were to look at it along the surface of the paper in a perpendicular direction to its length, it would generally be seen as an elongated object, pointed at one end. But, if we were to observe it along the plane of the paper at right angles to our former direction, it would be seen as a disk. This illustration is rough, but serves to show that the acquaintance derived through any specific means (e.g., vision) is also *partial* in another sense; it varies with the position., of any specified observer, Smith, or a camera.

Furthermore, any given means provides, for *different* observers, different acquaintances. Thus, vision shows the pencil to one observer, Smith, as a pointed rod, and to another observer, Jones, as a disk. Feeling, through other receptors, is just as dependent upon many conditions; and different observers receive different impressions. This is well illustrated by the familiar tale of the five blind men and the elephant.

Because of differences in sensitivity in the receptors of Smiths and Browns (partial colour-blindness, astigmatism, far-sightedness.,), any given means of acquaintance (e.g., vision) gives to different observers different reports of the one object. The acquaintance is thus personal and individual.

Again, the reports received through particular channels are influenced by the kind of reports that have already come through that channel. To one who has not seen trees frequently, a spruce and a balsam are not seen to be different. They are just 'evergreens'. With better educated seeing, this individual later differentiates, perhaps, four kinds of spruce.

Because of this factor of experience, the *response* of each individual to similar external stimuli is individual. We can only *agree* on colours, shapes, distances., by ignoring the fact that the effect of the 'same' stimulus is different in different individuals. Besides that, we have no accurate means of comparing our impressions.

The 'time' factor enters, in that we cannot become acquainted with our pencil *on all sides at once*. Nor can we observe the outer form and the inner structure at the 'same time'. We may even neglect to examine the inner structure entirely. Even more important is the fact that all our means together give us only a *partial* and personal acquaintance with the 'pencil'. Continually we invent extra-neural means which reveal new characteristics and finer detail. Nor is this process ever completed. No one can ever acquire a 'complete' acquaintance with even so simple an object as a pencil. The chemistry, the physics, the uses of the varieties., offer fields of acquaintance that can be extended indefinitely. Nature is inexhaustible; the events have infinite numbers of characteristics, and this accounts for the wealth and infinite numbers of possibilities in nature.

I used the word 'acquaintance' deliberately, because it seems vague, and, as yet, *el* gambling on words have not spoiled this term. I had to avoid the *el* terms 'senses' and 'mind' as much as possible in this analysis. If we recall the example of paper roses in the case of hay fever, we shall realize that the terms 'senses' and 'mind' are not reliable, particularly in humans. As a further instance, we have but to remember the experiment with newspaper headlines, also cited earlier.

We become better acquainted with the object by exploring it in manifold ways, and building for ourselves different pictures, all partial, and supplied by direct or indirect contact with different nerve centres. In these explorations, different nerve centres supply their *specific* responses to the different stimuli. Other higher nerve centres summarize them, eliminate weaker details, and so, gradually, our acquaintance becomes fuller while yet remaining *specific* and *partial,* and the semantic problems of *evaluation, meanings,* begin to be important.

If we try to select a term which would describe structurally the processes which are essential for our acquaintance with the object, we should select a term which implies 'non-allness' and the specificity of the response to the stimuli.

If we pass from such a primitive level to a level of 1933, and enquire what we actually know about an object and the structure of its material, we find that in 1933 we know positively that the internal structure of materials is very *different* from what we gather by our rough 'senses' on the macroscopic level. It appears of a dynamic character and of an extremely fine structure, which neither light, nor the nerve centres affected by light, can register.

What we see is structurally only a specific *statistical mass-effect* of happenings on a much finer grained level. We *see* what we see because we *miss* all the finer details. For our purpose, it is usually enough to deal only with sight; this simplifies writing, and the comments made apply to all other 'senses', though perhaps in different degrees.

In 1933, in our human economy, we have to take into account at least three levels. The one is the sub-microscopic level of science, what science 'knows' *about* 'it'. The second is the gross macroscopic, daily experience level of rough objects. The third is the verbal level.

We must also evaluate an important semantic issue; namely, the relative importance of these three levels. We know already that to become acquainted with an object, we must not only explore it from all possible points of view and put it in contact with as many nerve centres as we can, as this is an essential condition of 'knowing', but we must also not forget that our nerve centres must summarize the different partial, abstracted, specific pictures. In the human class of life, we find a new factor, non-existent in any other form of life; namely, that we have a capacity to collect all known experiences of different individuals. Such a capacity increases enormously the number of observations a single individual can handle, and so our acquaintance with the world around, and in, us becomes much more refined and exact. This capacity, which I call the time-binding capacity, is only possible because, in distinction from the animals, we have evolved, or perfected, extra-neural means by which, without altering our nervous system, we can refine its operation and expand its scope. Our scientific instruments record what ordinarily we cannot see, hear,. Our neural verbal centres allow us to exchange and accumulate experiences, although no one could live through all of them; and they would be soon forgotten if we had no neural and extra-neural means to record them.

Again the organism works as-a-whole. All forms of human activities are interconnected. It is impossible to select a special characteristic and treat it in a delusional *el* 'isolation' as the most important. Science becomes an extra-neural extension of the *human* nervous system. We might expect the structure of the nervous system to throw some light on the structure of science; and, vice versa, the structure of science might elucidate the working of the human nervous system.

This fact is very important, semantically, and usually is not sufficiently emphasized or analysed enough. When we take these undeniable facts into account, we find the results already reached to be quite natural and necessary, and we understand better why an individual cannot be considered entirely sane if he is wholly ignorant of *scientific method* and structure, and so retains primitive *s.r.*

For a theory of sanity, all three levels are important. Our 'senses' react as they do because they are united as-a-whole in one living structure, which has potentialities or capacities for language and science.

If we enquire what we *do* in science, we find that we 'observe' silently and then record our observations *verbally*. From a neurological point of view, we abstract whatever we and the instruments can; then we summarize; and, finally, we generalize, by which we mean the processes of abstracting carried further.

In our 'acquaintance' with daily objects, we do substantially a similar thing. We abstract whatever we can, and, according to the degree of intelligence and information we have, we summarize and generalize. From the psychophysiological point of view, the ignorant is neurologically deficient. But to 'know' or to 'believe' something which is false to facts is still more dangerous and akin to delusions, as psychiatry and daily experience teach us.[1] It is a neurological fallacy to treat science in 'isolation' and disregard its psychophysiological role.

In the building of our language, a similar neurological process becomes evident. If we were to see a series of different individuals, whom we might call Smith, Brown, Jones., we could, by a process of abstracting the characteristics, segregate the individuals by sizes or colours.; then, by concentration on one characteristic and disregarding the others, we could build classes or higher abstractions, such as 'whites', 'blacks',. Abstracting again, with rejection of the colour difference., we would finally reach the term 'man'. This procedure is general.

Anthropological studies show clearly how the degree of 'culture' among primitive peoples can be measured by the orders of the abstractions they have produced. Primitive languages are characterized particularly by an enormous number of names for individual objects. Some savage races have names for a pine or an oak., but have no 'tree', which is a higher abstraction from 'pines', 'oaks',. Some other tribes have the term 'tree', but do not have a still higher abstraction 'woods'. It does not need much emphasis to see that higher abstractions are extremely *expedient* devices. There is an enormous economy which facilitates mutual understanding in being able to be brief in a statement and yet cover wider subjects.

Let us consider a primitive statement 'I have seen tree$_1$', followed by a description of the individual characteristics 'I have seen tree$_2$', with minute individual description., where tree$_1$, tree$_2$., stand for names of the individual trees. If an event of interest had happened in a place where there were a hundred trees, it would take a long while to observe fairly well the individual trees and still longer to give an approximate description of them. Such a method is non-expedient, *fundamentally endless;* the mechanism is cumbersome, involves many *irrelevant* characteristics; and it is impossible to express in a few words much that might be *important*. Progress must be slow; the general level of development of a given race or individual must be low. It should be noticed that the problem of *evaluation* enters, at once implying many most important psycho-logical and semantic processes. Similar remarks apply to the abstracting of infants, 'mentally' deficient grown-ups, and some 'mentally' ill.

Indeed, as the readers of my *Manhood of Humanity* already know, the 'human class of life' is chiefly differentiated from 'animals' by its rapid rate of progress through the rapid rate of accumulation of past experiences. This is possible only when expedient means of communication are established; that is, when higher and higher orders of abstractions are worked out.

All scientific 'laws', and other generalizations of higher order (even single words), are precisely such methods of expediency, and represent abstractions of very high order. They are uniquely important because they accelerate progress and help the further summarizing and abstracting of results achieved by others. Naturally, this process of abstracting has also unique practical consequences. When chemical 'elements' were 'permanent' and 'immutable', our physics and chemistry were much undeveloped. With the advent of higher abstractions, such as the monistic and general dynamic theories of all 'matter' and 'electricity', unitary field theories., the creative freedom of science and the control over 'nature' have increased enormously and will increase still more.

Psychiatry also seems to give data indicating that 'mental' illnesses are connected either with arrested development or with regression to phylogenetically older and more primitive levels, all of which, of course, involves lower order abstractions. From the point of view of a theory of sanity, a sharp differentiation between 'man' and 'animal' becomes imperative. For with 'man', the lack of knowledge of this difference may lead to the copying of animals, which would involve semantic *regression* and ultimately become a 'mental' illness.

Although organisms have had acquaintance with objects for many hundreds or thousands of millions of years, the higher abstractions which characterize 'man' are only a few hundreds of thousands of years old. As a result, the nervous currents have a natural tendency to select the older, more travelled, nervous paths. Education should counteract this tendency which, from a *human* point of view, represents regression or under-development.

By now we know how important it is for a $\bar{A}$-system to abandon the older implications and adopt an actional, behaviouristic, operational, or functional language. On the neurological level, what the nervous system *does* is abstracting, of which the summarization, integration., are only special aspects. Hence, I select the term *abstracting* as fundamental.

The standard meaning of 'abstract', 'abstracting' implies 'selecting', 'picking out', 'separating', 'summarizing', 'deducting', 'removing', 'omitting', 'disengaging', 'taking away', 'stripping', and, as an adjective, not 'concrete'. We see that the term 'abstracting' implies structurally and semantically the activities characteristic of the nervous system, and so serves as an excellent *functional physiological* term.

There are other reasons for making the term 'abstracting' fundamental, which, from a *practical* point of view, are important. A bad habit cannot be easily eliminated except by forming a new semantic counter-reaction. All of us have some undesirable but thoroughly established *linguistic habits* and *s.r* which have become almost automatic, overloaded with unconscious 'emotional' evaluation. This is the reason why new 'non-systems' are, in the beginning, so extremely difficult to acquire. We have to break down the old structural habits before we can acquire the new *s.r*. The $\bar{E}$ geometries or the $\bar{N}$ systems are not any more difficult than the older systems were. Perhaps they are even simpler. The main semantic difficulty, for those accustomed to the old, consists in breaking the old structural linguistic habits, in becoming once more flexible and receptive in feelings, and in acquiring new *s.r*. Similar remarks apply in a more marked degree to a $\bar{A}$-system. The majority of us have very little to do *directly* with $\bar{E}$ or $\bar{N}$ systems (although indirectly we all have a good deal to do with them). But all of us live our immediate lives in a human world still desperately A. Hence a $\bar{A}$-system, no matter what benefits it may give, is much handicapped by the old semantic blockages.

In building such a system, this natural resistance or persistence of the old *s.r* must be taken into consideration and, if possible, counteracted. One of the most pernicious bad habits which we have acquired 'emotionally' from the old language is the feeling of 'allness', of 'concreteness', in connection with the 'is' of identity and elementalism. One of the main points in the present $\bar{A}$-system is first to remove entirely from our *s.r* this 'allness' and 'concreteness', both of which are structurally unjustified and lead to identification, absolutism, dogmatism, and other semantic disturbances. Usually, the term 'abstract' is contrasted with 'concrete', which is connected with some vague feeling of 'allness'. By making the functional term *abstracting* fundamental, we establish a most efficient semantic counter-reaction to replace the older terms which had such vicious structural implications. Indeed, it is comparatively easy to accept the term 'abstractions of different orders', and any one who does so will see how much clarity and how much semantic balance he will automatically acquire.

From a *non-el* point of view, the term 'abstracting' is also very satisfactory. The structure of the nervous system is in ordered levels, and all levels go through the process of abstracting from the other levels.

The term implies a general activity, not only of the nervous system as-a-whole, but even of all living protoplasm, as already explained. The characteristic activities of the nervous system, such as summarizing, integrating., are also included by implication.

If we wish to use our terms in the strictly *non-el* way, we must abandon the older division of 'physiological abstractions', which implies 'body', and of 'mental abstractions', which, in turn, implies 'mind', both taken in an *el* way. We can easily do that by postulating abstractions of different orders. We should notice that the above use of the term 'abstracting' differs from the old usage. The semantic difference is in uniting all the abstractions our nervous system performs under the one term, and in distinguishing between different abstractions by the order of them, which is functionally, as well as structurally, justified.

The term 'first order abstractions' or 'abstractions of lower order' does not distinguish between 'body' and 'mind'. *Practically,* it corresponds roughly to 'senses' or immediate feelings, except that by implication it *does not eliminate 'mind'*. Neither does the term 'abstractions of higher orders' eliminate 'body' or 'senses', although it corresponds roughly to 'mental' processes.

From the point of view of 'order', the term 'abstracting' has a great deal in its favor. We have seen what serious structural and semantic importance the term 'order' has, and how the activity of the nervous system has to be spoken of in terms of order. If we establish the term 'abstracting' as fundamental for its *general* semantic implications, we can easily make the meanings more definite and specific in each case by having 'abstractions of different orders'.

We have seen also that the terms we select should involve environment by implication: it is not difficult to see that the term 'abstracting' implies 'abstracting from something' and so involves the environment as an implication.

The term 'abstractions of different orders' is, in this work, as fundamental as the term 'time-binding' was in the author's earlier *Manhood of Humanity*. Hence, it is impossible to be comprehensive about it at this stage; more will be forthcoming as we proceed.

But we have already come to some important semantic results. We have selected our structural metaphysics, and decided that in 1933 we should accept the metaphysics of 1933, which is given *exclusively* by science. We have decided to abandon the false to facts 'is' of identity and to use, instead, the best available language; namely, an actional, behaviouristic, functional, operational language, based on 'order'. And, finally, we have found a term which is functionally satisfactory and has the correct structural and neural implications, and which represents a *non-el* term, and of which the meanings can be expanded and refined indefinitely by assigning to them different orders.

In passing on to the general scientific outlook, similar structural remarks upon a *non-el* point of view apply, and are semantically of importance. Because of the *non-el* character of the work of the writers on the Einstein and new quantum theories, much use is made of this material in the present work. There is a marked structural, methodological, *and semantic* parallelism between all modern *non-el* strivings, which are extremely effective psycho-logically. More material on this subject is given in Parts IX and X.

Now, returning to the analysis of the object which we called 'pencil', we observe that, in spite of all 'similarities', this object is unique, is different from anything else, and has a *unique* relationship to the rest of the world. Hence, we should give the object a *unique name*. Fortunately, we have already become acquainted with the way mathematicians manufacture an endless array of individual names without unduly expanding the vocabulary. If we call the given object 'pencil$_1$', we could call another similar object 'pencil$_2$',. In this way, we produce individual names, and so cover the *differences*. By keeping the main root word 'pencil', we keep the implications of daily life, and also of *similarities*. The habitual use of such a device is structurally and semantically of extreme importance. It has already been emphasized repeatedly that our abstracting from physical objects or situations proceeds by missing, neglecting, or forgetting, and that those disregarded characteristics usually produce errors in evaluation, resulting in the disasters of life. If we acquire this extensional mathematical habit of using special names for unique individuals, we become conscious, not only of the similarities, but also of the differences, which consciousness is one of the mechanisms for helping the proper evaluation and so preventing or eliminating semantic disturbances.

So we now have before us a unique object which we call by a unique name 'pencil$_1$'. If we enquire what science 1933 has to say about this object, we find that this object represents structurally an extremely complex, dynamic process. For our purpose, which is *intuitive,* it is of little importance whether we accept the object as made up of atoms and the atom as made up of whirling electrons., or whether we accept the newer quantum theory, as given in Part X, according to which the atom is formulated in terms of 'electrons' but the 'electron' is the region where some waves reinforce each other, instead of being a 'bit' of something. It is of no importance from our point of view whether the atoms are of a finite size or whether they extend indefinitely and are noticeable to us only in the regions of reinforcement of the waves. Naturally, this last hypothesis has a strong semantic appeal, since it would account, when worked out, for many other facts, such as 'fulness', in a *non-el* language; but probably it would necessitate a postulation of some sub-electronic structures.

What is important for our *s.r* is that we realize the fact that the gross macroscopic materials with which we are familiar are *not* simply what we see, feel., but consist of dynamic processes of some extremely fine structure; and that we realize further that our 'senses' are not adapted to register these processes without the help of extra-neural means and higher order abstractions.

Let us recall, in this connection, the familiar example of a rotary fan, which is made up of separate radial blades, but which, when rotating with a certain velocity, gives the impression of a *solid disk*. In this case the 'disk' is not 'reality', but a nervous integration, or abstraction from the rotating blades. We not only see the 'disk' (*b*) where there is no disk, but, if the blades rotate fast enough, we could not throw sand through them,

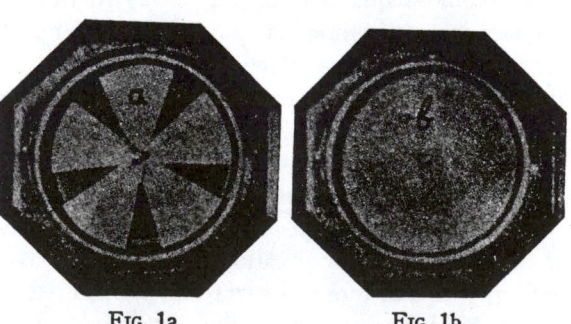

FIG. 1a   FIG. 1b

as the sand would be too slow to get through before being struck by one of the blades.

The 'disk' represents a *joint phenomenon* of the rotating blades (*a*) and of the abstracting power of our nervous system, which registers only the gross macroscopic aspects and slow velocities, but *not* the finer activities on subtler levels. We cannot blame 'the finite mind' for the failure to register the separate blades, because physical instruments may behave similarly. For instance, the illustrations (*a*) and (*b*) are photographs of a small fan which I use ·in lectures, and the photographic camera also missed the rotating blades and registered only a 'disk', in Fig. 1b.

Something roughly similar may be assumed for our purpose as going on in what we usually call 'materials'. These are composed of some dynamic, fine-grained processes, not unlike the 'rotating blades' of our example ; and what we register is the 'disk', be it a table or a chair or ourselves.

For a similar reason, we may assume that we cannot put our finger through a table, as our finger is too thick and too slow, and that, for some materials, it takes X-rays to be agile enough to penetrate.

The above analogies are helpful for our purpose only, but are over-simplified and should not be taken as a scientific explanation.

This neural process seems to be very general, and in all our daily experiences the dynamic fine structures are lost to our rough 'senses'. We register 'disks', although investigation discovers not 'disks', but rotating 'blades'. Our gross macroscopic experience is only a nervous abstraction of some definite order.

As we need to speak about such problems, we must select the best language at our disposal. This ought to be *non-el* and, structurally, the closest to facts. Such a language has been built, and is to be found in the differential and four-dimensional language of space-time, and in the new quantum mechanics. In practice, it is simple to ascribe to every 'point of space' a date, but it takes some training to get this *s.r.* The language of space-time is *non-el*. To the new notion of a 'point' in 'space-time', such a 'point', always having a date associated with it and hence never identical with any other point, the name of 'point-event', or simply 'event', has been given.

How to pass from point-events to extended macroscopic events is a problem in mathematical 'logic'. Several quite satisfactory schemes have been given, into the details of which we do not need to enter here. As the *non-el* structure of the language of space-time appears different from the older *el* language of 'space' *and* 'time', quite obviously the old term 'matter', which belonged to the descriptive apparatus of 'space' *and* 'time', should be abandoned also, and the 'bits' of materials we dealt with should be referred to by structurally new terms. In fact, we know that the old term 'matter' can be displaced by some other term connected with the 'curvature' of 'space-time'.

There is on record a striking example of what the structure of a form of representation means. In a paper printed in the Proceedings of the National Academy of Science, February, 1926, Professor G. Y. Rainich, the mathematician, tried· to introduce 'mass' into space-time, the terms belonging to forms of representation of different structure. He succeeded, but at the price of splitting space-time into the original space *and* time. This is, as far as my knowledge goes, the first proof of how intimately a form of representation is inwardly and structurally interconnected. This fact is of extraordinary semantic importance for psycho-logicians and psychiatrists, who always study symbolism of some sort. It would be of great interest to have such problems worked out by them.

As abstracting in many orders seems to be a general process found in all forms of life, but particularly in humans, it is of importance to be clear on this subject and to select a language of proper structure. As we know already, we use *one* term, say 'apple', for at least *four* entirely different entities; namely, (1) the event, or scientific object, or the sub-microscopic physico-chemical processes, (2) the ordinary object manufactured from the event by our lower nervous centres, (3) the psychological picture probably manufactured by the higher centres, and (4) the verbal definition of the term. If we use a language of adjectives and subject-predicate forms pertaining to 'sense' impressions, we are using a language which deals with entities *inside our skin* and characteristics entirely non-existent in the outside world. Thus the events outside our skin are neither cold nor warm, green nor red, sweet nor bitter., but these characteristics are manufactured by our nervous system inside our skins, as responses only to different energy manifestations, physico-chemical processes,. When we use such terms, we are dealing with characteristics which are absent in the external world, and build up an anthropomorphic and delusional world non-similar in structure to the world around us. Not so if we use a language of order, relations, or structure, which can be applied to sub-microscopic events, to objective levels, to semantic levels, and which can also be expressed in words. In using such language, we deal with characteristics found or discovered on all levels which give us *structural* data uniquely important for knowledge. The ordering on semantic levels in the meantime abolishes identification. It is of extreme importance to realize that the relational., attitude is optional and can be applied everywhere and always, once the above-mentioned benefits are realized. Thus, any object can be considered as a set of relations of its parts ., any 'sense' perception may be considered as a response to a stimulus ., which again introduces relations ,. As relations are found in the scientific sub-microscopic world, the objective world, and also in the psycho-logical and verbal worlds, it is beneficial to use such a language because it is *similar in structure* to the external world and our nervous system; and it is applicable to all levels. The use of such a language leads to the discovery of invariant relations usually called 'laws of nature', gives us structural data which make the only possible content of 'knowledge', and eliminates also anthropomorphic, primitive, and delusional speculations, identifications, and harmful *s.r.*

CHAPTER XXV

ON THE STRUCTURAL DIFFERENTIAL

You cannot recognise an event; because when it is gone, it is gone . . .
But a character of an event can be recognised. . . . Things which we thus
recognise I call objects. (573)   A. N. WHITEHEAD

When there is a judgment of identity or difference, it is because a particular associative reaction of the second order is occurring, conditioned
by the primary reaction, whether the same or different; this is a gain in
perceptive knowledge. (411)   HENRI PIÉRON

To some extent, the practice of thinking, deciding, feeling, appreciating,
and sympathizing molds the personality of the thinker. Presumably, the
stable patterns of cortical association are changed by the performance of
these acts just as on a lower plane muscles are changed by systematic
exercises. (222)   C. JUDSON HERRICK

Experimental analysis of the memory of forms insusceptible of symbolic
schematization has convinced me of the great importance of ocular kinaes-
thesia and the small part played by visualization in nearly all individuals,
with the general illusion of really visual representations, a very strong
illusion, especially when symbolic and verbal schematization is possible.
Ideas which are substituted for visual representation, and play the same
part, are easily mistaken for it. (411)   HENRI PIÉRON

The eyes of the dog give to him sometimes a more intelligent expression
than that of his master, and there is no doubt that he uses them to very
good advantage; but they are not our eyes. (221)   C. JUDSON HERRICK

Before I recapitulate, in the form of a structural diagram, what
has been said in the previous chapter, I must explain briefly the use of
the term 'event'. The introduction of new terms in a language always
represents initial difficulties to the student. It is always advisable, if
only possible, to introduce terms which are structurally close to our
daily experience. At present, in physics, we have a dual language; one
of 'space-time', in which 'matter' is connected somehow with its 'curvature', the other of the quanta. The structure of both languages is quite
different, and at present scientists have not succeeded in translating one
language into the other. Einstein, in his latest unified field theory, has
succeeded, by the introduction of new notions, in amalgamating the
electromagnetic phenomena with the general theory of relativity; but
even this new language does not include the quantum theory. For my
purpose, it is important to amalgamate both languages as an intuitive pictorial device, which, from a technical point of view, still awaits formulation. As the 'space-time' continuum is the closest to our daily experience,
I accept the language of 'events' as fundamental and add only a few

pictorial notions taken from the quantum theory. There is no doubt that the day is not far off when the unified field theory will be extended to include the new quantum theory, and so this anticipation does not appear illegitimate.

If we take something, anything, let us say the object already referred to, called 'pencil', and enquire what it represents, according to science 1933, we find that the 'scientific object' represents an 'event', a mad dance of 'electrons', which is different every instant, which never repeats itself, which is known to consist of extremely complex dynamic processes of very fine structure, acted upon by, and reacting upon, the rest of the universe, inextricably connected with everything else and dependent on everything else. If we enquire *how many characteristics* ($m.o$) we should ascribe to such an event, the only possible answer is that we should ascribe to an event infinite numbers of characteristics, as it represents a process which never stops in one form or another; neither, to the best of our knowledge, does it repeat itself.

In our diagram, Fig. 1, we indicate this by a parabola (A), which is supposed to extend indefinitely, which extension we indicate by a broken off line (B). We symbolize the characteristics by small circles (C), the number of which is obviously indefinitely great.

Underneath, we symbolize the 'object' by the circle (O), which has a finite size. The characteristics of the object we also denote by similar little circles (C′). The number of characteristics which an object has is large but *finite,* and is denoted by the finite number of the small circles (C′).

Then we attach a label to the object, its name, let us say 'pencil$_1$', which we indicate in our diagram by the label (L). We ascribe, also, characteristics to the labels, and we indicate these by the little circles (C″).

The number of characteristics which we ascribe by *definition* to the label is still smaller than the number of characteristics the object has. To the label 'pencil$_1$' we would ascribe, perhaps, its length, thickness, shape, colour, hardness,. But we would mostly *disregard* the accidental characteristics, such as a scratch on its surface, or the kind of glue by which the two wooden parts of the objective 'pencil' are held together,. If we want an objective 'pencil' and come to a shop to purchase one, we say so and specify verbally only these characteristics which are of particular *immediate interest* to us.

It is clear that the object is often of interest to us for some special characteristics of immediate usefulness or value. If we enquire as to the neurological processes involved in registering the object, we find that the

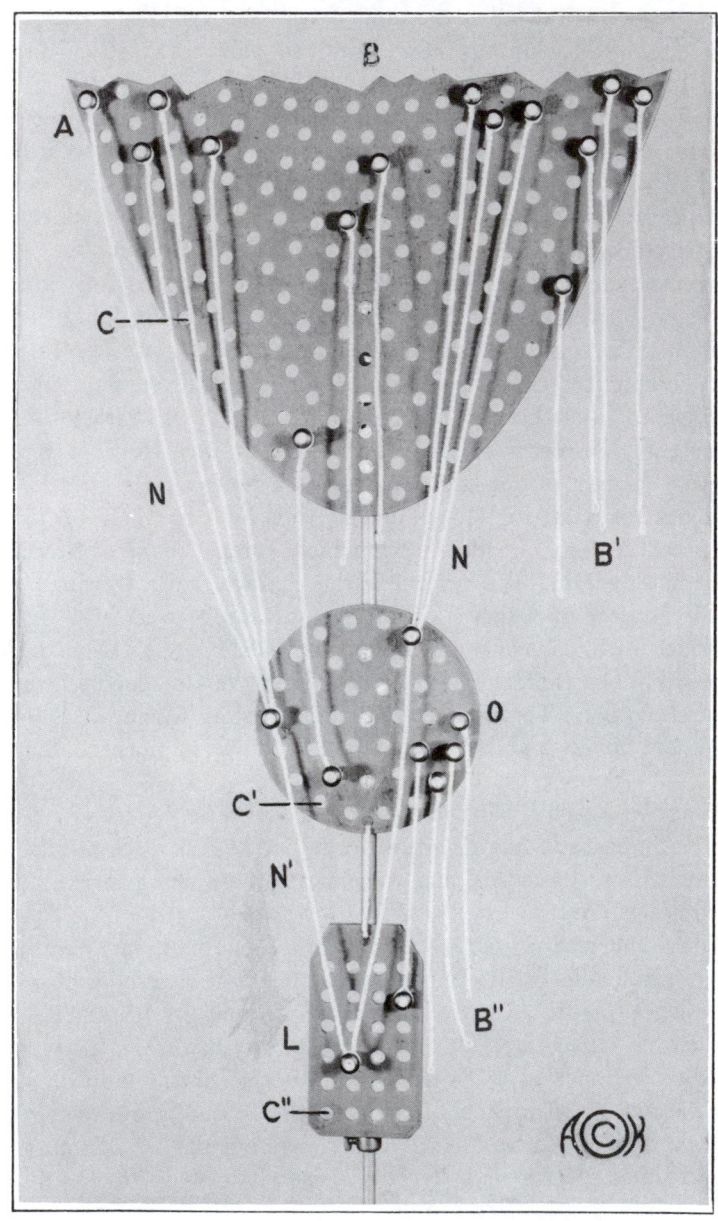

FIG. 1

nervous system has *abstracted,* from the infinite numbers of sub-micro-scopic characteristics of the event, a large but finite number of macro-scopic characteristics. In purchasing a 'pencil' we usually are not interested in its smell or taste. But if we were interested in these abstractions, we would have to find the smell and the taste of our object by experiment.

But this is not all. The object represents in this language a gross macroscopic abstraction, for our nervous system is not adapted for abstracting directly the infinite numbers of characteristics which the end-lessly complex dynamic fine structure of the event represents. We must consider the object as a 'first abstraction' (with a finite number of char-acteristics) from the infinite numbers of characteristics an event has. The above considerations are in perfect accord not only with the func-tioning of the nervous system but also with its structure. Our nervous system registers objects with its lower centres first, and each of these lower specific abstractions we call an object. If we were to define an object, we should have to say that an object represents a first abstrac-tion with a finite number of *m.o* characteristics from the infinite num-bers of *m.o* characteristics an event has.

Obviously, if our inspection of the object is through the lower nervous centres, the number of characteristics which the object has is larger (taste, smell., of our pencil$_1$) than the number of characteristics which we need to ascribe to the label. The label, the *importance* of which lies in its *meanings to us,* represents a still higher abstraction from the event, and usually labels, also, a *semantic reaction.*

We have come to some quite obvious and most important structural conclusions of evaluation of the *non-el* type. We see that the object *is not* the event but an abstraction from it, and that the label *is not* the object nor the event, but a still further abstraction. The nervous process of abstracting we represent by the lines (N), (N'). The characteristics *left out,* or not abstracted, are indicated by the lines (B'), (B'').

For our semantic purpose, the distinction between lower and higher abstractions seems fundamental; but, of course, we could call the object simply the first order abstraction, and the label, with its meanings, the second order abstraction, as indicated in the diagram.

If we were to enquire how this problem of abstracting in different orders appears as a limiting case among animals, we should select a defi-nite individual with which to carry on the analysis. For our analysis, which is deliberately of an extensional character, we select an animal with a definite, proper name, corresponding to 'Smith' among us. Such an animal suggests itself at once on purely verbal grounds.

It is the one we call 'Fido'. Practically all English speaking people are acquainted with the name 'Fido'. Besides, most of us like dogs and are aware of how 'intelligent' they are.

Investigations and experimenting have shown that the nervous system of a Fido presents, in structure and function, marked similarities to that of a Smith. Accordingly, we may assume that, in a general way, it functions similarly. We have already spoken of the event in terms of recognition; namely, that we can never recognize an event, as it changes continually. Whitehead points out the fundamental difference between an event and an object in terms of *recognition;* namely, that an event cannot be recognized, and that an object can be recognized. He defines the object as the recognizable part of the event. The use of this definition helps us to test whether Fido has 'objects'. Since experiments show that Fido can recognize, we have to ascribe to Fido objects by definition. If we enquire what the objects of Fido represent, the structure and function of his nervous system, which are very similar to ours, would suggest that Fido's objects represent, also, *abstractions* of some low order, from the events. Would his objects appear the 'same' as ours? No. First of all, the abstractions from events which we call objects are not the 'same', even when abstracted by different individuals among humans. An extreme example of this can be given in that limited form of colour-blindness which is called Daltonism, when an object which appears green to most persons appears red to the certain few who suffer from this disease. There is, at present, no doubt that the nervous abstractions of all organisms are individual, not only with each individual, but at different 'times' with one individual, and differ, also, for these higher groups (abstractions) which we call species. We can infer how the world appears to a particular organism only if its nervous structure is quite similar to our own. With species widely separated neurologically, such inferences are entirely unjustified. So, on general grounds, the 'objects' of Fido are not the 'same' as ours; on neurological grounds, they appear only similar. In daily experience, we know that we should have difficulty in recognizing our own glove among a thousand, but Fido could perform this detection for us much better. So the 'same' glove must have been registered in the nervous system of Fido differently from the way it has been in ours.

We indicate this similarity of the human object (O_h) and the animal object (O_a) by making the circle (O_a) smaller, and emphasize the difference between the objects by differently spacing the holes representing the characteristics. Whether we call the objects (O_h) and (O_a) 'first order' abstractions or '100th order' abstractions, or simply 'lower

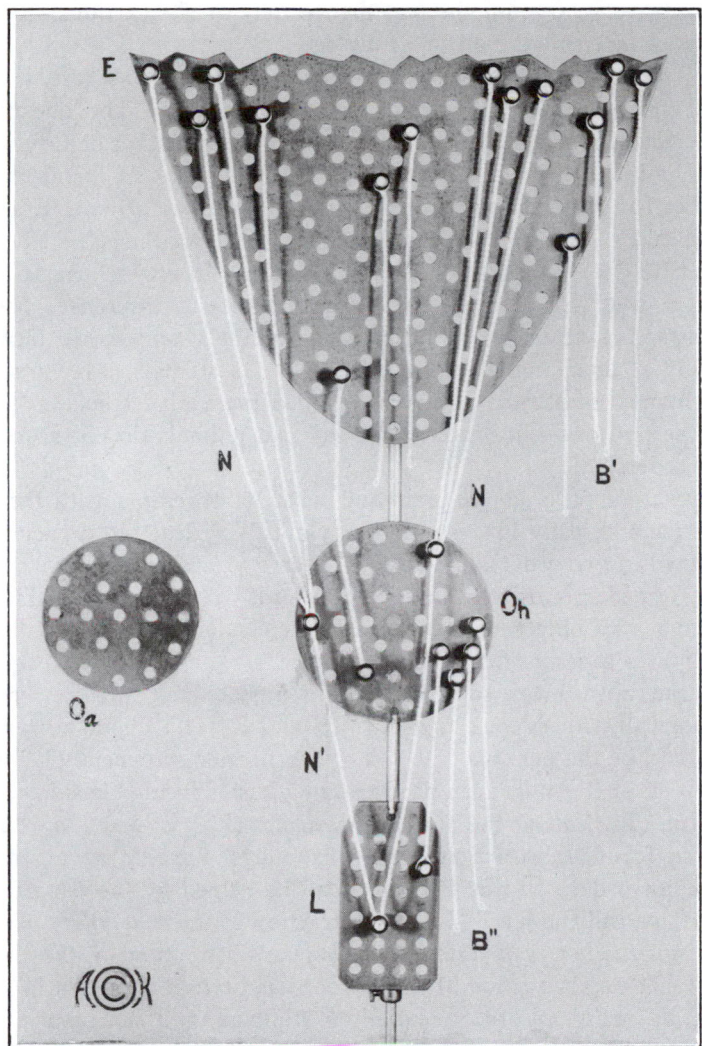

order' abstractions, is mainly optional. There is no neurological doubt that all 'objects' represent *low order abstractions* and the use of a number to indicate the order is simply a matter of convention and convenience. If we were to start with the simplest living cell, we might ascribe to its abstractions the term 'first order' abstractions. If we were to survey in this way all known forms of life, we might ascribe to Fido and Smith very large numbers as their orders of abstractions. But this is unnecessary, as we shall presently see.

We note that Fido does abstract from events, at any rate, in lower orders, 'has objects' (O_a) which he can recognize. The question is, does he abstract in higher orders? We might answer that he does within certain limits. Or, we might prefer to take the limits of his abstracting capacities for granted and to include them all as lower order abstractions. For the sake of convenience and simplicity, we select the last method and say that he does not abstract in higher orders. In our schematic representation, we shall discover some very important differences between the abstracting capacities of humans and animals, and so we introduce here only as much complexity as we need. As animals have no speech, in the human sense, and as we have called the verbal labelling* of the object 'second order abstraction' we say that animals do not abstract in higher orders.

If we compare our diagram and what it represents with the well-known facts of daily life, we see that Smith's abstracting capacities are not limited to two orders, or to any 'n' orders of abstractions.

In our diagrams, the label (L) stands for the *name* which we assigned to the object. But we can also consider the level of the first label (L) as a *descriptive* level or statement. We know very well that Smith can always say something *about* a statement (L), on record. Neurologically considered, this *next* statement (L_1) about a statement (L) would be the nervous response to the former statement (L) which he has seen or heard or even produced by himself inside his skin. So his statement (L_1), *about* the former statement (L), is a *new abstraction* from the former abstraction. In my language, I call it an abstraction of a higher order. In this case, we shall be helped by the use of numbers. If we call the level (L) an abstraction of *second order*, we must call an *abstraction from this abstraction* an abstraction of *third order*, (L_1). Once an abstraction of third order has been produced, it becomes, in turn, a fact on record, potentially a stimulus, and can be abstracted further and a statement made about it, which becomes an abstraction of the fourth order (L_2). This process has no definite limits, for, whenever statements of any order are made, we can always make a statement about them, and so produce an abstraction of still higher order. This capacity is practically universal among organisms which we call 'humans'. Here we reach a fundamental difference between 'Smith'

---

*In the present system the terms 'label', 'labelling'., are always connected with their meanings, and so, for simplicity, from now on the reference to meanings will be omitted.

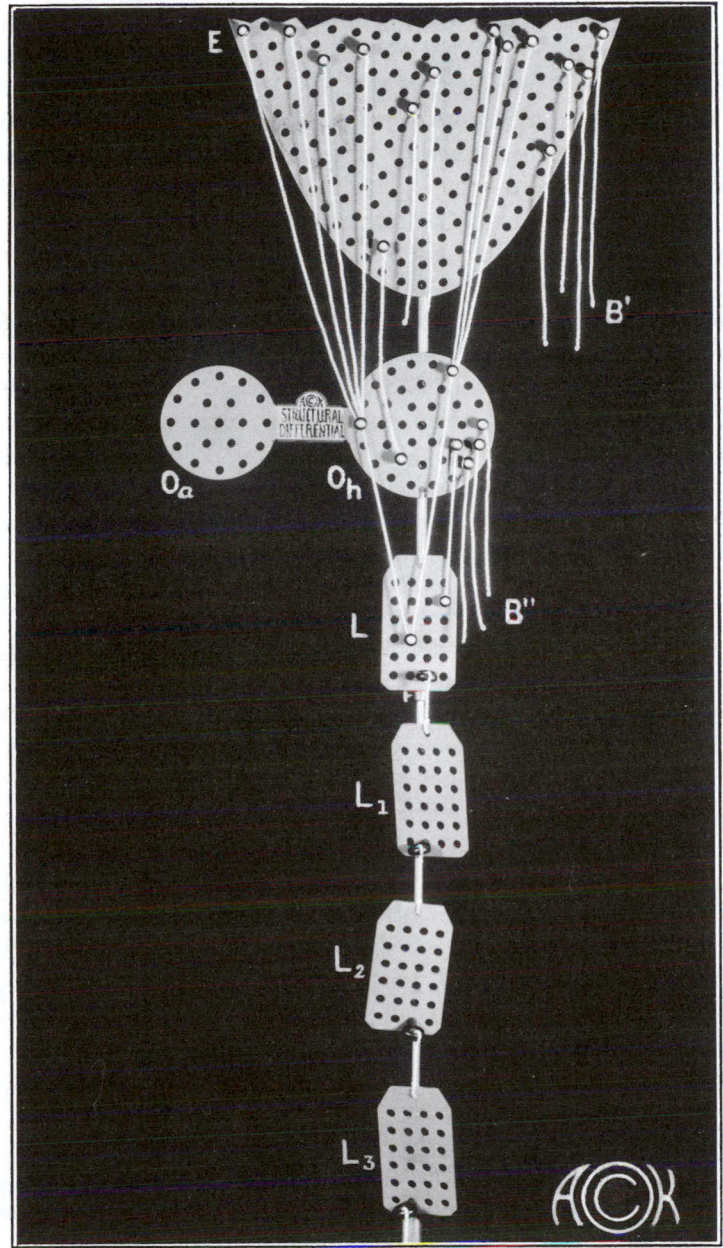

FIG. 3
THE STRUCTURAL DIFFERENTIAL

and 'Fido'. Fido's *power of abstracting stops somewhere*, although it may include a few orders. Not so with 'Smith'; his power of abstracting has no known limit (see Part VI).

Perhaps the reader is semantically perplexed by the unfamiliarity of the language of this analysis. It must be granted that the introduction of any new language is generally perplexing, and it is justified *only* if the new language accomplishes something structurally and semantically which the old languages did *not* accomplish. In this case, it has brought us to a new *sharp* distinction between 'man' and 'animal'. The number of orders of abstractions an 'animal' can produce is *limited*. The number of orders of abstractions a 'man' can produce is, in principle, *unlimited*.

Here is found the fundamental mechanism of the 'time-binding' power which characterizes man, and which allows him, in principle, to gather the *experiences* of all past generations. A higher order abstraction, let us say, of the $n+1$ order, is made as a response to the stimulus of abstractions of the nth order. Among 'humans' the abstractions of high orders produced by others, as well as those produced by oneself are stimuli to abstracting in still higher orders. Thus, in principle, we start where the former generation left off. It should be noticed that, in the present analysis, we have abandoned the structurally *el* methods and language, and the whole analysis becomes simple, although non-familiar because it involves new *non-el s.r.*

The preceding explanation justifies my former statement that the ascribing of absolute numbers to the orders of abstractions of 'animal' and of 'man' is unnecessary. In our diagram we could ascribe as many orders of abstractions to the animal as we please; yet we should have to admit, for the structural correctness of description of experimental facts, that the 'animal's' power of abstracting has limits, while the number of orders of abstractions a 'man' can produce has no known limits.

From an epistemological and semantic point of view, there is an important benefit in this method. In this language, we have discovered *sharp* verbal and analytical methods, in terms of the *non-el* 'orders of abstractions', by which these two 'classes of life', or these two high abstractions, can be differentiated. The terms 'animal' and 'man' each represent a name for an abstraction of very high order, and not a name for an objective individual. To formulate the difference between these 'classes' becomes a problem of *verbal structural ingenuity and methods*, as in life we deal only with absolute individuals on the un-speakable, objective levels. In our diagram, we could hang on the 'animal' object as many levels of labels, which stand for higher order abstractions, as we please; yet somewhere we would have to stop; but with 'man' we could continue indefinitely.

This *sharp* difference between 'man' and 'animal' may be called the '*horizontal difference*'. The habitual use of *our hands* in showing these different horizontal levels is extremely useful in studying this work, and it facilitates greatly the acquiring of the structurally new language and corresponding *s.r.* The solution of the majority of human semantic difficulties (evaluation), and the elimination of pathological identification, lie precisely in the maintenance, without confusion, of the sharp differentiation between these horizontal levels of orders of abstractions.

Let us now investigate the possibility of a sharp '*vertical difference*'. We have already come to the conclusion that Fido abstracts objects from events, and that, if his nervous system is similar to ours, his lower order abstractions are similar to ours. Here we may ask the question: Does Fido 'know', or can he 'know', that he abstracts? It seems undeniable that Fido does not 'know' and *cannot 'know'* that he abstracts, *because it takes science to 'know'* that we abstract, and Fido has no science. It is semantically important that we should be entirely convinced on this point. We do not argue about the kind of 'knowledge' animals may have or about the relative value of this 'knowledge' as compared with ours. Science was made possible by the human nervous system and the invention of extra-neural means for investigation and recording, which animals lack entirely. Whoever claims that animals have science should, to say the least, show libraries and scientific laboratories and instruments produced by animals.

We see that, although Fido has abstracted, he not only does not 'know' but *cannot* 'know' that he abstracts, as this last 'knowledge' is given exclusively by science, which animals do not have. *In this consciousness of abstracting, we find a most important 'vertical difference' between Smith and Fido.* The difference is sharp again.

If, in our diagram, Fig. 4, we ascribe to Fido more horizontal orders of abstractions, let us say two, (H_1) and (H_2), nevertheless, the 'animal' stops somewhere. This extended diagram illustrates that 'man' is capable of abstracting in higher and higher orders indefinitely. In this diagram, we symbolize the fact that Fido does not and cannot 'know' that he abstracts, by not connecting the characteristics of his object (O_a) by lines (A_n) with the event (E). *Without science, we have no event; Fido's gross macroscopic object* (O_a) *represents 'all' that he 'knows' or cares about.* We see that the *vertical difference* (V_1) formulated as consciousness of abstracting for Smith appears sharp, and completely differentiates Fido from Smith. In it, we find the semantic

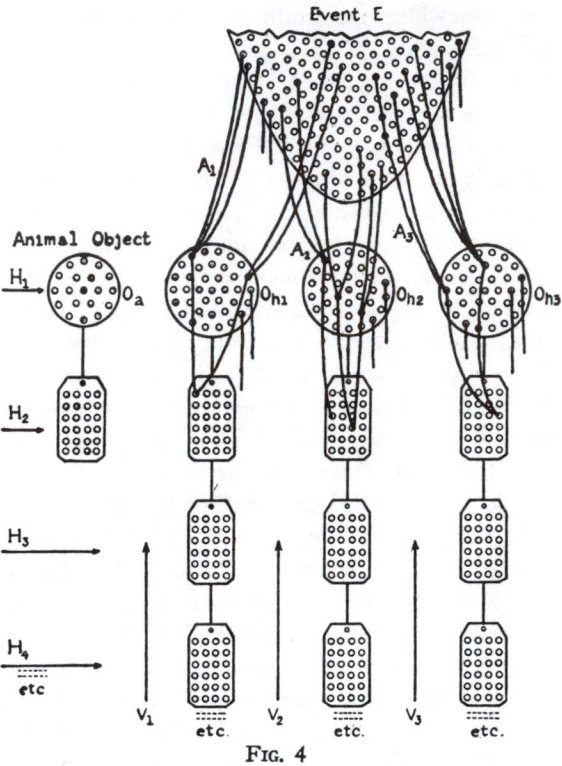

FIG. 4

mechanism of all proper *evaluation,* based on *non-identification* or the differentiation between orders of abstractions, impossible with animals.

In this diagram we have introduced more objects, because each individual abstracts, in general, from an event *different* objects, in the sense that they are *not identical* in every respect. We must be aware continuously that in life on the un-speakable objective level we deal only with absolute individuals, be they objects, situations, or *s.r.* The vertical stratification not only gives us representation for the sharp difference between 'man' and 'animal', but also allows us to train our *s.r* in the absolute individuality of our objects and those of different observers, and for the differences between their individual abstractions. What has been said here applies equally to all first order effects on the objective level, such as immediate feelings,.

The present theory can only be fully beneficial when the reader acquires in *his system* the habitual feeling of both the vertical and the horizontal stratifications with which identification becomes impossible.

In the experiments of Doctor Philip S. Graven with the 'mentally' ill, training in the realization of this stratification has either resulted in complete recovery or has markedly improved the conditions of the patient.

The diagram is used in *two* distinct ways. One is by showing the abstracting from the event to the object, and the applying of a name to the object. The other is by illustrating the level of statements which can be made about statements. If we have different objects, and label them with different names, say, A_1, A_2, A_3 . . . A_n, we still have *no* proposition. To make a proposition, we have to accept some undefined relational term, by which we relate one object to the other. The use of this diagram to illustrate the *levels or orders of statements* implies that we have selected some metaphysics as expressed in our undefined relational terms. We should be fully aware of the difference between these *two* uses of the one diagram for the structural illustration of two aspects of one process.

If we enquire: What do the characteristics of the event represent? We find that they are given only by science and represent at each date the highest, most verified, most reliable abstractions 'Smith' has produced.

Theory and practice have shown that the points illustrated by the above structural diagrams have a crucial semantic significance, as, without using them, it is practically impossible to train ourselves or others and to accomplish the psychophysiological re-education. For this reason, the diagrams have been produced for home and school use, separately, in the simplified form illustrated in Fig. 5. This structural diagram is called the 'Anthropometer' or the 'Structural Differential', as it illustrates the fundamental structural difference between the world, *and so the environment,* of the animal and man. If we live in such a very complex *human* world, but our *s.r*, owing to wrong *evaluation*, are adjusted only to the simpler animal world, free, to say the least, from man-made complications, then adjustment and sanity for *humans* is impossible. Our *s.r* are bound to follow the simpler animalistic patterns, *pathological for man.* All human experience, scientific or otherwise, shows that we still copy animals in our nervous reactions, trying to adjust ourselves to a world of fictitious, simple *animal structure,* while *actually* we live in a world of very complex *human structure* which is quite different. Naturally, under such conditions, which, ultimately, turn out to be delusional, *human* adjustment is impossible and results in false evaluations, animalistic *s.r*, and the general state of un-sanity.

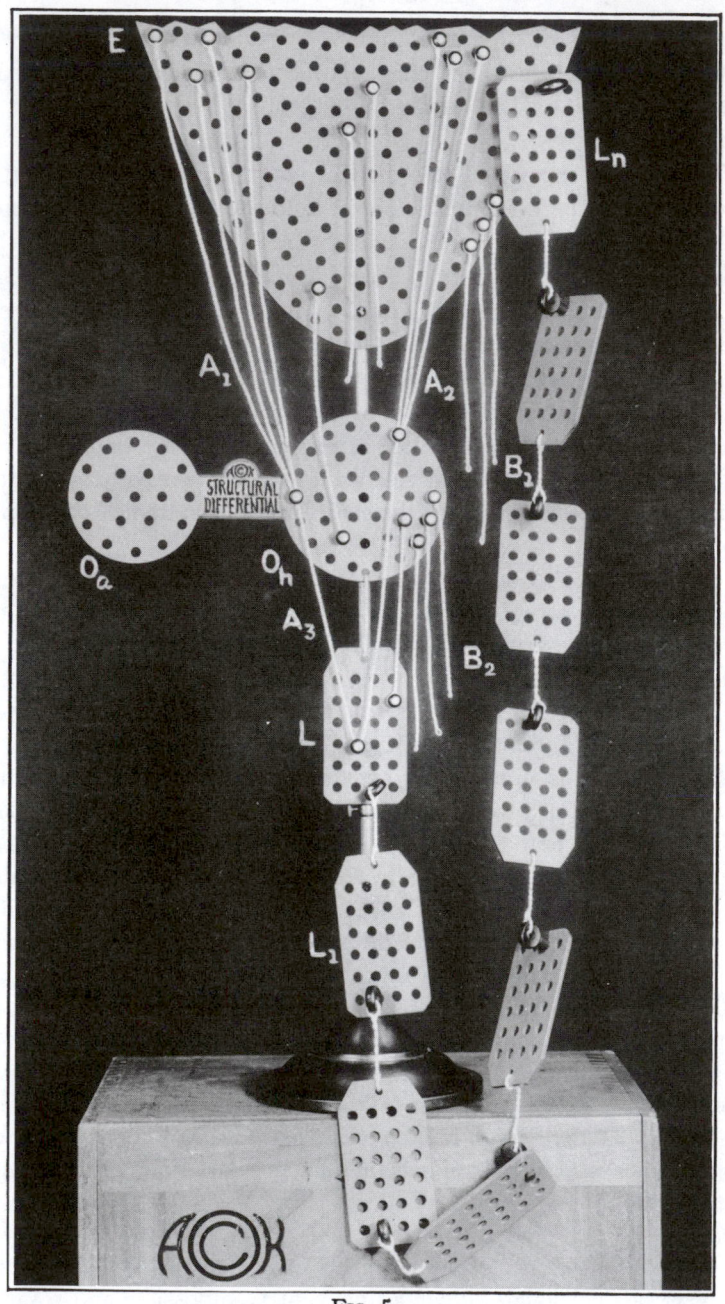

FIG. 5
THE STRUCTURAL DIFFERENTIAL

Any one who will work out the present analysis with the aid of the Differential will find clearly that the majority of human difficulties, the preventable or curable 'mental' or semantic disturbances included, are due to this fatal *structural* error, resulting in false evaluation due to identification or lack of differentiation.

The Structural Differentials are manufactured in two forms: (1) in a printed map-like scroll for hanging on the walls or black-board; (2) in relief form with detachable labels. As the main problem is to train and re-educate the *semantic psychophysiological reactions* in non-identity, the relief form is the most effective because of the freely hanging strings, detachable labels., which give means to engage more nerve centres in the training. I shall describe the latter type in some detail.

For the event we have a parabola in relief (E), broken off to indicate its limitless extension. The disk (O_h) symbolizes the human object; the disk (O_a) represents the animal object. The label (L) represents the higher abstraction called a name (with its meaning given by a definition). The lines (A_n) in the relief diagram are hanging strings which are tied to pegs. They indicate the process of abstracting. The free hanging strings (B_n) indicate the most important characteristics *left out*, neglected, or forgotten in the abstracting. The Structural Differentials are provided with a number of separate labels attached to pegs. These are hung, one to the other, in a series, and the last one may be attached by a long peg to the event, to indicate that the characteristics of the event represent the highest abstractions we have produced at each date. *The objective level is not words, and cannot be reached by words alone. We must point our finger and be silent, or we shall never reach this level.* Our personal feelings, also, *are not* words, and belong to the objective level.

The whole of the present theory can be illustrated on the Structural Differential by the childishly simple operation of the teacher pointing a finger to the event and then to the object, saying 'This *is not* this' and insisting on silence on the pupil's part. One should continue by showing with the finger the object and the label, saying again 'This *is not* this', *insisting on silence* on the objective level; then, showing the first and the second label, saying again 'This *is not* this',.

In a more complex language, one would say that the object *is not* the event, that the label *is not* the un-speakable object, and that a statement about a statement *is not* the 'same' statement, nor on one level. We see and are made to visualize that the $\bar{A}$-system is based *on the denial of the 'is' of identity,* which necessitates the differentiation of orders of abstractions.

The little word 'to be' appears as a very peculiar word and is, perhaps, responsible for many human semantic difficulties. If the anthropologists are correct, only a few of the primitive peoples have this verb. The majority do not have it and do not need it, because all their *s.r* and languages are practically based on, and involve, literal *identification.*[1] In passing from the primitive stage of human society to the present slightly higher stage, which might be called the infantile stage, or infantile period, too crude identification was no longer possible. Languages were built, based on slightly modified or limited identification, and, for flexibility, the 'is' of identity was introduced explicitly. Although very little has been done in the *structural* analysis of languages in general, and of those of primitive peoples in particular, we know that in the Indo-European languages the verb 'to be', among others, is used as an *auxiliary verb* and also for the purpose of positing false to facts identity. With the primitive prevalent lack of consciousness of abstracting, and the primitive belief in the magic of words, the *s.r* were such that words were identified with the objective levels. Perhaps it is not too much to say that the primitive 'psychology' peculiarly required such a fundamental identity. Identity may be defined as 'absolute sameness in all respects' which, in a world of ever-changing processes and a human world of indefinitely many orders of abstractions, appears as a *structural* impossibility. Identity appears, then, as a primitive 'over-emotional' generalization of similarity, equality, equivalence, equipollence., and, in no case, does it appear in fact as 'absolute sameness in all respects'. As soon as the structurally *delusional* character of identity is pointed out, it becomes imperative for sanity to eliminate such delusional factors from our languages and *s.r*. With the advent of 'civilization', the use of this word was enlarged, but some of the fundamental primitive implications and psycho-logical semantic effects were preserved. If we use the 'is' at all, and it is extremely difficult to avoid entirely this auxiliary verb when using languages which, to a large extent, depend on it, we must be particularly careful not to use 'is' as an identity term.

In 1933, the amount of knowledge we have about the primitive peoples is considerable. Anthropologists have gathered an enormous number of descriptive facts, on which they practically all agree, but the several schools of anthropology differ widely as to the interpretation of these facts. Roughly speaking, the British school tries to interpret the facts from the point of view of ascribing to the primitives the deficient 'psychology' and 'logic' of the white man. The French and Polish schools avoid these unjustified tendencies, and attempt to reconstruct the original primitive 'psychologies' and 'logics' which could be responsible for the developments, or the lack of developments, of the primitive peoples. All schools accept, as yet, the existing *el* 'psychologies' and two-valued *A* 'logic' as the standard, normal, and, perhaps, even as the final disciplines for an adult human civilization. No school suspects that an *A* stage of civilization appears to be built, to a large extent, on the slightly refined *primitive identifications* which produced only an *infantile period* of human development. They do not suspect that a future *Ā* society may differ as greatly from the present *A* society as the latter differs from the primitive society.

In my work, I prefer to follow the French and Polish schools of anthropology, as it seems to me that these schools are freer from semantic identification and aristotelianism than the others.

In 1933, it seems, beyond doubt, that *if* any single *semantic* characteristic could be selected to account for the primitive state of the individuals and their societies, we could say, without making too great a mistake, that it would be found in *identification,* understood in the more general sense as it is used in the present work. There is very little doubt, at present, that different physico-chemical factors, environment, climate, kind of food, colloidal behaviour, endocrine secretions ., are fundamental factors which condition the potentialities, as well as the behaviour, of an organism. It is equally certain that, as an end-result, these physico-chemical factors are connected with definite types of *s.r*. It is known that the reverse is also true; namely, that *s.r* affect colloidal behaviour, endocrine secretions, and metabolism. The exact type of dependence is not known, because too little experimenting on humans has been made. The present analysis is conducted from the semantic point of view, and its results, no matter how far-reaching, are limited to this special aspect.

Simple analysis shows that identification is a necessary condition which underlies the reactions of animals, of infants, and of primitives. If found in 'civilized' grown-ups, it equally indicates some remains of earlier periods of development, and can always be found in the analysis of any private or public difficulties which prevent any satisfactory solution. Identification in a slightly modified form represents, also, the very foundation of the *A*-system and those institutions which are founded on this system.

Mathematics gives us practically the only linguistic system free from pathological identifications, although mathematicians use this term uncritically. The more identification is eliminated from other sciences, the more the mathematical functional semantics and method are applied, and the further a given science progresses.

26

The best we know in 1933 is that the general structure of the world was not different in prehistoric times from what we find it today. We have no doubt that the materials in great antiquity consisted of molecules, molecules of atoms, and atoms of electrons and protons., or whatever else we shall be able to discover some day. We have no doubt that blood was circulating in the higher animals' and humans, that vitamins exhibited very similar characteristics as today, that different forms of radiant energy influenced colloidal behaviour., ., regardless of whether or not the given animal, primitive man or infant 'knew' or 'knows' about them.

How about the primitive *physical needs and wants* of an animal, a primitive man, and an infant? Besides all mystical and mythological reasons for identification, the structural facts of life *necessitated identification* on this level of development. *Without* modern knowledge, what a hungry animal, primitive man, or an infant 'wants' 'is' an 'object', say, called an 'apple'. He would 'define' his 'apple' the best he could as to shape, colour, smell, taste,. Was this what his organism needed? Obviously not. We could, at present, produce an undigestible synthetic apple which would satisfy his eventual objective definitions; he might eat it, many such 'apples', and eventually die of hunger. Is an abundant and pleasant diet free from unsuspected and unseen 'vitamins' satisfactory for survival? Again, no! Thus, we see clearly that what the organism needed for survival were the physico-chemical processes, not found in the 'ordinary object', but exclusively in the 'scientific object', or the event. Here we find the age old and necessary, on this early level, identification of the ordinary object with the scientific object. This form of identification is extremely common even in 1933, and, to a large extent, responsible for our low development, because, no matter what we 'think' or feel about an object, an object represents *only* an abstraction of low order, only a *general symbol* for the *scientific object,* which remains the only possible survival concern of the organism. But, obviously, such identification, being false to facts, can never be entirely reliable. If any one fancies that he deals with 'ultimate reality', yet that *m.o* reality represents only a shadow cast by the scientific object; he begins, with experience, to distrust the object and populates his world with delusional mysticism and mythologies to account for the mysteries of the shadow.

As any organism represents an *abstracting* in *different orders* process, which, again, the animal, the primitive man, and the infant cannot know, they, by necessity, identify different orders of abstractions. Thus, names are identified with the un-speakable objects, names for action with the un-speakable action itself, names for a feeling with the un-speakable feelings themselves ,. By confusing descriptions with inferences and descriptive words with inferential words, the 'judgements', 'opinions', 'beliefs', and similar *s.r*, which represent mostly, if not exclusively, inferential semantic end-products, are projected with varying pathological intensity on the outside world. By this method pre-'logical' primitive semantic attitudes were built. Mere similarities were evaluated as identities, primitive syllogisms were built of the type: 'stags run fast, some Indians run fast, some Indians are stags'. It is common to find among primitive peoples a kind of 'logic' based on the *post hoc, ergo propter hoc* (after this and, therefore, because of this) fallacy which obviously represents an identification of an ordinal description with an inference. The 'question begging epithets', which exercise a tremendous semantic influence on primitive and immature peoples and represent a semantic factor in many primitive as well as modern taboos, are also based on such confusions of orders of abstractions.

Identification is one of the primitive characteristics which cannot be eliminated from the animal or the infant, because we have no means to communicate with them properly. It cannot be eliminated from primitive peoples as long as they preserve their languages and environments. Identification is extremely wide-spread among ourselves, embodied strongly in the structure of our inherited language and systems. To change that primitive state of affairs, we need special simple means, such as a $\overline{A}$-system may offer, to combat effectively this serious menace to our *s.r*. It should never be forgotten that identification is practically never dangerous in the animal world, because unaided nature plays no tricks on animals and the elimination by non-survival is very sharp. It is dangerous in the primitive stage of man, however, as it prevents the primitive man to become more civilized, but under his primitive conditions of life his dangers are not so acute. It becomes only very dangerous to the infant if not taken care of, and to the modern white man in the midst of a very far advanced industrial system which affects all phases of his life, when his *s.r* are left unchanged from the ages gone by, and still remain on the infantile level.

The present $\overline{A}$-system is not only based on the complete rejection of the 'is' of identity, but every important term which has been introduced here, as well as the Structural Differential, is aimed at the elimination of these relics of the animal, the primitive man, and the infant in us.

Thus, the primitive 'mentality' does *not differentiate relations* enough; to counteract this, I introduce the *Structural Differential*. The primitive identifies; I introduce a system based on the denial of the 'is' of identity all through. The primitive man pays most attention to what is conveyed to him through the eye and the ear; I introduce the Structural Differential which indicates to the eye the stratification of human knowledge, which represents to the eye the verbal denial of the 'is' of identity. If we identify, we do not differentiate. If we differentiate, we cannot identify; hence, the Structural Differential.

The terms used also convey similar processes. Once we have *order,* we differentiate and have orders of abstractions. Once we abstract, we eliminate 'allness', the semantic foundation for identification. Once we abstract, we abstract in different orders, and so we *order,* abolishing fanciful infinities. Once we differentiate, differentiation becomes the denial of identity. Once we discriminate between the objective and verbal levels, we learn 'silence' on the un-speakable objective levels, and so introduce a most beneficial neurological 'delay'—engage the cortex to perform its natural function. Once we discriminate between the objective and verbal levels, structure becomes the only link between the two worlds. This results in search for similarity of structure and relations, which introduces the aggregate feeling, and the individual becomes a *social being.* Once we differentiate, we discriminate between descriptions and inferences. Once we discriminate, we consider descriptions separately and so are led to *observe* the facts, and only from description of facts do we tentatively form inferences,. Finally, the consciousness of abstracting introduces the general and permanent differentiation between orders of abstractions, introduces the ordering, and so stratifications, and abolishes for good the primitive or infantile identifications. The semantic passing from the primitive man or infantile state to the adult period becomes a semantic, accomplished fact. It should be noticed that these results are accomplished by starting with primitive means, the use of the simplest terms, such as 'this *is not* this', and by the direct appeal to the primitive main receptors—the eye and the ear.

The elimination of the 'is' of identity appears as a serious task, because the A-system and 'logic' by which we regulate our lives, and the influence of which has been eliminated only partially from science, represent only a very scholarly formulation of the restricted primitive identification. Thus, we usually assume, following A disciplines, that the 'is' of identity is fundamental for the 'laws of thought', which have been formulated as follows:

1) The Law of Identity: whatever is, is.

2) The Law of Contradiction: nothing can both be and not be.

3) The Law of Excluded Middle: everything must either be or not be.

It is impossible, short of a volume, to revise this 'logic' and to formulate a $\bar{A}$, ∞-valued, *non-elementalistic* semantics which would be structurally similar to the world and our nervous system; but it must be mentioned, even here, that the 'law of identity' is never applicable to processes. The 'law of excluded middle', or 'excluded third', as it is sometimes called, which gives the two-valued character to A 'logic', establishes, as a general principle, what represents only a limiting case and so, *as a general principle,* must be unsatisfactory. As on the objective, un-speakable levels, we deal exclusively with absolute individuals and individual situations, in the sense that they are not identical, all statements which, by necessity, represent higher order abstractions must only represent *probable* statements. Thus, we are led to ∞-valued semantics of probability, which introduces an inherent and general principle of uncertainty.

It is true that the above given 'laws of thought' can and have been expressed in other terms with many scholarly interpretations, but fundamentally the semantic state of affairs has not been altered.

From a *non-el* point of view, it is more expedient to treat the A-system on a similar footing with the E-system; namely, to consider the above 'laws of thought' as postulates which underlie that system and which express the 'laws of thought' of a given epoch and, eventually, of a race. We know other systems among the primitive peoples which follow other 'laws', in which identity plays a still more integral part of the system. Such natives reason quite well; their systems are consistent with their postulates, although these are quite incomprehensible to those who try to apply A postulates to them. From this point of view, we should not discuss how 'true' or 'false' the A-system appears, but we should simply say that, at a different epoch, other postulates seem structurally closer to our experience and appear more expedient. Such an attitude would not retard so greatly the appearance of new systems which will supersede the present $\bar{A}$-system.

In the present system, 'identification' represents a label for the semantic process of inappropriate evaluation on the un-speakable levels, or for such 'feelings', 'impulses', 'tendencies',. As in human life, we deal with many orders of abstractions, we could say in an ordinal language that identification originates *or* results in the confusion of orders of abstractions. This conclusion may assume different forms: one represented by the identification of the scientific object or the event with the ordinary object, which may be called ignorance, pathological to *man;* another, the identification of the objective levels with the verbal levels, which I call objectification; a third, the identification of descriptions with inferences, which I call confusion of higher order abstractions. In the latter case, we should notice that inferences involve usually more intense semantic components, such as 'opinions', 'beliefs', 'wishes'., than descriptions. These inferences may have a definite, objective, un-speakable character and may represent, then, a semantic state which *is not* words, and so objectifications of higher order may be produced.

When we introduce the ordinal language, we should notice that under known conditions we deal with an ordered natural series; namely, events first, object next; object first, label next; description first, inferences next,. This order expresses the natural importance, giving us the natural base for evaluation and so for our natural *human s.r.* If we identify two different orders, by necessity, we evaluate them equally, which always involves errors, resulting potentially in semantic shocks. As we deal in life with an established natural order of values which can be expressed, for my purpose, by a series decreasing in value: events or scientific objects, ordinary objects, labels, descriptions, inferences.; identification results in a very curious semantic situation.

Let us assume that the scientifically established value of any level could be expressed as 100, and the value of the next as 1. With the consciousness of abstracting we could not disregard, nor identify, these values, nor forget that $100 > 1$. If we confuse the orders of abstractions, this can be expressed as the identification in value and we have a *semantic* equation: (1) $100 = 100$, or (2) $1 = 1$, or any other number, say (3) $50 = 50$.

As we deal fundamentally with a natural, directed inequality, say, $100 > 1$, and, under some semantic pressure, 'want', 'wishful thinking', or ignorance, or lack of consciousness of abstracting, or 'mental' illness., we identified the two in value, we produce in the first and third cases an *over*-evaluation on the right-hand side, and, in the second and third cases, an *under*-evaluation. on the left-hand side. Thus, on the *semantic level*, any identification of *essentially different in value* different orders of abstractions, appears as the *reversal* of the natural order of evaluation, with different degrees of intensity. If the *natural* order of scientific evaluation would be $100 > 1$, and we would evaluate through identification as $2 = 2$, or $3 = 3$., $50 = 50$., $100 = 100$, we would be ascribing twice, or three times, or fifty times, or a hundred times., more *delusional* values to the right-hand side and under-evaluate the left-hand side, than the natural order of evaluation would require. Nature exhibits, in my language and in this field, an asymmetrical relation of 'more', or 'less' inaccessible to A procedure. Under the influence of aristotelianism, when, through identification, we ascribe to nature delusional values, adjustment becomes very difficult, particularly under modern complex life-conditions.

The above example indicates the degrees of intensity which we find in life in the reversal of the natural order of evaluation through identification, produced by, and resulting in, the lack of consciousness of abstracting. Un-sanity, which affects practically all of us, represents the reversal of lesser intensity; the reversal of greater intensity—the more advanced 'mental' ills.

We should realize that *experimentally* we find in this field a fundamental difference in value, which, on semantic levels, can be expressed as an asymmetrical relation of 'more' or 'less', establishing some natural order. If any one should claim a natural 'identity', the burden of proof falls on him. If 'absolute sameness in all respects' cannot be found in this world, then such a notion appears as false to facts, and becomes a structural falsification, preventing sanity and adjustment. If he accepts the fundamental, natural differences in value, but prefers to assume a different order of evaluation depending on his metaphysics, be it the *elementalistic* materialism, or equally *elementalistic* idealism, the semantic results are not changed, because identification in the second case would also ascribe delusional identity to essentially different orders of abstractions. It should be noticed that the $\bar{A}$ formulation applies equally to the older different, opposite doctrines and renders them illegitimate on similar grounds.

The status of the event, or the scientific object, is slightly more complex, because the event is *described* at each date by very reliable, constantly revised and tested, *hypothetical*, structural, inferential terms, exhibiting the peculiar circularity of human knowledge. If we should treat these inferential structures, not as hypothetical, but should identify them semantically with the eventual processes on the level of the submicroscopic event, we would have semantic disturbances of identification.

I have selected the above given order, not only for convenience and simplicity, but because of its experimental character. When we identify in values, we always exhibit in our *s.r* the reversed natural order, introduced here on space-time structural and evaluational grounds.

The above analysis represents a very rough outline, but is sufficient for my purpose. Any attentive and informed reader can carry it further as far as desired. The main point appears that different orders of abstractions exhibit different characteristics, and so any identification of entities essentially different in one or more aspects must introduce delusional semantic factors. I speak mostly about evaluation, because evaluation appears experimentally as an essential factor in all *s.r* and can be applied even profitably in those cases of 'mental' illness where no definite evaluation appears, the *absence* of evaluation being a form of evaluation (*m.o*). In training, it is of utmost importance to eliminate identification entirely, which invariably appears as a delusional semantic factor. To achieve these ends, all and every available means should be employed.

When one studies carefully the older disciplines, one is amazed to learn to what an extent the recorded 'thinkers' rebelled against the limitations and insufficiencies of aristotelianism, which system, naturally, became antiquated a short time after its formulation. One is amazed to find that 'everything has already been said', and that, to a large extent, these important, separated statements were *inoperative*. It is of little importance that some 'wise statements' had been made by some one, somewhere, if they had no influence on the great masses of the race. The reason for this tremendous public waste of private efforts is that aristotelianism, with its further elaborations and its delusional identification, elementalism., represents a co-ordinated *system* which moulded our *s.r*, languages, and institutions, and which influenced every phase of our lives. Under such conditions, isolated doctrines, no matter how wise, become powerless in the face of such a system, or, more correctly, a system of interlocked systems. Only a revision of the system and the tentative formulation of a $\bar{A}$-system can make many older fundamental clarifications workable, which, although known to a few specialists, appear generally unknown to the great masses and unavailable in elementary education, which alone can be generally effective. One is also amazed at the power of structurally correct terminology, and feels full of sympathy toward the primitive interpretation as the 'magic of words'! Happy, structural high abstractions really have a strong creative character. Since, for instance, the principle of 'least action', or the 'general principle of relativity' (the theory of the absolute)., have been formulated, all of our structural knowledge has been recast, clarified, and we constantly hear of some remarkable applications of the new knowledge. Similarly, if it is pointed out that our main private and public difficulties are due to infantilism produced by 'aristotelianism', in general, and, in particular, by identification and elementalism, we at once have practical means for a revision and applications. In such a first and novel attempt over-subtlety is impossible and even not desirable. It is preferable, as well as expedient, to formulate the general outline and, thereby, draw more men into the work for the details.

For thousands of years, millions upon millions of humans have used a great deal of their nervous energy in worrying upon delusional questions, forced upon them by the pernicious 'is' of identity, such as:

'What *is* an object?', 'What *is* life?', 'What *is* hell?', 'What *is* heaven?', 'What *is* space?' 'What *is* time?', and an endless array of such irritants. The answer, based on the human discrimination of orders of abstractions and so proper *human evaluation,* is definite, undeniable, simple, and *unique:* 'Whatever one might *say* something *"is", it is not.'* Whatever we might *say* belongs to the verbal level and *not* to the un-speakable, objective levels.

Let me repeat once more that the 'is' of identity forces us into semantic disturbances of wrong *evaluation.* We establish, for instance, the *identity* of the un-speakable objective level with words, which, once stated, becomes obviously false to facts. The 'is' of identity, if used as indicating 'identity' (structurally *impossible* on the objective levels), says nothing. Thus, the question, 'What *is* an object?', may be answered, 'An object *is* an object'—a statement which says nothing. If used in definitions or classifications, such as 'Smith *is* a man', a type of statement used even in the *Principia Mathematica,* or 'A *is* B or not B', as in the formulation of the law of 'excluded third' in the two-valued *A* 'logic', it always establishes an *identity,* false to facts. The first statement expresses the *identity* of a proper name with a class name which must lead to the confusion of classes (higher order abstractions) with individuals (lower order abstractions). This confusion leads automatically to disturbed evaluation in life, because the characteristics of a class are *not* the 'same' as, nor identical with, the characteristics of the individual. I shall not analyse in detail the 'A *is* B', because, obviously, it *is not.*

How about Fido? Fido has no science and, therefore, no 'event'. For him, the object is *not* an abstraction of some order, but *'is all'* he 'knows' and cares about. Smith not only abstracts in indefinite numbers of different orders, and does it automatically and habitually, but if he enquires he may also become *conscious of abstracting—'is not all'*, and *'this is not this'.* Now, Fido can *never* be conscious of abstracting, as his nervous system is incapable of being extended by extra-neural means, and this extension appears to be a necessary condition for the acquiring of *consciousness of abstracting.*

Although for Smith, 'This *is not* this', as illustrated on the Structural Differential, for Fido, that diagram would eventually mean 'this *is* this', the structure of his world being represented by the single disk (O_a). Fido cannot be conscious of abstracting, he *must* identify, because he 'knows' nothing of this process, and there is no means of informing him of these relations and structure.

If we *are not conscious* of abstracting, we must identify—in other words, whenever we confuse the different orders of abstractions, unavoidable if we use the 'is' of identity, we duplicate or copy the animal way of 'thinking', with similar 'emotional' responses. In the following chapters, this tragedy will be explained in detail, and it will be shown that practically all human difficulties involve this semantic factor of copying animals in our nervous reactions and evaluation as a component.

A theory which not only throws light on this serious problem, but which also gives means of replacing the old harmful *s.r* by more beneficial ones, may be useful, in spite of various temporary difficulties which are due to the old identity-reactions and the lack of familiarity with the new.

The old identity-reactions are extremely ingrained, particularly with grown-ups. Serious effort and permanent reminders are necessary to overcome them. The Structural Differential represents such a structural visual reminder, which we should keep constantly before our eyes until the pernicious disturbances of evaluation have been overcome. For Smith, the fundamental evaluation can be expressed in simple and quite primitive language—'This *is not* this'.

The above most vital semantic factors of evaluation indispensable for adjustment and sanity are conveyed to him whenever he looks at the stratification indicated on the Differential. The hanging free strings indicating the *non-abstracted* characteristics train his *s.r* to be aware of the non-allness of, and the lack of identity between, his abstractions.

Our old *s.r* were similar to Fido's; we were never *fully* conscious of abstracting. Through wrong evaluation we identified what is inherently different and longed for, or assumed some impossible 'allness' in our 'knowings'.

Practice has shown me, definitely, that to acquire these new reactions of *consciousness of abstracting* is difficult and requires 'time' and effort to accomplish, in spite of the exceptional, nearly primitive, simplicity of the means employed. The 'silence on the objective levels' sounds very innocent; yet it is extremely difficult to acquire, as it involves a complete checking of all semantic disturbances, identifications, confusions of orders of abstractions, habitual 'emotions', 'preconceived ideas'., practically impossible without the use of the *objective* Differential to which we can point our finger and be silent, to begin with. In fact, to disregard this point, actually means failure in accomplishing the desired semantic results. At present, as far as experience has gone, the main results were achieved when a given individual had conquered this first, simple, and obvious semantic obstacle. If the simple rules and conditions given in the present system for abolishing identification are followed persistently in the training with the Differential, a complete and very beneficial structural and semantic change in the character and 'mental' capacities of a given individual occurs, seemingly all out of proportion with the simplicity of the training. But if we consider the content of all knowledge as uniquely *structural,* and if the majority of us are semantically tied up, blocked, with antiquated, animalistic, primitive, infant-like, 'mentally'-ill and A structure and identity-reactions, owing to the lack of consciousness of abstracting, which we renounce *in toto* by acquiring the consciousness of abstracting, such remarkable transformation becomes intelligible.

The publication of the Structural Differential in separate, conveniently large copies has been forced upon me by experience and by various difficulties found in the re-educating of our *s.r*, without which a $\bar{A}$-system, adjustment, sanity, and all the desirable results which depend on them, are impossible.

CHAPTER XXVI

ON 'CONSCIOUSNESS' AND CONSCIOUSNESS OF ABSTRACTING

> But a felt 'contrary' is consciousness in germ. . . . Consciousness requires more than the mere entertainment of theory. It is the feeling of the contrast of theory, as *mere* theory, with fact, as *mere* fact. This contrast holds whether or no the theory be correct. (578)   A. N. WHITEHEAD

A language, to be most useful, should be similar in its structure to the structure of the events which it is supposed to represent. The language of 'abstractions of different orders' appears to be satisfactory in point of structure. It is a *non-el* language, since it does not discriminate between 'senses' and 'mind',. It is a functional language, since it describes, by implication, what is going on in the nervous system when it reacts to stimuli. It is a language which can be made as flexible and as sharp as desired, thus making it possible to establish sharp verbal differences, of both horizontal and vertical type, between the terms 'man' and 'animal'.

The last semantic characteristic of potential sharpness is extremely important for a theory of sanity. Evidence of 1933 leads us to conclude that, under the influence of external stimuli, the most primitive and simplest forms of life were moulded, transformed, and influenced in the process of survival, and, therefore, of adjustment. In this way, more and more complex structures evolved. It should be emphasized that organisms represent *functional* units, and that an additive change in structure does not necessarily involve a *simply additive* change in function. By physico-chemical, structural, colloidal necessity the organism works as-a-whole. Being a relative whole, any additive structural factor becomes a reactive and functional factor which influences the working of the whole. This is, perhaps, best illustrated by the boy who was born without a cortex, but with no other obvious defects. He was incomparably more helpless and unadjusted than animals who have no cortex, or even no nervous system at all. Although we could speak in *additive* terms of the difference between this boy and a normal boy, as one having no cortex and the other 'plus a cortex', yet the functioning was so different as not to be expressible in a 'plus' language.

Similar remarks could be generalized to all life. We must be very careful in building sharp distinctions, since the anatomical differences alone are unreliable. If we want to have more reliable differences, we should look for *functional* differences.

We have already discovered *functional* differences that are expressed by the horizontal and by the vertical differences between the abstracting capacities of Smith and Fido. The analysis of these differences is the subject of the present chapter.

'Thought' represents a reaction of the organism-as-a-whole, produced by the working of the whole, and influencing the whole. From our daily experience, we are familiar with what we usually denote as being 'conscious'; in other words, we are aware of something, be it an object, a process, an action, a 'feeling', or an 'idea'. A reaction that is very habitual and semi-automatic is not necessarily 'conscious'. The term 'consciousness', taken separately, is not a complete symbol; it lacks content, and one of the characteristics of 'consciousness' is to have some content. Usually, the term 'consciousness' is taken as undefined and *undefinable,* because of its immediate character for every one of us. Such a situation is not desirable, as it is always semantically useful to try to define a complex term by simpler terms. We may limit the general and undefined term 'consciousness' and make it a definite symbol by the deliberate ascribing of some content to this term. For this 'consciousness of something' I take 'consciousness of abstracting' as fundamental. Perhaps the only type of meanings the term 'consciousness' has is covered by the functional term 'consciousness of abstracting', which represents a general process going on in our nervous system. Even if this is not the only type of meanings, the term 'consciousness of abstracting' appears to be of such crucial semantic importance that its introduction is necessary.

The term 'consciousness', because of its hitherto undefined and traditionally *undefinable* character, did not allow us further analysis. Neither did we have any *workable,* educational, semantic means to handle the vast field of psycho-logical processes which this incomplete symbol indicated. If we now select the term 'consciousness of abstracting' as fundamental, we not only make the last symbol complete by assigning functional content to it, but we also find means to define it more specifically in *simpler terms.* Through understanding of the processes we gain educational means of handling and influencing a large group of semantic psycho-logical reactions.

Let us analyse this new term by aid of the diagram called the Structural Differential referred to in the previous chapter. Here the object (O_h) represents a nervous abstraction of a low order. In this abstracting, some characteristics of the event were missed or not abstracted; these are indicated by the not connected lines (B'). When we abstracted from our object further, by coining a definition or ascribing 'meanings' to the label (L), again we did not abstract 'all' the characteristics of the object into the definition; but some characteristics were left out, as indicated by the lines (B″). In other words, the number of characteristics

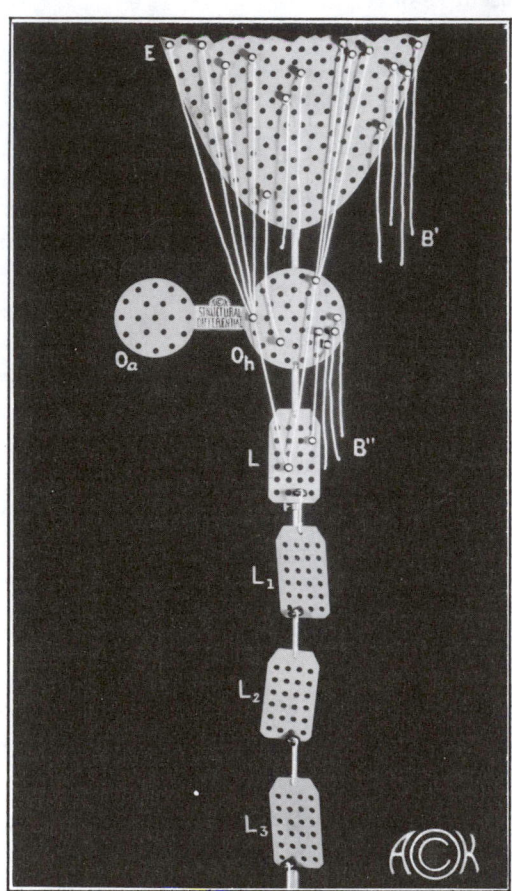

FIG. 1
THE STRUCTURAL DIFFERENTIAL

which we ascribe to the label, by some process of 'knowing', or 'wanting', or 'needing', or 'interest'., does *not* cover the number of characteristics the object has. The 'object' has more characteristics than we can include in the explicit or implicit definition of the label for the 'object'. Besides, the definition (implicit or explicit) of the 'object' *is not* the object itself, which always holds many surprises for us. The latter has the *'individuality of the object'*, as we may call it. Every one who uses a car, or a gun, or a typewriter, or who has had a number of wives, or husbands, or children, knows that well. In spite of the fact that these objects are, to a large extent, standardized, every individual object has individual peculiarities. With modern methods of physical, chemical, and astronomical investigation, scientists find that even their special materials and equipments have also peculiar individualities which must be taken into account in the more refined researches.

If we take any ordinary object and expect to find such and such characteristics, ascribed to the objects by *definition,* we may be disappointed. As a rule, we find or can find, if our analysis is subtle enough, these peculiar individualities. The reader can easily convince himself by looking over a box of matches, and by noticing the peculiar individuality of each match. But since, *by definition*, we expect that when we strike a match it should ignite, we may disregard all other characteristics as irrelevant for our purpose. A similar process is at work in other phases of life. We often live, feel happy or unhappy, *by what actually amounts to a definition*, and not by the empirical, individual facts less coloured by semantic factors. When $Smith_1$ marries $Smith_2$, they mostly do so *by a kind of definition*. They have certain notions as to what 'man', 'woman', and 'marriage' 'are' *by definition*. They actually go through the performance and find that the $Smith_1$ and his wife, $Smith_2$, have unexpected likes, dislikes, and particularities—in general, characteristic and semantic reactions *not included* in their definition of the terms 'man', 'woman', 'husband', 'wife', or 'marriage'. Characteristics 'left out' in the definitions make their appearance. 'Disappointments' accumulate, and a more or less unhappy life begins.

The above analysis applies to all phases of human life, and appears entirely general because of the structure of 'human knowledge'. Characteristics are discovered when it is *too late*. The *not knowing* or the *forgetting* of the relations explained above does the semantic havoc. On verbal, 'definitional', or doctrinal semantic grounds, we expect something else than what the experiences of life give us. The non-fulfillment of expectation produces a serious affective and semantic shock. If such shocks are repeated again and again, they disorganize the normal working of the nervous system, and often lead to pathological states. An indefinitely large number of experimental facts fully supports the above conclusions. Many of them have been supplied during the World War. Curiously enough, when the soldier *did* expect horrors, and later experienced them, he seldom became deranged 'mentally'. If he did not fully expect them, and yet had to experience them, he often broke down nervously.

The attack of hay fever at the sight of *paper roses,* referred to already, gives a similar semantic example. The attack followed from the semantic *'definition'* of 'roses', of 'hay fever', and from the situation as-a-whole, and was not due to *inspection* of the objective 'roses', or to the physico-chemical action of the 'roses'. If the patient had been blindfolded when the paper 'roses' were brought into his presence, no attack would have occurred.

We are now ready to define 'consciousness of abstracting' in *simpler terms;* namely, in terms of 'memory'. The term 'memory' is structurally a physico-chemical term. It implies that the events are interconnected, that everything in this world influences everything else, and that happenings leave some traces somewhere.

A similar analysis can be carried on in connection with the object and the event. Briefly, the object represents structurally an abstraction of some order, does not, and cannot, include all the characteristics of the event; and so, again, we have some characteristics *left out* as indicated by the lines (B').

Here we have the possibility of making a series of most general, and yet entirely true, *negative* statements of great semantic importance; that the label *is not* the object, and that the object *is not* the event,. For the number of *m.o* characteristics which we ascribe to the label by *definition* does not cover all the characteristics we recognize in the object; and the number of characteristics which we perceive in the object is also not equal to the infinite numbers of characteristics the event has. The differences are still more profound. Not only do the numbers of *m.o* characteristics differ, but also the *character* of these abstractions differs from level to level of the successive abstractions.

We can now define 'consciousness of abstracting' as '*awareness* that in our process of abstracting we have *left out* characteristics'. Or, consciousness of abstracting can be defined as '*remembering* the "*is not*", and that some characteristics have been *left out*'. It should be noticed that in this formulation, with the aid of the Structural Differential, we have succeeded in translating a *negative* process of forgetting into a *positive* process of *remembering* the denial of identity and that characteristics are left out. Such a positive formulation makes the whole system workable and available for the semantic training and education.

The use of the Structural Differential becomes a necessity for any one who wants to receive full semantic benefit from the present work. A book is, by necessity, *verbal*. Whatever any author can say is verbal, and nothing whatsoever can be *said* which is *not verbal*. It seems entirely obvious that in life we deal with an enormous number of things and situations, 'feelings'., which are *not verbal*. These belong to the 'objective level'. The crucial difficulty is found in the fact that whatever can be said *is not* and *cannot* be on the objective level, but belongs *only* to the verbal levels. This difference, being *inexpressible* by words, cannot be expressed by words. We must have *other means* to indicate this difference. We must show with our hand, by pointing our finger to the object, and by being silent outwardly as well as inwardly, which silence we may indicate by closing our lips with the other hand. The verbal denial of the 'is' of identity covers this point also when shown on the Differential. If we burst into speech based on the 'is' of identity, as we usually do, we find ourselves obviously on the verbal levels indicated by the labels L, L_1, L_2, . . . L_n, but never on the objective level (O_h). On this last level, we can look, handle., but *must be silent.* The reason that we nearly all identify the two levels is that it is impossible to train an individual in this semantic difference by *verbal means alone,* as all verbal means belong to the levels of labels and never to the objective un-speakable levels. With a visual and tactile *actual object* and labels on the Structural Differential, to point our finger at, handle., we now have simple means to convey the tremendously important semantic difference and train in *non-identity.*

We should notice that the consciousness of abstracting, or the remembering that we abstract in different orders with omission of characteristics, depends on the denial of the 'is' of identity and is connected with limitations or 'non-allness', so characteristic of the new non-systems.

The consciousness of abstracting eliminates *automatically* identification or 'confusion of the orders of abstractions', both applying to the semantic confusion on all levels. If we are *not* conscious of abstracting, we are bound to identify or confuse the object with its finite number of characteristics, with the event, with its infinite numbers of *different* characteristics. Confusion of these levels may misguide us into semantic situations ending in unpleasant shocks. If we acquire the consciousness of abstracting, and remember that the object *is not* the event and that we have abstracted characteristics fewer than, and different from, those the event has, we should expect many unforeseen happenings to occur. Consequently, when the unexpected happens, we are saved from painful and harmful semantic shocks.

If, through lack of consciousness of abstracting, we identify or con-fuse words with objects and feelings, or memories and 'ideas' with experiences which belong to the un-speakable objective level, we iden-tify higher order abstractions with lower. Since this special type of semantic identification or confusion is extremely general, it deserves a special name. I call it *objectification,* because it is generally the confusion of words or verbal issues (memories, 'ideas'.,) with objective, un-speak-able levels, such as objects, or experiences, or feelings,. If we objectify, we *forget,* or we *do not remember* that words *are not* the objects or feelings themselves, that the verbal levels are always different from the objective levels. When we identify them, we disregard the inherent differences, and so proper evaluation and full adjustment become im-possible.

Similar semantic difficulties arise from the confusion of higher order abstractions; for instance, the identification of inferences with

27

descriptions. This may be made clearer by examples. In studying these examples, it should be remembered that the organism acts as-a-whole, and that 'emotional' factors are, therefore, always present and should not be disregarded. In this study, the reader should try to put himself *'emotionally'* in the place of the Smith we speak about; then he cannot fail to understand the serious semantic disturbances these identifications create in everybody's life.

Let us begin with a Smith who knows nothing of what has been said here, and who is *not* conscious of abstracting. For him, as well as for Fido, there is, in principle, no realization of the 'characteristics left out'. He is 'emotionally' convinced that his words entirely cover the 'object' which *'is'* so and so'. He identifies his lower abstractions with characteristics left out, with higher abstractions which have all characteristics included. He ascribes to words an entirely false value and certitude which they cannot have. He does not realize that his words may have different meanings for the other fellow. He ascribes to words 'emotional' *objectivity* and value, and the verbal, *A* 'permanence', 'definiteness', 'one-value'., to objects. When he hears something that he does not like, he does not ask 'what do you mean?', but, under the semantic pressure of identification, he ascribes his own meanings to the other fellow's words. For him, words *'are'* 'emotionally' overloaded, objectified semantic fetishes, even as to the primitive man who believed in the 'magic of words'. Upon hearing anything strange, his *s.r* is undelayed and may appear as, 'I disagree with you', or 'I don't believe you',. There is no reason to be dramatic about any unwelcome statement. One needs definitions and interpretations of such statements, which probably are correct from the speaker's point of view, if we grant him his informations, his *undefined terms,* the structure of his language and premises which build up his *s.r.* But our Smith, innocent of the 'structure of human knowledge', has mostly a semantic belief in the one-value, absoluteness., of things, and thinghood of words, and does not know, or does not *remember,* that words *are not* the events themselves. Words represent higher order abstractions manufactured by higher nerve centres, and objects represent lower order abstractions manufactured by lower nerve centres. Under such *identity-delusions,* he becomes an absolutist, a dogmatist, a finalist,. He seeks to establish 'ultimate truths', 'eternal verities'., and is willing to fight for them, never knowing or remembering, otherwise forgetting, the 'characteristics left out'; never recognizing that the noises he makes *are not* the objective actualities we deal with. If somebody contradicts him, he is much disturbed. Forgetting characteristics left out, he is always 'right'. For him his statement is not only *the* only statement possible, but he actually attributes some cosmic objective evaluation to it.

The above *description* is unsatisfactory, but cannot be much improved upon, since the situation involves *un-speakable* affective components which *are not* words. We must simply try to put ourselves in his place, and to live through his experiences when he identifies and believes without question that his words *'are'* the things they only stand for. To give the full consequences of such identification resulting in wrong evaluation, I might add most tedious *descriptions* of the interplay of situations, evaluations., in quarrels, unhappinesses, disagreements., leading to dramas and tragedies, as well as to many forms of 'mental' illness effectively described only in the *belles-lettres*. Thus, Smith$_1$, who is *not* conscious of abstracting, makes the statement, 'A circle is not square'. Let us suppose that Brown$_1$ contradicts him. Smith$_1$ is angered; for his *s.r*, his statement 'is' the 'plain truth', and Brown$_1$ must be a fool. He objectifies it, ascribes to it undue value. For him, it 'is' 'experience', a 'fact'., and he bursts into speech, denouncing Brown$_1$ and showing how wrong he 'is'. From this semantic attitude, many difficulties and tragedies arise.

But if Smith$_2$ (conscious of abstracting) makes the statement, 'A circle is not square', and Brown$_2$ contradicts him, what would Smith$_2$ do? He would smile, would not burst into speech to defend *his* statement, but would ask Brown$_2$, 'What do you mean? I do not quite understand you'. After receiving some answer, Smith$_2$ would explain to Brown$_2$ that his statement is not anything to quarrel about, as it is verbal and is true only by *definition*. He would also grant the right of Brown$_2$ *not* to accept *his definition,* but to use another one to satisfy himself. The problem would then, naturally, arise as to what definition both could accept, or which would be generally acceptable. And the problem would then be solved by purely pragmatic considerations. Words appear as creatures of definitions, and optional; but this attitude involves important and new *s.r.*

This fact seems of tremendous semantic importance, as it provides the working foundation for a theory of 'universal agreement'. In the first part of the above example, Smith$_1$, according to the accepted standards, was 'right' ('a circle is not square'). Is he 'more right' than Brown$_1$, for whom the 'circle is square'? Not at all. Both statements belong to the verbal level, and represent only forms of representation for *s.r inside their skin*. Either may be 'right' by some explicit or implicit 'definitions'. Are the two statements equally valid? This *we do not know a priori;* we must investigate to find out if the noises uttered have meanings outside of pathology, or which statement structurally covers the situation better, carries us structurally further in describing and analysing this world,. Only scientific structural analysis can give the preference to one form over another. Smith and Brown can only produce their 'definitions' according to their *s.r*, but they are *not* judges as to which 'definitions' will *ultimately* stand the test of structure.

The moment we eliminate identification we become conscious of abstracting, and permanently and instinctively remember that the object is *not* the event, that the label is *not* the object, and that a statement about a statement is *not* the first statement; thus, we reach a semantic state, where we recognize that everybody 'is right' by his own 'definitions'. But any individual or unenlightened public opinion is not the sole judge as to what 'definitions' and what language should prevail. Only structural investigation (science) can decide which appears as the structurally more similar form of representation on the verbal levels for what is going on on the un-speakable, objective levels.

When it comes to 'description of facts', the situation is not fundamentally altered. Mistakes seem always possible and often occur. Besides, the semantic impressions which 'facts' make on us are also individual, and often in conflict, as comparison of the testimonies of eye-witnesses shows. But there is no need for permanent disagreement; more structural investigation of the objective and verbal levels will provide a solution. Once such an investigation is carried far enough, we can always reach a semantic basis where all may agree, provided we do not identify, do not objectify, and do not confuse description and inference, descriptive and inferential words,.

As our analysis is carried out from the structural and *non-el* point of view, we should not miss the fact that semantic components associated with words and statements are, outside of very pathological cases, never entirely absent, and become of paramount importance. In the older days, we had no simple and effective means by which we could affect painful, misplaced, or disproportionate evaluations, meanings., through a semantic re-education, which are supplied by the present analysis and the use of the Structural Differential. The means to eliminate identification consist of: first, an *objective* relief diagram to which we can *point our finger;* and second, a convincing explanation (pointing the finger to the labels) that the verbal levels, with their distressing and disastrous older *s.r, are not,* and differ entirely from, the levels of objects and events. Whatever we may say or feel, the objects and events remain on the un-speakable levels and cannot be reached by words. Under such natural structural conditions, we can only reach the objective level by seeing, handling, actually feeling ., and, therefore, by pointing our finger to the object on the Structural Differential and being silent—all of which cannot be conveyed by words *alone*.

In experiments with the 'mentally' ill in whom the semantic disturbances were very strong, it took several months to train the patients in non-identity and in silence on the objective levels. But, as soon as this was achieved, either complete or partial relief followed.

The main disturbances in daily life, as well as in 'mental' illness, are found in the affective field. We find an internal pressure of identifications, expressed by bursting into speech, and unjustified semantic over-evaluation of words, the ascribing of objectivity to words ,. In such cases, suppression or repression of words does not accomplish much, but often does *considerable harm* and must be avoided by all means. Under such conditions, the use of the relief diagram becomes a necessity in pointing to the difference between different orders of abstractions and inducing the semantically beneficial silence on the 'objective level' without repression or suppression.

With the use of the Structural Differential, we can eliminate identification, and so attain the benefits, avoiding the dangers. If any one identifies, and his *s.r* drive him into an outburst of speech, we do not repress or suppress him; we say, instead: 'At your pleasure [since it makes him feel better], but remember that your words occur on the verbal levels [showing with a gesture of the hands the hanging labels], and that they *are not* the objective level, which remains untouched and unchanged'. Such a procedure, when repeated again and again, gives him the proper semantic *evaluation of orders of abstractions, frees him from identification, yet without repression or suppression.* It teaches him, also, to enquire into alleged 'facts', and then to try to find structurally better forms of representation. If such results are not forthcoming, we may use the older forms, but by proper evaluation we do not semantically put 'belief' in these forms of representation. Such beliefs always appear as the result of identification somewhere.

The technique of training is simple. We live on the 'objective' or lower order of abstraction levels, where we must see, feel, touch, *perform* ., but *never* speak. In training, we must use our hands ,. It is very useful, after the Structural Differential has been repeatedly explained, stressing, in particular, the rejection of the 'is' of identity, not to interrupt the other fellow. Let him speak, but wave the hand, indicating the verbal levels; then point the finger to the objective level, and, with the other hand, close your own lips, to show that on the objective level one can only be silent. When performed repeatedly, this pantomime has a most beneficial, semantic, pacifying effect upon the 'over-emotionalized' identification-conditions. The neurological mechanism of this action is not fully known, but some aspects are quite clear.

The more elaborate a nervous system becomes, the further some parts of the brain are removed from immediate experience. Nerve currents, having finite velocity, eventually have longer and more numerous paths to travel; different possibilities and complications arise, resulting in 'delayed action'. It is known that the thalamus (roughly) appears connected with affective and 'emotional' life, and that the cortex, farther removed and isolated from the external world, has the effect of inducing this 'delay in action'. In unbalanced and 'emotional' 'thinking', which is so prevalent, the thalamus seems overworked, the cortex seems not worked enough. The results take on the form of a low kind of animalistic, primitive, or infantile behaviour, often of a pathological character in a supposedly civilized adult. It appears that the 'silence on objective levels' introduces this 'delayed action', unloading the thalamic material on the cortex. This psychophysiological method is very simple, scientific, and entirely general. The standard 'mental' therapy of today applies also a *method of re-education* of s.r, as if relieving the thalamus, and putting more of the nerve currents through the cortex, or eventually furnishing the cortex with different material, so that the thalamic material returning from the cortex could be properly influenced.

If we succeed in such a semantic re-education, the difficulties vanish. The older experimental data show that in many instances we have succeeded, and that in many we have failed. The successful cases show that we actually know the essential semantic points involved; the failures show that we do not know enough, and that our older theories are not sufficiently general. At present, only the more pronounced and morbid semantic disturbances come to the attention of physicians, and very little is done by way of *preventive* measures. Besides the pronounced disturbances in daily life, we see an enormous number of semantic disturbances which we disregard, and call 'peculiarities'. In the majority of cases, these 'peculiarities' are undesirable, and, under unfavorable conditions, may lead to more serious consequences of a morbid character. They usually involve a great deal of unhappiness for all concerned, and unhappiness appears as a sign of some semantic maladjustment somewhere, and so may be destructive to 'mental' and nervous health.

In advanced 'mental' illnesses, such as usually come to the attention of psychiatrists, there are certain psycho-logical symptoms which are generally present. The symptoms of interest to us in this work are called 'delusions', 'illusions', and 'hallucinations'. All of them involve the semantic identification or *confusions of the orders of abstractions,* the evaluation of lower orders of abstraction as higher, or higher as lower. It was explained already that some components of identification are invariably present there, and so identification may be considered as an elementary type of semantic disturbances from which all the other states differ only in intensity.

The main point is to find psychophysiological preventive means whereby this identification can be forestalled or eliminated. To date, experience and analysis show that all forms of identification may be successfully eliminated by training in *visualization,* if this semantic state can be produced. For this purpose the Structural Differential is uniquely useful and necessary. With its help we train all centres. The lower centres are involved, as we see, feel, hear. ; the higher centres are equally involved, as we 'remember', 'understand'. ; *with the result that all centres work together without conflict.* The 'consciousness of abstracting' is inculcated, replacing vicious *s.r* of confusion of orders of abstractions and identification.

This harmonious working of all centres on their proper levels has extremely far-reaching, practical consequences in 'mental' and physical hygiene. We become co-ordinated, adjusted, and difficulties which might otherwise occur in the future are eliminated in a preventive way. It must be remembered that, at present, it is impossible to foresee to how great an extent the elimination of identification on all levels will have a beneficial effect. At this stage we know even experimentally that the benefits are very large, but we may expect that they will become still more numerous when more experimenting has been done. Delusions, illusions, and hallucinations represent manifestations which occur in practically all 'mental' difficulties, and they only represent a semantic identification of orders of abstractions of different degrees of intensity. When this confusion is eliminated, we may expect general changes in the symptoms. But as the correspondence is probably not *one-to-one,* it is impossible to foretell theoretically what improvements may be expected in pronounced illness. In the slighter disturbances, which affect us in daily life, the results are much easier to foresee, and are *always* beneficial.

To how great an extent the consciousness of abstracting benefits semantically *the whole organism,* I may illustrate by one of my own experiences. Once I was travelling on a ship. A gentleman visited my cabin, and, seeing the Structural Differential, asked questions about it. After a short explanation, he asked about practical applications.

My guest was sitting on my berth; I was sitting on a small folding chair. I got up, went to the door, then pretended that I was coming in, and, at my suggestion, he said, 'Please have a seat'. I remained standing while explaining how, if I were not 'conscious of abstracting', to me his word 'seat' would be identified with the chair (objectification) and my *s.r* would be such that I would sit down with great confidence. If the chair were to collapse I would have, besides the bump, an affective shock, 'fright'., which might do harm to my nervous system. But if I were conscious of abstracting, my *s.r* would be different. I would re-member that the *word*, the *label* 'seat' is *not* the thing on which I am supposed to sit. I would remember that I am to sit on this individual, unique, un-speakable object, which might be strong or weak,. Accord-ingly, I would sit carefully. In case the chair should collapse, and I should hurt myself physically, I would still have been saved an affective nervous shock.

During all these explanations I was handling the little chair and shaking it. I did not notice that the legs were falling out, and that the chair was becoming unfit for use. Then, when I actually sat on the relic, it gave way under me. However, I did *not fall* on the floor. I caught myself in the air, so to say, and saved myself from a painful experience. It is important to notice that such physical readiness requires a very elaborate, nervous, unconscious co-ordination, which was accomplished by the semantic state of *non-identification* or *consciousness of abstract-ing*. When such a consciousness of abstracting is acquired, it works instinctively and automatically and does *not* require continual effort. Its operation involves a fraction of a second's *delay in action,* but this small delay is not harmful in practice; on the contrary, it has very important psycho-logical and neurological 'delayed action' effects.

It seems that 'silence' on the objective levels involves this psycho-physiological delay. No matter how small, it serves to unload the thal-amic material on the cortex. In a number of clinical cases, Dr. Philip S. Graven has demonstrated that the moment such a delay can actually be produced in the patient, he either improves or is entirely relieved. The precise neurological mechanism of this process is not known, but there is no doubt that this 'delayed action' has many very beneficial effects upon the whole working of the nervous system. It somehow balances harmful *s.r,* and also somehow stimulates the higher nervous centres to more *physiological* control over the lower centres.

A very vital point in this connection should be noticed. That this 'delayed action' is beneficial is acknowledged by the majority of nor-mally developed adults in the form of delay in action and finds its expression in such statements as 'think twice', 'keep your head', 'hold your horses', 'keep cool', 'steady', 'wait a minute'., and such functional recipes as 'when angry, count ten',. In daily life, such wisdom is acquired either by painful experience, or is taught to children in an A language, which, as practice shows, is rarely effective because of its inadequacy. It is seldom realized that the mechanism of these functional observations and familiar advices have very powerful and workable underlying *neurological processes,* which can be *reached and directly affected by psychophysiological,* ordinal, *non-el* methods in connection with *the structure of the language we use.* Thus, under an infantile, A, and prevailing system we use and teach our children a language involving the 'is' of identity, and so we must confuse orders of abstractions, preparing for ourselves and the children the harmful semantic predispositions for 'bursting into speech', instead of 'wait a minute', which, neurologically, means abusing our thalamus and keeping our cortex 'unemployed'. In a $\bar{A}$ ∞-valued system we reject the 'is' of identity; we cannot confuse orders of abstractions; we cannot identify words with the un-speakable objective levels or inferences with descriptions, and we cannot identify the different abstractions of different individuals,. This semantic state of proper evaluation results in discrimination between the different orders of abstractions; an automatic delay is introduced—the cortex is switched completely into the nervous circuit. The semantic foundation is laid for 'higher mentality' and 'emotional balance'.

We have already had occasion to mention the mechanism of projection in connection with identification, as a semantic state of affective ascribing of lower centre characteristics to higher order abstractions and vice versa, and in connection with the introverted or extroverted attitudes. Likewise, we have already reached the conclusion that a well-adjusted and, therefore, well-balanced individual should be neither of the extremes, but a balanced extroverted introvert. By training with the Differential this important semantic result may be brought about. By training with the 'object' on its level, we become extroverted, and we learn to observe; this results in semantic freedom from 'preconceived ideas', such as we have when we start with the evaluation, label first and object next, instead of the natural order, object first and label next. By passing to higher order abstractions and evaluating the successive ranks of labels, we train in introversion. The result, as a whole, is that we may achieve the desirable and balanced semantic state of the extroverted introvert.

That in the training with the Differential we use all available nerve centres is beneficial, because the lower centres are in closer connection with the vegetative nervous system than are the others.

CHAPTER XXVII

HIGHER ORDER ABSTRACTIONS

The characters which science discerns in nature are subtle characters, not obvious at first sight. They are relations of relations and characters of characters. (573)   A. N. WHITEHEAD

In this connection one should particularly remember that the human language permits the construction of sentences which do not involve any consequences and which therefore have no content at all—in spite of the fact that these sentences produce some kind of picture in our imagination; e. g., the statement that besides our world there exists another world, with which any connection is impossible in principle, does not lead to any experimental consequence, but does produce a kind of picture in the mind. Obviously such a statement can neither be proved nor disproved. One should be especially careful in using the words "reality," "actually," etc., since these words very often lead to statements of the type just mentioned. (215)   W. HEISENBERG

Section A. General.

In the previous chapters I demonstrated that there is a short cut which enables us to grasp, acquire, and apply what has been advanced in the present work. This semantic short cut is 'consciousness of abstracting'. It is a psycho-logical attitude toward all our abstracting on all levels, and so involves the co-ordinated working of the organism-as-a-whole.

The use of the Structural Differential is necessary, because some levels are un-speakable. We can see them, handle them, feel them., but under *no* circumstances can we reach those levels by speech alone. We must, therefore, have a diagram, by preference in relief form, which represents the empirical structural conditions, and which indicates the un-speakable level by some other means than speech. We must, in the simplest case, either point our finger to the object, insisting upon silence, or must perform bodily some activity and similarly insist upon silence, as the performing and feelings are also *not* words.

In such semantic training it is enough to insist upon the non-identity or the difference between the objective, *un-speakable* levels of lower order abstractions, (O_h), and the verbal or higher order abstractions, (L_n). When this habit and feeling are acquired, no one should have difficulties in extending the non-identity method to daily-life occurrences. To achieve these semantic aims, we must first emphasize the common-sense fact that an object *is not* the event. To do this, we start with the 1933 scientific structural 'metaphysics' about the event and stress the fact that the object, being a nervous abstraction of lower order, has fewer and different *m.o* characteristics than the event has. This is best accomplished by stressing the fact that in abstracting from the event to the object we left out some characteristics. We did not *abstract 'all'* characteristics; this would be a self-contradiction in terms, an impossibility.

We do not even need to stress a full understanding of the event. Common-sense examples, showing that what we recognize as a 'pencil' is not 'all', often suffice. No one will have difficulties, provided he trains himself in this direction, ,in *remembering* continually and instinctively the free hanging strings (B'), (B"), which indicate the non-abstracted or left-out characteristics

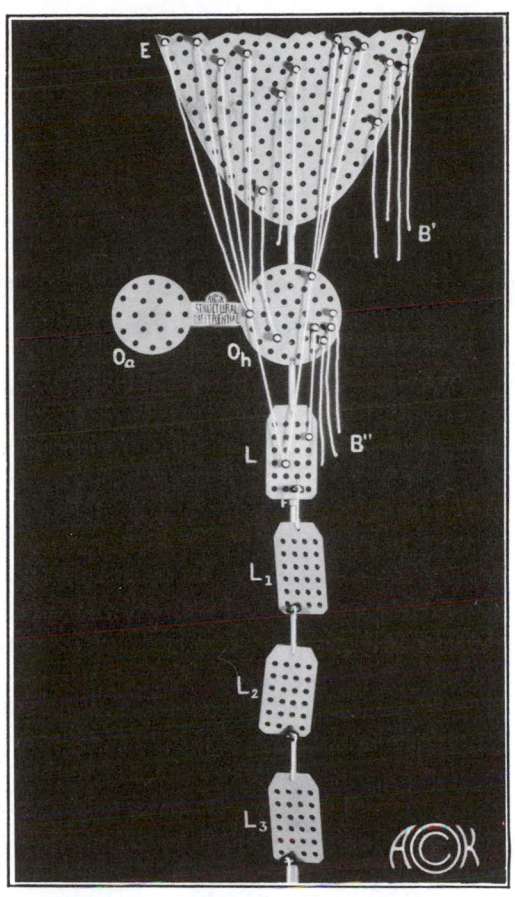

FIG. 1
THE STRUCTURAL DIFFERENTIAL

and which help to train in *non-identity*. With the relief diagram, the *s.r* of the student are trained *through all nervous centres*. He sees, he handles ., the hanging strings, and he also hears about them. This gives the *maximum probability* that the organism-as-a-whole will be affected. In this way an 'intellectual' theory engages the 'senses', feelings, and reflex mechanisms. To affect the organism-as-a-whole, organism-as-a-whole methods must be employed.

A similar structural situation is found when we deal with higher order abstractions. A word, or a name, or a statement is conveyed in spoken form or by writing, and affects first the lower centres and then is abstracted, and again transformed, by the higher centres. The order is generally not changed when the verbal issues are neither seen nor heard but originate in ourselves. Most 'impulses', 'interests', 'meanings', 'evaluations'., originate in lower centres and follow the usual course, from lower centres to higher. When 'experience' (reaction of lower centres) is transformed into 'memories' (higher centres)., the order is similar. Difficulties begin when the order is pathologically *reversed* and 'ideas' are evaluated as experience, words as objects,. In the building of language a similar process can be observed. We observe the absolute individuals with which we actually deal, we label them with individual names, say, $A_1, A_2, \ldots, A_{11}, A_{12}, \ldots, A_{21}, A_{22}, \ldots, A_{31}, A_{32}, \ldots,$. By a process of abstracting and *disregarding,* for instance, the characteristic subscripts '1', we would have only the ones which have the characteristic subscripts $2, 3, \ldots, 9, 22, 23, \ldots, 29,$. *Disregarding* the characteristic subscripts '2', we would have the ones with the characteristics $3, 4, \ldots, 9, 33, 34, \ldots, 39,$. Finally, if we should eliminate all *individual* characteristic subscripts, we would have a 'general' name A for the whole group without singling out individual characteristics.

All words of the type of 'man', 'animal', 'house', 'chair', 'pencil'., have been built by a similar process of abstraction, or *disregard* for individual differences. In each case of disregard of individual characteristics a *new* neurological process was involved.

Similarly, with 'statements about a statement'. When we hear a statement, or see it in a written form, such a statement becomes a stimulus entering through the lower centres, and a statement about it represents, in general, a new process of abstraction, or an abstraction of higher order.

It becomes obvious that the introduction of a language of 'different order of abstractions', although it is not familiar, yet structurally it represents very closely, in terms of *order,* most fundamental neurological processes going on in us. As we already know, a natural order has been established by evolution; namely, lower order abstractions first, higher next; the identifications of orders or the reversal of orders appears pathological for man and appears as a confusion of orders of abstractions, resulting in *false evaluation:* identification, illusions, delusions, and hallucinations.

Historically, the first to pay serious attention to the above problems in a consistent, yet very limited, way were mathematicians. In the investigation of the problems of the foundation of mathematics, mathematical 'logic', and the theory of aggregates, we came across self-contradictions which would make mathematics impossible. To avoid such a disaster, Russell invented what is called the 'theory of mathematical types'. The status of this theory is a very interesting and instructive one. The theory solves the mathematical difficulties, thus saving mathematics, but has *no* application to life. Practically all mathematicians, if I am not mistaken, the author of the theory included, somehow 'dislike' the theory and make efforts to solve the problems in a different way and possibly to abandon the theory altogether.

We have already shown that the introduction of a language of 'different orders of abstractions' is structurally entirely justified and physiologically natural, as it describes, in terms of order, the activities of the nervous system. Such facts are important; but if, in addition, the introduction of a language of a new $\overline{A}$ structure would give us further demonstrable advantages, then the introduction of such a language would become increasingly desirable.

Although the majority of mathematicians 'dislike' the theory of types, yet, at present, this theory is unconditionally necessary for non-self-contradictory mathematics. The author was pleasantly surprised to find that after his $\overline{A}$-system was formulated, this simple and natural, actional, functional, operational, *non-el* theory covers the theory of mathematical types and generalizes it, making the theory applicable not only to the solution of mathematical paradoxes but to the solution of the majority of purely human and scientific difficulties. One general rule of 'non-confusion of orders of abstractions', and the acquiring of the simple and workable 'consciousness of abstracting' based on the denial of the 'is' of identity, offers a *full* structural and semantic solution. The disregard of the issues involved leads fatalistically to the manufacture of endless and unnecessary human sufferings and unhappiness, the elimination of which is one of the main points in a theory of sanity. There is no mystery in 1933 that continuous small painful shocks may lead to serious semantic and physical disturbances. Psycho-logicians and psychiatrists will find it increasingly difficult to work at their problems if they disregard these semantic issues. Parents and teachers will find simple yet effective structural means for training the reactions of children *in sanity,* with all the ensuing semantic benefits to the individuals and to society.

When Whitehead and Russell were working at the foundations of mathematics, they came across endless paradoxes and self-contradictions, which, of course, would make mathematics impossible. After many efforts they found that all these paradoxes had one general source, in the rough, in the expressions which involve the word 'all', and the solution was found by introducing 'non-allness', a semantic forerunner of non-identity. Consider, for example, 'a proposition about *all* propositions'. They found that such totalities, or such 'all' statements, were not legitimate, as they involved a self-contradiction to start with. A proposition cannot be made legitimately about 'all' propositions without some restriction, since it would have to include the new proposition which is being made. If we consider a *m.o* term like 'propositions', which we can manufacture without known limits, and remember that any statement *about* propositions takes the form of a proposition, then obviously we cannot make statements about *all* propositions. In such a case the statement must be limited; such a set has *no total,* and a statement about 'all its members' cannot be made legitimately. Similarly, we cannot speak about *all* numbers.

Statements such as 'a proposition about *all* propositions' have been called by Russell 'illegitimate totalities'. In such cases, it is necessary to break up the set into smaller sets, each of which is capable of having a totality. This represents, in the main, what the theory of types aims to accomplish. In the language of the *Principia Mathematica,* the principle which enables us to avoid the illegitimate totalities may be expressed as follows: 'Whatever involves *all* of a collection must not be one of the collection', or, 'If, provided a certain collection had a total, it would have members only definable in terms of that total, then the said collection has no total'.[1] The above principle is called the 'vicious-circle principle', because it allows us to evade the vicious circles which the introduction of illegitimate totalities involve. Russell calls the arguments which involve the vicious-circle principle, 'vicious-circle fallacies'.

As an example, Russell gives the two-valued law of 'excluded third', formulated in the form that 'all propositions are true or false'. We involve a vicious-circle fallacy if we argue that the law of excluded third takes the form of a proposition, and, therefore, may be evaluated as true or false. Before we can make any statement about 'all propositions' legitimate, we must limit it in some way so that a statement about this totality must fall outside this totality.

Another example of a vicious-circle fallacy may be given as that of the imaginary sceptic who asserts that he knows nothing, but is refuted by the question—does he *know* that he *knows* nothing? Before the statement of the sceptic becomes significant, he must limit, somehow, the number of facts concerning which he asserts his ignorance, which represent an illegitimate totality. When such a limitation is imposed, and he asserts that he is ignorant of an extensional series of propositions, of which the proposition about his ignorance is not a member, then such scepticism cannot be refuted in the above way.

We do not need to enter into further details concerning the elaborate and difficult theory of types. In my $\bar{A}$ psychophysiological formulation, the theory becomes structurally extremely simple and natural, and applies to mathematics as well as to a very large number of daily experiences, eliminating an unbelievably large number of misunderstandings, vicious circles, and other semantic sources of human disagreements and unhappiness.

It should be noticed that, in the given examples, we always made a statement *about* another statement, and that the vicious circle arose from identifying or from the confusion of the orders of statements. The way out is found in the consciousness of abstracting, which leads to the semantic discrimination between *orders of abstractions*. If we have certain propositions, $p_1, p_2, p_3, \ldots p_n$, and make a new proposition about these propositions, say P, then, according to the present theory, the statement P about the statements $p_1, p_2.$, must be considered as an abstraction of higher order, and so different, and must not be identified as to order with the propositions $p_1, p_2, \ldots p_n$.

The above psychophysiological formulation is entirely general, yet simple and natural in a $\bar{A}$-system. To make this clearer, I shall take several statements concerning the theory of types from the *Principia Mathematica*, shall designate them by (Pr.), shall reformulate them in my language of *orders of abstractions*, and shall designate them as general semantics (G. S.).

Thus, 'The vicious circles in question arise from supposing that a collection of objects may contain members which can only be defined by means of the collection as a whole' (Pr.). Objects as individuals and 'collections of objects' obviously belong to different orders of abstractions and should not be confused (G.S.). A 'Proposition about *all* propositions' (Pr.). This involves a confusion of orders of abstractions, for if we posit propositions $p_1, p_2, \ldots p_n$, then a proposition P *about* these propositions represents a higher order abstraction and should not be identified with them (G.S.). 'More generally, given any set of objects such that, if we suppose the set to have a total, it will contain members which presuppose this total, then such a set cannot have a total. By saying that a set has "no total", we mean, primarily, that no significant statement can be made about "all its members"' (Pr.). A set of statements or objects or elements, or the like, and a statement *about* them belong to different orders of abstractions and should not be confused (G.S.). In the language of Wittgenstein: 'No proposition can say anything about itself, because the propositional sign cannot be contained in itself (that is the "whole theory of types").'[2]

In the language of the present general semantics a statement about a statement is not the 'same' statement, but represents, by structural and neurological necessity, a higher order of abstraction, and should not be confused with the original statement.

Similar reformulations apply to all cases given in the *Principia Mathematica,* and so it becomes evident that the present theory covers a similar ground as the theory of types, and also covers an endless list of daily-life applications which are of crucial semantic importance in a theory of sanity. We must stress here a simple, natural, and *single semantic law of non-identity which covers all confusions of orders of abstraction.* This one rule and training teach us not to confuse the higher orders with the lower, not to identify words with objects (not to objectify), as well as not to confuse higher abstractions of different orders. This generality and structural simplicity constitute an argument in favor of the present $\bar{A}$-system. It is easier to teach a single, simple, and natural rule which covers a vast field of semantic sources of human difficulties. For when the rule is explained, and the learner is trained with the Structural Differential, the semantic problem resolves itself simply into the showing with one's finger different orders of abstractions, and insisting that 'this *is not* this'.

If we consider the natural, structural, and *empirical* fact that our lives are lived in a world of non-identical abstractions of different orders, the discrimination between different orders becomes of paramount semantic importance for evaluation. Under such conditions we should become thoroughly acquainted with the mechanism of these different orders of abstractions. We should notice, first, that the language of the *Principia Mathematica* is A, and involves the 'is' of identity,. Such a language leads to identifications and to confusions, and makes simple issues difficult and perplexing. The term 'class' is very confusing. What do we mean by this term? In life we have, and deal with, *individuals* on *objective, un-speakable* levels. If we take a number of individuals, we have a number of them, yet they all remain individual. If we produce an abstraction of higher order, so that the individuality of each member is lost, then we have an abstraction of a higher order ('idea' in the old language), but no more the absolute individuals of our collection. The term 'class' in this respect is seriously confusing, as it tends to conceal a simple experimental fact, and leads to confusion of the orders of abstractions if the *multiordinality* of the term 'class' is not formulated.

Many critics and reviewers of the *Principia Mathematica* somehow feel this to be so, but their criticisms are not bold enough, and do not go to the roots of the A semantic difficulty. They do not pay attention to the A, 'logical', 'philosophical', and 'psychological' *elementalistic* method and language involving the 'is' of identity, in which the Introduction of the *Principia* is written. Doctor Alonzo Church is the first, as far as my knowledge goes, to suggest that, following Peano, numbers should be defined in the language of abstractions. He does not carry his analysis further, however, and does not state that it involves a language of entirely different $\bar{A}$ structure.[3] If we abandon the term 'class' and accept the language of 'abstractions of *different* orders', then we are led to the rejection of the 'is' of identity and to the present system, of which the theory of mathematical types becomes a necessary part. The problems of 'class' cease to be an 'assumption', as the different orders of abstractions are descriptions of experimental facts; and so the 'axiom of reducibility' becomes unnecessary. In my language, this axiom is also an *aristotelian description* of the experimental fact that we can abstract in different orders.

Section B. Multiordinal terms.

In the examples given in Section A, we used words such as 'proposition', which were applied to all higher order abstractions. We have already seen that such terms may have different uses or meanings if applied to different orders of abstractions. Thus originates what I call the *multiordinality* of terms. The words 'yes', 'no', 'true', 'false', 'function', 'property', 'relation', 'number', 'difference', 'name', 'definition', 'abstraction', 'proposition', 'fact', 'reality', 'structure', 'characteristic', 'problem', 'to know', 'to think', 'to speak', 'to hate', 'to love', 'to doubt', 'cause', 'effect', 'meaning', 'evaluation', and an endless array of the most important terms we have, must be considered as *multiordinal terms*. There is a most important semantic characteristic of these *m.o* terms; namely, that they are ambiguous, or ∞-valued, in general, and that each has a definite meaning, or one value, only and exclusively in a given context, when the order of abstraction can be definitely indicated.

These issues appear extremely simple and general, a part and parcel of the structure of 'human knowledge' and of our language. We cannot avoid these semantic issues, and, therefore, the only way left is to face them explicitly. The test for the multiordinality of a term is simple. Let us make any statement and see if a given term applies to it ('true', 'false', 'yes', 'no', 'fact', 'reality', 'to think', 'to love',.). If it does, let us deliberately make another statement *about* the former statement and test if the given term may be used again. If so, it is a safe assertion that this term should be considered as *m.o.* Any one can test such a *m.o*

28

term by himself without any difficulty. The main point about all such *m.o* terms is that, *in general,* they are *ambiguous,* and that all arguments about them, 'in general', lead only to *identification of orders of abstractions and semantic disturbances, and nowhere else.* Multiordinal terms have only definite meanings on a given level and in a given context. Before we can argue about them, we must fix their orders, whereupon the issues become simple and lead to agreement. As to 'orders of abstraction', we have no possibility of ascertaining the 'absolute' order of an abstraction; besides, we *never* need it. In human semantic difficulties, in science, as well as in private life, usually no more than three, perhaps even two, neighbouring levels require consideration. When it comes to a serious discussion of some problem, errors, ambiguity, confusion, and disagreement follow from confusing or identifying the neighbouring levels. In practice, it becomes *extremely simple* to settle these three (or two) levels and to keep them separated, *provided we are conscious of abstracting, but not otherwise.*

For a theory of sanity, these issues seem important and structurally essential. In identifications, delusions, illusions, and hallucinations, we have found a *confusion* between the orders of abstractions or a false evaluation expressed as a reversal of the natural order.

One of the symptoms of this confusion manifests itself as 'false beliefs', which again imply comparison of statements about 'facts' and 'reality', and involve such terms as 'yes', 'no', 'true', 'false',. As all these terms are multiordinal, and, therefore, ambiguous, 'general' 'philosophical' rigmaroles should be avoided. With the consciousness of abstracting, and, therefore, with a *feel* for this peculiar stratification of 'human knowledge', all semantic problems involved can be settled simply.

The avoidance of *m.o* terms is impossible and undesirable. Systematic ambiguity of the most important terms follows systematic analogy. They appear as a direct result and condition of our powers of abstracting in different orders, and allow us to apply one chain of ∞-valued reasoning to an endless array of different one-valued facts, all of which are different and become manageable only through our abstracting powers.

For further details about the theory of types, the reader is referred to the literature on the subject and Supplement II[4]; here I shall give only a few examples of the complexities and difficulties inherent in language, and show how simply they become solved by the aid of $\bar{A}$ general semantics and the resulting 'consciousness of abstracting'.

As an example, I quote Russell's analysis of the 'simple' statement 'I am lying', as given in the *Principia*. 'The oldest contradiction of the kind in question is the *Epimenides*. Epimenides the Cretan said that all Cretans were liars, and all other statements made by Cretans were certainly lies. Was this a lie? The simplest form of this contradiction is afforded by the man who says "I am lying"; if he is lying, he is speaking the truth, and vice versa. . . .

'When a man says "I am lying", we may interpret his statement as: "There is a proposition which I am affirming and which is false." That is to say, he is asserting the truth of some value of the function "I assert p, and p is false." But we saw that the word "false" is ambiguous, and that, in order to make it unambiguous, we must specify the order of falsehood, or, what comes to the same thing, the order of the proposition to which falsehood is ascribed. We saw also that, if p is a proposition of the nth order, a proposition in which p occurs as an apparent variable is not of the nth order, but of a higher order. Hence the kind of truth or falsehood which can belong to the statement "there is a proposition p which I am affirming and which has falsehood of the nth order" is truth or falsehood of a higher order than the nth. Hence the statement of Epimenides does not fall within its own scope, and therefore no contradiction emerges.

'If we regard the statement "I am lying" as a compact way of simultaneously making all the following statements: "I am asserting a false proposition of the first order," "I am asserting a false proposition of the second order," and so on, we find the following curious state of things: As no proposition of the first order is being asserted, the statement "I am asserting a false proposition of the first order" is false. This statement is of the second order, hence the statement "I am making a false statement of the second order" is true. This is a statement of the third order, and is the only statement of the third order which is being made. Hence the statement "I am making a false statement of the third order" is false. Thus we see that the statement "I am making a false statement of order $2n + 1$" is false, while the statement "I am making a false statement of order $2n$" is true. But in this state of things there is no contradiction.'[5]

Clearly, if we should apply the language of orders of abstractions to the above case, a similar outcome is reached more generally and more simply. If we should confuse the orders of abstractions, we might naturally have an endless argument at hand. This example shows how a confusion of orders of abstractions might lead to insoluble verbal problems, and how semantically important it is that we should not identify, and that we should be conscious of abstracting, with the resulting instinctive feeling for this peculiar structural stratification of 'human knowledge'. We should notice that with the confusion of orders of abstractions, and by the use of *m.o* terms, *without realizing their ∞-valued* character, we may always construct an endless array of such verbal arguments to befog the issues, but that as soon as we assign a definite order to the *m.o* terms, and so settle a specific single meaning in a given context for the many meanings any *m.o* term may have, the difficulties vanish.

As the above analysis applies to all *m.o* terms, and these terms happen to be most important in our lives, there is no use in trying to avoid these terms and the consequences of using them. Quite the contrary; often it is structurally necessary to build a *m.o* term—for instance, 'abstracting'—we must take for granted that it has many meanings, and indicate these meanings by assigning to the term the definite order of abstraction. Thus, such a term as 'abstracting' or 'characteristic'., might be confusing and troublesome; but 'abstracting in different orders'., is not, as in a given context we may always assign the definite order and single meaning to the term.

It has been repeatedly said that a *m.o* term has, by structural necessity, many meanings. No matter how we define it, its definition is again based on other *m.o* terms. If we try to give a *general* 'meaning' to a *m.o* term, which it cannot have, further and deeper analysis would disclose the multiordinality of the terms by which it is defined, restoring once more its multiordinality. As there is no possibility of avoiding the above structural issue, it is more correct and also more expedient to recognize at once the fundamental multiordinality of a term. If we do so, we shall not get confused as to the meaning of such a term in a given context, because, in principle, in a context its meaning is single and fixed by that context.

The semantic benefits of such a recognition of multiordinality are, in the main, sevenfold: (1) we gain an enormous economy of 'time' and effort, as we stop 'the hunting of the snark', usually called 'philosophy', or for a one-valued general definition of a *m.o* term, which would not be formulated in other *m.o* terms; (2) we acquire great versatility in expression, as our most important vocabulary consists of *m.o* terms, which can be extended indefinitely by assigning many different orders and, therefore, meanings; (3) we recognize that a definition of a *m.o* term must, by necessity, represent not a proposition but a propositional function involving variables; (4) we do not need to bother much about formal definitions of a *m.o* term outside of mathematics, but may use the term freely, realizing that its unique, in principle, meaning in a given context is structurally indicated by the context; (5) under such structural conditions, the freedom of the writer or speaker becomes very much accentuated; his vocabulary consists potentially of infinite numbers of words, and psycho-logical, semantic blockages are eliminated; (6) he knows that a reader who understands that ∞-valued mechanism will never be confused as to the meaning intended; and (7) the whole linguistic process becomes extremely flexible, yet it preserves its essential extensional one-valued character, in a given case.

In a certain sense, such a use of $m.o$ terms is to be found in poetry, and it is well known that many scientists, particularly the creative ones, like poetry. Moreover, poetry often conveys in a few sentences more of lasting values than a whole volume of scientific analysis. The free use of $m.o$ terms without the bother of a structurally impossible formalism outside of mathematics accomplishes this, *provided we are conscious of abstracting; otherwise only confusion results.*

It should be understood that I have no intention of condemning formalism. Formalism of the most rigorous character is an extremely important and valuable discipline (mathematics at present); but formalism, as such, in experimental science and life appears often as a handicap and not as a benefit, because, in empirical science and life, we are engaged in exploring and discovering the unknown structure of the world as a means for structural adjustment. The formal elaboration of some language is only the consistent elaboration of its structure, which must be accomplished independently if we are to have means to compare verbal with empirical structures. From a $\bar{A}$ point of view, both issues are equally important in the search for structure.

Under such structural empirical conditions the $m.o$ terms acquire great semantic importance, and perhaps, without them, language, mathematics, and science would be impossible. As soon as we understand this, we are forced to realize the profound structural and semantic difference between the A and $\bar{A}$ systems. What in the old days were considered propositions, become propositional functions, and most of our doctrines become the doctrinal functions of Keyser, or system-functions, allowing multiple interpretations.

Terms belong to verbal levels and their meanings *must* be given by definitions, these definitions depending on undefined terms, which consist always, as far as my knowledge goes, of $m.o$ terms. Perhaps it is necessary for them to have this character, to be useful at all. When these structural empirical conditions are taken into account, we must conclude that the postulational method which gives the structure of a given doctrine lies at the foundation of all human linguistic performances, in daily life as well as in mathematics and science. The study of these problems throws a most important light on all mysteries of language, and on the proper use of this most important human neurological and semantic function, without which sanity is impossible.

From a structural point of view, postulates or definitions or assumptions must be considered as those relational or multi-dimensional order structural assumptions which establish, conjointly with the undefined terms, the structure of a given language. Obviously, to find the structure of a language we must work out the given language to a system of postulates and find the minimum of its (never unique) undefined terms. This done, we should have the structure of such a system fully disclosed; and, with the structure of the language thoroughly known, we should have a most valuable tool for investigating empirical structure by predicting verbally, and then verifying empirically.

To pacify the non-specialist, let me say at once that this work is very tedious and difficult, although a crying need; nevertheless, it may be accomplished by a single individual. Because of the character of the problem, however, when this work is done, the semantic results have always proved thus far—and probably will continue so—quite simple and comprehensible to the common sense, even of a child.

One very important point should be noted. Since language was first used by the human race, the structural and related semantic conditions disclosed by the present analysis *have not been changed,* as they are inherent in the structure of 'human knowledge' and language. Historically, we were always most interested in the immediacy of our daily lives. We began with grunts symbolizing this immediacy, and we never realize, even now, that these historically first grunts were the most complex and difficult of them all. Besides these grunts, we have also developed others, which we call mathematics, dealing with, and elaborating, a language of numbers, or (as I define it semantically) a language of *two symmetrical and infinitely many asymmetrical unique, specific relations* for exploring the structure of the world, which is, at present, the most effective and the simplest language yet formed. Only in 1933, after many hundreds of thousands of years, have the last mentioned grunts become sufficiently elaborate to give us a sidelight on structure. We must revise the whole linguistic procedure and structure, and gain the means by which to disclose the structure of 'human knowledge'. Such semantic means will provide for the proper handling of our neurological structure, which, in turn, is the foundation for the structurally proper use of the human nervous system, and will lead to human nervous adjustment, appropriate *s.r*, and, therefore, to sanity.

Human beings are quite accustomed to the fact that words have different meanings, and by making use of this fact have produced some rather detrimental speculations, but, to the best of my knowledge, the structural discovery of the multiordinality of terms and of the psycho-physiological importance of the treatment of orders of abstractions resulting from the rejection of the 'is' of identity—as formulated in the present system—is novel. In this mechanism of multiordinality, we shall find an unusually important structural problem of human psycho-logics, responsible for a great many fundamental, desirable, undesirable, and even morbid, human characteristics. The full mastery of this mechanism is only possible when it is formulated, and leads automatically to a possibility of a complete psychophysiological adjustment. This adjustment often reverses the psycho-logical process prevailing at a given date; and this is the foundation, among others, of what we call 'culture' and 'sublimation' in psychiatry.

Let me recall that one of the most fundamental functional differences between animal and man consists in the fact that no matter in how many orders the animal may abstract, its abstractions stop on some level beyond which the animal cannot proceed. Not so with man. Structurally and potentially, man can abstract in indefinitely many orders, and no one can say legitimately that he has reached the 'final' order of abstractions beyond which no one can go. In the older days, when this semantic mechanism was not made structurally obvious, the majority of us copied animals, and stopped abstracting on some level, as if this were the 'final' level. In our semantic training in language and the 'is' of identity given to us by our parents or teachers or in school, the multiordinality of terms was never suspected, and, although the human physiological mechanism was operating all the while, we used it on the conscious level in the animalistic way, which means ceasing to abstract at some level. Instead of being told of the mechanism, and of being trained consciously in the fluid and dynamic *s.r* of *passing to higher and higher abstractions as normal,* for Smith, we preserved a sub-normal, animalistic semantic blockage, and 'emotionally' stopped abstracting on some level.

Thus, for instance, if, as a result of life, we come to a psycho-logical state of hate or doubt, and stop at that level, then, as we know from experience, the lives of the given individual and of those close to him are not so happy. But a hate or doubt of a higher order reverses or annuls the first order semantic effect. Thus, hate of hate, or doubt of doubt—a second order effect—has reversed or annulled the first order effect, which was detrimental to all concerned because it remained a *structurally-stopped or an animalistic* first order effect.

The whole subject of our human capacity for higher abstracting without discernible limits appears extremely broad, novel, and unanalysed. It will take many years and volumes to work it out; so, of necessity, the examples given below will be only suggestive and will serve to illustrate roughly the enormous power of the A methods and structure, aiming to make them workable as an educational, powerful, semantic device.

Let us take some terms which may be considered as of a positive character and represent the structure of 'culture', science, and what is known in psychiatry as 'sublimation'; such as curiosity, attention, analysis, reasoning, choice, consideration, knowing, evaluation,. The first order effects are well known, and we do not need to analyse them. But if we transform them into second order effects, we then have curiosity of curiosity, attention of attention, analysis of analysis, reasoning about reasoning (which represents science, psycho-logics, epistemology.,); choice of choice (which represents freedom, lack of psycho-logical blockages, and shows, also, the semantic mechanism of eliminating those blocks); consideration of consideration gives an important cultural achievement; knowing of knowing involves abstracting and structure, becomes 'consciousness', at least in its limited aspect, taken as consciousness of abstracting; evaluation of evaluation becomes a theory of sanity,.

Another group represents morbid semantic reactions. Thus the first order worry, nervousness, fear, pity., may be quite legitimate and comparatively harmless. But when these are of a higher order and identified with the first order as in worry about worry, fear of fear., they become morbid. Pity of pity is dangerously near to self-pity. Second order effects, such as belief in belief, makes fanaticism. To know that we know, to have conviction of conviction, ignorance of ignorance., shows the mechanism of dogmatism; while such effects as free will of free will, or cause of cause., often become delusions and illusions.

A third group is represented by such first order effects as inhibition, hate, doubt, contempt, disgust, anger, and similar semantic states; the *second order reverses and annuls* the first order effects. Thus an inhibition of an inhibition becomes a positive excitation or release (see Part VI); hate of hate is close to 'love'; doubt of doubt becomes scientific criticism and imparts the scientific tendency; the others obviously reverse or annul the first order undesirable *s.r.*

In this connection the pernicious effect of identification becomes quite obvious. In the first and third cases beneficial effects were *prevented,* because identification of orders of abstractions, as a semantic state, produced a semantic blockage which did not allow us to pass to higher order abstractions; in the second case, it actually produced morbid manifestations.

The consciousness of abstracting, which involves, among others, the full instinctive semantic realization of non-identity and the stratification of human knowledge, and so the multiordinality of the most important terms we use, solves these weighty and complex problems because it gives us structural methods for semantic evaluation, for orientation, and for handling them. By passing to higher orders these states which involve inhibition or negative excitation become reversed. Some of them on higher levels become culturally important; and some of them become morbid. Now consciousness of abstracting in all cases gives us the semantic *freedom* of all levels and so helps *evaluation* and selection, thus removing the possibility of remaining animalistically fixed or blocked on any one level. Here we find the mechanism of the 'change of human nature' and an assistance for persons in morbid states to revise by themselves their own afflictions by the simple realization that the symptoms are due to identifying levels which are essentially different, an unconscious jumping of a level or of otherwise confusing the orders of abstractions. Even at present all psychotherapy is unconsciously using this mechanism, although, as far as I know, it has never before been structurally formulated in a general way.

It should be added that the moment we eliminate identification and acquire the consciousness of abstracting, as explained in the present system, we have already acquired the permanent semantic feeling of this peculiar *structural stratification* of human knowledge which is found in the psycho-logics of the differential and integral calculus and mathematics, similar in structure to the world around us, without any difficult mathematical technique. Psycho-logically, both mathematics and the present system appear structurally similar, not only to themselves, but also to the world and our nervous system; and at this point it departs very widely from the older systems.

Let me give another example of how the recognition of order of abstractions clears up semantic difficulties.

I recall vividly an argument I had with a young and very gifted mathematician. Our conversation was about the geometries of Euclid and Lobatchevski, and we were discussing the *dropping* and *introduction* of assumptions. I maintained that Lobatchevski *introduced* an assumption; he maintained that Lobatchevski *dropped* an assumption. On the surface, it might have appeared that this is a problem of 'fact' and not of *preference*. The famous fifth postulate of Euclid reads, 'If a straight line falling on two straight lines makes the interior angles on the same side less than two right angles, the two straight lines, if produced indefinitely, meet on that side on which are the angles less than two right angles'. We should note, in passing, that a straight line is *assumed* to be of 'infinite' length, which involves a definite type of structural metaphysics of 'space', common to the A and older systems. This postulate of Euclid can be expressed in one of its equivalent forms, as, for instance, 'Through a point outside a straight line one, and only one, parallel to it can be drawn'. Lobatchevski and others decided to build up a geometry *without* this postulate, and in this they were successful. Let us consider what Lobatchevski did. For this, we go to a deeper level—otherwise, to a higher order abstraction—where we discover that what on *his level* had been the *dropping* of an assumption becomes on our deeper level or higher order abstraction the *introduction* of an assumption; namely, the assumption that through a point outside a straight line there passes *more* than one parallel line.

Now such a process is *structurally inherent in all human knowledge*. More than this, it is a unique characteristic of the structure of human knowledge. We can always do this. If we pass to higher orders of abstractions, situations seemingly 'insoluble', 'matters of fact', quite often become problems of *preference*. This problem is of extreme semantic importance, and of indefinitely extended consequences for all science, psychiatry, and education in particular.

The examples I have given show a most astonishing semantic situation; namely, that one question can sometimes be answered 'yes' or 'no', 'true' or 'false', depending on the order of abstractions the answerer is considering. The above facts alter considerably the former supposedly sharply defined fields of 'yes' and 'no', 'true' and 'false', and, in general, of all multiordinal terms. Many problems of 'fact' on one level of abstraction become problems of 'preference' on another, thereby helping to diminish the semantic field of disagreement.

It is interesting to throw some light on the problem of 'preference'. Which statement or attitude is preferable? The one claiming that Lobatchevski *dropped* a postulate, or the one claiming that Lobatchevski *introduced* a new postulate? Both are 'facts', but on different levels, or of different orders. The *dropping* appears as an historical fact; the *introducing* as a psycho-logical fact *inherent* in the structure of human knowledge. The preference is fairly indicated; the psycho-logical fact is of the utmost generality (as all psycho-logical facts are) and, therefore, more useful, since it applies to all human endeavours and not merely to what a certain mathematician did under certain circumstances.

Section C. Confusion of higher orders of abstractions.

We have already seen that Fido's power of abstracting stops some-where. If we are finalists of any kind, we also assume that *our* power of abstracting stops somewhere. In some such way the finalistic, dog-matic and absolutistic semantic attitudes are built.

If, however, by the aid of the Structural Differential we train the *s.r* of our children in $\overline{A}$ non-identity and the inherent stratification of human knowledge and power of abstracting, we *facilitate* the passing to higher order abstractions and establish *flexible s.r* of *full conditionality* which are unique for Smith and of great preventive and therapeutic value. We thus build up 'human mind' for efficiency and sanity, by eliminating the factors of semantic blockages, while, by engaging the activity of the higher nerve centres, we diminish the vicious overflow of nervous energy upon the lower nerve centres, which, if allowed, must, of necessity, make itself manifest in arrested or regressive symptoms.

The above issues are of serious semantic importance in our daily lives and in sanity. All semantic disturbances involve evaluation, doc-trines, creeds, speculations., and vice versa. Under circumstances such as described above, which appear inherent with us, it is dangerous not to have means to see one's way clear in the maze of verbal difficulties with all their dangerous and ever-present semantic components.

By disregarding the orders of abstractions, we can manufacture any kind of verbal difficulties; and, without the consciousness of abstracting, we all become nearly helpless and hopeless semantic victims of a primi-tive-made language and its underlying structural metaphysics. Yet the way out is simple; non-identity leads to 'consciousness of abstracting' and gives us a new working sense for *values,* new *s.r,* to guide us in the verbal labyrinth.

Outside of 'objectification', which is defined as the evaluation of higher order abstractions as lower; namely, words, memories., as objects, experiences, feelings., the most usual identification of different *higher order* abstractions appears as the confusion of inferences and inferential terms with descriptions and descriptive terms.

Obviously, if we consider a description as of the *n*th order, then an inference from such a description (or others) should be considered as an abstraction of a higher order $(n+1)$. Before we make a deci-sion, we usually make a more or less hasty survey of happenings, this survey establishing a foundation for our judgements, which become the basis of our action. This statement is fairly general, as the components of it can be found by analysis practically everywhere. Our problem is to analyse the general case. Let us follow up roughly the process.

We assume, for instance, an hypothetical case of an ideal observer who observes correctly and gives an impersonal, unbiased account of what he has observed. Let us assume that the happenings he has observed appeared as: ●, ◆, ♭, ♮, . . . , and then a new happening ⚵ occurred. At this level of *observation,* no speaking can be done, and, therefore, I use various fanciful symbols, and not words. The observer then gives a *description* of the above happenings, let us say $a, b, c, d, \ldots, x$; then he makes an inference from these descriptions and reaches a conclusion or forms a judgement A about these facts. We assume that facts unknown to him, which always exist, are not important in this case. Let us assume, also, that his conclusion seems correct and that the action A'' which this conclusion motivates is appropriate. Obviously, we deal with at least three different levels of abstractions: the seen, experienced . , lower order abstractions (un-speakable) ; then the descriptive level, and, finally, the inferential levels.

Let us assume now another individual, Smith$_1$, ignorant of structure or the orders of abstractions, of consciousness of abstracting, of *s.r.*; a politician or a preacher, let us say, a person who habitually identifies, confuses his orders, uses inferential language for descriptions, and rather makes a business out of it. Let us assume that Smith$_1$ observes the 'same happenings'. He would witness the happenings ●, ◆, ♭, ♮, . . . , and the happening ⚵ would appear new to him. The happenings ●, ◆, ♭, ♮, . . . , he would describe in the form $a, b, c, d, \ldots$, from which fewer descriptions he *would form a judgement, reach a conclusion, B*; which means that he would pass to another order of abstractions. When the new happening ⚵ occurs, he handles it with an already formed opinion B, and so his description of the happening ⚵ is *coloured* by his older *s.r* and no longer the x of the ideal observer, but $B(x) = y$. His description of 'facts' would *not* appear as the $a, b, c, d, \ldots, x$, of the ideal observer but $a, b, c, d, \ldots, B(x) = y$. Next he would abstract on a higher level, form a new judgement, about 'facts' $a, b, c, d, \ldots,$ $B(x) = y$, let us say, C. We see how the semantic error was produced. The happenings appeared the 'same', yet the unconscious identification of levels brought finally an entirely different conclusion to motivate a quite different action, C''.

A diagram will make this structurally clearer, as it is very difficult to explain this by words alone. On the Structural Differential it is shown without difficulty.

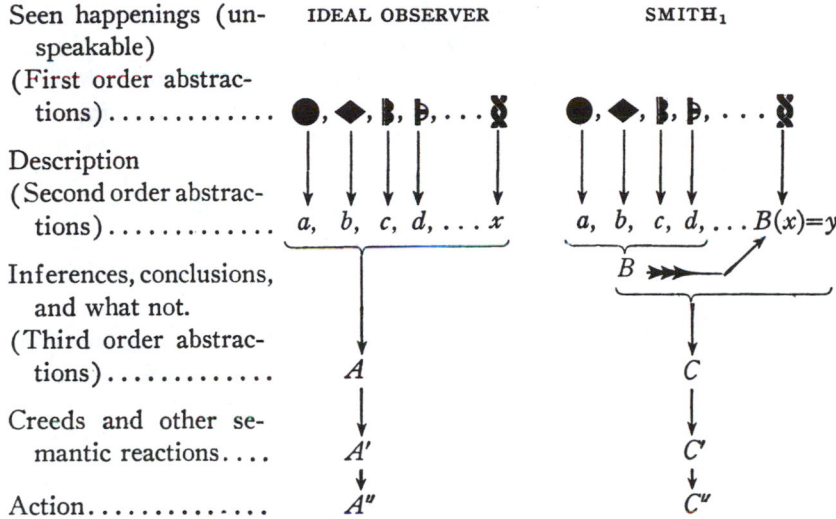

Let us illustrate the foregoing with two clinical examples. In one case, a young boy persistently did not get up in the morning. In another case, a boy persistently took money from his mother's pocketbook. In both cases, the actions were undesirable. In both cases, the parents unconsciously identified the levels, x was identified with $B(x)$, and confused their orders of abstractions. In the first case, they *concluded* that the boy was *lazy;* in the second, that the boy was a *thief.* The parents, through semantic identification, read these inferences into every new 'description' of forthcoming facts, so that the parents' new 'facts' became more and more semantically distorted and coloured in evaluation, and their actions more and more detrimental to all concerned. The general conditions in both families became continually worse, until the reading of inferences into descriptions by the ignorant parents produced a semantic background in the boys of driving them to murderous intents.

A psychiatrist dealt with the problem as shown in the diagram of the ideal observer. The net result was that the one boy was not 'lazy', nor the other a 'thief', but that both were ill. After medical attention, of which the first step was to clarify the symbolic semantic situation, though not in such a general way as given here, all went smoothly. Two families were saved from crime and wreck.

I may give another example out of a long list which it is unnecessary for our purpose to analyse, because as soon as the 'consciousness of abstracting' is acquired, the avoidance of these inherent semantic difficulties becomes automatic. In a common fallacy of *'Petitio Principii',* or 'Begging the Question' fallacy, we, by self-deceptive semantic evaluation, *assume the conclusion* to be proved. In other words, we confuse the orders of abstractions. Beside the wilful use of this fallacy by lawyers in courts to influence juries of low intelligence., a similar fallacy is widely committed in the reasonings of daily life and leads to many unnecessary semantic difficulties. Particularly vicious is the use of the so-called 'question-begging epithets'. We postulate the fact which we wish to prove, label it by another name, and then use the new higher order name in our premise. It represents clearly a confusion of orders of abstractions.

All such terms as 'un-patriotic', 'un-christian', 'un-american', 'pro-german' (during the World War), 'wet', 'dry'., fall into this group. It is probably no secret that a large part of the population of this world was swayed by such methods during the war. In times of peace, large countries are continually swayed by such use of terms which play upon the pathological *s.r* of the population, thereby facilitating the 'putting over' of different propagandas. Similar procedures lead to many semantic difficulties in daily life. It is easy to see that the difficulty is general; namely, 'the confusion of orders of abstractions'. The antidote is equally general, and is found in the elimination of the 'is' of identity, resulting in the 'consciousness of abstracting'. It should be noticed that these pathological reactions have long been known, and that they are extremely general. We are told about them in schools under the name of 'logical fallacies', disregarding their semantic character, and so it is practically impossible to eliminate them or to apply the wisdom we are taught. It is not difficult to see why this should occur. In the older days, all the 'wisdom' was taught to us by purely 'intellectual', 'verbal', classical *A* and *el* methods. We had no simple psycho*physiological* method of *complete* generality, which could be taught in a *non-el* way affecting *all* nerve centres. It is known how difficult it is to 'change human nature', which simply means that the older verbal educational methods could not properly affect the lower centres. It seems that the first step in developing a method to accomplish these ends is to use the Structural Differential, without which it is practically impossible to teach 'silence on the objective level' and 'delayed action' and to train through *all* centres in non-identity, 'stratification', natural order, and so in appropriate *s.r*. It appears that now, to begin with, we have acquired a workable and simple psychophysiological method for changing identification into visualization, and, in general, for the prevention or elimination of identification or confusion of orders of abstractions. We have now discovered a mechanism which involves and deals directly with the reactions of the lower centres, 'senses', affects, 'emotions',. The older, difficult 'change in human nature' becomes an easily accomplished fact in a structural, $\bar{A}$ semantic education. 'Human nature' can best be described, perhaps, as a complex of *s.r*, which *can* be educated and 'changed' to a large extent.

It seems unnecessary to enlarge further upon this subject. Every attentive reader can supply endless examples of this kind of semantic disturbances from his own observation or experience. Naturally, the generality, simplicity, and *physiological* character of the method proposed in this work become powerful assets, and instruction in the $\bar{A}$ methods can easily be given to, or acquired by, everybody. It can be taught in homes and schools. It gives a preventive psychophysiological method of training the *s.r* in the millions and millions of cases in which human life becomes wrecked through the lack of a *working structural educational theory* concerned with these reactions. But it is not enough to preach these 'platitudes'; they must be practised as well. If the parents and the boys mentioned above had been trained as children with the Structural Differential, it would have been an impossibility for the situation to have become so acute.

Let us follow our *daily experiences* by the aid of the Structural Differential. We find ourselves on at least five levels. The first represents the un-speakable event, or the scientific object, or the unseen physico-chemical processes on the sub-microscopic levels which constitute stimuli registered by our nervous system as objects. The second consists of the external, objective, also un-speakable, levels on which we see with our eyes,. On this level, we could make a moving picture, including actions., (writing a book is also behaviour). The third level represents the equally un-speakable psycho-logical 'pictures' and *s.r*. On the fourth level of abstractions we describe verbally our facts, that humans (*a*) eat, sleep.; (*b*) cheat, murder.; (*c*) moralize, philosophize, legislate.; (*d*) scientize, mathematize,. Finally, in the present context, our inferences belong to the fifth level.

Unfortunately, we usually abstract facts (*a*), identify the levels, and form a conclusion 'man is an animal',. From this *conclusion* we confuse the levels again and colour the description of the facts (*b*), (*c*), (*d*).; jump again to higher levels and build conclusions from descriptions (*a*) and from *distorted,* coloured descriptions (*b*), (*c*), (*d*), and so obtain the prevailing doctrines in all fields. These again lead us, in the field of action, to the mess we all find ourselves in. In this dervish dance between the levels we entirely *disregarded uncoloured facts* (*d*).

The ideal observer would observe *all* forms of human behaviour at a given date, *not leaving out facts* (*d*); then, without confusing his levels, and also without confusing descriptions with inferences, he would reach his higher order of abstractions properly, with very different resultant doctrines, which would produce entirely different semantic evaluation, and motivate equally different action.

We may understand now why we must constantly revise our doctrines, for the above analysis throws a considerable light on the fact that scientists need training with the Differential as much as other mortals (the author included). History shows that they have not officially checked themselves up sufficiently to become aware of this fatal habit of confusion of orders of abstractions through identification.

It might appear, at first glance, that all that has been said here is simple and easy. On the contrary, it is *not* for the grown-ups; it is easy only for children and the young. In all my studies and experimenting I have found that, for the reasons already given, the use of the Differential appears essential, and that it requires a long while and training to accomplish new semantic results. As a rule, unless they are very unhappy, people try to trust their 'understanding', and dislike to train repeatedly with the Differential. For some reason or other, they usually forget that they cannot acquire structural familiarity with, or reflex-reactions in, spelling, or typewriting, or driving a car., *by verbal means alone*. Similar considerations apply in this case. Without the actual training with the Differential, certainly the best results cannot be expected.

To gain the full benefit involves the uprooting of old habits, taboos, 'philosophies', and private doctrines, the worst being the structure of our primitive *A* language with the 'is' of identity, all of which are deeply rooted and work unconsciously. Only the semantic training with the Differential in *non-identity* can affect the 'habitual' and the 'unconscious'. Rationalization, lip-service to the 'understanding' of it, will be of no use whatsoever. Persistent training seems the only way to acquire this *special structural sense for proper evaluation,* and the habit of *feeling* when identification, or the confusion of orders of abstractions becomes particularly dangerous. This feeling, as it involves most important factors of evaluation, is difficult to acquire, as difficult, perhaps, as reflex-learning to spell or to typewrite. But, when acquired, it makes us aware of the continuous, necessary utilization of many levels of abstractions, which becomes dangerous only when we identify them or are *not conscious* of this fact. We can then utilize the different orders of abstractions *consciously,* without identification, and thus keep out of danger. Most of the important *terms* appear as multiordinal, and, although they *belong* to verbal levels, they *apply* often to all levels, an important structural fact impossible to avoid, and one which makes this special semantic sense uniquely necessary to acquire.

It seems unnecessary to repeat that everything that has been said above applies in the fullest extent to our ethical, social, political, economic, and international relations. Before any sanity can be brought into the analysis of these relations, before they can be rationally analysed, the investigators would have to be trained to observe correctly and to avoid verbal structural pitfalls. For the lack of such semantic training and re-education, the 'time-honoured' 'Fido' debates involving the 'is' of identity, continue on all sides, and lead to naught else but a waste of 'time' and effort.

I say waste of 'time', simply because there seems no end to the paradoxes which, with a little ingenuity, we can build up when we begin to gamble with confusion of orders of abstractions and disregard multi-ordinality. Any doctrine, no matter how structurally true or beneficial, can be defeated, confused, or delayed, by the use of such methods. These problems appear of crucial semantic importance, because our lives are lived in a *permanent* structural interplay between different orders of abstractions. All our achievements depend upon this interplay, yet the most acute and painful dangers also have their sources in the non-realization of this dervish dance between different orders of abstractions.

Since we cannot evade the passing from level to level, or the use of multiordinal terms, our wisdom should consist only in not abusing these semantic conditions of human life. As we must do that, let us do it, but let us not identify the orders, and thus let us evade the dangers. Consciousness of abstracting gives us the complete *psychophysiological* solution of this complex situation, as it allows us to have the psychological benefits and to avoid the dangers by the use of *physiological* means.

In conclusion, I must stress once more the importance of the structure of the language in which we analyse any given problem. In the $\overline{A}$-system I am proposing, the term *order* is accepted as one of its very foundations. In 1933, we know that as words *are not* the things spoken about., structure, and structure alone, becomes the only possible content of knowledge, and the search for structure, the only possible aim of science. If we try to define structure, we can do so in terms of relations and multi-dimensional order. The recent advances of science show, beyond doubt, that the day will come when all science will be formulated in terms of structure and, therefore, of physics, and physics formulated as a form of multi-dimensional geometry, based on multi-dimensional order, giving us, ultimately, multiordinal structure.

29

The application of the term *order*, which involves physiological, as well as semantic, mechanisms of evaluation, to the analysis of human behaviour, has led me to the present $\bar{A}$-system and the investigation of the structure of language. The discovery that some of the most important terms we use appear multiordinal, a character *concealed* by the 'is' of identity, has disclosed to us a most vital and inherent psycho-logical mechanism, responsible in humans for many most desirable, many undesirable, and many morbid human characteristics. It disclosed, also, the psycho-logical *structure* of these characteristics, and so we have obtained *physiological* means by which to enhance the development of desirable characteristics, and to prevent or transform the others.

Further analysis has disclosed a natural survival order in evaluation: the event first, the object next; the object first, the label next; description first, inferences next., in inherent importance. We have also found that the majority of human difficulties, 'mental' ills included, involve semantic disturbances and exhibit, *not* the natural survival order, but the identification of different orders, resulting in a reversed (patho-logical) order.

It is impossible in this book to review the data of psychiatry from the $\bar{A}$ point of view, as this would require a separate large volume, which, I hope, will be written some day; but any one can verify the statements made above by himself from clinical literature, and also by analysing his own or other persons' life-difficulties, quarrels, disagree-ments., which generally involve quite unnecessary sufferings. Psycho-therapeutic literature shows abundantly that the success of the physicians depends mostly on reversing the pathological reversed order in a given field, and so restoring the natural order in the *s.r.* It is easily verified that, in most cases, when 'mental' illness originated through different life experiences, these would have affected very little, if at all, a child or an adult who was conscious of abstracting, and whose nervous processes and corresponding semantic states followed the natural order.

With the aid of the Structural Differential and a $\bar{A}$ language of new structure, it is easy to train the *s.r* of an infant, a child, or a young person, and possible, although much more difficult, to train a grown-up in the natural order. Such a training becomes a potent preventive struc-tural *physiological* method, as it eliminates the psycho-logical states of identification or reversed order, both of which represent the raw seman-tic material out of which future nervous disorders are produced.

The *non-el* term 'order' is equally applicable in life and science; gives us, in 1933, the simplest structural common base, and allows us to attempt the formulation of a science of man, which ultimately becomes a theory of sanity, a consequence of a *non-aristotelian system.*

It should be recalled that order is accepted in the present system as undefined and fundamental; yet its use is easily explained by the aid of the term 'between', and can be shown and applied in reference to empirical structures.

If we can formulate a method which, through the application of a psychophysiological term such as *order,* and a simple device such as the training of *s.r* in the natural survival order, or reversing the pathological reversed order, includes the mechanism of non-identity and one of the most important human nervous functionings, such a method, because of its structural simplicity and physiological character, may be expected to prove very workable. I desire to stress most emphatically the very important *general, impersonal, preventive, semantic, reflex-character* of such a method.

In actual life, we deal, for the most part, with persons who are 'mentally' or nervously disturbed in different degrees. We could, eventually, divide them for our purpose into two groups: (1) those who do not want to improve or get well, but who somehow like their fictitious worlds and the maladjustments connected with them; (2) those who genuinely want to get over their difficulties.

In general, it is extremely difficult or impossible to achieve anything at all with the first group. The second group is greatly helped if we give them means to work by themselves at their problems. Very often it is most effective to explain to them this simple 'natural order', 'identification', and 'reversed order' mechanism, the multiordinality of terms., and so give them a *definite psychophysiological symptom* to struggle against. These symptoms of identification or reversed order, in their generality and structural neurological fundamentality, underlie the process of formation of practically all known semantic difficulties of evaluation.

The reader should not assume that it is always possible to eliminate identification and so achieve this coveted natural order, or the reversal of the reversed pathological order; but, whenever this *is* possible, the person is relieved in a great many psycho-logical fields. The simplicity and generality, the physiological and structural character of this method seems its main recommendation, particularly as a preventive measure or semantic training for sanity. The training is a laborious process, requiring great persistence; but, to my knowledge, very few trainings are easy, and, perhaps, none leads to more important results than does this one.

CHAPTER XXVIII

ON THE MECHANISM OF IDENTIFICATION AND VISUALIZATION

"Did you say 'pig,' or 'fig'?" said the Cat.

"I said 'pig'," replied Alice; "and I wish you wouldn't keep appearing and vanishing so suddenly: you make one quite giddy!"

"All right," said the Cat; and this time it vanished quite slowly, beginning with the end of the tail, and ending with the grin, which remained some time after the rest of it had gone.

"Well! I've often seen a cat without a grin," thought Alice; "but a grin without a cat! It's the most curious thing I ever saw in all my life!"*

LEWIS CARROLL

The significance of the paradoxical phase is not limited to pathological states such as those previously observed, and it is highly probable that it plays an important part in normal men too, who often are apt to be much more influenced by words than by the actual facts of the surrounding reality. (394)

I. P. PAVLOV

In the case of an imbecile, repetition without comprehension, psittacism, may prevail; the rôle of visual impressions is null or nearly so among the illiterate; the deaf from birth who have learned to speak have no auditory impressions to intervene. But, normally, it is feelings and ideas that appear in action, in the form of language. (411)

HENRI PIÉRON

The specific neurones necessary for sensation are also necessary for the associative reawakening of that sensation, which is called the image—a dynamic process and not a photographic negative resting miraculously in the nervous substance, where some subtle spirit might go to consult it. (411)

HENRI PIÉRON

It is none the less true that certain cultivated persons can use visual images, and can even use these images in preference to others. (411)

HENRI PIÉRON

Objectification and visualization are usually not differentiated. The first represents a very undesirable semantic process, whereas the second, visualization, represents one of the most beneficial and efficient forms of human 'thought'. From a $\bar{A}$ point of view, such a lack of differentiation between the two reactions appears as a very serious problem, requiring an analysis of the respective mechanisms.

To visualize, we must have such forms of representation as lend themselves to visualization; otherwise, we must fail. The A-system, which could not adequately handle asymmetrical relations, and could not be built explicitly on structure, necessarily involves identification. In the A period, we were able to visualize objects and a few objective situations, but all the higher abstractions were, in principle, inaccessible to visualization, making scientific theories needlessly difficult. A $\bar{A}$-system, free from identification, must be based explicitly on *structure* on all levels (structure defined in terms of relations and ultimately multi-dimensional

---

*Alice in Wonderland.

452

order), which can be easily visualized. It should be recalled that structure, relations, and multi-dimensional order supply us with a language which completely bridges daily-life experiences with all science, leading toward a *general theory of values*. Mathematics and mathematical physics then become the representatives and the foundation of all science; and in the human field a general theory of values will lead to adjustment or sanity and will some day include ethics, economics,.

For these reasons, the Structural Differential is uniquely useful, as, at a glance, it conveys to the eye structural differences between the world of the animal, the primitive man, and the infant, which, no matter how complex, is extremely simple in comparison with the world of the 'civilized' adult. The first involves a *one-valued orientation* which, if applied to the ∞-valued facts of life, gives extremely inadequate, wasteful, and ultimately painful adjustment, where only the few strongest survive. The second involves ∞-valued orientation, similar in structure to the actual, empirical, ∞-valued facts of life, allowing a one-to-one adjustment in evaluation with the facts in each individual case, and producing a semantic flexibility., necessary for adjustment. This flexibility is known to be the foundation for balanced semantic states, 'higher intelligence',.

Visualization requires a definite elimination, through differentiation, of harmful identification, which, as usual, is based on incorrect evaluation of structural issues. Thus, we have had endless, bitter, and futile arguments as to whether or not the 'mechanistic' point of view of the world and ourselves is legitimate, adequate,. The average person, as well as the majority of 'philosophers', identifies 'mechanistic' with 'machinistic'. Roughly, mechanics is a name for a science which deals with dynamic manifestations on all levels; thus, we have macroscopic classical mechanics, colloidal mechanics now being formulated, and the sub-microscopic quantum mechanics already being well-developed disciplines. In the rough, 'machine' is a label applied to a man-made apparatus for the application or transformation of power. But even machines differ greatly; thus, a dynamo is entirely different in principle, in theory, and in applications from a lathe or an automobile.

If we ask: 'Is the *machinistic* point of view of the world justified?', the answer is simple and undeniable; namely, that this point of view is grossly inadequate and should be entirely abandoned. But it is not so with the *mechanistic* point of view, understood in its modern sense and including the quantum mechanics point of view, which is entirely *structural*. In 1933, we know positively that even the gross macroscopic physico-chemical characteristics of everything we are dealing with depend on the sub-microscopic *structure* (see Part X). The details are not yet fully known, but the principles are firmly established. With a $\overline{A}$ understanding and evaluation of the unique importance of structure as the only possible content of 'knowledge', these 'firmly established' principles become *'irreversibly established'*. We may go further and say that the quantum mechanics point of view becomes the first structurally correct point of view and, as such, should be accepted fully in any sane orientation. If we stop identification, then we will differentiate between some simple facts. For instance, we will understand that any semantic state, reaction, or process has its corresponding sub-microscopic, structural, colloidal, and ultimately quantum mechanical processes going on in the nervous system; however, the *s.r*, or feelings of pain or pleasure., *are not* the sub-microscopic processes. These belong to different levels, but with ∞-valued semantics we can establish in principle a one-to-one correspondence between them. Thus, when we differentiate adequately, the older *mach*inistic objections disappear entirely; and, in its proper field, for structural reasons, we must preserve the *mechanistic,* and entirely abandon the too crude *mach*inistic attitudes. The *mechanistic* (1933) attitude is based on *structure* and so is indispensable for visualization; and *training in visualization automatically abolishes objectification,* which represents an important special case of all identification. From the point of view of a $\overline{A}$-system, adjustment and sanity in humans depend, to a large extent, on their 'understanding', which is entirely structural in character; therefore, we must accept a *mechanistic* (1933) *attitude,* which, in the meantime, can be visualized.

The finding of structural means of representation facilitates *visualization,* imagining, picturing,. In the adjustment trend we start with lower nerve-impressions, 'senses', 'feelings'., lower abstractions, and these are abstracted again by the higher centres. The higher centres produce the 'very abstract' theories, which cannot be visualized for a while. The lower centres, which are involved in visualization, can deal only with structures which can be 'concretely pictured'. So we always try to invent mechanistic or geometrical theories, such as can be handled by the lower centres.

Individual 'experiences', supplied by the lower centres of different individuals, do not blend directly. They are blended in the higher centres. In them manifold experiences, whether individual or accumulated by the race (time-binding), are abstracted further, integrated, and summarized. Once this has been accomplished, structural means are sought *and discovered* to translate these higher abstractions into lower, the only ones with which the lower centres can deal. Then we can 'visualize' our theories, and the higher centres not only influence the lower centres, but the lower centres have appropriate means by which to co-operate with the higher centres in their new *non-el* quests.

The lack of explicitly structural forms of representations is responsible, also, for the difficulties which arise when the higher order abstractions are translated into the reflex-reactions of the lower centres, which can deal with 'intuitions', 'orientations', 'visualization',. The so-called 'geniuses' have a very subtle nervous system in which the translation of higher order abstractions into lower and vice versa is easily accomplished. From the point of view of forms or representations, we can have two issues: (1) we may have *el* forms of representations which are not based on structure, visualization., and cannot efficiently affect the activities of the lower centres; (2) we may have a *non-el* system based on structure, visualization., which can be translated simply, easily, and efficiently into the terms of the lower centres. These problems are of educational importance and should be worked out more fully.

In my experience with grown-ups who have had only a *short* contact with my work, I find, in many cases, that, although they may have even given their complete verbal approval of the main point of the system, yet, invariably, in practice, the full application is lacking. Obviously, the semantic importance of the present findings is not in the verbal approval alone, when that approval is not applied, but in the consistent and permanent instinctive acquisition of the new semantic attitude which involves a complete elimination of identification, allness, elementalism,.

We can teach any one to repeat verbally, by heart, instructions for operating an automobile, a piano, or a typewriter; but no one could operate them satisfactorily by reflex-action after such verbal training alone. To operate effectively and skilfully any structural complex, we must become intimately familiar with its structural working through actual reflex-training, and only then can we expect the best results. In my experience, this is true with language, and, without the *visual* Structural Differential on which we can point our finger to the objective level and urge silence., such basic semantic *reflex-training* cannot properly be given.

If we ask a man: 'Do you know how to drive a car?', and he answers 'Yes', we assume that he has acquired the proper *reflexes*. If he answers 'No, but I know *about* it', he means that he has *not* acquired the proper reflexes, but that his 'knowledge' is on purely verbal levels, non-effective in application on *non-verbal* reflex-levels. This applies fully to *s.r*; we may 'know' *about* them, but we may never apply successfully what we supposedly 'know'. To 'know' represents a multiordinal process which involves equally the activities of the lower nerve centres and of the higher. In our *el* systems we had no such distinction, and so we confused them. The older 'knowledge', when presented in *el* language, could not have been absorbed easily by the *non-el* organisms-as-a-whole. As the main task, at present, is to unlearn the older *s.r*, the new reactions need a persistent training, particularly by the grown-ups. The *non-el*, $\bar{A}$ language and method prove to have psychophysiological importance.

Although the neurological mechanism underlying identification, objectification, visualization., is not well known (1933), neurology gives us evidence that in these states, as well as in delusions and hallucinations, the actual lower nerve centres are somehow engaged. We may assume that different 'resistances', 'blockages'., in some parts of the nervous system make the passage of nervous impulses more difficult, and it seems reasonable to suppose that, in such cases, the paths travelled by the nervous currents are different.

In Fig. 1, an hypothetical and over-simplified scheme of the different types of distribution of nervous currents, as is known functionally, is suggested. The ordering is not anatomical but functional in terms of degrees of intensity. In this scheme, we may consider that the nervous

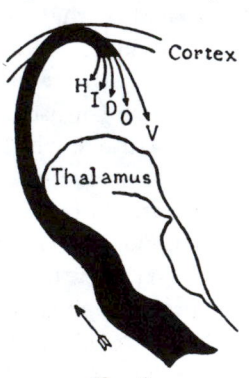

impulse (A) reaches the lower nerve centres, the brain-stem and the thalamus, passes through the sub-cortical layers and the cortex, continuously being transformed. Finally, in returning, it may take either the beneficial and adaptive semantic form of visualization (V), free from identification and semantic disturbances, or may involve identification, with semantic disturbances, such as objectifications of different orders (O), delusions (D), illusions (I), or, finally hallucinations (H).

FIG. 1

Identification, or confusion of orders of abstractions, consists of erroneous evaluation: that which is going on inside of our skin has objective existence outside of our skins; the ascribing of external objectivity to words; the identification in value of 'memories of experiences' with experience; the identification of our *s.r* and states with words; the identification of inferences with descriptions,. Identification is greatly facilitated, if not actually induced, by the *A* structure of language in which we have *one* *name* for at least *four* entirely different entities. Thus, the *A* 'apple' (without subscripts and date) is used as a label for the physico-chemical process; for an object, say, 'apple$_{1,\text{Feb.23,1933}}$'; for a 'mental' picture on the un-speakable semantic level, and for the verbal definition. Under such linguistic conditions, it is practically impossible, *without special training,* not to identify the *four* entirely different abstractions into *one*., with all the following sinister consequences.

Delusions represent incorrect notions and inappropriate *s.r* formed, not by insufficient knowledge or 'logic', but by affective pressure in a definite evaluational direction; as, for instance, delusions of grandeur; delusions of persecution; delusions of 'sin'; delusions of reference,.

Illusions appear more like real perceptions, but pathologically changed. For instance, anything may be semantically coloured or interpreted, or evaluated as an offense, or a threat, or a promise,.

Hallucinations consist of 'perceptions', with all their vividness, but *without* any external stimuli. Patients hear voices; see visions; feel pricks or burnings., when there is nothing to hear, or see, or to be pricked by.

In *visualization,* identification does not occur; orders of abstractions are not confused; semantic disturbances do not appear; the *evaluation is correct;* a 'picture' is evaluated as a picture and not as the events,. In other words, because of the consciousness of abstracting, the natural order of evaluation is preserved. But once, through identification, this natural order is reversed, it marks a pathological condition more or less morbid, and often of a non-adaptive character.

Identification represents, in affective tension, the mildest semantic disturbance, consisting of an error in meanings and evaluation. Objects are evaluated as events; 'ideas', or *higher* order abstractions, are evaluated as objects; as experience; as the un-speakable semantic states or reactions; otherwise, as *lower* order abstractions. The confusion in the field of higher order abstractions follows a similar rule. Inferences obviously represent higher order abstractions than descriptions; so, when they are not differentiated, higher order abstractions are again identified with the lower. We all know from daily-life experience the fantastic amount of suffering we can, and do, actually produce for ourselves and others with such identifications.

In delusions, a similar but more intense identification occurs, resulting in erroneous semantic evaluation; wishes, feelings, and other semantic states inside of our skins are projected into the external world, giving delusionally strong objective evaluation.

In illusions, we also ascribe to, or identify our complex semantic states with, different perceptions and evaluate our higher order abstractions as lower.

In hallucinations, this process of reversing the natural order comes to a culminating point: higher order abstractions are translated into, and have the full vividness and 'reality' of, lower order abstractions.

We see that the pathological processes of 'mental' illnesses involve identification as a generalized symptom; which means the reversal, in different degrees, of the natural order of evaluation based on the intensified confusion of orders of abstractions. The more intense this process of reversal becomes, the more non-adaptive and morbid the manifestations. It should be noticed that this analysis becomes a necessity once we decide to accept a *non-el* language. This analysis is far from exhaustive, but an analysis in new *non-el*, structurally correct terms, throws a new light on old problems.

Hallucinations which result from 'physical' illness do not represent a permanent danger, but when a patient seems 'physically' well, and his confusions of orders of abstractions, delusions, illusions, and hallucinations become completely 'rationalized', then these are unmistakable signs of serious 'mental' illness, suggesting sub-microscopic colloidal lesions. Now this 'rationalization' represents nothing else but a nervous disturbance and involves *identification* somewhere. In 'physical' ills the nervous system may be disturbed, but the illness does not usually originate in nervous disturbances, and so, as such, is not dangerous.

The distinction between visualization and objectification based on a A-system seems new; the difference is subtle, but when it is formulated we can discover a simple means whereby to control the situation. If we were to take a 'bone' made of papier-maché and smear it with fat or meat, Fido would, perhaps, *objectify* (identify) such a 'bone' from the smell and the form of the papier-maché with an edible one, and would fight for it. We do a similar sort of thing when we objectify. Religious wars, the 'holy inquisition', the persecution of science, which we are witnessing even at the present day in some countries and communities, are excellent examples.

We should notice that Fido was able to *trust* his natural, even 'objectified', instinct, for nature does not play such tricks on him, such as producing 'bones' of papier-maché. If nature did, dogs that objectify and persist in their liking for such 'food' would soon be eliminated. These particular objectifications would be dangerous and painful to those particular kinds of dogs with that particular nervous system, and would ultimately prove of no survival value. Thus identification, which represents an inappropriate evaluation, is harmful to all life, but is little noticed at present, because the main periods of the animal racial adjustment have been accomplished long ago. Experiments on flies show that the number of mutants which may be produced in a laboratory is large, but very few would survive outside of a laboratory. In unaided nature, these mutants probably occur, but seldom leave observable traces.[1]

However, even today, as Pavlov has shown in his laboratories, we can impose, by an interplay of a four-dimensional order of stimuli, such conditions upon animals for which their nervous survival structure was not naturally adapted, and so induce nervous pathological states. Wrong evaluation is, indeed, harmful to all life and accounts for such rigid survival laws in nature, which science teaches humans how to make more flexible. Practically word for word, this applies to ourselves. We are constantly producing more and more complex conditions of life, man-made, man-invented, and deceptive for the non-prepared. These new conditions are usually due to the application of the work of some genius, and the nervous system and $s.r$ of most of us are not prepared for such eventualities. In spite of inventions and discoveries of science, which are *human* achievements, we still preserve *animalistic* systems and doctrines which shape our $s.r$. Hence, life becomes more strained and increasingly more unhappy, thereby multiplying the number of nervous break-downs.

It is known that not all people are able to visualize equally well. In the older days this fact was taken for granted, and did not suggest further analysis. Under present conditions in many human beings and also in animals, as shown in the experiments of Pavlov, the visual stimuli are physiologically weaker than the auditory ones; in man, however, the visual stimuli should be physiologically stronger than the auditory. This difference does not affect the general mechanism of the cyclic nerve currents and orders of abstractions. In the auditory type the main returning currents are deviated into different paths. The division between 'visual' and 'auditory' types is not sharp. In life we deal mainly with individuals who have no more than a special inclination for one or the other types of reaction.

In the case of 'mental' processes, human adjustment has to be managed on higher, more numerous, and more complex levels. Obviously, then, the auditory types are more enmeshed by words, further removed from life than the visual ones, and so cannot be equally well adjusted. This fact should not be neglected, and on the human levels we should have educational methods to train in visualization, which automatically eliminates identification.

The auditory channels which connect us with the external world are much less subtle and effective than the visual ones. The eye is not merely a 'sense-organ'. Embryology shows that the eye is a part of the brain itself, and what is called the 'optic nerve' must be considered not a nerve but as a genuine nervous tract. This fact, of course, would assign to the eye a special semantic importance, not shared with other 'sense-organs' or receptors. We ought not to be surprised to find that the visual types are better adjusted to this world than the auditory types. In pathological states, such as identifications, delusions, illusions, and hallucinations, there seems to be involved a translation of *auditory* semantic stimuli into visual images. In these pathological cases the order of evaluation appears as label first and object next, while the adaptive order seems to require object first and label next,. There seems little doubt that visualization is very useful, and that identification is especially harmful. The *most effective means to transform the s.r of identification is found in visualization, which indicates its special semantic importance.*

The semantic *disturbance* of identification may have many sources, auditory included, but the only adaptive trend is in visualization, which involves in some way the optical neural structure. Some structural light is thrown on this subject when we realize that, physiologically, the eye is more closely related to the vegetative nervous system, which regulates our vital organs, than the ear is. In man the optic thalamus is greatly enlarged, so that the whole thalamus is often called the 'optic thalamus'. Actually, the thalamus has many functions, other than visual, and is connected with affective manifestations.

As most of our observations are accomplished by the aid of the eye, we should expect auditory types to be *poor observers,* and so racially, in the long run, not so well adjusted semantically. Observation shows that the auditory types often have infantile reactions—a serious handicap. From an adaptive point of view the 'normal', non-infantile, best-adjusted individual ought to be a visual type. Auditory types must also be further detached from actualities than the visual types, as auditory stimuli involve more inferences than descriptions, which is the opposite of the functioning of the visual types. If inferences, rather than descriptions, are involved, we naturally deal with higher abstractions first, and with the lower next, and so there is always a danger of the semantic confusion of orders of abstractions, which necessarily involves inappropriate evaluation, of which objectification is only a particular case.

Even to common sense it seems clear that there is a significant difference between 'knowing' this world by hearing and 'knowing' it by seeing. There is, likewise, a difference between the translation of higher abstractions into lower terms by the visual path, and the corresponding translation by the auditory path. In daily life we never say 'I hear' when we wish to convey that we understand; but we say 'I see'. When we say 'I hear', we usually wish to convey that we have heard something which we did not fully grasp or approve. The above relation is rather important, but has not been sufficiently analysed. The problems of introversion and extroversion are connected with it.

The relation between the problems of identification and the *number* of values *found* in the empirical world in connection with the *number* of values *ascribed,* or *assumed* ., by our semantic processes, is most important.

The following analysis is, by necessity, one-sided, over-simplified ., as a fuller analysis would require a separate volume. I consider many problems 'in principle' only ; this allows me a briefer treatment necessary for my purpose, but it must be realized that our language and general semantics, which, *in practice,* we use unconsciously, are *extremely complex* and involve one-, two-, three-, and ∞-valued components, never, as yet, sharply differentiated nor formulated. Investigation shows that the ∞-valued semantics is the most general and includes the one-, two- ., and few-valued semantics as particular cases. The one-valued semantics of literal identifications are found only among animals, primitive people, infants, and the 'mentally' ill, although more or less serious traces of some identification are found in practically all of us, because these are embodied in the structure of our language and prevent the acquisition of the ∞-valued systems necessary for sanity. For my purpose, it is enough to formulate the problems for the complete elimination of primitive identification, and then modern, ∞-valued, $\bar{A}$ semantics follow automatically. Under such conditions, I must concentrate on the vital problem of one-valued identification and treat the two- ., and few-valued systems sketchily, 'in principle', although we must realize that the last systems have been made more flexible by the use of many ingenious verbal devices which I do not even mention in the present work.

Let me repeat that the attitudes, flexibility, or fixity., of our *s.r* depend to a large extent on the structure of language used, which involves also its appropriate general semantics. The 'logic' of our schooldays represents a composite affair, in the main A, and we call it by this last name. This 'logic' can be considered as a two-valued 'logic' because of the fundamental 'law of the excluded third', expressed as 'A is B or not B', by which a third possibility is excluded. But even the traditional 'logic' had to admit in its scheme what was called 'modality' ; namely, some degrees of certainty or uncertainty with which a given statement is made. Lately, Łukasiewicz has shown that a three-valued 'logic' can be so formulated as to include modality. Later, he and Tarski generalized it to an *n*-valued 'logic'. When *n* tends toward infinity, this 'logic' becomes the 'logic' of probability. If these disciplines are made *non-el,*

we have what I call one-, two-, three- ., and ∞-valued *general semantics*
Theoretically, and in practice, we are interested mostly in the one-, two-,
three-, few-valued, and ∞-valued general semantics. For my purpose,
and for simplicity, I shall deal only with identification; that is, the
primitive one-valued semantics, the influence of which is found in both
the two- and three-valued semantics, *and may only be completely eliminated in an ∞-valued semantics.*

We live in a four-dimensional space-time manifold which, on all
levels, consists of absolutely individual events, objects, situations, abstractions., and we must conclude that structurally we live in an *indefinitely many-valued* or ∞-valued world, the possibilities of which follow
in principle the laws of combinations of higher orders. The above
statement represents a description of a structural observation about the
empirical world, independent of our pleasure, and can be contradicted
only by exhibiting empirically, actual 'identity' or 'absolute sameness'.,
of different events, objects, or situations., which exhibiting becomes an
impossibility if we decide to investigate facts more fully.

Under such empirical conditions, for adjustment and so for sanity,
we must have on semantic levels such theories, systems, methods.,
which would allow us in a given case, under given conditions, at a given
date., to evaluate the individual happenings *uniquely;* or which would
allow us to establish a one-to-one correspondence between the essentially ∞-valued facts of experience and our semantic states. It becomes
obvious that this can be accomplished only if we have ∞-valued and
non-el general semantics. We see that the two-, or three-valued, *el A*
'logic', 'psychology'., and, in general, the *A*-system, being structurally
different from the empirical world, will prevent, in principle, such an
adjustment and, therefore, sanity.

Identification may be considered as the remains of pre-human, or
primitive, or infantile, one-valued semantics, which establishes, or results
from, semantic states, by which the essentially ∞-valued facts of experience are not differentiated or evaluated properly, and so the indefinitely
many values of these facts are identified into a single value. Such
identification is always structurally unjustified and dangerous, and may
be the result of a great many factors, such as low development, ignorance, insufficient observation, 'wishful thinking', fears, pathological
states of our nervous system, different semantic disturbances, 'mental'
ills, infantilism in the grown-ups,. But among humans we cannot avoid
training, through the mechanism of language and its structure, in some,
most often unconscious, general semantics, and so a great deal depends

on what kind of semantics or methods of evaluation we impose on our children.

We should notice an important fact which is usually disregarded; namely, that a language, and often one word, involves a definite type of semantics. Thus, in primitive 'polysynthetic' languages, it is not a question of associations or superstitions; the mystic characteristics and the thing simply are not differentiated, but literally identified into one whole. Thus, we have one-valued semantics where the 'good' and 'evil spirits' actually participate in everything considered as a synthetic whole.[2]

A language of 'true' and 'false' involves two-valued semantics; the introduction of adverbs or their equivalents introduces modality and so three-valued semantics. The introduction of indefinitely many degrees between the 'true' and 'false' leads finally to ∞-valued semantics.

A diagram may help to make this clearer.

A, B, C., ∞-valued and different facts of experience, which, in a given case, have, by necessity, *indefinitely many, single, individual* values.

a, b, c., ∞-valued non-aristotelian orientation structurally *similar* to the empirical world which allows us, in a given case, to assign indefinitely many single, one-to-one corresponding values to the individual facts.

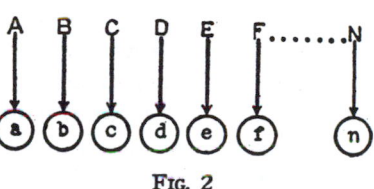

FIG. 2

A, B, C., ∞-valued and different facts of experience, which, in a given case, have, by necessity, *indefinitely many, single, individual* values.

Σ_1, Σ_2., two-, three-., and few-valued aristotelian orientation structurally non-similar to the empirical world, which compels us to ascribe two., or few values to the essentially indefinitely many-valued and different facts, resulting in identification of the many values into a few, which improper evaluation is projected on the facts.

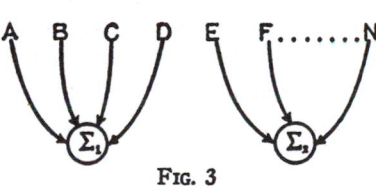

FIG. 3

A, B, C., ∞-valued and different facts of experience, which, in a given case, have, by necessity, *indefinitely many, single, individual* values.

Ω, one-valued, animal, primitive., orientation, structurally non-similar to the empirical world, which compels us to ascribe one value to the essentially indefinitely many-valued and different facts, resulting in *identification* of the many values into one, which improper evaluation is projected on the facts.

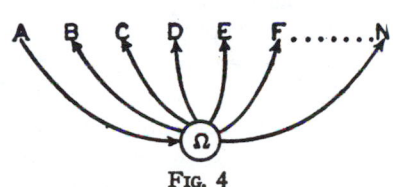

FIG. 4

In Fig. 2, the arrows Aa, Bb., indicate the $\bar{A}$ one-to-one correspondence between the ∞-valued individual facts of life, A,B,C., and the corresponding *s.r*, or a,b,c orientations., which ascribe single values to the different facts, establishing a foundation for structurally correct *proper evaluation* which helps adjustment and so sanity.

In Fig. 3, the A two-., few-valued orientation and type of correspondence is shown.

In Fig. 4, Ω indicates a single, say, proper evaluation of the *one* fact A. The arrows ΩB, ΩC, ΩD, ΩE, ΩF, . . . , ΩN indicate the *projection* of the one-valued semantic state, or orientation on the essentially unchanged ∞-valued facts A, B, C., *distorting them.* In other words, the ∞-valued facts, through the identification of many values into one, and by pathological projection, have been given wrong evaluation, thereby preventing, *in principle,* adjustment and sanity, particularly for a civilized human 1933.

If we train our children in one-, two-, three-, and more generally few-valued *el, A* reactions based on corresponding languages, 'logics'., the result must be that they will have great difficulty in adjusting themselves to a world of *non-el* ∞-valued facts, and that, even if they succeed, this would ultimately happen only after a great waste of efforts and unnecessary sufferings. If we approach the ∞-valued facts of life with one-, two-, or even few-valued semantic attitudes, we must identify some of the indefinitely many values into one, or a few values, and so approach the ∞-valued world with an orientation which *projects* ignorantly or pathologically our *restricted,* few-valued semantic evaluations on the ∞-valued individual facts of experience.

The above explanations apply in the fullest extent to the structure of language. The daily language, as well as our attitudes toward it, still reflects primitive structural *s.r* of the period before it was known that on the objective levels we deal *exclusively* with *∞-valued, four-dimensional processes.* The language in the A-system represents, *in principle,* what may be called a three-dimensional and one-, two-., more generally few-valued linguistic system structurally non-similar to the ∞-valued, four-dimensional event-process conditions. Let us analyse, for instance, the A term 'apple'. This term represents, *in principle,* a name for a verbal, one-valued, and constant intensional definition, in which space-time relations do not enter. What are the structural facts of experience? The object which we call 'apple' represents a process which changes continually; besides, every single apple that ever existed, or will exist, was an absolute individual, and different from any other objective 'apple'. In applying such a three-dimensional and one-valued language to essentially ∞-valued processes, we only make proper evaluation, and so adjustment and sanity, very difficult.

Yet the structural adjustment is simple in a $\bar{A}$-system. The A 'apple' was a name for a verbal *intensional definition;* in a $\bar{A}$-system, we manufacture indefinitely many names for the indefinitely many objective and different 'apples' by subscripts, 'apple₁', 'apple₂', 'apple₃'., supplementing the subscript with the date; thus, in 'apple₁ ₋Feb.₂₃,₁₉₃₃', we gain the possibility of considering 'apple$_{n,t}$' as ∞-valued, and so, in a given case., we are enabled to have a single name which we could relate to single values of the objective, absolute individuals, and absolute individual stages of the process. Similarly with multiordinal terms. Before the multiordinality of terms was discovered and formulated by me in 1925, these terms were silently assumed *in principle,* to be one-valued, and we were either prevented from using them in connection with ∞-valued orders of abstractions, or, if used by semantic necessity, we identified the indefinitely many values into one. Both results were undesirable; the first established semantic blockages to creative scientific work; the other promulgated semantic disturbances. But once the multiordinality of terms is established, we have ∞-valued terms to which, in a given context (by differentiating the different orders of abstractions which a context indicates), we can ascribe single values.

Such a pioneering analysis may appear difficult at first, but this is only due to the lack of familiarity and established pre-A and A one-, two-, three-, or few-valued *s.r,* all of which involve ultimately *identification* somewhere. Once identification is abolished, however, and this is childishly simple, although not easy and rather laborious for grown-ups, ∞-valued semantics become natural and *automatic,* evading very serious theoretical difficulties. In the present volume, I had to elaborate in detail upon different issues, simply because my readers will be mostly grown-ups with established *pre-A,* and A reactions, who must first be made to recognize the benefits of a $\bar{A}$ evaluation before they will be willing to undergo a laborious re-education of their older *s.r.* The procedure in the training of infants and children is extremely simple and entirely on their levels.

There is, however, one point that I wish to make entirely clear. From the older point of view, one might say that a $\bar{A}$-system may lead to 'over-rationalization' and, consequently, take 'all the joy out of life'. Such objections are entirely unjustified. First of all, the A-system leads to shallow, but often clever verbal interplay of definitions, mostly non-similar in structure to the world and ourselves, representing a species of apologetics, and usually called 'rationalization'. The $\bar{A}$-system leads to

30

structural adjustment of language and *s.r*, and a *structural* enquiry, resulting in *understanding*. It makes shallow infantile 'rationalization', 'wishful thinking', and apologetics of different brands impossible, but leads to a higher order of adult intelligence, based on *proper evaluation*. In mere 'rationalization', we often have clever, but shallow, infantile evaluation, based on the ignorance or disregard of structural facts, which alone make up the content of all 'knowledge'. In a $\bar{A}$-system, by eliminating the sources of infantile evaluation and reactions, we supply the nervous system of the infant with uniquely appropriate material, so that it may develop into a 'normal' adult. In the older system, instead of helping, we hindered the development of adult standards of evaluation, with well-known results. There is nothing wrong with 'human nature' or the majority of nervous systems as such, but there is something definitely wrong with our educational methods inside and outside our schools.

There is another point which is still more convincing and, perhaps, even more decisive. The above-mentioned older objections are due to *s.r* based on the play upon *elementalistic* terms and are a *neurological impossibility*. The organism works as-a-whole, and in the cyclic nerve currents it is impossible, by any known educational methods, to abolish 'emotions'. But what *can be* accomplished is this: By training in silence on the un-speakable objective levels and in differentiation between different orders of abstractions, we automatically abolish the infantile identifications and evaluations; we introduce a 'delay in action', which is the physiological means for getting our 'emotions' under control and for engaging the fuller co-operation of the cortex. Infantile 'over-emotionalism' is abolished in the adult. Infants would behave as infants, but this infantile behaviour would not be carried into the period when adulthood should begin. The 'emotions' are not abolished but 'sublimated'.

It is true that many standards would be changed. For instance, we might roughly say that an infantile type is often bored by a symphony and that jazz satisfies his infantile make-up. If we were to take such an infantile grown-up and compel him to listen only to symphonies, this would not be kind, nor would it transform his infantile *s.r* into adult reactions. But, if unhampered by inappropriate semantic and so neurological training, he would be free to develop normally into an adult, and his own preference would be toward a symphony rather than toward primitive throbbings, his enjoyment, then, would not be diminished, but, perhaps, made fuller.

Similar analysis could be made of all human interests, with the result that the forcing of adult standards on infantile types would remain unkind; but the sad part of it is that, in spite of repressions, impositions., these imposed standards remain largely ineffective and are abandoned as soon as compulsion ends. Not so, if, by proper semantic education, we allow the infant to develop normally into adulthood. The *new* standards are not imposed, but become his own. We do not then need compulsion from without, because the new standards act from within and become pleasurable and lasting.

A similar process is very obvious in the practice of psychotherapy. The standards of evaluation of the patients are usually inappropriate to the conditions of modern life and often clash sharply with the accepted standards. Moralizing without changing by *other means* his standards of evaluation never accomplishes any satisfactory therapeutic results; quite the contrary, it often does a great deal of harm. A physician would be very unwise to censure or condemn a symptom, as this would preclude any beneficial results. What physicians usually do is to treat any symptom, no matter how repulsive, with great sympathy and understanding. They *do not attempt* to change the symptom directly, but, by the understanding of its main mechanism, they try to *change the patient's standards of evaluation,* of which the symptom is only a consequence. If at all successful and the physician succeeds in changing the inappropriate standards of evaluation, the symptom then automatically disappears. In daily life, we usually attack only the symptoms, disregarding mostly the underlying structural foundations; this method accounts for the doubtful results.

Under infantile standards we apply similar methods to society. Many may want to abolish wars, revolutions, 'depressions'., but they do not investigate structurally deep enough. They attack the symptoms, instead of analysing the structural issues which produce these symptoms.

In conclusion, let us notice that the analysis of a semantic mechanism on a printed page requires new terms and the co-ordination of many details., which, at first, do not always appear so simple, although, once the theoretical side is mastered, the educational application is genuinely simple. Thus, the analysis of the one-, two-, three-., and ∞-valued semantics may appear difficult, yet, in practice, it only amounts to imparting through our educational systems a semantic flexibility, instead of fixity; to acquiring the inclination of starting with observations, followed by descriptions, from which we pass to inferences, in connection with awareness of these ordered processes,. In training, it is enough to abolish identification, and this is easily achieved once we have produced the proper method, based on a language of new $\bar{A}$ structure. This last actually consists of but a few, new, simple, and common-sense terms, the analysis of which helps us to discover a few simple and invariant psycho-physiological relations. Thus, identification is eliminated by starting with an *ordinal* language and method. Once we get the feel of the horizontal and vertical stratification, and learn to differentiate between orders of abstractions, identification disappears. Silence on the objective levels produces a 'delay', involves and trains the cortex; our reactions become more and more intelligent in the human sense., .; and the most important results are reached by the simplest means.

The training in visualization and the abolishing of objectification are the first and most important steps for a complete elimination of identification. When this first step is achieved, the rest is comparatively a very simple task.

But the reader may ask why we should have to use such unfamiliar and, therefore, seemingly difficult methods to achieve such obvious results. Do we really need a $\bar{A}$-system to achieve the results which, even in an A-system, are known to be desirable? The answer is weighty with consequences and should be taken very earnestly. In the A-system, these desirable results could not be attained generally, because the structure of our old languages and the method hampered rather than helped us. New theories, new systems., are built precisely with the aim to facilitate adjustment. Those questions which in the older days were supposed to be 'philosophical', 'metaphysical'., the application of which required a high grade of intelligence, knowledge., to start with, become, in the new way, simply *a problem of the structure of the language we use*. All the issues appear closely interrelated. We do not require 'high intelligence' nor 'higher education' to begin with, in order to obtain these desirable results, as the results follow *automatically* from the structure of the language we accept and teach our children. Thus, the older impossibilities are achieved simply and automatically, with the greatest possible efficiency and the most lasting results.

CHAPTER XXIX

ON NON-ARISTOTELIAN TRAINING

If the preliminary experiments described above should be fully upheld, an important fact in the physiology of the cortex will be disclosed—namely, that new connections can be established in the cortex, not only in the areas of optimal excitability, but also in those areas which are in one or another phase of inhibition. (394)   I. P. PAVLOV

That wretched monosyllable "all" has caused mathematicians more trouble than all the rest of the dictionary. (23)   E. T. BELL

. . . these observations . . . point to the view . . . that the mechanism of development of a conditioned reflex and the mechanism of external inhibition are somehow similar, and that the process of external inhibition bears some relation to the development of new connections between different cortical elements. (394)   I. P. PAVLOV
In particular the factor of duration of time was shown to act as a real physiological stimulus, and experiments were described in which definite time intervals appeared as effective stimuli. (394)   I. P. PAVLOV

The procedure for training in the present system by the aid of the Differential follows directly from the theoretical considerations which have been explained in the foregoing chapters. The contentions of the system have been verified experimentally in all cases where it has been consistently applied.

The main aim is to acquire the coveted 'consciousness of abstracting', on which non-delusional evaluation is based, and which becomes the foundation for non-pathological *s.r* and sanity. As we deal with different aspects of an organic process which inherently works as-a-whole, all these aspects appear strictly interrelated. We have found by analysis two main aspects which underlie the others. It appears that the A structure leads to semantic states which can be formulated as the feeling of 'allness', and that, through the 'is' of identity, it leads to the confusion of orders of abstractions. Thus, for training, the program is readily sketched: we must first eliminate the 'allness'; then we must impart this peculiar stratification of 'human knowledge' which follows from the rejection of the 'is' of identity; in other words, eliminate identification. It becomes also obvious that a theory of sanity cannot be separated from a $\bar{A}$-system.

Since the organism works as-a-whole, all nerve centres should be trained so as to impart a permanent, lasting, and ingrained feeling of abstracting. Once this has been achieved, the recognition of the vertical and the horizontal stratification of human knowledge becomes, also, a permanent semantic state. This gives us a kind of semantic co-ordinate system, in which we can represent any life situation or scientific situation, or any difficulty, with great clarity, and so evaluate them properly. In verbal theoretical explanations this procedure appears complex; in practice, it is not so. It is extremely simple, provided we persistently follow the instructions, which are based on theory and practice. Above all, we must not expect results too quickly.

For reasons already explained, students should not only hear and see the explanations, but should also *perform for themselves,* should handle the labels and indicate with their hands the different orders of abstractions. After preliminary explanations, the children should be called to the Differential, and, using their hands, they should explain it. This applies, also, to grown-ups and to patients. The Differential is not only a permanent structural and semantic reminder which affects many nerve centres; it is more, for, in training, it conveys the *natural order* through all centres. Any reader who refuses to use his hands in this connection handicaps himself seriously, because *ordering* abolishes identification.

Fundamentally, there is no structural difference between the use of language and the use of any other mechanical device; they all involve reflex-action. It is well known that any pianist, telegraph operator, typist, or chauffeur would not be a successful performer if he had to meditate about every move he makes. As a rule, *verbal* explanations of the working of the respective machines are necessary at first, yet the structural reflex-skill required is actually acquired by prolonged practice, where again all nervous centres are involved. We all know what amazing *unconscious* reflex-adjustments a good driver of a car can make in case of unexpected danger.

A similar semantic reflex-skill is required in handling our linguistic apparatus; and, in case of danger, of sudden turns and twists, our orientation should also work unconsciously. That is why the structural *feeling* for the working of the apparatus is required. All nerve centres should be trained to employ the most effective means to affect the organism and its working as-a-whole.

The semantic training of grown-ups and that of children do not differ in essentials. Children have fewer established habits, have more fluid *s.r* than adults, and, therefore, the results with children are achieved more quickly and last better.

I shall now explain how to train children. A similar method applies to adults, also; but an adult should not trust himself too much that he has completely acquired 'consciousness of abstracting'. He must train very thoroughly. I speak from personal experience. Although I have the Differential before my eyes practically always and am the author of the present system, yet every once in a while I catch myself with one of the old vicious semantic habits. Habits, and particularly linguistic habits, may be very pernicious and difficult to change.

We do not need to start with profound theoretical considerations; we may start with any familiar daily-life objects and a microscope or magnifying glass. We bring the Differential into the classroom, with labels (except one) detached, but do *not* proceed to explain it. We start with a little semantic *experiment* upon the subject of 'allness'. We take any actual object, an apple, a pencil, or anything else which is familiar to the children. The principles involved are entirely general and apply to all objective levels in a very similar way. We tell them that we will have some fun. Then we ask them to tell us 'everything' or 'all' about the object in question; in this case, the apple. When the children begin to tell us 'all' about it, we write the characteristics down on the black-board. *This*

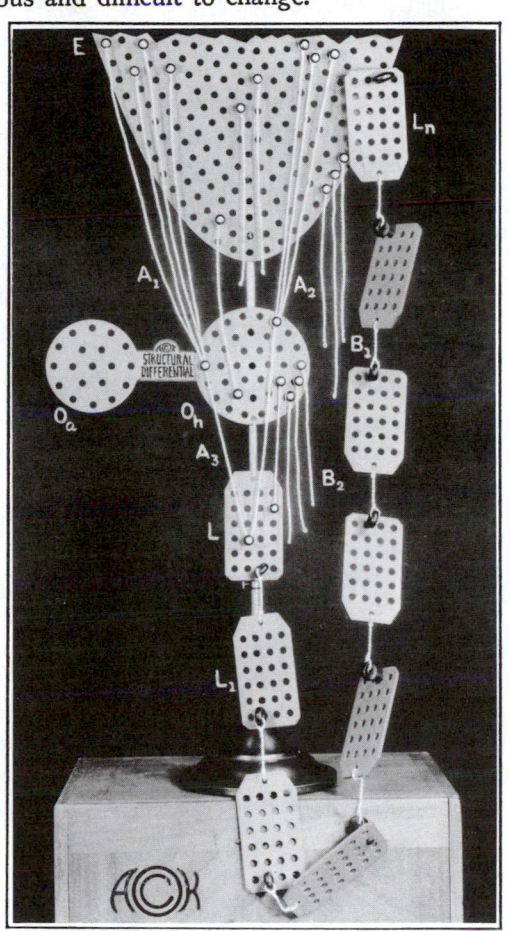

FIG. 1
THE STRUCTURAL DIFFERENTIAL

last is vital. We must have a visual and extensional record of the ascribed characteristics. When the children have exhausted their ingenuity in telling 'all' about the apple, we should *not* be satisfied. We should make them doubt, urging them that, perhaps, they did not tell 'all' about it, using the word 'all' continually. The term 'all' should be stressed and repeated to the point of the children's being thoroughly annoyed with the term. The more they learn to dislike this term, the better. We are already training a most important *s.r*.

We should *not* be satisfied with the best answers made by the most intelligent children. In a large class there may even be a child who tells us bluntly that it is impossible to tell 'all' about the apple. We should concentrate on the *less* intelligent children and deal particularly with them. There are many and important reasons for this. For one thing, the children become more eager and more interested in their own achievement. Then, too, they easily learn by example what a difference in intelligence means. This understanding of the shortcomings of others has an important semantic, broadening effect. In life, numerous serious 'hurts' occur precisely because we do not appreciate some natural shortcomings and expect *too much*. Expecting too much leads to very harmful semantic shocks, disappointments, suspicions, fears, hopelessness, helplessness, pessimism,.

The less bright children benefit also. The experiment is conducted on their level, so that they also have the maximum chance to benefit. Soon the children begin to argue about the new method and to explain it to each other by themselves; for we have touched very vital and complex semantic processes of 'curiosity', 'achievement', 'ambitions'., characteristics strongly represented in the child's life. We evade, also, the danger of taking clever, yet shallow, replies as standard. The last error would be fatal, as the issues are fundamental, and we should not rest content with mere verbal brilliancy.

When the subject seems exhausted, and the list of characteristics of the apple 'complete' (we repeatedly make certain that the children assume they have told us 'all' about it), we cut the apple into pieces and show the children experimentally, using eventually a microscope or a magnifying glass, that they did *not* tell us 'all' about the apple.

It may appear to some educators that such training might involve some undesirable psycho-logical results. But later, when consciousness of abstracting is acquired as a lasting semantic state, this fear appears entirely unjustified, as explained further on. The first step in dealing with 'reality' seems to demand that we abandon entirely the older delusional methods.

When the children have become thoroughly convinced of the non-allness and the *impossibility* of 'allness', we are ready to explain to them what the word abstracting means, using again the terms 'all' and 'not all'. We show them a small rotating fan and explain to them about the separate blades which when rotating we see as a disk. In such demonstrations we can go as far as desired. All science supplies data (e.g., the dynamic structure of seemingly solid materials). We must select the data according to the age of the children or the knowledge of the grown-ups. Everything said should be demonstrated empirically from a structural point of view.

The next step is to demonstrate practically that an object taken from different points of view has different aspects for different observers. We may use different objects or wooden geometrical figures painted with different colours on different sides. We may place the object in different positions and ask the children their descriptions, which should be written down. The descriptions will, of course, be different, and the children should be made thoroughly aware of this. In all these preliminary exercises the ingenuity of the teacher has a vast field for exercise, and we do not need to enter into details.

When all these results have been accomplished on the level of the *least developed* child, we then proceed to explain the Differential as a structural diagrammatic summary of the above results. It is a *positive condition* that the new language be used, and that an object be described as an *abstraction* of some order. If this vital structural point is disregarded, most of the psycho-logical semantic benefits of 'non-allness' are either lost or greatly lessened. We should make this term clear to the child, and should train him in its use, as it appears uniquely in accordance with the structure and the functioning of his nervous system. The child should be warned that the old languages are not structurally suitable for their future understanding and semantic adjustment. This warning should be repeated seriously and persistently.

Having eliminated 'allness', we begin to eliminate the 'is' of identity, which, at the primitive and infantile stages of racial human development, happens to be extremely ingrained in our *s.r*, embodied, as it is, in the structure of our daily language. As was explained before, identification is a natural reaction of the animal, the primitive man, and the infant, reflected and systematized in the A and older linguistic systems, which, through the ignorance or neglect of parents and teachers, is not counteracted and so is continued into the lives of children and grown-ups, until, finally, it becomes embodied in the structure of what we call 'civilization' (1933). In a theory of adjustment or sanity we must counteract this animalistic, primitive, or infantile *s.r* by building a $\bar{A}$-system, which entirely rejects the 'is' of identity.

In the A-system, through the use of this 'is', different orders of abstractions were unconsciously identified in values, in obvious contradiction to empirical facts. In other words, being identified in values, they were treated as of one order or on one level and so did not necessitate indefinitely many expanded orders of horizontal and vertical differences. Similarly with the objectively meaningless 'infinite velocity' of a process, it does *not* allow *order*. But once we have a finite velocity of a process, *order* makes its appearance as an indispensable aspect of a process. *The finite and known velocity of nerve currents on the physico-mathematical levels results in ordered series on physiological levels; in non-identity and proper evaluation on semantic levels, and in orders of abstractions and a non-aristotelian system and general semantics on verbal levels.*

Once we abolish in our language the always false to fact 'is' of identity, we automatically stop identifying different orders of abstractions. We do not assume that they represent one level, which becomes expanded into a natural ordered series of indefinitely many different orders of abstractions, with different values. Adjustment, therefore, sanity and adulthood of humanity, depend on proper evaluation, impossible under conditions of delusional identification of fundamentally different orders of abstractions. We must then train the *s.r* in the natural *physiological order* of the process of abstracting which, on the psychological levels, become non-pathological semantic evaluation.

In the case of training in the 'non-allness', it was necessary to start with the analysis of an ordinary object, to give the child a simplified theoretical explanation, and then to demonstrate it empirically. The child will be easily 'convinced', but this conviction is not enough, because it will not affect permanently his *s.r*. We explain this difficulty very simply, telling him that, although he 'agreed' with our presentation, he will very soon 'forget' it, and so we need a permanent visual reminder which is supplied by the strings, freely hanging from the event and from the object, and indicating those 'characteristics left out', or not abstracted.

In the elimination of the 'is' of identity, we have also structurally interconnected aspects. The rejection of this 'is' becomes an equivalent to the stressing of the *stratification* in the structure of 'human knowledge'. To facilitate training, we should *stress both aspects* by all available means, and should involve as many nerve centres as we can. Thus, through the ear we stress verbally the formula of the rejection of the 'is' of identity by indicating with our finger the different orders of abstractions, in the meantime, affecting the eye while we repeat 'this *is not* this'. We utilize the kinesthetic centres, not only by pointing the finger to different levels, but also by making broad motions with our hands, indicating the stratifications. We should train in both horizontal and vertical stratifications, always using the hands. The horizontal stratification indicates the difference, or ordering of different order abstractions; the vertical stratification indicates the difference between 'man' and 'animal' and the differences between the different absolute individuals. In both cases, the semantic effect of the 'is' of identity is counteracted.

The above procedure in training has an important neurological foundation. Besides what has been explained already, we find that a word has four principal characteristics with corresponding cortical representations. A word can be heard, seen, spoken, and written. Language, thus, involves many nervous functions; e.g., auditory, visual, and diversified motor nerve centres, interconnected in a most complex network of 'horizontal' and 'vertical' fibres. The use of the Differential involves all available nervous channels; we see, we hear, we speak, we move our hands, indicating stratification, 'non-allness'., engaging large cortical areas, and so have the maximum probability of affecting, through *non-el* methods, the organism-as-a-whole. The Differential gives us a special, simplified, yet advanced interracial *structural symbolism* (1933), which affects wide nervous areas of the illiterate, or nearly illiterate person, or of the infant., which otherwise could not be affected. It is known that extensive reading and writing, as well as speaking a number of languages, has a very marked cultural effect and helps visualization and consciousness of abstracting. The reason can be found, perhaps, in the fact that a learned polyglot, or a scholar, utilizes many nerve centres in co-ordination. In the older days, unless one became a scholar of some sort, it was extremely difficult to train these nerve centres in co-ordination. With the Differential we can train simply, and comparatively quickly, all necessary nerve centres, and so impart to children and to practically illiterate persons the cultural results of prolonged and difficult university training without any complicated technique. This last should always be regarded as a means and not as an end.

In my experience with children, and with men from the lowest 'mentality' to the highest, the non-identity of different orders of abstractions is usually taken lightly. It all seems so simple and self-evident that no one assumes that there could be serious, unconscious, structural, semantic, linguistic, and neurological delusional mechanisms involved, which cannot be reached without specially devised non-identity training. The delusional feelings of 'allness' and 'identity' are peculiar in that, like other pathological states, they tend to appear as all-pervading. It is the most difficult in daily, as well as in medical, experience to make a breach in this all-pervading tendency, but once this delusional state is even partially replaced by glimpses of *m.o* reality, the further elaboration and training in adjustment to 'reality' becomes comparatively simple. Thus, in practice, if we *start* with ordinary objects, feelings, and words, and train in the non-allness and non-identity, any child, or any grown-up, even an imbecile, can follow this easily. Once this feeling has been acquired, and in most cases it is only a matter of method and persistence to acquire it, the main *semantic* blockage has been eliminated, and the rest is comparatively easy. I have had no opportunity yet to verify it, but I am convinced that even a superior imbecile could be trained to differentiate between descriptions and inferences, after he has learned to differentiate between the objective levels and words. In such a training with imbeciles we can go in simplicity as far as desired ; thus, if the given individual is hungry and says he wants 'bread', we hand him a label which is attached to the objective bread, and he would be quickly made to realize that the symbol *is not* the thing symbolized.

It should be realized that in the training we should impart the obvious fact that words or labels represent conveniences, and *are not* the objects or feelings themselves. We should carry the labels in our pockets, so to say, as we carry our money, or checks for hats or trunks, and not identify them 'emotionally' with what they eventually stand for, because monetary standards change, and hats and trunks get exchanged, lost or burnt. To accomplish this, we must have *objective labels,* which we may handle and carry in our pockets, and also an objective something to which we can attach the labels. In the present $\bar{A}$-system the rejection of the 'is' of identity is complete and applies to all levels. Thus, the event *is not* the object ; the object *is not* the label ; description *is not* inference ; a proper name *is not* a class name . ; the characteristics ascribed to events, objects, or labels *are not* identical, an object, a situation, or a feeling *is not* identical with another object, situation or feeling ., ., all of which establishes a *structure of horizontal and vertical stratification.* At an early stage of the training, we must begin with what appears the simplest and most obvious to the child ; namely, the absence of identity between the word and the object, or that the word *is not* the object. We accomplish this by stressing that one cannot sit on the *word* 'chair', that one cannot write with the *word* 'pencil', or drink the *word* 'milk', . These simple facts should always be translated into the *generalized form,* indicating with the hand the two levels on the Differential, conjointly with the fundamental formula 'this *is not* this'. We should always tell the child that the formula is entirely general, but for the present we should not go into any further details.

At this stage we can advance one step further, still using *only* ordinary objects as examples, and explain the un-speakable character of the object; namely, that whatever we can see, taste, smell, handle., is an absolute individual (demonstrated empirically) and *un-speakable*. We then take the apple, bite it (actually performing), and explain that, although the object is *not words*., yet we are very much interested, and traditionally so, in this un-speakable level. Then we explain repeatedly and at length, emphasizing the important principle of evaluation, that to live we must deal with the objective level; yet this level cannot be reached by *words alone*. As a rule, it takes a few weeks, or even months, before this simple *s.r* is established, the old identification being psycho-physiologically very much ingrained. Once this is established, we stress the fact that we must handle, look, and listen., never speak, but remain silent, outwardly as well as inwardly, in order to find ourselves on the objective level. Here we come to one of the most difficult steps in the whole training. This 'silence on the objective level' involves checking upon neutral grounds of a great many 'emotions', 'preconceived ideas',. This step, in the meantime, appears as the first, the simplest, most obvious and most effective psychophysiological 'reality-factor' in eliminating the delusional identifications.

Once the child is thoroughly aware of the absence of identity between words and objects, we may attempt the expanding of the notion 'object' to the 'objective levels'. Such training requires persistence, even though it seems fundamentally simple. We demonstrate and explain that action, actual bodily performance, and all objective happenings, *are not* words. At a later stage we explain that a toothache, or demonstrate that the actual pain of a prick., *are not* words, and belong to the *objective un-speakable levels*. Still later, we enlarge this notion to cover all ordinary objects, all actions, functions, performances, processes going on outside our skin, and also all immediate feelings, 'emotions', 'moods'., going on inside our skins which also *are not* words. We enlarge the 'silence' to all happenings on the objective levels and the animalistic, 'human nature' begins to be 'changed' into quite a different *human nature*.

When this is accomplished the rest is much simplified, although much more subtle. We explain, as simply as we can, the problems of evaluation and *s.r*, stressing and making obvious the fact that our actual lives are lived *entirely* on objective, un-speakable levels. We illustrate this all the time by simple examples, such as our sleeping, or eating, any activities, or pain. or pleasure, or immediate feelings, 'emotions'., which *are not* words. If words are not translated into the first order un-speakable effects, with the result that we do not do something, or do not feel something, or do not learn or remember something., such words take *no effect* and become useless noises.

One fact should be stressed; namely, that the problem is *not* one of 'inadequacy of words'. We can always invent 'adequate words', but even the most ideal and structurally adequate language will *not be* the things or feelings themselves. On this point there is *no possible compromise.* Many people still utter quite happily, pessimistic expressions about the present *language,* based on silent assumptions connected with unconscious delusional identification, and believe that in an 'adequate' language the word by some good primitive magic would be identical with the thing. The more the denial of the 'is' of identity is driven home, and the sooner it becomes a part of one's *s.r,* the sooner the 'consciousness of abstracting' is acquired.

We are now ready to go further into the theory of *natural evaluation* based on natural *order.* As a preliminary step, we must show repeatedly the difference between descriptions and inferences, using simple examples. We must stress the fact that words, as such, must be divided into two categories: a first, of descriptive, in the main, functional words; and a second, of inferential words, which involve assumptions or inferences. Thus, 'A does not get up in the morning' may be considered as descriptive. If A explicitly refuses to get up, the statement 'A *refuses* to get up in the morning' may also be considered as descriptive. If A did not explicitly refuse, this statement becomes inferential, because A may be dead or paralysed. If we would say simply, 'A is lazy', such a statement represents an illegitimate inference of high orders based on ignorance, because in 1933 it is known that 'laziness' represents a symptom of physico-chemical, colloidal, or semantic disturbances. It should be stressed that this discrimination between descriptive and inferential words, although extremely important, is not based on any 'absolute' differences, but, to a large extent, depends on the context. I shall not analyse this problem further, because any parent or teacher who has acquired the consciousness of abstracting himself will find more examples at hand than are needed.

We should notice here a very vital, yet generally diregarded, structural fact—that *human* life is lived under conditions which establish a *natural order* of importance between different orders of abstractions. This natural order should be made the basis of natural adaptive evaluation and so survival *s.r.* As our lives are lived *entirely* on the *un-speakable level,* which includes not only scientific objects and ordinary objects, but, also, actions, functions, processes, performances, feelings, 'emotions'., this level is obviously first in importance, and the verbal level, which is only auxiliary, comes next in importance. The analysis of the relative evaluation between description and inferences appears extremely complex and would require a separate volume, beyond the scope of the present work. Here we may assume the generally accepted opinion that the reliability of inferences depends on the reliability of the descriptive premises, and that description is more reliable than inference. In importance and in temporal and neurological natural *order,* description comes first; inferences, next. If we consider different orders of inferences, or inferential words, inferences or inferential words of lower order are more reliable and so more important than inferences of higher orders (inferences from inferences of lower orders).

As science is a *racial* product and so represents structural descriptions and inferences of an enormous amount of constantly revised observations and formulations of past generations, this racial product, 'science', is more reliable and important *in principle,* particularly in its negative results, than the individual abstractions of individuals. If some individuals happen to be 'geniuses', who upset racial scientific abstractions, they are under the scrutiny of other scientists who, no matter how biased or slow, remain judges of their products. In 1933 the opinion of scientists is the most dependable opinion we have. We must accept, at a given date, the racial, particularly negative, abstractions as more reliable, establishing in evaluation the event (scientific object) first, and the ordinary object next. It should be stressed that the 'object' of daily experience, in human life, is by far not so reliable as that in the life of animals entirely without human interference. Thus, a high tension wire, or a third rail, or high explosives are not found in unaided nature and do not forewarn us as ordinary objects do. These 'objects' possess characteristics concealed or not obvious on the objective level of our ordinary inspection of, let us say, sight, hearing, or smell; yet these characteristics appear just as 'real' and dangerous as ever. It appears, then, that the 'scientific object', or the event, in contradistinction to the ordinary object, is more important than the daily object, no matter how important the latter might be. In fact, the only macroscopic importance of objects, outside of aesthetic and symbolic values, may be found in those not obvious physico-chemical, microscopic, and sub-microscopic characteristics. Thus the importance of food, or air, or a chair is found precisely in these physico-chemical effects which result from eating, from breathing, and from resting on a chair, and so again these hidden characteristics, revealed *only* by science, appear much more important than the gross characteristics manufactured by our nervous systems which we recognize as an object.

We come thus to a *natural scale* of a definite *natural order,* which also establishes the *natural order* of *genetic importance* and represents the *natural* base for *survival semantic evaluation.* For our purpose the relative order may be represented as the scientific object or the event first, ordinary object next; the ordinary object first, the label next; description first, inferences next, extended to descriptive and inferential words.

If we use the 'is' of identity and identify in value or importance the different, ultimately non-identical, levels, we nullify in principle the natural order of evaluation, which, by psychophysiological necessity, appears as a *reversal* of natural order in various degrees. We find many reasons for this curious fact, but, for our purpose, it will be enough to suggest that: (1) words are simpler and take less effort to handle than objects; (2) inferences being higher order abstractions than descriptions, are psycho-logically closer to our feelings and easier for any individual to manage than impersonal descriptions which require developed linguistic training, power of observation, self-mastery., and, in general, consciousness of abstracting. The reversal of the natural order must lead to non-adjustment and results in pathological symptoms in different degrees. The natural order consists of asymmetrical relations expressed by an ordered series, not only as to space-time, but as to values. All our experiences and all we know indicate definitely that ordinary materials ('objects') are extremely rare and very complex special cases of the beknottedness of the plenum; that the organic world and 'life' represent extremely rare and still more complex special cases of the material world; and, finally, so-called 'intelligent life' represents increasingly complex and still rarer special cases of 'life'. When we identify the members of these series, we disregard the asymmetrical character of this series and transform it into a fictitious, or delusional, or false to facts symmetrical relation of identity. It becomes also obvious why in the *A*-system, which did not allow asymmetrical relations, proper evaluation, adjustment, and sanity in general were, in principle, impossible.

Although the language used in this connection is not familiar, it is not entirely arbitrary. It appears experimentally that four-dimensional order has physiological importance on the one hand; on the other, that on the psycho-logical levels it involves the semantic factors of evaluation. In training in the *physiological natural order,* we train the *evaluation* or appropriate human and adult *s.r* on the *psycho-logical levels.*

In the difference between the un-speakable 'scientific object' and the ordinary object, the objective level and the verbal level, we find the precise spot at which we differ most radically from the animals. If we disregard these differences and retain the 'is' of identity, we must some-how copy animals in our nervous processes. Through wrong evaluation we are using the lower centres too much and cannot 'think' properly. We are 'over-emotional'; we get easily confused, worried, terrorized, or discouraged; or else we become absolutists, dogmatists,. The results of such copying of animals are usually tragic, as might be expected. Owing to wrong evaluation we add self-made semantic difficulties to the dif-ficulties which we actually find in nature. When we live in a *delusional* world, we multiply our worries, fears, and discouragements, and our higher nerve centres, instead of protecting us from over-stimulation, actually multiply the semantic harmful stimuli indefinitely. Under such circumstances 'sanity' is impossible.

It seems that here in the elimination of the 'is' of identity we have put our hands on an extremely powerful reflex-mechanism for the edu-cation, or re-education, of our 'emotional' life. As has already been said, suppressing or repressing our feelings is dangerous, and should be avoided. The old animalistic educational systems were built on repres-sion and suppression, with sad results. But since we had no other means of education, we had to use the older means or else abandon this special education altogether. Not so in the new $\bar{A}$ way with the Struc-tural Differential. We *do not* repress or suppress. We teach silence on the objective level *in general*, which is a most impressive *'emotional' edu-cation,* on perfectly *neutral* grounds, one of the consequences of the elimination of the 'is' of identity. Any bursting into speech is not repressed; a gesture of the hand to the labels reminds us that words are *not* objects, or action, or happenings, or feelings. Such a procedure has a most potent semantic effect. It gives a semantic jar; yet this jar is *not repression,* but the realization of a most fundamental, natural, struc-tural fact of evaluation in which we should all be well trained. Dis-turbing *s.r* subside, and no one is 'hurt'. It takes long and persistent training, but the results are most beneficial.

We must note an important difference between a statement involv-ing the 'is' of identity, that 'we are animals'—which has nothing to do with the actual facts; all of us (the animals, too) *are not* words, but rep-resent absolute individuals and all different—and the statement that we 'copy animals' in our nervous reactions. In the first case, nothing can be done. In the second case, although the results are *equally sad,* we can stop 'copying animals' the moment the mechanism is discovered and we begin to realize that we are doing so. Thus, the old hopeless becomes hopeful.

31

I have already mentioned that some educators may assume the eventual harmfulness of training in the consciousness of abstracting on the ground that children should be kept 'close to reality'. The answer to such an argument can be found in the recognition that what in the older days seemed 'reality' must now, in the light of new knowledge, be considered delusions, and the older training as preparatory for acquired un-sanity. The modern conditions of human life appear much more complex than those of animal life, or of the primitive man. Every year, perhaps even every month, new human 'realities' make their appearance; complexities arise and our educational systems do not equip the children semantically to meet these new conditions. After investigation one may find by himself that the older 'allness' and identifications represent delusional factors found nowhere in the empirical world, and thus have to conclude that if we train children in such delusions, adjustment to the actual world is made extremely difficult, if not impossible. It is true that some beneficial results do not appear at once, but only after the full consciousness of abstracting is acquired. Thus, at an early stage of the training, when the student begins to realize the delusional character introduced by the 'is' of identity, the general and well-known tendency may struggle hard to *retain* the *delusions*. His *first* reaction may be that of disappointment, with its many concomitants, depending on his temperament, metaphysics.; but when he acquires the freedom of the full consciousness of abstracting, all levels will be evaluated properly and he will be able to adjust himself to conditions of *m.o* reality described in the present work, which cannot be avoided by any one. 'Knowledge' or 'intelligence' is only possible with abstracting, and, therefore, it fundamentally involves 'non-allness'. 'Omniscience' would involve a 'knowledge' of every point-event. These are fundamentally different, and such a world would be one of chaos, where knowledge would be impossible. Life, *m.o* abstracting, and *m.o* intelligence start together and are *conditioned by the m.o process of abstracting*.

Among the many semantically beneficial results of such training, besides the training in sanity and, therefore, in adjustment, a few other benefits should be mentioned. Our life, our *m.o* mentation, the structure of our language, with its syllogisms, fallacies., consist for the most part of the constant utilization of the different levels of abstractions. This appears as an inherent characteristic of 'human knowledge', and, therefore, we cannot abolish it without abolishing *m.o* intelligence altogether. Intelligence requires the passing from level to level in both directions. All the benefits we possess follow from this; but also many semantic dangers are hidden in it. Similar remarks could be made about an automobile,. A great many beneficial results follow from the use of automobiles., but there are also great dangers involved. For instance, at present we have regulations for the driving of an automobile. A driver has to pass his examination, demonstrate his practical reflex-ability in driving., before he is allowed to drive in public. Similarly with our language; we find the greatest benefits in it, and we should utilize them. Proper training in the use of language should teach us how to avoid dangers. Obviously, 'consciousness of abstracting' teaches us how to avoid these dangers; likewise, once we become trained in the passing to higher and higher order of abstractions, we become capable of the performance of what we call 'high intelligence'. The difference between 'high intelligence' and 'low intelligence' consists in the fact that a 'high intelligence' has a larger outlook backwards as well as forward; a 'low intelligence', as suggested in Fig. 2, sees only a little backwards (ignorance) and foresees only a little. A 'high intelligence' has a larger span or field; it knows more about the past and looks further into the future.

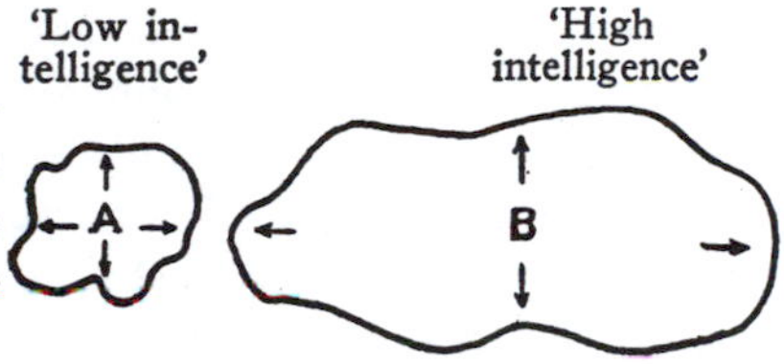

Fig. 2

It is no mystery that when we want to look further into the past and the future we need higher and higher order abstractions. By training in this passing to higher and higher abstractions we train the 'mind' to be more efficient; this 'mental' expanding should be the structural and semantic aim of every education.

Once we eliminate identification, we must accept *structure* as the only possible content of 'knowledge' and also realize that no 'knowledge' is ever free from some structural assumptions. Sometimes it is pathetic to watch the metaphysical performances of some otherwise very eminent scientists, who seem entirely innocent of these facts. They often attempt to divorce their metaphysics from science, and miss the point that primitive metaphysics represents 'science' or the structural assumptions of that period, whereas modern science represents structural assumptions or metaphysics of modern 'times', which cannot be reconciled with the older 'science'. The difference appears in dates, not in kind. The real problem before mankind presents itself in the selection of a structural metaphysics. If we select the primitive structural assumptions and have to live under present conditions, we must become a split personality which cannot adjust itself. If we accept modern structural assumptions called science, we may adjust ourselves. In no case can we free ourselves entirely from some structural assumptions. The problem becomes one *of dates,* and of un-sanity versus sanity. These problems appear of unusual importance, because the difficult scientific technique does not enter into this field at all, and the few structural data (1933) can be given in the simplest form to children and even to the feeble-minded. In the older days this problem was entirely misunderstood. We tried to 'popularize' science in the sense that we translated structurally correct language into the daily language of *primitive structure;* this resulted only in bewilderment; we did not analyse the structure of language and its role in our lives and *begin* with a structural linguistic revision. Once this revision is accomplished, and we build a $\bar{A}$ language, the semantic background is prepared for a natural acceptance of modern *structural* metaphysics (science) of each date and the older 'popularization' becomes unnecessary. Such procedure would help to integrate the individual, while the older methods only help to split him.

Let us recall that the animal stops somewhere in his abstracting. When we come to a *stop,* and consider it 'final' or that we 'know all about it', we copy animals in our nervous reactions. Training in passing from order to order of abstractions as such, trains the particularly *human,* fluid, non-blocked *s.r,* counteracts, and ultimately abolishes, the animalistic blockage. In a language of a given structure we can express ourselves in a definite way; and, if that way is incomplete, we must leave the field open, for in a structurally different language the issues may look entirely different.

It is fundamental to stress that the old 'unknowable' becomes entirely abolished. This 'unknowable' originated in the primitive identification and elementalism. Our ancestors could not miss indefinitely that identification was false to facts; yet somehow the emphasis, which the ecclesiastical authorities (for their purpose) laid on the importance of the *A-system,* prevented them from completely rejecting the 'is' of identity. The un-speakable was called the 'unknowable', a very gloomy term, indeed. The use of this term prevented them from discovering, long ago, that the only content of 'knowledge' appears as structural, with all the following *non-el* consequences. 'Knowledge' was expected to represent somehow more than 'knowledge'—a silent self-contradiction. On this foundation whole systems of delusions were built. With the newer realizations, we understand that the only possible content of 'knowledge' appears as *structural,* so that we can know all which belongs to the structural legitimate field of 'knowledge'. What does not belong to the field of 'knowledge' must be considered meaningless, and making noises about it, one way or another, will not help us at all; quite the contrary, it involves us in delusional states. Students of the history of 'philosophy' may realize, in this case, particularly, the drama and the dangers which the playing on such *m.o* terms as 'knowledge' may needlessly produce.

Through the semantic mechanism which it involves, 'consciousness of abstracting' abolishes many fears, despairs, and other disturbances which follow from the confusion of orders of abstractions. We become introverted extroverts; in other words, we become affectively well-balanced, and ready to deal with empirical first order effects on their levels, and with verbal problems on their different levels. We learn, also, to *observe,* as soon as we have learned 'silence' on the 'objective levels'. Realizing that we abstract in different orders, we slowly acquire the most creative structural feeling that human knowledge is inexhaustible; we become more and more interested in knowledge; our curiosity becomes aroused; our sporting spirit stimulated and our level of *m.o* intelligence raised.

It is well known that the higher intelligence is characterized by a critical attitude. By training with the Structural Differential until the memory of the characteristics left out and the non-identity become a permanent semantic acquisition with us, this critical attitude is also developed. No one who feels habitually these 'characteristics left out'—'this is *not* this'—will ever *take a word or a statement for granted.* He will enquire, investigate; will always ask 'what do you mean', a question which automatically leads to further investigation, and finally strikes the bottom of the undefined terms which divulge our silent structural creeds and metaphysics.

We should avoid the mistake of assuming that the average man, or a moron, does not 'think'. His nervous system works continually, as does that of a genius. The difference consists in that its working is not productive or efficient. Proper training and understanding of the semantic mechanism must add to efficiency and productiveness. By the elimination of semantic blockings, as in identification, we release the creative capacities of any individual. We release him from the primitive semantic bondage in the daily and constant use of a powerful instrument called language—full of benefits, but also full of dangers—the structure of which he totally misunderstands. Such misunderstanding must lead to inefficiency in the use, and so to the abuse, of this function. Instead of being a semantic slave of the structure of language, he becomes its master.

When we become more civilized and enlightened, no public speaker or writer will be allowed to operate publicly without demonstrating first that he knows the structure and semantic functioning of the linguistic capacities. Even at present no professor, teacher, lawyer, physician, or chemist., is allowed to operate publicly without passing examination to show that he knows his subject. The above statement does not mean control or censorship. Far from it. Our language involves a much more intricate, beneficial, or dangerous semantic mechanism than any automobile ever had or will have. We do not control the drivers in their destinations. They come and go as they please, but for *public safety* we demand that they should have acquired the necessary reflex-skill for driving, and so we eliminate unnecessary tragedies. Similarly with language, of which the ignorant or pathological use becomes a public danger of a very serious semantic character. At present public writers or speakers can hide behind ignorance (1933) of the verbal, semantic, and neurological mechanism. They may 'mean well'; yet, by playing upon .the pathological reactions of their own and those of the mob, they may 'put over' some very vicious propaganda and bring about very serious sufferings to all concerned. But once they would have to pass an examination to get their licence as public speakers or writers, they could not hide any longer behind ignorance. If found to have *misused* the linguistic mechanism, such an abuse on their part would be clearly a *wilful act,* and 'well meaning' would cease to be an alibi.

We must accept the obvious facts which make the older theoretical 'democracy' or the older theoretical 'socialism' a scientific impossibility. If, in 1933, 99% of the population of the globe appear as infantile or 'mentally' deficient, how can any one expect that the majority or the mass could ever have proper evaluation or non-pathological *s.r*? All history shows at present, and this evidence should not be taken lightly by *scientifically enlightened society,* that the majority appears 'always wrong', and that all that we call 'progress', 'civilization', 'science'., has been achieved by a very small minority. Such an understanding should guide our future conduct if we desire better results than we have at present. Under $\bar{A}$ conditions, not the state, nor different private societies, but professional scientific bodies would have to set the standards and perfect the technique of the linguistic structural examinations. They would also select members who would serve on the examination boards. It might seem that such a $\bar{A}$ innovation would not be important or far-reaching. This would be a mistake. It seems that most of those public writers and speakers may be considered privately as 'honest' men, who do not realize that under A conditions they often impose on defenceless masses delusional states which too often become of a pronounced morbid character. Once such an examination would force them to look into structural, semantic, and linguistic problems, we may take for granted that a great many of them would become able to evaluate properly their own activities and comprehend the harm they do. As a result, quite probably, a great amount of useless, befogging issues, delusional writings and speeches would *not* be produced, with great benefit to all concerned. No one would censure them. Consciousness of abstracting would accomplish that. They would become their own censors, aided, also, by the newly developed consciousness of abstracting on the part of some members of the audiences or readers.

It would be desirable to experiment and introduce parallel classes in schools for a while; one group continuing in the old A-system, the other being trained in the $\bar{A}$-system. We may expect that at the end of a year the results would be fairly tangible. The ones which acquire the 'consciousness of abstracting' should show a marked improvement in character, should behave better, and should also show better results in scholarship, not to mention, in addition, the *preventive* semantic benefits in their future life and adjustment. Experimenting under various conditions is very desirable, as we deal with such a tremendously broad and fundamental structural problem that it is impossible at present to foresee more than the main results and bearings.

In a school one three-dimensional Differential should be enough, but in each classroom a large printed diagram, which is published also, should be permanently exhibited on the walls and applied in all studies. This is necessary, not only because such a reminder makes the children thoroughly familiar with the 'characteristics left out', 'natural order'., but also because the children will discuss it and settle their educational and personal difficulties by its aid and so train themselves in $\bar{A}$ reactions. In my practice, I have found that one of the main difficulties of the learner, or in 'thinking', in general, consists in the fact that in any verbal discussion we must utilize different orders of abstractions and $m.o$ terms. If we do not realize this, the problem often seems very involved; once we are conscious of it, however, the problem becomes simple. In fact, it may be said that this special flexibility which is entirely absent in animals and little developed in the primitive man, represents the working mechanism of 'high intelligence', and that this special flexibility can be acquired through proper $\bar{A}$ training.

The dealing with reflex-reactions and with experimental theories in general has one very encouraging characteristic; namely, that no matter how difficult the theoretical side may be, the practical is invariably extremely simple. Thus, a theoretical treatise on the Einstein theory, or the new quantum mechanics, or on an automobile, radio, or piano, or on music, or the conditional reflexes of Pavlov., may present, and, in fact, generally does present, difficulties, because it is formulated on purely verbal and analytical levels. But these levels are most important, for we find that on these levels the *full evaluation,* and so the full realiza-tion, of existing or possible *relations* and *meanings* is accomplished. In these verbal levels we find also economical and effective means to analyse further developments on which the possible range of applications ulti-mately depends. Such a treatise can be produced by a single man and thereby becomes available for the rest of us.

A description of the application is, however, very simple; we label the related parts of some structure, describe, mostly in terms of order, their interrelations, and then give instructions how to act, push, pull, or turn a given part to get such and such results. These descriptions, al-though verbal, refer exclusively to some physical structure, so that men of very low 'mentality' can soon become acquainted with the practical problems concerned. When the reflex-handling of the physical structure is acquired, the experimental and behaviouristic aspects become child-ishly simple. A child can see the experimental results of any theory, or notice the ease and simplicity of the reflex-adjustments a good driver can make.

But what an infant, a savage. or an ignorant man can*not* do, is *appreciate* the *meanings* of given occurrences and *evaluate* them; in other words, they cannot *relate* the given occurrences to other occurrences which alone give the *significance* of the happenings. Thus, not only physicists, but even the average man, knew of the equality of gravi-tational and inertial mass; it took the genius of an Einstein, however, to *evaluate* properly, to have the proper *semantic reaction* toward this 'commonplace fact'. The present work shows clearly that all semantic disturbances exhibit a lack of proper evaluation; or in getting hold of the *meanings,* or *relations,* or *order* of different order abstrac-tions. Only a full *theoretical* understanding can supply us with those meanings and produce in us the proper *s.r* of evaluation—a necessary psychophysiological step for further advance, and for full application of the conquests already made.

As the present work is experimental throughout, and deals with verifiable subjects, such as the structure of languages on record, the natural order of development, the pathological reversal of order, which, if again reversed, restores the natural order., and, when applied, brings about most beneficial experimental results, all that has been said about experimental theories applies fully in our case.

Just as in other disciplines, the instructions are simple: 'push these', 'pull this', or 'turn that'; so, in our case, this simple descriptive rule which refers to the objective Differential is given as: 'this *is not* this'. Once the reflex-activities have been acquired, we can, for instance, enjoy the pleasure of an automobile trip, the music of a radio, or a semantic trip toward sanity in harmony with ourselves and others, very simply, in spite of the underlying theoretical complexities which are always means and not ends in themselves.

But here we must face an important difference. It is easy to demonstrate empirically to the majority of us the usefulness or pleasurableness of automobiles and radios, but it is very difficult to demonstrate the benefits of consciousness of abstracting to those who have not acquired it. Before the experimental data begin to accumulate and become common knowledge, the main evaluation will have to be made on theoretical grounds. Besides, before children can be trained by the simple and easy methods outlined above, adults must first re-train and re-shape their own *s.r*, which re-training and re-shaping are not easy, and require even more difficult theoretical considerations. Because of this, the present work had to be written in the form of a text-book for parents, teachers, physicians, and workers in 'mental' hygiene, and for future students and research workers in psychophysiology and semantic hygiene.

At the beginning, in the application of the method, a number of difficulties will be discovered which will have to be overcome. As a rule, the training in non-pathological *s.r* proves to be easiest and simplest with very young children. Most of it, or at least the laying down of the semantic foundations for such reactions, should be accomplished at home by specially trained teachers, if the parents are unable to do that themselves. In countries or communities where the national or local governments show an interest in the health of the population by providing, for instance, specialists in preventive vaccination, specialists to train in preventive measures against semantic disturbances will probably also be provided.

In elementary schools the teachers at first will have to train themselves as best they can with the help of specialists; but in high schools, colleges, and universities, special instructors will have to be employed.

The first concern, then, is to start the education and training of teachers. With this end in view, the present work has been written so as to form a fairly complete outline of the whole problem; reference literature has been indicated, so that any one who wants to specialize in the subject can find some suitable text-book as a primer. As to qualifications for the professional $\bar{A}$ instructors,—it is, at present, very difficult to foresee details, but, as the full consciousness of abstracting leads to *s.r* which, also, follow unconsciously, or which should so follow, from the study and acquiring the *feel* of the calculus,—students of mathematics would, perhaps, be the most desirable. Any specialist in a new line of work has to learn a great deal, and this cannot be avoided; but it makes a difference what kind of training one has had as a young person. Thus, it is simpler for a student of mathematics to learn about psychiatry, or psycho-logics, than for a psychiatrist or a psycho-logician to learn mathematics. However, for a person with university training, this is less important than a genuine willingness to master the subject. Once the consciousness of abstracting has been acquired by such a student, his semantic blockages will be eliminated. He will then have no difficulties whatsoever with details, or even in doing creative work along this line.

With very young children, in the beginning an hour a day for several months should be devoted to the subject. When they have acquired the consciousness of abstracting, the results should not be entirely trusted as to permanence, but, at least, once a week the problems should be recalled to them. How many hours a week should be devoted to it in high schools, colleges, and universities I shall not venture to suggest, because the working hours in these institutions are already very crowded. The training in consciousness of abstracting automatically eliminates an enormous amount of semantic blockages, and would facilitate the acquisition of learning in all branches of knowledge, and so save 'time' and effort—the more so, if the respective teachers themselves were to become conscious of abstracting.

The beneficial results which are to be expected may be found in better scholarship, more interest in studies, improved character, higher *m.o* intelligence, and a better general adjustment. All of this seems quite apart from the *preventive* character of the training as a protection against many semantic disturbances in the future. But when teachers of all subjects acquire consciousness of abstracting themselves, they will probably discover new means and methods of conveying more simply and more effectively what they wish to convey to their pupils. I am convinced that the hours spent on semantic training would actually turn out to be an important *economy of efforts*. Moreover, it would effectively give the children and students the highest grade of *cultural training,* which at present we acquire only occasionally and with difficulty, without the conscious co-operation of our teachers.

CHAPTER XXX

IDENTIFICATION, INFANTILISM, AND UN-SANITY VERSUS SANITY

Common sense, do what it will, cannot avoid being surprised occasionally. The object of science is to spare it this emotion and create mental habits which shall be in such close accord with the habits of the world as to secure that nothing shall be unexpected. (457)   BERTRAND RUSSELL

Medicine is to-day an Art or Calling, to whose exercise certain Sciences are no doubt ancillary; but she has forfeited pretension to be deemed a Science, *because* her Professors and Doctors decline to define fundamentals or to state first principles, and refuse to consider, in express terms, the relations between Things, Thoughts and Words involved in their communications to others. (122)   F. G. CROOKSHANK

Unless physiology like any other of the sciences basic to medicine will teach less fact and more method, it might as well be deleted from the catalogue.
Can anything be done to help the situation? Not, I think, without large vision. The student of medicine needs not more external but more internal discipline.*   MARTIN H. FISCHER

I wonder how soon we shall be far enough along to have the physician ask: How much and what, if anything, is *structural?* how much *functional, somatic* or *metabolic?* how much *constitutional, psychogenic* and *social?***
     ADOLF MEYER

Section A. General.

The name of Freud is usually associated with the term 'the unconscious'. This term appears as a general *descriptive* term standing for a great many psycho-logical semantic processes. In 1933 the work of Freud is generally accepted as important and very suggestive, although further experiments by many research workers and practicians have shown that the freudian formulations have not the exclusiveness formerly assumed for them.

It is useless to deny that the term 'unconscious' is fundamental and necessary. The use of the term is best shown in the study of hypnotic phenomena. Some patients do certain things under hypnotic influence, and then seemingly lose all memory traces of such doings upon emerging from hypnosis. Careful experiments showed that after prolonged efforts these recollections could be made accessible to the patients' waking consciousness. The difficulty in recalling was not ordinary 'forgetting'. What is 'forgotten' can also be spontaneously 'recollected'. Here the situation seemed different, in that these lost 'memories' required considerable work and effort for their reconstruction. The psycho-logical

---

*Teaching of Physiology. *Jour. Asso. Med. Colleges.* Apr. 1929.
**The "Complaint" as the Center of Genetic-Dynamic and Nosological Teaching In Psychiatry *New Eng. Jour. of Med.* Aug. 23, 1928.

state in such cases of perfected 'forgetting' was called 'unconscious', which, as a descriptive term, is very satisfactory.

The origin of the freudian theory of the unconscious was strictly scientific. The theory was a new generalization in a new structurally appropriate language to account for *experimental facts*. Subsequently, a large number of other experimental facts showed that the work of Freud was sound as far as it went. Different workers, from different sets of facts, amplified or reshaped the freudian theories. At present, there are several schools that differ widely in language, all of them based on the fundamental system-function of Freud, however.

In reading the literature of the subject, one finds most diversified material. Seemingly, there are facts which prove 'beyond doubt' each and every· theory, no matter how widely they differ among themselves. It is also easy to find experimental facts which can be accounted for by several different theories.

Such a situation appears unsatisfactory. This lack of generality conceals a very important and workable semantic mechanism, which, under A identification, 'allness', and elementalism, becomes pathological, resulting in arrested or regressive symptoms. Once we pass to a $\bar{A}$-system free from the above harmful semantic factors of delusional evaluation, the difficulties do not arise. The more my enquiry progressed, the more it became obvious that the underlying mechanism appears similar in all psychoanalytical theories. It seems that the general problem may be formulated as the need to discover methods for non-delusional evaluation affecting our *s.r*, and so be able to make the 'unconscious' 'conscious'.

The term 'consciousness' is an incomplete symbol, as it lacks content. If we use the term 'consciousness of abstracting', we ascribe content and also gain empirical means to bring under educational control a vast array of important psycho-logical processes. The *negative* term 'unconscious' does *not* imply specific content, and the main difficulty in its practical application is to find its content, or to ascribe content to it. Once this is achieved, the 'unconscious' becomes 'conscious'. A patient whose unconscious semantic difficulty is made conscious either improves or is entirely relieved. For a general theory, we must find general structural means of ascribing semantic content to the 'unconscious'. Different schools have elaborated different means of discovering this desired content. All schools agree that the behaviour difficulties are due to experiences hidden in the 'unconscious' and that bringing them to 'consciousness' seems the main goal. The diverse schools have an unduly bitter attitude toward one another, and have not attempted to analyse the problems at hand from a more general, more workable *non-el* structural, semantic, system-function, and linguistic point of view. Whether or not the term 'consciousness' has any content besides 'consciousness of abstracting' may be disregarded for the moment. At any rate, the term 'consciousness of abstracting' gives very vital and workable psychophysiological means of analysis, of an impersonal, and general structural, and semantic character. Enquiry into the clinical cases and literature shows that pathological cases, amenable to treatment, appear improved by a similar evaluational treatment; namely, the correction in some form or another of the semantic disturbance of the lack of 'consciousness of abstracting'.

'Mental' illnesses (infantilism included) appear as semantic arrested development or a regression to lower levels, to those of the primitive man, the infant, the animal. The animal *is not* conscious of abstracting; man can become so. Here we find the precise mechanism of a decisive nature which not only supplies us with *preventive* measures, but which should also become of therapeutic value.

All life exhibits conservative characteristics acquired during the long periods of its development. In the facts of heredity and embryology, we have an excellent evidence of this. In its development, the germ cell of an animal or man repeats in a very abbreviated way the structures of forms from which it descended. The ever-changing environmental conditions, although they affect each organism to a large extent, produce extremely few hereditary changes, which again may be considered an indication of the conservative characteristics of life.

As we have already explained, life, abstracting, and 'intelligence' started together, and are consequences of the physico-chemical colloidal structure of the protoplasm. Psychiatry also assumes that 'the unconscious', 'tendencies', and 'impulses' originated with life itself. From this point of view the past piled up structurally upon the past until the highly complex organism called Smith made its appearance. In this process of evolution the 'instincts' and 'impulses' have had an important role, not only conservative, but also compensatory and protective. In man, $s.r$ should be based on proper *evaluation* and so play both a stimulating and *protective* role. Under A conditions of delusional evaluation, the protective role appears practically non-existent; the human organism, under modern conditions, becomes over-stimulated, resulting often in pathological conditions. Consciousness of abstracting, or the elimination of delusional evaluation, abolishes man-made, artificial, and harmful irritants.

I shall borrow from Jelliffe an excellent diagram to illustrate the evolution of the periods of growth and shall follow closely his exposition.[1]

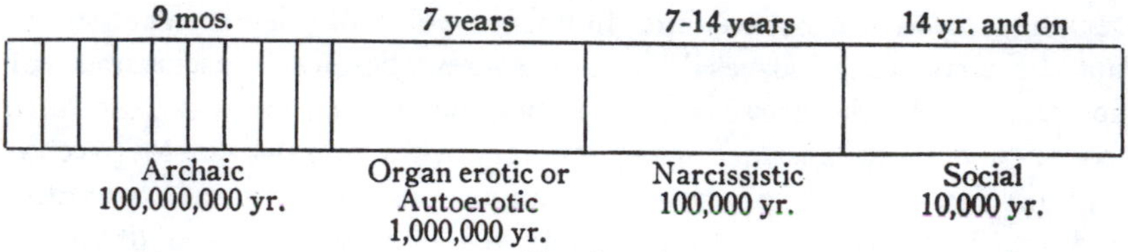

9 mos.	7 years	7-14 years	14 yr. and on

Archaic	Organ erotic or	Narcissistic	Social
100,000,000 yr.	Autoerotic	100,000 yr.	10,000 yr.
	1,000,000 yr.		

FIG. 1

Some such classification of the periods of growth has been forced upon psychiatrists by the study of 'mental' ills, and is justified by embryology and an endless chain of empirical observations.

The first period represents the archaic period and is of the greatest antiquity. In it the past is roughly recapitulated from, let us say, the beginning of unicellular life to the anthropoid ape. At the fertilization of the egg the hereditary constitution is established, and during the gestation period all prenatal influences are laid down. The life of the baby before birth may be described as a vegetative existence of full indolence with all needs supplied by the mother's body.

At birth, the first 'struggle for existence' begins, the struggle for air, as symbolized by the first cry. At this epoch the baby already appears as a self-running organism with some *s.r.* His vegetative nervous system is integrated and functioning. He begins to 'feel'. Pleasure and pain begin to be significant semantic factors. This period is called the organ erotic or autoerotic, since, as with animals, its main interests are *'sense' gratifications*. Many millions of 'sense' receptors suddenly have thrust upon them from the environment a mass of energy with which the organism has to deal somehow. At first there is a rivalry between different 'senses'. Later, co-ordination appears. Each group of receptors establishes its own semantic values for itself, depending upon its own cell growth. This period may be divided, schematically, from birth to seven years, and corresponds roughly to the evolution from the higher animals to primitive man. This period is extremely important from a semantic and educational point of view. At this stage in the human child the nervous system is not fully developed; and different environmental influences (language, doctrines included) may twist this development, so that irreparable harm can easily be done.

The narcissistic period is named after the Greek mythical figure, Narcissus, who, seeing his reflection in a pool of water, became so engrossed in self-adoration that he rejected the attentions of Venus and was killed. In another version of the myth his punishment was loss of sight. This period covers, more or less, from seven years to fourteen years. As the name indicates, it represents a semantic period of self-love.

The child has not entered, as yet, into a social stage of development. He remains egotistical, egoistical, self-centred, and *asocial*.

At about fourteen, the social semantic period begins, which leads, when 'normal', to the adult socialized individual.

We should realize that these semantic stages are 'normal' when they are lived through within the age limits indicated here. Even if children show some characteristics which are not desirable (organ erotic or autoerotic, narcissistic), this, in itself, does not constitute a danger, provided they outgrow these undesirable manifestations. The serious dangers, and even tragedies, begin when some of the infantile or narcissistic semantic characteristics are carried over into the life of the grown-ups.

Not only 'intellectual' growth but also 'emotional' development may be arrested on some earlier lower level. In such cases we speak of idiots, imbeciles, and morons on the 'mental' levels; and of moral imbecility, infantilism, narcissism, and, in general, of 'mental' illness on the 'emotional' levels.

Besides arrested growth, or under-development in some respect, cases of so-called *regression* are frequently encountered. Regression follows the general scheme as outlined in Fig. 1, but in a reversed order. The following diagram, Fig. 2, is also taken from Jelliffe, with slight modifications.

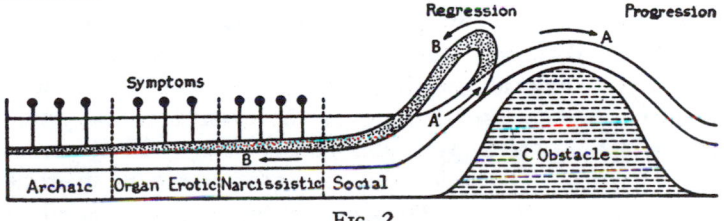

Fig. 2

As processes, life, development, or regression are best represented as 'vector quantities' which have direction and magnitude. In one type of cases of regression the progressive tendency or energy is strong; yet the obstacle is also very great, so that the progressive tendencies may not be strong enough to carry over the obstacle or to conquer it. Again, the progressive tendency or energy may be weak, and the obstacle correspondingly slight, yet strong enough to start the regressive movement.

In the healthy individual the progressive tendency is not easily diverted from its forward course. He conquers his obstacles (C) and goes on (arrow A). Weaker individuals (A') may surmount their obstacles with more difficulty or may start regression on smaller obstacles, as indicated by arrows (B). In such cases they may regress to different levels, developing a neurosis or a psychosis, in accordance with the degree of regression. It is extremely instructive to study these different phases in regression and to watch how the symptoms arrange themselves in a perfectly orderly manner. In some instances the regression goes so far as to bring the patients to the foetal level. Such a patient sits in a dark corner in the foetal position with the head covered with a rag. His 'mentality' and semantic responses are similar to those of the foetus, practically none.

Regressions to the archaic level are usually hopeless of improvement, so that I shall not analyse them in this work. We are mostly interested in *under-development,* or in regression not further than to the autoerotic or the narcissistic semantic levels, in which treatment in many instances yields curative results.

Jelliffe gives among others a very instructive diagram as a method of showing how the personal make-up of an individual can be plotted (psychogram). These diagrams afford excellent graphic means for orientation. One of them I reproduce on the opposite page (Fig. 3).[2]

The circular form of the diagram is particularly appropriate, as it shows clearly how the horizons, activities, and interests widen from the archaic (animal?) through the child and savage, to the adult socialized individual. The dips in the eye, stomach, and bladder sectors correspond to definite symptoms. In the eye sector the dip goes to the narcissistic semantic level. Whenever this patient is riding in an automobile and another car is coming close, so that a collision seems possible, the patient experiences a compulsory shutting of the eyes, a typical narcissistic semantic symptom which symbolizes that something which one cannot see cannot happen. The patient did not regress to the organ erotic level and become actually blind or deaf (psychic or rather *semantic* blindness or deafness); so the dips are not plotted to the organ erotic level. In the bladder and nutritive sectors we see that the curve sinks as low as the organ erotic level. These dips correspond to striking semantic symptoms. When the patient drives in her car and is held by the traffic, she has an involuntary passage of urine. The dip in the nutritive sector corresponds to the symptom that after eating certain foods the patient is able to bring them back into her mouth (selective rumination). Analysis by Doctor Jelliffe has revealed that in the case of the passage of urine when the patient is held up in traffic, her unconscious organ erotic semantic fantasy triumphs over the need for self-control, and she asserts her mastery through the early and necessary mastery acquired over the control of the bladder. Since she is prevented from doing one

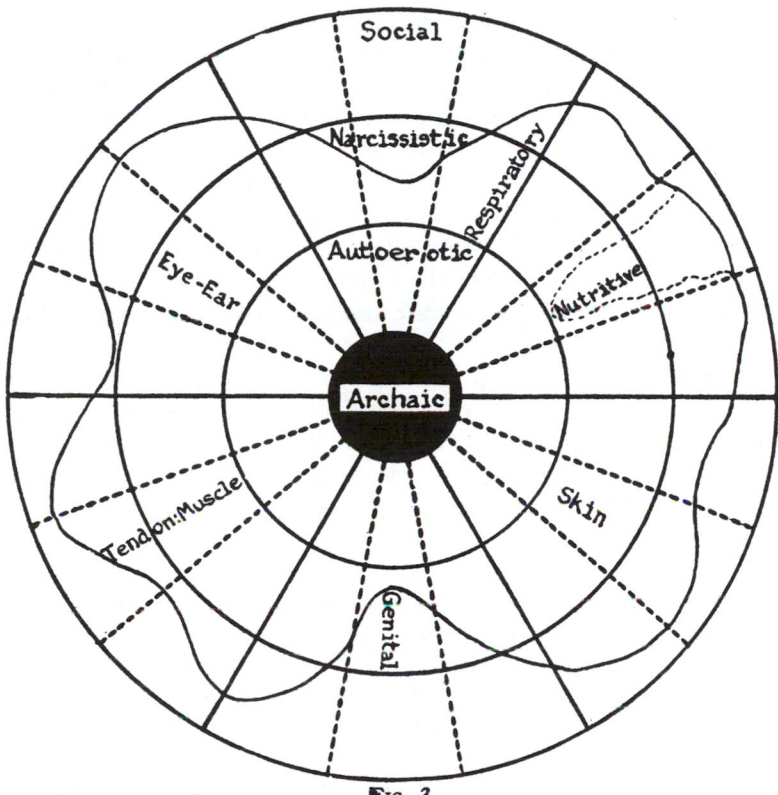

Fig. 3

Schematic representation of regressions and fixations in one patient. The dip in the respiratory sector represents a psychogenic asthmatic defence *s.r*; in the eye and ear sector, refusing to see or hear the 'truth' and 'reality'; in the genital, urinary eroticism. The deep incision in a fairly well socialized nutritive 'libido' represents an 'emotionally' conditioned *s.r* for selective rumination of individual ingredients in the stomach. The patient appears as a severely sick individual on the border of a psychotic reaction. Present nosological schemes would call this an anxiety-hysteria or a manic-depressive psychosis if the semantic compensation should break and further regression occur. (After Jelliffe.)

thing, her power of doing gets semantic expression through a substitute act which cannot be prevented by outside interference. The selective rumination symptom also goes back to the nursing semantic period. When suckling she would vomit after a full meal, and demand another nursing, to which the mother foolishly acceded. Vomiting then became her semantic way of controlling 'reality'. She used that as a weapon in getting her wishes with her family. She has amplified and refined the methods of semantic expression of this old mastery over 'reality', and the selective rumination seems one of the results.

32

The patient also exhibits other neurotic semantic symptoms. She is too impatient to read and does not remember what she has read. She can never remain quiet. She is very keen-sighted to find fault in others, very acute to hear the last verbal equivocations, and very neat and clean with reference to her bodily secretions. Here we see clearly the semantic mechanism of infantilism and the contradictions between the conscious performance and the unconscious fantasies.

In an hypothetically healthy individual, his make-up could be represented graphically by a circle on the social level of adjustment. He would have outgrown the passing semantic stages of the archaic, organ erotic, and narcissistic periods.

Failure in semantic adaptation to 'reality' might be represented by dips in the curve to such a level as the individual fixation or regression has put him. By such means we have an excellent method to represent clearly the weak spots and to show the focal semantic points of conflict in evaluation where energy is diverted to useless or harmful fantasy ends.

When the dips or deviations are few and slight, we call them idiosyncrasies; for instance, such a habit as the narcissistic tactile fantasy of toying with a button, a moustache, or eyebrow. When the number of failures is larger and the semantic symptoms go to lower levels of development, we speak of hysteria,. When the level of regression is still lower (organ erotic or archaic), we are usually entitled to speak of a psychosis.[3]

We have already emphasized over and over again that the organism works as-a-whole, and that, therefore, any *el* splittings cannot lead to satisfactory results. The verbal division of 'body' and 'mind' remains verbal, and also involves a language whose structure does not correspond to the structure and functioning of the organism. A language is like a map; it *is not* the territory represented, but it may be a good map or a bad map. If the map shows a different structure from the territory represented—for instance, shows the cities in a wrong *order,* or some places east of others while in the actual territory they are west.,—then the map is worse than useless, as it misinforms and leads astray. One who made use of it could never be certain of reaching his destination. The use of *el* language to represent events which operate as-a-whole is, at least, equally misguiding and semantically dangerous.

With this in mind, let us briefly analyse the 'obstacle' in Fig. 2. As we deal with 'obstacles' in a life sense, we can generalize the obstacle to some semantic factors involving meanings and evaluations which may arrest the development as well as result in regression.

From the *non-el* point of view every obstacle and difficulty involves semantic evaluation. Any and all reactions to lower order abstractions involve the cyclic chain of higher order abstractions, no matter how imperfectly. In ordinary language, a physical occurrence with which we become acquainted through lower nerve centres involves our 'mental' *attitudes,* doctrines., in general, *s.r* influenced by the activities of the higher centres. From this *non-el* point of view, *surprise,* fear, fright., enter, and usually do the harm. Physical pain seldom, if ever, leads to semantic disturbances, but fear, fright., and surprise usually do. Anticipation of danger, or proper evaluation of a situation, has a *protective* effect, as it usually tends to diminish or abolish the fear, fright., or *surprise.* The outside world is full of devastating energies, and an organism may only be called adapted to life when it not only receives stimuli but also has protective means against stimuli. Now such anticipation or expectation makes an organism *prepared,* and the difference between a prepared and an unprepared organism in the face of danger or pain may turn the scale of the outcome.

Section B. Consciousness of abstracting.

It is obvious that in the human organism the field for stimulations is vastly greater than in animals. We are subjected not only to all external stimuli but also to a large number of permanently operating *internal* semantic stimuli, against which we have had, as yet, very little protective psychophysiological means. Such structurally powerful semantic stimuli are found in our doctrines, metaphysics, language, attitudes,. These do not belong to the objective external world, and so the animals do not have them in a like degree. As our enquiry has shown, in practically all 'mental' ills, a confusion of orders of abstractions appears as a factor. When we confuse the orders of abstractions and ascribe objective reality to terms and symbols, or confuse conclusions and inferences with descriptions., a great deal of semantic suffering is produced.

Obviously, in such a delusional world, different from the actualities, we are not prepared for *actualities,* and then always something unexpected or 'frightful' may happen. The organism cannot adjust itself to such fictions; it is not prepared to face ∞-valued *m.o* realities and must suffer from constant surprises and painful semantic shocks, which do the harm.

As we have already seen, the general preventive psychophysiological discipline in all such cases of confusion of orders of abstractions is found in 'consciousness of abstracting'. When conscious of abstracting we cannot identify the symbol with the thing,. In the case just described, the difficulties of the patient were precisely in intensified *mis-evaluation* —the confusion of the symbol on the infantile semantic level with *m.o* reality—and this persisted, in spite of later serious inconvenience and difficulties when the symbol did not any longer produce the desired submission of others and became in itself a nuisance.

We could analyse from this semantic point of view all psychiatry, and we would find that the intensified mis-evaluation or confusion of orders of abstractions is always very prominent in 'mental' illnesses. This characteristic is very general, and the suffering these confusions produce is very acute. The general protective psychophysiological measure, however, is very simple; namely, 'consciousness of abstracting'.

A fundamental difference between 'man' and 'animal' is found in the fact that a man can be conscious of abstracting, and an animal cannot. This last statement could be reformulated: that animals are 'unconscious of abstracting'. Now consciousness of abstracting is not inborn as a rule, but becomes a *s.r* acquired only by education or through very long, and usually painful, experience in evaluation. If we are *un*conscious of abstracting, we obviously copy animals in our 'mental' processes and attitudes and cannot completely adapt ourselves to the structurally more complex human world (with higher order abstractions), so that some arrested or regressive processes are bound to result. In such a more *complex* world we need *protection* against semantic overstimulations, which the animals in their simpler world do not need. If, therefore, we copy animals in our 'mental' processes, we could, perhaps, live in their simpler world, but cannot adapt ourselves easily to a structurally more complex human world.

We see here the general semantic mechanism of human adaptation. Our human world is more complex; the number of stimuli is enormously increased. Against this excessive stimulation we need protection, which is found in the consciousness of abstracting. One adjusts oneself by increasing the field of 'consciousness', and by giving it properly evaluated content as against the vast 'unconscious' which covers the animal's life and our own past. In 'mental' ills we find the arrested or regressive stages, with a vast and harmful unconscious. 'Mental' therapy always has the semantic aim and method; namely, to discover the unconscious material and make it conscious, and so make proper evaluation possible.

It is quite remarkable that 'mental' therapy, which actually is a form of semantic, *non-el* re-education, is only successful when it succeeds in making the patient not only 'rationalize' his difficulties but also makes him 'emotionally' revive—live through again, so to say, and evaluate anew— his past experiences. This process can be compared with a glass of water in which some chalky sediment lies on the bottom. In semantic difficulties the different 'hurts'., may be compared to the water *and* the sediment. 'Rationalization', alone, is like throwing away the clean water and letting the sediment remain. No improvement follows; the semantic sediment of earlier evaluation is still there and does its work. But if we stir up the water *and* the chalk, then we can throw out both and a clearing up will follow. The *non-el* 'living through' of the past experiences is equivalent to this semantic stirring-up of meanings before eliminating the immature evaluations.

This semantic mechanism is well recognized, yet puzzling. It shows that it is more difficult to influence the affective than to affect the 'rationalization'. One may 'rationalize' perfectly well; yet his lower centres will not be affected sufficiently. It is possible that the confusion of orders of abstractions or identification is, in the main, responsible for it. With the use of the Structural Differential and with training in the orders of abstractions and in *silence* on the objective levels, we gain seemingly extremely powerful psychophysiological means of an entirely general character to influence directly the *affective* responses, in which we are aided by the utilization of all available nerve centres. It also shows once more how persistent is the working of the organism-as-a-whole. *The harm was done by organism-as-a-whole methods (affecting higher and lower centres); the protective semantic agencies should employ similar means.*

In the older 'mental' therapy we tried to bring the unconscious or buried material into the conscious, but each psychiatrist proceeded by a *private* method, and according to a special theory. Such procedure is obviously not general enough for simple *preventive* training on a large scale. The present system offers such general and effective semantic psychophysiological means. By making ourselves conscious of abstracting we prevent the animalistic *un*consciousness of abstracting, and so prevent arrested development or regression. We bring into consciousness some of the most fundamental human characteristics, of which animals are *un*conscious, and so prevent arrestment or regression to lower levels. The method is entirely general and simple, based on the elimination of identification, introducing natural and so adaptive evaluation which should not stir up resistance in the child.

In the freudian theory the famous Oedipus complex purported to explain the often unconscious hostility of the son toward the father and his excessive attachment to the mother. The researches of the anthropologist Malinowski show that in primitive matriarchal societies the biological father is not recognized as such and is only a kind of friend and nurse to his son. The other functions which the father has in patriarchal societies are here performed by the mother's brother. The taboos which apply to the mother in patriarchal societies apply to the sister in such matriarchal societies.

The results are quite interesting. Malinowski found that seemingly similar unconscious semantic mechanisms are at work. But the hostility is toward the uncle, and the excessive attachment is directed toward the sister.

Malinowski concludes that the freudian mechanisms are thus proven. According to the present theory these facts are very important and show clearly that 'sex', as such, has nothing or very little to do with these 'complexes', but that the active unconscious agents appear as semantic and *doctrinal*. Doctrines and their meanings to the individual, their applications, identifications., make the father in one case, and the uncle in the other, the dreaded ., member of the family. Because of lack of consciousness of abstracting the child reacts to such application of doctrines with some 'complex', or semantic state, based on identification, non-mature evaluation, involving non-mature *non-el* meanings, in spite of *el* theories and languages.

Similarly, some spanking or other pain in childhood may later result in a neurosis. A successful analysis can usually trace neuroses back to some such experiences. What did the harm? Was it the burning *physical* feeling? Obviously not; for any child has had in its childhood many more painful experiences, and yet no semantic harm has followed. So we must look in another direction, and the elimination of identification or of the confusion of order of abstractions at once offers a solution. The 'spanking' had many factors; some of them were 'physical', some 'mental'. If we consider among the 'mental' factors the objectifications of 'authority', 'hell', 'sin', and other terms of evaluation, these result in fright and other semantic shocks, which ultimately lead to the neurosis. We know from our own experience how little affected we are by an accidental hit. Such a purely physical experience which does not give a
* see page xii semantic shock cannot product a neurosis.

It is not difficult to see that an investigation of 'hurts', 'emotional shocks', 'fear', 'fright', 'surprise'., must lead to a more general enquiry into the structure of 'human knowledge', meanings, evaluation, *s.r.*, which must include the structure of science and mathematics.

In the disregard of the stratification of human knowledge, or in identification or the confusion of orders of abstractions, we find an ever-present and abundant semantic source of human suffering, which increases unnecessarily the internal stimulations and so disturbs the efficient working of the organism-as-a-whole.

Psychiatrists in purely pathological fields have also discovered different sources of human difficulties. They discovered that the 'unconscious' seems quite a dangerous affair, and that 'mental' ills exhibit symptoms of an arrested or regressive process.

This present investigation, as well as the psychopathological ones here referred to, although conducted on entirely different grounds, one more general than the other, has discovered very similar mechanisms; namely, the benefit in enlarging the field of 'consciousness', by bringing into 'consciousness' important factors of the 'unconscious' and thereby counteracting the semantic possibility for arrested development or regression.

With such divergence, both in methods and in material used, the similarity of results indicates strongly the soundness of the conclusions. Generality offers, also, a criterion of practical simplicity and workability; and on these grounds the more general semantic $\bar{A}$ discipline commends itself.

Because of this generality the present theory has not only a simple but also an *impersonal* character which makes it available as a preventive measure in elementary education. With the older theories we deal in practice with personal responses to meanings; in our case we deal with the *s.r* in *general* and with their psychophysiological mechanism in particular.

The present enquiry started with the search for a sharp difference between Fido and Smith. This was found in the fact that Smith functions as a 'time-binder', while Fido does not. Further investigation into the mechanism of the time-binding function revealed that its most important characteristic is found in its peculiar stratification into many orders of abstractions. The realization of this stratification eliminates identification and leads to the 'consciousness of abstracting', thus ascribing a permanent, *strictly human* content to 'consciousness', and so automatically eliminating such animalistic, and, therefore, arresting or regressive, '*un*consciousness'. It is found, also, that in the consciousness of abstracting we find a general and simple psychophysiological semantic method for the elimination of the majority of human difficulties. In the training in this consciousness of abstracting we find a workable physiological tool with which to integrate the functioning of the human nervous system. We use organism-as-a-whole methods and achieve organism-as-a-whole results. We find in the language of 'semantic reactions', 'non-elementalistic meanings'., psychophysiological means to integrate the 'emotional' with the 'intellectual', which was hampered, to say the least, by the old *el* languages and methods and systems. The organism by structural necessity acts as-a-whole, but the old elementalism with its psychophysiological effects prepared the semantic background for split personalities, which a *non-el* system helps to re-integrate. We find a rather astonishing result; namely, that the structure of human achievements corresponds to the principle of stratification with the resulting consciousness of abstracting, usually limited to the special field. The majority of individual and group difficulties are found, also, to be due to the very general disregard of this principle.

By the scientific data of 1933 it seems well established that the enlargement of the field of 'consciousness' is extremely desirable. With this aim, a more general enquiry into the character of the 'unconscious' may also be worth while. Let us investigate the structure of science (1933), and see if some 'unconscious' factors cannot be found there. We find a curious fact, that mathematicians, in addition to their other activities, make a business of unravelling hidden unconscious assumptions. Their enquiries have led to a thorough investigation of the structure of their language in two directions: one, investigation of the underlying assumptions; the other, working out the 'implications'.

Let us give a simple structural example of this. Two assumptions are said to be equivalent when each of them can be deduced from the other without the help of additional new assumptions. For instance: (*a*) The fifth postulate of Euclid—'If a straight line falling on two straight lines make the interior angles on the same side less than two right angles, the two straight lines, if produced indefinitely, meet on that side on which are the angles less than the two right angles', (*b*) 'Two straight lines parallel to a third are parallel to each other', (*c*) 'Through a point outside a straight line one and only one parallel to it can be drawn'. Each assumption silently, unconsciously, presupposes the other, so that they can be deduced from each other. They actually are different forms of the same propositional function.

Another case is equivalence *relatively* to a fundamental set of assumptions A, B, C, ... M. It might happen that, in diminishing the fundamental set, two assumptions which were formerly equivalent cease to be so. For instance, the following assumptions are mutually equivalent and also equivalent to the fifth postulate of Euclid. (*a*) 'The internal angles, which two parallels make with a transversal on the same side, are supplementary.' (Ptolemy.) (*b*) 'Two parallel straight lines are equidistant.' (*c*) 'If a straight line intersects one of two parallels, it also intersects the other.' (Proclus.) (*d*) 'A triangle being given, another triangle can be constructed similar to the given one and of any size whatever.' (Wallis.) (*e*) 'Through three points, not lying on a straight line, a sphere can always be drawn.' (W. Bolyai),.

But the following two assumptions are only equivalent to the *E* fifth postulate if we retain thé postulate of Archimedes*: (*a*) 'The locus of the points which are equidistant from a straight line is a straight line'; (*b*) 'The sum of the angles of a triangle is equal to two right angles.' (Saccheri.)[4]

The crucial point of this discussion is that all that has been said here is *not obvious* even to the attentive and intelligent reader, nor to many mathematicians. It took nearly two thousand years and some of the efforts of the best scientists of the world to discover these *connections and implications*. The above examples illustrate a *general* underlying structure of all our languages. They have inherent interconnection, underlying assumptions and implications, the analysis of which, outside of mathematics, is seldom, if ever, carried far enough. Now these structural assumptions and implications are inside of our skin when we accept a language—*any* language. If unravelled, they become conscious; if not, they remain *unconscious*. In the present work we have already had an opportunity to become acquainted with unconscious implications which are concealed in the structure of any language. We saw that we must start with undefined terms, which represent structural assumptions and postulates, as we have no means to explain them or define them at a given date. We found that these *undefined* terms represented our unconscious metaphysics, and that the way to make this unconscious metaphysics conscious was to start explicitly with undefined terms, produce a system of postulates., a procedure which is completely fulfilled only in mathematics.

It should be noticed (as this is very important) that the undefined terms, being undefined, are overloaded with '*emotional*' *values*. As the higher nervous centres cannot handle them, the lower nerve centres work upon them overtime. If we do not analyse our languages into their undefined terms and structural postulates, our strongest 'emotional' and semantic components, which made these languages, remain hidden and unconscious.

---

*The Postulate of Archimedes is stated by Hilbert thus: Let A_1 be any point upon a straight line between the arbitrarily chosen points A and B. Take the points A_2, A_3, . . . so that A_1 lies between A and A_2, A_2 between A_1 and A_3.; moreover, let the segments AA_1, A_1A_2, A_2A_3, . . . be all equal. Then, among this series of points, there always exists a certain point A_n, such that B lies between A and A_n.
This hypothesis is used by Saccheri in its intuitive form; namely, a segment, which passes continuously from the length a to the length b, different from a, takes, during its variation, every length intermediate between a and b.[5]

Here we are face to face for the first 'time' with a wider, more general, and impersonal 'unconscious', which underlies the structure of any language, and so is operative in every one who uses a language. We may call this general form the scientific, or public, or linguistic, or semantic, or by preference, as a term, the *structural unconscious*. It embodies the underlying *structural* assumptions and implications which are silently hidden behind our languages and their *structures*. These assumptions., may be called 'unconscious', because they are *totally unknown and unsuspected,* unless uncovered after painful research.

Any form of representation has its own structural assumptions at its basis, and when we accept a language we unconsciously accept sets of silent structural assumptions of which we become semantic victims. For a long while, the white race has been the victim of the unconscious assumptions and metaphysics which underlie the A, E, and N systems. It has needed a structural revision of these systems, culminating in $\bar{E}$, $\bar{N}$, and, finally, $\bar{A}$ systems. These non-systems are characterized not by the introduction of new assumptions, but by making the older unjustified, primitive, unconscious structural assumptions conscious, and so helping us in eliminating the semantically undesirable reactions. We have already seen how fallacies and taboos (1933) can be and were manufactured unconsciously by semantic processes; these start with more general, more natural, and more fundamental structural errors, such as the primitive 'identification', for instance, which are due to pre-human ways of 'thinking' and which result in semantic difficulties and regression even today.

Let us assume, as an illustration, that the fifth postulate of Euclid is a false assumption, seriously detrimental to human life and comparable to some of the false doctrines that underlie the morbid symptoms with which psychiatrists deal every day. Let us assume, further, that a doctor, ignorant of the structure of 'human knowledge', *s.r*, and the equivalence of assumptions, succeeds, after painful and laborious efforts, in eliminating from a patient this special vicious assumption. Yet, because of his oversight, he pays no attention to another assumption equivalent to the first, and does not eliminate it. In such a case, rationalization about the first false doctrine would probably make the treatment a failure, as the other unconscious and equivalent doctrine would, in virtue of the extremely formal, one-, and two-valued character of the unconscious, perform its task and make the treatment ineffective. The tangle of equivalent structural assumptions in daily life is still unanalysed. For instance, it is extremely difficult to attempt to impart 'proper evaluation' without eliminating identification ,.

The higher and lower order abstractions seem structurally and neurologically, as well as functionally, interconnected in a cyclic chain, and so can never be entirely divided. A language—any language—involves undefined terms which, with the structure of the given language, express the silent and unconscious metaphysics underlying it. A language, for its maximum serviceability, must, at least, have the structure of the events it attempts to describe; and so science must first discover the structure of events, for only then can we shape our languages and give them the necessary structure. Any advance in our knowledge of nature is strictly connected with new languages of similar structure which reflect the structure of the world. This last 'knowledge' at each date represents again 'modern metaphysics'. In all such enquiries we have to struggle with the older, mostly primitive structural metaphysics and unconscious linguistic semantic consequences. Enquiry into these subjects must throw new light on the unconscious processes, and so diminish the vast field of the unconscious.

The structural unconscious seems to be more general, more fundamental than the special, or individual, psychiatrical one, because analysis shows that the latter follows from the former. As the reader may recall, life, 'intelligence', and abstracting in different orders started together. Without abstracting, recognition, and, therefore, selection would not be possible. The world of the animal, as well as of man, represents nothing else than the structural results of abstracting, without which life itself would be totally impossible. Man alone has the power of extending the orders of abstractions indefinitely. When Smith has produced an abstraction of some order, perhaps by making a statement, he has the potential capacity of analysing and contemplating this statement, which has become a fact on record, and so he can abstract to a still higher order., without known limits. It is this capacity which crowds the world of Smith with endless 'facts' belonging to very different orders of abstractions. The animal's capacity for abstracting ceases on some level, and is never extended without a change in the nervous structure. So the animal's world is comparatively simple, the world structure of man being, by comparison, indescribably more complex. Man's problems of adjustment, therefore, also become more complex. Human medicine is much more complex than veterinary science, although the higher animals differ very little in their gross anatomical structure from humans. The structural *m.o* facts, which resulted from abstracting in different orders, differ in number as well as in complexity. The human capacity for expanding indefinitely the orders of abstractions brings about this peculiar stratification of 'human knowledge'.

It appears as a product of evolution just as stratified as rocks appear. This stratification appears as a crucial, structural *m.o* fact, though generally disregarded, except partially in mathematics and psychiatry. Its realization necessitates the elimination of the 'is' of identity and results in the consciousness of abstracting, so fundamental for sanity.

Section C. Infantilism.

As has already been mentioned, the main symptoms of physical and 'mental' illnesses are few and simple. This would suggest the possibility of simple and more general theories relating to the fundamental symptoms. The colloidal structure of protoplasm accounts for this peculiar simplicity and for the small number of the fundamental symptoms. In the 'mental' field these fundamental symptoms are accounted for by a simple structural, functional principle of 'copying animals' in our nervous processes, which must be harmful, and which is characterized by the lack of consciousness of abstracting, implying colloidal disturbances. Psychogalvanic experiments show clearly that every 'emotion' or 'thought' is always connected with some electrical currents, and that electricity seems fundamental for colloidal behaviour, and, therefore, for physical symptoms and the behaviour of the organism.

In the colloidal processes we find the bridge between the 'physical' and the 'mental', and the mutual link seems mainly electricity. It is more than mere coincidence that all illnesses, 'physical' or 'mental', have only a few fundamental symptoms; and we should no longer be surprised to find that physical ills result in 'mental' symptoms, and that 'mental' ills may also involve 'physical' symptoms.

If a simple symptom is completely *general*, it indicates that it is structurally fundamental, and we shall be repaid if we devote special attention to it. As a rule, in 'mental' ills we observe a striking appearance of symptoms which have a sinister parallel with the behaviour of infants. Arrested development or regression in grown-ups also exhibits these infantile characteristics. In other words, whenever infantile characteristics appear in grown-ups it indicates that the 'adult' has not grown up fully in some semantic respects, or has already started on the way of regression, implying some colloidal or *m.o* structural injury.

When we speak of 'infantilism' in 'adults', we include symptoms which belong to the period of childhood in its organ erotic or autoerotic and narcissistic stages. It should be recalled that in children these semantic phases are natural; they become pathological only when the individual does not outgrow them and exhibits them as a grown-up. The term 'infantilism' is a rather sinister one, and should never be applied to children. Children behave like children, and that ends the subject. But children have fewer responsibilities, their sex impulses are undeveloped ., and so their behaviour cannot be equally dangerous to themselves and others. But not so with grown-ups. They have responsibilities, duties, often strong sex impulses ., which make out of the infantile 'adult' an individual dangerous to himself and others. The term 'social' period, or 'socialized' individual, is sometimes wrongly interpreted. The fact that human achievements and capacities are accumulative and depend on achievements of others makes us, by necessity, a social time-binding class of life, which again involves more complex modes of adjustment. Whether we approve or disapprove the existing legal and police regulations has nothing to do with the fact that in a social class of life some restrictions are necessary. Our present commercial 'civilization' can be characterized as of an infantile type, governed mostly by structurally primitive mythologies and language very often involving primitive *s.r.* One need but read the speeches of different merchants, presidents, and kings to be thoroughly convinced of this. The rules and regulations are naturally antiquated, and belong to the period to which the underlying metaphysics and language belong. The 'adult' or scientific semantic stage of civilization would be precisely the 'social' stage of complete evaluation of our privileges and *duties*.

In speaking about infantilism, it should be remembered that the child has an advantage over the imbeciles, idiots, and 'mentally' ill who have stopped development or have regressed to the age of the infant or the child. The 'normal' child profits by experience and outgrows the semantic characteristics that are natural to a given age. In cases of arrested development or regression, the undesirable infantile characteristics persist in the grown-ups and are a source of endless difficulties and suffering to them and their associates. Thus, in our childhood we all have had experiences similar to that of the patient of Dr. Jelliffe, and are no worse off because of them. But, if the reader should imagine himself in the position of the patient with those infantile characteristics, he would realize that an enormous amount of suffering, fear, shame, bewilderment ., results for the patient. The worst feature in such cases is found in the fact that an infantile type usually cannot 'outgrow' or alter such characteristics by himself, and needs very wise and patient outside help in re-training, or medical assistance, if he is ever to overcome earlier inappropriate *s.r.* But if we *start* the education of an infant with appropriate *s.r*, such a procedure must play a most important preventive evaluational role.

We should remember that the nervous system of the human child is not finished at birth. The extension, the growth, and the multiplication of the appendages of the neurones., go on after birth. The nervous system of an adult shows striking differences in the length and complexity of the nerve cells over that of the infant. The researches of Hammarberg showed, in all the cases of idiocy he investigated, an arrest of development in a more or less large part of the cortex, at a stage corresponding to either an embryological period, or to the period of early infancy. Only a small number of cells had reached their full development during the growth of the cortex. The psycho-logical defects were in direct proportion to the defects of the development of the cells, and were the greater the earlier the period of arrest of development.[6] In extreme congenital imbecility the cortex is poorly organized, is thin and deficient in nerve cells. Bolton's second layer of pyramidal cells matures last; and its development in different mammals corresponds to the degree of their 'intelligence'. In humans, its degree of deficiency corresponds to the degree of 'mental' arrest or regression. In organic 'mental' diseases very diffuse cortical lesions, when present, impair 'intelligence'. Affective disturbances depend upon even smaller brain lesions, particularly when the thalamic regions are affected.[7] In general, lesions in the basal ganglia diminish the energy of the impulsive life (sleeping sickness). Lesions at the base of the frontal lobe, and some brain tumours, lead to euphoric excitement, which gives a feeble, stupid expression, facetiousness, and a tendency to teasing. Other lesions in the basal ganglia lead to a labile affectivity. All destructions of parts of the brain usually lead to irritability and moodiness. In different focal lesions of the brain the disturbances lead to anger and rage.[8]

Among other results of organic brain diseases we find semantic disturbances, absence of critical faculty, and a disturbance of judgement; complicated situations can no longer be grasped, the *evaluation of relations* is impaired,. In cases of labile affectivity the special 'emotion' dominates the patient completely. Trifles make him either very happy or desperate. Because of the *decrease in association* or deficiency in the process of relating, the patient often appears indifferent, although the defect is not primarily in the affective field. Similar difficulties in association or relating make many patients appear egotistic in their *s.r* and behaviour. Since the patients have lost their insight in, and evaluation of, different life situations, their actions appear un-ethical. Tenderness, consideration, tact, aesthetic sensibility, sense of duty, sense of right, feeling of shame., may all disappear at any moment, though they would otherwise naturally be present. Any kind of impulse may be translated into action without restraint.[9]   Hemorrhages in the thalamic region often result in marked lability of affects.   Diffuse nutritional disturbances of the cortex usually give similar symptoms, as in organic brain diseases , .[10]

It should be remembered that in the human nervous system the co-ordinated working of the higher and lower centres is a necessity for the optimum working of the whole.   In cats and dogs deprived of the association areas of the cortex, the difference is not so marked.   They still behave in a co-ordinated way, provided the thalamic regions are intact.   Even in a child without a cortex, we find facial grimaces if we give him something bitter, but higher adjustments are impossible.[11]   The general 'inhibitory' and regulative action of the higher centres increases with the differentiation of the nervous system; and in man this becomes of paramount importance.   This has been shown empirically.   For instance, in man and dogs strong negative action on the flow of gastric juices may be 'psychic' in origin.   This negative action is weak in the guinea pig, although it is discovered on decerebration.   With the tortoise, also, there is some acceleration of the movements of the stomach after decerebration, but in a frog we do not find any negative influence at all.[12]

The facts given above were established through anatomical and physiological structural examinations.   If the functioning of the nervous system is examined from the point of view of colloidal chemistry, the gross non-surgical lesions become interpretable as the result of changes in colloidal behaviour.   Thus Doctors Wilder D. Bancroft, J. Holmes Richter, H. Beckett Lang, John A. Paterson, Walter Freeman, and others demonstrated that it is possible to find a correlation between the functional psychoses and the state of dispersion of the nerve colloids.   For instance, in dementia praecox the nervous system appears in a state of colloidal over-dispersion; and in manic depressive psychoses, in a state of decreased dispersion.[13]   It is interesting to note that in infants the colloids appear more dispersed than in grown-ups and probably similar conditions will be found in cases of infantilism.   The above-mentioned scientists have found, also, that the colloidal behaviour of the nervous system can be altered by special chemical treatment with drugs, carbon dioxide, oxygen., with specific reactions on the psycho-logical level.   As s.r involve electrical occurrences fundamental in colloidal behaviour, similar symptoms on psycho-logical levels may imply corresponding submicroscopic colloidal states.   Taking into consideration the structural characteristics of colloidal behaviour and the elaboration of technical means, we may discover that semantic re-education must involve differences in electrical potentials., and result in differences in colloidal behaviour in different regions of the nervous system. Experiments should be made in combining chemical means, which effects are not lasting, but which might facilitate the semantic approach, with semantic re-education, which results, once achieved, often become lasting. Colloidal and psycho-galvanic investigations of a given patient before and after the semantic re-education should also be made.

Let me emphasize once more that from the colloidal point of view free from identification, the 'body-mind' problem ceases to be a puzzle, as we have a well-established electrodynamic, structural colloidal background which can account perfectly for the experimental facts of 'mind'. The subtleties of the sub-microscopic structure involve an endless array of possibilities. At present we lack detailed knowledge of this structure; for the colloidal developments are very recent, and in this special field very little experimentation has been done.

If we accept the *non-el* point of view, and all known evidence seems to demand it, we must conclude that if different macroscopic, microscopic, and sub-microscopic lesions of the nervous system *result* in quite definite psycho-logical symptoms, which on the semantic levels appear as a lack of *evaluation of relations;* then, vice versa, the use of linguistic systems, which systematically train the immature nervous system of the child and of the grown-ups in delusional evaluation, must result in *at least* colloidal disturbances of the nervous system. These functional colloidal disturbances become superimposed upon the inborn eventual deficiencies of the nervous system, and the end-results may be quite out of proportion to the seemingly slight induced discrepancy. The actual behaviour, adjustment, sanity., may be considerably impaired.

Before birth, the child can be considered as in ideal conditions. He floats comfortably in a fluid of a temperature equal to his own. All his wants are satisfied, as everything is supplied to him by the maternal body. At birth, the child must begin to breathe, and a little later he must take food, digest,. External influences begin to impinge on him, and he must begin to adjust himself. Very soon the average infant finds that he can get what he wants, within certain limits, by certain movements or by crying. For the infant, a cry or a word becomes semantic magic. In Pavlov's language, a word governs a conditional reflex. In psychiatry, a definite series of such conditional *s.r* of animalistic low order of conditionality is called a 'complex'. In Pavlov's experiments a dog is shown food and a bell rung simultaneously. At the sight of food, saliva and gastric juice flow. Associations soon *relate* the ringing of the bell and the food, and, later, simply the ringing of the bell will produce the flow. In another animal some other signal, a whistle, for instance, would produce similar effects. In different people, through experience, associations, relations, meanings, and *s.r* are built around some symbol. Obviously, in grown-up humans the identification of the symbol with the thing must be pathological. But in infancy the confusion of orders of abstractions must be considered as an entirely natural semantic period. The infant 'knows' nothing about science and events. Objects and 'sense perceptions' 'are' the only 'reality' he knows and cares about; so he does not and cannot discriminate between events and objects. By necessity, he identifies unknowingly two entirely different levels. As his symbol usually means a satisfaction of his wants, naturally he identifies the symbols with the objects and events. At this stage, also, he cannot know that the orders of his abstractions can be extended indefinitely, or that his most important terms have the multiordinal character. It is important to notice that objectification, and, in general, identification or confusion of orders of abstractions, are semantically *natural* for the infant. The more the child comes in touch with 'reality', the more he learns, and in a 'normal' child the 'pleasure principle', which was established as a method of adjustment on the infantile level, is slowly displaced by the 'reality principle', which thus becomes the semantic method of adjustment of the complete adult. Science alone gives us full knowledge of current 'reality'. But science represents a social achievement, and, therefore, a complete adult, in growing up to the social level, must become aware of the latest stages of *m.o* reality. These are given by the current scientific methods and structural notions about this world, and gradually become incorporated in the structure of the language we use, always deeply affecting our *s.r.*

It is important that in the twentieth century we should realize that the work of Einstein and the four-dimensional space-time continuum establishes a language of different structure, closer to the facts we know in 1933, and that it gives us a new semantic method of adjustment to a new 'reality' (see Part IX).

The semantic stages of the development of the child must naturally pass through the stages outlined above. When he begins to differentiate himself from the environment, he is self-centred and concentrated on his 'sensations' (autoerotic). Later he projects his own sensations on the outside events; he *personifies*. This semantic trait is often found in incomplete adults, when in anger they break dishes or furniture.

The child is interested, first, in himself (autoerotic); then in children like himself (homosexual). Slowly his interests turn away to persons less similar to himself, to the opposite sex, and so he enters the semantic period of the race development.

33

Similar semantic processes are to be seen in the racial developments as given by anthropology, and are reflected in the structure of the languages. In the archaic period of one-valued 'pre-logical thinking', which is found among primitive peoples, the 'consciousness of abstracting' is practically nil. The effect produced by something upon an individual inside his skin is projected outside his skin, thus acquiring a demonic semantic character. The 'idea' of an action or object is identified with the action or the object itself. Identification and confusion of orders of abstractions have full sway.

The paralogical stage is a little more advanced. In it the identification is based on *similarities,* and differences are neglected (not consciously, of course). Lévy-Bruhl describes this primitive semantic period by formulating the 'law of participation', by which all things which have *similar* characteristics *'are the same'*.[14] A primitive syllogism runs somewhat as follows: 'Certain Indians run fast, stags run fast; therefore, some Indians *are* stags'. This semantic process was entirely natural at an early stage and laid a foundation for the *building of language* and higher order abstractions. We proceeded by similarities, much too often considered as *identities,* with the result that differences were neglected. But in actual life, without some primitive metaphysics, we do not find identities, and differences become as important as similarities. The former primitive emphasis on identity, later enlarged to similarities, must, at some stage of human development, become semantically disastrous and the optimum adjustment an impossibility.

In building a $\bar{A}$-system, we have to stress *differences,* build a 'non-system' on 'non-allness', and reject identity. The older semantic inclinations and infantile or primitive tendencies were a necessary step in human evolution. For sanity, we must outgrow these infantile semantic fixations. Similarly, for civilization, we must grow out from primitive structural fixations, primitive metaphysics, taboos, and other primitive *s.r.* These primitive habits, languages, and structural metaphysics and reactions have been extremely ingrained in us through the ages, and it requires effort and new semantic *training* to overcome them.

In the 'mentally' ill we find sinister and very close parallels to the behaviour of the primitive man and the infant, not only in the 'mental' and 'emotional' responses, but even in physical behaviour, postures, drawings, and other modes of expression. These parallels are today recognized by practically all scientific workers and are analysed in many excellent volumes.

We should notice that in this maze of observational material, one general rule holds; namely, 'consciousness of abstracting' offers a *full semantic solution*. In it we find not only a complete foundation for a theory of sanity, but also the semantic, psychophysiological mechanism for the passing from the infantile, or primitive-man, level to the higher level of complete adulthood and civilized social man.

The following drawing is taken from Bleuler's *Textbook of Psychiatry* (p. 402) and was made by a very ill patient (chronic catatonic) *who could formerly draw well;* yet this drawing is obviously *childish.*

The literature of psychiatry abounds in such productions, and they confirm fully the processes as described in psychiatry.

Adult infantilism becomes usually a potent wrecker of individual lives, and, when viewed from a social, national, or international point of view, accounts, also, for the majority of our semantic difficulties in the social, economic, and political fields.

Although we treat infantile, arrested, and regressive symptoms together, it is important to realize that most of these characteristics are *normal* with the primitive man, and the infant, provided the child outgrows them. The difference between arrested development and regression seems somewhat like that between a poor man and one who has lost his fortune. In the first case, the functioning of the nervous system is unsatisfactory through some deficiency; in the second, the potentialities for efficient working were there, but some semantic obstacle originated the regression, or else some degenerative neural process set in.

FIG. 4

In the field of higher abstractions the train of 'ideas' of children, imbeciles, and idiots is restricted. Uncommon 'ideas' are left out, and only those which originated in immediate 'sense perception' are easily grasped. Until lately, even in science, such an attitude was noticeable, as, for instance, in gross empiricism, or in the case of the physicist already mentioned who was willing to 'fight' to prove that he 'saw' the 'electron',. He did not realize that inferential entities are just as good abstractions as those he 'sees'. The attitude of the 'practical man' who pooh-poohs science and the 'highbrows' may serve also as an example.

Children, idiots, and imbeciles cannot comprehend anything complicated; they see some elements, but miss the relative wholes. We have elaborated a racial language of 'senses' and elementalism. Similarly, in schizophrenics the relative whole is disregarded, while, on the other hand, a single semantically effective characteristic is sufficient to connect the most heterogeneous abstractions in an unnatural whole. Word-relations have a predominance over actualities (identification). Thus, a patient looks anxiously at a moving door and exclaims 'Da fressen mich die Thüren'* and refuses to pass through the door-way. Here we see the identification of words with objects carried to the limit. In general, the s.r of the schizophrenic seem such that he identifies intensely his higher abstractions with the lower.[15]

Much excellent material on infantilism can be found in Dr. Joseph Collins' *The Doctor Looks at Love and Life,* particularly in his chapter on Adult Infantilism, from which much of the following material is taken, and which I acknowledge gratefully.

Children and idiots live in the present only and do not concern themselves with the past and the future beyond their immediate gratification. Infantile types also want all the 'sense' enjoyment of the moment, never enquiring about the sufferings of others or of the consequences for themselves in the future. Indeed, their attitude is often hostile toward those who take into consideration a larger field. 'Après nous le déluge' represents their royal semantic motto. On national and commercial grounds, they devastate their natural resources, since they are interested only in some immediate and selfish advantage. They love praise and hate blame, not realizing that a *critical* attitude gives the foundation for proper evaluation and becomes a semantic characteristic of maturity and that, generally, it is *more beneficial* in the *long run.* They thrive and thrill on commendations and compliments, and shiver and shrink at disapproval. Such characteristics are found even in whole nations. They

---

*Animal = Thier, Door = Thür, so that the unconscious play upon words gives the meaning: 'Doors devour me', for 'Animals devour me'.

are self-satisfied, and keep aloof from others in international affairs, not realizing that this is impossible, and that the attempt is ultimately harmful to them. They assume, as an excuse, the superiority of their institutions., and the 'righteousness' of their own conduct.

Children and superior idiots appreciate resemblances more readily than differences. Simple generalizations are possible, but often they are hasty and faulty. A child's pride and self-respect are hurt if he is considered different from other children, or is dressed differently. Originality and individuality are tabooed among children. Because of semantic undevelopment, differences become a disturbing factor to them; they want everything standardized. On national grounds, the adult infants standardize all they can and have even a kind of hostility to anything which has an individual flavour. For instance, those who wear straw hats after an arbitrary date are attacked on the streets. Not wanting to 'think', or to bother about differences, they fancy that they can regulate life by legislation and they keep busy manufacturing 'laws', which are very often impracticable and self-contradictory. When they pass several thousand 'laws' a year, these become a maze and a joke. The ultimate semantic result of such over-legislation is a complete lack of justice or of any respect for 'law'. Not being able to 'think' for themselves, they leave that bothersome function to politicians, priests, newspapermen,. Under such conditions life is impossible without expensive lawyers.

Not having the critical semantic capacity for proper evaluation, their likes and dislikes are very intense. They cannot differentiate the essential from the unimportant. The immediate 'sense' perception or 'emotion' unduly influences their actions. Impulses to copy others dominate them. They are often prejudiced. This results in weak judgement, over-suggestiveness, 'emotional' outbreaks, exaggerated sensibility, variability of affective states., and, finally, in an attitude toward life devoid of proper evaluation. Their moods are changeable; their attention readily gained and as readily diverted. They become easily intimidated and frightened, and easily influenced by others.

The above semantic characteristics are sponsored by commercialism, and build up the kind of methods, advertisements, and business policies which we see about us. This also introduces a semantic factor of disintegration into human relationships, as it leads to methods of trickery, to 'putting something over' on the other fellow, and appeals to self-indulgence,. When such commercial tactics are national, their sinister educational effect is pronounced. Children, from the age when they begin to read, are impressed by such practices as *normal* and take them as semantic standards for their own further orientations. Unfortunately, even psychiatrists have not, as yet, analysed the semantic influence of such advertisements on the building and preserving of infantile characteristics.

Children lack moderation and a semantic sense of proper evaluation. Tolerance is not one of their characteristics. To them persons and 'ideas' are evaluated in extremes, either good, 'wonderful', or bad, 'terrible'. Their $s.r$ appear dogmatic and stubborn, as in all the unexperienced. They talk too much or are silent; they praise too much or blame too much; they work too hard or play too hard, and know no middle ground. The whole life of a nation may be coloured by such semantic attitudes. Nations become boastful of their own possessions and achievements, and happily borrow and forget the achievements of others. They pride themselves on having the largest airships, the largest cities, the highest buildings, the longest bridges,. They know no moderation in food or drink; they eat or drink too much or become total 'prohibitionists'. They exhibit quick friendships and quick dislikes. They are solemn in their games, like children who are playing father and mother, and make out of games a national event. The childish pleasure of defeating an adversary accounts for national crazes, like racing, boxing, football, baseball, and similar sports, which often overshadow in public attention all really important issues.

Children and many idiots are incapable of any choice which involves meanings and evaluation. When confronted with a situation in which they have to choose between two alternatives, they have difficulties, and often want both. Similarly with 'ideas'; they often keep sets of entirely self-contradictory 'ideas'. Even scientists of an infantile type do so, and then publish 'manifestos' in which they try to justify such behaviour and semantic attitudes. Merchants train salesmen especially to induce customers with such infantile $s.r$ to buy what they do not need. This attitude is often extended to marriage. Any man and woman may marry simply because they come across each other; then, when they meet somebody else, they soon change the object of their sentiments.

All classes of feeble-minded and children show marked credulity; they like fairy tales and fantastic stories. Free inventions, by a process of objectification, are taken as experience. Children and schizophrenics *pun and play on words*. They build up languages of their own. Perseveration and stereotypy in speech are also found among them. National commercialism utilizes this principle in advertisements and tries to run a country by verbal slogans and play on words.

Many children and feeble-minded show distinct acquisitiveness. Like some animals, they show a tendency for collection of objects, and value their collections highly. It is a well-known childish game to claim the best morsel of some food because one has put one's hand on it first. Acquisitiveness is made a national slogan and proclaimed a highest aim, which, of course, becomes a semantic source of endless wars and miseries. Infantile legalistic 'putting hands first', on a piece of paper as a title to land, or some such similar form of a 'claim', becomes a source of ridiculous fortunes for the few and of unbearable life-conditions for the many.

Children are gregarious and afraid to be alone. Similar tendencies are carried on by Rotary and other clubs and lodges. Infantile grown-ups are too empty in their heads to desire to be alone. Children seldom stick to anything for long. They hunt for new excitements, and the old toys are often soon forgotten. Similarly, grown-up infants hunt for new excitements, for new toys, whether they be a house or an automobile, a wife or a lover.

In children and the feeble-minded, we seldom find such feelings as shame, aesthetic sentiments, or appreciation of beauty. They like things bizarre, grotesque, glittering, and enormous, things which attract and hold their attention. Similar characteristics are found in incomplete adults. Children and the feeble-minded are usually untidy and noisy. Visiting a public park, or witnessing a 'celebration', will show an observer clearly how infantile grown-ups behave.

Children like to domineer over their younger brothers and sisters and to play the leading part in a game. Similar semantic characteristics are carried into adult life, sometimes taking the form of sadism. We often see infantile docility or resentment, as expressed in sentimental approval or bitter disillusionment, both generally unjustified.

Self-respect is little developed in the idiot, but plays an important semantic role in the life of imbeciles and children. The infantile adult also shows an exaggerated self-respect. Bus conductors and university professors label themselves with a title—even if it is only 'Mr.' John Smith, as if being called simply 'John Smith' would be offensive to him. An adult evaluates a man by what he has in his head or character, but the infantile type largely judges him by the *symbols* (money) which he has, or the kind of hat or clothes he wears. Since commercialism cannot sell brains, but can sell trousers or a dress, it establishes semantic standards whereby a man is evaluated by his clothes and hats.

In speaking of exaggerated self-regard based on improper self-evaluation, we touch the problems of infantile self-love and self-importance. Infantile grown-ups carry these even further, and are unable to make dependable attachments to other persons. The love of parents toward their child is largely because it is *their* child; and infantile A 'loves' B, only because B 'adores' A and gives up his individuality. The moment something changes in B, all the 'love' A had disappears. The unbelievable bitterness which appears in divorce-court scenes shows clearly the value of infantile 'love'. Such 'love' is often based on purely egoistic grounds. They 'love' what they represent to themselves, what they once represented, what they would like to represent. Infantile parents see all kinds of perfections in their babies, although a sober outsider does not share these opinions. An infantile mother treats her child like a doll, plays and is thrilled with it, but soon gets tired as the responsibilities become irksome. An infantile father sees in the child, first, a toy, and, later, a nuisance.

Infantile adults have little regard for, or endurance of, life responsibilities. They tire quickly, are easily discouraged and frightened. They are thus irresponsible, unreliable, and a source of suffering for those connected with, or dependent on, them. This permanent suspense for others produces, perhaps, one of the most serious sources of worries and unhappiness. Since it is persistent, it gives continual, painful nervous shocks, the cumulative effect of which is bound to be harmful.

The infantile individual himself cannot fail to notice that something is wrong, for life makes him quickly aware of it. But, in his self-love, exaggerated self-esteem., he overlooks his own shortcomings, and blames everybody and everything but himself. In the face of 'injustice', he becomes discouraged, timid, or bitter and pessimistic. He is unable to discharge his duties, and becomes a disappointment as a father, husband, friend, and, ultimately, as a human being and citizen. Bitterness, disappointment, and painful semantic shocks pile up on all sides under such conditions.

One of the important characteristics of infantilism of all degrees takes the form of exhibitionism, an impulse for showing off, even by crude display of himself, his body,. This tendency is very common, and leads to many results of a very undesirable social character. Infantile men and women are primarily in love with *themselves* and care only how pretty they are. They spend large portions of their income and life on dresses and grooming, which, of course, have no social value. Such types live in an infantile world and are socially useless, often parasitic on the social body. Often those who support them ruin their lives to satisfy these infantile semantic characteristics.

Infantile exhibitionism leads, also, very often to a selection of a career. Most diplomats, politicians, professional military men, preachers, actors, boxers, wrestlers, athletes, many lawyers and public speakers, to list only the more important professions, select their professions because of this infantile tendency. We should notice that in this list we find the most important professions which, as yet, have shaped our destinies. Royalty, hereditary potentates, and many plutocrats live under such infantile fairy-tale conditions that they necessarily become semantically twisted.

This pathological tendency probably accounts for our so-called civilization being at an infantile *a*social level, based, as it is, on selfishness, 'sense' gratification, might, brutal competition, acquisitiveness,. We should notice that whole 'philosophies', such as theism, the older ontology, teleology, materialism, solipsism, the Anglo-Saxon philosophy of selfishness, and different military and commercial philosophies, clearly display these infantile characteristics. Commercialism, the 'law of supply and demand', as a by-product, also follows from infantile world-outlooks. Those who are interested in problems of politics, economics, sociology, war, and peace., should investigate their problems from this semantic point of view. As Burrow well said, the problems of war are more the problems of psychiatry than of diplomacy.[16]

Many women at present are still infantile, very little developed as human beings; they are themselves exhibitionists and also *sponsor exhibitionism*. It should not surprise us to find that these characteristics, the lavishing of 'love' on shiny buttons and on regiments marching to their destruction, have favoured wars. During the Russian revolution of 1905, the czar's soldiers were on the streets. But women did not 'love' them then. Little children spat on them from behind corners. The result was that very soon the soldiers *refused to carry on this unapproved service*. I know many cases connected with the World War, where, in spite of unspeakable horrors, many regretted the ending of the war because of the infantile approval shown by their women for their 'glory' and the infantile thrill the soldiers themselves experienced because of the shiny buttons, martial music, and parades. In the old system, militarism, religionism, legalism, and commercialism are strictly interconnected by similar *s.r*. Eliminate any one of them for good, and the others would become obsolete or would disintegrate. Our infantile women, no doubt, have sponsored through the ages these infantile social cancers.

The future war will, perhaps, automatically bring these problems to the foreground. It will be an extremely devastating (and less picturesque) aerial war, in which women and children will not be spared.

Then, perhaps, some of these infantile women will begin to *face m.c reality,* and so will help to start a new era of human adulthood. Men will always depend in their standards on the wishes of women.

In infantile nations we witness, also, a great deal of exhibitionism, a craze for athletics, clothes, sumptuousness, noisy behaviour, parades, uniforms, 'military academies', military drills,. 'Serious', yet infantile, 'business men' love to parade on the streets dressed up like little boys or circus performers, give themselves some 'mysterious' high-sounding and empty high titles, play with swords which they do not know how to handle,. In international affairs, of course, a nation of a more pronounced infantile semantic tendency will seek to keep away from adult international associations. The attitude of the United States towards the League of Nations and that of Great Britain towards the project of a confederated Europe suggest themselves at once in this connection.

Infantilism has another serious and detrimental connection with race problems; namely, through the sex glands or gonads in their effect upon 'love' and other activities. We should realize and emphasize that the sex glands do not function only as 'sex' glands in the common meaning of the word, but even more as *internal secretion glands* with an enormous bearing on all life and 'mental' processes; a $\overline{A}$, *non-el* orientation should never forget that.

The various consequences of castration are well known, and need not be repeated here. But the interrelation of the gonads with the thymus and with thyroid glands is of interest to us. The term gonad means reproductive gland, which produces the egg cells or the sperms. The thymus is a term applied to a light pink gland situated in the superior and anterior part of the thorax. It extends up into the roots of the neck and comes close to the thyroid gland. The thyroid gland is a term applied to a deep red glandular mass consisting of two lobes which lie, one on each side of the upper part of the trachea and lower part of the larynx. In women and children the thymus is relatively larger than in the adult male.

In humans the thymus grows up to the second year of life and then rapidly diminishes, so that only traces of it are found at puberty. In certain cases of *arrested development* or of general weakness in young people the thymus has been found to be persistent. Castration at an early age leads to the persistence of the thymus gland. Normally, the gland atrophies before the gonads come to maturity and begin to function. In some of the lower mammals the gland does not disappear as early as it does in humans. The thymus of the calf is popularly called the 'chest sweetbread'.[17]

Atrophy of the thyroid in the adult is usually followed by blunting of 'mental' capacity. Speech is slow, and cerebration delayed. If the secretions are lacking in childhood we have what is known as cretinism. Excessive activity of the thyroid produces a condition known as exophthalmic goitre. In many females at each menstruation the thyroid is perceptibly enlarged. Extirpation of the thyroid before puberty brings about, among others, signs of cretinism, failure of development of the ovaries., so that puberty is delayed partially or completely.[18]

Even these few particulars are sufficient to make us understand that when we begin to deal with 'infantilism', 'arrested development', or 'regression', or 'adulthood', we deal with fundamental *non-el* semantic life-problems which are connected structurally with the organism-as-a-whole. Bleuler describes disturbances of affectivity thus: 'The so-called psychopaths are really nearly all exclusively or mainly *thymopaths*. Furthermore, since affectivity dominates all other functions, it assumes a prominent rôle in psychopathology generally, even in slight deviations, not only on account of its own morbid manifestations, but even more because in disturbances in any sphere, it is the affective mechanisms that first create the manifest symptoms. What we call psychogenic is mostly thymogenic. The influence of the affects on the associations produces delusions, systematic splittings of personality, and hysteroid twilight states; repressed pain is the source of most neurotic symptoms, while displacements and irradiations produce compulsive ideas, obsessive acts, and similar mechanisms.'[19]

The thymus appears not only as a childhood gland, but the adult gonads begin to function when the thymus ceases to function. When the thymus persists, we often find arrested development and psychopathological disturbances connected with infantilism. It should be remembered that in the organism not all 'cause and effect' sequences appear as *one-to-one*, but mostly as *many-to-one* relations. Therefore, no standard mechanism can be readily assigned to a semantic disturbance. But there are enough structural, functional, and colloidal mechanisms known to account for most disturbances, although the precise working is not known, 1933, in most cases.

Psychopathology and experience show that the 'self-love', 'self-sufficiency'., of infantilism are usually accompanied by marked sex disturbances, which, from a racial point of view, are just as important as the semantic disturbances.

Infantile types often have 'charming' qualities. The women 'are sweet', 'nice'; the men seem 'good mixers' and 'popular'. The opposite sex often likes these characteristics. In men a feeling of sympathy is aroused for the 'helpless little girl', or else pedophilic tendencies are released. (Pedophilia is used as the name of a 'mental' disturbance or desire for relations with children, often found among senile dementia and imbeciles.) In women, often a feeling of motherhood leads them to like infantile males. The charm of the child lies, to a great extent, in his narcissism, his self-sufficiency, and inaccessibility. Certain animals, such as cats and larger beasts of prey, fascinate us, as they do not concern themselves with us and are inaccessible. But this 'charm' has another and very tragic side. Such infantile types cannot stand responsibilities; their affections are shallow and unreliable; they know how to take, but do not know how to give,. In life, such connections lead invariably to great unhappiness, and often to disasters. The children produced by such infantile types are usually completely ruined by the lack of parental understanding or lack of care. Instead of liking such types, men and women of semantic maturity should either avoid them or suggest psychiatrical consultation.

Infantile types invariably show some sex disturbances, which also add greatly to family and social difficulties. The men are often impotent; the women, frigid. Onanistic and homosexual habits or tendencies persist, although such an adult infant is married and has an opportunity for normal life. A very important $\bar{A}$, *non-el* fact should be noticed. Since the organism works as-a-whole, 'mental' components should be considered in connection with sex life. An infantile type appears still in an organ erotic stage. He wants only sense gratification. From a theory of sanity point of view, *prostitution appears as a substitute for onanism.* In adult infants, we very often find either impotence, frigidity, onanism, homosexualism, as simple forms of arrested development or regression, or more extreme forms, like many cases of prostitution. Infantiles not only indulge in promiscuity, but build up fanciful rationalizations and represent their own infantile tendencies by 'theories' as 'normal' conduct. Many criminals, professional 'vamps', and professed 'heartbreakers' belong to this type. It is interesting to note that many 'mental' illnesses are connected with different onanistic rationalizations. Often excessive cleanliness, continual washing of hands., appears. If a schizophrenic onanist has a melancholic make-up, he rationalizes his problems that he is 'rotting because of his sins'. If he has a manic make-up, he feels that he 'is a saviour of mankind'.[20]

In all such cases family life is very unhappy, and the future of the children bred under such conditions is usually gloomy. Children need healthy family and semantic conditions to develop into healthy individuals.

The majority of professional criminals and prostitutes have an infantile make-up. No matter how cunning, they usually show little foresight. They appear egotistic, boastful, exhibitionistic,. Gangsters love pomp; their funerals are, as a rule, very expensive,—they want, even after death, to 'show off'. Criminals seldom become good fathers or mothers. They treat each other brutally and are generally promiscuous. Ethically, they behave usually as 'moral imbeciles', not realizing fully what they do. I am not advocating the abolishment of the death penalty on sentimental grounds, but an *enlightened* society should abolish any *penalty* on sick individuals. The 'mentally' ill criminal type should be either taken care of or else eliminated with some scientific benefit, but *not* as a *penalty*. Professional criminals can hardly ever become 'morally reformed' or useful members of society, unless the application of medical science can alter their pathological *s.r*. Without scientific attendance, they would practically always remain socially dangerous individuals. If we want to grow out of the present infantilism, experimentation on humans should be encouraged. Modern experimentation on animals is very humane, and suffering is eliminated. Criminals who are condemned to death should be given to science for experimenting. They would not suffer. Ultimately, they would probably die, but the benefits to the rest of mankind through scientific discoveries would be very important. Under present conditions, we 'take revenge on', 'punish'., mostly *sick* individuals, with seriously brutalizing semantic effects on the rest of us. There is not the slightest doubt that experimentation confined solely to animals, no matter how useful, will not solve many problems of Smith. Experimentation on humans is essential, and must be permitted. Most of the notorious criminals who go to the gallows appear at least infantile. How instructive it would be to make experiments on such individuals in respect to their thymus,. The list of experiments which science ought to make is very long, but material is lacking for such experimentation. Let me repeat that modern science can conduct its experiments without suffering to the individual, in spite of the fact that some of these experiments would be dangerous and might easily end in the painless death of the subject. The killing off of criminals (sick individuals) as a 'revenge' or 'punishment' or 'justice' is really too antiquated and too barbaric and *wasteful* for an enlightened society. If society wants to *eliminate* them, society can do it; but, at least, let us do it without such brutalizing morbidity, and with as great benefit to knowledge as possible.

The elimination of infantilism must be considered more than a personal issue with individuals; it becomes an *international semantic problem;* and such an international body as the League of Nations might originate a new era by starting a fundamental enquiry into this subject.

Infantilism in its national aspects is not equally distributed. Some countries are more infantile than others. In some countries even the university students show marked under-development for their age. Burrow reports that a questionnaire among students of a prominent university in the United States of America shows a surprisingly large percentage of onanism and homosexualism.[21]

We should notice that not even all scientists are free from infantilism. Many of them are childlike in that they do not really care for science, or civilization, or society, but are *asocial* and merely like to play with their toys. As an excuse (rationalization of tendencies and 'emotions'), they usually profess 'science for science' sake', not realizing that a complete adult must become a *socialized* individual and cannot keep aloof from general human interests, and that science represents a *public,* time-binding activity and concern, not the private pleasure or benefit of some one person.

Section D. Constructive suggestions.

As we have already seen, a young child cannot be 'conscious of abstracting', but he can acquire it gradually with experience. Racial, ordered experience is called science. Every one of us has the tendencies, and, to some extent, the capacities, for developing science. The main aim of such racial, ordered experience is to save effort and unnecessary experiences, so that a child may start where the father leaves off (time-binding). The problems of consciousness of abstracting should be formulated by science and made available for semantic training. This would fulfill the main requirements of science, to save experience and effort, and to predict the future, to help in the mastery over external and internal 'nature', and so to produce semantic and physical adjustment.

If we teach and train the children in the consciousness of abstracting, we save them an enormous amount of the effort which would be necessary to acquire it eventually by themselves, and we also eliminate a great deal of unnecessary sufferings and disappointments. There is no danger of taking 'the joy out of life'; the opposite is true. With the consciousness of abstracting, the joy of living is considerably increased. We have no more 'frights', bewilderments, or similar undesirable semantic experiences. We grow up to full adulthood; and when the body is matured for the taking up of life and its responsibilities, we accomplish that, and find joy in it, as our 'mind' and 'emotions' have also matured. Such a consciousness of abstracting leads to an integrated, semantically balanced and adapted adult personality. Joys, pleasures, and 'emotions' are not abolished, as this cannot be done, given the structure of our nervous system and 'mental' health, but they are 'sublimated' to higher adult human semantic levels. Life becomes fuller, and the individual ceases to act as a nuisance and a danger to himself and others.

In the racial aspects, if the development of the individual became normal, we should grow beyond infantile organ erotic fixations and *el* languages and infantile systems in all fields. A $\bar{A}$-system, in accordance with science 1933 ($\bar{E}$, $\bar{N}$ systems), would be the human link supplying scientific standards of evaluation to the affairs of Smith.

With the older infantilism and the practically general lack of full consciousness of abstracting, the fears, frights, painful 'emotional' shocks under which mankind lived were bound to have a marked, lasting, and sinister semantic and neurological effect upon the race. The race has never had an opportunity to develop in an adult way. What will be the results for the race of such a transformation it is impossible, at present, to foresee; but one thing is certain, that the results are bound to be very far-reaching.

To afford a better appreciation of what the consciousness of abstracting can accomplish, two more points should be explained. Most young fish do not know their parents, and, from the beginning, their life is independent of parental influences. The human child is helpless, and, for a comparatively long period, is under parental influence. His *s.r* are consequently moulded, 'mentally', 'emotionally', by the doctrines, taboos, structure of language., of the parents,. When we speak of a human child, we should never consider him in a fictitious isolation, which has nothing to do with *m.o* reality. *Both* parents and child should be made 'conscious of abstracting'. Only under such semantic conditions can the full benefit be reached. If parents are conscious of abstracting, and realize that their child represents, also, an abstracting in higher orders organism, which consciously or unconsciously registers in one form or another all happenings, the majority of the present unfortunate conditions, 'complexes'., could not possibly arise.

An important, yet usually disregarded, characteristic should be mentioned here. It is known that *repeated* 'emotional shocks' in childhood do harm. As the experiments of Watson show, the child is usually born without 'fears' and without 'frights'. Now 'fears' and 'frights' are not *simply additive* (a linear function) but follow some other more complex function of higher degree. If we denote the constitutional potentialities of the child by f, and the given event by x, the result of the impact of x on the life of the child would be a reaction $f(x) = F_1$. This F_1 designates what his make-up would make out of, or abstract from, x. When another event y happens, the reaction of the child is no longer $f(y)$, because this new event is usually taken by the child in the light of the former experience $f(x) = F_1$. Thus, the effect on the child would be different; namely, $F_1(y) = F_2$. If a new event z should happen, the child would react in his 'feelings'., as $F_2(z) = F_3$,. We see, then, that 'hurts' and, in general, *s.r* are *not simply additive* but may follow some other higher degree function. This process appears general, perhaps necessary, and yet it involves many dangers which can be completely eliminated *only* by the acquired consciousness of abstracting.

In practice, when we train a child in the consciousness of abstracting, we begin to check this devastating semantic process of piling up 'hurts' on 'hurts'. Let us assume that before we begin to train the child, the child has already had painful experiences. His memories are still fresh, still fluid; he has little difficulty in dwelling on them. With the consciousness of abstracting, and so proper evaluation, dawning upon him, further 'hurts' would 'hurt' less and less until the hurting process would stop altogether. In case some semantic harm had been done to the child before he became conscious of abstracting, the memories would be still fresh and he could apply his newly acquired semantic evaluational immunity to the harmful 'hurts'. New 'hurts' in practice are usually related or similar to the old ones; they would 'revive' the older hurts. Accordingly, he could not only 'live through' the older experiences but at once revise them, and after re-evaluation eliminate the harmful effects.

Semantic 'emotional pains' absorb nervous energy and prevent a full development of our capacities. Directly the consciousness of abstracting is acquired, the vast field of '*un*consciousness' is diminished, and the nervous energy which was engaged in fighting semantic phantoms is released. We should expect keener and sustained attention, strengthened interest, and other creative manifestations. Consciousness of abstracting, as it leads to proper evaluation, not only eliminates many unnecessary sufferings and semantic disturbances, but, by doing so, actually releases stores of energy for useful and creative purposes.

The human brain has vast areas which, at present, have no definitely known functions. Perhaps, with the older *lack* of consciousness of abstracting, the flow of nervous energy was misdirected or absorbed by the older ways of 'feeling' and 'thinking' in the lower centres. Thus, the available energy left was not sufficient to utilize the higher centres to the full extent.

Personal semantic difficulties always seem very personal, and no outsider can ever fully grasp the situation. One of the benefits of the present method of training in sanity consists in the fact that we do *not* dwell upon the personal affairs of the individual, but that we give, instead, a general structural semantic *method,* by the aid of which every one can solve his problems by *himself.*

We have established sharp differences between 'man' and 'animal'. These differences must be considered of higher order, as the terms 'man' and 'animal' are applied to abstractions of higher orders. We found that 'man' through ignorance and inappropriate *s.r* can copy animals in his nervous reactions. Such copying appears either as arrested development or as regression. In dealing with the terms 'conscious' and 'unconscious', we discovered a general and human content for *human* 'consciousness'; namely, the 'consciousness of abstracting'. The ascribing of a *general content* abolishes a vast field of 'unconsciousness', and so tends to prevent arrested development, infantilism, regression., whenever this is possible. The problem of making the *structure* of language similar at a given date to the structure of the events it symbolizes, is introduced. The conquests of science become incorporated into daily life by the use of the new *language.* The *structure* common to both science and language appears to be the intimate bond between science and *human* life. The masses gain simple structural and semantic means for adjustment.

A theory of sanity must draw attention to problems involving 'truth', 'falsehood', 'repressions',. Since the main usefulness of the theory is to help in attaining the most efficient working of the nervous system by the elimination of disturbing semantic factors, 'attitudes', doctrines., we must investigate the effect false (or repressed) statements may have on the working of the nervous system.

For instance, if we *see* that A, B, and C are given in the order A,B,C, such lower abstractions start cycles of nervous currents, which correspond to the *seen* order. If we *see* the order A,B,C, and *say untruthfully* that the order appears as C,B,A, this *statement* results, also, from some cyclic nerve currents. Obviously, we have *some conflict and disturbance in the working of the system.* If we make a mistake, the situation is different. Let us say that many observers definitely establish the given order as A,B,C. A new observer *sees by mistake* the order as C,B,A. His nervous currents correspond to his error, and when he makes a *truthful statement* that he has seen C,B,A, this statement also is connected with the appropriate nerve currents and there is *no conflict* or disturbance between the corresponding nerve currents. The seen and reported correspond to each other.

34

It is easy to conclude that mistakes and deliberate falsehoods have a different mechanism. A mistake, which leads to a subjectively true but objectively false statement, has no nervously disturbing factors. Deliberate false statements about facts involve semantic conflicts and disturbances in the functioning of the nervous system. Similarly, with 'repressed' material, permanently conflicting nervous currents are present. The nervous energy is spent on conflicts and struggles, while all of it is needed for constructive purposes.

In scientific work we have similar problems. We gather different abstractions of lower orders and then make higher abstractions about them. When these two different orders of abstractions fit nicely structurally, we are satisfied and enjoy the resultant harmony. If they conflict we feel restless. Often scientists spend years, or even a lifetime, formulating higher order abstractions which do not conflict structurally with the lower abstractions. Then they feel satisfied. Scientists know well the feelings of 'mental' pain and discomfort. Creative work is carried out because of such discomfort. Those who are *not creative* do not experience this, but they also do not produce important work.

The problems of structure, of correct symbolism, of evaluation, of the production of higher order abstractions which are structurally similar to the lower order abstractions., must have a neurological significance, and should be investigated from this point of view. Scientists should try to eliminate these unnecessary conflicts. Those who feel no conflicts may, nevertheless, be so involved in it that they have no free nervous energy left to overcome it. Some such attempt is being made empirically in semantic therapy. The psychiatrist tries to discover and eliminate the semantic conflict, thus freeing nervous energy which may then be spent on useful work.

In racial and national levels, systems of politics, economics., which are *based on falsehoods and repression of truth*, must unbalance the working of the nervous systems of the people. Since they are the result of the infantile $s.r$ of the race, they propagate the arrested or regressive development in the part of the race whose $s.r$ they influence. As usual, the vicious circle is working here also. A $\bar{A}$-system throws an entirely new light on the significance of science in *human* life. Radios, with their attendant possibilities of hearing jazz or a delusional 'revivalist', and the invention of bigger and better means for killing people, do not represent the main *practical* importance of science. *Generalized* science means scientific method and the discovery of the structure of events, to which the structure of our language must be adjusted if this daily tool of everybody is not to play dangerous semantic tricks.

A 'science of man' must follow science (1933) in its structure and method. Only by accepting the current 'scientific metaphysics' as given by science at a given date *is sanity possible*. The passing from an infantile 'civilization' to an *adult civilization* of fuller human life and happiness will come with the development of a scientific civilization which has scientific standards of evaluation. But the passing will not be so easy. As we have already seen, science contains affective factors, and many scientists still appear infantile. In order to enter upon an adult civilization, we must first have non-infantile leaders, who must be produced by appropriate training. This involves much research work along the lines sketched in the present work, and the establishment of chairs of general semantics and psychophysiology in universities. Educational methods must be radically revised, and experimentation encouraged in the widest sense.

In 1933 we know positively that in the physico-chemical and colloidal structures we find conditions of practically endless possibilities corresponding to the very large numbers of semantic states and reactions. Medical practice shows experimentally that a great many physical symptoms involving some colloidal states are produced by semantic disturbances; because, once these disturbances are eliminated, the physical symptoms vanish. The enormous numbers of observed and possible different *s.r* could not be accounted for by the older, still prevailing, *el*, *A*, and two-, or three-valued outlook, and the cumbersome, extremely limited, and necessarily slow chemical 'passing of different substances' through the nervous system.

It is true that every student of medical science is acquainted with colloidal behaviour, but this knowledge has been neither emphasized nor consistently applied, because colloidal behaviour represents physico-chemical processes involving electromagnetic, high pressure., manifestations which cannot be dealt with at all by *el*, *A* means. Thus, a physician who is not trained in $\bar{A}$ general semantics, cannot 'think' in colloidal and physico-chemical terms, which in 1933 are the only modern ways of dealing with the organism-as-a-whole. This is much more serious than the layman or even the physicians realize, and accounts for the fact that, in spite of different special achievements and different discoveries, the practice of medicine is becoming more and more unsatisfactory. It also explains why the average physician cannot grasp the importance of psychiatry for general medicine, and why some psychiatrists indulge in very unscientific and doubtful metaphysics. Thus, a general physician who 'thinks' uniquely in seriously antiquated chemical and physiological terms, deals with a non-existent, fictitiously isolated A and *el* 'body', and cannot grasp the necessity for a *non-el*, $\bar{A}$, and a physico-chemical, colloidal outlook, which integrates 'body' and 'mind'. The majority of psychiatrists in their turn, and for similar reasons, often have a highly metaphysical outlook, repulsive to the general physician. They do not seem to realize that they have at their disposal colloidal and physiological mechanisms as well as physico-mathematical formulations based on four-dimensional order, and that they, therefore, do not need any doubtful metaphysics. With $\bar{A}$ modern semantics, the only possible scientific outlook (1933) must be colloidal, physico-chemical and physico-mathematical, in which the long sought for *non-el* solution of the 'body-mind' is found. The difficulties I am dealing with are general and depend on fundamental principles, the disregard of which introduces semantic blocking factors, at present imposed on the medical students, and from which only a few exceptional, scientifically inclined individuals are capable of breaking away. From the present point of view the older reflexology is also unsatisfactory and requires a $\bar{A}$ reformulation.

The present system, although far from complete, already suggests many most important structural issues which should be *verified empirically*. Experiments *alone* can decide which verbal structures are similar to empirical structures, and experimentation should be encouraged in the widest sense. Some further theoretical work should also be done. Clinical literature describes many new and unexpected facts. These facts should be described anew in the new language, to see what relations survive the transformation of forms of representation. Thus, if it is found that *all* 'mental' ills in *all* different formulations indicate *improper evaluation*, we should be justified in concluding that *evaluation* represents an invariant general characteristic of the activities of the human organism-as-a-whole, and, consequently, must be of extraordinary importance for adjustment and sanity. When we reach this conclusion, we should investigate the *mechanism of evaluation,* starting with the simplest issues; namely, investigating those factors which make proper *evaluation impossible*. We should discover that identification in *all cases* makes proper evaluation impossible, and should then conclude that identification must be entirely eliminated before we can go one step further. In fact, once we have reached these rather obvious results, the rest of the $\bar{A}$-system follows. But this would not be enough; we must verify the conclusions *empirically*, and this suggests directly that a definite series of experiments should be undertaken.

In hospitals for 'mentally' ill two equally large groups of accessible patients exhibiting similar clinical symptoms should be selected, and isolated. A physician who himself has undergone a $\bar{A}$ training should attempt to re-train the *s.r* of one group. The other group should not be re-trained, but treated in the average passive and standard way,—it would be the control group. One physician should be in charge of both wards and keep a detailed record of the cases and treatment. It is to be expected that at the end of the year, in the ward trained in the $\bar{A}$ standards of evaluation, a larger number of unexpected and spontaneous recoveries would happen than in the untrained ward. It would be extremely instructive to have more than two groups, and to attempt a different group method, following some other medical school based on another system-function. The passive attitude toward the patients should be changed, as under the older methods physicians in 'mental' hospitals are more glorified keepers than medical men. This is what theory suggests. Experiments alone can show if these conclusions are correct. In special individual cases, the theory has already been confirmed, but it should be tried as a group method, and, if successful, only then would 'mental' hospitals become hospitals, and not mere places of detention.

A few words concerning psychotherapy will not be amiss. In a time-binding class of life, we must take into account the historical four-dimensional experiences of the race, which, even in individual cases, have sound neurological foundations, as it is known that the nervous reactions are influenced by past experiences. History teaches us that the work of some men has influenced great masses of mankind for many years, and that the works of others have had but little general, lasting effect. The considerations of doctrinal functions and system-functions explain this fact quite simply. The older an individual or a race grows, the more structural observations they gather, and the more they notice the structural dissimilarity of their forms of representation with the first order facts they encounter. As adjustment is generally useful, individuals, as well as groups, and particularly scientists, always attempt to discover more structural data about the world and themselves. This process requires, among others, the comparison of the structure of the forms of representation with the structure of the world and ourselves. All so-called 'progress', 'civilization', and science depend on this.

In this particular field, achievements are of two kinds:

1) Some individuals produce a *new* system-function, with a *new* structure, more similar to the world., (see Chapter XI). In the great majority of cases, the new system-function is *not* formulated *explicitly,* but is hidden implicitly behind some explicit particular and individual interpretation or particular system of the discoverer. The production of a new system-function is usually a most important event and is independent of the special value given to the variables in this function by the

originator. In such cases, the given originator has many followers, and there is a possibility of many doctrines or systems which have *one* doctrinal function or system-function. The content of the doctrines may be changed, but they all have *one structure*. The importance of the new doctrine or system was not in its particular interpretation, or in the assigning of a particular value to the variables, but in the underlying doctrinal function or system-function, *which alone has explicit structure,* and is given by the postulates which establish the function.

2) Some individuals do *not* produce explicitly or implicitly new doctrinal function or system-functions, with new structure, but simply assign a new and individual value to the variables in the *one* doctrinal function or system-function produced by others. These workers very often bitterly defend the private individual value they have assigned to a variable, and are often entirely innocent of the serious debt they owe to the originator of the new function which they utilize. But these works never mark a milestone in the progress of mankind and are usually soon forgotten.

Because of the disregard of the considerations explained here, the proper evaluation of different doctrines and systems is very difficult, and even in scientific circles the lack of orientation in this field is astonishing. It seems that it is not enough to produce a 'new theory' to have made an important contribution to knowledge; but it is essential to produce a new doctrinal function or system-function, because this only has a structural significance. This point of view, perhaps, solves the tremendous and, as yet, unsolved difficulties we have in reducing doctrines to sets of postulates, which is admittedly desirable, and yet so hard to produce. Thus, to find the doctrinal function or system-function which underlies a theory, *we must strip it of all accidental values privately ascribed to the variables, and formulate only the invariant relations which are posited between the variables.* The finding of this function is also equivalent to finding the *structure* of a given theory.

We may consider that the psychoanalytical and psychotherapeutic movement originated with the work of Freud. The epoch-making value of his work consists in the fact that, underlying his special theory, there can be discovered a new system-function. All other schools simply ascribe a different value to the variables, but do not produce, structurally, new system-functions; they represent different systems which have one freudian system-function.

It is here impossible to analyse this problem systematically or in detail, but a few structural hints may be useful.

First, we must discriminate permanently between the freudian particular system, which represents a particular interpretation of the freudian *system-function* without specific interpretations: say 'complex *x*', on semantic levels, which corresponds, let us say, to 'cluster *X*' on colloidal levels. Second, we must realize that the freudian system-function (not system) was scientific at the date of its production, but to be scientific 1933 it must be revised and reformulated, taking into consideration the newer physico-mathematical, physico-chemical, colloidal, $\bar{A}$., points of view. The struggle for a special 'complex *A*' or invention of a new 'complex *B*' is useless because in 1933 the colloidal, structural 'cluster *X*' which underlies the semantic 'complex *x*', which alone can legitimately be considered in a system-function, includes all of the 'complexes' in literary existence, and there are no assignable limits to their numbers. If we analyse in such ∞-valued terms as, for instance, 'cluster *X*', we evade an enormous amount of unnecessary and confusing metaphysics, and become scientific in the 1933 sense. From a modern, $\bar{A}$ system-function point of view, which means, when we recognize the necessity of ∞-valued semantics, structure., necessitating the reduction of a system to a postulate base, we readily see that the freudian *system-function* (not system) is a necessary and natural passing step between the *A* and $\bar{A}$ systems.

The postulates which are discovered in the freudian system-function can be divided into two main groups:

1) The observations of human behaviour and, in my language, of *s.r*, have to be formulated in a special language to fit the more structurally fundamental parts of the system.

2) The fundamental and revolutionary new postulates were, at the date of their introduction, quite scientific. In 1933 these postulates have to be reformulated and made to comply to modern physico-mathematical, physico-chemical, and general semantic $\bar{A}$ standards.

A satisfactory analysis of the above problems would require a special volume; therefore, I shall entirely disregard No. 1, and from No. 2 shall only suggest a few most important and new postulates. These can be expressed, roughly, as follows: (*a*) The postulation of an *active* 'dynamic' unconscious. This postulate departs widely from the older notions, although the word 'dynamic' is used in this connection in the vernacular, but not strictly scientific, sense. The methods of translating the dynamic into static and vice versa, are disregarded, owing to the innocence of modern science and the assumption by physicians, in general, of the permanent validity of *A* principles. (*b*) Once the *active* unconscious is postulated, some determinism follows according to the date. Freud, at his date, accepted the (in 1933 antiquated) two-valued determinism. Unfortunately, the great majority of physicians and the medical education still follow the antiquated notions. (*c*) As the past is taken into account and man is treated as a process in which the past experiences play an important role, we might say that the outlook is four-dimensional, but this statement is not entirely justified, because the notion of a consistent four-dimensional orientation carries us much further than physicians, who neglect physico-mathematical aspects, can possibly produce.

In a $\bar{A}$-system, the fundamental postulates of the system-function which underlies all psychotherapy have been accepted, although they have been vastly enlarged to comply with the known facts and scientific requirements of 1933. It is important to investigate independently, systematically, and in detail, the corresponding system-functions and to find to what extent they mutually intertranslate, but such an investigation cannot be carried out successfully if we confuse, through the habits of speech, the two different terminologies.

It seems that the older psychotherapeutic schools have been formulated as systems, and that the system-function, which underlies them, has not been explicitly stated, greatly hampering future creative work.

The special benefit of a generalized theory is in its fundamental simplicity and entirely general linguistic applicability, which, for prevention, plays a decisive role. Accidentally, we acquire psychophysiological means to influence the so difficult 'narcissistic' cases.

The present author has attempted to indicate the most important structural and semantic factors which would facilitate the future workers in the impending, necessary revision and co-ordination.

CHAPTER XXXI

CONCLUDING REMARKS

Pitiless indeed are the processes of Time and Creative Thought and Logic; they respect the convenience of none nor the love of things held sacred; agony attends their course. Yet their work is the increasing glory of a world,—the production of psychic light,—the growth of knowledge,—the advancement of understanding,—the enlargement of human life,—the emancipation of Man. (264)   CASSIUS J. KEYSER

Yet the barbarians, who are not divided by rival traditions, fight all the more incessantly for food and space. Peoples cannot love one another unless they love the same ideas. (461)   G. SANTAYANA

The individual whose brutish desire for personal profit is unrestrained by the needs and rights of his fellows reverts to barbarism. If a bandit he is outlawed; if a politician he is—usually reëlected, with resulting retrogression of the entire social organization. (221)   C. JUDSON HERRICK

. . . a "League of Sound Logic" is the best "League of Nations" because effective under the subtle inevitable laws of Logical Fate—Unified Doctrines Will Unify Man. (280)   A. K.

A little less worry over the child and a bit more concern about the world we make for the child to live in; an inclusion of the child in a life of which the aim is not merely to earn money so as to become independent of the job; more love for whole-hearted, creative work and progress that will make possible what we all can share in; with these conditions, the adult and the young both will have a better chance.*   ADOLF MEYER

The present remarks were originally written for the last chapter of the whole volume, but a final critical survey of the material suggested the newly ordered sequence of the present three main divisions. Book I gives a general preparatory introduction which will help the reader to differentiate between the A and $\bar{A}$ systems, and to evaluate properly the differences. Book II formulates the main $\bar{A}$ principles which constitute an organic interrelated whole, to which the present concluding remarks belong. Book III gives some additional structural data about mathematics and physics which usually are not treated from the present point of view, but which furnish the essential structural material needed.

The writing of Book I, and particularly Book III, was very laborious and difficult. I often had the temptation to omit Book III entirely, and to refer the reader to other authors. After months of search, however, I found, to my sorrow, that, in spite of many excellent volumes, there were no books written from a structural and semantic

---

*What Can the Psychiatrist Contribute to Character Education? *Rel. Educ.* May, 1930.

point of view. To refer the reader to other writers would necessitate the reading of a fair-sized library, because often from a whole book he would need only a few scattered paragraphs. This would involve a very expensive and laborious process of hunting, which very few would undertake; besides, it would not give a connected structural or semantic picture. I tried to induce some specialists to write a book on the structure and semantic aspects of mathematics, and another similar book on physics. I was told that it would be very laborious and difficult, if at all possible, and so I had no other choice than to try to write it myself.

I earnestly suggest the reading of Book III, so that the reader may, at least, become acquainted with the existence of such problems. I hope that even specialists may find some suggestions helpful, because the structural and semantic aspects of science and mathematics are usually neglected, the neglect of which introduces needless difficulties in the teaching. The elimination of identification helps to solve many scientific puzzles, besides eliminating semantic blockages and so helping creative activities.

The world affairs have seemingly come to an impasse and probably, without the help of scientists, mathematicians, and psychiatrists included, we shall not be able to solve our urgent problems soon enough to prevent a complete collapse. Now those who are professionally engaged in human affairs, economists, sociologists, politicians, bankers, priests of every kind, teachers., 'mental' hygiene workers, and psychiatrists included, do not even suspect that material and methods of great general semantic value can be found in mathematics and the exact sciences. The drawing of their attention to this fact, no matter how clumsily done at first, will stimulate further researches, produce better formulations and understanding, and ultimately create conditions where sanity will be possible.

Some of those who have seen my manuscript or with whom I have discussed the problems seemed to dislike the term 'copying animals in our nervous reactions' and also the explicit introduction of 'Fido'. As identification is found among animals, primitives, infants, and 'mentally' ill, it could be said that the introduction of 'Fido' was not necessary. I have given serious consideration to the eventual desirability of completely eliminating 'Fido' from my work and substituting the term 'primitive'.; but, after mature deliberation, I decided that it will be helpful to accentuate the distinction between the reactions of animal and man. The main justifications of this are as follows:

1) My whole work and the formulation of a $\bar{A}$-system started with an attempt to produce a science of man, this necessitating a mod-

ern, scientific *functional, non-elementalistic, sharp* definition of man. Such a definition was given in my *Manhood of Humanity* to the effect that man differs from the animals in the capacity of each human generation to begin where the former generation left off. This capacity I called the time-binding function. This definition cannot be denied, and it fulfills the modern requirements.

2) The present enquiry originated in the investigation of the mechanism of time-binding, and is a further analysis of the sharp differences between the reactions of animals and humans, which became the psychophysiological foundation of a $\bar{A}$-system and a theory of sanity.

3) The further the investigation advanced, it became increasingly evident that the issues involved are extremely complex, and that in this field, from a structural and *non-el* point of view, practically nothing has been done. In general, all existing 'logics' and 'psychologies' are structurally misleading, since they are still thoroughly *el* and pre-A or A; these conditions necessitate the elimination of them, as well as other dependent disciplines, to prevent their being accepted as structurally fundamental. It was then desirable, in my pioneering enterprise, to keep a simpler and more obvious contrast between 'Fido', whom we nearly all know quite well and usually like, and 'Smith', whom no one seems to know properly. This method has proved very useful to the writer and I am convinced that many readers will find it equally helpful. I frankly admit that if I had not followed this simplified method, I could not have produced the $\bar{A}$-system and discovered in this psycho-logical maze the blockages introduced into our *s.r* by identification, elementalism, lack of consciousness of abstracting, improper evaluation., and, in general, infantilism.

For these three main reasons it seemed advisable to retain 'Fido' as a most useful factor in my analysis, with all due apologies to 'Fido'.

I also admit that I did not realize the difficulties of the task and the magnitude of the undertaking. The last revision alone of the manuscript required more than a year. I am all too well aware to what extent the presentation falls short of my expectations and how much better it could have been written by some one more gifted, but the following rather unexpected developments sustained my courage.

1) Curiously enough, the principles involved are often childishly simple, often 'generally known', to the point that on several occasions some older scientists felt 'offended' that such 'obvious' principles should be so emphasized. Yet my experience, without any exception, was that no matter how much these simple principles were approved of verbally, *in no case* were they *fully applied in practice.* Slowly I understood that

we cannot train mankind in identification by all available means, which must prevent adjustment, and then live by non-identification. Thus, when non-identity is pointed out, even a moron will 'agree', or wonder at the silliness of an author who fusses about it; yet, because all of us were *trained* in a linguistic and semantic system based on identity, that infantile identification will unconsciously play havoc with all our *s.r* the rest of our lives, unless this semantic blockage is counteracted. Naturally, the 'simpler' a principle appears, to which we may pay lip service, but which we *never fully apply,* the more I became convinced that the discovery of new methods for the application of this simple yet neglected principle must be considered most important. Any reader may verify by himself to what extent identification introduces difficulties in his own life. In fact, the main difficulties we have can always be traced to some identification somewhere.

2) The experimental data of Doctor Philip S. Graven with the 'mentally' ill and those cases of semantic disturbances which, in the orthodox way, were not supposed to be un-sane, showed that a change from the A standards of evaluations involving identification to the $\bar{A}$ standards without identification often either brought about a complete semantic reconstruction of an individual, or semantic, expedient, and lasting 'cures'. This fact again impressed me with the genuine workability and so human importance of a $\bar{A}$-system. If the old 'impossible to change', 'human nature' can be 'changed' by the new simple psycho-physiological methods, this again suggests that this new system, no matter how imperfect, may be useful.

3) I was also very much impressed by the far-reaching power of the $\bar{A}$ methods. As a rule, only mathematicians and epistemologists fully appreciate what the power of a method means. Thus, the differential methods were invented, and later we found that these methods were structurally applicable to all processes. A tensor calculus was invented, and we found that it gave us absolute, invariant formulations applicable to all physics. Many other methodological innovations could be cited, and always the generality of the applications gave the value to those new formulations. The present $\bar{A}$-system was formulated in a way independent of other diciplines, as it was the direct result of structural semantic researches *free from identification.* This led to the formulation of fundamental *general* principles which underlie all human 'knowledge', such as non-identity, requiring the recognition of structure as the only possible content of 'knowledge' and so leading to the formulation of 'similarity of structure'; non-elementalism as a general principle; the general principle of uncertainty; ∞-valued general semantics,. It is

naturally very reassuring to find that the newest most important achievements of science have followed these principles unconsciously and have applied them *before* they were explicitly formulated.

From another point of view, a $\bar{A}$-system which could claim to be 'modern', should formulate general principles that all scientists in every field could follow. This was practically the case with the A-system until Francis Bacon. It is also the case with the present system except that different scientists have applied these new principles without having produced a *general formulation*. The fact that these principles had no general formulation was a retarding factor even in science and made the application of science to human affairs impossible. In the following examples, the different $\bar{A}$ aspects overlap, and I am emphasizing only the most marked features. Thus Einstein-Minkowski's space-time, Einstein-Mayer's new unified field theory, the newer quantum mechanics, the new physics of high pressure, piezochemistry., the tropism theory of the late Jacques Loeb, the physiological gradients of C. M. Child., ., exhibit clearly the application of non-elementalism. Heisenberg's restricted principle of uncertainty is also the result of the application of non-elementalism, based on the observation that the 'observer' and the 'observed' cannot be sharply divided. This principle becomes a particular instance of the general $\bar{A}$ principle of uncertainty, which again is based on the observation that we deal actually with absolute individuals and speak in more or less general terms, with the result that all statements are only probable in different degrees.

The absolute individuality of four-dimensional events, objects, situations, *s.r.*, necessitates an indefinitely flexible evaluation requiring ∞-valued $\bar{A}$ semantics. Outside of daily life, the best examples are given in science by the newer developments in vitamins, the effects of radiant energy on heredity, but particularly by the bewildering possibilities disclosed by the developments of physics, physics of high pressure, piezochemistry, polymorphism, colloidal behaviour, and the application of colloidal knowledge to psychiatry. The Polish school of mathematicians has produced the extension of the traditional two-valued A 'logic' to three-, and many-valued 'logic'; Chwistek has based a new foundation of mathematics and a new theory of aggregates on his semantic methods; but even these writers disregarded the *general* problems of non-elementalism, non-identity, and the necessity for a full-fledged $\bar{A}$-system before their formulations can become free from paradoxes, valid, and applied to life.

All these issues combined are of particular interest to mankind in general, and to the medical profession in particular, because, obviously, if mankind is to pass from an infantile stage of its development into an era of general sanity, this would require a serious collaboration of medical science. Unfortunately, medical science is one of the most laborious and difficult disciplines, and, of late, in spite of some specific advances, it is rapidly ceasing to be in general a modern science. Any one who attends medical congresses, scientific meetings, or follows up medical literature often wonders whether he listens to, or reads, scientific arguments, or sixteenth century religious disputes. Dr. F. G. Crookshank, in his chapter on *The Importance of a Theory of Signs and a Critique of Language in the Study of Medicine* in Ogden and Richard's *The Meaning of Meaning*, gives an excellent picture of the present sad state of affairs; but a $\bar{A}$ analysis discovers deeper foundations underlying the difficulties in medicine, which would have to be remedied by the revision of medical education. In this connection, $\bar{A}$ issues become very important. Organisms in general, and humans in particular, represent colloidal processes which involve tremendous pressure because of colloidal attraction for water. Dr. Neda Marinesco[1] has recently suggested that Ice VI constitutes an important factor in the human organism. Ice VI represents a new form of ice discovered by P. W. Bridgman[2] who found that water in bulk and at the temperature of the body may be found to crystallize by the application of high pressure. It is the notion of Dr. Marinesco that the forces of adsorption may be as high as the pressure used by Bridgman, so that in thin surface films the arrangement of water molecules may be much like that found in Ice VI. It may interest the reader to know that, among others, Professor Bridgman discovered that aceton becomes solidified at room temperature, albumen coagulates., under high pressure.

Although physicians in their university days are well acquainted with colloidal chemistry, yet somehow, in practice, they have great difficulties in 'thinking' in colloidal terms. With the newest discoveries of physics of high pressure and piezochemistry, with their bewildering variety of physical manifestations, which, under different pressure, change with every individual material, a modern physician will have to 'think' not only in terms of colloids, but of colloids in combination with the data of high-pressure physics and piezochemistry. Now such 'thinking' is humanly impossible under the traditional two-, or three-valued A disciplines and becomes only possible with ∞-valued $\bar{A}$ general semantics. One of the immediate results of the use of $\bar{A}$ disciplines is the elimination of the elementalism of 'body' *and* 'mind', 'intellect' *and* 'emotions'., and the introduction of the *non-el* point of view as given in the present work. This requires every physician to be acquainted with psychiatry, which acquaintance would eliminate many harmful cults. It should be fully realized that the older chemistry which dealt with different 'substances', having different 'properties', could have been treated by A subject-predicate and two-, or three-valued means. But not so in 1933; the older chemistry is gone, and today we deal only with a special branch of physics based on structure; the newer physics of high pressure show clearly that many of the older characteristics of 'substances' are only accidental functions of pressure, temperature, and what not, which vary in a bewildering way, requiring new semantic principles, new epistemologies.; in short, a new *non-el* and ∞-valued $\bar{A}$-system. In other words, whoever retains the A *s.r* is entirely unable to 'think' scientifically in the modern sense. If we want to have a science of man or a 1933 science of medicine., the first step is to revise thoroughly the *A-system*.

In fact, many more interconnections and interrelations could be shown which would make still more obvious how a $\bar{A}$-system results from, and leads to, modern scientific results, which can be extended and *applied to all human concerns* only after a *general* formulation *as a system.*

4) If the difference between the animal and man consists in the capacity of the latter to start where the former generation left off, obviously humans, to be humans, should exercise this capacity to the fullest extent. If we fail to do this, we again 'copy animals in our nervous reactions', which copying is the very thing we should struggle against. This 'where the former generation left off' would not only include all science, but also epistemology and the 'wisdom' which through painful experiences each former generation has accumulated., which, *in principle, should be given to every child.* Under the A conditions of our present education., systems, and evaluational systems, this is completely impossible and may sound visionary. Thus, to acquire scientific knowledge in all fields, one would have to spend a lifetime devoted to science, entirely free from financial worries., and even then he would only be able to acquire a small part of it. Before any older epistemological insight could be imparted, one should not only have special gifts, interests., but should also have an enormous amount of knowledge before such an education could be attempted. Similarly with 'wisdom'. The older and the younger generations, by colloidal necessity, cannot fully understand each other and, to a large extent, have mutual mistrust, which, as yet, is an entirely normal A *s.r*,.

In a *non-el*, $\bar{A}$-system this whole situation becomes radically changed. The impossible is made possible; I may say more, it is made simple and easy, and becomes a necessary and unavoidable factor in the life of any child. A *non-el, Ā*-system is based on the complete elimination of identification, from which it directly follows that the only link between the un-speakable objective levels and the verbal levels is found in *structure*. Structure, then, becomes the only possible content of all knowledge., and all scientific technicalities, admittedly laborious and difficult, become only a necessary tool in the search for structure, with little, if any, intrinsic value, and are unnecessary for 'knowledge' as soon as in a given case the structure is discovered. This structure is always simple and can be given to children.

It is meaningless and utterly useless to argue whether or not the world is 'simple'; as the world *is not* our understanding of it; but as our 'understanding' happens to be structural, our nervous system, through its abstracting capacities, makes it simple, once its structural content is discovered. As the search for structure involves similarity of linguistic and empirical *structures,* we readily understand that any language, which we cannot evade teaching our children, has structure and involves structural assumptions. In the *structural revision* of our language and the teaching of a few structurally appropriate terms, entirely abandoning a few structurally misleading ones, we directly impart all up-to-a-date fundamental knowlege to any child. We train him automatically in the appropriate linguistic structure, which builds up in him appropriate *s.r.* Mankind at large does not need scientific technicalities to absorb and thereby obtain semantic benefits from the structural results of science. These results are the only ones which really matter, and which can be given in an extremely simple way, automatically abolishing the primitive metaphysics, structural assumptions, and infantile *s.r.*

By abolishing the structurally false to facts one-valued identification, we automatically train in ∞-valued differentiation leading toward consciousness of abstracting, which results in all the wisdom that epistemology and private experience can give us, being structurally a total result of racial experience. As structure is based on relations and *order,* structural training, when done consciously, becomes a physiological method, working simply and automatically.

In the *A*-system these semantic mechanisms were not consciously recognized, although they worked fatalistically with us. We were imparting primitive psychophysiological reactions to our children, who had to spend a lifetime to learn by very painful experience that something was wrong somewhere. Now we understand that the origin of the difficulty was in the lack of scientific investigations which would have analysed, non-elementalistically, the structural aspects of language and connected *s.r.* All of which, let me repeat, works automatically, as experience and experiments abundantly show. Thus, an analysis of the mechanism of time-binding depends on the discovery of a sharp *non-el* difference between 'Fido' and 'Smith', and the formulation of means to make the time-binding characteristics of man fully effective with all except heavily pathological individuals.

By abolishing identification we generalize differentiation and so impart consciousness of abstracting, an indispensable factor in *proper evaluation,* and an absolute condition for adaptive and so survival behaviour. Thus a $\bar{A}$-system becomes a general theory of sanity and the general theory of time-binding, from which general semantics follow.

5) One of the most important features of the present $\bar{A}$-system consists of its *non-el* structural character. We may analyse problems in a scientific 'intellectual' way; yet this analysis, because *non-el*, structural, and semantic, appeals to, and affects, our 'feelings', 'intuitions'., involving psycho*physiological* factors based on order. Thus, the structurally necessary translations of one level of abstractions into the others and vice versa, is enormously facilitated, while in *el* systems these translations were hampered by unavoidable semantic blockages. Accordingly, 'intellect', 'emotions', 'body', and 'mind'., are not divided. The organism is affected *as-a-whole,* because structurally correct *non-el* means are employed, making many benefits of the system accessible to children, morons, and, perhaps, even superior idiots. The last results are to be foreseen, although they have not, as yet, been verified empirically.

6) But the most workable feature of the system consists in the fact that, being based on such fundamental principles as non-identification, non-elementalism., it has an organic unity. The main issues are all strictly interrelated and apply to 'body', 'mind', 'emotions'., in a *non-el* way, all working *automatically,* no matter from what angle we approach the training.

Thus, if we start with order, we are led to relations and structure; these establish differentiation and stratification, eliminating identification and 'allness', which result in consciousness of abstracting, necessitating ∞-valued general semantics, indispensable to proper evaluation and adjustment. If we start with non-identity, we are led to order, relations, structure, differentiation, stratification, non-allness, consciousness of abstracting, ∞-valued semantics, proper evaluation, and adjustment. If we start with differentiation or stratification, we are led to order, relations, structure, non-identification, non-allness, consciousness of abstracting, and proper evaluation.

35

It should be noticed that consciousness of abstracting and proper evaluation are complex end-results which cannot be imparted directly, but which become automatically lasting semantic states only after we have eliminated one-valued identification, or introduced order, ∞-valued differentiation, stratification,. The *non-el* benefit of the system consists in engaging the organism-as-a-whole. Thus, four-dimensional *order* plays the role of a potent *physiological* factor in the process and becomes the foundation for psychophysiology. Non-identity is a term applied on the verbal levels, which, on visual and intuitive levels, involves differentiation, ordering, and stratification. This system thus involves all necessary nerve centres and operates in a *non-el* way, as reactions on one level are easily and organically translated into the terms of other levels, making psychophysiology possible.

7) Finally, it is significant that many publications in the last ten years have shown efforts in a similar direction, which have received scientific and public approval. As I am more interested in creative work, rather than critical, I shall not analyse these strivings except to make one general remark that, because they are not based on order, structure, *non-el s.r*, the complete elimination of identification., they are valuable and useful to the selected few, but under no conditions could a psychophysiology or a theory of sanity be based on these works which could be applicable in general elementary linguistic and semantic education. If I am not mistaken, in this respect the present work differs radically from the others with which I am acquainted.

From a *non-el* point of view we can never disregard the effect the 'body' or 'emotions' have on the 'mind', and vice versa the effect that the 'mind' has on the 'emotions' and the 'body',. Identification and all its consequences involve seriously disturbing semantic factors with corresponding colloidal disturbances, and it seems that, as yet, the human race, outside of very exceptional cases, has never been free from these disturbances. What effect the elimination of such disturbances will have on the human race it is impossible to foretell at this stage, beyond expressing the expectation that the consequences must be highly beneficial.

We have already become acquainted with the terms 'conditional' and 'unconditional' reactions. In the example of the patient and the paper roses, we have seen that the pathological symptoms were 'unconditional'. They were compulsory, as in the case of the dogs mentioned in Part VI. In a healthy individual they would have been fully conditional reactions, under semantic control. The above terminology may be extended so as to apply to all 'mental' ills, for here that which in the 'normal' person is a fully conditional reaction becomes unconditional, or a *reaction of lower order conditionality* (compulsory) beyond conscious control. Here we differ from animals and hospital cases. When our conditional reactions are not fully regulated by proper *s.r* and become unconditional, we copy animals, and so are in a state of arrested development or of regression.

The general therapeutic and *preventive* measures are clearly indicated by such considerations. Conditional reactions in man should become *fully* conditional and not fixed as *un*conditional, or conditional of lower orders. In other words, instead of 'fixation', we should have means and methods to preserve and foster *semantic flexibility. This last is accomplished by acquiring the semantic reactions connected with the consciousness of abstracting.* I recommend this last point to the attention of specialists, as it is impossible in the present work to go further into details. Flexibility is an important semantic characteristic of healthy youth. Fixation is a semantic characteristic of old age. With the colloidal background, the imparting of permanent semantic flexibility which every one acquires who becomes conscious of abstracting might prove to be a crucial neuro-physico-chemical colloidal factor of, at present, unrealized power. The colloidal behaviour of our 'bodies' is dependent on electromagnetic., manifestations, which, in their turn, are connected with 'mental' states of every description. If the colloidal ageing, which brings on old age, 'physical' and 'mental' symptoms, and, ultimately, death, is connected with such 'mental' fixity, we may expect some rather startling results if we impart a permanent semantic flexibility. The 'ageing' involves electrical changes in the colloidal background, which must be connected with the older semantic states. The new fluid semantic states should have different electrical influences, which, in their turn, would bring about a difference in the colloidal behaviour on which our 'physical' states depend.

From a $\bar{A}$ point of view, a new era of human development seems possible, in which, by mere structural analysis and a linguistic revision, we will discover disregarded semantic mechanisms operating in all of us, which can be easily influenced and controlled; and we will discover, also, that at least a great deal of prevention can be accomplished.

It seems, also, that we will discover more about the dependence of 'human nature' on the structure of our languages, doctrines, institutions., and will conclude that for adjustment, stability., we must adjust these man-made and man-invented semantic and other conditions in comformity with that newly discovered 'human nature'. This, of course, would require a thorough scientific 1933, physico-mathematical, epistemological, structural, and semantic revision of all existing human interests, inclinations, institutions., to be made by those specialized in a 'science of man'. If such a revision is produced soon enough, it will, perhaps, help to adjust peacefully the standards of evaluation and prevent the repetition of bloody protests of unenlightened blind forces against *equally blind* forces of existing powers and reactions.

The forces of life, humanity, and time-binding are at odds; in modern slang, a 'show-down' is imminent; it *will happen,* and no one can prevent it. To a $\bar{A}$ understanding the only problem of importance is whether this 'show-down' will be scientific, enlightened, orderly, and peaceful, with minimum suffering; or whether it will take a blind, chaotic, silly, bloody, and wasteful turn with maximum suffering.

The problems of structure, language, and 'consciousness of abstracting' play a crucial semantic role. To be modern, one must accept modern metaphysics and a structurally revised modern language. As yet, these semantic problems have been *completely* disregarded as far as general education is concerned. This is probably due to the fact that in an infantile and commercial civilization we encourage engineering and applied sciences, medicine, biology., to increase private profits., and preserve or increase the ranks of buyers. But we do not encourage to an equal extent branches of science like mathematics, mathematical philosophy, linguistic, structural, and semantic researches., which would not directly increase profits or the numbers of customers, but which would, nevertheless, discover structural means for more happiness for all.

Accidentally—and this is recommended to the attention of economists—the classical law of 'supply and demand' is structurally and semantically an *animalistic law,* which in an adult human civilization must be reformulated. In fact, an adult human civilization cannot be produced at all if we preserve such fundamental animalistic 'laws'. In the animal world the numbers of individuals cannot increase beyond what the given conditions allow. The animals do not produce artificially.

Not so with our human world. We produce artificially because we are time-binders, and all of us stand on the shoulders of others and on the labours of the dead. We can over-populate this globe as we have done. Our numbers are not controlled by unaided nature, but can be increased considerably. In the animal world the numbers are regulated by the supply of food., and not by conditions imposed by the animals on that food supply. The animal law of 'supply and demand' is strict. In a human class of life, which does produce artificially, production should satisfy the wants of all, or their number should be controlled until the wants can be filled. The application of animalistic laws to ourselves makes conditions very complicated, and detrimental to most, if not all of us. It is also easily understood why it should be so. Ignorant and *A* handling of powerful symbols has proved to be dangerous when we do not realize the overwhelming semantic role and the importance of symbols in a symbolic class of life.

Another interesting application of the consciousness of abstracting is given in our attitude toward money, bonds, titles to property,. Money represents a symbol for all human time-binding characteristics. Animals do not have it. No doubt bees produce honey, but these products of the bees do not constitute wealth until man puts his hands on them. Money is not edible or habitable. It is worthless if the other fellow refuses to take it. The *m.o reality* behind the symbol is found in *human agreement*. The *value* behind the symbol is *doctrinal*. Fido does not discriminate between the different orders of abstractions. If we copy him, we worship the symbol alone. 'In gold we trust' becomes the motto, with all its identifications and destructive consequences. Smith should not identify the *m.o* reality behind the symbol with the symbol. It is amusing, when not tragic, to see how the so-called 'practical man' deals mostly with fictitious values, for which he is willing to live and die. When he has the upper hand and ignorantly plays with symbols, disregarding the *m.o* realities behind them, of course, he drives civilization to disasters. History is full of examples of this.

We see the utter folly of racing to accumulate symbols, worthless in themselves, while destroying the 'mental' and 'moral' values which are behind the symbols. For it is useless to 'own' a semantically unbalanced world. Such ownership is a fiction, no matter how stable it may look on paper. Commercialism, as a creed, is a folly of this type. Some day even economists, bankers, and merchants will understand that such 'impractical' works as this present one on structure, *s.r.*, lead to the revision of standards of evaluation and are directly helping the stabilization of an economic system. Meanwhile, in their ignorance, they do their best to keep the economic system unscientific, and, therefore, unbalanced. History shows clearly how the rulers have generally made life unbearable for the rest of mankind, and what bloody results have followed. Since the World War certain conditions are becoming increasingly more difficult, and the infantile and animalistic systems drive us fatalistically toward further catastrophes. Whether these disasters will occur, the unknown future shall decide; but out of this unknown, one fact remains a certainty; namely, that this will depend on whether or not science can take hold of human affairs; I hope it can, but the blind forces of identification are so strong and powerful that perhaps such hopes are premature. Perhaps a new race can accomplish it after this one is extinct, with the exception of a few remnants in museums.

The problems of determinism and indeterminism are not purely 'academic' but influence, to a large extent, our theories and behaviour, and so are fundamental for adjustment. Historically, science has utilized determinism of the two-, and three-valued variety, which has lately, in the case of the newer quantum mechanics, proved insufficient. The lack of the formulation of ∞-valued semantics, necessary for ∞-valued determinism, seemed to indicate that even science tends to drift toward indeterminism, a tendency which was rather baffling and disturbing to many scientists.

Different 'ethics' and 'morals' have fought determinism throughout all our past on the ground that in a deterministic world all 'morals' and 'ethics' would be impossible. If a man is compelled to do something, then, we are told, he is not responsible. They state that the result would be undesirable licence, forgetting that determinism implies quite the opposite of licence.

We have already become acquainted with infantile self-love and self-importance. These infantile characteristics have not only shaped our semantic attitudes, but also our 'scientific' theories. Smith and this little earth were in many ways postulated as the centre of the universe. Scientific discoveries showed that such statements did not cover the facts at hand, and Smith was displaced from this primitive and infantile self-centred position. The Polish astronomer Copernicus was the first to give this rude shock. The little earth was no longer *the* 'centre of *the* universe'. Next came Darwin with another shock to such infantile pride. Smith was no longer a 'special creation', but belonged to the general series of living forms, none of which were 'special creations'. Finally, Freud developed the notion that even in semantic processes, determinism prevails. All our actions, psycho-logical and semantic states., have very definite conscious and unconscious psychophysiological 'causes' which activate us.

An infantile society had difficulties in abandoning their pleasing delusions, and these three men were duly persecuted, criticized, and bitterly attacked and hated by many.

The present situation may appear baffling because science discovers facts which would seem to lead to an 'undesirable' indeterminism in science, and to a determinism in 'mental' processes. The reader, by now, I hope, realizes that both 'undesirable' results are only undesirable because of identification and the confusion of orders of abstractions, which resulted in the ascribing of *undue generality and uniqueness* to the *A* two and three-valued 'logic'. But once we realize that in a $\bar{A}$, ∞-valued, more general system, the two-, and three-valued aspects are only particular instances, which apply to some instances but not to others, all our difficulties vanish. From a $\bar{A}$ structural point of view we also understand that ∞-valued determinism becomes a necessity of our *s.r* in the search and comparison of structures.

The result seems to be that the problem of determinism or indeterminism is not primarily a problem of the outside world, but simply one of our *s.r* and ignorance versus 'knowledge'. Abandoning elementalism and identification, we stop arguing '*is* the *world* deterministic or not'.; but, by analysis, we find which semantics better fit, structurally, the facts and our abstracting capacities. The results we reach are not entirely new, but the semantic conflict is eliminated.

Science employs determinism because of the structure and function of our nervous system. We cannot do otherwise than preserve ∞-valued determinism and step by step supply the missing links in our structural adjustments of language to the structure of empirical data.

Let us again repeat that the older problems of 'determinism' *in general* were the results of elementalism and identification and of a complete misunderstanding of the role of structure. Once these undesirable afflictions are eliminated, the artificial problems which they create are also eliminated. Structural considerations show clearly that determinism is a neurological necessity. If empirical facts lead to linguistic indeterminism, it is an unmistakable sign that the language used is not similar in structure to the structure of the world around us, and that we should simply produce a language of different structure. *Such determinism is a vital condition in the search for structure, and cannot be abandoned.*

Shall we, then, preserve the deterministic attitude in our 'mental' processes? Are the objections on 'moral' and 'ethical' grounds serious enough to induce us to reinstate in our semantic attitudes the older structurally misleading 'indeterminism'?

Let us, first, recall the facts. In our old *el* and infantile attitudes with identification we analysed a child or an adult in *isolation*. Determinism was applied to such a fictitious *non-existent* individual, and the old objectified and *el* speculations followed. If any one is inclined to challenge the above statement, let him perform an experiment and immediately after birth isolate a child 'completely'. He will find that this cannot be done with a human baby without destroying the child. Therefore, the old speculations deal with structurally *fictitious* conditions. The facts are that a baby is, from the first, subjected to a treatment based on the semantics, structure of language, doctrines, understanding, knowledge, attitudes, metaphysics., of his parents or their substitutes, which *shape his semantic reactions.*

If we abandon the problem of the two-valued 'determinism' in connection with such a fictitious, isolated individual, and apply ∞-valued determinism to an actual, non-isolated individual, we see at once that the whole situation is different. If parents and society accept ∞-valued determinism, they realize their own *responsibilities* toward the individual, and understand that the actions of parents, society., are, to a large extent, responsible for the future development of the child on quite deterministic psychophysiological grounds. If an individual behaves in a way detrimental to others and to himself, and an enlightened society decides to do this or that with him, that is a different proposition. The main point is that, if we were to accept an indeterministic attitude, a great deal of harm would be done by parents, teachers, preachers, and society in general; harm which could be prevented. This is, to a large extent, unrealized, and in the old way no one was supposed to be *responsible* except the poor victim of 'free will'. Under such A conditions, we sponsor bitterness, cruelty., under the labels of 'sin', 'justice', 'revenge', 'punishment', or whatever it may be. On deterministic grounds, when society and educators realize fully their own responsibilities, we should blame the individual less, and should more and more investigate structure, language, our systems, metaphysics, education, conditions of living,. Instead of a holy frenzy for 'justice', 'punishment', 'revenge'., we would try to improve conditions of life and education, so that a new-born individual would not be handicapped from the day of his birth.

Since the organism operates as-a-whole and *no one* is free from higher order abstractions and structural assumptions, we see that the *keeping of savage-made metaphysics* must involve us individually and collectively in an arrested or regressive development. From the organism-as-a-whole point of view structural ignorance must result in some semantic defectiveness.

The objection that there are cases of great 'mental' brilliancy accompanied by very vicious tendencies is easily answered by the fact that the problem is formulated in an *el* way. 'Mental' brilliancy does not tell the whole story of the organism-as-a-whole. One may be 'mentally' brilliant, yet infantile or a 'moral imbecile'. In life, we deal with the whole non-isolated individual, who may be pathological in a great many ways. If it is objected that science is so complicated that it would be impossible to impart such knowledge to the masses, the answer is that, as this enquiry shows, science involves some structural metaphysics and semantic components which, once discovered, are childishly simple, and can be given in elementary education.

Science represents the highest structural abstractions that have been produced at each date. It is a supreme abstraction from all the experiences of countless individuals and generations. Since the lower centres produce the raw material from which the higher abstractions are made, and these higher abstractions again influence the working of the lower centres, obviously *some means can be devised to put back into the nervous circuits the beneficial effects of those highest abstractions*.

The above statement may appear visionary, and many are likely to say, 'It cannot be done'. Now, the main contention of the present theory, verified empirically, is that it *can* be done in an extremely simple way, provided we study the neglected *non-el* aspects of mathematics and science; namely, their structural and semantic aspects. Such study has helped us to discover in a $\bar{A}$-system the means for affecting lower centres by the products of the higher centres of the best men we had. We have already discovered that all advances in science and mathematics supply us with an unbelievable amount of purely psycho-logical and semantic data of extreme simplicity, which, without any technicalities, can be imparted to the masses in elementary structural education. Such education allows us to give very simply to children the 'cultural results', or to impart the *s.r*, which are the aim of university training, in a relatively short period and without any technicalities. These benefits, under an A education, are too rarely acquired even by university graduates, and impossible to impart to the masses, who are left helpless with archaic, delusional structural assumptions.

From one point of view the $\bar{A}$ issues are childishly simple and obvious, but from another, because of the power of old established habits and *s.r.* are quite difficult for the grown-ups to apply. It seems evident that an infant must be under the influence of the standards of evaluation of those who take care of him, automatically connected with the structure of the language he is taught. Under such unavoidable conditions, it is obvious that to give the full benefit of a $\bar{A}$-system in the training of children, parents and teachers should, themselves, have entirely absorbed these new standards.

A $\bar{A}$ civilization will require a unification of all existing human disciplines on the base of exact sciences. This unification will require all scientists, mathematicians, physicists, and psychiatrists included, to become acquainted and *fully* to practice $\bar{A}$ standards of evaluation. A $\bar{A}$ revision would have an international and interracial application, requiring a very thorough revision of all doctrines, a better acquaintance of specialists in one field with the accomplishments in other fields, and an *up-to-date epistemology*. If we try to disregard epistemology consciously, we delude ourselves, as we cannot eliminate *some* epistemology as a foundation for our methods of evaluation, and, therefore, unconsciously retain some primitive epistemology which through inappropriate standards of evaluation, introduces semantic blockages.

Mach said long ago: 'Not every physicist is an epistemologist, and not every one must or can be. Special investigation claims a whole man, so also does the theory of knowledge.' The influence of Mach on modern science is well known; men such as the late Jacques Loeb, Einstein, the younger quantum pioneers., were deeply influenced by the writings of Mach, because Mach was a deep student of epistemology. But in a $\bar{A}$ society his statement must be slightly reworded; namely: 'Not every individual knows or realizes the importance of, or seemingly consciously cares for, epistemology; yet every one unconsciously has one and acts and lives by it. Each individual has his own special problems, the solution of which always claims the whole man, and no man is complete, unless he consciously realizes the permanent presence in his life of some standards of evaluation. Every one has thus *some* epistemology. There is no way of parting with it,—nor with air, nor with water,—and live. The only problem is whether his standards of evaluation are polluted with primitive remains of bygone ages, in a variety of ways; or sanified by science and modern epistemology.'

The present work shows how any system involves a special epistemology which we accept unconsciously, once we accept the system. To evaluate a system is practically equivalent to formulating its epistemology. This is strictly connected with linguistic and structural investigations.

To centralize and co-ordinate the $\bar{A}$ efforts, an *International Non-aristotelian Library* has been originated, which field embraces, ultimately, all known doctrines and human interests, the first publication being the present handbook. To facilitate the application of $\bar{A}$ disciplines and to stimulate further researches, an *International Non-aristotelian Society* has been incorporated with headquarters in New York City and branches to be established in all cities of the world which have educational institutions. The main aims of the Society are scientific and educational for the study, by means of papers and lectures followed by discussions, of the $\bar{A}$ aspects necessary for a revision and, therefore, a co-ordination of all existing sciences and concerns of man. As the aspects of science which are of interest to the Society would be *structural* and *semantic*, from the point of view of a *general theory of values*, the lectures would be of a general non-technical character on the level of intelligent laymen.

Science would not be 'popularized' but analysed from a fundamental $\bar{A}$ epistemological point of view, compelling the speakers and authors of papers to analyse the deeper *non-el*, structural, and semantic foundations of experience, as well as of theories. The layman would benefit because he would be given a structural education readily understood, without being led astray by the older A 'popularization'. Later, if economically feasible, it is intended to issue a monthly *International Non-aristotelian Review,* and also to organize *International Non-aristotelian Congresses.*

The A-system was the result of the *s.r* of the white race of more than two thousand years ago; it built up the doctrines, institutions ., appropriate to this system. In those days, knowledge was very scanty; the interconnection of different peoples, vague; the means of communications, very primitive,. It may be considered that science, and particularly mathematics, began a $\bar{A}$ revolution by explicitly searching for structure and adjusting the structure of the scientific languages, which we usually call 'terminology', 'theories',. Modern conditions of life are, to a large extent, affected by $\bar{A}$ science but exploited by the thoroughly A doctrines of the commercialists, militarists, politicians, priests, lawyers., which results in a bewildering chaos, resulting in needless, great, and imposed suffering for the great masses of mankind, as exemplified by such cataclysms as wars, revolutions, unemployment, different economic crises,.

$\bar{A}$ disciplines, or science *as such,* are thoroughly beneficial to mankind at large; but an A exploitation and use of these $\bar{A}$ products are, and must be, a source of endless sufferings to the enormous majority of mankind, leading automatically to every kind of break-down. It is impossible to give a fuller analysis of this complex interrelation, as this would require a separate volume; I shall, therefore, tabulate only a few overlapping suggestions.

NON-ARISTOTELIAN, SCIENTIFIC, ADULT STANDARDS OF EVALUATION	ARISTOTELIAN, INFANTILE STANDARDS OF EVALUATION OF COMMERCIALISM, MILITARISM ,.
Biological Sciences.	
Particularly medicine has discovered means of how to keep or restore health, how to eliminate suffering, how to save and to prolong life .; but medicine also gives means whereby we may prevent the overpopulation of this globe and so teaches us how to avoid great sufferings through overcrowding, which results in bitter struggle for food, shelter, employment., in turn, leading to wars, revolutions, unemployment,.	Commercialized medicine unavailable to great masses of poor people. Commercialism, militarism, infantile dreams of 'world empires' sponsor unintelligent breeding in vast numbers which runs up the prices of land and houses, lowers the price of labour, and supplies cannon-fodder,. The intelligent control of numbers of population is prevented by jailing and otherwise persecuting scientific workers, which persecution affects only the poor and uneducated.

NON-ARISTOTELIAN, SCIENTIFIC, ADULT STANDARDS OF EVALUATION	ARISTOTELIAN, INFANTILE STANDARDS OF EVALUATION OF COMMERCIALISM, MILITARISM ,

Chemistry.

Antiseptics, fundamental for medicine and the control of the increase or decrease in the population.	Sponsoring of overpopulation by forcibly withholding knowledge from the masses.
Drugs, fundamental for medicine.	Use of drugs in war to allow a soldier to suffer twice. Commercialized drug pedlars.
High explosives, necessary for agriculture, mining ,.	High explosives used for killing in the struggle for ultimately futile 'world empires'.
Production of food products. Alcohol, wine, beer,.	Food destroyed to keep up prices. Commercialization of drinking. The saloon, crimes, 'prohibitions', the financing of gangsters, corruption of government and justice ,.
Poisonous gas, necessary for the elimination of insects.	Poisonous gas in wars.

Linguistics.

Law, as expressing some standards of evaluation.	Interpretations by commercialized lawyers to evade law, and the influencing of the wording of the law so as to make evasion possible. Lobbyists.
Newspapers, magazines., giving most powerful educational means.	Commercialized newspapers.., controlled by profits and advertisements, supplying stultifying, controlled material, stimulating the morbid potentialities of the mob, to increase circulation ,.
Other public prints, giving necessary or useful informations.	Commercialized advertisements. Schizophrenic play on words, the promoting of infantilism ,.
Religions represent primitive structural rationalizations, or primitive 'science'; intended, also, as guides for conduct and adjustment, under the structural assumptions of the epoch of their primitive origin.	Commercialized religions. Religions having outlived their usefulness, often become priestcraft as a source of income and control. The sponsoring of primitive and delusional standards of evaluation often for private gains. The imposition of primitive standards of evaluation involves pathological factors.

Physics and Related Sciences.

Aeroplanes, as means of communication and scientific exploring.	Militarized aeroplanes, as means of destruction and murder.
Automobiles, as means of transportation and pleasure.	Militarized automobiles, as means of destruction and suffering.

NON-ARISTOTELIAN, SCIENTIFIC, ADULT STANDARDS OF EVALUATION	ARISTOTELIAN, INFANTILE STANDARDS OF EVALUATION OF COMMERCIALISM, MILITARISM, .
Physics and Related Sciences (continued).	
Machinery and tools, as means to eliminate avoidable efforts and to benefit fully through natural resources; to give greater comfort and sanitary conditions; to allow greater leisure for cultural pursuits, .	Commercialized machines, as means for larger individual profits and suffering for the increasing numbers of unemployed and starving masses. Mass production of guns, ammunitions., for mass extermination and destruction, and larger profits for the manufacturers and investors.
Moving pictures, as powerful educational means.	Commercialized moving pictures, stimulating and satisfying the erotic and morbid crowds, and for private propaganda.
Radio, as powerful means of communication and education.	Commercialized radio, advertisements, private propaganda, often stimulating morbid inclinations of the mob.
Railroads, as means of public transportation.	Commercialized railroads, as means for private gains and the control by a few of large areas and many people.
Tractors, as means of transportation and sources of power in agriculture.	Tractors, as tanks and means of destruction and murder in wars.
Public Servants.	
Judges, as guardians of some standards of evaluation.	Commercialized politicians, as judges, corruption, lack of justice, .
Lawyers, as assistants in the administration of justice.	Commercialization and corruption of the legal profession. Means to evade or pervert justice.
Police, as an executive, regulating, safety-force.	Commercialization of police by politicians, corruption, combinations with the underworld, .
Sports, as means for preserving and building up health, co-ordinated orientation, fair play, and in a lesser degree as recreation.	Commercialization of sports, elimination of benefits. Gambling. Lowering of educational standards, .

From a $\bar{A}$ point of view, which eliminates primitive *s.r*, it becomes obvious that mankind represents an interdependent time-binding class of life, and any group of people who possess physical means for destruction and still preserve infantile standards of evaluation become a menace to the culture of the whole race. Under such conditions we must have agencies for an international exchange and evaluation of our standards, as well as methods which would help us in adjusting these standards.

At present, we must admit that with the modern, rapid, and international advancement of science we have fairly well-established international $\bar{A}$ standards of scientific values. International Scientific Congresses are not only necessary for the advancement of science, but they also explicitly prove science to be entirely international.

The latest, most important $\bar{A}$ institution is found in the League of Nations, which embraces practically the whole civilized world, with the exception of a very few nations who display infantile and A aloofness, using different self-deceptive excuses, and, to a large extent, handicapping the power and usefulness of the League.

As we have learned lately, not only human achievements, but also human disasters, are mostly interrelated and international, and are becoming more so every year. Obviously, with A narrowness, selfishness, shortsightedness, infantilism, commercialism, militarism, nationalism., rampant, mankind, to prevent further major A disasters, would have to produce a special international body which would co-ordinate various structural achievements, strivings., formulate and inform the great masses of mankind of the modern scientific $\bar{A}$, adult standards of evaluation.

At present, we already have the necessary agencies; but, as yet, they are inefficient and non-co-ordinated. These are to be found in International Scientific Congresses, and the League of Nations. The weak spots of these organizations are found in the fact that the Scientific Congresses are too cumbersome, expensive, non-co-ordinated and only periodic. The League of Nations, although a $\bar{A}$ body in structure, is mostly made up of men who do not know any other standards of evaluation than A, and so they often lack the means to present a scientific, or $\bar{A}$ argument, and usually do not realize the tremendous power they would have in a $\bar{A}$-system. In human affairs, for instance, there cannot be a neutral and innocent absentee. One such absentee with guns and battleships becomes a powerful blocking and so, ultimately, disrupting factor for the rest of civilization. Such an absentee is not, then, guilty by omission; but, from a $\bar{A}$ point of view, becomes guilty by commission. The League, when definitely and fully allied with international science, will some day have the pluck to make such a declaration and act accordingly. A consciously $\bar{A}$ League of Nations will not limit itself to the thankless and very often useless task of adjusting inevitable clashes of A standards of evaluation, but will, with the full co-operation of scientists, undertake the much more important, constructive, and unique duty of a guardian and leader of human culture. Such a League would become a scientific, professional, international, co-ordinating, cultural, time-binding advisory organization for all nations. National A governments, instead of only *instructing* their A representatives, would first *consult* with the new $\bar{A}$ *specialists*.

Many politicians and their followers often become all but hysterical at the mention of the League of Nations, which, in some mysterious way, they associate with a 'super-state' or 'control',. Let me say, at once, that a symbolic, or human class of life, is very largely controlled by ignorant, hidden, often pathological., factors *beyond public control,* of which the majority are entirely innocent. In the human symbolic class of life no one is entirely free, but all our lives are entangled in an interdependence of human relations. The dependence on those powers which are now hidden, and *beyond public control,* constitutes a grave danger to all. Not so with a scientific, enlightened public opinion with adult standards of evaluation as formulated by a future co-ordination of science and the League of Nations. Such great majority-opinions will remain opinions, or statements of standards of evaluation, which any one member of the League may accept or reject; but, then, it would be necessary for him to state publicly his standards of evaluation and to decide consciously to act with or against, or to enlighten further the opinion of the human race. There is, of course, no question of 'super-states' or 'control' except the unified demand for a conscious and explicit stand on any important subject by any nation. Public opinion will do the rest, once it is convoked to act.

I am not a pacifist in the accepted sense. In an animalistic, infantile, or A society, this would be not only impossible but downright silly. Quite the contrary, I am disgusted with the infantile standards under which wars are conducted. Thus, our rulers and war lords, sponsored by commercialism, like little boys make a sort of game out of wars, and thus help to preserve them as an institution. In a consistent society, wars should be as ruthless as possible *to all.* If any one wants a war, he should consistently take all the consequences. But this would not suit our infantile rulers; they know that when little boys play at war, and one group becomes too rough, the other group refuses to play, and so the play at war ends.[3] All these perverted 'humanitarianisms' only sponsor wars because, in an unlimited modern warfare, the people would soon come to their senses and would refuse to suffer for the benefit of the very few. So I am far from being an A pacifist.

But why our destinies should be dependent on the accidental and primitive structure of the language we use., is beyond my comprehension. I grant that if we accept such and such postulates, two-valued, *el*, structurally false-to-fact A 'logics', 'psychologies'., all the old, too familiar consequences follow, which we, in our ignorance, have forced upon human life.

But if we put all systems and all 'logics'., on new $\bar{A}$ foundations, which are structurally closer to the facts of life (1933), all the older conclusions may even be reversed. The problem now before mankind is whether or not the new $\bar{A}$-system is more similar in structure to the world and our nervous system than the old. On the answer to this question, the future of civilization depends.

From the present point of view, we should establish with the League of Nations a permanent $\bar{A}$ or scientific department, composed of a few of the best scholars from all countries, who would keep in touch, not only with the developments of their specialties, but would also co-ordinate them on general structural and epistemological foundations. This department would be the international authority on modern revised and co-ordinated standards of evaluation, which would be published in special proceedings. The present discrepancy and lack of co-ordination between different branches of knowledge becomes genuinely alarming and detrimental to mankind, because in 1933 it is humanly impossible for a single individual to attempt such a co-ordination. Members of this group would be selected by the universities of each country. In their researches, joint studies, and results, mankind at large would find the most reliable scientific and $\bar{A}$ opinions produced at each date, and would have some definite and conscious standards of evaluation by which to orient themselves.

The modern 'voting' has some benefits in local affairs, but when its very limited validity is not understood, it becomes a serious danger to mankind. Thus, when we are ill, or when we want a bridge built, we ask specialists for their scientific co-operation; we would hardly depend on ignorant voters. Similarly, in a $\bar{A}$ scientific civilization, the major problems of mankind would be analysed by scientific specialists, recommendations offered to be accepted or rejected, as the case may be; but the ignorant voter would have at his disposal an unbiased, impersonal, and responsible opinion of international scientific specialists to compare with the equivocations of some local ignorant politician.

To facilitate such future $\bar{A}$ activities, the International Non-aristotelian Society has been established. It is hoped that soon the scientific, educational, 'mental' hygiene., workers will begin to unite on a local and national $\bar{A}$ basis. Later, International Congresses will unite the local societies, which ultimately will be embodied as a permanent institution, most probably in the League of Nations.

In the process of formulating the above system a curious observation has been forced upon me; namely, that statements which are, for instance, quite legitimate for the English language, even though they probably apply in general to all Indo-european languages, do not apply in a similar degree.

I am intimately acquainted with six languages, two Slavic, two Latin, and two Teutonic, and also with the psycho-logical trends of these groups. I have been led to suspect strongly that the finer differences in the structure of these languages and their use are connected with the semantics of these national groups. An enquiry into this problem, in my opinion, presents great semantic possibilities and might be the foundation for the understanding of international psycho-logical differences. Once formulated, this would lead us to a better mutual understanding, particularly if a $\bar{A}$ semantic revision of these different languages is undertaken. To the best of my knowledge, this field of enquiry is entirely new and very promising.

It must be obvious to the reader that such a vast program is beyond the power of a single man to carry out, and the present author hopes for public interest in this enterprise.

If the $\bar{A}$-system has accomplished nothing more than to draw the attention of mankind to some disregarded problems; if it has done nothing more than point the way, not to panaceas, but to suggestions toward an expedient, constructive, and unified scientific program whereby future disasters may be avoided or lessened—the writer will be satisfied.

BOOK III

ADDITIONAL STRUCTURAL DATA
ABOUT LANGUAGES AND
THE EMPIRICAL WORLD

The every-day language reeks with philosophies . . . It shatters at every touch of advancing knowledge. At its heart lies paradox.

The language of mathematics, on the contrary, stands and grows in firmness. It gives service to men beyond all other language. (25)

ARTHUR F. BENTLEY

Nothing is more interesting to the true theorist than a fact which directly contradicts a theory generally accepted up to that time, for this is his particular work. (415)   **M. PLANCK**

It is not surprising that our language should be incapable of describing the processes occurring within the atoms, for, as has been remarked, it was invented to describe the experiences of daily life, and these consist only of processes involving exceedingly large numbers of atoms. Furthermore, it is very difficult to modify our language so that it will be able to describe these atomic processes, for words can only describe things of which we can form mental pictures, and this ability, too, is a result of daily experience. (215)   **W. HEISENBERG**

PREFATORY REMARKS

In re mathematica ars proponendi quaestionem pluris facienda est quam
solvendi. (74)   GEORG CANTOR

We cannot describe substance; we can only give a name to it. Any attempt to do more than give a name leads at once to an attribution of structure. But structure can be described to some extent; and when reduced to
ultimate terms it appears to resolve itself into a complex of relations . . .
A law of nature resolves itself into a constant relation, . . . , of the two
world-conditions to which the different classes of observed quantities forming the two sides of the equation are traceable. Such a constant relation
independent of measure-code is only to be expressed by a tensor equation.
(148)   A. S. EDDINGTON
We have found reason to believe that this creative action of the mind
follows closely the mathematical process of Hamiltonian differentiation of
an invariant. (148)   A. S. EDDINGTON

The only justification for our concepts and system of concepts is that
they serve to represent the complex of our experiences; beyond this they
have no legitimacy. I am convinced that the philosophers have had a harmful effect upon the progress of scientific thinking in removing certain fundamental concepts from the domain of empiricism, where they are under our
control, to the intangible heights of the *a priori*. (152)   A. EINSTEIN

In writing the following *semantic* survey of a rather wide field of mathematics and physics, I was confronted with a difficult task of selecting source-
books. *Any* mathematical treatise involves conscious and many unconscious
notions concerning 'infinity', the nature of numbers, mathematics, 'proof',
'rigour'. , which underlie the definitions of further fundamental terms, such as
'continuity', 'limits', . It seems that when we discover a universally *constant
empirical relation*, such as 'non-identity', and apply it; then all other assumptions
have to be revised, from this new point of view, irrespective of what startling
consequences may follow.

At present, neither the laymen nor the majority of scientists realize that
human mathematical behaviour has many aspects which should never be
identified. Thus, (1) to be somehow aware that 'one and one combine in some
way into two', is a notion which is common even among children, 'mentally'
deficients, and most primitive peoples. (2) The mathematical '1 +1 = 2' already
represents a very advanced stage (in theory, and in method . ,) of development,
although in *practice* both of these *s.r* may lead to one result. It should be
noticed that the above (1) represents an individual *s.r*, as it is not a general
formulation; and (2) represents and involves a generalized *s.r*. Does that
exhaust the problem of '1 +1 = 2'? It does not seem to. Thus, (3), in the
Principia Mathematica of Whitehead and Russell which deals with the meanings
and foundations of mathematics, written in a special shorthand, abbreviating
statements perhaps tenfold, it takes more than 350 large 'shorthand' pages to
arrive at the notion of 'number one'.

565

It becomes obvious that we should not identify the manipulation of mathematical symbols with the semantic aspects of mathematics. History and investigations show that both aspects are necessary and important, although of the two, the semantic discoveries are strictly connected with the revolutionary advances in science, and have invariably marked a new period of human development. In Chapter XXXIX, the reader will find a very impressive example of this *general* fact. Thus, what is known as the 'Lorentz transformation', *looks like* the 'Einstein transformation'. When manipulated numerically both give equal numerical results, yet the meanings, and the semantic aspects, are different. Although Lorentz produced the 'Lorentz transformation' he did not, and *could not* have produced the revolutionary Einstein theory.

It is well known that when it comes to the manipulation of symbols mathematicians agree, but when it comes to the semantic aspects or meanings . , they are *admittedly* hopelessly at variance. In a prevailingly A world we have had no satisfactory theory of 'infinity', or a $\bar{A}$ definition of numbers and mathematics. This necessarily resulted in the fact that the semantic aspects of practically all important mathematical works by different authors often involve *individual semantic presuppositions*, or orientations concerning fundamentals. My presentation intends to be primarily semantic and elementary, and is only remotely concerned with the manipulation of symbols. A $\bar{A}$-system, which rejects 'identity', differs very widely from A attitudes, and introduces distinct $\bar{A}$ requirements. I had, therefore, to select from many works, with their *individual* presuppositions, those which were less in conflict with $\bar{A}$ principles than the others.

A survey of important mathematical treatises shows that although the majority of modern mathematicians explicitly abjure the 'infinitesimal', yet, in some presentations, this notion persists. In my presentation I reject the 'infinitesimal' explicitly and implicitly, although the formulae are not altered. 'Modern' calculus is based *officially* on the theory of limits, but as the theory of limits involves the unclarified theory of 'infinity'. , nothing would be gained semantically and for my purpose, had I stressed these formal possibilities of the calculus. Quite the opposite, if I had done so, I would have failed to stress the most fundamental $\bar{A}$ principle and task of establishing the *similarity of structure between languages and the un-speakable levels and happenings as the first and crucial consequence of the elimination of identity*. For these weighty reasons, in my presentation, I followed some older textbooks, particularly Osgood's, which, from a $\bar{A}$ point of view, are sounder than the newer, largely A rationalizations and apologetics.

However, it should be realized that practically all outstanding and creative mathematicians have had, and still have, $\bar{A}$ attitudes, yet, these private beneficial attitudes, not being formulated in a $\bar{A}$-system, could not become *conscious*, simple, workable, public, and educational assets. We can be simple about this point. With the elimination of identity, structure becomes the *only possible* content of 'knowledge'—and structure of the un-speakable levels has to be *discovered*. Discovery depends on the finding of *new*, and therefore *different* characteristics. In the *formulation* of the last sentence, we cannot make the 'training in discovery' an *educational discipline*. The opposite is true in a $\bar{A}$-system, based on non-identity, as *we can train simply and effectively in non-identity*, which ultimately leads to differentiation, and so discovery.

Because of the elementary, and purely semantic character of the following pages, I have often restrained myself from giving technical, supposedly 'rigorous', and often A rationalizations, which we occasionally call definitions. In a semantic and $\bar{A}$ treatment, at this pioneering stage, stressing old definitions would be seriously confusing; and I wished to avoid such witty wittgensteinian 'definitions' as 'A point in space is a place for an argument'. In a number of instances, and for my purpose, I often avoided unsatisfactory formal definitions, preferring to depend upon the ordinary meanings of words.

For the reader who wishes to acquaint himself with an elementary theory of limits and corresponding sets of definitions, I would suggest the book of the late Professor J. G. Leathem, *Elements of the Mathematical Theory of Limits* (London and Chicago, 1925). This theory is based on Pascal's *Calcolo Infinitesimale*, Borel's *Théorie des fonctions*, and Godefroy's *Théorie des séries*. Leathem's book has been printed under the supervision of Professor H. F. Baker, F.R.S., of the University of Cambridge, and Professor E. T. Whittaker, F.R.S., of Edinburgh. I give these names for professional mathematicians, to indicate the semantic trend which underlies this particular treatment of limits and which does not greatly conflict with a $\bar{A}$ outlook. This outlook may be summarized in part, in the words of Borel somewhat as follows: 'To the evolution of physics should correspond an evolution of mathematics, which, without abandoning the classical and well-tried theories, should develop however, with the results of experiments in view'. This statement implies vaguely the 'similarity of structure'. , and so requires as a *modus operandi* the rejection of identity.

There seems to be little doubt that a complete and radical revision of the semantic aspects of human mathematical behaviour is pending. Such a revision appears to be laborious and difficult, and should be undertaken from the point of view of the theory of the unique and specific relations, called numbers. I doubt if a single man could accomplish this revision. Such an undertaking will probably be the result of group activities, and may, in the beginning, be *unified* by the formulation of one fundamental $\bar{A}$ principle of non-identity, the disregard of which, from science down to 'mental' ills, can be found at the bottom of practically all avoidable human difficulties.

The problems are very complicated and extremely difficult, and need to be treated from many angles. At present, we have many scientific societies, grouped by their specialties; but we do not have a scientific society composed of *many different specialists* whose work could be unified by some *common and general principle*. There can be no doubt that the principle of 'identity', or 'absolute sameness in *all* aspects', is invariably false to facts. The main problem is *to trace* this semantic disturbance of improper evaluation in all fields of science and life, and this requires a new *co-ordinating scientific body* of many specialists, with branches in all universities. Each group would meet, say monthly, to discuss their problems, and give mutual technical assistance in tracing *this first general semantic disturbance*. Such meetings would stimulate enormously scientific productivity. In fact, without such a co-ordinating body, the present enormous technical developments in each branch of science preclude the revision of general principles, on which, in the last analysis, all other of our activities greatly depend. The first task then, is to find a co-ordinating principle, and present it to the scientific world.

Psychiatry, and common experience, teach us, that in heavy cases of dementia praecox we find the most highly developed 'identification'. $\bar{A}$ considerations suggest that *any* identification, no matter how slight, represents a dementia praecox factor in our semantic reactions. The rest is only a question of degrees of this maladjustment. From this point of view, we will find dementia praecox factors even in mathematics. In physics, only since Einstein has this factor of un-sanity been eliminated, and this elimination has already produced an ever-growing crop of 'geniuses', which merely means, that some inhibitions of mis-evaluation have been eliminated from these younger men, and that they are humanly more 'normal' than the others.

In mathematics, from a $\bar{A}$ point of view, we must first of all *not identify* different aspects of our mathematical behaviour, nor try to cover up these identifications of endless aspects by the one very old term 'mathematics'. This word, 'mathematics', in its *accepted sense* covers a non-existing fiction. What does exist, and the only thing we actually deal with, is *human mathematical behaviour*, human *s.r*, and the *results* of human *mathematical behaviour* and *s.r*. A treatise, say, on a new quantum mechanics, has no value to a monkey or a corpse, and only human *mathematical behaviour* and *s.r*, have any actual *non-el* existence, and is the *only* thing which actually matters. So we see that 'mathematics' covers a non-existent fiction *if elementalistically* separated from human mathematical behaviour and *s.r*. I use the term 'mathematics' in the *non-el* sense, and attempt to signalize some of the difficulties non-elementalism involves at this transitory stage.

From a $\bar{A}$, non-identity, structural, *non-el* point of view, human mathematical behaviour must be treated uniquely as a physico-mathematical discipline, and postulational methods to be used exclusively as a most valuable *checking* method. To *base* mathematical behaviour and *s.r* on postulational methods exclusively, is to introduce dementia praecox factors into science, which only induces the spread of semantic maladjustment in life.

Our main task in producing a $\bar{A}$ revision of mathematical *s.r*, is in the elimination of identification from our *s.r* about 'infinity' and in the formulation of a $\bar{A}$ definition of numbers in terms of relations. This would enable us to rebuild human mathematical *s.r* from a theory of numbers point of view, as a *physico-mathematical* discipline. The intrinsic, or internal theory of surfaces, and the tensor, or absolute calculus, are methodologically our most secure epistemological guides.

I would suggest that mathematical and scientific readers who are interested in a $\bar{A}$ revision should, at first, in their special fields, sketch in technical papers, presented before the local International Non-aristotelian Societies, A pitfalls and $\bar{A}$ problems and outlooks. Only after this is done, shall we be able to begin a co-ordination of their findings, and thereby initiate a revised and unified $\bar{A}$ science, mathematics, and perhaps ultimately a saner *scientific civilization*.

The scientific achievements dealt with in Book III, are developing so rapidly, and the technical points of view alter so often, that on a static printed page it is impossible to do them justice. The writer has spared no efforts to keep informed of these scientific developments until two weeks before the appearance of this book; yet because these new developments do not represent new and fundamental semantic factors, I deliberately do not include them here. In some instances, a given author may seem to change his opinions, but, from a $\bar{A}$ point of view, it sometimes appears that the original notions were more justified, and so I preserved them without change.

The following pages are written exclusively from a semantic point of view, an undertaking which is far more difficult than dealing with a restricted technical physico-mathematical problem, because it involves *second order* observations, of the first order observations, of the first order observer, and of the relations between them , . When it came to a final revision of the manuscript, and reading of the proofs, I found that dealing with so many varied fields, languages, and symbolisms at one period, was no small task, and I only hope that I have not over-looked too many errors or misprints.

If we must have slogans, a $\bar{A}$ motto readily suggests itself—'Scientists of the world unite'. Perhaps this motto may prove more constructive and workable than the familiar A *elementalistic* slogans which have mostly led to the dismembering of human society. Protests against any misrule should not be confused with the proclaiming of disrupting *general principles*. Let me repeat once more, that the most lowly manual worker is useful *only* because of his human nervous system, which produced all science, and which differentiates him from an animal, and not primarily for his hands alone; otherwise we would breed apes to do the world's work.

In the explanations of some geometrical notions, and some parts of the theory of Einstein, I have followed often very closely the *Einstein's Theory of Relativity* by Max Born, which is easily the best elementary exposition I have read, and also the books of Eddington. In the quantum field I have followed mostly the books by Biggs, Birtwistle, Bôcher, Haas, and Sommerfeld, and I wish to acknowledge my indebtedness to the above authors.

I am also under heavy obligations to Professors E. T. Bell, P. W. Bridgman, B. F. Dostal, R. J. Kennedy, and G. Y. Rainich, who were so kind as to read the MS. and/or proofs, and whose criticism and suggestions were invaluable to me. However, I assume entire responsibility for the following pages, especially since I have not always followed the suggestions made.

PART VIII

ON THE STRUCTURE OF MATHEMATICS

Being myself a remarkably stupid fellow, I have had to unteach myself the difficulties, and now beg to present to my fellow fools the parts that are not hard. Master these thoroughly, and the rest will follow. What one fool can do, another can. (510)   SILVANUS P. THOMPSON

Besides the theory of surfaces is the model on which all the higher theories are built and must be built, and it is well to master it completely before attempting generalizations. (425)   G. Y. RAINICH

To find such relations Einstein has applied a mathematical method of great power—the calculus of tensors—with extraordinary success. The calculus threshes out the laws of nature, separating the observer's eccentricities from what is independent of him, with the superb efficiency of a modern harvester. (21)   E. T. BELL

CHAPTER XXXII

ON THE SEMANTICS OF THE DIFFERENTIAL CALCULUS

The principle of gaining knowledge of the external world from the behaviour of its infinitesimal parts is the mainspring of the theory of knowledge in infinitesimal physics as in Riemann's geometry, and, indeed, the mainspring of all the eminent work of Riemann, in particular, that dealing with the theory of complex function. (547)   HERMANN WEYL

The conception of tensors is possible owing to the circumstance that the transition from one co-ordinate system to another expresses itself as a **linear** transformation in the differentials. One here uses the exceedingly fruitful mathematical device of making a problem "linear" by reverting to infinitely small quantities. (547)   HERMANN WEYL

Section A. Introductory.

In the first draft of this book written in 1928, the following pages preceded Part VII. In a final revision in 1932, it seemed advisable to transfer pages which to laymen look 'mathematical', to the end of the volume, because the majority of even intelligent readers have a sort of 'inferiority complex' about anything 'mathematical'.

The patient reader knows by now, I hope, that on neurological grounds, he must for the sake of sanity, be able to translate the dynamic into the static, and the static into the dynamic; and also that he must know at least about the modern structure of 'space', 'time', 'matter'. These conditions seem essential for sanity, and so I had no choice but to give the minimum of a structural and semantic outline, and to acquaint the reader with the existence of modern scientific problems and vocabularies. It is not my aim to teach the reader mathematics or modern physics. I must limit myself to structural and semantic issues, for there are excellent elementary books which will give him the necessary informations.

The following pages should in no way intimidate the intelligent reader. Elementary structural statements and definitions are given in simple language, followed by illustrations to render their meanings more understandable. The pages are less technical than they look, as each example is carried through in the most elementary way in all of its details, so as to make easy reading. A real difficulty for some readers may come from the semantic blockage created by the use of apparently strange, and, to them, unknown terms, or from a feeling of fright or abhorrence of anything mathematical, due to deplorably faulty introduction to some branch of mathematics at the hands of some teacher innocent of the broader epistemological aspects of science. I am acquainted with scientists of very considerable mathematical gifts, who have had to overcome this phobia of mathematics. Once the word 'mathematics' was mentioned to them, they became 'mentally' paralysed. An 'emotional' fright seized them and it took some months to overcome this undesirable childish s.r. I use the subject of mathematics as an illustration of this difficulty, because I want to contrast the comparative simplicity of mathematical notions with the complexity of human problems and language. For when we have understood the *simplest* notions, which happen to be mathematical, then only shall we be able better to understand our human problems, which are in comparison so difficult and so confused.

Any reader who has a distaste for mathematics will benefit most if he overcomes his semantic phobia and struggles through these pages, even several times. As a result of so doing he will find it simple although not always easy. It is always semantically useful to overcome one's phobias; it liberates one from unjustified fears, feelings of inferiority, . The main point of this whole discussion is to evoke the semantic components of a living Smith, when he habitually uses the method which will be explained herewith. This method is so simple and so fundamental that in the form given by a $\bar{A}$-system and further simplified according to the gifts of the teacher, it will some day be introduced into *elementary* schools without technicalities, as a *preventive* semantic method against 'insanity', un-sanity and other nervous and semantic difficulties, and as a foundation for a training in *sanity* and adjustment.

Section B. On the Differential Calculus.

1. GENERAL CONSIDERATIONS

As we have already seen, the structural notion of a function is strictly connected with that of the variable. The variable on one level does not 'vary'; it is a selection by Smith of a definite value from a given set. As these processes are going on inside of the skin of Smith he might experience on a different level a feeling of 'change'. The method of dealing with such problems is given by the mathematical differential and integral calculus.

The beginnings of methods dealing with 'change' are to be found even among the ancients. Galileo, Roberval, Napier, Barrow, and others were interested in 'fluxional' methods, before Newton and Leibnitz.[1] The epoch-making discoveries of the last two mathematicians consisted not only in perfecting the knowledge they had and in inventing new methods, but also— and this is perhaps the most important—they formulated a *general* theory of these structural methods and invented a new notation suitable for their purpose. The definite abandonment of the old tentative methods of integration in favour of methods in which integration is regarded as the inverse of differentiation was especially the work of Newton. Leibnitz' main work was in the field of precise formulation of simple rules for differentiation in special cases and the introduction of a very useful notation.

It is not an exaggeration to say that the calculus is one of the most inspiring, creative, structural methods in mathematics. There is little doubt that the analysis of the foundations of mathematics, and their revision, was suggested by a study of the methods of the calculus. *It is structurally and semantically the 'logic' of sanity* and, as such, can be given ultimately without technicalities by the present $\bar{A}$-system and semantic training, with the aid of the Structural Differential.

The application of the differential calculus to geometry produced differential geometry. This prepared the way for the notions of Einstein and Minkowski.

The whole of modern physics becomes possible through the calculus, and it will probably be correct to say that the achievements of the future also will be dependent on it.

The present work is also to a large extent inspired by it, and develops simple non-technical methods by which the psycho-logical structural *s.r* necessitated by the calculus can be given to the masses in elementary education without any technical knowledge of it. This statement does not include teachers, who should be acquainted with at least the rudiments of the calculus.[2]

It is true that in the beginning we did not suspect that the semantics of the calculus are indispensable in education for *sanity*. It is the *only* structural method which can reconcile the as yet irreconcilable higher and lower order abstractions. Without such a reconciliation, at our present level of development, sanity is a matter of good luck quite beyond our conscious or educational control.

Let us recall the rough definition of a function: y is said to be a function of x if, when x is given, y is determined. In symbols we write $y = f(x)$ which we read 'y is equal to a function of x' or 'y is equal to f of x'. If y is a function of x, or $y = f(x)$, then x is called the independent variable, being the one to which we arbitrarily assign *any* value we choose out of a given set of values. The y is called the dependent variable as its value depends on the value we assign to x.

A function may have more than one independent variable; in which case we have a function of several variables. It happens frequently that to one value of the independent variable there may correspond several values of the dependent variable. Then y is said to be a multiple-valued function of x.

Roughly speaking, a function is said to be continuous if a small increment in the variable gives rise to a small increment of the function.

A theory of functions can be developed without any references to graphs and geometrical notions of co-ordinates and lengths; but in practice (and in this work), it is extremely useful to introduce these geometrical notions, as they help intuition. A modern definition of an analytic function is technical and unnecessary for our purpose. Suffice it to say that it is connected with derivatives and power series, *which means structure*.

Geometry is a very remarkable science. It may be treated as pure mathematics, or it may be treated as physics. It may therefore be used as a link between the two or as a link between the higher and lower order of abstractions. This fact is of tremendous psycho-logical and semantic importance. It is not by pure 'chance' that the most important writers on mathematical philosophy, authors who have generalized their knowledge of mathematics to include human results, were mostly geometers.

Indeed, Whitehead, in his *Universal Algebra* (p. 32), says, and justly so, that a treatise on universal algebra is also a treatise on certain generalized notions of 'space'. 'Space' should be understood as 'fulness', 'fulness of something', a plenum. Naturally coherent speech, like universal algebra, must be coherent speech about *something*. 'Generalized space' becomes generalized plenum, and so it belongs to *two* realms. One is contentless and formal, hence generalized algebra; the other, in that it refers to a generalized plenum, becomes generalized geometry, or generalized physics.

The main importance, perhaps, of geometry is in the fact that it can be interpreted *both ways*. One way appears as pure mathematics, and therefore as the study of sets of numbers representing co-ordinates. The other takes the form of an interpretation, in which its terms imply a connection with the empirical entities of our world. Obviously if speech is not the things spoken about, we must have a special discipline which will translate the coherent language of pure mathematics, which is contentless by definition, into another way of speaking which uses a different vocabulary capable of *both* interpretations.

Again, the different orders of abstractions, which our nervous structure produces, are perfectly reflected in the very structure and methods of mathematics. The possibility of the use of the 'intuitions' of lower order abstractions, is extremely useful in pure mathematics. This fact makes geometry also *unique*. It allows us to apply to the development of geometry both orders of abstractions—the 'intuitions', 'feelings', of the lower order of abstractions, and the static, 'quantum' jump methods of pure analysis. This is also why the einsteinian physics becomes four-dimensional geometry; which, because it can be treated on both levels of abstraction, gives tremendously powerful and important psycho-logical means for sanity and nervous co-ordination of the individual. Since Einstein, many far-sighted scientists have said that although they do not know in what respect the Einstein theory will affect our lives, yet they feel that it will have a tremendous influence. I venture to suggest that the bearing of the Einstein theory and its development on the problems of sanity, as explained in this work, is a new and unexpected semantic result of the application of modern science to our lives. As the Einstein theory could have been formulated more than two hundred years ago when the finite velocity of light was discovered, so the present theory is also several hundred years overdue. The only consolation we have left is that it is better late than never.

The scope of this work allows us to go but a little beyond these simple remarks, and permits only a very brief explanation of the most fundamental and elementary beginnings of the calculus. In this presentation I shall appeal very often to intuition (lower order abstractions), as this will help the reader.

The notion of differentiation of a continuous function is the process for measuring the rate of growth; that is to say, the evaluation of the increment of the function as compared with the growth or increment of the variable. We may describe this process as follows: If y is a function of x, it is helpful not to consider x as having one or another special value but as flowing or growing, just as we feel 'time' or follow the ripples made by a stone thrown into a pond.

The function y varies with x, sometimes increasing, sometimes decreasing. We have already defined the variable as *any* value selected from a given range. Let us consider our x as given in the interval between 1 and 5. We are now interested in all values which our x may take between these two values, or, as we say, in this interval. Obviously, we can select a few values, or, in other words, take big steps; as, for instance, assigning to x the successive values $x_1 = 1$, $x_2 = 2$, $x_3 = 3$, $x_4 = 4$, $x_5 = 5$. In such a case we would have few values and the difference between two successive values would be rather large, for instance, $x_3 - x_2 = 1$. But such large differences are not of much interest to us here. We may, if we choose, select smaller differences; in other words, assign more values to our variable in the given range.

Let us take, for instance, for our x the series of values: 1, $1\frac{1}{2}$, 2, $2\frac{1}{2}$, 3, $3\frac{1}{2}$, 4, $4\frac{1}{2}$, 5. Here we see that the difference between two successive values is smaller than 1, it is $\frac{1}{2}$. So we already have nine, instead of the former five, values which we may assign to our x. Thus we have selected smaller steps by which to proceed. Let us select still smaller steps; for instance, $\frac{1}{4}$. Our extensional manifold of values for x in the interval between 1 and 5 would then be: 1, $1\frac{1}{4}$, $1\frac{1}{2}$, $1\frac{3}{4}$, 2, $2\frac{1}{4}$, $2\frac{1}{2}$, $2\frac{3}{4}$, 3, $3\frac{1}{4}$, $3\frac{1}{2}$, $3\frac{3}{4}$, 4, $4\frac{1}{4}$, $4\frac{1}{2}$, $4\frac{3}{4}$, 5. We see that in the interval between 1 and 5, we have already 17 values which we may assign to our variable, but we have followed the 'growth' of our x by smaller steps; namely, by steps of $\frac{1}{4}$. If we choose to diminish the steps to $\frac{1}{10}$, we would have for our extensional manifold of values: 1, 1.1, . . . , 1.9, 2, 2.1, . . . , 2.9, 3, 3.1, . . . , 3.9, 4, 4.1, . . . , 4.9, 5: in all, 41 values for x, any two succeeding values differing by $\frac{1}{10}$. If we select still smaller steps—let us say, $\frac{1}{100}$—we have 401 values for x and the difference between two successive values is still smaller; namely, $\frac{1}{100}$. This process may be carried on until we have as many numbers between 1 and 5 as we choose, since we may make the difference between successive numbers in the sequence as small as we please. In the limit, between any two numbers, let us say, 1 and 2, or any two fractions, there are infinite numbers of other numbers or fractions. It is obvious that in a given interval, let us say, between 1 and 5, we can have an indefinitely large number of intermediary numbers arranged in an increasing progression, such that the difference between two successive numbers can be made smaller than any assigned value, which is itself greater than zero.

The above may be made clearer by a geometrical illustration. Let us take a segment of a line of definite length, let us say, 2 inches. Let us designate the ends by numbers 1 and 3. In figure (A) we divide the segment into 2 equal parts of one inch each, and see that to reach 3 starting with 1 we have to proceed by two large jumps from 1 to 2, 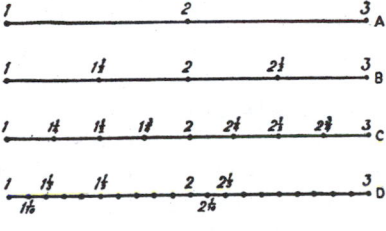 and from 2 to 3. In figure (B) we have more steps in the interval and therefore the steps are smaller. In figures (C) and (D) the steps are still smaller and their number greater. If the number of steps is very large, the steps are very small. In the limit, if the numbers of steps become infinite, the length of the steps tends

37

toward zero and the aggregate of such points of division represents (in the rough only) a *continuous* line.

It is important that the reader should become thoroughly acquainted with the above simple considerations as they will be very useful in *any* line of endeavour. Here we already have learned how, somehow, to translate discontinuous jumps into 'continuous' smooth entities. Because of the structure of our nervous system we 'feel' 'continuity', yet we can analyse it into a smaller or larger number of definite jumps, according to our needs. The secret of this process lies in assigning an increasing number of jumps, which as they become vanishingly small, or tend to zero, as we say, cease to be felt as jumps and are felt as a 'continuous' motion, or change, or growth or anything of this sort.

An excellent example is given by the motion pictures. When we look at them we see a very good representation of life with all its continuity of transitions between joy and sorrow. If we look at an arrested film we find a definite number of *static* pictures, each differing from the next by a measurable difference or jump, and the joy or sorrow which moved us so in the play of the actors on the *moving* film, becomes a static manifold of static pictures each differing measurably from its neighbour by a slightly more or less accentuated grimace. If we increase the number of pictures in a unit of 'time' by using a faster camera and then release this film at the ordinary speed, we get what is called slow motion pictures with which we are all familiar. In them we notice a much greater smoothness of movements which in life are jerky, as, for instance, the movements of a running horse. They appear smooth and non-jerky, the horse looks as if it were swimming. Indeed we do swim no less than fishes, except that our medium; namely, air, is less dense than water, and so our movements have to be more energetic to overcome gravitation. The above example is indeed the best analogy in existence of the working of our nervous system and of the difference between orders of abstractions. Let us imagine that some one wants to *study* some event as presented by the moving picture camera. What would he do? He would first see the picture, in its moving, dynamic form, and later he would arrest the movement and devote himself to the contemplation of the static extensional manifold, or series, of the static pictures of the film. It should be noticed that the differences between the static pictures are finite, definite and *measurable*.

The power of analysis which we humans possess in our higher order abstractions is due precisely to the fact that they are *static* and so we can take our 'time' to investigate, analyse , . The lower order abstractions, such as our *looking* at the moving picture, are shifting and non-permanent and thus evade any serious analysis. On the level of *looking* at the *moving* film, we get a general *feeling* of the events, with a very imperfect memory of what we have seen, coloured to a large extent by our moods and other 'emotional' or organic states. We are on the shifting level of lower order abstractions, 'feelings', 'motions', and 'emotions'. The first lower centres do the best they can in a given case but the value of their results is highly doubtful, as they are not especially reliable. Now the higher order abstractions are produced by the higher centres, further removed, and not in direct contact with the world around us. With the finite velocity of nerve currents it takes 'time' for impulses to reach these centres, as the cortical pathways offer higher neural resistances than the other pathways.[3] So there has to be a survival mechanism in the production of nervous means for arresting the stream of events and producing *static* pictures of permanent character, which may allow us to investigate, verify, analyse , . It must be noticed that because of this higher neural resistance of higher centres and the static character of the higher abstractions, these abstractions are less distorted by affective moods. For, since the higher abstractions persist, if we care to remember them, and the moods vary, we can contemplate the abstractions under different moods and so come to some *average* outlook on a given problem. It is true that we seldom do this, but we *may* do it, and this is of importance to us.

As one of the aims of the calculus is to study relative rates of change we will consider a series of successive values of our variable which differ by little from each other. If we have $y = f(x)$ we can consider the change in x for a short interval, let us say, from x_0 to x_1, so that we assign to our x two values, $x = x_0$ and $x = x_1$. The corresponding values of our function or y will be $y_0 = f(x_0)$ and $y_1 = f(x_1)$. In general, small changes in y will be almost proportional to the corresponding changes in x, provided $f(x)$ is 'continuous'.

Denoting the small increment of x by Δx, so that $x_1 - x_0 = \Delta x$ or $x_1 = x_0 + \Delta x$, function y receives the increment $y_1 - y_0 = \Delta y$ or $y_1 = y_0 + \Delta y$. Since $y_1 = f(x_1)$ and $x_1 = x_0 + \Delta x$ we have:

$$y_0 + \Delta y = f(x_0 + \Delta x); \quad \text{if we subtract}$$

from both sides
$$y_0 = f(x_0) \qquad \text{we would have}$$

$$\Delta y = f(x_0 + \Delta x) - f(x_0); \text{ dividing both sides}$$

by Δx we have
$$\frac{\Delta y}{\Delta x} = \frac{f(x_0 + \Delta x) - f(x_0)}{\Delta x} \tag{1}$$

The above ratio represents the ratio of the increment of the function to the increment of the variable. In the limit when the increment in the variable becomes vanishingly small or when Δx tends toward zero, and our function is continuous, the limit of this ratio gives us the law of change or growth of our function.

The limit which the ratio (1) approaches when Δx approaches 0,

$$\lim_{\Delta x \to 0} \frac{\Delta y}{\Delta x} = \lim_{\Delta x \to 0} \frac{f(x_0 + \Delta x) - f(x_0)}{\Delta x} \tag{2}$$

is called the derivative of y with respect to x and is denoted by $D_x y$, which we read 'D_x of y', in symbols,

$$\lim_{\Delta x \to 0} \frac{\Delta y}{\Delta x} = D_x y \tag{3}$$

Let us illustrate this by a simple numerical example. Take the equation $y = x^2$ and assume that $x = 100$, whence $y = 10,000$. Suppose the increment of x, namely, $\Delta x = 1/10$. Then $x + \Delta x = 100.1$ and $(x + \Delta x)^2 = 100.1 \times 100.1 = 10020.01$. The last 1 is $1/100$ and only one millionth part of the 10,000, and so, we can neglect it and consider $y + \Delta y = 10{,}020$; whence $\Delta y = 20$ and $\Delta y / \Delta x = 20/0.1 = 200$. In the general case, if $y = x^2$ and instead of x we take a slightly larger value, $x + \Delta x$, then our function y also becomes slightly larger; thus,

$$y + \Delta y = (x + \Delta x)^2 = x^2 + 2x\Delta x + (\Delta x)^2$$

If we subtract $y = x^2$ from the last expression we have

$$\Delta y = 2x\Delta x + (\Delta x)^2, \text{ dividing by } \Delta x, \text{ we have}$$
$$\Delta y / \Delta x = 2x + \Delta x.$$

In the limit as Δx approaches zero, the value of the above ratio, or the rate of change of our function, would be $2x$, as Δx would disappear. If our $x = 100$, the above ratio would be 200, as determined above in the case of the numerical example. Another way of symbolizing the derivative is $D_x y = dy/dx$, but this requires a short explanation.

In Chapter XV we have already discussed the problem of the 'infinitesimal' and we have seen that 'infinitesimal' is a misnomer and that there is no such thing at all. Yet this word is very often uncritically used by mathematicians and is therefore often confusing. By an 'infinitesimal' mathematicians mean a *variable* which approaches zero as a limit. The condition that it should be a variable is essential. It would probably be better to call an 'infinitesimal' an *indefinitely* small quantity or 'indefinitesimal', and that is what the reader should understand when he sees anywhere the word 'infinitesimal' or 'infinitely small quantity'.

These indefinitely small quantities are in general neither equal, nor even of one order. Some by comparison are indefinitely smaller than others, and hence are said to be 'of higher order'. Usually several quantities are considered which approach zero simultaneously. In such a case one of them is chosen as the principal indefinitely small quantity. Let us recall that if we take any number, for example, 1, and divide it by 2 we have $1/2$. If we divide 1 by 4 we have $1/4$ which is smaller than $1/2$; if we divide 1 by 10 we have $1/10$ which is still smaller. If we carry this process on indefinitely, taking larger and larger denominators, the results are fractions of smaller and smaller values. In the limit, as the value of the denominator becomes indefinitely large the value of the fraction approaches zero. This simple consideration will help us in the classification of indefinitely small quantities.

Let us take a as the principal indefinitely small quantity and b another indefinitely small quantity. If the ratio b/a approaches zero with a we say that b is an indefinitely small quantity of higher order with respect to a. In other words, although a approaches zero in the limit yet it is infinitely larger than b and so the ratio b/a also approaches zero.

If the ratio b/a approaches a limit k different from zero as a approaches zero, then b is said to be of the 'same order' as a and $b/a = k + \epsilon$ where ϵ is indefinitely small with respect to a. In such a case $b = a\,(k + \epsilon) = ka + a\epsilon$, and ka is called the principal part of b. The term $a\epsilon$ is obviously of a higher order than a.

We may say in general that if we have a power of a, for instance a^n, such that the ratio b/a^n approaches a limit different from zero, b is called an 'infinitesimal' (indefinitesimal) of order n with respect to a.

Let us give a numerical illustration. We know that there are 60 minutes in one hour, 24 hours in a day, or that there are 1440 minutes in a day, and by multiplying 1440 by 7, that there are 10,080 minutes in a week. Our forefathers called this $1/10,080$ part of a week a 'minute' because of its minuteness. It is obvious that a minute is very small as compared with a week. But if we subdivide a minute into 60 equal parts we have a still smaller quantity, a quantity of second order smallness and so we called it a second. Indeed there are 3600 seconds in one hour, 86,400 seconds in a day and 604,800 seconds in a week. If we decide that for some purpose a minute is as short a period of 'time' as we need to consider, then the second, $1/60$ of a minute, is relatively so small that it could be neglected. In a calculation where $1/100$ of some unit is the smallest value which needs to be considered, we may define this $1/100$ as of first order

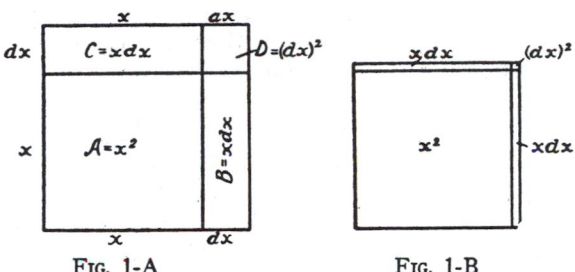

FIG. 1-A   FIG. 1-B

smallness. Then $1/100$ of $1/100$, or $1/10,000$, of that unit, which is relatively of second order smallness, is entirely negligible. The fractions whose smallness we are considering here are comparatively large, and we usually deal with much smaller quantities, but the smaller a quantity is, the more negligible the correspondingly smaller quantity of higher order becomes.

Let us consider a geometrical interpretation of the above. If we represent a quantity x by a line segment, and a slightly greater quantity, $x+dx$, by a slightly longer line segment; then the quantities x^2 and $(x+dx)^2 = x^2+2xdx+(dx)^2$ may be represented by squares where sides are the line segments which represent the quantities x and $x+dx$ respectively.

If we denote the areas by A, B, C, D, we see that $A = x^2$ and that $A+B+C+D = x^2+2xdx+(dx)^2$. If we select our dx smaller and smaller the areas $B = C = xdx$ diminishing in one dimension only, become also smaller and smaller, but $D = (dx)^2$ is vanishing much more rapidly as it is diminishing in each of two dimensions, whence it is said to be a quantity of second order smallness, which for all purposes at hand may be neglected.

If we take $\qquad\qquad y = f(x) \qquad$ and its derivative

$$\lim_{\Delta x \to 0} \frac{\Delta y}{\Delta x} = D_x y.$$

Then $\qquad\qquad \dfrac{\Delta y}{\Delta x} = D_x y + \epsilon, \qquad$ where ϵ is an indefinitesimal,

and $\qquad\qquad \Delta y = D_x y \Delta x + \epsilon \Delta x.$

In the above expression $D_x y \Delta x$ represents the principal part and $\epsilon \Delta x$ appears as an indefinitesimal of higher order. This principal part is called the differential of y and is denoted by dy. If we choose $f(x) = x$ we have $dx = \Delta x$ and so, $dy = D_x y \, dx$.

So we see that the differential of the independent variable x is equal to the increment of that variable. This statement is not generally true about the dependent variable, as ϵ does not generally vanish.

The derivative is also sometimes denoted as $f'(x)$ or y' and this notation is due to Lagrange; all three notations are used and it is well to be acquainted with them.

The derivative of a function $f(x)$ is in general another function of x, let us say $f'(x)$. If $f'(x)$ has a derivative, the new function is the derivative of the derivative or the *second derivative* of $f(x)$ and is denoted by y'' or $f''(x)$. Similarly the third derivative y''' or $f'''(x)$ is defined as the derivative of the second derivative and so on. In the other notations we have:

$$D_x(D_x y) = D_x{}^2 y \text{ or } \frac{d}{dx}\left(\frac{dy}{dx}\right) = \frac{d^2 y}{dx^2}$$

Having introduced these few definitions it must be emphasized that the main importance of the calculus is in its central idea; namely, the study of a *continuous function* by following its history by *indefinitely small steps*, as the function *changes* when we give indefinitely small increments to the independent variable. As was emphasized before, the whole psycho-logics of this process is intimately connected with the activities of the *nervous structure* and also with the structure of science. In this work we are not interested in calculations, complications, or analytical niceties. Mathematicians have taken excellent care of all that. We need only to know about the structure and *method* which help to translate dynamic into static, and vice versa; to translate 'continuity' on one level, or order of abstraction, into 'steps' on another.

To illustrate what has been said and to give the reader the *feel* of the process, let us take for instance a simple equation $y = 2x^3 - x + 5$ where y represents the function of the variable x expressed by a group of symbols to the right of the sign of equality.

To determine the relative rate of growth of this function, that is, to differentiate it, we replace x by a slightly larger value; namely, $x + \Delta x$, and see what happens to the expression. $2x^3$ becomes $2(x + \Delta x)^3 = 2x^3 + 6x^2\Delta x + 6x(\Delta x)^2 + 2(\Delta x)^3$; $-x$ becomes $-x - \Delta x$ and the constant 5 remains unchanged. In symbols, $y + \Delta y = 2x^3 + 6x^2\Delta x + 6x(\Delta x)^2 + 2(\Delta x)^3 - x - \Delta x + 5$, where Δy represents the increment of the function and Δx represents the increment of the independent variable.

Subtracting the original expression $y = 2x^3 - x + 5$ we get the amount by which the function has been increased, namely:

$$\Delta y = 6x^2\Delta x + 6x(\Delta x)^2 + 2(\Delta x)^3 - \Delta x.$$

To determine the *relation*, or *ratio*, of Δy, the increment of the function, to Δx the increment of the independent variable which produced Δy, we divide Δy by Δx, and obtain the equation

$$\frac{\Delta y}{\Delta x} = 6x^2 + 6x\Delta x + 2(\Delta x)^2 - 1.$$

Then as Δx approaches 0 the terms in the right-hand side of the equation which contain Δx as a factor also approach 0 and replacing the left-hand side by $\frac{dy}{dx}$ we obtain the equation $\frac{dy}{dx} = 6x^2 - 1$ which means, that as Δx approaches 0, the ratio of the increment of the function to the increment of the independent variable approaches $6x^2 - 1$, true for any value we may arbitrarily assign to x.

It should be noticed that in our function the left-hand side represents the 'whole' as composed of interrelated elements which are represented by the right-hand side. When instead of x we selected a slightly larger value; namely, $x + \Delta x$, we performed upon this altered value *all* the operations indicated by our expression. We thus have in mathematics, because of the self-imposed limitations, the first and only example of *complete* analysis, impossible in physical problems as in these there are always characteristics left out.

An important structural and methodological issue should also be emphasized. In the calculus we introduce a 'small increment' of the variable; we performed upon it certain indicated operations, and in the final results this arbitrary increment disappeared leaving important information as to the rate of change of our function. This device is structurally extremely useful and can be generalized and applied to language with similar results.

It has been noticed already that the calculus can be developed without any reference to graphs, co-ordinates or any appeal to geometrical notions; but as geometry is an all-important link between pure analysis and the outside world of physics, we find in geometry also the psycho-logical link between the higher and lower orders of abstraction. But the appeal to geometrical notions helps *intuition* and so is extremely useful. For this reason we will explain briefly a system of co-ordinates and show what geometrical significance the derivative has

We take in a plane two straight lines $X'X$ and $Y'Y$, intersecting at O at right angles, so that $X'OX$ is horizontal extending to the left and right of O, and YOY', is vertical, extending above and below O, as a frame of reference for the locations of point, lines, and other geometrical figures in the plane. We call this a two-dimensional rectangular system of co-ordinates. This method may be extended to three dimensions, and our points, lines, and other geometrical figures referred to a three-dimensional rectangular system of co-ordinates consisting of three mutually perpendicular and intersecting planes.

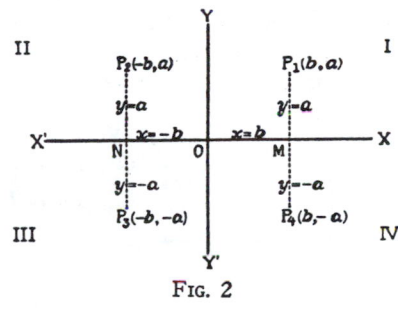

Fig. 2

As we see in Fig. 2, we have four quadrants I, II, III, IV, formed by the intersecting axes $X'X$ and $Y'Y$. The co-ordinates of a point P, by which we mean the distances from the axes determine the position of the point uniquely. We call $X'X$ and $Y'Y$ the *axis* of X and the *axis* of Y respectively, and O the *origin*. If we select a point P_1 in the plane of $X'X$ and $Y'Y$ and draw a line P_1M perpendicular to $X'X$ then OM and MP_1 are called the co-ordinates of P_1; OM is called the *abscissa* and is denoted by $x = b$; and MP_1 is called the ordinate and denoted by $y = a$. We speak of P_1 as the point (b,a), or, in general, of any point as the point (x,y).

Let us draw $ON = OM = b$ and draw lines P_1P_4 and P_2P_3 through M and N respectively perpendicular to $X'X$, making $MP_4 = NP_2 = NP_3 = MP_1 = a$. We then have four points P_1, P_2, P_3, P_4, in each one of the four quadrants and all of them by construction would have equal numerical values for their abscissas and ordinates. To be able to discriminate between the four quadrants, and so avoid ambiguity, we make the convention that all values of y above $X'X$ are to be positive and below $X'X$ negative; and all values of x to the right of $Y'Y$ positive, to the left negative. Thus we see that by such conventions the point P_1 would have both b and a positive; P_2 would have b negative and a positive; P_3 both b and a negative, and finally P_4 would have b positive and a negative, or in symbols $P_1(b,a)$; $P_2(-b,a)$; $P_3(-b,-a)$; and finally $P_4(b,-a)$.

It is obvious that for any point on the X axis (for instance M) the ordinate $y = 0$. If our point is on the Y axis the abscissa $x = 0$ and the co-ordinates of the origin O are both zero $(0,0)$.

From the above definitions we see at once how to plot, or locate, a point. To plot the point $(-4,3)$, since the abscissa x is negative and the ordinate y is positive we locate N on $X'X$, 4 units to the left of O. At N we erect a perpendicular upon which we locate the point $(-4,3)$, 3 units above N. The symbol $(-4,3)$ represents a particular case of the general symbol (x,y) and is accordingly plotted as a particular point as just shown. If instead of the pair or relations expressed by two equations $x = -4$, $y = 3$, we have a single relation expressed by one equation, for example, $y = x - 2$, we have y expressed as a function of x, whence by assigning to x different values, corresponding values of y are determined, and a set of points may be plotted where abscissas and ordinates are corresponding values of x and y respectively. Thus, when $x = 0$, $y = -2$, when $x = 1$, $y = -1$, when $x = 2$, $y = 0$, when $x = 3$, $y = 1$, when $x = 4$, $y = 2$, .

We may now plot the points $A(0,-2)$; $B(1,-1)$; $C(2,0)$; $D(3,1)$; $E(4,2)$; or as many more points as we may choose by giving x additional different values.

If we give to x successive values with smaller differences our points would be closer together, for instance for

$x = 0$	$y = -2$	(A)
$x = 0.5$	$y = -1.5$	(A')
$x = 1$	$y = -1$	(B)
$x = 1.5$	$y = -0.5$	(B')
$x = 2$	$y = 0$	(C)
$\cdots$	$\cdots$	

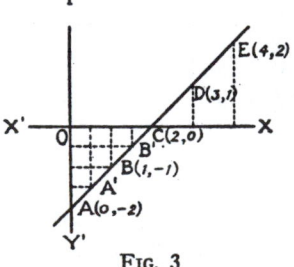

Fig. 3

As we plot larger and larger numbers of points closer and closer together, in the limit, if we take indefinitely many such points, we approach a smooth line. It can be proved that an equation of the type given in this example; namely, where both variables are of the first order, always represents a *straight* line. Such equations are called therefore *linear* equations, as they represent straight lines.

The problem of linearity and non-linearity is of extreme importance, and we will return to it later on. Here we are interested only in the definition and meaning of linearity of equations.

Let us consider next a simple equation of second degree, $y = \dfrac{x^2}{2}$. In assigning arbitrary values to x, we note that x^2 is always positive (by the rule of signs) whether x is positive or negative. Hence, we may tabulate values of x with the double sign $\pm$ meaning either $+$ or $-$.

$x = 0$	$y = 0$	(O)
$x = \pm 1$	$y = \frac{1}{2}$	(A) ,
$x = \pm 2$	$y = 2$	(B)
$x = \pm 3$	$y = 4\frac{1}{2}$	(C)
$x = \pm 4$	$y = 8$	(D)
$\cdots$	$\cdots$	

We see for each value of y we have two values for x which differ only in sign This means that we have points on two sides of the Y axis with numerically equal abscissas and, since for $x = 0$, $y = 0$, the beginning of our curve is at the origin of co-ordinates and the curve is symmetrical with respect to the Y axis.

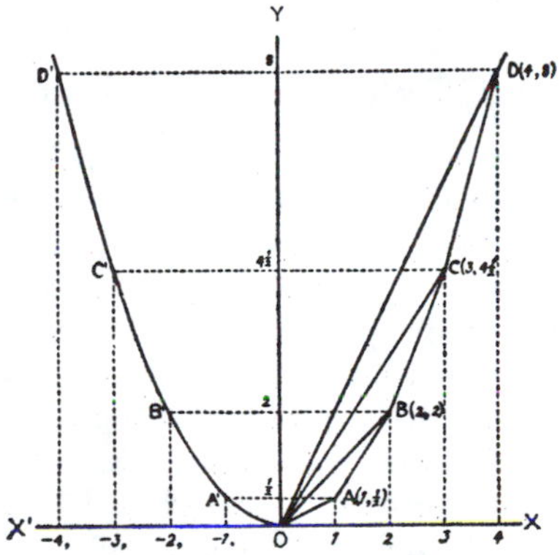

If we connect the points D', C', B', A', O, A, B, C, D, with straight lines we have a broken line. But if we choose smaller and smaller differences between the successive values of x, the broken line becomes smoother and smoother, and, in the limit, as we take increasingly smaller steps, or, in other words, plot indefinitely larger numbers of points in one interval, we approach a smooth, or continuous curve.

Fig. 4

It must be noticed that in equations of higher orders the ratio of changes in the function y to corresponding changes in the variable x vary from point to point, and so we have a *curve* instead of a straight line. It is necessary to become quite clear on this point so we may better compare the two different types of equations as to the law of their growth.

Let us write down in two columns the successive values for the two types of equations. Let us take the equation $y = \dfrac{x^2}{2}$ with the graph shown in the preceding diagram (Fig. 4) and the equation $y = 2x$ as shown in Fig. 5.

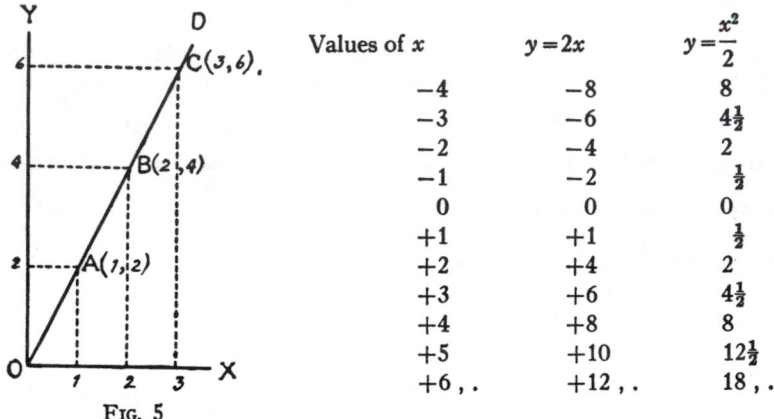

Values of x	$y = 2x$	$y = \dfrac{x^2}{2}$
-4	-8	8
-3	-6	$4\frac{1}{2}$
-2	-4	2
-1	-2	$\frac{1}{2}$
0	0	0
$+1$	$+1$	$\frac{1}{2}$
$+2$	$+4$	2
$+3$	$+6$	$4\frac{1}{2}$
$+4$	$+8$	8
$+5$	$+10$	$12\frac{1}{2}$
$+6$, .	$+12$, .	18, .

FIG. 5

The equation $y = 2x$ involves the variables in the first degree and we see that the ratio of changes in the ordinates to corresponding changes in the abscissas remains constant (proportional). The triangles in Fig. 5, are either equal or similar, which necessitates the equality of angles and so the line $OABCD$ is of necessity a straight line. In this case as $x = 0$ gave us $y = 0$ the line passes through the origin of co-ordinates.

The picture is entirely different in the case of the higher degree equation, $y = \dfrac{x^2}{2}$, illustrated in Fig. 4. From the table of values of the function we see that the value of the function increases increasingly more rapidly than the values of the independent variable and so the ordinates are *not proportional* to the abscissas. If in Fig. 4 we connect O with A, O with B, O with C, O with D, respectively, we see that the lines OA, OB, OC, and OD have *different* angles with the axis $X'X$; the respective triangles are not similar, and so there is no proportionality. The lines OA, OB, OC, OD . , do *not* represent a straight line as they have all different angles with the axis XX' and so the points A, B, C, D . , cannot lie on a straight line but represent a *broken* line which, in the limit, when the points plotted become sufficiently near together, becomes a smooth and continuous curve.

The fact that equations in which the variables are only of the first degree, represent straight lines, and that equations of higher degrees represent curved lines is very important, as will appear later on. We must notice also that the problem of *linearity* is connected with *proportionality*.

These few simple notions concerning the use of co-ordinates will allow us to explain the geometrical meaning of the derivative and the differential.

Consider P_1 and P_2, (Fig. 6) two points on the curve, $y=f(x)$, referred to the axes OX and OY. Drop perpendiculars P_1M_1 and P_2M_2 from P_1 and P_2 to OX. These are the ordinates $y_1=f(x_1)$ and $y_2=f(x_2)$ of the points P_1 and P_2, and OM_1 and OM_2 are the abscissas x_1 and x_2 of the points P_1 and P_2. Through P_1 draw the secant P_1P_2, the tangent to the curve P_1T, and the line P_1Q parallel to OX. Then P_1Q represents $\Delta x=x_2-x_1$ the change in the variable x, and P_2Q represents $\Delta y=y_2-y_1=f(x_2)-f(x_1)$ the change in the function y.

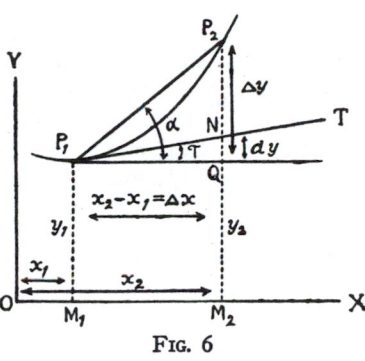

In the right triangle P_1QP_2 the ratio P_2Q/P_1Q is a measure (the tangent) of the angle P_2P_1Q ($=\alpha$) that is, $\tan \alpha$

$$= P_2Q/P_1Q = \Delta y/\Delta x = \frac{f(x_2)-f(x_1)}{\Delta x} \quad \text{or,}$$

Fig. 6

since $x_2=x_1+\Delta x$

we may write
$$\tan \alpha = \frac{f(x_1+\Delta x)-f(x_1)}{\Delta x}$$

As P_2 approaches P_1 along the curve, the secant P_1P_2 rotates about P_1 approaching P_1T as its limit, and the tangent of α approaches the tangent of τ, τ being the angle which P_1T, the tangent to the curve at P_1, makes with P_1Q. But as P_2 approaches P_1, $\Delta x=x_2-x_1=M_1M_2$ approaches zero or symbolically as $\Delta x \rightarrow 0$; $(\Delta y/\Delta x) \rightarrow \tan \tau$, that is $\tan \tau = \lim\limits_{\Delta x \rightarrow 0} (\Delta y/\Delta x)$.

We see that the $\lim\limits_{x_2 \rightarrow x_1} \dfrac{y_2-y_1}{x_2-x_1} = \lim\limits_{\Delta x \rightarrow 0} \dfrac{\Delta y}{\Delta x}$ represents nothing more or less than the derivative of the function representing the curve. In other words, the geometrical interpretation of the analytical process of differentiation is the finding of the slope of the graph of the function. The increment Δy of the function is represented by P_2Q; the differential dy is equal to NQ and $\Delta x = dx = P_1Q$; $\tan \angle TP_1Q = \dfrac{dy}{dx}$.

From the above considerations we see that the differential calculus gives, by the application of some extremely simple structural principles, a method of analysis by which we can discover a tendency at a particular stage rather than the final outcome after a definite interval. From such fundamental yet simple beginnings the whole calculus is developed. Most of these developments are not needed for our purpose, but we will explain one specially important theorem. The theorem in question is that the derivative of the sum of two functions is equal to the sum of their derivatives. In symbols
$$D_x(u+v) = D_xu+D_xv.$$

Let us symbolize $u+v=y$ and select a special value

$$y_0 = u_0 + v_0 \tag{4}$$

then $y_0 + \Delta y = u_0 + \Delta u + v_0 + \Delta v.$ By subtracting (4),

we have $\Delta y = \Delta u + \Delta v.$ Dividing by Δx,

we have $\dfrac{\Delta y}{\Delta x} = \dfrac{\Delta u}{\Delta x} + \dfrac{\Delta v}{\Delta x}.$ When Δx approaches zero the

left-hand side approaches $D_x y = D_x(u+v)$; and the first term of the right-hand approaches $D_x u$, while the second term approaches $D_x v$ and so,

$$D_x(u+v) = D_x u + D_x v.$$

The symbol D_x means also that certain operations are to be performed upon our function; namely, to find its derivative. When used in this sense it is called an operator. The operator D_x can be also written in its differential form as $\dfrac{d}{dx},$ and similarly for higher derivatives.

2. MAXIMA AND MINIMA

It will be useful to have some applications of the differential calculus explained.

If a function $y = f(x)$ is continuous in an interval $a < x < b$ and has larger (or smaller) values at some intermediate points than it has at or near the ends, then it has a maximum (or minimum) at some point $x = x_0$, inside this interval. If Fig. 7 represents the graph of the function, it is obvious that at the maximum (or minimum) the tangent to the curve is parallel to the axis and therefore the slope of this tangent is zero. As this slope is given by the derivative and the slope is zero we have a simple method of finding the maximum (or the minimum) of a function by equating the first derivative to zero; namely, $D_x y = 0$ when $x = x_0$.

It is useful to be able to discriminate between the maximum and the minimum of a function. Fig. 7 shows that this can be done by finding means to discriminate between the two cases when our curve is concave upwards or concave downwards. The slope of a curve for a particular value of x

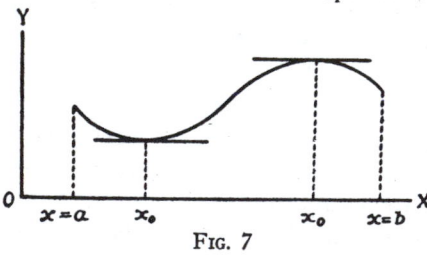

FIG. 7

is given by the value of $D_x y$, corresponding to that value of x. If the value $D_x y$ is positive, y increases as x increases, and the curve slopes up as we move to the right; if the value of $D_x y$ is negative, y decreases as x increases, and the curve slopes down as we move to the right.

If we consider the curve $y = f(x)$ which has its concave side turned upward (Fig. 8), the slope of the curve itself is a function of x, $\tan \alpha = f'(x)$. If we consider a variable point P on a curve $y = f(x)$, together with the tangent to the curve P, as following the curve in the direction of increasing values of x, the curve is concave upward whenever the slope is increasing algebraically, that is when $D_x \tan \alpha = 0$. In other words, the curve is concave upwards for those values of x for which $D_x \tan \alpha$ is positive, or since $\tan \alpha = D_x y$ for those values of x for which $D_x \tan \alpha = D_x(D_x y) = D_x^2 y$

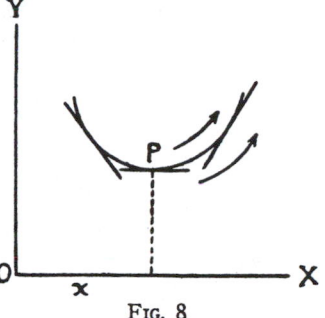

FIG. 8

is positive. Similarly a curve is concave downwards for those values of x for which $D_x \tan \alpha = D_x(D_x y) = D_x^2 y$ is negative. These results can be expressed thus:

A curve $y = f(x)$ is concave upward when $D_x^2 y > 0$, or, in words, when the second derivative is positive, and the curve is concave downward when the second derivative is negative, or, in symbols, when $D_x^2 y < 0$.

From Fig. 7, we see that for a maximum we must have our tangent parallel to the $X'X$ axis and our curve concave downwards, hence for these conditions the first derivative $[D_x y]_{x=x_0} = 0$, and the second derivative $[D_x^2 y]_{x=x_0} < 0$. For a minimum the first derivative must again be zero and the second derivative positive, whence the concave side of the curve is turned upwards. It should be noticed that the problems of maxima and minima play an extremely important structural psycho-logical and semantic role in our lives. All theories, somehow, are built on some minimum or maximum principle involving evaluations which are fundamental factors of all semantic reactions. In daily life we apply these structural and semantic notions continually. In science this tendency made its appearance quite early. The problem of maxima and minima was treated seriously as far back as the second century B.C. In the eighteenth century Maupertuis formulated a 'supreme law of nature', that in all natural processes the 'action' (energy multiplied by 'time') must be a minimum. Euler and Lagrange gave an exact basis and form to this principle; and finally Hamilton, in 1834, established this principle structurally as a *variational* principle, known as the hamiltonian principle, which appears to be of extreme generality and usefulness. It facilitates the derivation of the fundamental equations of mechanics, electrodynamics and electron theory. It has also survived, in a generalized form, the einsteinian revolution, for it contains nothing whatever which would connect it with a definite co-ordinate system; it involves only pure numbers and so is invariant to all transformations. It is structurally one of the most important invariants ascribed to nature, being independent of the systems of reference of the observers.

It is very desirable that this problem should be investigated further from the structural psycho-logical semantic and neurological point of view, as the very foundations of human psycho-logics are fundamentally connected with such a principle, which itself is an *invariant* in human psycho-logics.

Its importance is still increasing, and the hamiltonian principle plays a most remarkable role in all the newest advances of science. Any reader need only look attentively at his daily life to realize that there too this principle plays a predominant role.

3. *CURVATURE*

In modern scientific literature we hear often the fundamental term 'curvature' mentioned, and a few words about it will not be amiss. If we take two perpendicular lines $X'OX$ and OY and select on OY a number of points A, $B, C, D.$, further and further away from O

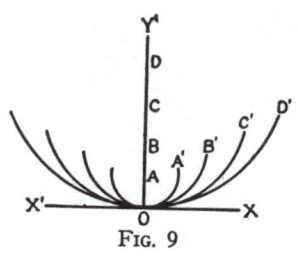

Fig. 9

and describe arcs of circles with these points as centres with radii AO, BO, CO, DO ., respectively Fig. 9, we find each successive arc flatter and closer to the line $X'X$ than its predecessor. In other words, the larger the radius of our circle, the flatter its arc is. In the limit as the radius of the circle becomes indefinitely large, the arc approaches a straight line by intuition and by definition. We notice also that the curvature of each circle is uniform, that is, one-valued at every point; but that when we pass from one circle to another of different radius, the curvature changes.

If we consider a curve and two points on it, M_1 and M_2, (Fig. 10) and draw two tangents at these points; then the angle between these two tangents will depend on two factors, the sharpness of the curve and the distance between

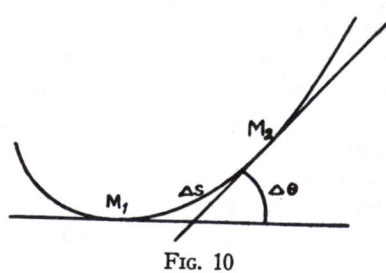

Fig. 10

the points M_1 and M_2. If we take the points near enough and designate the length of the arc between them by Δs, the angle between the two tangents by $\Delta\theta$, then the limiting value of the ratio $\Delta\theta/\Delta s$, as M_2 approaches M_1, becomes $d\theta/ds$, and is a measure of the rate of change of the direction of the tangent at M, as M moves along the curve. Let us designate the rate at which the tangent turns where the point describes the curve with unit velocity as the curvature, or $k = \pm d\theta/ds$, but as k is essentially a positive number or zero we accept only the absolute value of this ratio. To find $d\theta/ds$ we notice that $\tan\theta = dy/dx$

or $\theta = \tan^{-1}\dfrac{dy}{dx} = \tan^{-1} y'$, whence $d\theta = \dfrac{dy'}{1+y'^2} = \dfrac{y''dx}{1+y'^2}$.

But $k = \dfrac{d\theta}{ds}$ where $ds = \sqrt{(dx)^2 + (dy)^2} = \sqrt{1 + y'^2}\,dx$

whence $k = \dfrac{y''}{(1+y'^2)^{3/2}}$.

The reciprocal of the curvature is called the radius of curvature. The radius of curvature of a circle is its radius. The curvature of a curve is measured by the radius of the osculating circle, that circle which fits the curve the most closely in the neighbourhood of our point.

4. *VELOCITY*

Until now we have treated our independent variables as *any* quantity: but there are many problems where the independent variable represents 'time'. For instance, if we travel by railroad the distance increases as 'time' increases, plants and animals grow with 'time', . By the average velocity with which a given point moves for a given length of 'time' we mean the distance s traversed divided by the 'time' elapsed. If, for instance, a train makes 5 miles in 10 minutes we say that its average velocity is 30 miles per hour, or, in symbols: Velocity, $v = \frac{s}{t}$. In this case we were considering uniform velocity, but very often we have to deal with velocities which are not uniform and which might be increasing or decreasing. In such a case we can describe the velocity approximately at any given moment if we take a short interval of 'time' immediately after the moment in question and take the average velocity for this short interval.

For instance, the distance a stone falls is according to the law, $s = 16\,t^2$. We want to find how fast it is going after t_1 seconds when $s_1 = 16\,t_1^2$, and a short interval after we have, let us say, $s_2 = 16\,t_2^2$. Obviously the average velocity for the interval $t_2 - t_1$ is $\frac{s_2 - s_1}{t_2 - t_1}$ feet per second. If we take $t_1 = 1$, $s_1 = 16$, and the difference $t_2 - t_1 = 0.1$ of a second then $s_2 = 16\,t_2^2 = 16 \times 1.21 = 19.36$ and

$$\frac{s_2 - s_1}{t_2 - t_1} = \frac{19.36 - 16}{0.1} = \frac{3.36}{0.1} = 33.6 \text{ ft. per second.}$$

If we take the interval of 'time' smaller, for instance, $1/100$ of a second we would have $\frac{s_2 - s_1}{t_2 - t_1} = 32.2$ feet per second, and if we take the intervals as $1/1000$ of a second the average velocity would be 32.0 feet per second. We see that we could determine the speed of the stone at any instant with any degree of accuracy by direct calculation, but this is not necessary. If we regard the interval $t_2 - t_1$ as an increment of the variable t, that is as Δt, and $s_2 - s_1 = \Delta s$ which represents the increment of the distance considered as a function of the 'time' we would have the average velocity $= \Delta s / \Delta t$. As Δt approaches zero in the limit, the average velocity approaches a limit and this limit is the velocity v at the instant t_1, or in symbols

$$v = \lim_{\Delta t \to 0} \frac{\Delta s}{\Delta t} = \frac{ds}{dt}.$$ In words, the velocity of a point is the 'time' derivative of the space traveled.

If the velocity is not uniform, the rate at which the velocity is increasing is called the acceleration and may be written as $a = \frac{dv}{dt}$, but as we have already seen dv is itself $d\left(\frac{ds}{dt}\right)$, hence $a = \frac{d^2s}{dt^2}$. In words, the acceleration is the second derivative of the distance with respect to 'time'.

In the above notes we have not attempted to give the reader more than some structural and methodological notions, and what amount really to short structural explanations of definitions which will be useful later on. The reader can find many excellent books which give all the additional information he may want.

Section C. On the integral calculus.

So far, we have been studying a method by which to find the variation of a given function corresponding to an indefinitely small variation of our variable. We saw that the *rate of change* of our function was given by the first derivative, which in turn was also a function (usually different) of our independent variable and so could itself vary and have a rate of change, and so give us a second derivative , .

And now we must explain briefly the inverse problem; namely, given the derivative to find the function. In symbols, given $u = D_x U$, find U.

The function U is called the integral of u with respect to x, or, in symbols, $U = \int u dx$.

To integrate a function $f(x)$ is to find a function $F(x)$ which when differentiated gives again the function $f(x)$ with which we started. As in this work we are not interested in computations, but only in the structural, methodological, and semantic aspects, the inverse problem of differentiation; namely, integration, is less important for us here, and I will explain only a single example. We have already differentiated the function $y = 2x^3 - x + 5$ and found its derivative $dy/dx = 6x^2 - 1$. Just as the derivative of the sum of a number of functions is equal to the sum of their derivatives, a similar rule holds for the integrals; namely, that the integral of the sum of a number of functions is equal to the sum of their integrals. Hence we can take in our example only the first term of our equation. In symbols $D_x(2x^3) = 6x^2$; in words, the derivative of $2x^3$ is $6x^2$.

In a problem in integration we would have $6x^2$ given and we would have to find the original function from which $6x^2$ was obtained by differentiation. In our case the solution is already given; namely, $\int 6x^2 dx = 2x^3$.* In general

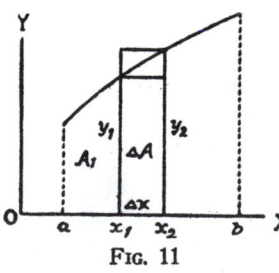

FIG. 11

the solution of problems of integration is largely dependent on the ingenuity of the solver, although we have a number of standard formulae and methods. The geometrical meaning of integration is much more interesting for us and we will give a short explanation of it.

If we consider the curve given by an equation $y = f(x)$ and the area bounded by the X axis, the two ordinates whose abscissas are $x = a$ and $x = b$ and the curve, we may find the area as follows:

---

*The constant of integration is omitted so as not to confuse the reader.

If we select an arbitrary value $x = x_1$ for which $y = y_1 = f(x_1)$, denoting the corresponding value of the area A by A_1 (Fig. 11) and give to x_1 an increment Δx, then the area A_1 would receive the increment ΔA. We can approximate ΔA by the help of two rectangles, one of height $y_1 = f(x_1)$, the other of height $y_2 = y_1 + \Delta y = f(x_2) = f(x_1 + \Delta x)$.

We see that ΔA is larger than the smaller rectangle. In symbols
$$y_1 \Delta x < \Delta A < (y_1 + \Delta y)\Delta x,$$

hence
$$y_1 < \frac{\Delta A}{\Delta x} < (y_1 + \Delta y).$$

As we pass to the limit and let Δx approach zero, we have

$\lim\limits_{\Delta x \to 0} \dfrac{\Delta A}{\Delta x} = y_1$. That is, $D_x A = y_1 = f(x_1)$ when $x = x_1$; which means that the

ordinate of the curve at any point is equal to the x derivative of the area at that point. In general, $D_x A = y$, and hence, $A = \int y\,dx$.

The consideration of what is called the definite integral is still more instructive. Let us take the curve in Fig. 12 represented by an equation $y = f(x)$ and a pair of ordinates which intersect the X axis at the points $x = x_0$ and $x = x_n$.

Let us divide the interval $x_0 x_n$ into n equal parts and erect ordinates at each point of division. Let us construct a set of pairs of rectangles with these ordinates as we constructed the single pair of rectangles in Fig. 11. By inspection of the figure we see that the area under the curve is slightly greater than the sum of the areas of the included rectangles and slightly less than the sum of

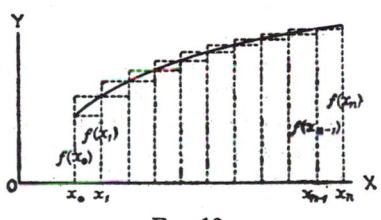

FIG. 12

the areas of the including rectangles. When n is allowed to increase without limit the sum of the areas of either set of these rectangles approaches the area bounded by the curve, the X axis, and the end ordinates. In symbols, the area of the first rectangle beneath the curve is $f(x_0)\Delta x$, where Δx denotes $x_1 - x_0 = \dfrac{x_n - x_0}{n}$. The area of the second rectangle is $f(x_1)\Delta x$, . The sum of

these areas is $f(x_0)\Delta x + f(x_1)\Delta x + \ldots + f(x_{n-1})\Delta x = \sum\limits_{i=0}^{n-1} f(x_i)\Delta x$.

If we allow n to increase without limit we have the area under the curve:
$$A = \lim_{n \to \infty} [f(x_0)\Delta x + f(x_1)\Delta x + \ldots + f(x_{n-1})\Delta x] = \lim_{\Delta x \to 0} \sum f(x)\Delta x$$
$$= \int_{x=x_0}^{x=x_n} f(x)dx = \Big[F(x) \Big]_{x=x_0}^{x=x_n} = F(x_n) - F(x_0).$$

In words, the above formula indicates the fundamental process of the integral calculus; namely: Let $f(x)$ be a continuous function of x throughout the interval $x_0 \leqq x \leqq x_n$. If we divide this interval into n equal parts by the points $x = x_0, x_1, \ldots, x_n$, and form the sum $f(x_0)\Delta x + f(x_1)\Delta x + \ldots + f(x_{n-1})\Delta x$,

38

as we let n increase without limit, this sum will approach a limit, which can be found by integrating the function $f(x)$, that is, by finding the function $F(x)$ of which $f(x)$ is the derivative, and by taking the integral between the limits $x = x_0$ and $x = x_n$; that is, by taking the difference between $F(x_n)$ and $F(x_0)$.

It must be noticed that in our first example, the case of the indefinite integral, we considered integration as the inverse of differentiation; in the second example, we considered the definite integral as the limit of a sum.

The symbol of the integral, $\int$, had its origin in the letter S from the latin word 'summa', the integral being historically understood as the definite integral, or the limit of a sum.

Section D. Further applications.

1. PARTIAL DIFFERENTIATION

When we have more than one independent variable, for example, two, we have to become acquainted with what is called partial differentiation. This process is important, as in practice we usually deal with several independent variables. It presents very little that is new from a structural and methodological point of view, but we give it here, simply to explain the meaning of the term, as the reader may find it used in other works.

If we have a function z of two independent variables x and y, $z = f(x,y)$ which geometrically represents a surface, we may differentiate with respect to one of the variables, let us say x, and hold the other variable y fast, that is, treat it as a constant. In this way we should then have a partial derivative of z with respect to x. Similarly, if we treat x as a constant and differentiate in respect to y, we should have the partial derivative of z with respect to y. The above definitions give us the rules for partial differentiation—that is, following the ordinary rules, considering each variable individually and treating all the other variables as constant.

The notation for partial derivatives is similar to the ones explained before, except that the lower case letter d is replaced by the script form ∂ or a subscript is used to indicate the variable with respect to which the differentiation is performed; for instance, $\dfrac{\partial f}{\partial x} = \dfrac{\partial z}{\partial x} = f_x' = z_x' = D_x f = D_x z$, . Higher derivatives are obtained without difficulty in like manner. If $z = f(x,y)$ and $\dfrac{\partial z}{\partial x} = f_x'(x,y)$ and $\dfrac{\partial z}{\partial y} = f_y'(x,y)$, the partial derivatives themselves are in general also functions of x and y and can in turn be differentiated. Thus, $\dfrac{\partial}{\partial x}\left(\dfrac{\partial z}{\partial x}\right) = \dfrac{\partial^2 z}{\partial x^2} = f_{xx}''(x,y)$, or $\dfrac{\partial}{\partial y}\left(\dfrac{\partial z}{\partial x}\right) = \dfrac{\partial^2 z}{\partial x \partial y} = f_{xy}''(x,y)$. The order in which we differentiate is immaterial provided that the derivatives concerned are continuous. The *total differential* of a function of two variables, for example, $f(x,y)$, $df = d_x f + d_y f = \dfrac{\partial f}{\partial x}dx + \dfrac{\partial f}{\partial y}dy$ is equal to the sum of the partial differentials of the first order if we neglect terms of higher orders, whose values are indefinitely small quantities relative to the first order differentials. In symbols, $df = d_x f + d_y f = \left(\dfrac{\partial f}{\partial x}\right)dx + \left(\dfrac{\partial f}{\partial y}\right)dy$. In words, the total differential of $f(x,y)$ is found by finding the partial derivatives with respect to x and y, multiplying them respectively by dx and dy, and adding.

2. *DIFFERENTIAL EQUATIONS*

A natural development of the invention of the calculus was the introduction of differential equations. Differential equations differ from the ordinary equations of mathematics in that in addition to variables and constants they contain also derivatives of one or more of the variables involved. Differential equations are of extreme importance, and arise in many problems. Newton solved his first differential equation in 1676 by the use of an infinite series, eleven years after his discovery of the calculus in 1665. Leibnitz solved his first differential equation in 1693, the year in which Newton first published his results. From this date on, progress in the development and application of differential equations was very rapid, and today the subject of differential equations occupies in the general field of mathematics a central position from which important and useful lines of development flow in many different directions.

To integrate or solve a differential equation means, analytically, to find all the functions which satisfy the equation. In geometry, it means to find all the curves which have the property expressed by the equation. In mechanics it means to find all the motions that may possibly result from a given set of forces , . The *degree* of the differential equation is defined as the degree of the derivative of the highest order which enters the equation. The *order* of the differential equations is the order of the highest derivative it contains.

Equations in x and y, of the first degree in y and its derivatives with respect to x, y', y''. , are called *linear equations*. The main equations of physics are *linear* differential equations of the second order, since y, the primitive function, y', the first derivative, and y'', the second derivative, appear only in the first degree. For instance the equation $\dfrac{d^2y}{dx^2} + a_1\dfrac{dy}{dx} + a_2y = X$, or $y'' + a_1y' + a_2y = X$, when X represents a function of x alone is such an equation. It is linear, or of the first degree, because the second derivative, y'', appears only to the first degree. It is of the second order because that is the highest order derivative in the equation. As we may recall, the derivative of a function gives us the *rate of change* of the function when we give successive values to the independent variable. When we study the rate of change of the rate of change of our function, we study the rate of change of the first derivative which expresses the rate of change of the function, whence we obtain the derivative of the second order, and so on. If we equate our derivatives to zero, or choose a value of the variable for which our derivative becomes zero, the rate of change of our function becomes zero. In other words, the value of our function is momentarily constant, it has a stationary value.

Quite naturally, differential equations which involve derivatives involve implicitly and explicitly the whole fundamental structural framework of the calculus as explained in this chapter by expressing the 'rate of change' of some natural process. If the rate of change is zero, it might express some 'natural law', or some uniformity as found in nature. In other words, differential equations express differential laws, which in turn express the momentary tendencies of processes whose outcomes are given by the process of integration.

From what has already been said here, it is obvious that differential equations and the differential laws which they express are of extreme structural importance. They formulate not only the uniformities and tendencies found in nature, but also of necessity somehow involve causality. Besides which, they are also in accord with the physical structure and function of the nervous system. We shall return to this most important subject in the next chapter, in which we shall analyse the physical significance and aspects of what has been explained here.

3. *METHODS OF APPROXIMATION*

In discussing the above fundamental notions of the calculus we considered a portion AB, of the curve given by the equation $y = f(x)$, (Fig. 13) and two points on this curve P_1 with co-ordinates (x_1, y_1) and P_2 with co-ordinates

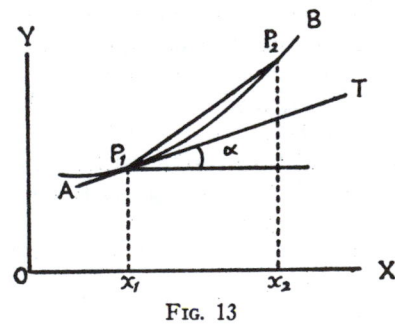

FIG. 13

(x_2, y_2) moving along the curve, the secant, or chord, P_1P_2 rotates about P_1, its length steadily diminishing, and in the limit as the length of the chord P_1P_2 tends toward zero, the slope of the secant approaches the slope of the tangent P_1T. We saw that the slope of this tangent was given by the value of the first derivative of the function which represented the curve. We were trying to get some knowledge of the direction of our curve at a given point by considering the slope of a *straight line* of smaller and smaller length. When we studied the curvature of our curve we considered the rate of change of the slope of our tangent and so, by the aid of a second derivative, we found the curvature. In this case we approximated our curve to a circle of radius equal to the radius of curvature of the curve at a given point.

In attempting to determine the length of a portion of our curve a point cannot be regarded as a piece of the curve but only as marking a position on it. For the purpose of determining the length of an arc it is convenient to replace each small element of the arc by its chord, a *lineal element*. By definition the length of an arc of a curve is the limit, if such limit exists, toward which the sum of the lengths of the chords of its small subdivisions tends as the number of chords increases indefinitely and their individual lengths all approach zero uniformly. For example, the circumference of a circle is the limit approached by the perimeter of an inscribed polygon as the number of its sides increases indefinitely, the lengths of the individual sides all approaching zero.

Similarly the length of a curve may be approximated by the sum of the lengths of segments of tangents at successive arbitrarily chosen points, merely by choosing the points nearer and nearer together. For example, the circumference of a circle is the limit approached by the perimeter of a circumscribed polygon as the number of its sides increases indefinitely, the lengths of the individual sides all approaching zero. In either case, a point on a curve taken with a vanishingly small portion of the tangent to the curve at that point may be called the *lineal element* of the curve.

The above definitions apply equally well in either two or three dimensions. The lineal element in two dimensions may be defined by three co-ordinates x, y, p, of which x and y are the co-ordinates of the point through which the lineal element passes and p is the slope of the element. This slope, as we already know, is to be found by differentiation, and is given by the formula $p = dy/dx$. In geometrical problems which relate the slope of a tangent to that of other lines, it is not the tangent that is of real importance but the *lineal element*. From this point of view a curve is made up of infinite numbers of vanishingly small lineal elements which are tangent to it, which is the point of view of the differential calculus. Or the curve is composed of infinite numbers of vanishingly small chords which are the sides of an inscribed polygon, which is the point of view of the integral calculus.

Obviously, in the limit, both points of view are equivalent, although as a matter of convenience they may be different. In any case, it must be obvious to the reader that using *straight lines* instead of pieces of a curve, or using as closer approximations arcs of circles, facilitates our study of the curves, indeed renders such study possible at all, and in practice we can carry our work to any degree of approximation we choose. But in theoretical work we require precision, hence we think in terms of infinite numbers of vanishingly small steps. The differential and integral calculus supply the only perfect technique for these processes of analysis and synthesis.

4. *PERIODIC FUNCTIONS AND WAVES*

We have already said that the most important relations of physics are represented by linear differential equations of the second order. It is important to know the connection of these equations with the general theory of waves or oscillations.

If on a circle of unit radius, as shown in Fig. 14, we take several points P_1, P_2, P_3, P_4, and connect these points by straight lines with the centre O, we get angles XOP_1, XOP_2, . In trigonometry we define certain functions of these angles and a unit of measurement. For our purpose we will only define the so-called sine and cosine, as we have already met the definition of tangent

$\tan \theta = \dfrac{M_1 P_1}{OM_1}$. The angles XOP_1, XOP_2., may be specified by the ratios $M_1 P_1 / OP_1$, $M_2 P_2 / OP_2$.., respectively each of which ratios has a definite value. This ratio in any case is called the *sine* of the angle, and is written in abbreviated form $\sin \theta$. If the radius of our circle is taken as unity then simply $M_1 P_1 = \sin \angle P_1 O M_1 = \sin \theta$, since $OP_1 = 1$. The ratio OM_1 / OP_1 is called the *cosine* of the angle XOP_1, and is written $\cos \theta$.

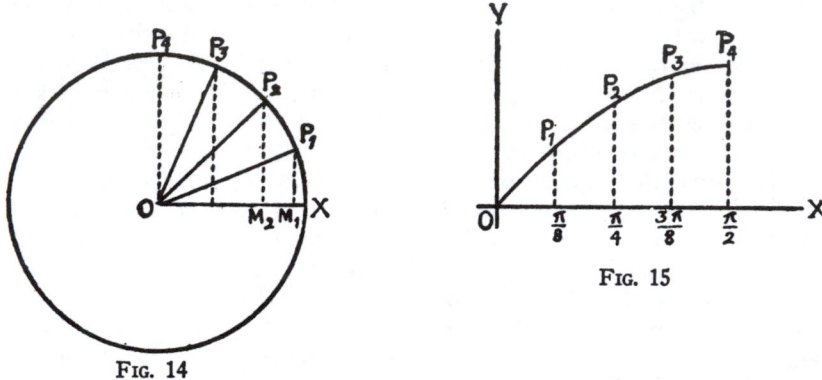

FIG. 15

FIG. 14

There are two units of measurement of angles. In ordinary, or sexagesimal, measure, the unit angle is the degree, $1/360$ of the entire angle about a point, $1/180$ of a straight angle, or $1/90$ of a right angle. The degree is divided into 60 equal parts called minutes. The minute is divided into 60 parts called seconds. In circular measure the unit angle is the radian, the angle at the centre of a circle whose arc is equal to the radius of the circle. This angle is a constant whether the circle be large or small, due to the fact that the cir-cumferences of circles vary as their radii, and, in one circle, angles at the centre are proportional to their arcs. The constant ratio of the circumference of the circle to its radius is given by the number $\pi = 3.14159 \ldots$, this number being 'incommensurable' with unity. As the length of the circumference of a circle with radius R, is $2 \pi R$ we see that the entire angle about the centre, which in degrees is 360, is in radians 2π; that a straight angle equals 180 degrees or π radians; and that a right angle equals 90 degrees or $\dfrac{\pi}{2}$ radians.

Thus 1 radian $= \dfrac{180°}{\pi} = 57°17'44''$. 806 ... which, as it depends on the value of π, is itself an 'irrational' number. The 'incommensurability' of the radian with right and straight angles makes its practical use inconvenient. One of the main uses of the radian is in theory as it introduces a marked simplification in that the ratio of the sine of an indefinitely small angle to the angle itself is 1, when the angle is measured in radians. In other words, the equivalence of an indefinitely small arc and chord becomes apparent numerically when the angle and sine are expressed in one unit.

The following table gives the ordinary and radian measures, the sine, cosine and tangent of angles of 0, 1, 2, 3, and 4 right angles.

Angle in Right Angles	Angle in Degrees	Angle in Radians	Sine	Cosine	Tangent
0	0	0	0	1	0
1	90	$\pi/2$	1	0	$\pm\infty$
2	180	π	0	-1	0
3	270	$3\pi/2$	-1	0	$\mp\infty$
4	360	2π	0	1	0

From Fig. 14 and from this table, it follows that the values of the trigonometric functions are equal for the angles 0 and 2π, or in the language of degrees, for the angles $0°$ and $360°$. We see also from Fig. 14 that the angle XOP_1, or any other angle, has one measure as expressed by its trigonometric functions if we add to it $360°$ or 2π radians.

The structural importance of the trigonometric functions in analysis lies in the fact that they are the *simplest* singly *periodic* functions and are therefore adapted for the representation of undulations. As we have already seen the sine and cosine have the single real period 2π, which means that they are not altered in value by the addition of 2π to the variable. The tangent has the period π.

Besides the three functions defined above, we usually define three others, the secant, the cosecant and the cotangent as reciprocals respectively of the cosine, the sine, and the tangent. These last three we may disregard in our present discussion.

Let us consider the function $y = \sin x$, and construct the curve which this equation represents. If we draw a circle of *unit* radius, Fig. 14, the ordinates corresponding to the different angles XOP_1, XOP_2 . , give the values of y, while the angles measured in radians, give the corresponding values of the abscissa x.

Plotting corresponding values of x and y as thus obtained in Fig. 14 we get in Fig. 15 the partial graph of the function $y = \sin x$. Proceeding again around our circle in Fig. 14, that is, adding $360°$ or 2π, to each of our angles, hence to their abscissas of the curve in Fig. 15, we add to the graph a second complete wave. We may thus proceed either forward or backward obtaining as many complete waves, or undulations, as we please, as in Fig. 16.

The curve represented by $y = \cos x$ is obtained in like manner and is quite similar to the sine curve. (See Fig. 17.)

To differentiate $\sin x$ we give to x the arbitrary values x_1, and $x_1 + \Delta x$ and compute for y the corresponding values $y_1 = \sin x_1$ and $y_1 + \Delta y = \sin (x_1 + \Delta x)$.

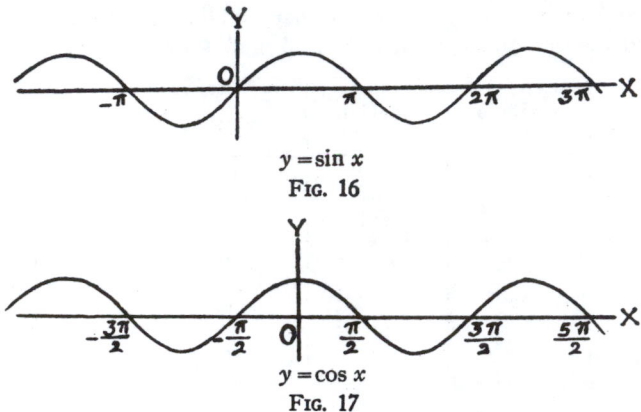

$y = \sin x$

FIG. 16

$y = \cos x$

FIG. 17

Subtracting y_1 from $y_1 + \Delta y$, we have $\Delta y = \sin (x_1 + \Delta x) - \sin x_1$. Dividing by Δx we have
$$\Delta y / \Delta x = \frac{\sin (x_1 + \Delta x) - \sin x_1}{\Delta x}.$$

To express the meaning of the above geometrically we may take a circle of unit radius and construct the angles x_1 and $(x_1 + \Delta x)$, (Fig. 18). Then $M_1 P_1 = \sin x_1$, $M_2 P_2 = \sin (x_1 + \Delta x)$, $QP_2 = \sin (x_1 + \Delta x) - \sin x_1 = \Delta y$, and arc $P_1 P_2 = \angle P_1 P_2 = \Delta x$.

The limit approached by the ratio $\Delta y / \Delta x = QP_2 / P_1 P_2$ as $P_2 \to P_1$, or as $\Delta x \to 0$, is by previous definitions the sine of the angle $M_2 P_2 P_1$, since *in the limit* the arc $P_1 P_2$ becomes a straight line, the hypotenuse of the right triangle $P_1 Q P_2$, which is similar to the right triangle $P_1 M_1 O$, whence angle $\angle M_2 P_2 P_1 = \angle x_1$.

In other words, $\lim\limits_{\Delta x \to 0} \dfrac{\Delta y}{\Delta x} = \lim\limits_{\Delta x \to 0} \dfrac{QP_2}{P_1 P_2} = \dfrac{OM_1}{OP_1} = \cos x$, or $D_x \sin x = \cos x$.

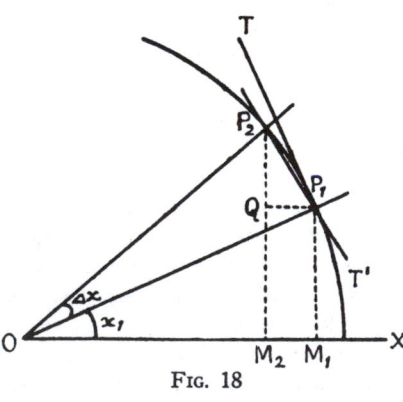

FIG. 18

It may be easily and similarly shown that $D_x \cos x = -\sin x$. Our main interest is in the second derivative. We see that the derivative of $\sin x$ is $\cos x$, and that the derivative of $\cos x$ is $-\sin x$.

Differentiating again we obtain the second derivative of $\sin x$ as $-\sin x$, and the second derivative of $\cos x$ as $-\cos x$.

The sine and cosine and their linear combinations are the only functions which when differentiated twice give us the second differential coefficient equal and of opposite sign to the original function. In symbols
$$\frac{d^2 (\sin x)}{dx^2} = -\sin x; \text{ and } \frac{d^2 (\cos x)}{dx^2} = -\cos x.$$

In physics we deal with many processes which are structurally periodic, which means that a definite physical condition constantly recurs after equal intervals of 'time'. The number of seconds or fractions of seconds within which the process runs its course is called the period. We know already that the simplest periodic functions are sine and cosine functions of the type

$$\sin (x+2n\pi) = \sin x, \text{ and}$$
$$\cos (x+2n\pi) = \cos x, \text{ where } n \text{ may have any in-}$$

tegral value. Furthermore we have already seen that the first derivatives, therefore all derivatives of such functions are likewise simple sine and cosine functions. In particular, the second derivatives of the sine and cosine functions are likewise sine and cosine functions taken with the opposite algebraic sign.

If we express the variability of a process as a function of 'time', that is, by an equation of the form $S = F(t)$, then in a periodic process, $F(t_1 + nT) = F(t_1)$, where T is the period and n any integer. If the process repeats itself, as in a periodic process, we must have

$$\left| \frac{dF}{dt} \right|_{t=t_1+nT} = \left| \frac{dF}{dt} \right|_{t=t_1} \text{ and } \left| \frac{d^2F}{dt^2} \right|_{t=t_1+nT} = \left| \frac{d^2F}{dt^2} \right|_{t=t_1} \cdot \text{,}$$

but, as we have already seen, the sine and cosine functions satisfy these conditions.

A process which can be described by an equation of the type $S = A \sin \dfrac{2\pi t}{T}$

is called a *harmonic vibration* or, a 'pure sine vibration', or simply a 'vibration' or 'oscillation'. The constant A, which represents the maximum value of the displacement on either side, is called the amplitude. The period T is called the 'time of vibration', its reciprocal value which gives the number of vibrations in a unit of 'time' is called the vibration number or *frequency*.

As the second derivatives of sine and cosine functions are equal to the original functions taken with the opposite signs, we can describe harmonic vibrations by differential equations of the first degree (linear) and of the second order of the special type $\dfrac{d^2S}{dt^2} = -a^2S$, where $S = A \sin \left(\dfrac{2\pi t}{T} + \epsilon \right)$, A representing the amplitude, T the period, ϵ the phase of the vibration. The factor of proportionality a is taken as the square of any arbitrary real quantity to indicate that the right-hand side must always have the opposite sign to that of S.

The propagation of a vibration is called an advancing plane wave which has both velocity and direction.

Fourier has shown that any given form of wave may be represented by the superposition of a series of sine-waves, which gives sine-waves great theoretical and practical importance.

In writing this chapter I had two main aims. One was to briefly indicate the essential semantic factors involved in the differential methods. The other, to make the general reader and even specialists who are not mathematicians acquainted with some terms and rudiments of method which will be necessary for further discussion.

The differential methods involve semantic factors essential for a $\bar{A}$-system, the ∞-valued semantics of probability and for sanity and cannot be longer disregarded.

The main pressing issues are twofold. One, to formulate methods which would impart the $\bar{A}$ semantic reactions of the calculus, which need not involve any technicalities, and can be imparted in the most elementary home or school education. The other is to draw the attention of specialists to these semantic problems so that they will work them out.

An attempt to solve the first issue has been undertaken in the present volume. The second task will probably be accomplished in the not too distant future.

It is earnestly suggested to all scientists, professional men and teachers, who are not mathematicians, to become familiar with differential methods and so acquire the appropriate semantic reactions. Experience, in many cases, has shown that this will assist them in acquiring semantic balance and 'mental' efficiency. Teachers and physicians in particular, would be greatly helped in their efforts to train children and patients in the $\bar{A}$ reactions. The benefit is not in any 'calculations' whatsoever, but in the method and the related psycho-logical reactions.

There is an excellent, short, most elementary and amusing account of the calculus by Sylvanus P. Thompson *Calculus Made Easy* (Macmillan) which, for the present, is all that is needed for this purpose.

CHAPTER XXXIII

ON LINEARITY

The conception of linear transformation thus plays the same part in affine geometry as congruence plays in general geometry; hence its fundamental importance. (547)   HERMANN WEYL

It is instructive to compare the mathematical apparatus of quantum theory with that of the theory of relativity. In both cases there is an application of the theory of linear algebras. (215)   W. HEISENBERG

This "perturbation theory" is the complete counterpart of that of classical mechanics, except that it is simpler because in undulatory mechanics we are always in the domain of *linear* relations. (466)   E. SCHRÖDINGER

As a result of experimental research on association, in 1904, I was led to show the complexity of the factors governing evocation. . . . And I have often insisted since then on this essential idea, in opposition to the simple schema of linear associative connection. (411)   HENRI PIÉRON

We have already had several occasions to mention the 'plus' or additive issues as connected with linearity. This problem is of structural and linguistic as well as empirical and psycho-logical semantic importance. It is sufficient for our purpose at present that we should notice two facts; namely, (1) That in one dimension, linearity expresses the *relation* of proportionality; (2) That the problems of linearity are dependent on the *relation* of additivity.

The structural notion of additivity is of great antiquity. Being the simplest of such notions, it naturally originated very early in our history. The earliest records show that the Babylonians and the Egyptians used the additive principle in their notation. Our primitive ancestors, long before any records were written, had similar structural conditions present, open for investigation and reflection, that we have today. That this was the case is not a mere guess. Otherwise we would still be at their stage of development. Some beginning had to be made somewhere. There is little doubt that the men of remote antiquity presented many types of make-up, as we do today. Some, for instance, were more curious than others; some more inventive, some more reflective . , which, as we know today, is found even among animals. These more gifted individuals were, as usual, the inventors, discoverers, and builders of systems and language of their period. They could not long fail to recognize the fact that a stone *and* a stone, or a fruit *and* a fruit are *different* from *one* stone or *one* fruit. For instance, the two stones might have saved the early observer's life in defence, or the two fruits might have satisfied his hunger or thirst, where one would not have done so. An accumulation of objects was obviously somehow different from a single object. As these problems were often of vital importance to their lives, names for such accumulations of objects began to be invented, and one and one was called two, two and one was called three , . Number and mathematics were born as a structural semantic life-necessity for a time-binding class of life. They were an expression of the neurological structure and function and of the tendency toward induction.

In the beginning, names and generalizations were made from the simplest brute facts of life, and our primitive ancestors did not realize, that these crude generalizations might not have a structural validity, which they seldom doubted that they possessed, even as we today, seldom doubt. Those primitive scientists, (and we today differ very little from them), having produced *terms*, objectified them, and began to speculate about them. Let us examine some examples of such primitive mathematical speculations. Addition, of course, by which we *generate numbers*,—one and one make two, two and one make three . , was all-important. They could not miss the simple fact that three, which is equal to two and one, by definition, is more than two or one. A primitive generalization; namely, that the sum is always more than the summands taken separately, was still further generalized to a postulate that a part is smaller than the whole. This generalization has hampered mathematics almost up to our own day, and for many thousands of years it prevented the discovery of the notion of mathematical infinity, which we have already discussed in Chapter XIV.

It must be noticed that such generalizations involve *s.r*, which are objective and un-speakable. If verbally formulated they should have a structure similar to that of the facts, otherwise they are fanciful and vicious, because not properly formulated. When formulated they become public structural facts (*s.r* are personal, individual, non-transmittable, and un-speakable) and so they may be criticized, improved, revised, rejected , . All human history shows that the correct structural formulation of a problem is usually as good as the solution of it, because sooner or later a solution always follows a formulation.

After many thousands of years—in fact, practically only the other day—it was found that these primitive generalizations were in general not valid. Negative numbers were invented, and two plus minus-one was no more three but one, $2+(-1)=1$. The sum was *no* longer greater than its summands. The usual tragedy takes place here also. A few people know the facts, but the old primitive structural *s.r remain* in some of these few, as well as in the great majority of us who did not even know the facts. That such structural *s.r* do not vanish quickly, or generally, is proven again and again throughout history. We see it very clearly in the problems of 'infinity', or $\bar{E}$ geometries, or $\bar{N}$ physics. But the most pathetic sight is to see scientists who have *rationalized* the technique without a deeper re-education of their *s.r*. This is most clearly seen in the case of many writers on the foundations of mathematics, the Einstein theory or on the newer quantum mechanics. They *feel* in the old structural way, they *rationalize* in the new, hence their works are full of self-contradictions. Readers and students alike feel how 'difficult' and messy the whole subject is. As a matter of fact, the new theories are neither messy nor difficult. They are really much simpler and easier than the old theories, *provided* our structural *s.r* are purged of the primitive structural tendencies to which every one of us is heir. When this semantic re-education of our structural feelings is accomplished it is the old that becomes 'unthinkable' and incomprehensible, because it gives such a structural mess.

Something similar might be said about a feeling deep-rooted in all of us namely, the 'plus' feeling. In all the advances of science we struggle against it. For instance, the example of the green man-made leaf given previously shows clearly that man-made affairs may with some plausibility be considered as 'plus' affairs, but not so with non-man-made natural leaves, which appear not as 'plus', but functional affairs, where the greenness was structurally *not added* but happened, or became. As a structural fact, the world around us is *not* a 'plus' affair, and requires a functional representation. In chemistry, for instance, does hydrogen 'plus' oxygen produce water, H_2O? If we mix the two gases, two parts of hydrogen with one of oxygen we do not get water. We must first pass a spark through the mixture, when an explosion occurs and the result becomes *water*, a *new* compound quite *different* from its elements or from a mere mixture of them. Does one gallon of water and one gallon of alcohol make two gallons of a mixture? No, it makes less than two gallons. Does light added to light make more light? Not always. The phenomena of interference show clearly that light 'added' to light sometimes makes darkness. Four atoms of hydrogen, of atomic weight 1.008, produce, under proper conditions, one atom of helium, not of atomic weight 4.032, but of atomic weight 4. The 0.032 has somehow mysteriously vanished. Such examples could be quoted endlessly. They show unmistakably that structurally this world is not a 'plus' affair, but that *other* than additive principles must be looked for.

The struggle against this 'plus' feeling is quite evident, but often unsuccessful, in scientific literature. Man 'is' an animal 'plus' something. Life 'is' 'dead matter', 'plus' some 'vitalizing principle', . In scientific literature we find curious expressions: as for instance, 'It is impossible to express the conduct of a whole animal as the algebraic sum of the reflexes of its isolated segments'; or, 'The individual represents heredity *plus* environment'; or, 'That the abstraction does not merely take away from a number of engram groups some components and combine the rest into one sum, but forms thereby a new psychic structure is self evident and is in no way peculiar to the psyche. Thus a clock work is as little the mere *sum* of its little wheels as a human being is the *sum* of his cells and molecules'; and later on, 'to be exact the ego consists of the engrams of all our experiences *plus* the actual psychism'. There is endless material that might be quoted, but for our purpose these few samples will suffice. We do not give them with the purpose of citing authoritative examples of the need of non-plus considerations. Far from it. We do it to emphasize the astounding fact that, although the best men in their fields have vaguely felt this necessity, yet even they become a prey to this very old structural linguistic semantic tendency. In all three cases quoted the authors were of the best we have. They have fought all their lives against the 'plus' tendency and methods; and yet, if they succeed in eliminating this tendency from one part of their subject, they plant it quite obviously somewhere else. We see that we are dealing here with an ingrained psycho-logical tendency which can be remedied only by a fundamental, $\bar{A}$, structural, semantic investigation.

Let us analyse these quotations. In the second case, we hear, after a successful attack on *plus* tendencies, a statement that the 'individual represents heredity *plus* environment'. Is this statement true? Let us take examples. There are certain fishes which are heliotropic and swim toward the light, but if we change the temperature of the water they become negatively heliotropic and swim *away* from the light. Is this most complex activity of the organism-as-a-whole a 'plus environment' fact, or does the change of temperature produce some fundamental functional changes? When, for instance, a good mother rat, having been put on a different though still abundant diet, which is deprived of some minute amount of special vitamins, begins to eat her litters, is this again a 'plus' reaction, or is it a most complex functional change of the organism-as-a-whole? Or when a human being, because he received in childhood an 'emotional' shock through outside events (action or language of parents, for instance) develops a functional disorder, or even a physical ailment, is this again a 'plus environment' problem? Or, when chickens fed on eggs laid by hens kept without sunlight or violet rays, or which have only received sunlight through a glass window, develop rickets and soon die, though they do not do so when the glass windows are removed and the sunlight is allowed to operate directly upon the hens. Is this again a 'plus environment' example?

One 'Smith' and one 'Smith' make two 'Smiths', as far as theatre or railway tickets are concerned, but in life, under proper conditions, they form a family and very often many more than two 'Smiths' come out of such 'addition'. How about their work? Is it a mere sum? In the case of inventors who may have been influenced by one or many men directly or indirectly, do their inventions produce a sum of the work of as many men? Surely the steam engine or the dynamo produces more work than not only the inventors, but the series of other men who have been indirectly responsible for the inspiration of the inventors, could ever have produced. So again it is not a 'plus' affair.

In the third case we see the author attacking the 'plus' tendency on one page, and planting another 'plus' a few pages further on, which implies at once some objectified *additional* entity. In this respect it must be noticed that this *additive* tendency represents a partial and important structural and semantic *mechanism of identification*, and to deal successfully with it, we must clear up the problem connected with the *additive* tendency.

The numberless and endless 'philosophical' volumes, for instance, which have been written about the 'body-soul' problems, show the tremendous structural and semantic importance of the clearing up of this 'plus' versus 'non-plus' issue. The reader may recall that the $\bar{A}$, the $\bar{E}$, and the $\bar{N}$ systems have one underlying structural metaphysics. The $\bar{E}$ systems deal with non-linear equations and with curved lines, of which the linear equation and the straight line, (one of zero curvature), are only particular cases. And the general theory of Einstein, which is the foundation of $\bar{N}$ systems, also introduces non-linear equations. Ought we to be surprised to find that a $\bar{A}$-system must also solve this difficult structural and semantic problem of linearity versus non-linearity, of additivity versus non-additivity?

Indeed the problems demanding our attention are extremely baffling and difficult. Even in such a perfected science as physics, we have great difficulties in using non-linear equations, and are still at the stage where we solve few equations other than linear ones. To make any progress at all we must start with the *simplest* available problems in this field; namely, mathematical problems. The main point at this stage is not a solution of the problem but its formulation. When formulated and brought to the attention of mankind, there is no doubt that it will be eventually solved.

To better understand the additive principle, let us consider a group of elements, the individuals of which we denote by letters a, b, c, d , . Let us take two or more of these elements and produce a synthesis which results in a third or n-th entity. Let this synthesis be of such a nature that the characteristics ascribed to the elements are also present in the resultant synthesis, in other words, let them have the so-called group characteristic. If our elements are, for instance, numbers, the new synthesis is also a number and belongs to the original group. We must notice that the problem of *order* is important in the formulation of the additive principle. If a and b are the two elements the synthesis of which we define, we must be clear that a first and b second, or b first and a second, must be recognized in the synthesis. Let us assume also that only the two alternative orders a and b, or b and a, are of importance in this case. The commutative law asserts that a plus b is equal to b plus a, $a + b = b + a$, which means that the two possible alternative orders give equivalent results. We must notice that this does not mean that order does not enter into this synthesis; in such a case the above mentioned commutative law would make no assertion at all. It is of importance that order should be involved in the synthesis. It is a matter of indifference only as far as equivalence by a commutative law is concerned.

We should notice for our purpose that the synthesis has the 'same' characteristics as the elements had. In other words, if we know the characteristics of the elements we know the characteristics of the result. For instance, if the elements were numbers, the result will be a number, and no characteristic absent in the elements will appear in the result. This predictability from the characteristics of the elements to those of the result is perhaps one of the most striking characteristics of additivity. On the one hand, it allows us to foretell the future; on the other hand, it limits considerably the applicability of the additive principle. It is obvious that when we combine elements, and the results have *new* characteristics absent in the original elements, the new problems are structurally no more of an additive character, and the synthesis must be different.

Only a few of the simplest entities in physics possess additive characteristics. If we take, for instance, 'weight' or 'length' or 'time', we see that these units are additive. One pound, or inch, or second, if added respectively to one pound, or inch, or second, gives us two pounds, or two inches, or two seconds.

Not so, however, with temperature, or density, or many other derived magnitudes, as we call them. If we have a body of temperature of one degree and combine it with another body of equal temperature the synthesis will not have a temperature of two degrees, (as in the case of weight), but of one degree. This applies to density . , two bodies of density one each will not give us a body of density two, but of density one.

Before further analysis of the problems of linearity and additivity, it will be well to consider a few definitions.

If an entity u is changed into an entity v by some process, the change may be regarded as the result of an operation performed upon u, the operand, which has converted it into v. If we denote the operation by f, then the result might be written as $v = fu$. The symbol of the operation f is called the *operator*. We are familiar with many such; indeed the symbols for all mathematical operations may be treated as operators. So for instance the symbol $\sqrt{}$ indicates the operation of extracting the square root. If we deal with a range of values for a variable x, what we have defined as the function symbol $f(x)$ may be treated as an operator whose operation on x may be indicated by the symbol fx. The operation of differentiation may be symbolized by D, the result of whose operation on the variable u, Du is the derivative of u. The sign of the definite integral $\int_b^a$ may be taken as indicating an operation which converts a function into a number , .

It is important to know that many of the rules of algebra and arithmetic when defined in this way, give rise to a calculus of operations. The fundamental notion in such a calculus is that of a product. If u is operated upon by f the result v is indicated by fu, or symbolically, $v = fu$. If v in turn is operated upon by g the result w is indicated by gv, or symbolically, $w = gv = gfu$, whence the operation gf which converts u directly into w is called the product of f and g. If this operation is repeated several times in succession the usual notation of powers is used, for instance $ff = f^2$, $fff = f^3$, . Not applying the operator at all, which we would denote by f^0, leaves u unchanged, which we indicate symbolically by the equation $f^0 u = u$. The operator f^0 is equivalent to multiplication by 1, $f^0 = 1$, whence f^0 may be called the *idem operator*. We see also that the law of indices holds; namely, the $f^m \times f^n = f^{m+n}$.

For our purpose we will analyse only one special case; namely, where we have u, v and $u + v$ as operands, and such an operator, f, that $f(u + v) = fu + fv$. Expressed in words, this means that the operator applied to the sum of the two operands gives a result equal to the sum of the results of operating upon each operand separately. Such a special operator is called a *linear*, or distributive, operator.

In terms of functions we would have $f(x + y) = f(x) + f(y)$ which may be called a functional equation. It has been proved that such a functional equation has one type of solutions; namely, when f is equivalent to a multiplication by a *constant*, or $fx = cx$. This fact is of great importance for us. Many problems in science are stated in terms of variation. For purposes of analysis a statement that 'x varies as y' is written $y = kx$, where k is called a factor of proportionality, which enables us to convert a statement of variation into an equation. If y varies inversely as x, we write $y = k(1/x)$ or $y = k/x$. A multiplication by a *constant* thus introduces a relation of proportionality, hence the importance of proportionality in a world where constants are present.

It must also be noticed that the two fundamental operations of the calculus are *linear* without being equivalent to multiplication by a constant. These are: 'the derivative of the sum is the sum of the derivatives', that is, $D(u+v) = Du + Dv$; and 'the integral of the sum is the sum of the integrals', that is $\int (u+v)dx = \int u dx + \int v dx$. But as the fundamental notion of the calculus is to substitute for a given function a *linear* function, in other words, to deal with curves as the limits of vanishingly small straight lines, this linearity underlies structurally all fundamental assumptions of the calculus, and one might say with Weyl that 'one here uses the exceedingly fruitful mathematical device of making a problem *linear* by reverting to infinitely small quantities'.[1]

A vector is defined roughly as a line-segment which has a definite direction and magnitude, and any quantity which can be represented by such a segment is defined as a vector quantity.

The addition of vectors is defined by the law of the parallelogram, as in the case of two forces. It should be noticed that because of this definition the sum of two vectors *differs* in general from the arithmetical sum of the lengths, and only collinear, or parallel vectors obey the arithmetical summation law.

The introduction by definition of mathematical entities which obey different laws from the usual arithmetical laws is an important structural and methodological innovation. It gives us the useful *precedent* of defining our *operations* to suit our needs. The vector calculus accepted as the definition of the sum of two vectors the law established *experimentally* in physics for the sum of two forces; and so the vector calculus from the beginning was structurally a particularly useful language in physics. Only since Einstein has the value and importance of the vector calculus for physics become generally appreciated.

If we have two vectors, **a** and **b**, starting from a common origin O and complete the parallelogram as in Fig. 1, then the diagonal of the parallelogram will be the required sum, **a**+**b**, by *definition*.

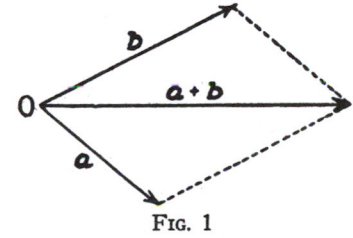

Fig. 1

If we choose two co-initial vectors of unit length, one on the X axis, and the other on the Y axis, and call them **i** and **j**, we can always represent any vector **x** as the sum of two vectors, one of which is the projection of **x** on the X axis, and the other the projection of **x** on the Y axis. (See Fig. 2.) Let us call these vectors **x**' and **x**'' respectively. Then **x** = **x**' + **x**'', by definition. But **x**' differs from **i** in length only, hence it can be obtained by multiplying **i** by

39

an appropriate number, say a. Similarly, $\mathbf{x}''$ can be obtained from $\mathbf{j}$ by multiplying $\mathbf{j}$ by b, and so, in symbols, $\mathbf{x}' = a\mathbf{i}$, $\mathbf{x}'' = b\mathbf{j}$, and $\mathbf{x} = a\mathbf{i} + b\mathbf{j}$. All vectors of the plane can be obtained from $\mathbf{i}$ and $\mathbf{j}$ in this form. The numbers a and b are called *components of* $\mathbf{x}$.

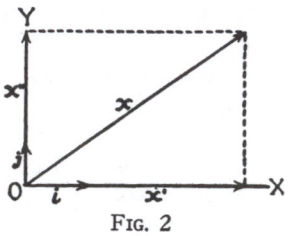

Fig. 2

Now that we know how to express a vector in terms of its components; namely, $\mathbf{x} = a\mathbf{i} + b\mathbf{j}$, let us consider a vector function $f(\mathbf{x})$ which satisfies the equation $f(\mathbf{x} + \mathbf{y}) = f(\mathbf{x}) + f(\mathbf{y})$. We may take $a\mathbf{i} = \mathbf{x}$ and $b\mathbf{j} = \mathbf{y}$ and $\mathbf{x} + \mathbf{y} = \mathbf{z}$ then we have $f(\mathbf{z}) = f(\mathbf{x} + \mathbf{y}) = f(a\mathbf{i}) + f(b\mathbf{j})$. But since a and b are numbers, we have $f(a\mathbf{i}) = af(\mathbf{i})$, and likewise, $f(b\mathbf{j}) = bf(\mathbf{j})$; so that $f(\mathbf{z}) = af(\mathbf{i}) + bf(\mathbf{j})$. But $f(\mathbf{i})$ is itself a vector and therefore expressible in the form $a'\mathbf{i} + b'\mathbf{j}$, and, $f(\mathbf{j}) = c\mathbf{i} + d\mathbf{j}$. Hence, $f(\mathbf{z}) = a(a'\mathbf{i} + b'\mathbf{j}) + b(c\mathbf{i} + d\mathbf{j}) = (aa' + bc)\mathbf{i} + (ab' + bd)\mathbf{j}$. In general, the components are the coefficients accompanying $\mathbf{i}$ and $\mathbf{j}$, and so we have the components of $f(\mathbf{z}) = f(\mathbf{x} + \mathbf{y})$ in terms of the components of $\mathbf{z}$; and we see how the components of a vector are changed into the components of the linear vector function of the vector.

In general terms, a continuous vector function of a vector is said to be a *linear* vector function when the function of the sum of any two vectors is the sum of the functions of those vectors; that is, the function f is linear if $f(\mathbf{r}_1 + \mathbf{r}_2) = f(\mathbf{r}_1) + f(\mathbf{r}_2)$, whence, if a be any positive or negative *number* and if f be a *linear* function then the function of a times $\mathbf{r}$ is a times the function of $\mathbf{r}$; $f(a\mathbf{r}) = af(\mathbf{r})$.

Linear vector operators are also defined by a similar equation; namely, $L(\mathbf{a} + \mathbf{b}) = L\mathbf{a} + L\mathbf{b}$.

Let us recapitulate. If we take the functional equation $f(x + y) = f(x) + f(y)$, which might be used as a definition of *linearity*, and which is based on *additivity*, and take $x = y = 1$, then we have $f(1+1) = f(2)$ and also $f(1) + f(1) = 2f(1)$; and so our original equation becomes by substitution $f(2) = 2f(1)$. It is obvious that the original equation, $f(x + y) = f(x) + f(y)$, is the source of indefinitely many such relations for particular numbers. For instance, $f(3) = f(2+1) = f(2) + f(1)$; but, in accordance with what we obtained before, $f(2) = 2f(1)$; so that $f(3) = 2f(1) + f(1) = 3f(1)$, and in general, $f(x) = xf(1)$. So, if we have an equation $f(x + y) = f(x) + f(y)$ for numbers, we know that we can obtain the value of this function for any x if we know it for 1. If we denote the function of 1, which is a *constant*, by $f(1) = k$, we have the general form of the function which satisfied $f(x + y) = f(x) + f(y)$ expressed by $f(x) = kx$. In words, a functional equation of the above type; namely, a function of the sum equal to the sum of the functions, has only one possible type of solution; namely, when f is equivalent to a multiplication by a constant, or, $f(x) = kx$. But this last means proportionality. The values of the function are proportional to the arguments (variables). In fact, let us consider two arguments, that is, two values of the independent variables x and y. We have, as shown before, $f(x) = kx$ and

$f(y) = ky$. Dividing the first by the second, we obtain $f(x)/f(y) = kx/ky = x/y$; or, in another form, $f(x)/x = f(y)/y = k$.

Let us consider an example. We know from elementary geometry that if we take an angle α and draw parallels which cut the sides at AA', BB', CC', DD' . , the corresponding intercepts are proportional. In general, the lengths of the segments on the left side are not equal, neither are they equal to the segments on the right side. If we designate the segment AB as x and BC as y, the corresponding segments $A'B'$ and $B'C'$ we may designate as $f(x)$ and $f(y)$, respectively, which means function of x and function of y. $A'B' = f(x)$, $B'C' = f(y)$. But the above intercepts are proportional, which

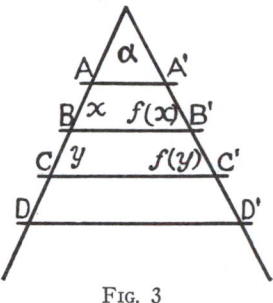

FIG. 3

means that $\dfrac{AB}{A'B'} = \dfrac{BC}{B'C'} = \dfrac{AB+BC}{A'B'+B'C'}$.

We easily see from Fig. 3, that $AB+BC = AC = x+y$; and $A'B'+B'C' = A'C'$ and so $A'C'$ on the one hand is $f(AC) = f(x+y)$ and on the other hand it is $f(x) + f(y)$ and therefore $f(x+y) = f(x) + f(y)$. We could multiply examples by taking relations between central angles in a circle and arcs of its circumference. In fact, any problem of *measure* in E geometry could be used as an example.

In our development we started with definite *additive* natural tendencies, not only in our highest, yet undeveloped, mathematics, which we call our daily and scientific language, but also in our lowest, but perfected, language which we call mathematics. In this perfected language the notion of *additivity* is connected with *linearity*, and the *methods* of *approximation* are also founded on additivity and linearity.

Yet the world around us in its more fundamental structural aspects is not additive; and for *adjustment* we must find means of passing from additive tendencies and formulations to non-additive tendencies and formulations. Modern mathematics has developed these methods, and modern physics is beginning to apply them. Let us repeat: the importance of linear functions implies the importance of 'straight' lines. They are important on two counts: first, because they are simpler than all other curves, so that naturally we want to study them before we study other curves, such as, for instance, circles or the other conic sections in elementary geometry; and secondly, because all curves can be approximated by straight lines. This point is very important, as approximation is the most powerful method we have of handling complicated situations.

There are two methods of approximating a curve in the vicinity of a point. If we are interested in the immediate vicinity of a point we approximate the curve by its tangent, as the tangent approximates the curve in the vicinity of a point better than any other straight line. If we want to decrease the error which we make in this approximation, we have only to decrease the vicinity in which we consider it. If we do not want to restrict ourselves to a small neighbourhood we have to use more complicated methods of approximation. We inscribe into the curve a broken line which consists of segments of straight lines. The beginnings of the study of curves consist in reducing the study of curves to: (1) The study of straight lines connected with the curves' tangents, which is the point of departure of the differential calculus; and (2) The study of the inscribed broken lines, which is the point of departure of the integral calculus.

Now curves represent only the simplest dependences. In other cases we have more complex kinds of functions; for instance, vector functions; but in every case we have *linear* functions, the simplest of their type; and other functions are studied by approximating them in one way or another by linear functions. In using the term 'function', we mean not merely numerical functions but also *operators*, which are to the ordinary functions what ordinary functions are to numbers. A general definition of linearity can be connected with that of proportionality in the following manner. If two variables are proportional, one to another, then to the sum of any values of the first corresponds the sum of the corresponding values of the second.

The simplest part of any field is the consideration of linear, additive questions; linear equations (equations of first degree in algebra), linear differential equations, linear integral equations, linear matrices, linear operators , . But sooner or later we come to the more difficult and more interesting non-linear problems. Perhaps the main importance of the General Theory of Einstein lies in the fact that the equations of physics become *non-linear*. Now, although non-linear equations can be approximated by linear equations, the character of a world determined by non-linear equations must be entirely different from a world determined by linear equations. In a linear world electrons would not repel each other but would travel independently of each other, and there could be no relation between the charges of different electrons. But we know that electrons do repel each other, and attract protons, and that their charges are equal. In physics, if a system can be described by linear differential equations, the causal trains started by different events propagate themselves *without interference*, with simple *addition* of effects.

The properties of systems which can be described by linear differential equations have, as we have already seen, the property of *additivity*. This means that the result of the effects of a number of elements is the sum of the effects separately, and no new effects will appear in the aggregate which were not present in the elements. In such a universe there is 'continuity', fields are superposable, wave disturbances are additive, 'energy' and 'mass' are indestructible , . In such a universe we can have two-valued *causality*, as causal trains started by different events propagate themselves without *interference*, and with simple addition of effects, and the present can be analysed backwards into the sum of elementary events, that is, a two-valued causal analysis is possible.

If our equations are not linear, the effects are not additive and a two-valued causal analysis is not possible.

The joint effect of *two* causes working together is *not* the *sum* of their effects separately[2], and we need ∞-valued causality.

Analytically, if we have *linear* differential equations and we have one solution $y_1 = f(x)$ and another solution $y_2 = F(x)$ then their sum is also a solution; namely, $y_3 = f(x) + F(x)$. If the differential equations are non-linear and if $y_1 = f(x)$ and $y_2 = F(x)$ are two solutions, then $f(x) + F(x)$ is *not* a solution.

Linear problems and linear equations play a very important structural role in science and there is little doubt that linear equations preponderate enormously, although many fundamental events cannot be described by such equations. A universe which can be described by linear differential equations of the second order has definite structural characteristics—in the main in rough accord with observation. As such differential equations give us the tendency of a process, we may use them to describe large-scale phenomena by integration, or the statistical phenomena of great numbers.

Unfortunately, the study of non-linear problems is structurally very difficult and largely a problem of the future.

There is one very important point which we should not miss. We know already that there is a fundamental difference between different orders of abstractions. Physical abstractions have always characteristics left out, and our higher order abstractions are further removed from life, but they have all characteristics included. The problem of sanity being a problem of adjustment, we must somehow correlate these abstractions in which characteristics are *left out* with those which include all characteristics, and so *must* proceed *by approximations*. Mathematical methods, particularly those of the differential and integral calculus, have evolved the best technique of *approximation* in existence today, which, as we have seen, is strictly connected with *linearity* or *additivity*.

A similar urge which prompted us in the expression of our additive tendencies and methods in the structure of language, has led to the production of the calculus. For organisms which abstract in so numerous and such different orders, the methods of the calculus are therefore fundamental psycho-logical devices, conditioning sanity.

In conclusion, we should notice two quite important facts. One of these is that the nervous system, being in a state of nervous tension, cannot structurally be a simple additive affair in all its functions, a fact which every one of us has experienced. Too many stimulations dull, or abolish, or change reaction in an enormous variety of ways. Piéron, as a result of experimenting in association, has not only shown the complexity of these processes, but also reaches the conclusion that the associative connections are non-linear.[3] The other most important point is that structurally the term 'and' implies addition. When we confuse orders of abstractions or levels of analysis, the 'and' additive implications falsify the issues. Thus for instance two atoms of hydrogen *and* one atom of oxygen *and* a spark produce water. The second 'and', at least, is used illegitimately, as it applies to an entirely different level (the spark) from that of the atoms. Linguistically we introduced additive implications, while empirically we are dealing with most complex non-additive, non-linear higher-degree functions. When we confuse orders of abstractions, as we all do, the 'and' is bound to introduce structurally false implications, which it is very difficult to avoid—the more so since these semantic problems are generally entirely neglected.

CHAPTER XXXIV

ON GEOMETRY

At the same time it will not be forgotten that the physical reality of geometry can not be put in evidence with full clarity unless there is an abstract theory also. . . . Thus, for example, while the term electron may have more than one physical meaning, it is by no means such a protean object as a point or a triangle. (259)   OSWALD VEBLEN

Euclidean space is simply a group. (417)   HENRI POINCARÉ

It is only in Euclidean "gravitationless" geometry that integrability obtains. (551)   HERMANN WEYL
The fundamental fact of Euclidean geometry is that the square of the distance between two points is a quadratic form of the relative co-ordinates of the two points (*Pythagoras' Theorem*). *But if we look upon this law as being strictly valid only for the case when these two points are infinitely near, we enter the domain of Riemann's geometry.* (547)   HERMANN WEYL
. . . *parallel displacement of a vector must leave unchanged the distance which it determines. Thus, the principle of transference of distances or lengths* which is the basis of metrical geometry, carries with it a *principle of transference of direction;* in other words, **an affine relationship is inherent in metrical space.** (547)   HERMANN WEYL

But before dealing with the brain, it is well to distinguish a second characteristic of nervous organization which renders it an organization in levels. (411)   HENRY PIÉRON

Section A. Introductory.

The main metrical rule in geometry is the familiar pythagorean theorem. In 1933 this rule is no longer considered as generally valid outside of the euclidean system, as its proof depends on the doubtful postulate of parallels. It is considered as an empirical generalization in which the relative error decreases when the distances become smaller. Indeed the small element of length, ds, given by the pythagorean rule is considered convenient and reliable in our exploration of the world.

The pythagorean rule states that in any right triangle, ABC, the square of the side opposite the right angle (the hypotenuse) is equal to the sum of the squares of the two other sides (the legs). In symbols, $AB^2 = AC^2 + BC^2$. If we build squares on all three sides of the triangle ABC and denote the areas of the squares by C', A', and B' then we have $C' = A' + B'$.

The above rule is also the main metrical rule for co-ordinate geometry, which gives us the length of the line segment joining any two

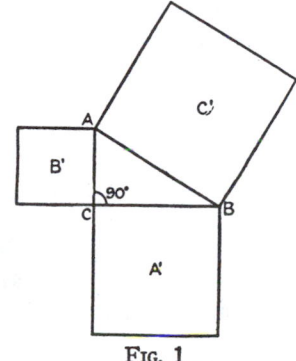

FIG. 1

points. Consider, for example, two points in two dimensions, P_1 and P_2, whose co-ordinates referred to a pair of axes in the plane are (x_1, y_1) and (x_2, y_2). By drawing the lines P_1Q and P_2Q parallel to the X and Y axes respectively, a right triangle P_1QP_2 is formed whose legs P_1Q and P_2Q are equal to $x_2 - x_1$ and $y_2 - y_1$ respectively, whence P_1P_2, the hypotenuse of the right triangle, or the distance s between the points, is equal to $\sqrt{P_1Q^2 + P_2Q^2}$, or $s = \sqrt{(x_2 - x_1)^2 + (y_2 - y_1)^2}$. If we pass to indefinitely small quantities and choose to deal with differentials we have $ds^2 = dx^2 + dy^2$ where $dx = x_2 - x_1$ and $dy = y_2 - y_1$. Usually the physicists treat their differentials as very small quantities and we may do likewise, although this is not precisely what a differential represents.

In three dimensions similar formulae appear; namely, $s^2 = x^2 + y^2 + z^2$ for the distance of a point from the origin and $s = \sqrt{(x_2 - x_1)^2 + (y_2 - y_1)^2 + (z_2 - z_1)^2}$

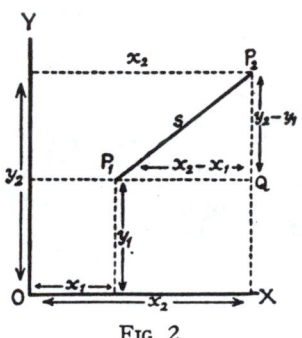

FIG. 2

for the distance between two points P_1 (x_1, y_1, z_1) and P_2 (x_2, y_2, z_2), and also $ds^2 = dx^2 + dy^2 + dz^2$, for the indefinitesimally small distance between two points.

In referring our geometrical entities to co-ordinate axes, or frames of reference, as they are called, we are interested in the properties of our geometrical entities and not in the accidental characteristics of our frames of reference, or the accidental characteristics of the form of representation we are using. Mathematicians discovered long ago that the form of representation is not of indifference to the results they obtain. Speaking roughly they have discovered that in one form of representation, they obtained characteristics $a, b, c, d, \ldots m, n;$ in another form, characteristics $a, b, c, d, \ldots p, q;$ and in still another form, characteristics $a, b, c, d, \ldots s, t, .$ In cases where direct inspection was possible they find by checking up predicted characteristics, that such characteristics as a, b, c, d in our example actually belong to the subject of our analysis, whereas the characteristics $m, n, \ldots p, q, \ldots s, t, \ldots$, do *not* belong to our subject at all, but *vary* from one form to another depending on the form of representation. Such facts make mathematicians distinguish between characteristics which are *intrinsic*, which actually belong to the subject independently of the form of representation; and those which are *extrinsic*, which do not belong to the subject, but are accidental and vary with the form of representation we happen to use.

If we mix intrinsic and extrinsic characteristics we have a structurally distorted knowledge of our subject. Obviously we are interested in methods by which these two types of characteristics can be separated and distinguished.

Such methods are found in what we call the transformation of co-ordinates, which means the passing from one form of representation to another, from one system of co-ordinates to another which corresponds to translation from one language to another. Obviously those characteristics which are intrinsic to our subject are and must be *independent* of the accidental selection of our form of representation, and therefore should remain unchanged when we pass from one frame of reference to another. Any characteristic which is changed by such a transformation of our systems of reference is clearly an extrinsic characteristic injected by the form of representation and not belonging to our subject; and so the transformation of co-ordinates is precisely the test we need and use.

Let us take for instance the line segment P_1P_2, as in Fig. 3. We may refer P_1P_2 to a system O, or to a system O'. Obviously the length of the line P_1P_2 is independent of the axes of reference used, and the formula for the length of a line is not altered, although the values for the x's and y's are different in the two systems. In other words, the sum of the squares of the differences of the co-ordinates remains invariant. In symbols,

$$s = \sqrt{(x_2-x_1)^2+(y_2-y_1)^2} = \sqrt{(x_2'-x_1')^2+(y_2'-y_1')^2}.$$

Such expressions, however, as x_1+x_2 or $y_1y_2.$, are *not* characteristics of our subject but characteristics of the particular frame of reference used, and so are mostly of no interest to us.

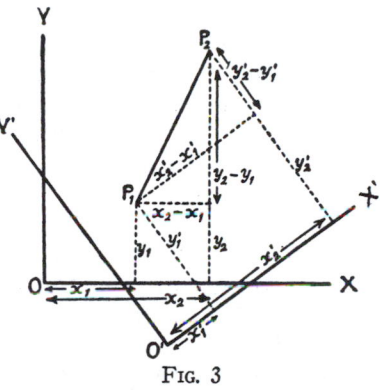

In such an elementary example as given here we are directly acquainted with our entities, and so we can inspect them directly and check for intrinsic and extrinsic characteristics. But when we deal with geometries of more than three dimensions, such checking becomes very difficult, if at all possible, and so new methods have to be invented.

FIG. 3

If we wish to eliminate the unit by which we measure our lines, this can be done by using a relation called a *ratio*. Let us, for instance, select 3 points A, B, C, and write the invariant formulae for the distance AB and AC in the form $\sqrt{(x_2-x_1)^2+(y_2-y_1)^2}$ and $\sqrt{(x_3-x_1)^2+(y_3-y_1)^2}$

then the *ratio*

$$R = \frac{\sqrt{(x_2-x_1)^2+(y_2-y_1)^2}}{\sqrt{(x_3-x_1)^2+(y_3-y_1)^2}}$$

is *independent of our unit of measurement.* If, for instance, this ratio $R=1$, we conclude that $AB=AC$, a characteristic which belongs to our lines and which is *independent not only of our system of reference but also of the unit which we have used.*

A great step forward in the formulation of methods which lead to invariant and intrinsic formulations was made in the invention of what is called the vector calculus and its extension in the modern tensor calculus. A few explanations of this principle will be of interest.

A vector is roughly a directed segment of a straight line on which we distinguish the initial and the terminal points. A vector has thus magnitude and direction. In practice we deal with two kinds of entities; some are purely numerical, establishing a specific, mostly asymmetrical, relation and have no direction, as for instance, mass, density, temperature, energy, electrical charge, population, mortality , . These quantities which do not involve direction are called scalar quantities.

Such quantities as velocity, acceleration, electric current, stresses, flow of heat or fluids . , which involve not only magnitude but also a definite direction, are called vector quantities, and have given rise to a special calculus called the vector calculus.

The invention of the vector calculus was a most revolutionary and beneficial structural and methodological step. It was originated independently by Hamilton and Grassmann. The benefits of this method are manifold, but we are interested mainly in but two of them. The first is that vector equations are simpler and fewer in number than co-ordinate equations. The second, and most important, is that the language of vectors is independent of choice of axes, and of frames of reference. It is naturally invariant for any transformations of axes. If axes are needed we can easily and simply introduce them, but we always have means to discriminate between intrinsic and extrinsic characteristics. The modern tensor calculus which made the general theory of Einstein possible is simply an extension of the vector calculus.

The above methodological and structural remarks are of fundamental semantic importance to us in all our affairs. Human life and affairs are never free from linguistic issues. Their role is similar to that of *mathematics*, that is to say, a form of *representation* gives us not only the characteristics which are intrinsic in our subject, but also introduces extrinsic characteristics which do not belong to the subject of our analysis but are due to the particular language we use and its *structure*. The analysis of these linguistic issues is much belated and extremely difficult because of the structural complexity of our language. These issues were discovered first in mathematics because of its clear-cut structural *simplicity*; and it is important that we should be aware of such new and unexpected fundamental semantic problems. We will not enlarge upon this phase of the problem here, except to mention that the whole of the present work, which uses a different language, of a different *structure*, already shows the usefulness of the new method. Sometimes we discover new characteristics, and sometimes we are led to emphasize characteristics which are known but have not yet been sufficiently analysed.

To carry our linguistic analogy further, we may take, for instance, the statement, 'knowledge is useful'. We could translate this statement into any other language and it would preserve its meaning. But if we make the statement, 'knowledge is a word which has six consonants and three vowels', such a statement may be false when translated into another language. Mathematics, being a language, has difficulties similar to ordinary language, but in mathematics it is often much more difficult to separate from other statements those which are purely about the language used. The so-called tensor calculus attempts to perform this last task.

The tensor calculus is an extension of the vector calculus, which has become famous since Einstein. It gives us formulations independent of any special frame of reference. In using it we are automatically prevented from ascribing to the events around us characteristics which do not belong to them. The tensor equations give us absolute formulations, absolute being understood as relative, no matter to what. Obviously the only language fit to express the 'laws of nature' should be independent of the particular point of view or language of some observer. It should give us formulations invariant for any and all systems of reference, although we might use preferred systems of reference, as, for instance, the principal axes of an ellipse, without any danger. The reader should not miss the point that such an ideal should be considered as the highest ideal in science. It is the mathematical species of a theory of 'universal agreement'. The above *sounds* simple and innocent; but, when actually applied, plays havoc with most of our old 'universal laws'. These laws do not survive this important and uniquely valid test, and so become mere local gossip instead of being the 'universal laws' that they claim to be. We will return to the structural problem of invariant formulations later. At present we must explain some other simple considerations.

On any surface we need two numbers or 'co-ordinates' to specify the position of a point, and so a surface is called a two-dimensional manifold. Points in three-dimensional manifolds require three numbers; points in four-dimensional manifolds four numbers; and similarly for any number of dimensions.

For our purpose, it is enough to speak in two dimensions, as our statements can easily be generalized to any number of dimensions. If we want to localize a point on a surface it is enough to divide the surface into *meshes* by any two line-systems which cross each other. By labeling the lines of each system with consecutive numbers, two numbers, one from each system, will specify a particular mesh. If the meshes are small enough we will be able to locate any point accurately.

These specifying labels or numbers require that we know what kind of mesh we are using. *Distances* between points are *independent* of mesh systems.

For the above reasons it is important to have more data about the mesh system we are using, which means that we have formulae which express the *distance* between two points, which is independent of the mesh systems, in terms of the mesh system.

We have already seen, in our study of the differential calculus, that, as a rule, it is simpler to deal with very short distances, and that it is easy to pass to larger distances by the process of integration. As yet we have used only plane rectangular systems of meshes in our illustrations, but this restriction is not necessary. If we use oblique co-ordinates (Fig. 4), the formula for the elemental distance is $ds^2 = dx_1^2 - 2k dx_1 dx_2 + dx_2^2$, where $k =$ cosine of the angle between the lines of partition.

The polar co-ordinates (Fig. 5) of the point P are the distance $r = OP$, of the point from the origin O, and the angle, $\theta = \angle XOP$, between the line OP

and the axis OX. The formula for the elemental distance in polar co-ordinates is $ds^2 = dr^2 + r^2 d\theta^2$.

Fig. 6 shows the co-ordinates frequently used by geographers; namely, geographic longitude and latitude, where the distance $ds^2 = d\beta^2 + \cos^2\beta d\lambda^2$.

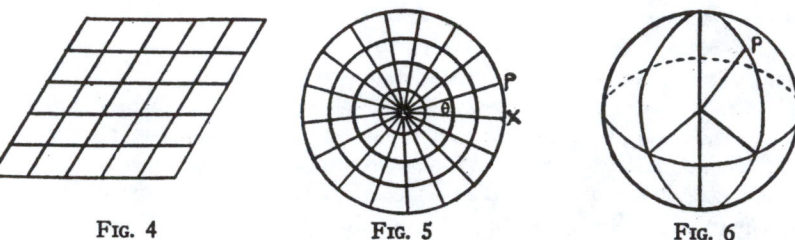

FIG. 4   FIG. 5   FIG. 6

It should be noticed that these formulae for different systems of co-ordinates are *different*. To make it still more obvious to ocular inspection, we will tabulate them in one lettering, thus:

$$
\begin{aligned}
ds^2 &= dx_1^2 + dx_2^2 && \text{for rectangular systems} \\
ds^2 &= dx_1^2 + x_1^2 dx_2^2 && \text{for polar systems} \\
ds^2 &= dx_1^2 - 2k\,dx_1 dx_2 + dx_2^2 && \text{for oblique systems} \\
ds^2 &= dx_1^2 + \cos^2 x_1 dx_2^2 && \text{for latitude and longitude systems.[1]}
\end{aligned}
\tag{1}
$$

It must be noticed that the values for the variables are not equal in these different equations. It is not necessary for the reader to know in detail how these formulae are obtained, but it is necessary to see that they are *different*, that they have different structure. The numbers of different co-ordinate systems we can use are infinite, but in practice we use only a few well-known types. There are also definite and simple formulae for passing from one system of co-ordinates to another.

We should not assume that in practice we always know what system of co-ordinates we are employing. For instance, before we learned that our earth is 'round', we did not know whether in our measurements we were employing the flat co-ordinates of a plane or spherical co-ordinates. We made some measurements and then we had to discover what kind of formulae would fit these measurements.

To find out what kind of co-ordinate system we are using, we select two points, let us say (x_1, x_2) and $(x_1 + dx, x_2 + dx)$ very close together, make our measurements of ds, and then test our ds to find which formula it fits. If we find for instance that our ds^2 is always equal to $dx_1^2 + dx_2^2$ we may assume for simplicity and our purpose that our co-ordinate system is plane and rectangular.

If our measurements fit any of the first three formulae (1), we may assume for simplicity and our purpose, that we are dealing with a plane surface, as each of these systems belongs to the plane. But if we find that the actual measurements of ds^2 are such that they never fit these first three formulae, but only the fourth one, we know, that our surface is not plane but curved like a sphere. Try as we may, we shall be unable to build on a plane any co-ordinate system which will fit the last formula. Thus we arrive at an important conclusion; namely, that from *measurements* we have a *structural* hint as to the *kind of world* we are in.

Section B. On the notion of the 'Internal Theory of Surfaces'.

Let us imagine some two-dimensional beings confined to their surface and unable to have a look at that surface from our third dimension. For them our third dimension would be 'unthinkable', and therefore the surface of a sphere like our earth which is curved in the third dimension would also be 'unthinkable' or 'beyond them'. Should they conduct some measurements in their 'world' and find that these measurements did not fit any of the first three formulae but only the fourth, they would have to reconstruct radically their 'world conception' and conclude that their world was a spherical surface. Our own situation does not differ radically from the situation of the inhabitants of this hypothetical two-dimensional world.

If we find ourselves so restricted as not to know whether we are finally dealing with a flat or spherical surface, we can select a point O and with a definite radius R describe from this point a circle ABC. Then we can measure the circumference of this circle. Now we know from geometry that in *the plane* the circumference of the circle $L=2\pi R$

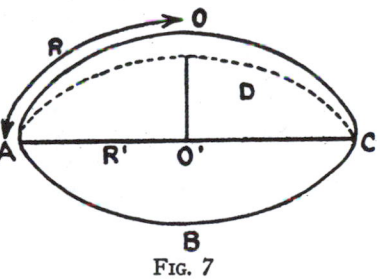

where R is the radius of the circle and $\pi = 3.1415\ldots$ If our surface is flat ($ABCD$), our measurement of L and R will satisfy the relation expressed in the formula. But if the surface is curved, our $R = OA$ will be larger than $R' = AO'$, and we shall find that our π is not $3.1415\ldots$, but smaller. We see once more that the *metrical properties of our world throw some light on its structural character.*

FIG. 7

We should notice also that the curvature of a two-dimensional surface is in the third dimension and that it is the means of giving us data about the surface without our leaving the surface and going into a third dimension. It is easy to convince oneself about these facts by taking 12 wires or strings of equal length and constructing the figure shown in Fig. 8. If we build it on a flat surface the 12 equal wires will fit exactly. But if we try this experiment on a curved surface, for instance on a pillow-, or saddle-shaped surface, the

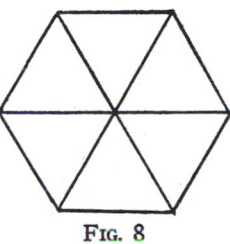

FIG. 8

last closing wire will not fit, and will be too short or too long depending on the kind of surfaces we have.

The formulae (1) have been generalized to

$$ds^2 = g_{11}dx_1{}^2 + 2g_{12}dx_1dx_2 + g_{22}dx_2{}^2 \tag{2}$$

for two dimensions and to

$$ds^2 = g_{11}dx_1{}^2 + g_{22}dx_2{}^2 + g_{33}dx_3{}^2 + g_{44}dx_4{}^2 + 2g_{12}dx_1dx_2$$
$$+ 2g_{13}dx_1dx_3 + 2g_{14}dx_1dx_4 + 2g_{23}dx_2dx_3 + 2g_{24}dx_2dx_4 + 2g_{34}dx_3dx_4 \tag{3}$$

for four dimensions. It is easy to see that $ds^2 = dx_1{}^2 + dx_2{}^2$ is obtained from (2) by taking $g_{11} = 1$; $g_{12} = 0$ and $g_{22} = 1$. This applies to formula (3) out of which we can have any of the other formulae by equating some of the g's to zeros, or to one or to other values. Formula (3) is called 'the generalized pythagorean rule', of which the ordinary form as given previously is only a particular case. We see, by comparing the formulae (1) with (2) and (3), that these g's are not equal for different systems of co-ordinates, and that they are factors in measure-determination which represent the geometry of the surface considered. It is customary to write the above formulae in an abbreviated form: thus, $ds^2 = \sum\sum g_{mn}dx_m dx_n$, where we give to m and n the values 1, 2, 3, 4, or, $(m, n = 1, 2, 3, 4)$ and where the symbol $\sum$ means summation.

We will now explain briefly the above generalizations and the meaning of the g's given in the expressions.

In the beginning of the nineteenth century the mathematician Gauss formulated the *internal* theory of surfaces without reference to the plenum in which they are embedded. This theory perhaps is and will remain a model on which all theories should be built. He introduced also a new kind of co-ordinates which have become of paramount importance, and which since Einstein are called gaussian co-ordinates. Gauss investigated the theory of surfaces, which are in general curved, embedded in three-dimensional 'space'. In 1854 the great mathematician Riemann generalized the two-dimensional gaussian theory to a continuous manifold of any number of dimensions. Historically, both Gauss and Riemann can be considered as the precursors of Einstein.

Let us imagine a surveyor to have the task of mapping a thickly wooded hilly region. Because of the conditions of his work, he can not use optical instruments, and he has no 'straight lines' to deal with. So euclidean geometry will, in general, not be applicable to the region as a whole. It can be assumed, however, that euclidean geometry may be applied to *very small* regions which can be considered flat. What we know already about the differential and integral calculus shows us that such approximations, when taken on a very small scale, are perfectly reliable and justifiable.

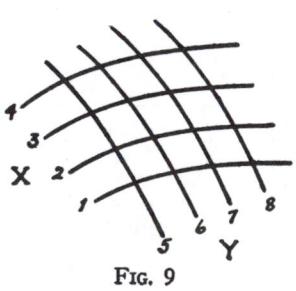

FIG. 9

The surveyor would lay out on his ground a network of smoothly curving lines, in two families, an X family and a Y family. (Fig. 9.) All the curves of the X family would intersect all the curves of the Y family but no X curve would intersect another X curve, nor a Y curve another Y curve.

Let us take the surveyor's network and label the curves by consecutive numbers in each family. The essential point is that these numbers, (let us call them the X and Y numbers) do *not* represent either lengths, or angles or other measurable quantities, but are simply labels for the curves, much as when we label streets by numbers.

But such numbering does not lead us far. We must introduce some *measure relations*. We have at our disposal a measuring chain and the *arbitrary* meshes of the network which we have introduced. The next step is to measure the small meshes one after another and plot them on our map. When this is done we have a complete map similar in structure to our region. Because of the smallness of the meshes we can consider them as small parallelograms, and such parallelograms can be defined by the lengths of two adjacent sides and one angle.

We may, however, proceed differently, as shown on Fig. 10.

Let us select one mesh, for instance the one bounded by the curves, 3 and 4, and by the curves 7 and 8. Let us consider a point P within the mesh, and let us denote its distance from the point O ($x=3$, $y=7$) by s. This distance could be directly measured. Let us draw from the point P parallels to our mesh lines and label the intersections with mesh lines by A and B, respectively. Let us also draw PC perpendicular to the x axis.

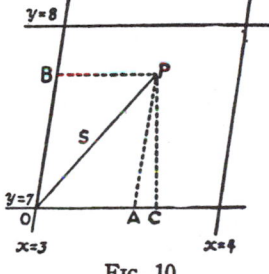

Fɪɢ. 10

The points A and B then also have numbers or labels or gaussian co-ordinates in our network. The co-ordinate of A may be determined by measuring the side of the parallelogram on which A lies and the distance of A from O. We can regard the relation called the *ratio* of these two lengths as the *increase* of the x co-ordinate of A towards O. We shall denote this increase itself by x, choosing O as the origin of the gaussian co-ordinates. Similarly, we determine the gaussian co-ordinate of y of B as the ratio in which B cuts the corresponding side. We see that these two *ratios*, which for brevity we call x and y, are the co-ordinates of our point P.

As x and y are *ratios* they of course do not give us the lengths of OA and OB but the lengths are given, for example, by ax, and by, where a and b are definite numbers, to be found by further measurements. If we move the point P about, its gaussian co-ordinates change but the *numbers* a and b which give the ratio of the gaussian co-ordinates to the true lengths *remain unchanged*.

We find the length, s, which is the distance of the point P from O, from the right triangle OPC by the pythagorean rule: $s^2 = OP^2 = OC^2 + CP^2$. But $OC = OA + AC$ and therefore by substituting and squaring we have $s^2 = AO^2 + 2OA \cdot AC + AC^2 + CP^2$. From the right triangle APC we have $AC^2 + CP^2 = AP^2$, whence substituting again we have $s^2 = OA^2 + 2OA \cdot AC + AP^2$. But $OA = ax$, $AP = OB = by$, and as AC is the projection of $AP = by$, it also has a *fixed ratio* to it, whence we may put $AC = cy$, and so we obtain the important formula $s^2 = a^2x^2 + 2acxy + b^2y^2$, in which a, b, c are *ratios* given by *fixed numbers*. Usually this formula is represented differently, a^2 is designated by g_{11}, ac by g_{12}, and b^2 by g_{22}; whence our formula becomes $s^2 = g_{11}x^2 + 2g_{12}xy + g_{22}y^2$ in which the numbers 11, 12 and 22 are simply ordering labels without quantitative values, mere subscripts, labels, indices ., which indicate that the different g's have different values. We see that the above formula is the one which was given previously by (2).

The g's with different labels serve just as sides or angles for the determination of the actual sizes of the parallelograms and we call them the *factors* of the measure *determination*. They may have different values from mesh to mesh, but if they are known for every mesh, then, by the last formula, the true distance of an arbitrary point P, within an arbitrary mesh from the origin can be calculated.[2]

The procedure by which we can locate any point on the surface is simple. If our point P is between the two curves $x = 3$ and $x = 4$ we can draw nine curves between these two curves and label them 3,1; 3,2; . . . ; 3,9. If P now lies between curves 3,1 and 3,2 we can draw nine curves between these two curves and label them 3,11; 3,12; . . . ; 3,19 , . We could do similarly with the y curves and in this way we would succeed in assigning to any point as accurate a pair of numerical labels as we pleased, and so finally have the gaussian co-ordinates of any point. We used nine curves simply to get the very convenient decimal method of labeling. The cartesian co-ordinate systems which we use in plane geometry obviously represent only *special cases* of gaussian systems.

As we have already seen, our g's are *ratios*, and so represent numbers. Such numbers may be regarded as tensors of zero rank for convenience of the mathematical treatment; and the quantities g_{xx}, g_{xy}, g_{yy}, may be treated as components of a tensor. Since this tensor determines the measure relations in any particular region, it is called the *metric fundamental tensor*. Its value must be given for the region in which we want to make our calculations. It determines the full geometry of the surface in a given region; and, conversely, we can also determine the fundamental tensor in a given region from measurements made in that region, without any previous knowledge of how our curved surface is embedded in 'space' at the place in question. The fundamental tensor in general varies continuously from place to place, and so every geometric manifold may be regarded as the field of its metric fundamental tensor.

Purely mathematical investigations show that the fundamental tensor defines a number called the 'Riemann scalar', which is completely independent of the co-ordinate system and leads to the definition of the *curvature tensor*, which can be connected with the 'matter tensor'.[3]

The main importance of the introduction of such arbitrary curves is to produce formulae for the surfaces which remain unaltered for a change of the gaussian co-ordinates—in other words, which remain invariant. This was achieved by the introduction of the relations called *ratios* which are pure numbers, and so the geometry of surfaces becomes a theory of invariants of a very general type.

On curved surfaces there are in general no straight lines—there are *shortest lines*, which are called 'geodetic lines'. To find them, we divide any arbitrary lines joining two points into small elements, which we measure, and select the line for which the sum of these elements is less than for any other line between the two points.* Analytically we can calculate them, when the g's are given, by the aid of the generalized pythagorean theorem. The geodetic lines, and also the curvature, are given by invariant formulae, which represent intrinsic characteristics of the surface, independent of any co-ordinates. All higher invariants are obtained from these invariants.[4]

We shall not attempt to give an explanation of the tensor calculus, as at present there is no elementary means of presenting a brief explanation; short of a small volume—at least the writer does not know of any.[5]

The name 'tensor' originally came from the Latin word *tendere* = to stretch, whence *tensio* = tension. Nowadays, however, it is used in a more general way; namely, to express the *relation* of one vector to another, and not necessarily to imply stress or tension. As an example, we can give the representation for stresses occurring in elastic bodies, which originally led to the name.[6]

As we have already seen, when we deal with *relations* of vectors our expressions become additionally independent of units. Such equations, independent of the measure code, are called tensor equations.[7]

As we are interested in equations which are invariant under arbitrary transformations, certain functions, called tensors, are defined, with respect to any system of co-ordinates by a number of functions of these co-ordinates, called the components of the tensor, from which we can calculate them for any new system of co-ordinates. If two tensors of one kind are equal in one system, they will be equal in any other system. If the components vanish in one system, they vanish in all systems. Such equations express conditions which are independent of the choice of co-ordinates. By the study of structural laws of the formation of tensors we acquire means of formulating structural laws of nature in generally invariant forms. Obviously, such methods and language are uniquely appropriate for physics and the formulations of the laws of nature. If a law cannot be formulated in some such form, there must be something wrong with the formulation and it needs revision.

The tensor calculus is also peculiarly fitted to describe processes in a *plenum*. We do not use it to describe the metrical conditions but to describe the *field* which expresses the physical states in a metrical plenum.

Eddington gives an excellent example of the fact that it is definitely necessary to look into the way we build up our formulae (structure) and the method of handling them.

---

*More generally, the geodetic represents a track of minimum or maximum interval-length between two distant events, either of them being unique (one-valued) in a given case.

40

The problem is to determine whether a particular kind of space-time is possible. We must investigate the different g's which give us different kinds of space-time, and not those which distinguish different kinds of mesh systems in one space-time. This means that our formulae must not be altered in any way if we change the mesh system.

The above condition makes an extraordinarily simple test of laws that have been or may be suggested. Among others, Newton's law is swept away. How this happens can be shown in two dimensions.

If in one mesh system (x,y) we have $ds^2 = g_{11}dx^2 + 2g_{12}dxdy + g_{22}dy^2$, and in another system (x',y') $ds^2 = g_{11}'dx'^2 + 2g_{12}'dx'dy' + g_{22}'dy'^2$, one law must be satisfied if the unaccented letters are replaced by accented letters. Let us suppose that the law $g_{11} = g_{22}$ is assumed. We change the mesh system, for instance, by spacing the y lines twice as far apart, that is, we take $y' = y/2$ and keep $x' = x$. Then $ds^2 = g_{11}dx^2 + 2g_{12}dxdy + g_{22}dy^2 = g_{11}dx'^2 + 4g_{12}dx'dy' + 4g_{22}dy'^2$. We see that $g_{11}' = g_{11}$ and $g_{22}' = 4g_{22}$. Whence if g_{11} is taken equal to g_{22}, g_{11}' *cannot* be equal to g_{22}'.

A few examples would convince us that it is extremely easy to change a formula entirely by the mere change of mesh systems. It seems unnecessary to emphasize the fact that 'universal laws', to be 'universal', should not depend structurally to such an extent on the accidental and, after all, unimportant, choice of reference systems.[8]

To remedy such a state of affairs, impossible in mature science, the tensor calculus was invented. The whole general theory of Einstein seems to demand that the equations of physics should ultimately be expressed in tensor forms; in other words, that 'universal laws' should cease to be 'local gossip'; a demand which must be granted, and *on this point* the Einstein theory is beyond criticism and is an epochal methodological advance of an irreversible structural linguistic character.

Section C. Space-time.

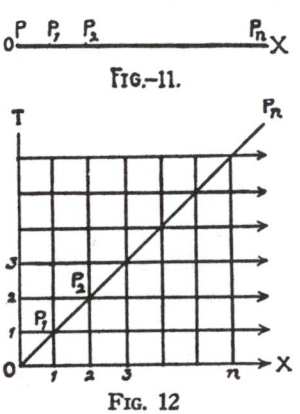

FIG.–11.

FIG. 12

In dealing with co-ordinate systems we have heretofore used them to represent only 'spatial' entities, spreads of different dimensions. It is desirable to become acquainted with a different use of co-ordinates, in which one of them will represent 'time'. The last use is just as simple as the former, but the graphs which we obtain are different.

Let us take the simplest example, of a point P moving uniformly along a straight line OX with the velocity of one inch per second. We could represent its movement in one dimension, as in Fig. 11, and *say* that our point P is at P_1 after one second ($t=1$), at P_2 after two seconds ($t=2$) . ; at the point P_n after n seconds ($t=n$).

But we could also represent this movement in a different way. We could choose two mutually perpendicular axes OX and OT as in Fig. 12, OX representing the 'spatial' actual direction of the movement and OT, which we have heretofore used to represent a second 'spatial' co-ordinate, now representing the 'time' co-ordinate.

We would lay off on the X axis our inches, 1, 2, 3, ... n, and on the T axis our seconds 1, 2, 3, ... n. In our two-dimensional *space-time* our point P would be at the point O ($x=0$, $t=0$). After one second it would be at the point P_1 ($x=1$, $t=1$), after two seconds at the point P_2 ($x=2$, $t=2$) . ; after n seconds at the point P_n ($x=n$, $t=n$). We see that the position of our *point* P in two-dimensional *space-time* would be represented by a series of points each given by two data: one 'spatial', the other a corresponding 'time'. If the intervals are taken indefinitely small, in the limit our *'moving' point would be represented by a static line inclined to the X axis*. We could then speak either of our 'moving' point, or else *not use* the term 'moving' but speak of infinitely many *static points*, each given by two numbers, one representing a distance, the other 'time'. Our 'moving' *point* would become a *static world-line*. The reader should notice that in this case we have structurally changed our *language* from dynamic to static, and raised the dimension. Our mathematical 'moving' 'point', which had no dimension in our one-dimensional 'space', is in our two-dimensional *space-time* represented by a *static* one-dimensional *line*.

In this example we had uniform translation. We did not introduce acceleration. The distances were proportional to the 'times', hence our line was 'straight' and inclined to the X axis at a constant angle.

Using such space-time representation we see that a point when it is not 'moving', but is stationary, is represented by a line parallel to the 'time' axis T, as shown at A on Fig. 13. Our point A is getting older, so to speak, but

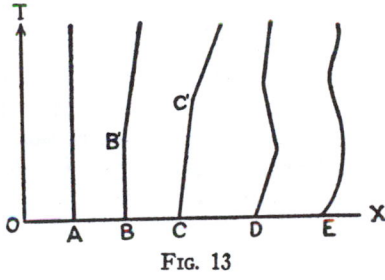

FIG. 13

does not 'move'. In the next case, the point B does not 'move' until it is some seconds old, when at B' it begins to 'move' with constant velocity. Point C 'moves' in the beginning at one constant velocity until C' where it acquires a certain different velocity and the direction changes.

In Fig. 13, D represents a point experiencing a series of sudden changes of velocities. The graph is a succession of short straight lines forming a broken line or open polygon. As the changes of velocity occur more and more frequently the sides of our polygon become smaller and smaller; and in the limit, as the changes of velocity become continuous, our broken line becomes a smooth curve E.

Motion with continuously changing velocity is called accelerated or retarded motion. The rate of change of velocity is called acceleration and is represented by the second derivative of the distance with respect to the 'time'; symbolically, $A = dv/dt = d^2s/dt^2$.

It is important to notice that in space-time an *accelerated* motion is represented by a *curved* line. In uniform (constant velocity) motion the distances are proportional to the 'times', and the line is straight and its equation is of the first degree. In accelerated motion the distances are *not* proportional to the 'times', the lines are curved and the 'time' element dt enters in the second degree at least; namely, as dt^2.

For example, let us study the graph of the motion represented by the equation $x = At^2/2$ which means that the distance x is proportional to the square of the 'time'.

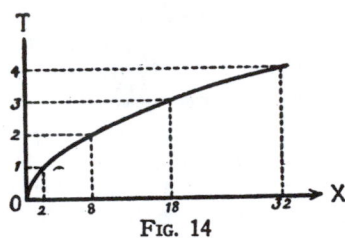

FIG. 14

Let OX (Fig. 14) be the 'spatial' axis, and OT the 'temporal' axis. We lay off on our T axis equally-spaced points, representing the seconds 1, 2, 3, 4 . , and calculate the distances x for each of these values from the equation $x = At^2/2$ where A represents a constant acceleration.

Let us assume that the constant acceleration A is given as 4 metres per second per second. The equation $x = 4t^2/2$ becomes $x = 2t^2$. Corresponding to the values $t = 0$, 1, 2, 3, 4 . , we have the values $x = 0$, 2, 8, 18, 32 , . If we plot these points, and assume that the change is continuous, we may join the points by a continuous *curve*, which represents the motion of the point as a *curved* world-line.

Similarly, in three-dimensional space-time, a point moving uniformly in the *plane XY* would be represented in the *plane XY* by the *line AB*, and in three-dimensional space-time by the static line AB', where the 'times' are proportional to the distance.

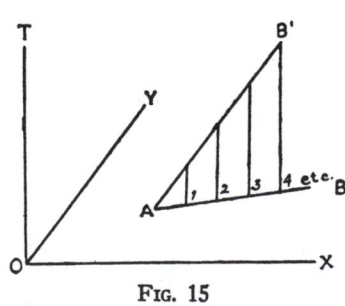

FIG. 15

As we have already seen, non-rectilinear motion may be considered as accelerated motion. We will generalize the above to the case where any *curved* path is traversed with *constant* velocity. In this case the direction of the velocity is changed. If we take the motion of a point which describes a circular orbit with *constant* velocity, it is easy to find its acceleration, which is called in this case centripetal acceleration.

Let us consider a point P, moving in a circular orbit with a *constant* velocity v, as given in Fig. 16. If at a certain 'time' it is at A, after a short interval t, it will be at B. The direction of the velocity will be changed from AA' to BB'.

If we construct the triangle DCE by drawing CD parallel and equal to AA', and CE parallel and equal to BB', we see that the angle $\angle DCE$ is equal to the angle $\angle A'A''B'$ because the sides are parallel, and it is also equal to the angle $\angle AOB$ whose sides are perpendicular to AA' and BB'. The triangles ABO and CDE are similar be-

cause they are isosceles and the angles between the equal sides are equal. Clearly the side $DE = w$, represents the supplementary velocity which transforms AA' into BB'. We know that in similar triangles the sides are proportional so we can write $DE/CD = AB/OA$. By inspection of our figures we see that $DE = w$; $CD = v$; $OA = r$, the radius of the circle. The chord AB may be taken as the arc AB of the circle, provided

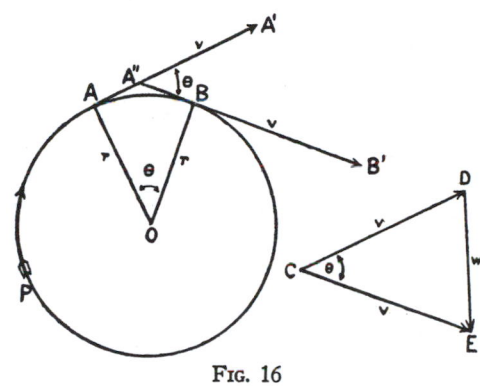

FIG. 16

the 'time'-interval is taken sufficiently small. Let us write chord $AB = s$. We have $w/v = s/r$ or $w = sv/r$. If we divide both sides of our equation by t we have $w/t = sv/tr$. But $w/t = A$, the acceleration, and $s/t = v$ hence $A = v^2/r$. In words, the centripetal acceleration is equal to the square of the velocity in the circle divided by the radius.

The above formula is of structural importance because it is the foundation for the empirical proof of Newton's law of gravitation. For our purpose it is important for other reasons, to be stated later.

There are two more diagrams which should be considered, in this connection. Fig. 17 represents the plane circular motion of a point P whose orbit in the plane XY is the circle PAB. In three-dimen-

sional space-time the plane circular orbit of motion would be represented by the static cylindrical helix (or screw-line) with axis parallel to the 'time' axis T. (Fig. 17.) We should note that the motion is *dynamically circular* in the XY plane, yet a three-dimensional space-time representation gives us a stationary helix.

Similarly for vibrational movements which could be represented in one dimension by to-and-fro movements on the X axis from A to B and from B to A. (Fig. 18.) If we introduce our space-time form of representation by introducing the T axis, our vibrational world-line would be represented structurally

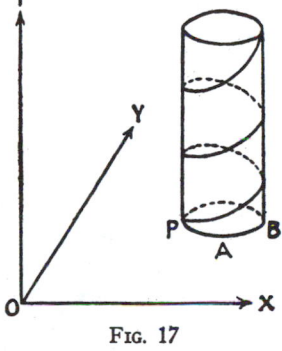

FIG. 17

by a wave-line along the T axis. In particular, if the vibrational motion is simply harmonic, a proper choice of the 'time' unit makes the wave-line a sine curve.[9]

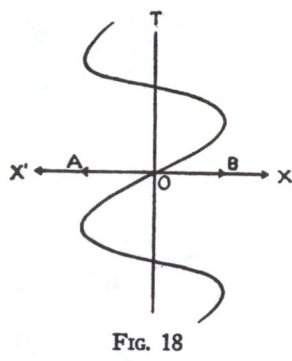

FIG. 18

Becoming thoroughly familiar with these few simple examples takes away a great deal of mystery from the Minkowski-Einstein and the new quantum world. We see that after all there is nothing extraordinary in the fact that in languages of different structures we get different forms of representations and pictures, and that in a world where accelerations abound we may very profitably use the term 'curved'.

When we come to speak about the Einstein theory, the four-dimensional space-time world of Minkowski, and the new quantum mechanics, we shall have considerable use for these few notions and illustrations.

Section D. The application of geometrical notions to cerebral localization.

In the present work we are dealing in the main with structure and the adjustment of the structure of our languages to empirical structures, and at this point it will be of use to suggest some of the consequences which follow from what has been said.

The question of cerebral localization is a difficult and vital problem. In former days it was supposed that the brain had individualized centres with strictly defined functions. Attempts were made to ascribe to definite cerebral parts definite functions such as 'memory', 'intelligence', 'morality', 'talents' , . In the meantime, experimental facts disproved such structural views, and as a reaction another tendency appeared; namely, to deny any localization.

Modern researches show unmistakably that both of these extreme tendencies are at variance with experimental structural facts. It appears that the lower centres play a more important role according as the terminal, or higher centres, are less developed and that there is considerable variability, at least in man, not only from the morphological and histological aspects but also from the functional aspect. It was found impossible to generalize from the particular development of centralization and functional distribution in one species to the distribution in another species. Localization may vary even in one individual under different circumstances.[10] Metabolism, and slight disturbances in the functioning of a neuron, were also found to have a most far-reaching influence, shown in its relations to other groups of neurons. The problems of localization are far too complex to attempt even an account of them, the more so since the reader will find excellent accounts of them in the large literature on the subject. The general conclusion reached by practically all investigators is that some localization of nervous function does exist, yet it has a certain variability which depends on an enormous number of factors.

The methods explained in this chapter will enable us to suggest a method by which we can orient ourselves in the bewildering complexity of the functioning of the nervous system.

One of the main difficulties is that the structure of this world is such, that it is made up of absolute individuals, each with unique relationship toward environment (in the broadest sense); and we have to speak about it in terms of generalities. 'Laws' . , formulated in the *old two-valued ways*, can never account adequately for the facts at hand, being only approximations. The mathematical methods which have already been explained give us at once a great advantage. We have seen that if we have a function, $y = f(x)$, let us say, and take the graph of this function, to every point of the graph there corresponds a pair of values x and y. We have seen also that each of the four quadrants I, II, III, IV has a characteristic pair of signs. In quadrant I, both x and y are positive; in II, x is negative and y positive; in III, both x and y are negative; and finally, in IV, x is positive and y negative. We can easily see that the value of the variables may be thought of as variable conditions different for each individual, and that definite *localizations* correspond to them. In our example we had to do with a function of one independent variable, and we had a one-dimensional line, curved in two dimensions. When we had a function of two independent variables we had a surface, which in general was curved in a third

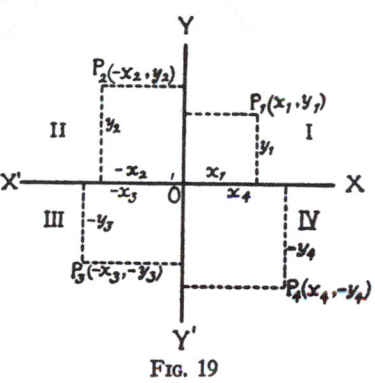

Fig. 19

dimension. By analogy we may pass to any number of dimensions, where by dimension we do not mean anything mysterious, but roughly the number of variables involved in the problem.

We see that if we think of the activity of the nervous system in terms of a mathematical function with an enormous number of variables, we shall not only have place for the uniqueness of each individual, determined by the value of the variables and the character of the function, but that this would also imply a *localization*, which is *permanent* in a given individual at a given 'time'; which again implies the totality of 'circumstances' , . Our function would be $N = f(x_1, x_2, x_3, \ldots x_n)$.

In fact it is hard to see how it is possible to analyse the activities of the nervous system in any other way. The facts are, that every organism is an individual, distinct and different from others, and so we must have means to take this individuality into account. Different values for different variables take care of this point. Similarities are accounted for by the general structural character of the functions. For instance, any quadratic equation with two unknowns gives us a conic section. An equation of the type $y^2 = ax$ represents a parabola, the graph of any equation of the form $xy = a$ represents a hyperbola , . For every definite set of values of our variables the implied localization is also definite, which corresponds to the fact that in a given individual at a given 'time' . , the localization is definite. One value for the whole function can be reached by giving different sets of values to the different variables. For instance, in the function $z = 5x - 2y - 1$, if $x = 1$ and $y = 1$, then $z = 2$; but we can have $z = 2$ by taking $x = 2$ and $y = 3.5$. Or if one of the conditions be non-existent, which means that the value of one of our variables is zero, for instance, $x = 0$, we still could have the value $z = 2$ by taking $y = -1.5$. This fact accounts for the *many-to-one* correspondences of causal factors, typified, for instance, by sunshine *or* cod liver oil producing a similar effect.

It should be understood that in what is said here, the numerical values do not matter. In most of the cases we are not advanced enough to be able to deal with such numerical values. What is to be emphasized is the structure of the language we use. The method should enable us, instead of dealing with generalizations in the old language, which somewhere have to be contradicted, to use a language of mathematical structure which shall account for the facts and leave room for the great individual varieties of organisms in structure and function.

After all, we should not be surprised that the theory of functions and language of functions is structurally appropriate in expressing, and so in understanding, the *functioning* of the nervous system, or any other system. Personally I have benefitted greatly through this method; and many baffling structural complexities have been much simplified.

Structurally, when we use the language of functions, variables., we automatically introduce *extensional* structure, as already explained, and we have at our disposal methods of translation of different orders of abstractions—dynamic into static, and vice versa—which is a neurological structural necessity for being rational and sane. And surely science should try to be rational. It should be stressed again that in our problem numerical values matter very little, but structure and method, for the many reasons already explained, are of paramount importance. Perhaps even the value of numbers is due mainly to the structural fact that it has forced upon us extensional and relational methods. It is the only language which is in accordance with the structure and functioning of the nervous system, and so helps to co-ordinate these activities instead of disorganizing them.

That these simple structural dependences have been discovered so late is really astonishing. The only explanation I can give of this is that we have been so engrossed in generalizing and generalizations that we lost sight of the fact that in life we deal structurally with *absolute individuals*, and that the only *language* which preserves the *extensional structural individuality* for its elements is found in mathematics—specifically, in numbers.

It may be that a study of mathematical structure and the psycho-logics of mathematics will give results of unparalleled human values, particularly for our sanity. The problems of sanity are problems of adjustment, and no means of adjustment should be disregarded. It may also be that the main importance of mathematics will be found some day to be more in the mathematical *methods* and structure which it has originated, methods forced upon the mathematician by the relational character of the entities he has to deal with, than in the possible combinations of these entities themselves.

At any rate, we must sadly admit that the problems of mathematical methods and structure and the psycho-logical values of mathematics have so far received very little attention, since we have failed to realize their human importance. In the future this problem will be further, and thoroughly, investigated.

PART IX

ON THE SIMILARITY OF EMPIRICAL
AND VERBAL STRUCTURES

The theory of relativity has resulted from a combination of the three elements which were called for in a reconstruction of physics: first, delicate experiment; secondly, logical analysis; and thirdly, epistemological considerations. (457)   BERTRAND RUSSELL

The essence of Einstein's generalization is its final disentanglement of that part of any physical event which is contributed by the observer from that which is inherent in the nature of things and independent of all observers. (21)   E. T. BELL

Even Leibniz formulated the postulate of continuity, of infinitely near action, as a general principle, and could not, for this reason, become reconciled to Newton's Law of Gravitation, which entails action at a distance and which corresponds fully to that of Coulomb. (547)   HERMANN WEYL

This limitation to what is directly observable is ultimately based on Mach's philosophy and, directly inspired by Mach, led three decades ago to the propagation of the so-called theory of "Energetics," which sought to recognise only quantities of energy as physically given and observable quantities. (481)   A. SOMMERFELD

CHAPTER XXXV

ACTION BY CONTACT

> The difficulty involved is that the proper and adequate means of describing changes in continuous deformable bodies is the method of *differential equations*. . . . They express mathematically the physical conception of contiguous action. (45)   MAX BORN

The analysis of 'matter', 'space', and 'time' from the point of view of structure and of orders of abstractions has led us to far-reaching conclusions. Let us summarize the semantic results, and consider some of the immediate consequences.

We may begin by recalling the difference between the lower order and higher order abstractions. The lower order abstractions are given to us by the lower nerve centres. They are 'dynamic', 'continuous', non-permanent, shifting, unreliable, and above all *un-speakable*.

They have a character of immediacy, because, structurally in terms of order, they are closest to outside events. They come first in order in the functioning of the nervous system. We always associate with them some 'objectivity' as, by necessity, the eventual definition of an 'object' starts at this level.

It should be emphasized over and over again that, speaking correctly, *on this level* we cannot define anything, since abstractions on this level are fundamentally *un-speakable*. We may look, listen, handle, feel . , but *cannot speak* and therefore cannot define. The moment we *define* our objects, we are no longer on the level of lower order abstractions. By neurological structural necessity we have passed to the higher nerve centres (speech), and higher order abstractions. This is what is meant when we say that this lower level is *un-speakable*.

Because these lower order abstractions are closer to the outside events, and because they come *first in order*, they have a special character of immediacy, with which we *must start*. The struggle begins when, through some primitive-made doctrines or structural assumptions (metaphysics), we try to avoid going any further than these lower order abstractions. As a matter of fact, this is an impossibility, because of the very structure of our nervous system. However intensely we believe that it is possible to do so, and however 'emotionally' we attempt to do it, we are cherishing delusions, which easily become *morbid* identifications, delusions, illusions, and often hallucinations.

This level being *un-speakable*, the only way to function on this level is to look, listen . , but to be *silent* outwardly as well as *inwardly*. This last condition represents a most beneficial semantic state, really difficult, perhaps impossible, to acquire without training.

The higher order abstractions appear to be products of the activity of the higher nerve centres, further removed from the external events and lacking, therefore, in immediacy. But these higher abstractions are static and so may be analysed. They have a separable unit 'quantum' structure, which can be treated individually. It should be carefully noticed that the static character of these higher order abstractions is the origin of their separable quantum character, conditioned by the human nervous structure. They are, if properly treated, reliable and are uniquely responsible for our being time-binders.

Again, by the structural necessity of our nervous system, we deal first with lower order abstractions, and next with higher order abstractions. It must be noticed that *no one*, unless he is (pathologically) entirely deprived of the higher nerve centres, is, or can be, an exception. We all deal with lower abstractions first, and with the higher next, no matter how perfect or imperfect these abstractions may be.

The general confusion of orders of abstractions, the lack of theories, and therefore of structural understanding of the entirely different characters of these distinct orders of abstractions, leads to, and must result in, identification or confusion of orders of abstractions. As the different processes are going on, whether we will it or not, in every single one of us, they may result in the *delusional* ascribing of the characteristics of the higher order abstractions to the lower order abstractions, as for instance, permanence, immutability . , somewhere involving 'infinities'. When objectified, we have such semantic disturbances as fanaticism, absolutism, dogmatism, finalism . , which often become morbid semantic states.

A similar confusion may lead to the delusional ascribing of the characteristics of the lower order abstractions to the higher ones. Under such delusions we ascribe to the higher abstractions fluidity, shiftiness, non-permanence, 'non-knowability'. , which results in pessimism, cynicism, disregard for science, bitterness, fright, hopelessness and other equally vicious semantic disturbances. These in turn affect by structural necessity the proper working of the entire organism, which always works as-a-whole.

The A-system and other older systems were not only built before these facts became structurally known but were actually based on such confusion. Hence their viciousness. By building a language and a method of this nature they perpetuated and made effective *mechanically* through the structure of language, a harmful confusion. This language being not in accordance with the structure and functioning of the nervous system and the world, must produce pathological results somewhere.

We have already seen that the use of the 'is' of identity is unconditionally delusional. Naturally, attitudes (affective, lower order abstractions) which can assert, (higher order abstractions) that so and so on objective levels 'is' so and so, must lead to pathological results. In science this is a profoundly unsatisfactory state of affairs and needs structural revision.

Mathematicians, though in the main unconscious or innocent of the structural, semantic, and neurological issues involved, nevertheless have solved this problem by producing methods of passing from one order of abstractions to another, from dynamic to static, and vice versa. The influence of these discoveries has also affected the other sciences *unconsciously*. Without consciously recognizing it, the modern trend of science is to banish from its habits and methods the application of the 'is' of identity.

So in science we have to use an actional, 'behaviouristic', 'functional', 'operational' language, in which we do *not* say that this and this 'is' so and so, but where we describe *extensionally* what happens in certain *order*. We describe how something *behaves*, what something *does*, what we *do* in our research work , . If one asks, for instance, what *is* 'length', what *is* 'space', what *is* 'time', what *is* 'matter'. , the only correct answer would be, 'As you asked the question verbally, and I answer it *verbally*, the above *terms remain terms*, which beside structure, have no connection whatsoever with the external world'. Yet undoubtedly we are interested in this external world and we should like to use a language which would help us in understanding this world better. What shall we do? It seems that if we produce a language which is *similar in structure*, to the external world, somehow, as a map or picture is similar in structure to the region it pictures, we should have a uniquely appropriate language. How can we do it? It is quite simple the moment we discover the principle. First of all, abandon completely the *A* 'is' of identity, and, instead, describe ordered happenings in an actional and functional language. Such a language shares with the external world at least the multi-dimensional order of happenings, and it gives us a solution.

It is easy to see that arguments (verbal) about 'matter', 'space', 'time'. , will never become anything else than verbal. All uses of the 'is' of identity, must lead to delusional evaluation. The situation is radically changed when we use an actional or functional language, when we describe what a physicist does when he finds his 'length' or 'second' or any other entity he is interested in.

We should notice here that the above procedure involves extremely far-reaching structural and semantic consequences. First of all, we abandon the vicious use of the 'is' of identity, and eliminate the semantic disturbance called identification. We introduce automatically the full psycho-logical working mechanism of *order*, *extensional* methods and discrimination between the orders of abstractions. We introduce the four-dimensional and differential methods, we build up static units, 'quanta', and so introduce *measurement* and its language called mathematics, which leads to structure and so to knowledge at each date.

It will be useful to recall why mathematics and measurements are somehow so important in our lives. Our nervous system, as we have seen, exhibits different activities on different levels. On one level the abstractions are shifting, non-permanent; on the other static and permanent in principle. This is expressed in our lives in a longing for some permanency, some security, some 'absolutes'. Mathematics formulated this tendency first and with *full success*. Mathematics has not only formulated full and successful theories of 'change', as, for instance, the theory of functions and the different calculi, but also full-fledged and remarkable theories of *invariance* under transformations. These new theories of invariance are actually *absolute* formulations in the only sense in which the term 'absolute' has a meaning; namely, relative, no matter to what; all of which leads to the only content of knowledge—structure.

The whole Einstein theory should, in this sense, be called the 'theory of the absolute', and can be expressed as the simple demand that 'universal laws' should be formulated in an invariant form, a most revolutionary demand and yet so *structurally natural* that no one can deny it.

When we mathematize or speak about potential or actual measurements, we are dealing with *ordered, extensional*, actional, behaviouristic, functional and operational entities, and so we build up a language which at least has a similar *structure* to the external events. Numbers imply units, quanta, but also order. It seems that number is the only abstraction upon which we all must agree. We never doubt that a statement, such as that 'I have in my pocket five pennies', may be perfectly definite and ascertainable for all. The specific and unique relations called numbers seems to have absolute significance. It must be added that the existence of non-quantitative branches of mathematics does not alter what is said here. In these branches, the asymmetrical relation of order remains paramount and we may treat numbers from either of their two aspects, the cardinal or the ordinal.

The epoch-making significance of the Einstein-Minkowski work consists precisely in the fact that they were the first to *apply* the above, though without, it is true, formulating the general principle. The lack of such a general, $\bar{A}$, epistemological formulation retards considerably the understanding of their work, and so laymen miss the enormous structural, and semantic beneficial effect upon the proper working of our nervous system and our sanity.

Before giving a short methodological account of the Einstein theory it will be well to recall some structural and semantic conclusions which the differential calculus suggests.

When we were dealing with the notion of a variable, we saw that the variable might be *any* element selected out of an ordered aggregate of elements. We can select elements relatively widely separated from each other, as, for instance, the numbers 1 and 2, or points, let us say, an inch apart. It is obvious that if we choose, we can make the gaps smaller, and postulate an infinity of intermediate steps. When we make our gaps smaller, the elements are ordered more densely and closer together. In the limit, if we choose indefinitely many elements between any two elements, our series become compact, if we still have a possibility of gaps; or they eventually become what we call *continuous*, when there are no more gaps.

Without legislating as to whether the entities we use in physics are 'continuous', 'compact', or 'discontinuous', we may grant that the maximum elucidation of the above terms in mathematics is very useful. We can easily see that in terms of *action* a continuous series gives us *action by contact*, since consecutive elements are indefinitely near each other. As the differential and integral calculus were built on the structural assumption of *continuity*, the use of the calculus brings us in touch not only with our x but also with its indefinitely close neighbour $x + dx$. We see that the calculus introduces a most important structural and semantic innovation; namely, that it is a language for describing *action by contact*, in sharp contradistinction to the structural assumption of action at a distance.

Let us illustrate the above by a structural example. Consider a series of equal small material spheres connected with each other by small spiral springs as shown on Fig. 1.

FIG. 1

These little spheres all have inertia, because of which, and because of the little springs, they resist displacement. If we displace the first of our spheres either in the transverse or longitudinal direction, it acts upon the second sphere, which in turn acts upon the third , . We see that the disturbance of equilibrium of the first little sphere is transmitted like a wave to the next sphere and so along the whole series. The most significant point in the analysis of such a wave of excitation is that it is not transmitted with some 'infinite velocity', or 'infinitely quickly' or in 'no time'. The action of each sphere is slightly delayed owing to its inertia, that is, it does not respond 'instantaneously' to an impulse. It must be noticed that the displacement is not due to a velocity, but to an acceleration, which is a change of velocity and requires a short interval of 'time'. The change in velocity again requires an interval of 'time' to overcome inertia and produce displacement. Similar reasoning applies to a long train just being started by the engine. The cars being coupled together by more or less elastic means, the engine may be moving uniformly and some of the last cars still be stationary. The pull of the engine is *not* transmitted instantaneously but with a *finite velocity*, due again to the inertia of the cars.

We see that the only structurally adequate means of describing changes in continuous, deformable materials is to be found in differential equations which express a method of dealing with *action by contact*.

We have already seen that this action by contact involves also the *finite velocity* of propagation, a fact of crucial structural and semantic importance. In the history of science we can distinguish three periods. The first was naturally the period of action at a distance, the best exemplified by the work of two great men, Euclid and Newton. In it we find of course, a superabundance of 'infinities'. With the advent of the differential calculus, and the introduction of differential equations in the study of nature, the notion of action at a distance became more and more untenable. We had a period of pseudo-contiguous action, which indeed involved differential equations; but the *velocity of propagation* was not introduced explicitly, and so there remained an implicit structural assumption of 'infinite velocity' of propagation. As an example of such pseudo-contiguous action we can cite the older theories of potential, which give differential equations for the change in the intensity of the field from place to place,

41

but which do not contain members that express a change in 'time', and hence do not take into account the transmission of electricity with finite velocity.[1]

The modern theories, as for instance, the Maxwell theory of electromagnetism, and the Einstein theory, are based on *action by contact*. These theories not only use the differential method, but they also introduce explicitly the *finite velocity* of propagation.

The invention of differential geometry with the recent contribution of Weyl, which we have already mentioned, transforms the geometry of Euclid from a language of action at a distance into contact geometry, or a language of indefinitely near action.

It should be mentioned perhaps that the riemannian differential geometry is more general than all the $\bar{E}$ geometries which preceded it, and includes them, as well as the E geometry, as special cases. Perhaps, as Weyl points out,[2] the investigation of the famous fifth postulate, which was the beginning of $\bar{E}$ geometry, was accidental in importance and the main structural value of the $\bar{E}$ geometries lies precisely in the application of the differential methods to geometry which was originated by the great work of Riemann. This work, we see, has carried us from metaphysical action at a distance to a physical action by contact. In passing from the older mechanics to electromagnetic events a very striking analogy appears, which explains the finite velocity of propagation.

In mechanics, when we have waves in an elastic medium, the finite velocity of propagation is due to the delay which occurs due to the inertia of materials. Now inertia is determined by acceleration (d^2s/dt^2), which represents the rate of change of the velocity ($v = ds/dt$), velocity itself being a rate of change of displacement. We see that this retardation, or negative acceleration, is represented by a double differentiation.

Something analogous occurs in electromagnetic events. The rate of change of the electric field (de/dt) determines the magnetic field; and then the rate of change (dh/dt) of the latter determines the electric field at a neighbouring point. The advance of the electric field from point to point is thus conditioned by two differentiations with respect to 'time', which is quite analogous to acceleration.

It is due to this double differentiation with respect to 'time' that the formulation of electromagnetic waves are structurally possible. If the partial effects were to occur without loss of 'time', no propagation of the electric waves would occur. The maxwellian 'field equations' not only express the above-mentioned relations, but introduce structurally the finite velocity of propagation which makes the Maxwell's electromagnetic theory structurally a contact theory.

The Einstein theory is also structurally a contact theory, and it may be said that it was originated by this contact tendency, and has carried it to the limit, as we shall see later. The gaussian theory of surfaces, whose extension to any number of dimensions was made by Riemann, also represents action by contact. This theory does not state the laws of surfaces on a large scale, but only their differential properties, the coefficients of the measure determination, the invariants which we can form, and the curvature and its measure. The form of a surface and its characteristics can then be calculated by a process similar to the solution of differential equations in physics.

We are now in a position to understand why the newer physics and the $\bar{N}$ systems, which are built entirely on the foundations of action by contact, found the E-system unsatisfactory. The E-system was built on the structural assumption of action at a distance, and we had to select the $\bar{E}$ geometries as originated by Gauss, Lobatchevski, Riemann, and others, which gave to physics the necessary geometry of action by contact.

But the question of action at a distance versus action by contact has also an experimental aspect which makes the latter theory more satisfactory.

Faraday (1791-1867) was not a learned academician, and he was much freer from scientific prejudices than any of his contemporaries. From a bookbinder's apprentice he became through his genius one of the founders of modern physics. His method of experimenting was to try every possible experiment and note what happened.

In 1838 Faraday made an important structural discovery; namely, that the mutual action between two electrically charged bodies depends upon the character of the intervening medium. Faraday established by this experiment that the capacity of a spherical condenser changes when another material is used as the separating medium, rather than air. He found that the capacity became twice as large when the medium was paraffin, three times as large for shellac, six times as large for glass, and about eighty times as large for water.

This experiment became the foundation of the new theory. The old 'action at a distance' theory postulated that the electrostatic field was merely a geometrical structure without physical significance, while this new experiment showed that the field had physical significance. Every charge acts first upon its immediate surroundings, and it is only through the medium of these that the action is propagated. The discovery of displacement currents necessitated an extension of his point of view to all distances.[3]

Faraday was so impressed by this discovery that he abandoned the older theories of action at a distance and formulated a structurally new theory of contiguous action for electric and magnetic events. Any one

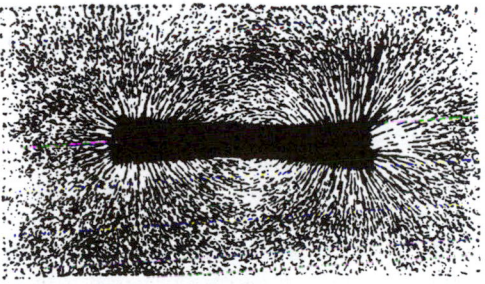

FIG. 2

can convince himself of the fact that the fields represent very actual physical conditions by taking a sheet of paper, sprinkling some iron dust upon it and putting a magnet under the paper. He will find that the particles of iron dust arrange themselves in a very definite structure as shown in Fig. 2.

Faraday also discovered that the forces between two magnetic poles likewise depend on the medium that happens to be between them. He concluded correctly that the electric as well as the magnetic forces are produced by a state of tension in the intervening medium.

These two examples will suffice as illustrations, but it can be said in general that all modern physics gives ample proofs of the correctness of Faraday's structural point of view. Some physicists, for instance Helmholtz, built special devices to test the correctness of this theory. As a matter of fact the success of the whole electromagnetic theory of Maxwell, which is structurally built as a contact theory, in which the velocity of propagation is considered finite, is in itself one of the best proofs of the correctness of the theory.

The finite velocity of light was discovered by Olaf Römer in 1676 and has since been repeatedly verified. This velocity is usually denoted by c and is known to be approximately 300,000 kilometres per second, that is, $c = 3 \times 10^{10}$ cm./sec., or 186,000 miles per second.

In 1856 Weber and Kohlraush calculated a certain constant which appears in the electromagnetic theory, and discovered that the constant had the dimension of velocity, $[c] = \left[\dfrac{L}{T}\right]$, and that its numerical value was 3×10^{10} cm./sec., which is the exact value for the velocity of light. This fact led Maxwell to associate light with electromagnetic waves, a view justified by experiments. In 1888 Hertz not only established once more the interrelation of optics and electrodynamics but found that the velocity of propagation of electromagnetic waves is finite and exactly equal to the velocity of light.[4]

Outside the exact sciences the principle of action by contact is making but slow progress, perhaps because of A s.r and the lack of structural formulations of the general issues at hand. We are happy to find a notable exception in the biological work of Professor C. M. Child, who has laid down a foundation for $\bar{A}$ biology and his system is structurally based on action by contact. This $\bar{A}$ biology has been applied to neurology by Professor C. J. Herrick. This present work, being a $\bar{A}$-system, must follow the methodological and structural advances explained here, and the $\bar{A}$ biology and neurology founded by Child and Herrick.

It is interesting to follow up the structural merging of geometry and physics. There are certain Smiths and Browns who call themselves physicists. There are some rooms with various instruments, which are called physical laboratories. The activities of the physicists which interest us are twofold. First, these scientists come to their laboratories, manipulate their instruments, note the positions of some pointers, manipulate the instruments again, note positions again , . This represents the *un-speakable* level of activity. Whatever happens happens, but there is no speaking to be done on that level.

Later the scientist describes his experiments in words. Obviously there are two entirely different stages in building physics, which usually we do not distinguish.

Quite obviously the *un-speakable* level cannot be called 'physics', and so we must apply the term to the higher order abstractions on the *verbal level;* namely, to the reasoned verbal account of what the experimenter saw, or felt, or experienced, in general abstracted on the lower levels; summarized, generalized . , in higher orders.

Physics represents then a verbal discipline. Being verbal, it needs a language. What language shall we select? As we want to have a science called physics, we shall naturally try to use the most structurally correct language in existence, so by necessity we must look in the direction of mathematics.

In mathematics we find originally two entirely different disciplines. One we may call arithmetic; the other, geometry. Becoming acquainted with these two originally separated languages, we find that the actual experiments and the stimuli for many experiences of importance to us, are outside our skins; so we try to choose the one of these two languages which is the more closely related in structure to the lower abstractions—that is, to what we see, feel , . Naturally we have an inclination toward the geometrical languages, dealing with 'lines', 'surfaces', 'volumes' . , terms for which we find immediate and quite obvious applications. By further investigation we find that of late both languages have become so developed in structure, that either can be translated perfectly into the other. This fact makes geometry the link between the higher order abstractions and the lower order abstractions. We have seen that *physics,* as well as geometry, must be considered *verbal* disciplines and their fusion becomes a very natural fact.

It is true that, as yet, 'time' appears as the bothersome factor, but 'time' may very well be represented geometrically, except that our diagrams and figures look a little different. For instance, a *flat* circular orbit in two-dimensional 'space' becomes a helix in three-dimensional space-time, a vibrational motion in one-dimensional 'space' becomes a wave-line in two-dimensional space-time , . 'Time', when properly represented, becomes simply another geometrical dimension.

It should not be forgotten that mathematicians obtain most of their structural inspirations from physics and build up mathematical theories to supply the structural needs of the physicist. We see an excellent example of this in the $\bar{E}$ geometries. In the days of Euclid, when physics hardly existed, we had 'emptiness', 'action at a distance', and such notions as were quite satisfactory for the needs of surveyors and builders. With the development of astronomy and physics, curved lines became more and more structurally important, and the haziness of the definition of 'straight line' also became apparent. The notion of 'emptiness' also became slowly structurally untenable. Such geometers as Gauss, Lobatchevski, and others, began to demand that the axioms of geometry be tested by experiment. With the introduction of 'curvature', the 'straight line' became only a special case of a curve with zero curvature.

The invention of the differential calculus also had a tremendous structural influence. It introduced continuity as a basic assumption in the vast structure of science, and cleared the way for psycho-logically trained scientific workers in structural continuity, and therefore in *action by contact*.

The discovery that light appears as electromagnetic waves and the finite velocity of both, made the notion of 'absolute emptiness' structurally untenable; and so E geometry with its action at a distance, 'emptiness' and neglect of gravitation and electricity, became very unsatisfactory. Indeed if our universe were E, light could not reach us.

Leibnitz, who invented the differential calculus independently of Newton, formulated a postulate of action by contact, and therefore could not become reconciled to Newton's Law of Gravitation which was structurally a law of action at a distance, corresponding fully to Coulomb's law in electricity. The latter law states that the force exerted by two electrically charged bodies upon each other is inversely proportional to the square of the distance between them, and acts in the direction of the line joining them.[5]

The introduction by Faraday of the structural notion of a 'field', instead of the notion of electrical charges acting at a distance, introduced the notion of a strain of the electrical field, which appears structurally as 'lines of forces'. Here we already have a 'fulness' of 'lines' and a big step toward the structural fusion of physics with geometry has been taken.

The transition from E to riemannian geometry corresponds structurally to the transition from physics based on action at a distance to physics based on action by contact. The fundamental metrical theorem of E geometry is the pythagorean rule, which expresses the fact that the square of the distance between two points is a quadratic form of the co-ordinates of the points. If we regard this theorem as strictly valid only in the case of points which are very near each other, we pass at once from E geometry to differential geometry. By doing so we gain a notable structural advantage, as we dispense with the necessity of defining our co-ordinates more precisely; because the pythagorean law, when expressed in differential form, is invariant for arbitrary transformations.[6]

Semantically, Riemann was the immediate predecessor of Einstein, although Einstein was not directly influenced by him. In differential geometry we ought to start with indefinitesimally near points, and depend for the analysis of greater distances, areas and volumes, on integration. The difficult notion of 'straight line' has to be replaced by the notion of the shortest line (geodetic), which is easily defined by differential methods and found empirically. In the older method, the length of a curve was to be found, in general, by the process of integration. The length of a 'straight line' between two points was supposed to be defined as a whole, and not as the limit of a sum of indefinitely little bits. Riemann considered that a 'straight line' does not differ in this respect from a curve. Measurements which are always performed by means of some instrument are *physical* operations, and their results depend for their interpretation upon the theories of physics. Dealing with geodetics is therefore preferable to dealing with 'straight lines'.[7]

We see that the problem was ripe for a final stroke of genius. Einstein's structural discovery of the dependence of 'space' and 'time', and Minkowski's success in giving a geometrical interpretation to the Einstein theory accomplished the probably irreversible fusion.

Three-dimensional kinematics becomes four-dimensional geometry, *three-dimensional dynamics* can be considered as *four-dimensional statics*.

We see immediately the human, psycho-logical, semantic and neurological importance of this fact. Our nervous system by its structure produces abstractions of different orders, dynamic on some levels, static on others. The problems of sanity and adjustment become problems of translation from one level to another, for which the structural advances in science supply us with methods of solution.

It should be noticed that the semantic gain due to the above facts is considerable, and that being structural, it is practical as well as theoretical. The fact that geometry has lost its old restricted status, which formerly applied principally to what could be 'intuitively visualized' and has been further abstracted to apply to what can be 'conceived', has merged geometry with the rest of mathematics. This merging represents a great structural and semantic step forward, and makes possible the treatment of geometrical problems by purely analytical means. It liberates geometry from the restrictions of lower order abstractions. By using 'geometrical intuition' (lower order abstractions) we find again a great help in analysis.

In the cyclic nerve currents, our so-called 'intuitions' (lower order abstractions) are not structurally isolated from our 'conceptions' (higher order abstractions), but both are intimately connected and influence each other. Modern advances are not only in perfect accord with the 'organism-as-a-whole' principle, but indeed give us excellent proofs that this principle is sound. 'Psychologists' miss a great deal by disregarding this important and unique form of human behaviour which we call mathematizing.

CHAPTER XXXVI

ON THE SEMANTICS OF THE EINSTEIN THEORY

It is precisely here, in an improved understanding of our mental relations
to nature, that the permanent contribution of relativity is to be found.
We should now make it our business to understand so thoroughly the
character of our permanent mental relations to nature that another change
in our attitude such as that due to Einstein, shall be forever impossible.
(55)   P. W. BRIDGMAN

It is not my aim to expound the Einstein Theory as such. There are many
excellent and competent books written on this subject. I have already explained
and stressed several structural points which in the last analysis are the foundation of Einstein's work. Many 'thinkers' through the ages have felt vaguely the
dangers of the structure of language and the viciousness of objectification, that
is, of the delusional ascribing of objective values to verbal forms. This vague
feeling, of course, is useful in individuals, but it is a private benefit, which
cannot be made public without some sort of formulation. The stroke of genius
of Einstein was that he produced a *non-elementalistic*, linguistic system of new
structure. Einstein, being a physicist, decided rightly, as we understand now,
to be entirely actional, behaviouristic, functional, and operational, and to stop
gambling on words. The older *el* linguistic problems of 'matter', 'space', and
'time' were in such a mess, due to the objectification of verbal structures, that
it was useless to talk any more in the old way. He decided to describe what a
physicist *does* when he measures 'space' and 'time', and to abandon, perhaps
unconsciously, the 'is' of identity.

It seems unnecessary to stress the simple fact that when we measure a
piece of wood, for example, we mark it off with another piece of material which
we have accepted arbitrarily as our 'unit of length'. The coincidence of our
'unit' with the intervals between the marks is again judged by an extremely
complex electromagnetic-neural process, which was quite disregarded until
Einstein. Our judgement is conditioned by the light rays travelling with finite
velocity which excite our nervous system through the retina, this excitation in
turn also travelling with finite velocity. We see that the apparently simple
measurement of a 'length' is really an extremely complex process, in which the
finite velocity of light and of the nerve currents plays a very important role.
Naturally, if we were to assume an 'infinite' velocity of the propagation of light,
our verbal speculations about 'space' and 'time' might be perhaps entertaining,
but they would nevertheless be fundamentally and structurally wrong.

Similar remarks apply to the measurement of 'time'. What do we mean
when we say that a train has arrived at the station at 9 o'clock? We mean no
more and no less than that the arrival of the train coincided with the arrival
of the pointer of a clock at a point marked 9 on the clock face. In other words,

648

we saw 'simultaneously' the arrival of the train and the pointer of the clock reaching the mark 9.

Our judgement about the results of measurements of 'time' depends on the *seen* coincidence of events—in this case, of the arrival of the train with the arrival of the pointer of the clock at the mark 9. Similar considerations, which applied to the measurements of 'lengths', apply also to the measurements of 'time'.

We see with Einstein that if we want to make any headway we shall have to investigate the two key terms; namely, 'velocity' and 'simultaneity'.

The newtonians take a particular delight in accusing Einstein of being a 'psychologist' and not a physicist. We have already stressed the physical subjectivity of physical instruments. What is said there applies, not only to the retina of the eye, but also to a photographic camera, or to a microscope or telescope, or any other instrument. Before an energetic packet, be it a light-impulse or a bullet, is able to accomplish any result it must first reach its mark, and so the finite velocity of propagation must be taken into consideration, which is a hard, established, empirical structural fact. So the criticisms of the newtonians are simply shallow and unscientific (1933). They disregard most important empirical physical facts, and so simply defend a semantic disturbance without aiding science (1933).

With the einsteinians, we treat the eye on the same footing as we would treat the camera or any other physical instrument. Even the newtonians must admit that when they photograph some happening on the sun, for example, the happening actually occurred (approximately) eight minutes *before* the photographic plate was affected. The eight minutes is the 'time' taken by the light to reach the earth from the sun.

Let us analyse the term 'velocity' first. We find ourselves here, as in any other human problem, on two distinct levels of abstraction, and we must discriminate between them.

Let us take up the verbal level first. We see that before we can talk about our terms 'space' or 'time', 'length' or 'seconds', we have to know a great deal about the term 'velocity'. How do we define the term 'velocity'? We define it as 'space divided by time', $v = s/t$. We see that on the *verbal* level the situation is perfectly hopeless and no result can be expected from verbal gambling. It may be added that older notions were based on objectification, or confusion between the two levels of abstraction, and the affective belief in the magic of words, identification playing most of the structural havoc.

How about the instrumental level, the silent level of the lower order abstraction? On this level, we find that physicists in their actions, behaviour, operations . , have elaborated a fairly definite technique for finding the data they require. So we see that there is no choice, we must *start* on this level.

But starting on this level is not all, and not enough. We must somehow *talk* about these doings and operations. Hence we must select a language which in its *structure* will reflect the structure of these actions and operations. Therefore we must abandon the 'is' of identity and *describe* in the asymmetrical language of order the happenings recorded by an instrument or by our lower nerve centres.

Without going into details we may summarize the results as achieved by the physicists. Experiments by the physicists, as indicated by the coincidences of pointers on different instruments, have seemingly established the fact that the 'velocity' of light, as defined by *behaviouristic, operational* instrumental means, is a *constant, $c = 3 \cdot 10^{10}$* cm./sec., independent of the relative velocity of the observers. By the 'observers' we mean again the readings on the instruments which the observer carries with him. Now this result contradicts flatly the established *verbal expectations* which we reached on verbal levels through the elementalistic structure of language and the semantic disturbance, of ascribing 'objective' existence, to the *terms* 'space' and 'time'.

The situation is acute. Shall we follow our semantic disturbances and reject hard empirical structural facts, or shall we accept the experimental facts and eliminate semantic disturbances?

As usual, the answer is implied by the method of putting the question. We accept the experimental facts and revise our semantic disturbances. In this case a psychiatrist might be a useful co-worker with the physicist.

The einsteinian revolution is structurally and semantically so fundamental, that every intelligent person should be acquainted with it. It will therefore be as well to consider some of its details.

In classical mechanics we had the classical mechanical principle of relativity; namely, that all mechanical equations have one form for two co-ordinate systems moving uniformly with respect to each other. The above has a very simple empirical meaning. If we travel in a train, let us say at a velocity of 50 miles an hour, all our activities in the train have one familiar relative velocity as if the train were at rest. If we throw a ball with a velocity of 20 miles an hour to another passenger on the train in the direction of the movement of the train, the ball will not reach the other passenger with the velocity of 20 miles an hour *plus* the additional 50 miles an hour velocity of the train but will reach him with the velocity as if the train were standing still. Not so, however, if the ball were thrown to an observer, standing on the tracks. The ball might hurt him, because it would have, relative to him, the velocity of 20 miles an hour of the ball, plus the velocity of 50 miles an hour of the train, or in all, a velocity of 70 miles an hour.

Quite probably, even our remote ancestors who used artificial means of transportation on land or water did not overlook the structural fact that mechanical events happen in just one way, whether the system is at rest or in relative motion. With the advent of *verbal* formulations of physics and mechanics, such happenings were formulated verbally; and so, slowly, the language of old structure with its consequent objectifications was built.

Now on *verbal grounds*, which seemed to be justified by experimental, macro-mechanical facts, we concluded that one law should prove valid in the case of electrodynamic and optical events.

To reformulate the above in simple symbols, let us imagine two parallel co-ordinate systems, O' and O'', of which the second moves with a velocity u relative to the first in the common x direction. If we denote the co-ordinates of the first system by single primes, and the co-ordinates of the second system by double primes, then, as usual, the co-ordinates of a point, P, in the second system would be connected with its co-ordinates in the first system by the equation $x'' = x' - ut$, which means that the x'' co-ordinate is less than the x' co-ordinate by the amount that our second co-ordinate system has moved; namely, by $a = ut$. We gave the diagram in two dimensions because it is simpler,

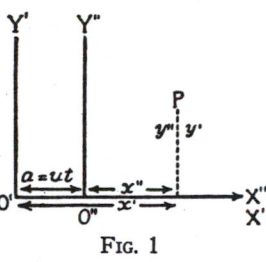

FIG. 1

and, as we have assumed that the displacement is parallel to the x axis, the other co-ordinates remain unaltered, $y'' = y'$, $z'' = z'$. 'Time' by the older assumptions, being 'objective' and 'absolute' would be 'the same'; namely, $t'' = t'$ ('absolute time'). The classical law of relative motion states that if the equation of motion in the first system is $f'(x', y', z', t) = 0$, this function must also be zero when x' is replaced by its new value; namely, $(x' - ut) = x''$ so that $f'(x'', y'', z'', t) = 0$.

Let us see if the above conditions hold true when we deal with the *propagation of light in spherical waves.*

$$OB^2 = OA^2 + AB^2$$
$$OB^2 = x^2 + y^2$$
$$OP^2 = OB^2 + BP^2$$
$$OP^2 = x^2 + y^2 + z^2$$
$$c^2t^2 = x^2 + y^2 + z^2$$

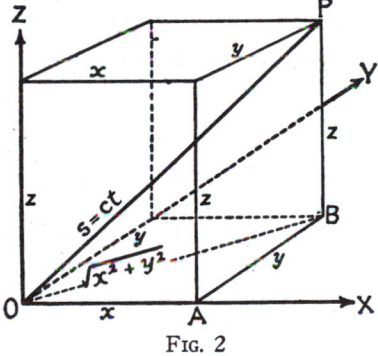

FIG. 2

If we select a three dimensional co-ordinate system O, the distance s of the point P from O is equal, by the pythagorean rule, to $s^2 = x^2 + y^2 + z^2$. If we assume that a light ray is travelling from O to P, the distance s could then be represented by the product of the velocity of light c by the 'time' or ct. The square of this distance would then be $c^2t^2 = s^2$. We have $x^2 + y^2 + z^2 = c^2t^2$, or $x^2 + y^2 + z^2 - c^2t^2 = 0$.

We can easily convince ourselves that if the last equation for light-waves holds good in the first co-ordinate system it cannot hold in the second.

Writing the last equation in our primed letters, we have $x'^2 + y'^2 + z'^2 - c^2t^2 = 0$. If we pass to our second system of co-ordinates moving uniformly in the X direction with the velocity u relative to the first system, our y', z', t, do not alter by assumption, but only $x'' = x' - ut$. We would have by substituting $x' - ut$ for x'', and retaining the primed values for y'', z'', t:

$$(x' - ut)^2 + y'^2 + z'^2 - c^2t^2 = x'^2 - 2x'ut + u^2t^2 + y'^2 + z'^2 - c^2t^2 \qquad (1)$$
$$= x'^2 + y'^2 + z'^2 - c^2t^2 + (u^2t^2 - 2x'ut).$$

But, by assumption, $x'^2+y'^2+z'^2-c^2t^2=0$, and therefore equation (1) cannot be zero unless $(u^2t^2-2x'ut)=0$. This last condition would mean that our second system of co-ordinates is also at rest. We see that for *light-waves* the older mechanical principle of relativity does *not* hold, as the equations *are altered* when we pass from one system of co-ordinates to another which moves with uniform velocity relative to the first.

To indicate this more obviously, we will express it in formulae. Consider two co-ordinate systems O' and O'', in which the second moves with a uniform velocity in the X direction relative to the first. If for the *light-waves* the equation $x'^2+y'^2+z'^2-c^2t^2=0$ holds in the first system, a similar equation for the second moving co-ordinate system, $x''^2+y''^2+z''^2-c^2t^2=0$, cannot be true. In other words, $x'^2+y'^2+z'^2-c^2t^2 \neq x''^2+y''^2+z''^2-c^2t^2$; whence we have an *inequality*, fundamentally contradicting the classical principle of relativity.

This extraordinary and unexpected inequality, because it contradicted structurally the classical mechanical principles of relativity, which apparently had been well established experimentally, created a baffling semantic situation which was profoundly unsatisfactory.

What could we do about it? Should we abandon the older principle of mechanical relativity; or should we have two different laws, one for the older gross macroscopic mechanical relativity, and another for optical and electrodynamic events; or should we investigate the fundamental structural assumptions which underlie our formulae, and see if the discrepancy is not due to some prejudice or some structural dogma which we have overlooked for centuries?

If a solution of the last kind should be found it would naturally be most satisfactory. The admission of two relativities, one for the mechanical events, the other for the optical events, would be against the whole trend of science, which requires the unification of theories.

Such a structural revision; namely, the rooting out of the old unjustified dogma which made all the trouble, was the work of Einstein's genius. In this epoch-making discovery he was *assisted* in the beginning by the famous Michelson-Morley experiment, since performed repeatedly with similar results, *seemingly* proving that the velocity of light is a constant no matter what the relative motion of the observer. If we take the equation for the spherical propagation of light-waves $x^2+y^2+z^2-c^2t^2=0$ or $x^2+y^2+z^2=c^2t^2$;

then $$c = \frac{\sqrt{x^2+y^2+z^2}}{t}$$ in one system of co-ordinates,

and $$c = \frac{\sqrt{x'^2+y'^2+z'^2}}{t}$$ in another system of co-ordinates,

are equal, which stated in another form would mean that
$$x^2+y^2+z^2-c^2t^2 = x'^2+y'^2+z'^2-c^2t^2.$$
The experiment says this relation is true; the arguments already advanced show it cannot be true. So we have to hunt for some error or compensation.

With Einstein's explanation, the finding of the error is simplicity itself. In the older mechanical relativity our 'space' and 'time' were *objectified*, we endowed them with objective values of definiteness and rigidity, we dealt with 'absolute space' and with 'absolute time', which was 'unchanging' and 'the same for all'. In the older assumptions our *velocities varied*. If A had a velocity of 5 centimetres per second, for instance, and B was overtaking A with a velocity of 7 centimetres per second, the relative velocity between A and B would be $7-5=2$ cm. per second. The *units of 'space' and 'time' were definite, immutable and did not and could not vary*, which followed directly from the assumptions of an 'objective' 'absolute space' and 'absolute time'.

In the case of light, we came in contact with a velocity which did *not vary* for any observer no matter what his relative motion. The velocity c was found to be constant, so the natural assumption to make is that our *'space' and 'time' vary* for different observers.

In the above equations as they stand, 'absolute time', $t=t$, the 'same' for all observers is assumed, which made such equality impossible. Assuming different 'times' for different observers, t for the first, and t' for the second, such a compensation transforms our inequality into an equality, as demanded both by the experiment and by the theory. Instead of writing

$$x^2+y^2+z^2-c^2t^2 = x'^2+y'^2+z'^2-c^2t^2, \ (t=t),$$

which *cannot be true*, we write,

$$x^2+y^2+z^2-c^2t^2 = x'^2+y'^2+z'^2-c^2t'^2 = 0, \ (t \neq t')$$

which can be true. We should notice that in the first equation we have on both sides t, which makes the equation impossible, whereas in the second equation we have on the left-hand side t and on the right-hand side a different t; namely, t'.

The above considerations mean that there is a definite structural discrepancy between the old language and the empirical world, requiring a fundamental structural linguistic revision. This revision has been accomplished, and is known as the Einstein theory. It is not implied that Einstein's work is final, but that it shows clearly the structural errors of the old elementalism to which we can never return.

In other words, in the older mechanics we had definite and permanent 'time' (absolute) and varying relative velocities. Dealing with *light-waves* we find experimentally that the velocity, c, of light does *not vary* with the relative motions of the observers and we must assume a *variable time* to preserve our equations.

An obvious objection can be raised to this: why alter our habitual notions of 'time'? Can we not keep the old *s.r* and find some other method of compensation, less bothersome and less revolutionary? The older physicists and Einstein give a long and convincing list of perfectly sufficient reasons for such a change; yet their arguments always leave us somehow in doubt, with the feeling of a lurking possibility that the old can be preserved.

What has already been said in this work about structure and semantic disturbances and the fact that the *terms* 'matter', 'space', and 'time' *are not objects*, which they cannot be, removes perhaps for good and all, the last doubt as to the revolutionary and epoch-making significance and value of the structural linguistic discoveries of Einstein. On these grounds *alone* the return to the old is impossible. The old is due to objectification of the structural peculiarities of the old *el* language, and to semantic disturbances, which at the present low level of our development is inevitably the result of *copying lower animals in our 'thinking'*, a *pathological* process for 'man'.

It will be well to explain at this point why I said that the Michelson-Morley experiment only *assisted* Einstein, and only *seemingly* proved the constant velocity of light. Historically, there is no doubt that the beginning of the theory of Einstein was suggested by, and had its physical basis in this experiment. In reality, as the whole of this present work about structure shows, the two issues are quite independent. The fact of the *finite* velocity of light has never been challenged, on the contrary it is becoming more and more solidly established, both empirically and theoretically, simply because an 'infinite velocity' has no meaning.

With the structural results of this present work, and the establishment of the fact of the finite velocity of light, the whole Einstein *theory* has a perfectly solid structural, linguistic foundation (1933). Nevertheless it is extremely gratifying that the latest, very important, and painstaking work of Doctor Roy J. Kennedy seems once more to add fundamental experimental support to the correctness of the Einstein theory.* From the point of view of structure, Einstein merely eliminated some primitive, perhaps even animalistic, remains of objectification which still lingered in the structure of our language of 'matter', 'space', and 'time'. These, being animalistic, were unfit for humans; vitiating not only our daily lives but science as well. (Eddington in *The Mathematical Theory of Relativity*, p. 196, uses the term 'pre-human' in a similar connection.)

It must be recalled that the definition of velocity is connected in a *circular* way with 'space' and 'time'. That is, in the definition of the relation of velocity, ($v = s/t$), 'space' and 'time', the definition of any one of our three terms depends upon our definition of the other two; whence there are many possible ways of verbal adjustment.

As we saw, the mechanical verbal principle of relativity with which we are all familiar was not structurally able to account satisfactorily for a similar relativity of optical and electrodynamic events. The older formulae of transformations were, as already given, $x' = x - ut$, $y' = y$, $z' = z$, $t = t$. These formulae are called the Galileo transformations in honour of the founder of mechanics; and, as we have seen, structurally they are not general enough.

If we consider the equation $x^2 + y^2 + z^2 - c^2 t^2 = 0$ and $x'^2 + y'^2 + z'^2 - c^2 t'^2 = 0$, we find that the galilean transformations do not satisfy them. Lorentz and Einstein have found another set of transformations which satisfies uniquely the above equations. These formulae of new structure are called the Lorentz-Einstein transformation, and are given by the following equations: $x' = \beta (x - vt)$, $y' = y$, $z' = z$, $t' = \beta (t - vx/c^2)$; where v is the relative velocity of one system with

---

*See 'The Velocity of Light', in *Nature*, Aug. 20, 1932, by R. J. Kennedy, and his latest paper (No. 261) in the Bibliography.

respect to the other, c, as usual, represents the velocity of light, and the factor $\beta = 1/\sqrt{(1 - v^2/c^2)}$.

The most striking characteristic of these formulae is that if we assume that c, the velocity of light, is '*infinite*', all the expressions containing c^2 would become zero, c^2 entering only in the denominators of fractions. In such a limiting case $\beta = 1/\sqrt{(1 - 0)} = 1/1 = 1$ and $x' = (x - ut)$, $y' = y$, $z' = z$, $t' = t$ which are the older galilean transformations.

Thus there appears the astonishing fact that all the pre-einsteinian physics and mechanics which involved the structural assumption of the galilean transformation, had a *tacit structural assumption* of the infinite velocity of light. This assumption, *known since 1676* to be false as to facts, remained unnoticed before Einstein.

As $c = 3 \times 10^{10}$ cm./sec., $c^2 = 9 \times 10^{20}$ is a very large number, whence the fractions vx/c^2 and v^2/c^2 are very small, and β differs very little from unity.

If we apply the Lorentz-Einstein transformation instead of the older galilean transformation to mechanical problems, the changes are so small that they can hardly be detected by experiments, the terrestrial velocities v^2 or vx being so small in comparison with the square of the velocity of light.

The galilean transformations are *experimentally* shown to be structurally invalid for optical and electrodynamic events. The Lorentz-Einstein transformations satisfy structurally the optical and electrodynamic events, and also apply to the older mechanical problems. We see that the Lorentz-Einstein transformations are *more general*, as they include the galilean transformations as a particular case when we assume $c = \infty$.

In a few instances, where we deal with large velocities, the values of the fractions containing the square of the velocity of light become appreciable and allow experimental testing. As yet all such experiments have verified the Einstein theory.

We should repeat again that the achievement of Einstein was the building of a linguistic system similar in structure to the world, which eliminated a pathological pre-human factor of objectification of terms. Such structural elimination was bound to bring some sanity to our theories; and this fact is independent of experiments in physical laboratories. However, it is gratifying to find that experiments support (1933) the Einstein theory. It was particularly gratifying in the beginning, when physicists and Einstein himself believed that his theory would stand or fall by experiment. Today we see that this theory represents such an enormous general, structural, epistemological, psycho-logical, and methodological *non-elementalistic* advance, that no matter what the experiments show or may show in the future we cannot return to a language of the old, *el*, obviously wrong, structure of the pre-einsteinian days. As usual, the negative results are the important ones. No matter what experiments may show we shall never again accept the silent structural assumption of 'infinite' velocity for light, when we know positively that the velocity is finite. We shall never again treat *terms* of 'matter', 'space', and 'time' as objects—lower order abstractions, when we know that they represent *terms*—higher order abstractions. When once this is realized, we cannot ascribe 'finiteness' or 'infiniteness', 'definiteness', 'rigidity'. , to *terms*, verbal forms, forms of representation. From this point of view we may consider the Einstein theory as an irreversible gain. If it had achieved only the elimination of various structural prejudices and dogmas, it has done well; and at least this much Einstein has already achieved.

The structural, verbal, cortical quest for invariance in our formulations also becomes apparent. The older mechanics were invariant under the galilean transformation, equations preserved their form in different systems of co-ordinates. In the special theory of relativity the new laws are invariant under the Lorentz-Einstein transformation. In this special, or restricted, theory of relativity only uniform relative motion was taken into account. If we generalize the principle of relativity to *any* kind of relative motion we pass from the restricted to the *general* theory, which demands that the laws of physics should be formulated in a generally invariant form for any arbitrary transformations.

For this structural, cortical reason it is necessary to express all the laws of physics in tensor equations, which satisfy such conditions of general invariance. If this cannot be done, there must be something wrong with our language, as such, and with our verbal laws. We require structural revision of those laws so as to be able to express them in tensor equations. The newtonian law of gravitation and the older form of the law of conservation of energy are perhaps the most remarkable examples. They do not survive such minimal, and yet entirely justified, structural requirements as those of the general theory of Einstein, and therefore they cannot be structurally satisfactory.

We have already seen that the equation $x^2 + y^2 + z^2 = c^2 t^2$ or $x^2 + y^2 + z^2 - c^2 t^2 = 0$ represents the equation of the spherical *propagation of light* with the finite velocity c. The discovery that the velocity of light is a universal constant for all observers, and the above equation, led historically to the re-discovery by Einstein of the Lorentz transformation which, as we have seen, has assumed such overwhelming structural importance. The meaning of these facts is worth considering.

In Chapter XVII we analysed briefly the elementalistic language of 'matter', 'space', and 'time', and came to the conclusion that to eliminate objectification we must abandon the semantic disturbance and the use of the term 'is' of identity. Instead, we must use an actional functional language to describe ordered functioning, behaviour, or operations. By necessity we were led to a 'contact' method. We also discovered that in accepting the above structural methods we were compelled to discriminate between different orders of abstraction, since what we see, feel, and experience is *not* what we say about it. We found that on the 'objective' level of our actual activities, (manipulating instruments . ,) which represent the silent *un-speakable* level, we could never find a situation in which the old language of 'matter', 'space', and 'time' could be used without coming violently into conflict with the properly analysed facts. We came to the conclusion that this language was not structurally satisfactory, for *verbally*, 'space', 'time', and 'matter' were supposed to be quite clear-cut and *separate* entities, while in actual experience we *never could find such separated* objective entities. It became obvious that the *structure* of the old language of 'matter', 'space', and 'time' was *different* from the structure of the outside world as we now know it. We found ourselves in a situation where we had to choose either to keep the old language which, as it differed from them in structure, could never give a coherent account of facts at hand, or else to build up a new language with *structure* similar to that of the outside world, in order to have the possibility of coherent conversation about it.

The invention of such a new language is of course an extremely difficult undertaking. In fact, it requires some genius to invent new, more structurally similar, forms of representation for the old facts. Lorentz, Einstein, and Minkowski prepared and finally produced such a structurally new language. The difficulty was that verbally we had already separated what empirically could not be separated. The problem was to amalgamate somehow the old structurally *elementalistic* language of 'space' *and* 'time' into a *non-elementalistic* language. The key to such an amalgamation is found in the light-wave equation which gives us the structural information about the world, $x^2 + y^2 + z^2 = c^2 t^2$.

This equation represents an equality. The left-hand side is expressed in 'spatial' terms only—the *distance* between two points O and P. The right-hand side expresses the 'spatial' length, but in a 'temporal' term. We see that here we have means of translation, and a possibility of amalgamation of two *elementalistic* languages, which were not supposed to be intertranslatable.

The Lorentz-Einstein transformation formulae are $x' = \beta\ (x - vt)$, $y' = y$, $z' = z$, $t' = \beta\ (t - vx/c^2)$ where v is the relative velocity of the two systems of coordinates, c, the constant velocity of light, and $\beta = 1/\sqrt{(1 - v^2/c^2)}$.

The formulae for x' and t' which typify, on the left-hand side, x' a 'spatial' *length* and t' a 'time', are of particular interest. We see that on the right-hand side of the expressions the value of the 'spatial' x' is given by $\beta\ (x - vt)$ which involves *'time'*. The value of 'time', t' is given by $\beta\ (t - vx/c^2)$, which involves the 'spatial' *length* x. So we see that our amalgamation is complete, and separation impossible. The above formulae express structurally the *simple experimental* fact that 'space' and 'time' cannot be separated. At this point we are not ready to discuss 'matter'. This will be considered further on in this work (see Chapters XL and XLI).

The above formulae have also a very important physical and experimental meaning, as they introduce the 'contact' methods into our language. Our actual measurements of 'space' and 'time' are strictly connected with readings on some instruments, and involve therefore coincidences between pointers and 'simultaneity'. In all instances the finite velocity of propagation of signals must be taken into consideration. When our instrument, or the eye, is affected by signals there is always a delay due to the finite velocity of the propagation of the signals. These delays are part and parcel of our experiment, and so our formulae must contain terms explicitly involving this finite velocity of propagation. This innovation involves not only a most profound structural epistemological and semantic revolution but supplies the very factor that enables us to formulate more structurally satisfactory languages (theories), which Lorentz, Einstein, and Minkowski have produced.

42

We have been contrasting finite and 'infinite' velocities. Let us say frankly that 'infinite' velocity is a polite way of speaking about blunders of observation. 'Infinite' velocity is meaningless. Velocity is defined as $v = s/t$ and if t is taken as zero or in other words, if one of the fundamental factors in our *definition* is lacking, our *definition ceases to define the term in question*—in this case, velocity. So when the term 'time' is lacking, we have *no velocity*, by definition; so, to speak or speculate about 'infinite' velocity is simply making noises, and not saying anything. The *negative* of this noise; namely, saying that velocity is *not* 'infinite', or in a positive sense, that velocity is 'finite', is on a different verbal footing, although it remains a polite invitation to stop talking non-sense.

It should be noticed carefully that the *general* theory of Einstein is a high structural generalization of the special theory; and that both of them are generalizations of the classical mechanical principle of relativity. It is founded, not on the introduction of any extraordinary structural assumptions, but on the elimination of some unjustified and false-as-to-facts structural assumptions, such as that of the 'infinite' velocity of light.

Both the theory of Einstein, and the theory presented in this work are long overdue. The Einstein theory could have been formulated as soon as we discovered the *finite* velocity of light, in 1676. It should be noticed that this last discovery was also overdue, as it did *not* require experiments to establish the finite velocity of light. It was sufficient to establish the meaningless character of 'infinite' velocity, which on symbolic grounds, could have been accomplished much earlier, and to conclude, that the velocity of light *must* be finite. This example shows the hampering, blocking, semantic effect which different meaningless verbal structures have on us. To express this high and satisfactory structural generalization, Einstein had to select the most general and structurally appropriate language in existence. He chose at some stage of his work the language of $\bar{E}$ and four-dimensional geometries in general and that of the differential geometry and the tensor calculus in particular. In the latest field theory, Einstein and Mayer introduce a new *more general* and very revolutionary mathematical language where vectors and tensors in an n-dimensional spread may have m components.

At present it appears that two other very general mathematical disciplines will be used increasingly in the future. One of them is the *theory of groups;* the other is *analysis situs*. In the latter we study only these characteristics of figures that are unaffected (invariant) by continuous deformation produced without tearing. Two structural points are relevant for us in this connection: namely, that the analysis situs is fundamentally a *differential* and also an *ordinal* discipline, based on asymmetrical relations. In the next chapter, as an illustration of the actional, behaviouristic, functional, operational, differential, contact method a short account will be given of the way Einstein structurally treated 'simultaneity'. The elimination of the old structural dogma about 'simultaneity' resulting from the semantic disturbance of objectification of 'time', is one of the outstanding achievements of Einstein and is historically the beginning of his theory.

CHAPTER XXXVII

ON THE NOTION OF 'SIMULTANEITY'

So we see that we cannot attach any *absolute* signification to the concept of simultaneity, but that two events which, viewed from a system of co-ordinates, are simultaneous, can no longer be looked upon as simultaneous events when envisaged from a system which is in motion relatively to that system. (155)   A. EINSTEIN

In the older days we accepted as self-evident the structural assumption that there is sense in such a statement as that an event A on the sun was 'simultaneous' with an event B on the earth. We assumed also that the 'moments of our consciousness' had a universal 'meaning'. We tacitly assumed, for instance, that when we saw or photographed an event on the sun, that it happened just the moment we saw it. Such structural assumptions were rudely disturbed by the discovery of the *finite* velocity of light. Today we know that when we see or photograph an event on the sun, that event happened approximately eight minutes earlier, as it takes about eight minutes for the light from the sun to reach our earth. We begin to realize that the moments of our perceptions have no universal significance.

We inquire first what we mean structurally by simultaneity. We do not need to go into details. The application of functional and contact methods, even in the rough, will assist us. We can speak in terms of instruments. For instance, we can build a special, very fast moving picture camera, C, with two lenses D and E, at two opposite sides, and a calibrated film, F, running rapidly through the middle of the camera as shown in Fig. 1. If we focus our double camera on two flashes, A and B, occurring at 'equal distances', L, from the film, we say that the flashes occur simultaneously by definition if the pictures, a and b of the flashes A and B, appear exactly opposite each other on the film, or if we have *one* picture. If, under the conditions of the experiment, where the distances between the origins of the flashes and the film are equal, and our film is moving very rapidly, the pictures of the flashes do not occur exactly opposite each other, but one picture is separated from the other, then we have two pictures, and conclude, by *definition*, that the flashes are *not* simultaneous.

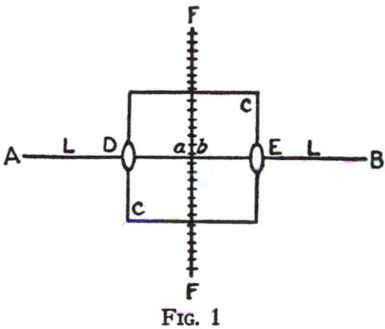

Fig. 1

We introduce this hypothetical instrument to show that, in discussing physics, and the theory of Einstein in physics, we do not speak of 'psychology' or personal 'subjectivity', but that we do deal with the inherent physical subjectivity of the instruments and the finite velocity of propagation. When we discuss the psycho-logical or methodological or semantic significance of science and scientific method, we deal with different subjects.

When we use the term 'observer' we mean an observer so equipped that he can do whatever is demanded of him.

What was said about the definition of 'simultaneity' by the aid of the camera, applies also to ourselves.

The problem of prime importance before us is to find out if 'simultaneity', as defined, has an 'absolute' and universal significance, or if it is perhaps a private and relative notion.

We will carry out the analysis in two ways, the first by example, which will be instructive, though perhaps not completely conclusive; the other, by the use of the Lorentz-Einstein transformation.

Let us perform our last experiment, which, with modern methods seems to be feasible, in a slightly more complicated form.

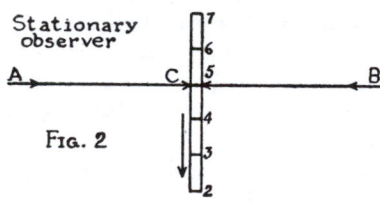

Fig. 2

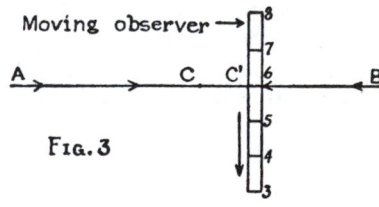

Fig. 3

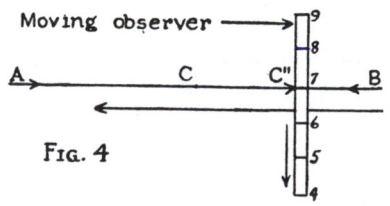

Fig. 4

We can select a dark night in which flashes will photograph well even at considerable distances. We can place powerful projectors at A and B, and we place our camera so that the film will come exactly at C, midway between A and B. We can start the mechanism of the rapidly moving film and, by an electrical contact made at C, we can produce a short flash from each of the two projectors. Because of the assumptions, $AC = CB$, and equal velocity of the propagation of electrical currents and light-waves in all directions we shall have by the structural definitions which condition the experiment, *one* picture in Fig. 2, say at the spot of our moving film marked by 5. The rays of light from A and B would arrive 'simultaneously'— that is, 'at the same time'—and would affect our moving film in *one spot*. Our definition was for a stationary observer, and under the conditions, the experiment was fairly definite—all the underlying structural assumptions, of course, being taken for granted.

Now consider an observer, as shown in Fig. 3, moving uniformly in the direction from A to B.

Let us assume that he is also equipped with a similar sort of moving picture camera as the stationary observer, and that just before he passes the point C the electrical impulse to the projectors is sent. Let us assume further that the mark 5 on his moving film is exactly at the focal point of the camera as C is passed. The electrical impulses travelling from C to A and B would travel the distance AC = BC, produce the flashes A and B which again would travel with finite velocity in all directions. During the interval these impulses and light-waves are travelling, our observer is moving from A toward B, and spot 5 on his moving film is no more at the focus of the camera. Obviously he will meet the light-wave from B first, at C', let us say, when mark 6 on his film is at the focus (Fig. 3). After another short interval when he reaches C'' and mark 7 on his film is at the focus, the light-wave from A overtakes him (Fig. 4).

So we see that what was 'simultaneous' (by definition) and produced *one* impression on the moving film of the stationary observer, was not 'simultaneous', (again by definition), for the moving observer, as *his* film registers *two* pictures.

As both observers use similar instruments and one set of definitions, obviously both are entitled to claim that their records on the film are conclusive. So the first can claim that the flashes were 'simultaneous', the second can claim that they were not 'simultaneous'. The reverse is equally true. If the moving observer had *one* picture, and claimed 'simultaneity', the stationary observer would have *two* pictures, and deny 'simultaneity'.

But when two observers are *equally justified* in making *two* opposing claims where, by their very meanings, there is only one possible, we must conclude that the claim itself is meaningless. We see that 'absolute simultaneity' is a fiction and impossible to ascertain, as it would depend on some impossible 'absolute motion', or 'infinite velocity' of propagation of signals.

The analytical form of showing the impossibility of 'absolute simultaneity' is very simple, and follows directly from the Lorentz-Einstein transformation.

Let us imagine two observers, one in an S system of co-ordinates (x, y, z, t) and another in an S' system of co-ordinates (x', y', z', t') moving relatively with the velocity v.

Let us assume two events happening in the unprimed system at the point (x_1, y_1, z_1) at the 'time' t_1, and the other at the point (x_2, y_2, z_2) at the 'time' t_2. According to the Lorentz-Einstein transformation the 'times' at which the two events occur relatively to the primed system are given by the formulae:

$t_1' = \beta (t_1 - x_1 v/c^2)$, $t_2' = \beta (t_2 - x_2 v/c^2)$, where as usual $\beta = 1/\sqrt{1 - v^2/c^2}$.

If we assume that in our unprimed system S the two events were 'simultaneous', which means that they 'occurred at the same time', t_1 would be equal to t_2, that is, $t_1 = t_2$, or $t_1 - t_2 = 0$. Let us find the difference between the two primed 'times' in the moving system S', and see if this difference is zero, which would mean that the primed 'times' are equal.

Returning to our formulae which give us the values for the primed system 'times', we express their difference as

$$t_1' - t_2' = \beta (t_1 - x_1 v/c^2) - \beta (t_2 - x_2 v/c^2) = \beta (t_1 - t_2 + x_2 v/c^2 - x_1 v/c^2). \qquad (1)$$

But we assumed $t_1 - t_2 = 0$; therefore $t_1' - t_2' = \beta (x_2 v/c^2 - x_1 v/c^2)$.

This last formula shows clearly that $t_1' - t_2'$ cannot be zero; or in other words, t_1' cannot be equal to t_2' unless $x_1 = x_2$.

The two events which, for an observer in the unprimed system, happen 'simultaneously', ($t_1 = t_2$, or $t_1 - t_2 = 0$) at different places, (or x_1 not equal to x_2, $x_1 \neq x_2$), cannot be 'simultaneous' for the moving observer in the primed system S', but will happen at different 'times' (t_1' is not equal to t_2', or $t_1' - t_2' \neq 0$).

It is extremely instructive to consider further what happens in measuring 'times' and 'lengths' in systems which are moving relatively to each other.

If, in the equation (1) above, we assume $x_1 = x_2$, this means that both events occur at one place in the stationary unprimed system S.

By changing the signs and cancelling the terms with x_1 and x_2, which are equal and of opposite signs, we have $t_2' - t_1' = \beta \, (t_2 - t_1)$ whence, substituting for β its value $1/\sqrt{1 - v^2/c^2}$, we obtain

$$t_2' - t_1' = \frac{t_2 - t_1}{\sqrt{1 - v^2/c^2}}$$

This last formula brings out a few remarkable issues. In terrestrial velocities the square of the velocity of the motion v^2 of the observer in the primed system S' is very small as compared with the square of the velocity of light c^2, so the fraction v^2/c^2 is small, $\sqrt{1 - v^2/c^2}$ differs very little from unity but the whole denominator is *less* than unity, and so $t_2' - t_1'$ is *not* equal to $t_2 - t_1$, but greater.

In other words, the interval of 'time' between the two events appears *larger* to the observer in the primed moving system than to the observer in the stationary unprimed system. In general, among all systems in a state of uniform relative motion, that one in which two events occur at *one place*, is characterized by the fact that the 'time' interval between the two events appears *shortest* to an observer in this system. The shortest interval means that to an observer in the system, the events run their course most rapidly. A process which, with reference to a given system, occurs in *one* place, appears to run its course most rapidly to an observer in that system, but more slowly to a moving observer in any other system.

The more rapid the relative motion, the slower the process will appear, and, in the limit, if an observer could move with the velocity of light, $v^2 = c^2$, the denominator of our equation would become $1 - 1 = 0$ and $t_2' - t_1'$ would become 'infinite' and all events would be at a standstill.

As the formulae for length, x and x', involve the 'times' and, as we see, the intervals of 'time' are dependent on the relative velocities, by a similar process of reasoning we find that the standards of length are also relative, and that the length L' in the primed system is represented by $L' = L\sqrt{1 - v^2/c^2}$. In other words, to an observer who sees the rod in motion, it will appear 'shortened', and among all systems in a state of uniform relative motion, the one in which the rod is at rest is distinguished from all others by the fact that in it the rod appears longer than in any other system. For instance, a metre rod lying on the earth in the direction of its motion would appear to an observer on the sun to be shortened by 5×10^{-7} cm. In the limit, when $v = c$, the fraction $v^2/c^2 = 1$, $1 - 1 = 0$ and $L' = 0$, which means that to an observer moving with the velocity of light, a three-dimensional body would appear as two-dimensional, or a two-dimensional figure as one-dimensional. The co-ordinates y and z, as we have seen, do not enter into consideration as they are equal in both systems moving relatively in the X direction, and the 'time' co-ordinates are independent of them.

If a body at rest appears to the observer in the unprimed system as a sphere, it will appear as an *oblate spheroid* to an observer in the primed system.

We see that structurally not only 'simultaneity' and 'time' are not absolute but also that length, and therefore *shape*, is relative.

We have seen that the 'shortest' and 'longest' values are important characteristics of the motion. This suggests why in the general theory of Einstein we are interested in, and introduce, geodetics.

It should be mentioned here that the Lorentz transformation has been reached by difficult considerations involving Maxwell's electromagnetic field equations, unrelated to the Einstein theory. Einstein found the Lorentz transformation by the *simplest consideration* connected with his theory. The finding of such important equations by two methods, entirely different structurally, must be considered as a convincing proof of the fundamental importance of such formulae, the more so since they follow from very simple and fundamental structural principles which in themselves cannot be denied because they are negative in character. Negative statements are on a different footing in the new systems; they follow structurally from a $\bar{A}$ orientation, just as the older positive dogmas were the structural results of aristotelianism and the delusional results of identification.

The facts mentioned concerning the measures of length and the behaviour of clocks do not present any paradoxes. They simply say that these discrepancies are mutual and inevitable, as any measurement is only a measurement when it can be registered by an instrument, or seen, or recorded in some way. If the measuring rods and clocks are moving relatively to us, what we see or what our instruments record is *not* what is happening on the moving system, which no one can see or record from outside the system. What reaches us is simply what the light-waves or other signals moving with finite velocity (and therefore retarded by a motion away from us) bring to us. As all existing methods of communication and all known signals have *finite velocities*, these structural differences which are conditioned by the inherent characteristics of the world should be taken into consideration in modern science.

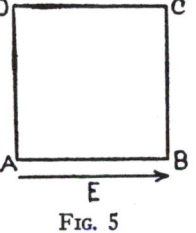

Fig. 5

If we draw a square $ABCD$ (Fig. 5) and an aviator E were to pass this square sign with a velocity of 161,000 miles a second* in the direction AB,

---

*I deliberately select such a velocity so as to make the contraction given by the formula $L' = L\sqrt{1 - v^2/c^2} = 1/2$. With this aim we must make the fraction represented by $v^2/c^2 = 3/4$, then $1 - 3/4 = 1/4$ and $\sqrt{1/4} = 1/2$. We find the square of our velocity v by taking 3/4 of the square of the velocity of light $v^2 = \frac{3}{4}c^2$ and find $v = \frac{c}{2}\sqrt{3} = 161,000$ miles a second.

he would see—and any instrument carried by him would register it—the sides of *our square* ($AB = BC = CD = DA$) in the direction of his flight, namely AB and CD, as 'contracted' to half their length. If he turned at right angles, the sides AB and CD would 'expand' and the other sides, which are at right angles, BC and DA, would 'contract'. For us the sides AB and BC are *equal*, for him one appears twice the other. To him our square appears oblong.

Under such *natural structural* conditions it is a fundamental fallacy to ascribe to 'lengths' or 'shapes' or 'times' any 'absolute' significance. If we grasp the structural fact that 'length' and 'duration' are not *things* inherent in the external world, nor are 'matter', 'space', and 'time', but that they appear as relations between events and some specified observer, and forms of representations, then all paradoxes would disappear.

A suggestion which concerns visualization may be helpful. If we realize the structural fact that words *are not* the objects they represent, we shall always discriminate automatically between what we *see*, *feel*. , on the level of lower order abstractions, and what we *say* on the level of higher order abstractions. When we have conquered that single difficulty we could never then identify the two different orders of abstractions. We would evaluate the *terms* 'matter', 'space', and 'time' as forms of representation, and non-objects, and we would describe events in a functional, operational, behaviouristic language of order. If we realize and feel the *finite* velocity of propagation of all processes, we may visualize all that has been explained here. Diagrammatizing and even following with one's hand, the *visualized order* of occurrences, helps enormously. Try to visualize how the aviator in the last example is flying away and how much more slowly the light impressions from the earth are reaching him or his instruments, and the difficulties will soon vanish.

We shall also be greatly helped in our power of visualization when we become acquainted with the structure of the Minkowski four-dimensional world. An explanation of this appears in the next chapter.

CHAPTER XXXVIII

ON THE 'WORLD' OF MINKOWSKI

> Moreover, the really fundamental things have a way of appearing to be simple once they have been stated by a genius, who was in this case Minkowski. (431)
>
> G. Y. RAINICH

We have already freely used the structural term 'dimension' and only hinted at its meanings. Before we approach the Minkowski world we must summarize roughly what for our purpose we should know about dimensions.

There is nothing mysterious about the term 'dimension'. First of all, the dimensionality of a manifold is not inherent in the manifold as such. It is a characteristic of *order* and so of structure. A manifold can be ordered in different ways, so that it follows that one manifold may have different dimensionality, depending on how we *order* it.

A manifold which has linear order and structure is called one-dimensional. A two-dimensional manifold is then a linearly ordered manifold of linearly ordered manifolds , .

Usually we speak about our 'space' of daily experiences as a three-dimensional manifold, but this is true only with reference to points, and not true with reference to lines or spheres. The manifold of all spheres in 'space' is, for instance, a four-dimensional manifold; so also is a manifold of lines.

Let us explain the line-dimensionality of our 'space' in terms of lines. A line can be given by two points—one, let us say, in the floor of our room, the other in the ceiling. Each of these points is given by two co-ordinates; it has two degrees of freedom; and so our 'space' is a four-dimensional (2×2) manifold *in lines*. This means that to distinguish any line in our 'space' from any other line we would have to have *four data*. Similarly, if we deal with spheres, a manifold made up of spheres requires four data, three for locating the centre and one giving the radius of the sphere. The above examples, of course, do not exhaust the structural possibilities.[1]

The term 'dimension' does not apply solely to what we call 'space'. The term applies to any manifold which we can order in some particular way. Manifolds or aggregates abound everywhere in our lives. The domain of colours, for instance, is a manifold; and so is the domain of tone, or of remembrances , . No manifold in itself has any dimensionality. To ascribe dimensionality to the manifold we must first order it and the number of its dimensionality, or its ascribed or discovered structure, may differ according to the principle of ordering used.

In discussing dimensionality we have two purposes. First, to dispel the semantic fright about this simple term; and, second, to suggest *means for visualization*, which for our purpose are of great neurological importance.

When we say that the world is structurally a four-dimensional manifold, we mean only that according to our experience and the structure of our nervous system, the world of our experience is represented by a fourfold order. We can order the events as to the right and to the left, forward and backward, up and down, and sooner and later. In our experience this fourfold order is completely united, and cannot be separated unless we deliberately choose to *neglect* some of these orderings.

Nor does it mean that all these dimensions are 'identical'. We are accustomed, for instance, to consider the three dimensions of 'space' as 'identical', or at least equivalent. Is this true in life? Can we disregard, for instance, the structural difference between vertical and horizontal? If we did, quite probably, as Eddington remarks, we should come to an untimely end, and break our necks.

Obviously if we *visualize* our *plenum*, as made up of lines or particles, by necessity we visualize structurally a four-dimensional manifold. It should be noticed that a four-dimensional 'absolute void', or 'absolute nothingness', besides being non-sense, cannot be visualized at all, because it could have *no structure*.

We see that all metaphysical 'fourth dimensions' are not only non-sense, but usually indicate a pathological semantic disturbance. The intensity of such disturbances is often high, because it is entirely impossible for a sane person to deal with such meaningless noises. The victim is obsessed with attempts to do the impossible,—a semantically hopeless and painful task.

Such objectifications of terms are very dangerous and science should try, by proper emphasis, to eliminate them. Outside of science the term 'dimension' has *no meaning* and ought to be definitely abandoned in our speculations, for the sake of sanity.

The notion of 'time' as a 'fourth dimension' is by no means new. It appeared in a vague form centuries ago. The notion however was not formulated properly, and therefore was unworkable. Instead of helping science, it only hindered it.

Inspired by Einstein's work, the mathematician Minkowski, whose work had been mainly in the theory of numbers, began to work at the theory of manifolds of any number of dimensions. In 1908 he delivered his famous and semantically epoch-making address on *Space and Time* which fused geometry and physics structurally. In this address he insisted that the connection between 'space' and 'time' as given by the Lorentz-Einstein formulae is not accidental but exhibits that inner connection or structure to which we had not paid enough attention.[2]

In our *experience*, 'space' and 'time' can never be entirely separated, as already explained, and so Minkowski combined them into a higher entity which is called the 'Minkowski world'. In the world of experience the datum appears to be, not a place and a point of 'time', but the *event* or the *world point*—that is, a place at a definite *date*.

The graphic picture of a moving point is a *world-line*. Rectilinear uniform motion corresponds then to a straight world-line; accelerated motion, to one that is curved.

The *event* is the most elementary notion. We shall use it from now on in this work in the sense of a four-dimensional volume of space-time which is small in all four dimensions. We do not posit whether events themselves have structure or not, but it is preferable to assume that they have no space-time structure, which means that the event has no parts which are external to each other in space-time. The order of events is fourfold, as previously shown.

The aggregate or manifold of all point-events is then called the world. The point-events are given by four numbers representing the co-ordinates, three giving the 'space' co-ordinates, and the fourth the 'time' co-ordinate.

The term 'space-time continuum' or 'space-time manifold' is used often and implies that the numbers x, y, z, t, are to vary continuously.

In such a space-time continuum all happenings are structurally the intersections of world-lines, and if we could describe the world-lines of all points of the universe we would have a full account of the universe, 'past' and 'future'. We see that all physics, with the *rest of our problems*, must then be considered as a chapter of the *general* structural and semantic study of continuous manifolds of four dimensions.

But we are already acquainted with such theories. For instance, the internal theory of surfaces may be considered as a part of the subject in two and three dimensions. We have seen that different surfaces are characterized by the expression for the line element $ds^2 = g_{11}dx_1^2 + 2g_{12}dx_1dx_2 + g_{22}dx_2^2$, or by that group of transformations which leaves the line element invariant. We know already that in the E as well as riemannian geometries we have similar expressions and characteristic transformations.

If physics is to be considered a branch of the theory of four-dimensional manifolds, we should naturally look for some such transformations. The manifold represents the world, the *generalized** theory of relativity gives the desired answer. Minkowski proposed a postulate, which he calls the *postulate of an absolute world*, or the world-postulate which asserts the *invariance of all the laws* of nature in relation to linear transformations, for which the function $x^2 + y^2 + z^2 - c^2t^2$ is invariant.

The reader is already familiar with the expression $x^2 + y^2$, which gives the invariant length in E geometry in two dimensions, and $x^2 + y^2 + z^2$ which gives it in three dimensions. It would be natural to expect that in four dimensions we should have an expression of the type $x^2 + y^2 + z^2 + t^2$ but in this case our expression is $x^2 + y^2 + z^2 - c^2t^2$. It should be noticed that the above different types of expressions have different origins. The first two arise in pure geometry, and the last has its roots in physics. The problem was to bring an experimental expression into harmony with a familiar geometrical expression. Minkowski introduced the expression $ict = u_4$, where i is as usual the square root of minus one, $(i = \sqrt{-1})$. Then of course $-c^2t^2$ becomes $(ict)^2 = u_4^2$.

---

*I use the term 'generalized' to embrace the unified field theory and eventually the quantum theory, although, for our purpose, I utilize only the special and general theory.

If we change the lettering, and denote $x=u_1$; $y=u_2$; $z=u_3$ then our expression $x^2+y^2+z^2-c^2t^2$ becomes $u_1^2+u_2^2+u_3^2+u_4^2$, a simple formula for distance in four-dimensional geometry. It would not be profitable for us to speculate upon this substitution; it was introduced merely for the sake of mathematical, verbal treatment and can be easily translated back into the usual terms of c and t.

We have already seen that the expression $x^2+y^2+z^2-c^2t^2 = x'^2+y'^2+z'^2-c^2t'^2$ is invariant under the Lorentz-Einstein transformation. This fact of invariance is fundamental, and it is well to convince ourselves that it is so. The Lorentz-

' see page xii Einstein transformation was $x'=\beta(x-vt)$, $y'=y$, $z'=z$, $t'=\beta(t-vx/c)^2$ where $\beta=1/\sqrt{1-v^2/c^2}$. As the co-ordinates y and z are equal in both systems, we can disregard them, and verify that $c^2t'^2-x'^2=c^2t^2-x^2$. Let us substitute for t' and x' the values given by the Lorentz-Einstein transformation. We have then:

$$c^2t'^2-x'^2=c^2\beta^2(t-vx/c^2)^2-\beta^2(x-vt)^2=c^2t^2\beta^2+\frac{c^2\beta^2v^2x^2}{c^4}-\frac{2c^2\beta^2tvx}{c^2}-\beta^2x^2-\beta^2v^2t^2$$

$$+2\beta^2xvt=c^2t^2\beta^2(1-v^2/c^2)-x^2\beta^2(1-v^2/c^2)=c^2t^2\frac{1-v^2/c^2}{1-v^2/c^2}-x^2\frac{1-v^2/c^2}{1-v^2/c^2}=c^2t^2-x^2;$$

since $\beta^2=\dfrac{1}{1-v^2/c^2}$. Similarly it is easy to show, if we take an event-particle,

as, for instance, a momentary spark, which has the co-ordinates x_1, y_1, z_1, t_1, in one system of co-ordinates, say S, and let another event-particle occur in that system at x_2, y_2, z_2, t_2, that the formulae remain invariant. If we designate the distance between the two events by r, its value would be given by $r^2=(x_2-x_1)^2+(y_2-y_1)^2+(z_2-z_1)^2$.

In a different system, S', moving uniformly relatively to S, r'^2 would in general not be equal to r^2, but the expression $r^2-c^2(t_2-t_1)^2$ would be equal to $r'^2-c^2(t_2'-t_1')^2$, $r^2-c^2(t_2-t_1)^2=r'^2-c^2(t_2'-t_1')^2$.

The above expression is called the *interval* and expresses a most fundamental structural characteristic; namely, that the interval is invariant for all systems in uniform relative motion. This result is quite general and independent of the relative orientation of the axes, or of the angle the velocity has to the axes. The interval plays in the theory of Einstein a similar role which the pythagorean rule played in the E geometry.

Because of the finite velocity of our measuring signals, our formulae must involve finite velocity. Therefore the interval is the only actual measurement which we can ever make in practice. Hence its fundamental semantic importance.

Eddington gives a very fine

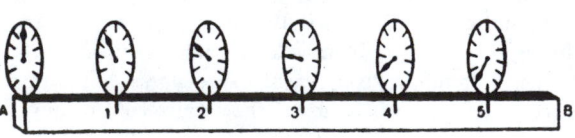

FIG. 1

diagram in explaining how intervals are measured. I reproduce it herewith (Fig. 1).

The expression ct, where c is the velocity of light, 300,000 kilometres per second, gives us the *distance* travelled by light in the 'time' t. It is natural to regard the velocity of light, which is a constant and translates easily into the language of length, as a unit of 'time'. In the Minkowski world it is customary, because of its convenience, to regard 1 second as the equivalent of 300,000 kilometres and measure lengths or 'times' in seconds or kilometres indiscriminately.

Let us imagine a scale graduated in kilometres, and clocks whose faces are also graduated in kilometres (1/300,000 of a second). If the clocks are set correctly and we look at them from A the sum of the reading of any clock and the scale division beside it is one for all because the scale reading gives the correction for the 'time' taken by light, travelling with unit velocity, to reach A.

If we lay the scale in line with the two events and note the clock and scale readings, t_1 and x_1, of the first event, and the corresponding readings, t_2 and x_2, of the second event, then $s^2 = (t_2 - t_1)^2 - (x_2 - x_1)^2$ where s represents the 'interval' mentioned above.

If we set the scale moving in the direction AB then the divisions would have advanced to meet the second event and the difference $(x_2 - x_1)$ would be smaller. But this is *compensated*, because $(t_2 - t_1)$ also becomes altered. When A is advancing to meet the light coming from any of the clocks on the scale the light arrives too quickly, and the reading of the clock appears smaller.

The net result is, roughly, that it does not matter what uniform motion is given to the scale, the final results for the interval s are always equal.[3]

We can now understand the vital importance of the minus sign with the 'time' co-ordinate. In fact, if in our equations all the signs were plus, using the 'space' and 'time' of one observer, one value of s would be obtained; but using the 'space' and 'time' of another observer, a different value would be obtained. With the minus sign for the 'time' co-ordinate, we see that we can have values of s which are equal for all observers. If the distances increase, the 'time' element increases also, and so the *difference* may not be changed, but with the positive sign this would not be the case.

We see that the interval s represents something which concerns only the events under consideration. The corresponding entity in ordinary geometry is *distance*, which is independent of the accidental choice of co-ordinates. The minus sign makes the geometry of space-time *non-euclidean*.

To familiarize ourselves with what has been already explained about simultaneity and the geometry of space-time, we will work it out once more, but now by the Minkowski method.

It will be enough to use two dimensions, one represented on the X axis, the other on the T axis. Let us consider three points A, B, C, at rest in our system O on the X axis (Fig. 2). In our space-time they will be represented by three parallels to the T axis. Let C be midway between A and B so that $AC = CB$. Let us assume that light signals are sent in both directions from C at the moment $t = 0$. We assume that the system is 'at rest', which means that the light signals propagate themselves to the right and to the left with equal velocities. Hence we can represent them by straight lines equally inclined to the X axis. These lines are called 'light-lines'.

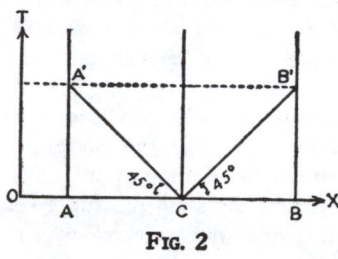

FIG. 2

The points A', B' which are the intersection of the 'world-lines' of the points A and B with the light-lines give us the 'times' at which the signals arrive. It follows from the drawing that $A'B'$ is parallel to the X axis, which means that A' and B' are 'simultaneous' (equal 'times').

Let us now take another case in which our points A, B, C, move uniformly with an equal velocity (Fig. 3). Their world-lines will also be parallel to each other but *inclined* to the axis. In the drawing the light-lines will be represented by similar lines but their intersections with the world-lines of A and B will not be on a parallel to the X axis, and so they will not be simultaneous.

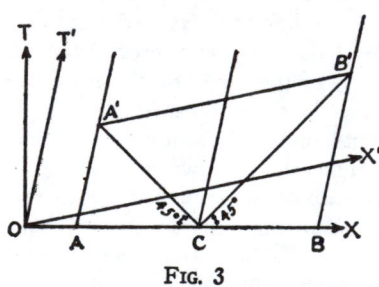

FIG. 3

We should notice that an observer who moved with the system in the direction OX' would be perfectly entitled to claim that A' and B' are *simultaneous to him*. His co-ordinate system would be $OX'T'$, in which the points A' and B' are on a parallel to *his X' axis* as he is *at rest* in his system $OX'T'$. The world-lines A, B, C, are parallel to the T' axis because the points are supposed to be at rest in this system and hence the x's have equal values for all t's.

An important point should be noticed; namely, that we have only one space-time and that the indefinitely numerous ways different observers partition their 'space' and 'time' represent merely the indefinitely many ways in which it can be partitioned. If we keep the whole of it under consideration we see that we cannot divide it into 'space' *and* 'time', as any subdivision has both aspects.[4]

The Minkowski method of representation makes the change in our measurements of length, as given by the Lorentz-Einstein transformation, very obvious.

A measuring rod is not purely a 'spatial' configuration, as in the actual world such a thing does not exist, but it is a space-time configuration.

Every point of the rod exists at each moment of 'time'. We see that in space-time we cannot represent our rod as a segment on the X axis but must represent it structurally as a *strip* in the XT plane. We assume here for simplicity that the rod is one-dimensional (Fig. 4).

A rod which is at rest in a system is represented by a strip parallel to the T axis. If it is moving, its strip is inclined to the T axis. The 'contraction' does not affect the strip at all but it is rather a section cut out of the X axis. In actual experience, it is only the strip as a manifold of world points which has *physical reality*, and not the cross sections, which, as we see, are not equal on different axes. The 'contraction' is not a change in 'physical reality' but merely a consequence of our way of regarding things. We see that the notorious argument as to whether the 'contraction' is 'real' or 'apparent' is based on a misunderstanding. Born gives an excellent example. If we slice a cucumber in different directions it is fallacious to argue that the smallest slice which is perpendicular to the axis is the 'real' one and the larger oblique slices only 'apparent'. Similarly in the Einstein theory, a rod has various lengths according to the motion of the observer.

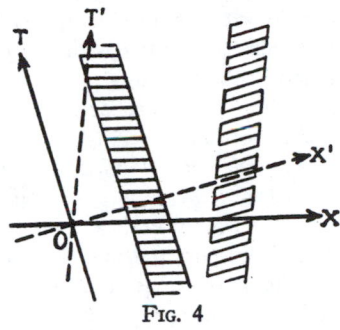

FIG. 4

One of these lengths, the static length, is the greatest, but it is no more 'real' than any other. Similar remarks can be made about 'time'.

Attention should be given to one extremely important semantic point concerning the Minkowski four-dimensional world. We already know that for our nervous systems the passing from dynamic to static, and vice versa, is a most vital structural problem. The first step of this translation has already been given in the notion of the 'variable'. The calculus carried it a step further. In the Minkowski world we reach the complete solution of the problem.

As Keyser points out in his *Mathematical Philosophy*, we had two verbal methods of dealing with 'time'. One was the method of Newton, the method of the structural importation of 'time'. From the objective dynamic world of the lower order abstractions, 'time' is imported into the static world of the higher order abstractions. We import it with 'motion', we say things 'move'. Such language is structurally unsatisfactory, even on the earlier level of our development. It hampers analysis, and is contrary to the structure and function of the *human* nervous system. It breeds tremendous metaphysical impasses, and is ultimately based on semantic disturbances due to identification.

If we introduce dynamic, shifting entities into static higher order abstractions, rationality is impossible and we drift toward mysticism.

A very real semantic problem appears here. We want to give the best possible account of the structurally *dynamic* world around us; yet our higher order abstractions are structurally *static*, and for their proper working they must use static means. Here seemed to be an impasse which for milleniums had defied solution. 'Philosophers' of different schools were preaching and teaching that we should never be able to be 'rational' and understand this world and ourselves. Anti-intellectual schools began to flourish, to the bewilderment of all.

The issue, after all, was simple, the moment some one discovered and stated it. We did not need to change either the world around us or ourselves;

we simply had to discover a structurally new method of dealing with the old problems without changing them.

The new method is given by Einstein and Minkowski. Instead of making the static world of higher order abstractions dynamic, which cannot be done at all without producing semantic disturbances, they invented structural methods for dealing with the dynamic world by static means. The key was found in the handling of the troublesome factor, 'time'. Minkowski decided to put 'time' in its proper place by introducing the structurally new four-dimensional world.

In the case of particle P, we habitually used to say that the point P at an instant t was at a 'space' point (x, y, z). At the instant t' it was at the point (x', y', z') , . We always needed four numbers, which gave us the where and when in respect to some frame of reference.

As we have seen, Minkowski decided to look structurally at this tetrad of four numbers (x, y, z, t) 'as-a-whole'. In other words, he placed himself on a higher level of abstraction. He took under consideration the older results, combined them, and called the combination by one single name, the 'world point'. Such a world point has also four numbers (not 3 plus 1, but just 4). A world made up of such points is a four-dimensional world in which all the points co-exist. The flux of the lower order abstractions and 'time' is abolished. There is no more 'motion' in a 'flow of time'. In such a world the term 'where' has completer structural meanings, it has absorbed the *when*. If we ask, where in such a world the particle P is, we answer, at the point (x, y, z, t). Where is the particle P'? At the point (x', y', z', t').

We see that the particles of such a world *are never 'the same'*; they do not 'change' or 'pass'; they co-exist, and all is *static*. In this way the three-dimensional dynamics become four-dimensional statics.

It should be noticed that we are now dealing with a language of new structure, uniquely befitting the structure and function of our nervous system. Of course we have altered nothing in the world around us. The example of the moving picture and the static film as given before is an excellent example of this structural innovation.

The fact that in this new world nothing repeats itself because it has a different date, unless the time-lines are closed, has very far-reaching consequences, of which we have already spoken and which we will analyse in more detail later on.

CHAPTER XXXIX

GENERAL REFLECTIONS ON THE EINSTEIN THEORY

For, beyond the bounds of science, too, objective and relative reflection is a gain, a release from prejudice, a liberation of the spirit from standards whose claim to absolute validity melts away before the critical judgment of the relativist. (45)　　　　　　　　　　　　　　　MAX BORN

It is extremely instructive to follow the elements of identification as they appear in the evolution of the Einstein theory.

We know that the results of the Michelson experiment which disclosed that light is propagated in all directions with equal velocities for all observers, irrespective of their relative velocities, could not be reconciled with the older mechanics. These results disturbed the physicists profoundly and attempts were made to solve this apparent impasse. In what follows we shall have to analyse incidentally the activities of some of our great scientists, men who have added enormously to our knowledge, and this fact should be appreciated. What we have to say is not intended as criticism—far from it—but simply as a structural and a semantic analysis.

The feeling that we objectify unduly and that we should not use a language of 'is' of identity, but that we should use an actional, behaviouristic, functional, operational language and methods, is not new in science, although the need was not formulated structurally, it is true, and therefore it never became a workable foundation. The main successes in these fields were rather accidental, and were the personal prerogatives of those few men whose psycho-logical make-up urged them to achieve. Objectification, which as we know, is a semantic ascribing of *objective* existence and values to terms, was bound to make its appearance somewhere.

This struggle against identification is apparent in all science, but it will suffice to point out the most striking example in the relation of the works of Lorentz and Einstein. Lorentz objectified, Einstein did not. We here come across a tremendous semantic fact which has to do with the *interpretation* of mathematical formulae. Lorentz on elaborate and difficult grounds, connected with Maxwell's field equations, produced what is usually called the Lorentz transformation. He gave it an *objectified* interpretation. Einstein introduced an entirely different fundamental interpretation of the structural *principle* involved. The formulae look alike but they now have different and very simple meanings.

Hertz, whose epoch-making discoveries made wireless possible, advocated long ago what is termed the phenomenological point of view, which in our language is approximately the actional, behaviouristic, operational, functional language and method. In his writings he implicitly refused to use the vicious

673

term 'is' of 'identity', and so to objectify his terms, which refusal in picturesque language he expressed as a refusal to legislate about 'essences'.

The old E and N language of 'absolute space' and 'absolute emptiness' were for a long while structurally unsatisfactory. Physicists felt that somehow they could not deal with it, but it never occurred to them that this 'absolute nothingness' is objectively meaningless, and that therefore no one can possibly deal with it. Not knowing that, they politely called this non-sense a 'metaphysical question' and evaded issues by leaving the solution in the hands of 'philosophers', never to be solved.

By now I hope that the reader is quite aware that meaningless problems cannot be solved by any one, and that there are no such things as *meta*-physical questions. There may, however, be a question about *enlarging* the domain of physics.

Being forced to abandon this 'absolute emptiness', physicists went to the other extreme and postulated some kind of 'material' ether. Let us note that such a postulate involves structurally the 'is' of identity and objectification. Lorentz in opposition to Hertz postulated an 'ether' which was 'motionless' in 'absolute space'. Note that here we have a perfect example of structural *objectification of terms*. 'Absolute space' is for him semantically some kind of 'absolute emptiness', which, not being satisfactory for the physicist, is filled with some 'material', 'motionless' ether. 'Motionless' is itself an objectification of language, as such a term has here *no* physical or objective meanings at all.

In pursuing the speculations on *objectified terms* (semantic disturbances) it was natural to expect, as the earth is not at rest with respect to the sun, the other planets . , that some 'ether wind' or 'ether drift' should appear which would make the constant velocity of light impossible for observers moving with different relative velocities. But these structural expectations were not fulfilled. The velocity of light, as shown by many experiments, was a constant for all observers. The 'motionless material ether' also became structurally impossible, as might be expected, if we stop objectifying terms.

In 1892 FitzGerald suggested an *objectified* theory, assuming 'absolute' 'length' and 'time' *superior to measurement*, which involve identification and do not allow the use of the actional, behaviouristic, operational, functional attitudes, language, and methods. FitzGerald assumed that every body 'moving' with the velocity v in the 'ether' is shortened in the direction of motion. It should be noted that every mention of 'shortening' or 'contraction', *presupposes some 'absolute'* standards of 'rest' or 'motion' or 'length', which do not, and cannot, exist outside of our skin, but are only semantic disturbances, inside our skin, which occur when we identify and ascribe *objective* existence and value to *terms*.

How deeply and completely these objectifications permeate our daily and scientific lives is best shown again in the case of Lorentz. Even in 1917, in his Haarlem lectures, he expressed structural hopes that a 'material', 'substantial' ether can be preserved, that 'space' and 'time' can be sharply separated, and that 'simultaneity' can have an absolute meaning.

In the *Theory of Relativity* of Whitehead, and in some others writers who deal with the theory of Einstein, and particularly in all critics of Einstein, we find a similar objectification of terms.

They still *feel* the older E and N 'absolute emptiness', 'absolute space', 'absolute time', to which *terms* they ascribe structural objectivity. In such works the term 'contraction' is used frequently.

Let me recall the mechanism of objectification. If we do not reject explicitly and implicitly the 'is' of identity, we automatically identify different orders of abstractions and ascribe objective characteristics to terms. Thus the term 'time' which represents a label for a feeling inside our skin, is given an objective evaluation. If 'objective' it must have a 'property' of 'simultaneity', a semantic process taken over from comparing two objective sticks when the two ends are made to coincide. On the objective external level, we never deal with 'time' but we simply *compare processes*. When we select an arbitrary unit-process on the objective level, whatever we might *say* that it 'is', well, it *is not*, and the difficulty is found exclusively in the use of the 'is' of identity.

If we abandon entirely the 'is' of identity, we stop objectification, we do not ascribe objective existence and values *outside* our skin, to terms and semantic reactions *inside* our skin. But then of course we have to change the *structure* of our language; as otherwise the old *s.r* will continue to play tricks on us. An actional, operational, functional language of *order* is the structural solution of our semantic difficulty.

If we objectify 'space' into 'absolute space', we must objectify it as 'absolute emptiness' for only such an 'absolute space' can be at 'absolute rest', that is, static in the E or N sense. Similarly only objectified 'time' can have the 'property' of 'absolute simultaneity'.

If we realize that these 'absolutes' are only the semantic objectifications of terms, (where the activities of the lower nerve centres are structurally ascribed to the activities of the higher nerve centres and vice versa), we begin to differentiate between different order abstractions, and to keep them differentiated. In terms of our structurally new language we become 'conscious of abstracting', and then habitually *and unconsciously* use the behaviouristic language and methods of *order*.

If we picture this 'absolute emptiness' or 'absolute nothingness' (which cannot be done successfully, as it has no meaning), and try to compare it with a plenum, or 'fulness' (a cloud of smoke, for instance), we see at once that only this 'absolute emptiness' can be static, homogeneous . , a condition that is impossible with a dynamic fulness.

Perhaps we can now appreciate the tremendous semantic significance of the Einstein theory, which introduces structurally a *non-objectified*, human, sane attitude of proper evaluation toward this world. We should not be surprised to find that a $\bar{A}$-system which is an inevitable general structural concomitant of the $\bar{E}$ and $\bar{N}$ systems of geometry and physics should *formulate as a general structural and semantic issue* what the $\bar{E}$ and $\bar{N}$ systems have *done* in their special fields, without such general formulation.

From our structural point of view there is no retreat; the Einstein work is irreversible. In the younger scientists of today the non-objectified attitude toward *terms* of 'space' and 'time' is already an accomplished semantic fact, entirely independent of what future experiments may show. For experiments can never justify identification, and so can have no detrimental effect upon this fundamental and most beneficial structural, linguistic, and semantic revolution. Our $\bar{A}$ task was to formulate these issues *in general* so as to make us conscious of them; and I assume that it is at this semantic point that the tremendous value of Einstein's work will manifest itself in life. Indeed we shall see later on in this volume that the newer quantum mechanics, which have begun to spring up rather rapidly, is made possible only by the semantic background imparted *unconsciously* (as yet) to younger physicists by the Einstein theory. It is my hope that the present work may make the above issues *conscious*, and so enable us not only to impart this semantic attitude more easily and with less labour but also to benefit by them more universally in our *daily life*. The problems of science and life do not differ in this respect. In both we are equally hampered by semantic disturbances, 'emotional stupors', identification, and similar difficulties, the elimination of which means better adjustment for all of us, as well as swifter progress in science.

A study of the history of science shows how slow and painful scientific progress has been. Now we begin to see why. 'Geniuses', as history shows, are men who at least in some fields are freer from identification and false evaluation than others. They are not hampered to a similar extent by 'emotional stupor'; hence they can evaluate the old *anew*. Lorentz, for instance, produced the formulae, but his objectifications *prevented* him from evaluating properly the new formulae. As a fact of history the formulae of Lorentz were discovered by Voigt a number of years before, but identification made impossible the evaluation of these formulae, and so *delayed* the discovery of the Einstein theory. This factor of identification can be found all through recorded history as a retarding semantic blockage.

If we could find methods of eliminating these semantic disturbances, an extremely hampering, paralysing psycho-logical factor would be eliminated, and 'geniuses' could be made the rule rather than the exception. Let me say again: in the old days morons were made and geniuses were born; in the new days, perhaps, this can be reversed, and *morons will be born but geniuses made*. We witness something of this kind among the younger post-einsteinian physicists, where the number of 'geniuses' is growing rapidly, in spite of the fact that the above structural issues are not as yet consciously applied in general education. The secret of creative work is freedom from structural bondage, and particularly the structural semantic bondage of words.

The reader should not assume that the few simple structural explanations given in this book exhaust the Einstein theory. I have not even attempted to summarize the theory; I have only given a few semantic facts, which belong to general semantics and to the theory of knowledge. The Einstein theory is indeed such a tremendous structural linguistic achievement that quite probably its full semantic significance and meanings will not be worked out for many years to come. We have given here only the minimum of explanation necessary for our special purpose.

The historical development of a theory has usually little to do with the semantic importance of the theory or its deeper meanings. The constancy of the velocity of light for all observers, which started the ball rolling, was an historical beginning and it served its purpose well, though the objectified 'contractions' and formulae of FitzGerald and Lorentz also did their share, as they helped Einstein and Minkowski to produce their epoch-making structural challenge to old prejudices such as 'absolute space' and 'absolute time', which were semantic remains of a primitive, perhaps pre-human, remote past. Once this is accomplished, no matter how, there is no return possible. Of physical structural facts, all that we need is the *finite* velocity of the propagation of events,* which as we already know involves far-reaching structural and semantic issues. Of the psycho-logical issues involved, we need only to eliminate semantic disturbances which still occur when we copy animals in our nervous processes and do not discriminate between different orders of abstractions—which animals do not recognize. This elimination can be done by training in the $\overline{A}$ methods explained before, with the net result that we become 'conscious of abstracting' on different levels and so can instinctively and by feeling discriminate habitually between orders of abstractions, which structurally and semantically could not be done by the old disciplines.

The theory of Einstein has manifold applications but we need only mention a few, which we shall utilize later on.

First, and above all, there are no possible 'absolute' meanings to 'space' and 'time', beyond the relations established by measurements. The structure of our language involving 'space' and 'time' should be similar to the structure of experimental facts, which ultimately show the impossibility of sharply dividing them.

If any one challenges this statement, he could not *a priori* be criticized. Such criticism would be entirely against the whole tendency of the present work. But such a person might be approached with no little curiosity and expectation. He could be asked: 'You claim that you can absolutely divide

---

*'But,' some reader may ask, 'though you assume a finite velocity of propagation, may it not happen that some day an "infinite" velocity will be discovered?'

Such a question would show that the reader has missed the point in the present work. We are confident in saying that an 'infinite' velocity has no meaning, and that no matter what we discover, this will never be discovered. This becomes still clearer if we use the differential *definition* of 'velocity'. Velocity is defined as the 'time' derivative of 'space' travelled. If 'time' is taken as zero, or if we have 'no time', there can be no 'time derivative' by our very assumption, and, therefore, no 'velocity'. There is, therefore, no danger that we shall ever discover in the actual world an 'infinite' velocity.

"space" and "time" on the objective level. That would be an epoch-making structural discovery. *Please demonstrate how to do it.*'

The fact is, of course, that he cannot demonstrate the process, because he refers to identifications inside his skin; yet he is claiming to be able to show it *objectively* outside his skin. That ends this problem.

While speaking of Einstein's Theory, it will be well to mention a few of the many structural differences between the older newtonian and the new einsteinian mechanics.

In the N-system, relative velocities were simply added $W_N = v + v'$. In the einsteinian system which we will denote by $\bar{N}$, it is not so structurally simple. We must introduce the finite velocity of propagation of our signals, which *alone* give us the data, and so

$$W_{\bar{N}} = \frac{v + v'}{1 + vv'/c^2}$$

The above formula involves the remarkable constant, c, the velocity of light. If we assume in the above formula that our velocity v' is equal to the velocity of light, c, we would have

$$W_{\bar{N}} = \frac{v + c}{1 + vc/c^2} = \frac{v + c}{1 + v/c} = c.$$

This means that the addition of some velocity to the velocity of light does not alter the velocity of light, which thus appears as a *limiting* velocity.

This applies to the difference of velocities where

$$W_{\bar{N}} = \frac{v - v'}{1 - vv'/c^2}$$

Let us here give an example of Eddington's. Let us assume two relative velocities each differing by only 1 km./sec. from the velocity of light. Let us say that one is 299,999 km./sec. and the other 300,001 km./sec. Now let us calculate the relative velocity. This relative velocity will be found to be 180,000,000,000 km./sec. For in our formula $v - v' = (c+1) - (c-1) = 2$, and

$$(1 - vv'/c^2) = 1 - \frac{(c+1)\ (c-1)}{c^2} = 1 - \frac{c^2 - 1}{c^2} = 1 - 1 + 1/c^2 = 1/c^2,$$

whence $W_{\bar{N}} = \dfrac{2}{1/c^2} = 2c^2 = 2 \times 300{,}000 \times 300{,}000 = 180{,}000{,}000{,}000.$

We see that a particle which might try to overtake light by having a velocity of one km./sec. greater than the velocity of light could never succeed. When the velocity 299,000, for example, was reached, the particle would find itself further away from its goal than when it started.[1]

Similar general considerations apply to mass. If we designate the mass of a particle at rest by m_0 its mass in motion

$$m_{\bar{N}} = \frac{m_0}{1 - v^2/c^2}$$

As the denominator is smaller than unity the mass in motion, $m_{\bar{N}}$ is larger than m_0, the mass at rest. In the limiting case, when the velocity would become equal to c the denominator would become zero and our mass $m_{\bar{N}}$ would tend toward infinite values, which is another way of saying that it is physically impossible.

In the N-system we had two kinds of energy; one was called *vis viva*, or kinetic energy, and was represented by $T = mv^2/2$; the other was called potential energy, or capacity for work, and was denoted by U. The law of conservation of energy in the N-system was expressed by the statement that the sum, $T + U = E$, or the total mechanical energy of a system remains constant (zero variation) during the motion of the body.

We see that as the above formula involves the terms m and v, the older formulae for energy must be altered, especially since they do not survive a Lorentz-Einstein transformation. It is found that
$$T_{\widetilde{N}} = c^2 (m - m_0), \text{ or } m = m_0 + T_{\widetilde{N}}/c^2,$$
which formula appears rigorous as a definition of kinetic energy even if members of order higher than the second are taken into account. In words, the mass in motion differs from mass at rest by the kinetic energy divided by the square of the velocity of light.

This expression suggests immediately that the static mass, m, is similarly related to the energy content in the body at rest. Generalizing our results we would have $m = E/c^2$, an equation which holds generally between mass and energy. This fact has been called by Einstein the law of the *inertia of energy*. It has been verified repeatedly by experiments, and is one of the most striking structural results of Einstein's theory. The above statement means that the two fundamental notions of 'mass' and 'energy' are equivalent and thus we have a clearer vision of the structure of 'matter'. The two older structural laws of 'conservation of matter' and of 'conservation of energy' become fused into one. Mass becomes structurally and verbally nothing else than energy concentrated at a point, and it appears as a form of energy manifestation.[2]

The above considerations have also led to a revision of our structural notions about 'energy' which we do not need to explain here. Suffice it to say that the old 'potential' energy is not associated structurally any longer with any features of this world. It can be made to vanish by a proper selection of co-ordinates, hence it is no longer considered as energy of any kind.[3]

With the Minkowski world we became acquainted with a new *language* which represents structurally more nearly the facts of experience (lower order abstractions) and shares the structure of our higher order of abstractions. So we have the language of 'space-time'. How about 'matter'? The bumping against something hard is not to be disregarded. True, *we need a language of new structure, but that is all*. In the Einstein theory, 'matter' of course is not treated separately as such. It is an offspring of the field, and is connected with the curvature of the world. The reader should not be surprised to find that the Minkowski world, which has accelerations, must be curved in this structurally new form of representation.

We have already defined a most fundamental entity called 'action'. Naturally in a space-time manifold, energy multiplied by 'time' should be a more fundamental entity than energy, and we call it 'action'. When we speak about some continuous material present in 'space' and 'time' we speak in terms of density. Density multiplied by a three-dimensional volume of 'space' gives us mass, or what appears as its equivalent—*energy*. From a four-dimensional, or space-time point of view, density multiplied by a four-dimensional volume of space-time gives us action. We see that the multiplication of density by the three dimensions of 'space' gives us mass or energy. A fourth multiplication by the dimension of 'time' gives us mass or energy multiplied by 'time' which becomes action by definition. It is obvious that, structurally, action must be more fundamental than the older quantities.

In terms of curvature, action represents the curvature of the world, because where we find 'action', we also find 'matter', acceleration, gravitation , .[4]

'Action' is fundamental, because structurally in a four-dimensional metrical manifold it takes the form of the simplest integral *invariant* that can exist at all. On this form of action Maxwell's electromagnetic theory is built. The quantity action appears as a pure number,[5] a unique, specific relation which conditions structure.

We should expect that the action represented by the number 1 would be most interesting and would eventually represent the indivisible atom of action. The modern quantum theory seems to favour such a point of view.

When we encounter a pure number having such crucial significance in this world we should not wonder that such a number intrigues us. As yet it is impossible to state that action cannot have fractional numbers. What, then, would the action represent?

Eddington suggests that the number may represent a *probability* or some function of a probability.

We combine probabilities by multiplication, but we combine actions in two regions by addition. We see, therefore, that the logarithm of a probability gives the function indicated and Eddington suggests the provisional equivalence of action with the negative of the logarithm of the statistical probability of the state of the world around us. Such a suggestion is extremely appealing and important because the principle of *Least Action* can be stated as the principle of *greatest probability*. The laws of nature appear to be such that the actual state of the world is represented by that which is statistically the most probable![6]

That such structural conclusion can be drawn at all is of tremendous semantic importance for us because, as we are *abstracting* in different orders all through, the only appropriate language in which we can eventually hope to speak correctly, is the language of probabilities, statistical averages , .

Action is one of the terms of pre-einsteinian physics which has survived unmodified, the only other one being entropy. The law of gravitation, the laws of mechanics, and the laws of electromagnetism, can all be, not only summed up, but also deduced, from a single principle of least action. This important structural unification was accomplished even before the advent of the einsteinian theory, and only the addition of gravitation to this list is new.[7]

In this brief structural and semantic survey we have had neither the opportunity nor the necessity of analysing the general theory of Einstein, which embodies and unifies most of the laws of mechanics, that of gravitation included.* In this unification lies the unrivalled grandeur of the theory. As we shall see later, the newer quantum theories have been already very much influenced by the Einstein theory. As all possible theories are dependent on *human ingenuity* and never can be the events themselves, we can rest assured that once freed from 'emotional stupors' and semantic disturbances, the world will not be long in producing a whole structurally unified system of science.

In our discussions we deal with 'apparent', 'real', 'actual', and similar *m.o* terms. We should recall that mathematics is *exclusive* in one respect; namely, that it has no content. It is entirely a product of higher abstractions created by definition from undefined terms. We have seen that mathematics must be considered as a language of special structure which is, however, similar to the structure of the world around us.

Our daily A language, among others, being based on the 'is' of 'identity', can never give a structurally satisfactory picture of this world or ourselves, but actually prevents such an achievement. Having abandoned a language which leads to identification, we shall be able to apply a new language, with new structure, by which we achieve better means for representing the events around us. From this point of view, mathematics and our daily language do not differ. Terms, being *not* the things they represent, must by necessity be creatures of definitions and undefined terms. The solution of many baffling semantic problems is found in the *structure* of a language which involves different semantic and unconscious attitudes.

---

*In fact, a few months ago, Einstein and Mayer succeeded in reducing the laws of gravitational and electromagnetic fields to a single basis. This was accomplished by the aid of a very revolutionary mathematical discovery that it is possible to introduce into a 'space' of n dimensions, vectors with m components. Although at present the results of the quantum theory are not included in this theory, there is no doubt that shortly, because of this mathematical discovery, these will be included in a *generalized* theory of relativity.

PART X

ON THE STRUCTURE OF 'MATTER'

Rather against my better judgment I will try to give a rough impression of the theory. It would probably be wiser to nail up over the door of the new quantum theory a notice, "Structural alterations in progress—No admittance except on business", and particularly to warn the doorkeeper to keep out prying philosophers. (149)
                                                        A. S. EDDINGTON

CHAPTER XL

THE OLDER 'MATTER'

And yet when I hear to-day protests against the Bolshevism of modern science and regrets for the old-established order, I am inclined to think that Rutherford not Einstein, is the real villain of the piece. (149)

A. S. EDDINGTON

Micro-mechanics appears as a refinement of macro-mechanics, which is necessitated by the geometrical and mechanical smallness of the objects, and the transition is of the same nature as that from geometrical to physical optics. (466)

E. SCHRÖDINGER

From the dawn of history, man has had to deal with different bits of materials, some hard and solid like stones, some soft like fruit or flesh, some liquid. In remote antiquity air and gases were not considered as 'matter'.

In those days 'matter' was structurally only what could be seen, or felt, or touched . : anything else was some kind of 'spirit', and everything 'existed' in an 'absolute void'. But even in remote antiquity our primitive ancestry could not miss the fact that the bits of materials they dealt with could be divided into smaller bits. Naturally, if we can subdivide bits into smaller bits, an interesting question arises: How far can this division be carried on? It seems that Democritus (about 460-360 B.C.) was the first man on record to formulate an atomistic theory. He already postulated structurally a subjective world picture, to be contrasted with an 'absolute' or objective world in which 'motion' was all important. This theory started us on the mechanistic road formulated for macroscopic events, and also on the road of individualization, the study of smaller and smaller bits of materials and the search for some unit bricks out of which this world appeared to be built; all of which was already a search for $m.o$ structure.

With the advent of chemistry some further fundamental structural light was thrown on the problem of individualization. It was found that certain materials, as, for instance, iron, copper . , remain one material, no matter how far we carry our subdivision. These were called 'elements'. At present we recognize 92 elements, a number which is supposed to represent all possible elements. Out of these a remaining few were at first predicted theoretically, and just the other day discovered experimentally. All other materials do not stand division so well. At some stage they decompose into their elements. The smallest bit of one of these last materials which still has the characteristics as the bulk, is called a molecule. The molecule is found to be built up from atoms of the elements. For instance, the molecule of water still has the characteristics of water, and consists of two atoms of hydrogen and one atom of oxygen, which are no longer water but elements of entirely different characteristics.*

Electrochemistry taught us in the meanwhile an important structural lesson; namely, that definite electrical charges are combined with the atoms. Such electrified atoms are called 'ions' (Greek for traveller). For instance, a

---

*The above statements are over-simplified, but satisfactory for my purpose.

molecule of water is broken up into a positively charged hydrogen ion consisting of two hydrogen atoms, and a negatively charged oxygen ion consisting of one oxygen atom.

But electricity had more structural surprises in store for us. About 1880 new facts were discovered. One of them was that a moving electrical charge has the effect of an electrical current; namely, it can deflect a magnet just as a current does. Such moving electrical charges were called convection currents, and the fact that they produce effects similar to an electric current led J. J. Thomson to a surprising conclusion. According to the Maxwell theory of electromagnetism a certain amount of energy must be associated with every electric or magnetic field. If an electrical charge in motion can produce magnetic effects, hence energy, it was concluded, and verified by experiment, that energy was required to set an electrical charge in motion. From which it follows structurally that an electrical charge possesses a characteristic in common with other materials; namely, inertia, which can be overcome only by the application of energy. This inertial mass of the electric charge was called electromagnetic mass.

Here we see two fundamental structural issues involved. One is that electricity seemingly has an inertial mass similar to that of 'matter'. The other is, that in convection currents, we find means to study electromechanical parallelism, and so discover the relationship between electrical and mechanical theories.

In the year 1895 Lorentz proposed the electron theory. He assumed structurally that moving molecules contain electrical charges and so produce convection currents. These charges are further assumed to be one electrical quantum and were called electrons. The electron theory proved to be enormously fruitful and all further advance in our structural knowledge is intimately connected with it.

As knowledge advanced, the more convincing became the structural evidence for some electronic theory. As already noted, a moving electrical charge produces magnetic effects. So also should a moving electron, whence we succeeded in accounting for magnetic effects in terms of moving electrons which in this case represent molecular convection currents.

Granting this, a revolving electron should represent also a small magnet and a mechanical gyroscope as well. Einstein in 1915 verified this assumption by an experiment. If these structural assumptions were true, then, by a quick reversal of magnetism, a soft iron rod should turn by a slight but definite amount. The reverse effect has also been verified; namely, a soft iron rod when rotated rapidly around its axis becomes magnetized.

The discovery of radioactive materials was also of enormous structural importance, as it gave us a means of studying directly the rays emitted from these materials. The rays were found to be of three kinds and were called by the first letters of the Greek alphabet. The α-rays have been found to be similar to positive rays, the β-rays similar to cathode rays, and finally the γ-rays to Röntgen-rays. Further investigation revealed that the α-rays were atoms of helium charged with a double positive charge of electricity, and that the β-rays were negatively charged particles with the charge of an electron.

These few remarks already make apparent the structural fact that electromagnetic phenomena exhibit characteristics quite similar to those of 'matter', some of their processes are atomic, they have inertia , .

In the older days we tried to apply macroscopic mechanical structural laws to electromagnetic phenomena, but were not very successful. The laws which applied to those sub-microscopic levels were seemingly different from those which applied to gross macroscopic levels, just as the psycho-logics of the individual differ from the psycho-logics of the mob.

An epoch-making semantic step was taken by Rutherford when he formulated the electromagnetic theory of the structure of 'matter'. In this theory the atoms represent complex structures built up of positive and negative electrons, and their number and arrangements (structure) determine the chemical and physical characteristics of the atom in question. The old structural dogma of the immutability of the elements became untenable; and today, theoretically, and in a few instances experimentally, it has been established that a transmutation of elements is not only a possibility but a rather well-established structural fact of 1933.

It should be noticed that once more one of the fanciful linguistic, structural, 'infinities' has been abolished. The elements appear as transitory processes with a 'life' of a *limited* span of years. The experimental structural evidence which physicists and chemists have gathered is overwhelming and, though the *positive* theories (verbal structures) may not always be satisfactory, the *negative* results leading to the rejection of the older theories are conclusive. This point is of supreme structural importance to us.

A short description of the different atomic models comes later in this chapter, but first something must be said about the old quantum theory, which represents at present the central problem in science and out of the solution of which the most revolutionary consequences are bound to follow.

The main problems of the quantum theory may be described and contrasted somehow as follows. If we take a line $X'X$ and select a point O as the origin, we may fix the position of a point P on this line by the co-ordinate x. In practice we find the values of x by measurement. If we assume that x varies *continuously* we can expect that by refined measurements we can find values of x as close together as we please.

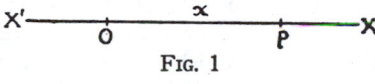

Fig. 1

Experiments show that for the processes going on 'inside of the atom' the structural conditions are somehow radically different. In comparing them with the above example, we should have to give our point only the freedom of occupying certain discrete points, let us say 1, 2, 3 . , but all fractional values, such as 1/2 or 2/3, would be impossible. If the only possible values of x are whole numbers, then the possibility of finding values of x as close together as we choose in order to make more precise measurements is excluded. If we find, for instance, that our x is neither 1 nor 2 nor 4, and cannot be larger than 4, the only solution is that our x must be 3.

These conditions, which are actually found in the atomic mechanics, represent an entirely new and unexpected structural state of affairs. The whole of the older quantum mechanics can be summarized in the statement, that its peculiarity consists in the fact that structurally characteristic *discrete numbers* make their appearance, and that the processes 'inside the atom' are to be described by *discrete* numbers.

The usual classical quantum mechanics demands that x be allowed to take all possible continuous values, but that the *integral values* of x represent the so-called stationary states through the quantum conditions. Under such conditions the intermediate fractional values had no meanings.

It is safe to say that, when in 1900 Planck formulated his quantum theory along the structural lines sketched here, it meant a complete and revolutionary structural and semantic departure from all accepted standards in life and science, for studying this world.

Planck has shown that it is impossible to explain the spectral distribution of the energy radiated by a black body under the older assumptions that energy can be divided indefinitely into smaller and smaller parts, but that it may be explained on the structural assumption that the energy exists in quanta of finite size $h\nu$, where ν is the frequency of the radiation and h is a constant ($h = 6.54 \times 10^{-27}$ erg sec.).

These observations lead to the revolutionary structural conclusion that the emission of radiation occurs *discontinuously*, and so the characteristic discrete numbers make their appearance.

It seems natural that because of this peculiar appearance of whole numbers, periodic processes such as rotations or oscillations should be closely related structurally with the quantum theory. As a matter of fact the most important structural and semantic reconciliation of continuous differential equations of the older mechanics with the appearance of discontinuous whole numbers has been solved by the newer quantum mechanics on this basis as explained in the following chapter.

The kinetic theory of heat and the atomistic theory of electricity have shown an enormous productivity. It is quite natural that these theories (verbal structures) should be highly workable, considering the structure of our nervous systems, as explained in the foregoing chapters. It was this principle of individualization which helped so greatly. The quantum theory is a structural attempt to extend this method of individualization, or the atomistic principle, to processes themselves.

As in the older days we introduced units or elementary quanta of mass, and later, an elementary quantum of electric charge, so in our newer knowledge we have need for an elementary quantum of action. Action is defined as energy multiplied by 'time', or $A = Et$.

Naturally such a product as energy multiplied by 'time' must play an extremely important structural and semantic role in this world of space-time, where nothing happens 'instantaneously', but all action requires 'time'. If we could discover some unit of action, we could change from the language of 'energy' and 'time' to the language of 'action' and 'times'. This language, by the way, is much more satisfactory and structurally closer to experience than the old languages. 'Action' as structurally defined (product of 'energy' by 'time') is one of the two fundamental entities of pre-relativity physics which have survived the Einstein revolution. It is really a universal term which we can apply without danger of degrading science into private gossip. From the neurological standpoint, as it deals with definite units and times, such a term has all the structural earmarks of an abstraction of highest order and of being really semantically important. Energy in space-time must by necessity be reformulated as 'action'. The quantum theory posits structurally that the action of physical processes is built up of a number of elementary quanta of action.

From the fact that electromagnetic waves and light-waves have one velocity, Maxwell concluded that light-waves are of an electromagnetic character, a conclusion which further experiments have fully justified. Einstein, in 1905, successfully applied the quantum structural principle to the theory of light, and in 1907 to the theory of heat of solid bodies.

The evolution of our theories concerning the internal structure of atoms has, until lately, closely followed our astronomical theories, but with the newest quantum mechanics this structural analogy seems less useful. The first atomic model on an electrical basis was proposed by J. J. Thomson. He assumed the atom to consist of a uniformly dense spherical volume charge of positive electricity, within which electrons described circular orbits. But the discovery of radioactivity and the fact that the alpha-rays could pass through several centimetres of atoms (which means penetrating through many thousands of atoms), without their direction being altered made these assumptions structurally untenable.

The Wilson photographs (Fig. 2) show clearly that a single atom can deflect the alpha-particle by a large angle which makes it clear that the nucleus of an atom must be considered as a very small part of the volume of the atom. The large deflections of the alpha-particles show also that the mass of the nucleus must be much larger than the mass of the deflected particles. Observations show also that the deflections increase with the atomic weight of the deflecting materials. These and similar facts led to the structural assumption that the mass of the atom is principally concentrated in the nucleus and that the mass of

FIG. 2

the electrons must be very small in comparison with that of the nucleus.

As the atoms are in general electrically neutral, we had to assume that the positive charges of the nucleus are compensated by the negative charges of the

44

electrons. These and other structural considerations led Rutherford to propose a different atomic model, which was much more successful for a while.

The Rutherford atom is supposed to be composed of a nucleus of positive electricity surrounded by negative electrons. The simplest atom is that of hydrogen, and was assumed to consist of one electron revolving around the simplest positive nucleus or proton, each having a charge, $e = 4.77 \times 10^{-10}$ e.s.u., and different masses, the mass of the nucleus or proton being 1845 times the mass of the electron. The other atoms represent more complex structures built up of protons and electrons, into the details of which we need not enter here. But this theory encountered difficulties of theoretical as well as of experimental character. Niels Bohr eliminated most of them by the application of the quantum theory to the atom.

For simplicity of writing in that which follows, I will use a descriptive language omitting in each statement 'we assume' . , but the reader should be continuously aware, that when we deal with the sub-microscopic levels we deal only with inferential units the representation of which involves a great many assumptions. For my purpose it is enough to stress: (1) the *negative* fact, that the structure of materials is definitely different from that which was assumed before the advent of the quantum theories, (2) that *in science*, inferential units represent abstractions of higher order and are as reliable as the lower order abstractions which we gather on the macroscopic levels, *if* they are treated semantically as *hypothetical* units. The layman should realize that his 'world-outlook' appears as full of assumptions as any scientific one, except that his assumptions are not conscious and cannot be verified, whereas most of the scientific assumptions are conscious and are *continually verified*.

In the older theory the orbits of the electrons were supposed to be arbitrary; in the Bohr theory the orbits have precedence, for which a definite magnitude, a whole number multiple of the elementary quantum of action is specified. We posit one-quantum orbits, two-quantum orbits . , to which definite values of the orbits, the velocity, the number of revolutions, and the energy correspond. In a one-quantum orbit, for instance, the velocity is supposed to be equal to

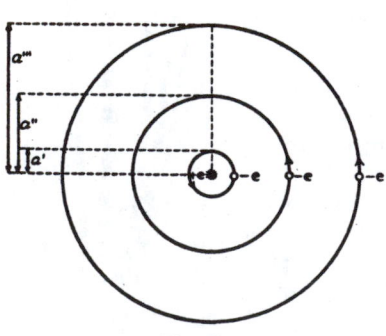

FIG. 3

$c/140$, that is one 140th of the velocity of light, and the number of revolutions equal to 6000 billions a second.

Bohr later modified his atomic model structurally by taking into account the movement of the nucleus. The electron was not supposed to revolve any longer around the proton, but both proton and electron were assumed to revolve around their common centre of gravity. In the simplest form the Bohr atomic model is shown in Fig. 3, representing the atom of hydrogen, which we assume to consist of a nucleus with one positive charge and of an electron that revolves about this nucleus. The nucleus is designated by a star, the three circles represent the possible orbits for the electron. The orbit of radius a' is the most stable, and usually the hydrogen electron is supposed to be found there, but through the action of heat or electric fields or collisions . , the electron may be removed to one of the outer orbits a'' or a'''. Such a condition is not so stable, and sooner or later the electron is assumed to return to the orbit a'. During these transitions of the electron, energy is radiated. This structural model is similar to the copernican planetary system, the planet-electrons revolving around the sun-nucleus.[1]

The above diagrams show schematically the supposed structure of some of the simpler atoms. Fig. 4 represents the hydrogen atom, consisting of one proton and one electron revolving around the proton. The mass of the proton is about 1845 times the mass of the electron, and we assume that the proton effectively gives us the mass of the atom.

Fig. 5 represents the neutral atom of helium. Its nucleus consists of four protons and two electrons, and it has two revolving electrons; in all, four protons with four positive charges and four electrons with four negative charges, the eight charges just neutralizing each other.

Fig. 6 represents a helium atom which has lost one electron. It has, therefore, four positive charges and only three negative charges. Such an atom has a resultant positive charge, and is denoted by He_+. If the helium atom loses two electrons it is doubly charged with a positive charge (He_{++}).

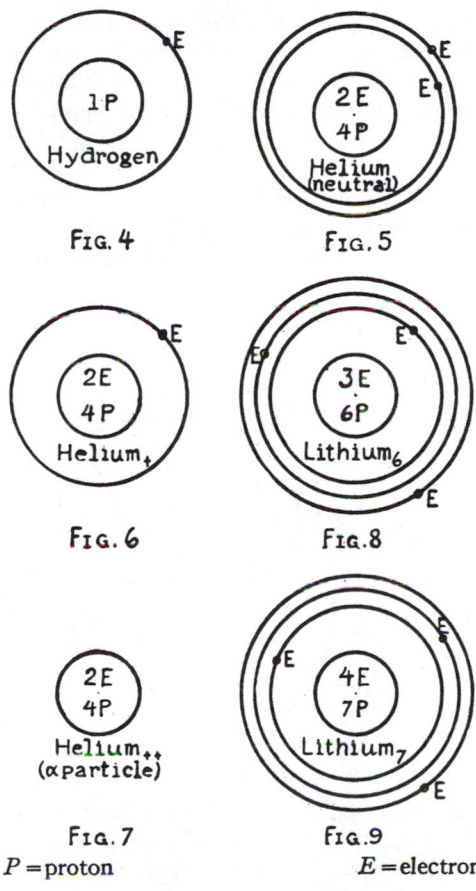

FIG. 4

FIG. 5

FIG. 6

FIG. 8

FIG. 7

FIG. 9

P = proton

E = electron

The helium nucleus He_{++}, as shown in Fig. 7, represents the particle emitted from radio-active materials.

Lithium consists of two isotopes; that is, two elements which appear extremely similar to each other in physical and chemical characteristics, but differ from each other in the number of electrons and protons. In Fig. 8 is shown lithium$_6$, with 6 protons and 3 electrons in the nucleus, and 3 revolving electrons. In Fig. 9 is shown lithium$_7$, with 7 protons and 4 electrons in the nucleus and 3 revolving electrons.[2]

In general terms, Bohr tried to account for all other atoms on the base of the structure of the hydrogen atom. The next important generalization and extension of the Bohr theory was accomplished by Sommerfeld about 1915. The achievement of Sommerfeld can be compared with the advance which Kepler made over the copernican theory of planetary motions. Copernicus considered the planetary orbits as circular. Kepler* considered them as elliptical, and thus introduced a tremendous structural advance in astronomy. Sommerfeld replaced the circular orbits of Bohr by elliptical ones. The theory became much more complicated, because a circle is given by one magnitude; namely, its radius, while an ellipse needs two data, its major and minor axes, and so two quantum numbers for the specification of an orbit. Sommerfeld also introduced some of the results of the Einstein theory; for example, that the mass of a body also depends on its velocity. Since the velocity of the negative electrons in the atom is supposedly very large it was quite probable that the relativity considerations should be appreciable. According to the Einstein theory, the faster a body is moving the greater is its mass. In an elliptical orbit the electron should have a larger mass at the perihelion than at the aphelion, and so the orbit would not be exactly an ellipse but the perihelion would advance slightly at every revolution.

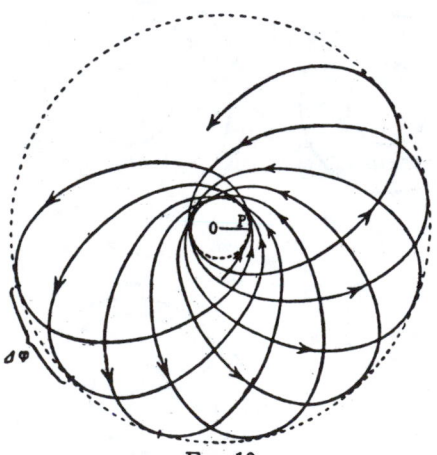

Fig. 10 gives us the relativistic Kepler orbit as introduced by Sommerfeld. O is the fixed focus in which the nucleus is situated and P is the initial position of the perihelion. The motion of the perihelion occurs in one sense with that of the orbit.[3]

FIG. 10

The last analogy in the structure of the atom taken over from astronomy was introduced in 1925, when Goudsmit and Uhlenbeck proposed their theory of the spinning electron. The electron was supposed to be spinning about its axis like a planet or a top. A similar notion was used by Compton in 1921, in connection with the magneton, but the notion of using the spinning electron for

---

*Kepler's first law states: 'The planet moves in an ellipse, at one focus of which the sun is situated. Perihelion is that point of a planet's orbit at which it is nearest the sun. Aphelion is that point of a planet's orbit at which it is farthest from the sun.'

the solution of a structural difficulty in the quantum theory, and thus assigning a fourth degree of freedom to the electrons, originated with Goudsmit and Uhlenbeck independently of the work of Compton.[4] It is not necessary for our purpose to follow all the further refinements of the classical theories. Suffice it to say that scientists work under uniquely severe mutual supervision, and that any theories advanced in science are taken under consideration only when the new theories agree better with experiments, and when they also prove structurally fruitful in predicting new experimental facts, which again must stand the test of experiment.

The Sommerfeld orbits have proved to be an advance over the older Bohr orbits, but they had also to be refined to take into account that the electron does not seem to revolve around a simple nucleus but around a *core* consisting of the nucleus and one or more electrons; and so again we had more complex orbits.

For our semantic purpose it is enough to say that to the best of our knowledge (1933), this world appears entirely *different* from what our primitive ancestors knew thousands of years ago, and perhaps from what the average layman knows today. As the problems of 'sanity' represent problems of semantic *adjustment*, and adjustment means adjustment to something—in this case to the structures of the world around and in us—it appears imperative that we should take into account the best knowledge we have of these structures.

The few remarks given above about the structure of 'matter' already show unmistakably that the old 'matter' is not so very 'material', so very 'solid', so very definite, as we once assumed it was; but it represents a *process*. We see that our nervous system, because of its gradual growth and evolution, has developed different levels or strata; our 'knowledge' also has different levels or strata, operative as-a-whole, although different aspects of it can be analysed in terms of order. The reader should realize that because of the old *s.r* we still 'need' some 'bits' of something to speak about. It is a *linguistic* semantic consequence of our pre-scientific, *el* language, which posits absolute 'matter', 'space', and 'time'. Thus, through a process of identification, we ascribe to these terms objective existence. In the old manner of speaking the term 'is' of identity played the main semantic havoc.

In the older days electrons were often taken as 'bits' of something or other. For the layman a 'bit' was identified with 'matter'; and here a great deal of confusion comes to light. Even a 'bit' of something is not necessarily material. Materials, by *definition*, are supposed to exhibit colour, temperature, hardness , . A 'bit' which did not have these characteristics would *not* be material by definition.

Although the 'electron' is defined as an electrical charge, in the older days we had the habit of considering the electrons as some definite 'bits' of something, some kind of 'matter'. Through a process of objectification we made them revolve in definite 'orbits', with definite 'velocities'. , which implies the definite application of *terms* such as 'space' and 'time', derived from *macroscopic gross experience*, but not necessarily applicable to the sub-atomic levels.

What has been said here about the structure of 'matter' is quite sufficient for our purpose. Here, as always, the *negative*—the 'is not'—results count. We are in a position to realize by now that the overwhelming evidence which science gives and which would be impossible to repeat here, shows us a structural picture of the world of tremendous complexity, beauty, and mystery of a structure undreamed of by our primitive ancestors who formulated the current mythological structures which moulded our older *s.r* and languages.

We can sum up, for our purpose, what we know about the structure of 'matter' somewhat as follows. The bits of materials visible and invisible to the unaided eye seem to be less simple than we assume them to be and to the best of our knowledge (1933) represent extremely complex processes of a dynamic structure. It appears also that our usual forms of verbal representation which were built by our primitive ancestors are not similar in structure to the world and so are not fit to represent the happenings going on on the un-speakable levels. As all our knowledge is due to the structure and function of our nervous system, which represents an abstracting mechanism, all our knowledge therefore, appears as some kind of abstractions of different orders, on different levels, of different character, and of varying precision and intensity, resulting in various definite general or individual *s.r.*

To bring what is said here to the lower level of abstracting; namely, to the level of structural visualization and feeling, we may use the rough analogue of an electric or mechanical fan. When such a fan rotates we *see a disk*, simply because our nervous system was evolved under natural conditions necessitating integration, and so does not discriminate between the rotating blades. The separate rotating blades are visually abstracted by us as a single solid disk, although there is no disk present.

To the best of our knowledge atoms represent very minute energetic configurations or dynamic structures where extremely rapid processes are going on, which our nervous system abstracts as 'solid'. Judging by our present standards in science and the amount of knowledge we have we may consider that science in the days of Newton (1643-1727) was in its infancy. During that period we knew a little about the shining specks we see in the skies, and more about the rough macroscopic facts of our daily experience. The genius of Newton not only advanced the detailed knowledge of his day in many branches of science, but also formulated two general theories. One was the differential and integral calculus, which he discovered independently of his contemporary Leibnitz, the other was what we call mechanics.

In Newton's era the problems of macroscopic, microscopic, and sub-microscopic levels of investigation had not yet arisen in the modern sense, although in formulating the differential and integral calculus a theoretical structural step was taken toward the analysis of the processes on the subtler levels. Quite naturally we applied the wisdom we derived from Newton to all phases of life and knowledge. With the advent of more detailed structural knowledge of electromagnetic phenomena which occur on sub-microscopic, as well as macroscopic levels, difficulties began to appear. It seemed as if the newtonian mechanics were not entirely applicable to these new and smaller scale phenomena. Finally Maxwell (1831-1879) produced his famous theory of electromagnetism. This theory appears structurally at variance with the classical mechanics. Attempts were made to reconcile both kinds of phenomena in one theory. The problems of macroscopic, microscopic, and sub-microscopic structure and levels came to the foreground.

With the advent of the quantum theory further difficulties made their appearance. It became quite obvious that neither the classical 'continuous' mechanics, nor the classical electromagnetic theory could fully account for the 'discontinuous' quantum facts. The situation became acute and bewildering. The Einstein theory with its profound structural semantic and methodological revolution liberated us from our semantic delusions of the uniqueness, absoluteness, and 'objectivity' of 'matter', 'space', and 'time'. It built up a new semantic attitude in the younger generation of scientists already educated on this new structure, and therefore unhampered by the old prejudices. New theories are now being formulated along increasingly more constructive and creative lines.

It is true that as yet neither 'psychologists' nor 'philosophers' have paid enough attention to the subjects discussed here, and so have not made us conscious of the structural and semantic problems involved. However, the Einstein theory has had a profound structural influence on the semantic attitudes of the younger scientists, though in the main they are unconscious of this fact.

The main issues at hand are twofold. One is semantic; namely, to inculcate the permanent structural feeling that words are *not* the things they stand for. If applied habitually, this leads to the rejection of the term 'is' of identity. The other is to replace old languages and methods by structurally new languages and new methods, in which when we describe ordered happenings, we describe the functioning, the behaviour . , by speaking more in a language of what something 'does' than in the old language of what something 'is', which as we have seen *must* be always structurally fallacious and semantically dangerous.

The reader should not take lightly these most general structural and semantic issues. They are unusually important for sanity. When they are formulated we can pass them on, and train children in the new *s.r* quite easily. It is much more difficult, after training a child thoroughly in the *old* vicious structural semantic habits of identification, eventually to have to appoint a guardian angel to watch him day and night to remind him that a word *is not* an object , . Such a procedure would lay a terrific strain on us and on the guardian angel. It would probably be also very expensive, judging by our present earthly substitutes for the 'heavenly powers'.

Once this is realized and applied, the second issue becomes a purely structural linguistic one. There is no *a priori* reason why a language which applies to one level should apply to another.

With these two main issues in view, it is readily understood why modern science tries so hard to develop functional languages and methods in order to be able to describe in terms of order happenings and processes which are observed. Something similar could be said about *all* theories which postulate too much of a definite mechanism, usually involving some identification somewhere.

The slightest discrepancy between such a theory and observation eliminates the theory as structurally unsatisfactory; while theories which succeed in not postulating mechanisms, and so are formulated in a functional language, last much better. One of the enormous advantages of the Maxwell electromagnetic theory is the fact that it describes the behaviour of electricity and magnetism while hardly positing any mechanism at all. A similar statement applies also to the Einstein theory.

The above general remarks are extremely well illustrated by the newer quantum mechanics.

The classical theories, as usual with scientific theories, were very satisfactory in many respects, but not in all; which is an unattainable ideal always demanded from a good scientific theory. They also postulated too much of a definite mechanism, which was the result of, and led to, the semantic disturbance called identification. Indeed, I have read an address by a prominent physicist in which he claims to have 'seen', and invites everybody else to 'see', an 'electron'. He challenges his critics, and seems to feel like fighting—a quite usual result of identification. Electrons represent *inferential entities*, and as such cannot be 'seen', but only inferred, which does not detract at all from the importance of the 'electrons'. The 'seeing' business was good enough in the infancy of science, but not in 1933. We 'see' the stick broken in water, the camera records it as broken, and yet it is not broken. We 'see' the fan as a disk, the camera records it so, but there is no disk. We 'see' a 'solid' piece of wood or stone, which under the microscope proves to have a very different structure , .

In the older days the electrodynamics of moving bodies presented difficulties quite similar to the difficulties encountered in the quantum mechanics. Einstein by an epoch-making stroke of genius solved the problem by observing that, in the languages in question, we operated with a notion of 'simultaneity' which did not correspond to any observable structural phenomenon in the physical world. He discovered that it is impossible to establish the simultaneity of two events occurring at different places, and that a thorough revision of our old theories is necessary in this connection. Einstein formulated a procedure, a method for measurements, taking into account the known laws of the propagation of light and electromagnetic phenomena. He once more established the most important semantic thesis that the laws of nature are relations which are discovered between events which are actually observed, or which are *fundamentally observable*.

It appears that in the older quantum mechanics there were introduced some objectified entities which were never observed, as, for instance, the positions, velocities, and periods of 'electrons' inside the atom. How indeed could we find lengths and 'times' *inside* the atom? Such a procedure requires the introduction of rods and clocks, which themselves consist of atoms; so that *inside* the atom such a procedure cannot be applied. We see clearly that all such conclusions are of an indirect character; but of course such conclusions should be based on some observable facts, and not only on our freedom to use words in any manner whatsoever. It follows that we must give up a language that speaks in terms of the 'position' of an electron at a given 'time'. , and use instead a language that describes observable characteristics, as, for instance, energy levels which are directly measurable by electron impacts and the frequencies which are derivable from them, the intensity and polarization of the emitted waves . ; instead of electronic 'motions' inside the atom, which never are and never *can be* actually *observed*. It is structurally indispensable to look for such data which are actual or at least can be observed.

As words are not the things we speak about, and structure is the only link between them, structure becomes the only content of knowledge. If we gamble on verbal structures that have no observable empirical structures, such gambling can never give us any structural information about the world. Therefore such verbal structures are structurally obsolete, and if we believe in them, they induce delusions or other semantic disturbances.

CHAPTER XLI

THE NEWER 'MATTER'

The twofold nature of light as a light-wave and as a light-quantum is thus extended to electrons and, further, to atoms: their wave-nature is asserting itself more and more, theoretically and experimentally, as concurrent with their corpuscular nature. (481)   A. SOMMERFELD

The concepts of wave amplitude, electric and magnetic field strengths, energy density, etc., were originally derived from primitive experiences of daily life, such as the observation of water waves or the vibrations of elastic bodies. (215)   W. HEISENBERG

The problem of quantum theory centers on the fact that the particle picture and the wave picture are merely two different aspects of one and the same physical reality. (215)   W. HEISENBERG

To me it seems extraordinarily difficult to tackle problems of the above kind, as long as we feel obliged on epistemological grounds to repress intuition in atomic dynamics, and to operate only with such abstract ideas as transition probabilities, energy levels, etc. (466)   E. SCHRÖDINGER

. . . for visualization, however, we must content ourselves with two incomplete analogies—the wave picture and the corpuscular picture. (215)
                                                         W. HEISENBERG

Not every physicist is an epistemologist, and not everyone must or can be one. Special investigation claims a whole man, so does the theory of knowledge. (326)   E. MACH

The following chapter was written in 1928 and since then the newer quantum mechanics has been developed much further, proved enormously fruitful, and has been repeatedly supported by experiments. The literature on this subject is steadily accumulating, the most important classical memoirs by the originators of this new scientific trend have been collected into book form and are now easily accessible. There is also a large number of excellent technical, as well as non-technical presentations. On reading in December 1932 what I had written in 1928, I find that although from some aspects the presentation may be considered unsatisfactory and antiquated, yet the epistemological side of the older presentation remains valid. So it seems advisable to retain this chapter and add only a few further $\bar{A}$ suggestions.

It is known that practically all creative and constructive physicists, who have produced revolutionary and lasting works, were interested in epistemology. There are many physicists who know as much physics as an Einstein, for instance, yet Einstein remains quite unique and his work is to a large extent responsible for the present revolutionary developments of physics. The reason is simple. Einstein has corrected a long established epistemological fallacy, which can be expressed in my language as the rejection of the structural fallacy of elementalism in a limited yet very important field of physics. He also established and applied new fundamental epistemological principles, which is another way of saying that he established new standards of evaluation in physics, as for instance, that we should never postulate entities which cannot possibly be observed, that the 'laws of nature' should be formulated in terms of generally invariant relations expressed in tensor equations , .

The weakness of the system of Einstein, resulting in many futile criticisms, lies in the fact that he eliminated elementalism in one vital region of physics, but he did not formulate the *general epistemological principle* of *non-elementalism*, which should be applied everywhere, daily life included. This he could not have accomplished without a still deeper enquiry into the mechanism of time-binding, which produces all science, and which leads to the discovery of the fundamental fallacy in the use of the 'is' of identity. Only after the elimination of this remainder in us of the primitive man, does *structure* become the only possible link between the objective and verbal worlds, and becomes also the only possible content of 'knowledge'. 'Similarity of structure' then demands the complete and general elimination from science and life of any elementalism.

The strength of the newer quantum mechanics lies in the fact that the younger physicists have accepted the new epoch-making einsteinian standards of evaluation or epistemological principles; the weakness lies in the fact that the scientists do not realize that the fallacy of elementalism is entirely general and vitiates *all* scientific outlooks. No one can produce satisfactory theories, nor evaluate, nor interpret them properly as long as he continues to use the few-valued and *elementalistic* 'logics' and 'psychologies', which are at present always found at the bottom of any 'evaluation' or 'interpretation'.

The latest work of Dirac goes very far in the direction of building $\overline{A}$ physics by establishing his language of transformations, states, observables . , ascribing *structure* to protons, magnetic poles . , but even Dirac does not seem to realize the general fundamental $\overline{A}$ issues involved. Dirac says: 'The description which quantum mechanics allows us to give is *merely* a manner of speaking which is of value in helping us to deduce and to remember the results of experiments and which never leads to wrong conclusions. *One should not try to give too much meaning to it.*'[1] (Italics are mine.) The italicized words show that even Dirac does not realize fully the mechanism of identification, as otherwise he would not have used these words in this form. If we entirely abandon identification, then a theory or a book, being verbal, represents nothing else but special language; there is no 'merely' about that either, structure being the only possible link between the non-verbal and verbal worlds. Instead of warning the reader 'that one should not try to give too much meaning to it', we must simply *insist* that the *only* 'meaning' should be looked for in structure . , the 'too much meaning' always indicating inappropriate evaluation and ultimately semantic disturbances.

Current physical literature shows that the main problems of 'interpretation' depend on the *solution* of the *m.o* problems of 'observation', 'reality', 'fact'. , and border on the *scientific* solution of the problems of pathological 'delusions', 'illusions', and 'hallucinations', all of which involve the fundamental issues of the elimination of *identification*. But once the $\overline{A}$ issues are structurally formulated and applied in practice, they result in a $\bar{A}$, *non-el*, ∞-valued orientation which involves the recognition of the *multiordinality* of terms . , which also solves the problems of quantum 'interpretations', the details of which I cannot enter into here.

Originally the quantum writers had an inclination to ascribe 'physical meaning' to waves. The present tendency of specialists is to regard the waves as 'purely symbolic', forgetting that experimentally something else besides the symbols 'bends around the corners'. From a $\bar{A}$ point of view, when the problems of the multiordinality of such terms as 'observation', 'fact', 'reality'. , are understood we will have to ascribe 'physical reality' to the waves, ascribe finer *structure* to the 'electron', . We would also have to abandon the A 'particle'-orientation and treat the 'electron', 'proton'. , in a $\bar{A}$, ∞-valued way as minute *fields*, which under the present experimental conditions behave as 'particles'. This $\bar{A}$ field-orientation suggests a great many possible interpretations, impossible in the A 'particle'-orientation.

Mathematically, the geometry of 'space filling' curves would have to be elaborated further so that we would better understand the *structure of plenums*, and this knowledge should be applied to physics.

We should also perform a direct series of experiments with a more elaborate Faraday box. A small wooden laboratory should be isolated from the rest of the world by every available *energetic screen* and physical experiments repeated under such new dynamic conditions. Technically the winding of, say, a foot thickness of insulated wires for different currents would not present any difficulties except for the door which should be also in the circuit. The eventual probable results of experiments in such a laboratory under different conditions could be calculated in advance, and it may be fairly well anticipated that at least significant discrepancies between the calculations and actual experiments would appear, throwing new light on the structure of the space-time plenum, the eventual connection between gravitation and electromagnetism , . Arguments alone will not help in this field and only experiments will point the pathway.

Because of the extensive literature dealing with the new quantum mechanics it seems unnecessary to dwell upon it any further, except by expressing the hope that some mathematicians and physicists will master the $\bar{A}$ ∞-valued orientation and will revise the theories now existing.

Section A. Introductory.

The new researches in the structure of the materials of the universe proceeded under unique conditions. On the one hand, since Planck in 1900 originated the quantum theory, which earlier elaboration we now call the classical quantum theory, the amount of experimental facts pointing in the direction of *some* quantum theory had become very convincing, yet, on the other hand, the lack of a structurally satisfactory theory to co-ordinate these new experimental facts was becoming distressing.

It appeared as though there were either a lack of 'geniuses' able to produce the required new theories, or the geniuses existed, but were unable to function properly. They seemed to suffer from a semantic blockage due to some identification which successfully prevented broader and unhampered vision.

Presently the co-ordinating theories *were* produced. In the production of the new quantum mechanics we see at work the unconscious semantic liberating influence of the $\bar{E}$ and $\bar{N}$ systems which had been developed and which we have already analysed. The classical theories had come to a structural impasse, but with the advent of younger scientists, who were educated in the theoretical semantic freedom of these new systems, the semantic blockage due to ascribing 'objectivity' to 'matter', 'space', and 'time' was removed, creative forces were released, and these young scientists proceeded to construct the structural formulations needed. They are now rightly hailed as geniuses.

It is astonishing how these young post-einsteinian scientists of various lands and different inclinations produced, independently and practically simultaneously, various new quantum theories, using different methods and different mathematical, as well as national, languages. When these different theories were studied and compared they were found to amount practically to one theory, but expressed by different mathematical languages. Now the use of different mathematical languages to express one group of experimental facts brings an additional benefit in that it gives diversified verbal structural information about the problems at hand. Since these developments are very recent and progress has been extremely rapid it is hard to keep track of the status of the problem.

In this account of the newer quantum mechanics I will emphasize only the structural and semantic side, treating the different theories as behaviour of their respective authors, and as illustrating the issues above mentioned. From this point of view, we are not interested in discussing to what extent the given theories are 'true' or 'false', which means no more than similar or dissimilar to the world in structure. We are interested directly in those semantic aspects of human behaviour which have been neglected. When 'Smith' puts a black mark on white paper, it is to be called *human behaviour*, and behaviour *unique* for man. Our analysis does not involve the question of the validity of his doings. But since he did it, let us analyse his doings.

In speaking about theories as complex and technical as the newer quantum mechanics, it is practically impossible to give a satisfactory account in a non-technical way. Such theories have not yet been worked out thoroughly enough, many points are still not clear, and no proper evaluation is possible at present. These difficulties are really immaterial to our present purpose, because these theories are *empirical facts on record*, and they throw significant light on human behaviour. It is not here proposed to make reflections on the world around us, but to analyse a certain linguistic, structural form of human behaviour.

From the point of view of structure and *s.r*, the classical quantum mechanics would be quite enough for *sanity* and adjustment; it is enough to realize that the still older theories of 'matter', 'space', and 'time' are *el, structurally fallacious* and represent only primitive identifications.

The advent of such a crop of geniuses and of several theories expressed quite differently, yet nearly equivalent, is an event of deep human semantic significance. It helps to understand the working of the human nervous system, and is in accord with the present general theory.

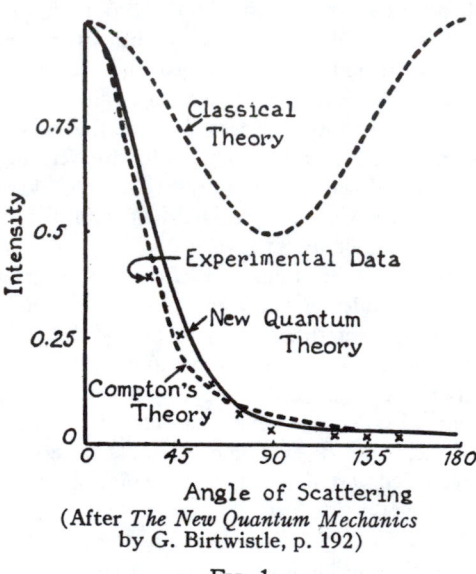

Angle of Scattering
(After *The New Quantum Mechanics*
by G. Birtwistle, p. 192)

Fig. 1

From the point of view of the physicist, these new theories are a marked structural improvement over the classical theory, a fact which can best be illustrated by a diagram of a special case. In Fig. 1, the crosses indicate the experimental data, the curves indicate the results as predicted by the classical theory, by the Compton theory, and by the new quantum mechanics. It should be noticed that the new quantum theory appears much more in accordance with the experimental data than the older theories. This fact is of great structural and semantic importance to us as well as to the physicist.

Section B. The nature of the problem.

At this point we may explain briefly the nature of the problem that was demanding solution.

We have become familiar with the use of co-ordinates. This procedure has been generalized and has given rise to 'generalized co-ordinates'. These are defined as *arbitrary* variables which represent not merely lengths but may also represent angles, surfaces, volumes . , though they must be capable of representing the $3n$ orthogonal co-ordinates. If, in a special case, we make the number of generalized co-ordinates equal to the number of degrees of freedom which the system has, these s generalized co-ordinates can be regarded as independent of each other. If we denote by q_i the generalized co-ordinate, q_i ($i = 1$ to s), then the orthogonal co-ordinates of any of the n particles can be represented as definite functions of the generalized co-ordinates, so that

$$x_h = f_h (q_1, q_2, q_3, \ldots q_s) . \tag{1}$$

We know that kinetic energy is represented by $mv^2/2$ where m represents the mass and v the velocity, or the 'time' derivative of the 'space' travelled.

If we want to find the value for the energy we must differentiate each of the $3n$ equations (1) with respect to 'time', which gives the components of the velocity, square them, multiply them by the corresponding masses and add them together to find the double value for the energy.

For simplicity we will denote the 'time' derivatives by the chosen letter, but with a dot over it (newtonian method). Thus,

* see page xii

$$\dot{x}_h = \frac{\partial x_h}{\partial q_1}\dot{q}_1 + \frac{\partial x_h}{\partial q_2}\dot{q}_2 + \cdots + \frac{\partial x_h}{\partial q_s}\dot{q}_s \qquad (2)$$

Squaring (2), we have

$$\begin{aligned}
\dot{x}_h{}^2 &= \left(\frac{\partial x_h}{\partial q_1}\right)^2 \dot{q}_1{}^2 + \left(\frac{\partial x_h}{\partial q_2}\right)^2 \dot{q}_2{}^2 + \cdots + \left(\frac{\partial x_h}{\partial q_s}\right)^2 \dot{q}_s{}^2 \\[2mm]
&\quad + 2\frac{\partial x_h}{\partial q_1}\frac{\partial x_h}{\partial q_2}\dot{q}_1\dot{q}_2 + 2\frac{\partial x_h}{\partial q_1}\frac{\partial x_h}{\partial q_3}\dot{q}_1\dot{q}_3 + \cdots \\[2mm]
&\quad + 2\frac{\partial x_h}{\partial q_2}\frac{\partial x_h}{\partial q_3}\dot{q}_2\dot{q}_3 + \cdots
\end{aligned} \qquad (3)$$

The last expression (3) can be simplified:

$$\dot{x}_h{}^2 = \sum_{i=1}^{i=s}\sum_{k=1}^{k=s}\frac{\partial x_h}{\partial q_i}\frac{\partial x_h}{\partial q_k}\dot{q}_i\dot{q}_k . \qquad (4)$$

It is easy to see that if in (4) we put $k = i$ we will have square members, and when $i \neq k$, every term will occur twice and so the above abbreviation (4), covers the formula (3).

If we write similar expressions for all of the $3n$ orthogonal co-ordinates, multiply by the corresponding masses and then add them together we obtain twice the value for the kinetic energy $2L$.

$$2L = \sum_{i=1}^{i=s}\sum_{k=1}^{k=s} c_{ik}\,\dot{q}_i\,\dot{q}_k, \text{ where}$$

$$c_{ik} = \sum_{h=1}^{h=n} m_h \left[\frac{\partial x_h}{\partial q_i}\frac{\partial x_h}{\partial q_k} + \frac{\partial y_h}{\partial q_i}\frac{\partial y_h}{\partial q_k} + \frac{\partial z_h}{\partial q_i}\frac{\partial z_h}{\partial q_k}\right].$$

The coefficients in the expansion of c_{ik} depend only on the values of the generalized co-ordinates and are independent of the value of the time-derivatives. The time-derivatives can be properly called *generalized velocities*, and we may denote them by q_i.[2]

In establishing formulae for the quantum theory we want to be as general as possible and not restrict ourselves to vibrational energy only. But we want to take into consideration *any* arbitrary point-mass, independently of whether we assume this point to be charged or not.

We define the momentum or impulse as the product of the mass and the velocity, or, $p = mv$. If, instead of denoting our co-ordinates by x, y, and z, we use the generalized co-ordinates q_i, we would have for the magnitude and direction of the velocities the time-derivatives of the co-ordinates; namely, $\dot{q}_i$, where $\dot{q}_1 = \dot{x} = dx/dt$, $\dot{q}_2 = \dot{y} = dy/dt$, .

If p_1, p_2, p_3, represent the corresponding components of the momentum or impulse, then we would have $p_i = m\dot{q}_i$. $\qquad (5)$

We should notice that the dynamical triplet of impulse co-ordinates occurs conjointly with the geometrical triplet of the co-ordinates of position. The second law of motion tells us that 'the change in momentum is proportional to the impressed force and takes place in the direction in which that force acts'. If we assume that the force K is derivable from the potential energy E_{pot}, (a function of q_i), then we have $\dot{p}_i = K_i = \dfrac{-\partial E_{pot}}{\partial q_i}$. (6)

The kinetic energy (E_{kin}) is represented by

$$E_{kin} = \frac{m}{2}\,(\dot{q}_1{}^2 + \dot{q}_2{}^2 + \dot{q}_3{}^2) = \frac{p_1{}^2 + p_2{}^2 + p_3{}^2}{2m}$$

where by (5), $\dot{q}_i{}^2 = p_i{}^2/m^2$. We call the total energy, which is represented as the sum of the kinetic and the potential energy, as expressed in terms of the generalized co-ordinates and momenta, the hamiltonian function H. Then we have:

$$H\,(q,\,p) = E_{kin} + E_{pot},\quad \frac{\partial H}{\partial q_i} = \frac{\partial E_{pot}}{\partial q_i},\quad \frac{\partial H}{\partial p_i} = \frac{\partial E_{kin}}{\partial p_i} = \frac{p_i}{m}. \qquad (7)$$

From (5), (6), and (7), we get the fundamental equations of motion,

$$\frac{dq_i}{dt} = \frac{\partial H}{\partial p_i},\ \text{and}\ \frac{dp_i}{dt} = -\frac{\partial H}{\partial q_i}.$$

The above hamiltonian, or canonical, form of the equations is remarkable because it preserves its form if any arbitrary co-ordinates are introduced; it is invariant under the transformation of co-ordinates. The equations hold not only for an individual point-mass but also for any arbitrary mechanical system. For arbitrary co-ordinates and systems the momentum or impulse p is defined by

$$p_i = \frac{\partial E_{kin}}{\partial \dot{q}_i},$$
so that the

kinetic energy is expressed as a function of the q_i's and their derivatives the $\dot{q}_i$'s.

To help visualization we can construct and consider the p and q as rectangular co-ordinates in two dimensions in the phase plane of our system. In this plane the sequence of those graph-points that correspond to the successive states of motion of the system represent the phase paths or phase-orbits. The characteristic structural feature of the quantum theory is that it selects a discrete family of phase-orbits from the infinity of possible orbits.

We next consider a point-mass m that is bound elastically to its position of rest, and which can move to either side of the central position only in the direction $x = q$, or its reverse, when experiencing a restoring force. We call such point-mass a linear oscillator. If we wish to emphasize that our oscillator is capable only of definite vibrations, on account of its elastic attachment, we call it a 'harmonic oscillator'. If the vibration number, or the frequency of the oscillator, which is represented by the number of its free vibrations per unit of 'time', is denoted by ν, then the vibration is represented by $x = q = a \sin 2\pi\nu t$.

The impulse becomes $p = mv = mq = 2\pi\nu ma \cos 2\pi\nu t$. The phase-orbit is represented by an ellipse in the p-q-plane and is given by the equation $\dfrac{q^2}{a^2} + \dfrac{p^2}{b^2} = 1$, where the minor axis $b = 2\pi\nu ma$.

In our family of orbits the phase area between two orbits is equal to the quantum of action h. Sommerfeld regards h as an elementary region or element of the phase area, and considers it as the definition of the Planck quantum of action h. If W_n represents the energy of the oscillator when it describes the n-th orbit, then $W_n = nh\nu$. In these orbits the energy appears as a whole multiple of the elementary quantum of energy; $\epsilon = h\nu$, and $W_n = n\epsilon$.

We call *stationary states* of the oscillator those states which the oscillator may pass through without cessation and without loss of energy, or, without radiation.

When an oscillator retains its stationary state, its energy is constant and its graph appears as an ellipse of the family in the phase plane. However, when the energy of the oscillator changes and jumps over to a smaller orbit, it emits energy. When it passes to a larger orbit it absorbs energy. The emission and absorption of energy occurs in multiples of the energy quantum, ϵ.

The graphs of the system in the phase plane are restricted to certain 'quantised' orbits. Between each orbit and its successor there is an elementary region, of area h. The n-th orbit, if closed, has as area nh. Or, expressed symbolically, $\int p\,dq = nh$. This integral is called the phase integral and is taken along the n-th orbit.

The quantum hypothesis can be structurally formulated so that the phase integral must be a whole multiple of the quantum of action h. This form of the classical quantum postulate is more general than the original formulation of Planck, although it includes the latter as a particular case.

In case of a rotating point-mass, a similar analysis gives us $E_{kin} = \dfrac{p\dot{q}}{2}$ and when $\nu = \dfrac{q}{2\pi}$, $E_{kin} = \dfrac{nh}{2}\dfrac{\dot{q}}{2\pi} = \dfrac{nh\nu}{2}$ where ν represents the rotation frequency of the rotator, or the number of full revolutions per unit of 'time', and takes the place of the vibration number of the oscillator.

In the classical theory, the quantised states were distinguished from all other possibilities by the characteristic whole numbers, and so we had a network. In a quantum orbit the 'electron', if undisturbed, was supposed to move permanently without resistance and not to emit radiation. The phase-space, representing the manifold of the possible states, including non-stationary states, is crossed, mesh-like, by the graph curves of the stationary orbits. The size of the meshes is determined by Planck's constant h.[3]

Section C. Matrices.

The older quantum mechanics forms an elaborate system, and we have a large accumulation of numerical data on record. Some of these data corroborated the older theories nicely, but some data were in contradiction to the classical

45

theory. The problem was not to discard the numerical data, which, whatever they mean structurally, represent quite solidly established data, but to find new equations which would be satisfied by these facts. Now 'new equations' really mean *languages of new structure*, and therefore new formulations had to be discovered.

Dealing with tables which give special theoretical data, it was natural to start with a calculus which deals with such numerical special tables. Such a calculus had been developed long ago, and was called the matrix calculus. Later on, when matrices themselves were treated as complex quantities, and still later, as operators, we were enabled to pass to the more developed calculi which use ordinary differential equations. The new quantum theories give us a unique case, in which several mathematical methods have been used at once and of which the results are fairly in accord.

At this point it is advisable to give a few structural explanations of these mathematical notions, including the matrix calculus. If we have two equations of the first degree with two variables; namely, $a_1x+b_1y=c_1$ and $a_2x+b_2y=c_2$, the solution of these equations takes the form: $x=\dfrac{b_2c_1-b_1c_2}{a_1b_2-a_2b_1}$; $y=\dfrac{c_2a_1-c_1a_2}{a_1b_2-a_2b_1}$.
The common denominator of the two solutions can be written in a two-dimensional table

$$\begin{vmatrix} a_1, & b_1 \\ a_2, & b_2 \end{vmatrix} \qquad (1)$$

which is understood as the product of the upper left-hand number and the lower right-hand number, minus the product of the lower left-hand and upper right-hand numbers.

Similarly, the numerators of these solutions can also be represented in the form of two-dimensional tables; namely,

$$b_2c_1-b_1c_2 = \begin{vmatrix} c_1, & b_1 \\ c_2, & b_2 \end{vmatrix} \quad (2) \text{ and } \quad c_2a_1-c_1a_2 = \begin{vmatrix} a_1, & c_1 \\ a_2, & c_2 \end{vmatrix} \qquad (3)$$

to which the above-mentioned rule applies. Expressions like (1), (2), (3), are called determinants of the second order.

The numbers in the first, second . , horizontal lines are called the first, second . , *rows*, respectively; the vertical lines are called first, second . , *columns*.

The above definitions and method can be applied to any number of equations with an equal number of variables, and in each case our determinant would have n^2 numbers, n rows and n columns.

We may use another notation which employs one letter for the coefficients of our variables, with indexes or suffixes to indicate that their values are different. Let us consider n^2 elements in the table:

$$\begin{vmatrix} a_{1,1}, & a_{1,2}, & a_{1,3}, & \ldots, & a_{1,n} \\ a_{2,1}, & a_{2,2}, & a_{2,3}, & \ldots, & a_{2,n} \\ \cdot & \cdot & \cdot & \cdot & \cdot \\ a_{n,1}, & a_{n,2}, & a_{n,3}, & \ldots, & a_{n,n} \end{vmatrix} \qquad (4)$$

The expression (4) is called a determinant of the n-th order.

The notation by suffixes is very convenient and is very much used these days. The first suffix denotes the row, the second the column in which the element is situated. Usually the comma dividing the two numbers in the index is dropped and the coefficients are written simply a_{11}, instead of $a_{1,1}$. In general the element $a_{i,k}$, or a_{ik}, represents the element in row i and column k.

The elements lying in the diagonal joining the upper left-hand to the lower right-hand number are called the principal diagonal. In our example we notice that the elements in the diagonal are such that $i = k$.

We have definite rules by which we can arrive at the solution of our equations, once the coefficients, which are the elements of the determinant, are given. In general, the determinants are treated as a functional form.

If m and n are positive integers, a manifold, or system of mn ordered quantities or elements arranged in m horizontal and n vertical rows, will be called a *rectangular matrix* and we may use the notation $A = (a(mn))$:

$$\begin{pmatrix} a(11) & a(12) & a(13) & \cdots \\ a(21) & a(22) & a(23) & \cdots \\ a(31) & a(32) & a(33) & \cdots \\ \cdots & \cdots & \cdots & \cdots \end{pmatrix}, \text{ or } (a(nm)).$$

The numbers m and n are called the orders of the matrix. If $m = n$ the matrix is called a *square* matrix. Without loss of generality we can treat any rectangular matrix in which $m \neq n$ as a square matrix by supplementing the missing rows and columns with zeros.

A matrix of the type,

$$1 = \begin{pmatrix} 1 & 0 & 0 & 0 & \cdots \\ 0 & 1 & 0 & 0 & \cdots \\ 0 & 0 & 1 & 0 & \cdots \\ 0 & 0 & 0 & 1 & \cdots \\ \cdots & \cdots & \cdots & \cdots & \cdots \end{pmatrix} \text{ or } (\delta(nm)),$$

where $\delta(nm) = 1$ for $n = m$, and $\delta(nm) = 0$ for $n \neq m$, is called a unit matrix. The matrix

$$\begin{pmatrix} a(11) & 0 & 0 & \cdots \\ 0 & a(22) & 0 & \cdots \\ 0 & 0 & a(33) & \cdots \\ \cdots & \cdots & \cdots & \cdots \end{pmatrix} \text{ or } (a(nm)\,\delta(nm)),$$

is called a diagonal matrix. In the new quantum mechanics a diagonal matrix is independent of t and represents a constant of the classical theory. The reverse is not necessarily true. The operation of differentiation can be expressed in terms of multiplication of matrices with the aid of the unit matrix.[4]

Equations in which matrices are equated are called matrix equations. If the equations involve only one unknown matrix, which does not occur more than once as a factor, such equations are called matrix equations of the first degree.

708   X. ON THE STRUCTURE OF 'MATTER'

The m scalar equations

$$a_{11}x_1 + a_{12}x_2 + \ldots + a_{1n}x_n = c_1$$
$$a_{21}x_1 + a_{22}x_2 + \ldots + a_{2n}x_n = c_2$$

. . . .

. . . .

$$a_{m1}x_1 + a_{m2}x_2 + \ldots + a_{mn}x_n = c_m$$

are together equivalent to one single matrix equation. There are several ways in which the notation can be simplified.

The difference between a determinant and a matrix is subtle, but important. By a determinant we understand, by definition, a certain homogeneous polynomial of the n-th degree, in the n^2 elements a_{ij}. Accordingly, a determinant gives a definite number when calculated.

But in many instances we are interested in the *table*, or the n^2 elements arranged in a certain order but *not* combined into a polynomial. Such an array, or table, is called a matrix. Thus, from this point of view, a matrix does not represent a definite quantity, but a system of quantities, and so a matrix is *not* a determinant.

We can illustrate this difference by an example. If we take a determinant of the second order

$$\begin{vmatrix} a_1, & b_1 \\ a_2, & b_2 \end{vmatrix}$$

and change the rows into columns, or vice versa, thus:

$$\begin{vmatrix} a_1, & a_2 \\ b_1, & b_2 \end{vmatrix}$$

the value of both determinants will be equal; namely,

$$a_1 b_2 - a_2 b_1,$$

by the definition rule already given; yet the *matrices* of the two determinants are *different*.

Although different, a determinant nevertheless defines a matrix, called the matrix of the determinant; conversely, a matrix defines a determinant, called the determinant of the matrix.

We have said that a matrix does not represent a quantity, while a determinant does. At this stage, and from this point of view, we may say so legitimately. However, we might eventually treat a matrix as a quantity also; but for this purpose we should have to enlarge the meaning of the term 'quantity'.

In our present use of the term 'quantity' we mean the real and complex quantities of ordinary algebra.

It may be said that mathematicians have had a peculiar tendency, which has proven of great value in the development of mathematics, gradually to extend the meaning of terms in order to embrace new notions as they arise. For instance, we have enlarged the primitive meaning applied to positive integers to embrace negative numbers, which formerly would not have been considered as quantities. Similarly, if we here use the ordinary notion of algebraic quantity, then a matrix is not a quantity but a system of quantities. The problem is, how shall we enlarge this meaning to include the matrices?

Mathematics recognizes that this generalization of *mathematical* notions is extremely useful and *legitimate*. This structural issue appears to be of very general application, as all of us exhibit a tendency towards it. It is a purely mathematical and useful tendency in mathematics, but it leads to disastrous results when applied to daily-l.. abstractions, as explained in Part VII. In this connection we should recall the difference between the mathematical contentless abstractions and the abstractions with physical content, with which we are generally concerned in science and life.

Let us now follow up the method by which a matrix can be considered as a quantity. If we have objects of two or more kinds which can be counted or measured, and if we consider an aggregate of such objects, say 5 horses, 3 cows and 2 sheep, we could denote such a complex quantity by the symbol (5,3,2). In this case, the first place in our symbol would be reserved for horses, the second for cows, and the third for sheep.

In mathematics, we do not specify horses, or cows, or sheep, but consider sets of quantities, and distinguish them by the position which they have in our symbolism. We may denote such a complex quantity by a single letter, $A = (a,b,c,)$. (For instance, we denote a fraction by a single letter, although a fraction is specified by *two* numbers.)

In such an instance, we should call a complex quantity equal to another when, and only when, the components are respectively equal. And a complex quantity is said to vanish only in case all the components vanish.

Ordinary mathematical operations can be applied to such complex quantities. For instance, we may define a sum or difference of two complex quantities $A' = (a_1, b_1, c_1)$ and $A'' = (a_2, b_2, c_2)$ as

$$A' \pm A'' = (a_1 \pm a_2, b_1 \pm b_2, c_1 \pm c_2) \, . \, ,$$

a definition which is entirely satisfactory theoretically, and also practically, as can be verified from our example.

From this point of view we may consider a matrix as a complex quantity with mn or m^2 components. A matrix would then represent a complex quantity, as a special case under the general method sketched above.

We could then define our further operations. A matrix would be said to be zero when all elements are equal to zero. Two matrices would be said to be equal when they have equal numbers of rows and columns and every element of one is equal to the corresponding element of the other.

By setting up some such rules we could develop a calculus of matrices, and matrices would be considered as complex numbers. In general, the algebraic rules would be found to be applicable to matrices, which would further justify us in treating matrices as complex numbers.

One of the notable exceptions in our operations would be found in the application of the classical operation of multiplication and its dependencies. In ordinary algebra and arithmetic, multiplication is what is called 'commutative' which means that $2 \times 3 = 3 \times 2 = 6$, or $a \times b = b \times a$.

In defining the multiplication of matrices we have no *a priori* grounds for determining why one definition or restriction should be preferable to another.

Only practice can show which definition is more workable or more fruitful in results. In the matrix calculus the definition of Cayley is generally accepted, as it has led to the most workable results. It was based on considerations of the composition of *linear* tranformations.

The definition is approximately as follows: The product **ab** of two square matrices of the n-th order gives a square matrix of the n-th order in which the element which lies in the i-th row and j-th column is obtained by multiplying each element of the i-th row of **a** by the corresponding element of the j-th column of **b** and adding the results.

If we denote by a_{ij} and b_{ij} the elements in the i-th row and j-th column of **a** and **b** respectively, then by definition the element (i,j) of our product **ab** would give,

$$a_{i1}b_{1j} + a_{i2}b_{2j} + \cdots + a_{in}b_{nj}, \tag{5}$$

and the (i,j) element in the matrix **ba** would be

$$a_{1j}b_{i1} + a_{2j}b_{i2} + \cdots + a_{nj}b_{in}. \tag{6}$$

In general, the quantities (5) and (6) are not equal and therefore we see that the multiplication of matrices is, in general, *not commutative*. The *order* in which we perform our multiplication is of importance and **ab** *is not* generally equal to **ba**, (**ab** $\neq$ **ba**).[5]

It should be noticed that the vector calculus has made us familiar with new operations which differ from arithmetical operations. For instance, the sum of two vectors differs in general from the arithmetical sum and is defined by the law of the parallelogram (see Chapter XXXIII). This definition is more general, and the arithmetical definition expresses only the particular case in which the vectors have one direction. Similarly, the non-commutative law of multiplication corresponds more closely to vector multiplication than to arithmetical multiplication.

We will not go further into the details of the matrix calculus, which is a well-developed mathematical discipline with a large literature, but will emphasize some methodological points of importance.

One of the main applications of the theory of matrices is found in the subject of *linear transformations*.

In mathematics, instead of using the given variables, we very often introduce new variables which are functions of the old. Such transformations, or change of variables, are particularly simple and important when the functions in question are homogeneous and *linear*.

If $x_1, x_2, \ldots, x_n$ represent the original variables, and $x_1', \ldots, x_n'$ the new variables, we have, by definition, the formulae of transformations:

$$x_1' = a_{11}x_1 + \cdots + a_{1n}x_n$$

$$\cdot \quad \cdot \quad \cdot \quad \cdot \quad \cdot \quad \cdot \quad \cdot \quad \cdot$$

$$x_n' = a_{n1}x_1 + \cdots + a_{nn}x_n.$$

The square matrix made up of the coefficients is called the matrix of the transformation and the determinant is called the determinant of the transformation and is completely determined by the matrix. We have already seen the importance of linear equations and linear transformations in physics and therefore in the investigation of the world around us. The theory of matrices is connected with such transformations, hence the importance of the theory of matrices for physics.

For our purpose another characteristic of the matrix calculus is of interest and that is the fact that in physics we usually have a large number of empirical numerical data which enter as coefficients in equations and which can always be put in the form of a two-dimensional array of numbers, or a table, which we have just called a matrix.

It appears that every physical quantity, however complicated, can be represented by such a table giving the values of the parameters which determine its character. From the definition of the term 'variable' as *any* value out of a possible range of values, we might treat our variables in two distinct ways, one from the point of view of *function* or operations, the other from the *extensional* point of view, when the function or operations are unknown, although the particular values of the variable are given. The matrix calculus takes this last point of view.

In physical research work we deal for the most part with arrays of numbers or unique and specific, mostly asymmetrical, relations which the experiments give us. Our usual problem is to find the structure, the function, and the operations which are satisfied by the given experimental relations.

We see that the dual approach to our solutions is due entirely to the definition which we have accepted for the variable. We have two issues: Either to find the values of the variable which satisfy the given function and operations or, having particular values of the variable (experimental), to find the function and operations.

Obviously every physical quantity can be represented by a matrix, which may be a sequence, and every mathematical theorem can be reduced to a property of matrices. Once the proper mathematical theories are worked out it will be always possible to pass from one form of representation to the other.[6]

In the older mechanics the functions were rather obvious and so the use of the matrix calculus was not so imperative. In the newer mechanics, the opposite is the case. We have a large amount of numerical experimental relations, but the functions and operations connecting these variables are unknown and the problem is to find them. From this point of view the matrix calculus represents an *extensional* calculus, a calculus of *observation*. In using the descriptive term 'observation' we must add that some objections have been advanced to such a use of the term. The answer is that the term 'observation', like most of our most important terms, must be considered a *multiordinal term*. Once this is understood the objections to the use of the term do not hold.

There is no limitation as to what the elements of a matrix may represent; they may be functions, functions of functions , .

Section D. The operator calculus.

The use of the *operator calculus* is interesting, structurally and psychologically, in that attention is concentrated, not on the numerical quantities, but on the semantic *operations of combining* them. The calculi used in the newer quantum mechanics are peculiar, because, while they retain the numerical data, and as far as possible, the classical equations, they alter the operations by which these quantities are combined, or the interpretation of the equations.

As an illustration of such a procedure we can take two different formulae for the addition of velocities; one from the classical mechanics, where

$$V_{13} = V_{12} + V_{23},\qquad(1)$$

and the other, the formula for velocity as given by the Einstein theory; namely,

$$V_{13} = \frac{V_{12} + V_{23}}{1 + \dfrac{V_{12} \times V_{23}}{c^2}}.\qquad(2)$$

In these formulae V_{12} represents the velocity of body 1 relative to body 2 , . In formula (1) the sign '+' symbolizes the ordinary arithmetical operation of addition. As we already know, this formula has proven too simple to represent accurately the experimental data, and Einstein has replaced it by the more elaborate formula (2).

The above statement is the usual way of speaking about the modification in formulation which has taken place in physics since Einstein. But we could equally well say that the formula has *not* been altered except that the '+' has no longer the old meaning and does not now represent the arithmetical operation of addition. Both points of view lead ultimately to one value, V_{13}, and the *computations* are similar in both cases.

We should notice especially the great freedom with which we can treat mathematical entities. Our voluntary selection of the point of view becomes important. A similar freedom of selection of interpretation appears to a still larger extent in all *verbal* problems, a fact of considerable structural and semantic importance in any theory of sanity, as we have already seen.

Further illustration of this freedom can be seen in the way in which the ordinary notion of multiplication is re-interpreted in the operator calculus. Let us denote by q and f, two numerical quantities, and qf as their product. But we could view this problem differently. We could say that *qf results* from a semantic *operation* q performed on f, or a semantic *operator* $(q \times)$ acting upon f which transformed f into qf. We could denote the *operation* of multiplying by q or the operator $(q \times)$ by a single symbol Q. Quite obviously the operator Q is not the number q; in other words, the semantic operation of multiplying, say by two, is *not* the number 2.

The operation of multiplying integer 1 by integer 2 gives the result 2. Similarly, the operation of multiplying 1 by q, or, in our new language, the application of the operator Q to the integer 1, gives q. In symbols, $Q1 = q$. If we take any arbitrary function f, the result of the operation of Q upon f is written $Qf = qf$. If we were to follow the operation Q by the differential operation d/dx the result would be*:

$$\frac{d}{dx}(Qf) = \frac{dq}{dx}f + q\frac{df}{dx} = \left[\frac{dq}{dx}\times\right]f + Q\frac{df}{dx}.$$

But as f is arbitrary, it may be omitted from the equations and the result written in the operator form as

$$\frac{d}{dx}Q - Q\frac{d}{dx} = \left[\frac{dq}{dx}\times\right] = \frac{dQ}{dX}.$$

The symbol $\frac{dQ}{dX}$, defined by this equation, should be read as 'the operation of multiplying by dq/dx'. Similarly, $\frac{d(QP)}{dX} = \frac{dQ}{dX}P + Q\frac{dP}{dX}$.

In translating the ordinary equations into the language of operators nothing new is introduced. This translation involves only a change of *'mental' focus*. Instead of concentrating our attention on the numerical values we concentrate on the operations of combining them. Since the great problems of the quantum mechanics consist in finding new methods of computation, or of combining numerical values, such a change of attitude may prove to be structurally useful.

It should be noticed here that once matrices are considered, and treated, as quantities, or unique and specific relations, by similar reasoning they can be treated as operators. This problem is of fundamental structural and semantic importance because in the quantum theory we deal with matrices which have infinite numbers of terms and since this complexity presents great technical difficulties it is of enormous advantage to be able to pass to some more developed methods of calculation.

At the preliminary stage in the operator calculus we have assumed that multiplication was commutative, that $QP = PQ$; but in the further and more general development of the theory, this does not hold.

In general we must assume in the operator calculus that multiplication is not commutative, that $QP \neq PQ$. For instance, if $Q = (q\times)$ and $P = d/dq$, the two operations are certainly not commutative. Naturally, the validity of $2\times 2 = 4$ is not doubted, but generalized non-commutative multiplication has a definite asymmetrical and so structural geometrical interpretation, to be found in the vector calculus. When we associate with each numerical quantity its own operation of multiplication we thus obtain a more general calculus. Operators may be regarded as compound, or built up from the elementary arithmetical operations of addition and multiplication. They represent, so to say, functions of these operations.[7]

---

*The operator D_x, also written d/dx, is called a differential operator. If applied to a product (uv) the results are given by the formula $D_x(uv) = uD_xv + vD_xu$.

Section E. The new quantum mechanics.

The main problem of the quantum theory is to determine these functions of the operations, so that the solution of certain equations (hamiltonian) may represent the experimental facts. The original equations of the new mechanics of Heisenberg, Born, and Jordan were frankly founded on an empirical basis. As Dirac puts it, in seeking for the new equations, the classical equations were to be retained as far as possible and only the operations by which these quantities are combined were to be altered.

To gain this freedom to alter multiplication the data were first interpreted as matrices. Then it was found, by Born and Wiener, that the matrices could be interpreted as a special kind of operator, which furnished means to calculate the matrices. Carl Eckart independently developed a simple operator calculus for the solution of the quantum problems. In this present work I follow closely the paper of Eckart.[7]

The origin of the new quantum mechanics was an epoch-making paper by Werner Heisenberg, in July, 1925. The older quantum theory had postulated the existence of stationary states of the atom, which were calculated with the aid of the older mechanics. In the new mechanics the equations have similar form as in the classical theory, but the variables no longer obey the commutative law of multiplication. In general pq is not equal to qp, ($pq = qp$ and $pq - qp = 0$ in the classical theory) but $pq - qp = \dfrac{h}{2\pi i}1$, where h represents the Planck constant, q the generalized co-ordinate, p the momentum, 1 stands for the unit matrix, and π and i have the usual meaning. The fact that multiplication is not commutative in our calculus allows us to give a definite value to the above difference and by introducing the Planck constant h, we are enabled to introduce the quantum conditions in our calculations.

The quantum conditions of the older theory led to an algebraic equation. By using the classical equation with a non-commutative multiplication law for the variables it is possible to perform calculations in the new and wider scheme of dynamics. The difference between pq and qp is expressed in terms of the Planck constant h. When h is made to approach zero, pq approaches qp, and so we pass to the classical mechanics. Thus we see that the classical mechanics appear only as a particular case of this more general theory.

In introducing his theory, Heisenberg pointed out that the older mechanics uses quantities which are *never observable*, and *can never be observed*, such as, for instance, orbital frequencies and amplitudes, or the position and 'time' of revolution of an 'electron'. , which, as such, have no physical meaning. He proposes to use *observable* data, such as the frequencies and intensities of the radiations , . Now these frequencies are always differences between two terms given by integers. If T_n and T_m are two such terms, the observable frequency is theoretically represented by $\nu_{nm} = T_n - T_m$. Such numbers as ν_{nm} characterize the atom as far as it is observable. It was natural that such a collection ('sum' in this case has no longer any physical meaning) of terms could best be represented by a matrix. In the classical theory a dynamical quantity was represented structurally by a trigonometrical Fourier series; in the new, it is represented by a two-dimensional table of values; that is, by a matrix giving the frequencies and the intensities of radiations.

An important and interesting structural issue now appears. It is that the Heisenberg theory gives a new formulation for the hamiltonian equations of motion, whereby their form is preserved, yet they are made applicable both to periodic and to *non-periodic motion*. It becomes possible to fuse the classical mechanics with the quantum mechanics. The distinction between 'quantised' and 'unquantised' motion loses all meaning, and a fundamental equation,

$$pq - qp = \frac{h}{2\pi i} 1,$$ is formulated which is valid for *all motion.*

The Heisenberg theory is also characterized by its thoroughly behaviouristic, actional, functional, and operational character. The number of unjustifiable assumptions is the lowest in existence and most of the identifications are eliminated. According to Heisenberg, electrons and atoms do not have the 'same' kind of 'reality' as ordinary *objects* of lower order abstractions. This conclusion, which underlies his whole work, is of particular importance structurally. As we know, differences in character separate different orders of abstraction and since the quantum phenomena belong to a higher order of abstraction they must differ from objects which belong to a lower order of abstraction. In this theory the feelings of 'space' and 'time' are no longer applicable to the 'inside' of the atom—as might be expected.

The distinction between 'inner' and 'outer' electrons in an atom becomes meaningless, since it is impossible to recognize a particular entity among a series of similar entities. In accordance with the new 'space-time' outlook we gain a physical basis for the absolute individuality of some eventual unit.

Because of its structure, the Heisenberg theory is a very fundamental one and there is little doubt that the Heisenberg *methods* will be elaborated further and will be kept as a permanent *checking method* in physics. A theory which is thoroughly behaviouristic, with a minimum of assumptions, will probably remain both a most important instrument of research and an inspiration to physicists and mathematicians.

The Heisenberg theory, again, because of its structure and method, does not lend itself easily to visualization. This is not against the theory. The pictorial representations of lower order abstractions are not to be relied upon. Besides, visualization depends on the *lower centres* and therefore must be represented by a *macroscopic* representation of a continuous (rather than a discrete) character, such as waves , .

If we were to try to describe the Heisenberg theory pictorially, which is obviously difficult to do, we would have to give a *negative* description. We should have to say that what we observe must be considered only as radiations from the location which the atom was supposed to occupy.[8]

There remains but to mention some more characteristics of the Heisenberg theory which seem to have very far-reaching structural and semantic bearings

This theory appears frankly statistical and introduces fundamental probability assumptions. The moment we realize that the human organism is essentially an *abstracting* affair and that the abstracting is performed on different levels, or in different orders, it becomes obvious that statistical methods and probability notions become fundamental.

In the earlier days we used to assume that statistical laws were laws with exceptions. Such an outlook was conditioned by our dealings with macroscopic events. Now, we analyse such macroscopic events in terms of microscopic and sub-microscopic events, the statistical laws become accurate laws, not for individuals but for *groups* of individuals. Because we abstract in different orders, we deal *only* with statistical data, mass effects of different 'packets' of nervous excitement, as is best illustrated by different thresholds in different nervous tissues.

The processes in the higher centres, *more remote from the external world*, deal with a special material, no more with statistical data of 'packets' and averages, but with what we used to call 'inferences', 'inductions'. , which give only the *probability* of happenings. But as we have already seen, probability has become a well-developed structural mathematical discipline, which has not yet made much effect upon our primitive-made macroscopic metaphysics and language. It should be noticed that the highest activities of the nervous centres are based on statistical data furnished by the lower centres. So we see that, to the best of our knowledge about ourselves and the world around us, a modern structural and semantic outlook, in science or in life, must be based on statistical and probability methods.

In space-time every point has a date, and therefore in the language of space-time all points are different and do not repeat themselves. Such structural outlook is, of course, again conditioned, and leads toward the statistical and probability methods. The main psycho-logical importance of the new methods is to diminish affective tension, which is always wasteful and harmful. Inferences may involve belief. When *belief* is *too strong*, although this is never justified according to the best modern knowledge, we very easily fall into identification, delusions, illusions, and the like. It should be emphasized that the last-mentioned pathological states are always compound. They involve at least two components. One of these consists of some ignorance somewhere; the other of strong affective belief in the 'truth' of our mistaken notions. The stronger the affective tension is, the more dangerous the semantic disturbance becomes.

The Heisenberg theory has succeeded in formulating (verbal) structural methods which are best suited to represent the experimental facts which underlie physics, as well as being structurally in accord with the working of the human nervous system. That is why I venture to assert that this theory will never be abandoned as a checking and research instrument.

In an hypothetical experiment in the quantum field, we may assume what may be called a gamma-ray microscope. If we were to illuminate an 'electron' by gamma-rays, the rays would disturb the experiment, and in our fundamental equation, $pq - qp = \dfrac{h}{2\pi i}$. 1, by which the 'position' was to be determined, the 'momentum' would thereby be disturbed. This change of 'momentum' would be greater the shorter the wave-length of the rays used; and the shorter the wave-length of the ray, the more accurate the determination of 'position' would be. Hence, the more exactly a co-ordinate q could be found, the less exactly could its momentum p be found, and vice versa.

So we have to introduce corrections for errors, and have to introduce 'mean values' and 'probability functions', which we can develop and compute. Lately, Bohr has further developed the probability aspects of the newer quantum mechanics but I have not seen this work. Heisenberg introduces 'probability packets' which correspond to the 'wave packets' of Schrödinger.

It is difficult to speak briefly, and yet in a satisfactory way, about these new developments, and particularly difficult to give credit properly to different authors. All their works are interwoven and at present they all really work together in spite of the fact that historically some of these theories have been developed independently.

What we call today, for the sake of brevity, the Heisenberg theory, because of its originator, has been further developed by Heisenberg, Born, Jordan, and others. Later, when the wave-mechanics appeared, all the new theories were finally fused into a very elaborate and impressive structure.

Historically, P. A. M. Dirac worked at the theory from a different mathematical point of view, utilizing what is called the 'Poisson bracket' method. In this treatment the difficulties of the matrix calculus were avoided. He introduced dynamical variables which he called the q numbers. These do not obey the commutative law of multiplication although the c numbers (classical) do. He also considered the difference of the non-commutative products $xy - yx$, where x and y are functions, respectively, of the co-ordinates $q_1 \ldots q_s$, and of the momenta $p_1 \ldots p_s$ of a multiple periodic system with s degrees of freedom.

Dirac has generalized the matrix theory and the Schrödinger equations. His work seems to be most important, in physics and mathematics, but it is not possible for us to consider it here in any detail.[9]

Let us recall once more that there are fundamental differences between the different orders of abstraction, and that we all have to abstract in different orders. From this point of view it is natural that *every* theory, even if expressed at present in a form which cannot be *visualized*, like the Heisenberg theory or the original Dirac theory, has sooner or later to be expressed in a structural form which can be visualized. These problems have really nothing, or at most very little, to do with the world around us. They are concerned with the neurological structure which produces *all* theories.

Theories of a structure such as that of the Heisenberg theory are extremely important, as already explained, but in them we lose the help of 'intuition'. Now 'intuition' (lower centres) has two quite different effects—sometimes it leads us astray, but on other occasions it helps greatly.

An 'intuitive' theory has a creative aspect, but always ought later to be revised and scrutinized by non-intuitive means. In fact, because of our nervous structure, we should always strive to produce *both aspects* of theories—strive *consciously*—for thus we facilitate progress. Historically, we can never completely avoid producing both types of theories, as they are inherent in our nervous structure and in the different orders of abstractions we produce.

It is precisely in the newer quantum mechanics that a typical example of this simple neurological fact is found. The non-intuitive handling of data was introduced by Heisenberg: the translation of the matrix calculus into operational and 'Poisson brackets' methods; and, finally, the new 'wave mechanics' of de Broglie, Schrödinger, and others, gives us a perfect translation into intuitive methods.

It should be noticed that according to the old notions such two methods, the intuitive and the non-intuitive, were not supposed to be a *neurological necessity*. We still assumed that they were separated 'absolutely', and even today in many quarters we argue as if they were absolutely separable. If we accept the principle of non-elementalism, we realize that this distinction is verbal only and that the invention of verbal means has little or nothing to do with the world around us, but that it depends on human structural ingenuity.

Investigation of the ordered cyclic nerve currents shows unmistakably that such sharp differentiation is unjustifiable; and we must conclude, in accordance with historical experience, that translation from one method to the other must be a necessity, and so will be accomplished some day in every field. It is true that at present the Einstein theory has not been translated with entire success into terms of lower order abstractions. This is a task which is facilitated by this present work. The newer quantum mechanics gives us an unparalleled example of such translation, and hence our main interest should be concentrated on this structural aspect.

On neurological grounds it seems certain that visualization involves in some way the lower nerve centres, which again, by evolutionary necessity, involve macroscopic forms of representation. Our macroscopic experience led us to *geometrical intuitions*. These were framed three-dimensionally in '*absolute emptiness*', and were impossible in higher dimensions. The old structure represented a static empty 'space' in which nothing could happen and which was thus unfit to represent this world around us where something is going on everywhere.

The new structure represents '*fulness*' or a plenum. We can visualize it as a network of intervals or world-lines and then, by the notions of the differential calculus, as already explained, we pass easily to the visualization of the many-dimensional space-time world of Minkowski-Einstein. Now, in such a world, the curves are represented by functions, and, vice versa, functions represent curves. Thus it is obvious that analytical 'non-intuitive' methods have 'intuitive' structural geometrical counterparts. From this point of view the method of operators represents a passing step from the non-intuitive to intuitive methods. As soon as we have functions, we can represent a functional calculus as an operational calculus. This involves a more behaviouristic semantic attitude and so leads to the possibility of translation of either method into the other. Ultimately these are psycho-logical transformations and translations.

We should not be surprised to find that in the development of the newer quantum mechanics the operational calculus plays just such a role.

Now our older macroscopic experience, which affected our lower nerve centres, gave rise to the elementary geometrical structural notions of 'lines', 'surfaces', 'volumes', . For the building of physics we had to introduce 'time', 'motion', . In older days we did not realize that these give us forms of representation and that it is optional with us which forms we will accept as fundamental or use as a starting point.

The old descriptive apparatus posited structurally an absolute and immutable 'space' (emptiness), 'time', 'matter' out of which we built up a verbal definition for 'motion'. The semantic attitude of all of us, scientists included, depended upon identification. We ascribed lower-centre significance to higher-centre abstractions. We did not discriminate enough between the macroscopic and small scale events. So we had 'geometrical optics' in which we 'perceived' a ray of light (in a dusty room, let us say) as a 'straight line'. Further investigation disclosed that the 'rays' on one level of abstraction were *waves* on another, but they were not perceived as waves by the lower centres.

But through our lower centres we had acquaintance with some waves, such as in water; so representation for waves was developed. A wave-theory still remained intuitively workable, even when we dealt with waves which we could not see. Now the equations of waves are well known. It is then possible to translate a non-intuitive matrix mechanics, when we treat matrices as operators, into a functional calculus which has an intuitive geometrical wave representation.

This is precisely what has happened and now, perhaps for the first time in human history, we have all the aspects of a theory being worked out simultaneously, with mutual co-operation of all workers and the use of methods which are mutually complementary from the neurological side. There seems little risk in predicting that because of these neurological factors the newer quantum mechanics will give extremely rapid and far-reaching results. When scientists become aware of the structural semantic and neurological issues involved, perhaps such achievements will be multiplied *consciously*, instead of being a kind of coincidence.

Personally, I am convinced that these new achievements are not simply coincident. It seems that the abolishment of the old, *el*, static 'absolute space' and 'absolute time' *has relieved the younger scientists from a semantic blockage.* This release was due to the bold stroke of genius of Einstein in refusing to use the vicious aristotelian 'is' of identity. As soon as we realize that words are not the objective levels, we gain an *unconscious semantic freedom* in handling words, as words. At once this freedom is bound to produce many different forms of representation for events, according to the personal make-up of the individual workers. And of course these forms can be translated into one another.

Section F. The wave mechanics.

We have not sufficient space at our disposal to discuss more fully the new wave mechanics. I found that short of a small volume, no explanations, readily intelligible, could be made.

In mathematics and physics, which represent the most developed sciences, we consciously and unconsciously strive for more and more general formulations. The work of Einstein, showing that the classical mechanics was only a particular case of a more general mechanics, has given a healthy stimulation to such a fruitful line of work.

As the quantum phenomena could not be accounted for by the old mechanics, it was natural that physicists should try schemes of new mechanics which included classical mechanics as a particular case. Thus, Sommerfeld, through his methods of the application of the Einstein theory to the quantum mechanics and his generalizations of phase-space, and his treatment of the relation of wave-optics to ray-optics and of the relation of mechanics to ray-optics, came close to the discovery of the wave mechanics.[10]

The new wave mechanics originated in 1924 in Paris, with the thesis of Louis de Broglie, published in 1925, and republished in a book form in 1926.

The controversy between the corpuscular light theory of Newton (emission theory) and the wave theory of Huygens is well known. The emission theory had its support in the 'rectilinear' propagation of light, which followed from the inertia of light particles. Also, it explained the reflection and the refraction . , of light, but failed in other respects. It is true that the wave theory also had its weak structural spots. In it the 'rectilinear' propagation of light remained a complete mystery and it completely failed to account for the dispersion of light, until this was explained on the electron theory.

Both theories assumed the periodicity of light phenomena, but the acceptance of one theory was generally held to mean the rejection of the other. It did not occur to many that both theories might be correct but only partial structural aspects of a more general theory.

With the advent of the quantum theory of Planck (1900) new methods were found. In 1905 Einstein propounded his theory of 'light quanta' successfully. He assumed that radiation occurs in discrete quanta of energy $h\nu$, where ν represents the frequency. From this point of view the quantum had the characteristic discreteness of a corpuscle, and yet the frequency characteristic of a wave. We see that the new theory involved a kind of a blend of the two older theories.

De Broglie generalized still further the above notions. His theory is in a way the result of the theory of Einstein. As we already know, Einstein shows the connection of mass and energy, so that the conservation of mass becomes also the conservation of energy, and vice versa. Starting with these premises de Broglie concluded that if any element, in the most general sense, be it an electron or proton or light quantum or what not, has energy W, there must be in the system a periodic phenomenon of frequency ν, defined by $W = h\nu$. From this point of view all forms of energy, radiation included, must have an atomic structure and the atoms of energy must be grouped around certain points, forming what we call 'electrons', 'light quanta', .

Applying the Lorentz-Einstein transformation, he finds a rather startling fact—that the frequency associated with any assumed mass m_0; namely, $\nu_0 = \dfrac{m_0 c^2}{h}$, represents no more and no less than a *periodic* phenomenon, analogous to a stationary wave, which spreads around the point of which the mass is a singularity.[11]

In other words, a 'mass particle' at rest is the centre of a pulsation throughout the spread, or otherwise it is a singularity of the pulsation. The quantity which pulsates is called ψ and $\psi\bar{\psi}$ is interpreted as the *electric density*, where $\bar{\psi}$ is the conjugate of the complex quantity ψ.

In the Minkowski representation, the above astonishing result becomes quite simple, and we can see clearly how simultaneous pulsations become travelling waves. In Fig. 2, we give a two-dimensional diagram of space-time. OX_0 is the 'space' co-ordinate, OT_0 the 'time' (ict_0) co-ordinate. ψ_1, ψ_2, ψ_3, represent the traces of the surfaces of constant phase which are perpendicular to OT_0. The Lorentz-Einstein transformation is equivalent to the transformation from rectangular system $X_0 O T_0$ to the rectangular system XOT, forming an angle θ. In the new system the lines ψ_1, ψ_2, $\psi_3 \ldots$, are no longer parallel to the X axis and so represent a moving *wave front*. In this new system the ψ-lines represent the moving wave front for different points, P, P'. , on one phase line and have *different* values of t. The smaller the velocity (v) of the particle, the smaller the angle θ and the greater the distance PQ travelled by the phase in a given 'time', $P'Q$ or Δt, which means the greater the phase velocity u.

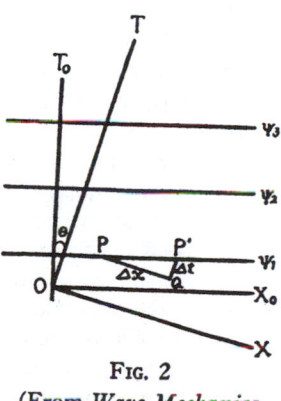

Fig. 2

(From *Wave Mechanics*, by H. F. Biggs)

The frequency of these pulsations or waves is the 'total energy of the particle' divided by Planck's constant, h. in symbols, $h\nu = mc^2 +$ potential energy, where

$$m = \frac{m_0}{\sqrt{1 - \beta^2}}, \quad \beta = \frac{v}{c}.$$

The problem before us is to connect structurally, the waves with two observations, one of the radiations, the other of what we call the 'material particles'.

Our interest at present lies only in the connection with the latter. According to this theory, the 'region occupied by the particle' is only a region

46

where a set of ψ-waves, which vary continuously in direction and in frequency in a small range, reinforce each other to give a *wave-group* travelling with what we usually call 'the velocity of the particle'.

As the waves have different frequencies, they travel with different velocities and so we have to face the problem of the dispersing medium, where, according to classical theories, the region of reinforcement has a velocity different from that of the phase. The ordinary expression for 'group velocity' gives, on the wave interpretation, the magnitude of the 'velocity of the particle'.[12]

That which, in the classical theories, we called the 'motion of the particle' is represented by the motion of the region of reinforcement of ψ-waves. The direction of motion is represented by the direction of a ray-, or wave-normal. The particular ray selected as the position of the 'path of the particle' is represented by the ray cut in coinciding phase by a set of ψ-waves with slightly varying directions. The 'position of the particle' is given by the small region occupied by the *group* of waves of slightly different frequencies and velocities.

In this connection we should notice an important point which is necessitated by the methods of generalization and of translation from macroscopic to sub-microscopic events and vice versa. Schrödinger, by formulating the differential equations for the ψ-waves, has brought out clearly the important structural point that wave mechanics bears a similar relationship to classical mechanics of particles as wave-optics bears to ray-optics.

Here again we have the macroscopic phenomena capable of being treated by mathematical methods different from those used for the small-scale phenomena.

In classical mechanics the state of a system whose co-ordinates were $q_1 \ldots q_s$ and whose momenta were $p_1 \ldots p_s$, was represented by a point in a $2s$-dimensional q-spread and the changes in the system were represented by the passage of the point along some curve, a 'ray', so to say.

Schrödinger regards the classical mechanics as only an approximation, while rigorous treatment must be made by the aid of wave mechanics.

The large-scale, or macroscopic, mechanical processes correspond to a wave signal in the q-spread and can be regarded as a point in comparison with the geometrical structure of the path. In small-scale phenomena, such as the atomic processes, a rigorous wave formulation must be used.

This analysis can be carried further and Hamilton knew well and used the analogy between mechanics and geometrical optics. The Hamilton variation principle, $\delta \int L dt = 0$, is Fermat's principle for a wave motion; the Hamilton-Jacobi equation expresses Huygens' principle for wave motion; and the new wave mechanics expresses the Kirchhoff analysis of physical optics. As Huygens' principle could deal with the problems of physical optics up to a certain point, so the Hamilton-Jacobi equations could deal with atomic problems up to a certain point. At the exact wave analysis of Kirchhoff was needed to clear up the finer points of physical optics, so the new wave mechanics is required for the exact solution of the atomic problems.[13]

A detailed analysis shows that classical mechanics was associated with geometrical optics (ray-optics). Obviously, a more exact system of mechanics would be one associated with wave-optics, which would give the classical results in all cases where the wave-length was negligible in comparison with the dimensions of the path. Schrödinger suggests that a correct extension of the analogy would be to regard the wave system as *sine-waves*. In this connection it should be recollected that Fourier has shown that any given form of waves can be represented by the superposition of *sine-waves*, and that therefore a *sine-wave* (see Chapter XXXII) may be considered as a general formulation.

The ray methods in physics worked only to some extent, that is to say, in the cases where the radii of curvature of such rays and the dimensions of the spreads were large in comparison with the wave-length. When this is not the case we have to consider waves and not rays. Naturally, dealing with atomic dimensions, which are very small, instead of using the paths of the particles or the ψ-rays, we have to use the ψ-waves. It appears that this difference was the main, rather puzzling distinction between the classical mechanics and the quantum mechanics, between the macroscopic theories and the sub-microscopic ones.

The above realization, and its formulation into a mathematical theory, seems to be an important and extremely fruitful generalization, which probably will be retained as a method.

One of the puzzling features of the quantum theory was the structural appearance of the whole-number laws of the 'orbits'. That some such whole-number relation is justified seems to be well established, yet it contradicted the older 'continuous' mechanics. A new theory, to be at all satisfactory, should be able to fit these whole-number empirical data. The first test of the new wave mechanics, and also its first success, was precisely in this field.

If a ray of the ψ-wave was supposed to run around in a circle for a stationary state, the circumference must be a *whole multiple* of the wave-length, or

$$2\pi r = n\lambda = n\frac{h}{mv}, \text{ and } mvr = n\frac{h}{2\pi}$$

where n is an integer. We see that the quantum condition of the Bohr theory, that the angular momentum must be a whole multiple of $h/2\pi$, is only the result of the requirement that the wave-function ψ shall be single-valued, which is another way of saying that the circumference ($2\pi r$) must contain a whole number of wave-lengths. It may be compared perhaps with waves travelling around a circular loop of string. If they travelled both ways we would have stationary waves.

At this point a very important structural feature of the new wave mechanics makes its appearance. In the above interpretation the 'velocity of the electron' has lost its physical meaning, it becomes simply the *wave length* of the ψ-waves. In the wave mechanics, as well as in the matrix mechanics, there is no meaning to the older 'position of an electron on its orbit'. So the wave mechanics again embodies the advantages of the matrix mechanics by not postulating entities which can never be observed. The whole numbers, as Schrödinger remarks, 'appear as naturally as do "integers" in the theory of vibrating strings'. In the theory of string vibrations these whole numbers are determined by certain boundary conditions which have to be satisfied by the solution of a differential equation. In the new wave mechanics there is also a differential equation representing the Schrödinger wave equation.[14]

Section G. Structural aspects of the new theories.

We should notice the important distinction between the two structural types of these theories. The extensional matrix theory can hardly be visualized, with all the consequent advantages and disadvantages. The wave mechanics can be visualized. From what we already know of the structure and working of the nervous system, we see that the wave mechanics will have a *creative* element and the matrix mechanics will remain an important *checking* method.

At present, all these new theories seemingly have blended or perhaps it would be better to say that they have been translated from one language to another and all the workers in this field work from all angles.

It should be mentioned also, that Einstein, Bose, Jordan, and others, work from the point of view of *statistics*, and that these methods, too, are being retranslated and connected with the rest of the new theories.

The new wave mechanics evades the difficulties of the matrix calculus and brings the new mechanics within the scope of the highly developed analysis of the theory of differential equations. It also enlists the creative aspects of 'intuition', 'visualization', .

Concluding our consideration of the subject, three remarkable aspects of the wave mechanics must be referred to. We are already acquainted with the term 'action'. It appears that the main point of the passing from the old mechanics to the new was the stroke of genius of de Broglie, when he divided action by the fundamental constant h with some definite numerical factor which then gives us the *phase*. In the expression for ψ, the energy appears as the 'time' component of a space-time vector whose 'space' components are those of the momentum. When this vector is divided by h, its components become the *frequency*, or the number of waves which each axis cuts per centimetre.

These are the methods by which we can use differential equations, whereby the older discontinuities disappear and the particle is represented as a group of reinforcing waves.[15]

From this point of view we also come to the conclusion that the 'conservation of energy', which was very valuable in the old days, is perhaps only a gross macroscopic generalization and will give place to a newer and more fundamental notion of the conservation of *frequency* or 'times'.[16]

It has been already mentioned that the newer mechanics must be represented in accordance with *statistical* data, *probabilities*, with due attention paid to the theory of errors , . While these requirements have very little to do with the world around us, they are unconditionally required by our nervous structure, which is, after all, the general author of all our 'knowledge' and 'theories'. Let us be candid about it; there is no such thing as 'knowledge' outside of a nervous system, and therefore the neurological requirements, as already mentioned, become paramount. The newer theories brilliantly satisfy this requirement.

As an example, we may perhaps mention an aspect of the wave mechanics theory which is not at present settled, but which remains just as interesting.

Schrödinger shows that in highly excited states a suitably chosen group of waves represents a *'wave packet'*, which behaves like a point-mass of the ordinary mechanics. It oscillates with frequency v_0 in a rectilinear path. The number and breadth of the waves which form the packet vary with the 'time', but the width of the packet remains constant. The remarkable part about the shape of the curve is that it represents the *Gauss error curve*.

Heisenberg has shown that this result is only accidentally true, but for our purpose of illustration this is quite enough.

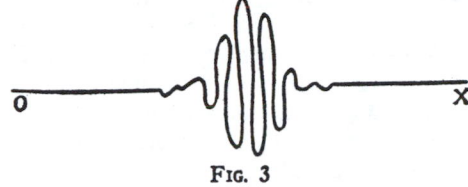

FIG. 3

(From *The New Quantum Mechanics*, p. 115, by George Birtwistle)

The newer quantum mechanics has shown once more the necessity for a re-analysis of our fundamental notions. From a space-time point of view, which seems to be a permanent acquisition of science, since it is a language and method of structure closer related to the external world than the older languages and methods, it seems beyond dispute that even macroscopic phenomena are the results of *repeated observation*. Now, such a point of view, although it is extremely plausible, and close to neurological and physical data, necessitates a complete reconstruction of our describing apparatus, which is not adapted to such an outlook.

The problem of 'observation' enters. Bohr suggests, and rightly, that this vocabulary is strictly connected with the older 'causal' vocabulary. One of the main points of the present work is to draw attention to the *multiordinal mechanism and terms* and to show that the analysis of these problems cannot be even attempted without first analysing the structure of our languages, of our 'knowledge', and the neurological and semantic aspects which such analysis involves.

When this analysis is carried through we see that the problems of 'continuity' and 'discontinuity' lose their absolute character. They become *verbal* problems, to be solved through the ingenuity of some one who will suggest the solutions.

The newer quantum mechanics give us ample material for work on these problems, but they also illustrate a much more general and important problem, which is the subject matter of the present book; namely, that all 'knowledge' is *structural*, strictly dependent on the nervous structure and functioning, and the language we use. 'Method' is that aspect of the search for structure which deals with the most expedient means for finding structure. Since words *are not* the things we speak about, the study of linguistic structure becomes a most important research method. The more languages (theories) we have for analysis and structural comparison, the more glimpses do we get at the structure of the world. The newer quantum mechanics give us enormous material of a linguistic, structural, and semantic character.

It is natural that this wisdom can come only from the study of the structure of the highest developed languages in existence, which are mathematical languages. If we want progress in any line of human endeavour, this progress is always dependent on the languages we use, since what we call 'progress' is a co-operative affair and therefore dependent on means of communications and languages.

From the point of view of structure, we deal with a world of *absolute individuals* and therefore our languages must be such as to reflect such individuality. We already know that this involves an extensional attitude and methods, which historically have produced mathematics as the only language which as yet reflects in structure the world around us.

With the newer quantum mechanics, the old 'discontinuity' resolves itself into an essential *individuality*, as noticed by Bohr, perfectly foreign to the older theories.

History proves that we were slow in arriving at that point. Our tragedies began when the 'intensional' biologist Aristotle took the lead over the 'extensional' mathematical philosopher Plato, and formulated all the primitive identifications, subject-predicativism . , into an imposing system, which for more than two thousand years we were not allowed to revise under penalty of persecution. Mathematics was not particularly encouraged, but at least, not persecuted, so that it was developed into the present day great linguistic system. The theory of function involves semantic factors of non-identification.

The invention of the differential and the integral calculi, represents the two great structural and psycho-logical aspects of analysis and differentiation, versus synthesis and integration.

The application of these methods led us to differential geometries, to methods of treating 'fulness', and to 'contact' methods. 'Fulness' necessitated geometries of higher dimensions, impossible in 'absolute emptiness', and so the fusion of geometry with physics became possible. The four-dimensional world of Minkowski and the theory of Einstein finally achieved this fusion. The next step was the invention of the new quantum mechanics, where all these important, nay, all-important, structural, semantic, and linguistic achievements find their culmination. The old primitive metaphysics become too 'materialistic' for an enlightened age.

Without legislating about the 'truth' or 'falsehood' of the newer mechanics, as a matter of *human behaviour* these theories are the best indications and examples of the structure of human 'knowledge', which I have attempted to formulate in this work as a general theory.

The A-system was strictly interconnected with primitive-made structural assumptions or metaphysics, reflected in the structure of the older languages and in the *el* notions about language, 'psychology', 'logic', and the pre-scientific anthropomorphic astronomy, physics, and other disciplines. Of late, science has developed in spite of all handicaps and persecutions, and has begun to depart structurally and semantically from the path of aristotelianism and the dark ages. Every science has had to build its own language and this fact completely condemns the A language, which, it is shocking to notice, we continue to preserve in our daily life.

Should we wonder that we have shown hardly any progress at all in our purely human affairs and notions? We should wonder, rather, that we have been able to survive until now—though with needless difficulties and suffering. More wars, more revolutions, more insanity, more morons, more struggle and competition, and more unhappiness are what we are entitled to *expect and predict* as the outcome of this structurally and semantically impossible situation.

As the organism works as-a-whole, such things as 'pure intellect' or 'pure emotions' represent structurally *el* fictions and scientists should realize that their professed detached scientific attitude is profoundly and fundamentally unjustified. All science has 'emotional' components, which play most important roles in life. If we live in a modern world, but keep the 'emotional attitudes' of primitive bygone days, then naturally we are bound to be semantically unbalanced, and cannot be adjusted to a fundamentally primitive 'civilization' in the midst of great technical achievements.* When scientists understand that, then the layman will have a different attitude toward science. He will understand that science is not a privilege of the few, something without effect upon all and every one. He will realize that while he lives in a modern world, *made so by science*, *structural ignorance* of the fundamentals as discovered by science leaves him with primitive structural assumptions or metaphysics, which by necessity build for him a delusional world leading to semantic unbalance and ultimately to 'mental' and nervous ills.

From the *non-elementalistic* point of view, the only escape is to realize that ignorance in an adult is, and must be, pathological, because 'knowledge' is to be considered as a normal characteristic of human *nervous tissue*.

A special structural and methodological brief and simplified account of scientific achievements, such as I have attempted in this work, must be a part of a theory of *sanity*. Sanity means *adjustment* and without the minimum of the best structural knowledge of each date concerning this world, such adjustment is impossible.

It is not necessary that the reader should fully understand all technical details of a theory, to be *aware* (instinctive, affective . ,) of the *existence* of the structural and semantic problems and to realize that some of the most competent and skilled professionals are working at these theories. Such awareness has great pacifying semantic influence; it eliminates the older affective tensions which were due to identifications, absolutism, dogmatism, flights into mysticism, and other similar pathological disturbances.

---

*See my *Manhood of Humanity*.

A *non-aristotelian system* must deal with all these structural and semantic issues. It is hoped that once a $\overline{A}$-system is presented to the public, scientists and laymen will become more interested in the structural and semantic issues emphasized here, and that new and wider researches will be undertaken.

History shows that such hopes are not illusory. The greatest men of science have always had wide human aims and interests. From the *non-elementalistic* point of view, they probably became productive geniuses because of this broad human urge. From the point of view of psychiatry, it is well known that 'mental' ills involve usually anti-social affective attitudes. When we see men with distinctly anti-social tendencies, no matter how they rationalize them, they are invariably ill in some way. A fully healthy individual is never anti-social.

That science should include structural and semantic factors of sanity may be a startling notion, but only at first! In the present analysis this turns out to be, rather unexpectedly, a necessity. But on second consideration we should rather expect it. Science and mathematics show the working of the 'human mind' at its best. Accordingly, we can learn from science and mathematics how this 'human mind' should work, *to be at its best*. Then we should make an analysis of science and mathematics from some wider structural and semantic point of view,—the task which has been undertaken in the present work.

At this early stage it is, of course, of comparatively little importance to what extent this analysis turns out to be satisfactory. The main point is that it has been *originated*. If the present author fails, others, perhaps even because of his failure, may be stimulated to do it better. The great and vital thing is that it should be done, by someone.

SUPPLEMENT I

THE LOGIC OF RELATIVITY

BY R. D. CARMICHAEL

In order to be able to deal with such quantities as are involved in the measurement of motion, time, velocity, etc., or indeed in the quantitative analysis of any physical phenomena, it is necessary to have some system or systems of reference with respect to which measurements can be made. Let us consider any set of things consisting of objects and any kind of physical quantities whatever, as electric charges or magnets or light-sources or telescopes or other objects and instruments, each of which is at rest with respect to each of the others. Let us suppose that among the objects are clocks, to be used for measuring time, and rods or rules to be used for measuring length, and that time and length may be measured at any desired instant and any assigned place. Such a set of objects and quantities and instruments, including the equipment for measuring time and length, all being at rest relatively to each other, we shall call a system of reference. Such a system we shall denote by S. In case we have to deal at once with two or more systems of reference we shall denote them by S, S', S_1, S_2, ...

In this definition of systems of reference nothing specific has been said about the units of length and of time. If we were dealing with our usual principles of mechanics we might pass over such a matter without any feeling of difficulty about it; it would be sufficient to proceed in accordance with our intuitive conceptions of time and length. But in the theory of relativity these appear in a new light. We can not proceed with confident dependence upon our intuition. On the other hand we shall not attempt to give explicit definitions of units of time and length. We shall proceed from certain principles or postulates, presently to be stated, to an analysis of time and length and so arrive at a suitable precision of these conceptions by means of certain guiding principles. It will be seen that it is not far from the truth to say that our fundamental terms are defined implicitly and indirectly by means of the statements made about them and accepted initially as valid and that they may mean anything which is consistent with the truth of these fundamental principles and postulates.

The restricted principle of relativity may now be stated in the following form:

RESTRICTED PRINCIPLE OF RELATIVITY. *If S_1 and S_2 are two systems of reference having with respect to each other a uniform unaccelerated motion, then natural phenomena run their course with respect to S_2 in accordance with precisely the same general laws as with respect to S_1.*

This principle says nothing about the suitability of any particular system of reference for the convenient expression of the laws of nature; but it does say that if either S_1 or S_2 is suitable the other is equally suitable, the relative motion of the two being unaccelerated.

In order to bring into suitable relations the measurements made on one system of reference and those made on another it is necessary to have some agreement as to the correspondence of units on the two systems. Accordingly we shall make the following assumption concerning the correspondence of units:

PRINCIPLE OF CORRESPONDENCE OF UNITS. *The units of any two systems S_1 and S_2 are such that the same numerical result will be obtained in measuring with the units of S_1 a quantity L_1 and with the units of S_2 a quantity L_2 when the relation of L_1 to S_1 is precisely the same as that of L_2 to S_2.*

We shall agree that the restricted principle of relativity is to be understood in a sense which implies this assumption concerning the correspondence of units; that is, the latter will be taken as a more precise formulation of a part of the content of the former. It is clear that the possibility of realizing this latter is taken for granted in the Galileo-Newtonian mechanics; it is often passed over without remark although it is a profound fact and is a part of the essential basis of any theory of motion.

It is a grave question whether the restricted principle of relativity can be maintained in the interpretation of natural phenomena. Indeed in the more general theory of relativity, to be taken up later, it is treated merely as a sort of approximation to a more comprehensive principle—an approximation strictly valid only in the absence of a gravitational field but very close to the truth for a wide variety of phenomena including most of those which are purely terrestrial.

There are two particular characteristic postulates, or 'laws of nature', lying at the base of the restricted theory of relativity. These may be stated as follows:

POSTULATE M. *The unaccelerated motion of a system of reference S can not be detected by observations made on S alone, the units of measurement being those belonging to S.*

POSTULATE R. *The velocity of light, in free space, measured on an unaccelerated system of reference S by means of units belonging to S, is independent of the velocity of S and of the unaccelerated velocity of the light-source.*

For these two particular postulates there is the strongest possible experimental evidence. Everything known points toward their truth, and there is nothing known which in any way seems to be in disagreement with them. It is to be observed that they apply only to the ideal case, that is, the case in which there is supposed to be no gravitational field.

For the development of the restricted theory of relativity there are three additional necessary postulates, or 'laws of nature;' those that theory shares in common with the Galileo-Newtonian mechanics. Such assumptions in some form are essential to the initial arguments and to the conclusions which are drawn by means of them. To the present writer it seems to be preferable to have these assumptions explicitly stated. They may be put into the following form:

POSTULATE V. *If the velocity of a system of reference S_2 relative to a system of reference S_1 is measured by means of the units belonging to S_1 and if the velocity of S_1 relative to S_2 is measured by means of the units belonging to S_2 the two results will agree in numerical value.*

POSTULATE T. *If two systems of reference S_1 and S_2 move with unaccelerated relative velocity and if a body moves relatively to one of the systems in a straight line with unaccelerated velocity then it also moves in a straight line relatively to the other and with unaccelerated velocity.*

POSTULATE L. *If two systems of reference S_1 and S_2 move with unaccelerated relative velocity and if a line segment l is perpendicular to the line of relative motion of S_1 and S_2 and is fixed to one of these systems, then the length of l measured by means of the units belonging to S_1 will be the same as its length measured by means of the units belonging to S_2.*

We now have before us the logical basis upon which may be built the restricted theory of relativity in all its details. It has been put in essentially the same form as that employed in my "Theory of Relativity" (published by Wiley and Sons, New York) and in my earlier articles in "The Physical Review". Reference may be made to the book named for the detailed development of the theory. Here we shall attempt to sketch only the progress of ideas and to indicate the main conclusions.

The first thing to be done in developing the theory on this basis is to consider carefully the relation between the time units of the two systems. The following remarkable conclusion is reached by a process of reasoning which is fully cogent in character:

If two systems of reference S_1 and S_2 move with a relative velocity v and β is the ratio v/c of v to the velocity c of light as measured on either system, then to an observer on S_1 the time unit of S_1 appears to be in the ratio $\sqrt{1-\beta^2}:1$ to that which is described to him as a unit by an observer on S_2 while to an observer on S_2 the time unit of S_2 appears to be in the ratio $\sqrt{1-\beta^2}:1$ to that which is described to him as a unit by the observer on S_1.

Thus we have the extraordinary conclusion that the time units of the two systems of reference S_1 and S_2, not at rest relatively to each other, are of different lengths in such a way that an observer on either system thinks that the time unit of the other system is greater than his own. It is evident that no simple change of the unit on either system (or both) will bring the units into agreement for observers on both systems. As postulates V and L and T are generally accepted and have not elsewhere led to such strange conclusions it is natural to suppose that the strangeness here is not due to them. In the argument the restricted principle of relativity needs to be used only in so far as it is involved in the conclusion that the units of any two systems of reference S_1 and S_2 are such that the same numerical result is obtained in measuring with the units of S_1 a quantity L_1 and with the units of S_2 a quantity L_2 when the relation of L_1 to S_1 is precisely the same as the relation of L_2 to S_2. But this principle is accepted in the classical mechanics and has not elsewhere led to strange results. The conclusion in postulate M appears to be demanded by the strongest experimental evidence; it is generally accepted; if the strange element in the result concerning units of time is due to this postulate, it appears that we must accept it as being required by such experience as has already been tested with due care. Hence the conclusion seems to be inevitable that the strangeness in our result is due principally to postulate R.

We shall presently see that the same basis of postulates leads to the conclusion that corresponding units of length in the two systems are also different when taken in certain directions. From the transformations of time and space which result from the conclusions thus obtained the whole restricted theory of relativity may be deduced (as is shown in the book mentioned). Therefore this theory depends essentially on the principle of correspondence of units in two systems of reference and on the propositions set forth explicitly in the postulates; and all of these are either generalizations from experiment or statements of laws which have usually been accepted. Hence we conclude: *The restricted theory of relativity may be developed by logical processes from the generalized results of certain experiments together with certain laws which have for a long time been accepted.*

The main result concerning the relation of units of length may be put in the following form:

If two systems of reference S_1 and S_2 move with a relative velocity v and if β is the ratio v/c of v to the velocity c of light as measured on either system, then to an observer on S_1 the unit of length of S_1 along the line of relative motion appears to be in the ratio $\sqrt{1-\beta^2}:1$ to that which is described to him as a unit by an observer on S_2 while to an observer on S_2 the unit of length of S_2 along the line of relative motion appears to be in the ratio $\sqrt{1-\beta^2}:1$ to that which is described to him as a unit by the observer on S_1.

These remarkable conclusions concerning units of length in two systems of reference rest on just those postulates which led to the strange results as to the units of time.

What often impresses one as the most remarkable conclusion in the theory of relativity is one which implies that the notion of simultaneity of events happening at different places is indefinite in meaning until some convention is adopted as to how simultaneity is to be determined. In fact, *there is no such thing as absolute simultaneity of events happening at different places.* With respect to the measured time and space of physics we must conclude that time does not run its course independently of space. Measured time and space are indissolubly bound together. The theorem which sets this forth most concretely may be stated in the following way:

Let two systems of reference S_1 and S_2 have an unaccelerated relative velocity v. Let an observer on S_2 place two clocks in the line of relative motion of S_1 and S_2 and adjust them so that they appear to him to mark simultaneously the same time. Then to an observer on S_1 the clock on S_2 which is forward in point of motion appears to be behind in point of time by the amount

$$\frac{v}{c^2} \cdot \frac{d}{\sqrt{1-\beta^2}},$$

where c is the velocity of light, $\beta = v/c$, and d is the distance between the two clocks as measured by the observer on S_1.

By means of the foregoing theorems we may readily obtain the formulae for the celebrated Lorentz transformation of space and time coordinates. (The non-mathematical reader may omit the remainder of this paragraph.) Let two systems of reference S and S' have the relative velocity v in the line l. Let systems of rectangular coordinates be attached to the systems of reference S and S' in such a way that the x-axis of each system is in the line l and that the two x-axes have the same positive direction, and let the y-axis and the z-axis of one system be parallel to the y-axis and z-axis respectively of the other system and have their positive senses in the same directions. Let these two systems of axes coincide at the time zero. Furthermore, for the sake of distinction, denote the space and time coordinates on S by x, y, z, t, and those on S' by x', y', z', t'. Let us suppose that S' moves with respect to S in the direction of increasing values of x. Then it turns out that the foregoing theorems imply the following relations between the two systems of coordinates:

$$t' = \frac{1}{\sqrt{1-\beta^2}} (t - \frac{v}{c^2}x),$$

$$x' = \frac{1}{\sqrt{1-\beta^2}} (x - vt),$$

$$y' = y,$$
$$z' = z,$$

where $\beta = v/c$ and c is the velocity of light.

The foregoing theorems, or (in more compact language) the foregoing equations of transformation, furnish the effective means for developing the whole of the restricted theory of relativity. Our purpose does not require us to follow that development further in detail. But we may mention a few of the remarkable conclusions which now emerge readily. If two velocities, each of which is less than c, are combined the resultant velocity is also less than c. The mass of a body increases with an increase in its velocity relative to the system on which the mass is measured. The mass of a body at rest appears to be the measure of its internal energy. Mass and energy in general appear to be essentially convertible terms. The velocity of light is a maximum which the velocity of a material body may approach but can never equal or exceed.

The development by Einstein in 1905 of the foregoing restricted theory of relativity led to a fresh analysis of the whole foundations of physics. This was made inevitable by its effective attack upon such fundamental notions as those of length and time and mass and velocity. Einstein himself succeeded in 1915 in greatly extending the range of his theory, developing what has since been called the general theory of relativity. We shall now speak briefly concerning the foundations underlying the latter.

Already in the restricted theory time and space had become essentially blended so that we could no longer speak of a three-dimensional space as separate and apart from the one dimension of time. A sort of combination of the two came into our conception and we began to realize that they can not be disentangled by the measurements of physics. We are forced to consider a four-dimensional continuum of space and time. It is with this space-time extension of four-dimensions that the general theory of relativity has essentially to do; and its problems are intimately connected with the relations of two systems of reference of the generalized sort which this makes necessary. The Lorentz transformation was a great psychological (and even logical) aid in the formation of the new theory.

Let us consider a four-dimensional extension in which space and time are intimately connected and blended so that each point P in these four dimensions represents a definite place A at a definite time t at which A is to be considered. In the course of time a material particle is represented by a succession of these points P. All these points for a given material particle lie on what is called the "world-line" of that particle; and this world-line represents the state of motion (or eventually the state of rest) of the material particle. If two objects come into coincidence at an instant their world-lines have a corresponding intersection. The things which the physicist deals with ultimately are these intersections of world-lines.

In order to deal with them he finds it necessary to introduce certain reference numbers which we may call the coordinates x_1, x_2, x_3, x_4. These numbers change in such a way that their variation along any world-line is continuous and that no two points ever have the same ordered set of four numbers assigned to them. This gives us a very general set of coordinates. It is clear that coordinates can be set up in an immense variety of ways so as to have these few very general properties. One of the first problems in the general theory of relativity is that of the character of the transformation by means of which we can pass from a given choice x_1, x_2, x_3, x_4 of coordinates to a second one ξ_1, ξ_2, ξ_3, ξ_4. It is clear that we must have relations of the form

$$\xi_i = f_i(x_1, x_2, x_3, x_4), \quad i = 1, 2, 3, 4,$$

where the functions f_i of the variables x_1, x_2, x_3, x_4 are rather general functions of these four arguments and are indeed to a large extent arbitrary. Now suppose that the laws of nature are expressed in terms of the coordinates x and also in terms of the coordinates ξ; the question arises as to what relation one ought to expect between these two forms of the law. Now there are no coordinates in nature. These have been inserted by us for our convenience. What is more natural, then, than the demand that we shall formulate our statements of these laws so that they shall have the same form in these two systems of reference, and indeed in all possible systems of reference? This is precisely one of the fundamental basic requirements upon which Einstein insists. The corresponding principle he has called the principle of covariance. In detailed and precise form, it may be stated somewhat as follows:

PRINCIPLE OF COVARIANCE. *The laws of nature can be (and are to be) expressed in such mathematical form in terms of the space-and-time coordinates x_1, x_2, x_3, x_4 that they shall remain invariant under every transformation of the form*
$$\xi_i = f_i (x_1, x_2, x_3, x_4), \ i = 1, 2, 3, 4,$$
where the functions f are subject to the following conditions:

1) *They are (apart from exceptional points or regions of fewer than four dimensions) finite and continuous and indefinitely differentiable;*

2) *They are such that the transformation is uniquely reversible, the inverse transformation having the properties demanded for the direct transformation;*

3) *They are such that in both the transformation and its inverse the fourth variable has the character of a time variable while the other three have the character of space variables.*

This principle demands the attainment of an ideal which is admittedly mathematical in its character. By means of it alone one could not come to grips with phenomena. One needs some additional hypotheses. One of these is to the effect that the restricted theory of relativity is valid in free space, that is, in space free of a gravitational field. The other is the celebrated law of the equivalence of gravitational forces and the apparent forces due to acceleration. This may be set forth as follows:

PRINCIPLE OF EQUIVALENCE. *For an indefinitely small region of the world (that is, a region so small that the variation of gravitation in it in both time and space is negligible) there exists a coordinate system S_0 (X_1, X_2, X_3, X_4) with respect to which gravitation has no influence either upon the motions of mass particles or upon any other physical phenomena whatsoever.*

Such is the logical basis from which the general theory of relativity proceeds. We can not here follow it in its high enterprise of conquest over the laws of nature. The road (at present and perhaps for a long time to come) can be followed only by one who is willing to give serious and long-continued attention to the study of certain branches of mathematics. In the earlier parts of the argument the reasoning is rather technical and abstruse in character and the general steps are intelligible only to those who have a considerable acquaintance with a certain range of mathematical ideas. After a time the exposition comes down, if not to earth, at least to the solar system and cases begin to appear in which it is possible to find means for choosing between the theory of Newton and that of Einstein.

Three crucial phenomena have been brought to light by means of which to test between the two theories. We shall now speak briefly of each of these.

For a long time astronomers have known that there is a certain forward advance in the perihelion position of the planet Mercury which can not be accounted for on Newton's theory. It amounts to about 42 seconds of angular measure per century. This is well accounted for by Einstein's theory.

Einstein predicted, on the basis of his theory, that a ray of light from a star which is seen apparently close to the edge of the sun would be found to be bent out of a straight path and that the deflection thus caused would turn out to be 1.74 seconds of angular measure, the bending being in such a direction that the star could actually be seen when just behind the edge of the sun. The prediction has been verified with a good degree of precision, observations having been taken at two eclipses of the sun.

A third crucial phenomenon is associated with the vibrations of an atom in a gravitational field. Since the periods of an atom furnish a sort of natural clock, it should give an invariant measure of an interval of time. Proceeding from this hypothesis one concludes that an atom vibrates more slowly on the sun than on the earth, due to the influence of the larger gravitational field of the sun. Hence the lines of the spectrum should be displaced towards the red. For the part of the spectrum usually observed this amounts to about .008 tenth-meters (a tenth-meter $= 10^{-10}$ meters). For a long time there was grave doubt whether this phenomenon is actually existent; but the evidence for its existence now (1933) seems to be conclusive.

Moreover in recent years it has come to be recognized that the stars known as white dwarfs have masses which are comparable with that of the sun, while their radii are much smaller. The companion of Sirius is a star whose radius is about 1/35 of that of our sun. Computation shows that the shift in the lines of the spectrum produced by light passing near this star should be about .30 tenth-meters. This matter was put to the test at Mount Wilson Observatory and an actual shift of .32 tenth-meters was found. One would conclude then that it is now hardly possible to doubt the actual existence of the spectral shift predicted by the Einstein theory.

Whatever may be the final verdict concerning the validity of the theory of relativity as a whole, it has certainly made a fundamental and permanent contribution to astronomy in developing a modification of Newton's law of gravitation. It has been checked experimentally in three very different ways and is thus established on a rather secure basis. Three such conquests as those just recorded have probably never before been made so nearly simultaneously by a single theory developed from one point of view consistently maintained throughout.

SUPPLEMENT II

THE THEORY OF TYPES[1]

by Paul Weiss[*]

It would seem from the interpretation that Whitehead and Russell put on the theory of types, that it is impossible or meaningless to state propositions which have an unrestricted possible range of values, or which, in any sense, are arguments to themselves. Thus on the acceptance of the principle that statements about all propositions are meaningless,[2] it would be illegitimate to say, "all propositions are representable by symbols," "all propositions involve judgment," "all propositions are elementary or not elementary," and if no statement could be made about all the members of a set,[3] it would be impossible to say, "all meanings are limited by a context," "all ideas are psychologically conditioned," "all significant assertions have grammatical structures," etc., all of which are intended to apply to themselves as well. The theory seems also to make ineffective a familiar form of refutation. General propositions are frequently denied because their enunciation or acknowledgment depends on the tacit supposition of the truth of a contradictory or contrary proposition. Such refutations assume that the general proposition should be capable of being an argument of the same type and to the same function as its own arguments, so that according to Whitehead and Russell, they fallaciously refute "by an argument which involves a vicious circle fallacy".[4]

That these limitations on the scope of assertions or on the validity of refutations are rarely heeded is apparent even from a cursory examination of philosophical writings since 1910. Thus Russell, apropos to Bergson's attempt to state a formula for the comic says,[5] "it would seem to be impossible to find any such formula as M. Bergson seeks. Every formula treats what is living as if it were mechanical, and is therefore by his own rules a fitting object of laughter." The characterisation of all formulæ, even though it refers to a totality, seems to Mr. Russell to be of the same type as the formulæ characterised.

---

[1] Chap. II., *Principia Mathematica.*

[*] [*Reprinted from* Mind: *a Quarterly Review of Psychology and Philosophy.* Vol. XXXVII., N.S., No. 147; with minor corrections.]

[2] P. 37, *ibid.* (second edition).   [3] P. 37, *ibid.*   [4] P. 38, *ibid.*

[5] "Prof. Guide to Laughter," *Cambridge Review*, Vol. 32, 1912, and Jourdain's *Philosophy of Mr. B*tr*nd R*ss*ll*, pp. 86-7.

If the theory were without any embarrassments of its own, and were indispensable for the resolution of the so-called paradoxes[1] (which no one seems to believe), there would be nothing to do but to acknowledge the impossibility of cosmic formulations, as well as the inadequacy of philosophic criticisms, and to pass charitably over such remarks as Russell's as mere accidents in a busy life. However, the statement of the theory itself involves the following difficulties in connection with (1) its scope, (2) its applicability to propositions made about it, and (3) its description.

1. It is either about all propositions or it is not.

A. If it were about all propositions it would violate the theory of types and be meaningless or self-contradictory.

B. If it were not about all propositions, it would not be universally applicable. To state it, its limitations of application would have to be specified. One cannot say that there is a different theory of types for each order of the hierarchy, for the proposition about the hierarchy introduces the difficulty over again.

2. Propositions about the theory of types (such as the present ones, as well as those in the *Principia*) are subject to the theory of types, or they are not.

A. If they were, the theory would include within its own scope propositions of a higher order, and thus be an argument to what is an argument to it.[2]

B. If they were not, there would be an unlimited number of propositions, not subject to the theory, that could be made directly or indirectly about it. Among these propositions there might be some which refer to a totality and involve functions which have arguments presupposing the function.

3. The statement of the theory of types is either a proposition or a propositional function, neither or both.

A. If it were a proposition, it would be either elementary, first order, general, etc., have a definite place in a hierarchy and refer only to those propositions which are of a lower order. If it were held to be a proposition of the last order, then the number of orders would have a last term, and there could not be meaningful propositions made about the theory. The *Principia* should not be able to say, on that basis, just what the purpose, character and application of the theory is.

B. Similarly, if it were a propositional function, it would have a definite place in a hierarchy, being derived from a proposition by generalisa-

---

[1] Paradoxes, though contrary to common opinion, may be and frequently are true. Paranoumena, violating principles of logic or reason, if they are not meaningless, are false, and it is only they which are capable of logical analysis and resolution. What the *Principia* attempts to do is to solve apparent paranoumena with a real paranoumenon.

[2] P. 39, *Principia Mathematica.*

tion. It could not refer to all propositions or propositional functions, but only to those of a lower order.

C. If it were neither it could not be true or false, nor refer to anything that was true or false. It could not apply to propositions, for only propositions or propositional functions, in a logic, refer to propositions.

D. If both at once, it would be necessarily self-reflexive.

    a. If as function it had itself as value, it would refer to itself. But the theory of types denies that a function can have itself as value.

    b. If as function it had something else as value, it would conform to the theory, which insists that functions have something else as values. The theory then applies to itself and is self-reflexive, and thus does not apply to itself. As, by hypothesis, it is a value of some other function, there must be propositions of a higher order and wider range than the theory of types.

It is no wonder that the perpetrators of the theory have not been altogether happy about it! What is sound in it—and there is much that is—is best discovered by forgetting their statements altogether, and by endeavouring to analyse the problems it was designed to answer, without recourse to their machinery. The result will be an acknowledgment of a theory of types having a limited application, and a formulation of a principle which will permit certain kinds of unrestricted general propositions.

To do this we shall deal in detail with two apparent paranoumena dealt with in the *Principia*, where the difficulty is largely *methodological*. We shall then treat of Weyl's "heterological-autological" problem, where the difficulty is due to a confusion in *meanings*. Those problems which cannot be dealt with under either heading will be those which need a theory of types for their resolution.

1. *Epimenides.* The proposition "All Cretans are liars" must be false if it applies to Epimenides as well, for it cannot be true, and only as false has it meaning. If it were true, it would involve its own falsity. When taken as false, no contradiction, or even paradox, is involved, for the truth would then be "*some* Cretans tell the truth". (The truth could not be "all Cretans tell the truth" for Epimenides must be a liar for that to be true and by that token it must be false). Epimenides himself would be one of the lying Cretans, and one of the lies that the Cretans were wont to make would be "all Cretans are liars". Thus if Epimenides meant to include all his own remarks within the scope of the assertion, he would contradict himself or state a falsehood. If it be denied that a contradictory assertion can have meaning, he must be saying something false if he is saying anything significant. Had he meant to refer to all other Cretans there is, of course, no difficulty, for he then invokes a kind of theory of types by which he makes a remark not intended to apply to himself. All difficulty disappears when it is recognised that the formal implication, "all Cretanic statements are lies" can as a particular statement be taken as one of the values of the terms of this implication. Letting Ep / p represent "Epimenides once asserted p"; ϕ represent "Cretanic" and p represent a statement or

proposition, then for "All Cretanic statements are false (or lies)," we have:

$$1.\ \phi p \mathbin{.} \supset_p \mathbin{.} \smallfrown p.$$

And as Epimenides is a Cretan, for any assertion he makes we have:

$$2.\ Ep\ !\ p \mathbin{.} \supset_p \mathbin{.} \phi p.$$

As No. 1 is an argument to the above—it being Epimenides' present remark—we get:

$$3.\ Ep\ !\ \{\phi p \mathbin{.} \supset_p \mathbin{.} \smallfrown p\} \mathbin{.} \supset \mathbin{.} \phi\{\phi p \mathbin{.} \supset_p \mathbin{.} \smallfrown p\}$$

No. 1, as a Cretanic statement, is an argument to No. 1 as a formal implication or principle about Cretanic statements, so that:

$$3A.\ \phi\{\phi p \mathbin{.} \supset_p \smallfrown p\} \mathbin{.} \supset \mathbin{.} \smallfrown \{\phi p \mathbin{.} \supset_p \mathbin{.} \smallfrown p\}$$

No. 3 and No. 3A by the syllogism yield:

$$3B.\ Ep\ !\ \{\phi p \mathbin{.} \supset_p \mathbin{.} \smallfrown p\} \mathbin{.} \supset \mathbin{.} \smallfrown \{\phi p \mathbin{.} \supset_p \mathbin{.} \smallfrown p\}$$

so that in this instance Epimenides lied.

It is important to note that No. 1 states a formal implication, and that No. 3, No. 3A and No. 3B employ No. 1 as a particular assertion or specific argument to their functions. No. 3A is an instance of the implication expressed by No. 1, and is this instance because of the particular argument it does have. It states the fact that " 'all Cretanic statements are false' is a Cretanic Statement," implies that " 'all Cretanic statements are false' is false". Substitution of another argument would give a different instance; though of course of the same implication. The implication contained in its argument does not have instances. " 'Some Cretanic statements are false' is a Cretanic statement" or " 'This Cretanic statement is false' is a Cretanic statement" are not instances of " 'All Cretanic statements are false' is a Cretanic statement," but of "P is a Cretanic statement". These three propositions have different subjects; they are different values of the same propositional function. That these subjects have relations to one another is of no moment. "My wife loves me" and "my mother-in-law is old (or loves me)" are two distinct and logically independent propositions, even though there is a relationship between the two subjects.

It is because any considered general proposition is at once an individual fact, and a formal implication or principle, with many possible arguments, that it is capable of being taken as an argument to itself. All propositions about words, logic, truth, meaning, ideas, etc., take arguments which fall in these same categories, and in so far as such a general proposition is stated in words, determined by logic, etc., it should, as such a fact, be an argument to itself as a formal implication. The principle must be false if this cannot be done, for it is sufficient, in order to overthrow a proposition of this kind, to produce one argument for which it does not hold. One may limit the principle by asserting that it holds for "all but . . .", in which case it is a *restricted* general proposition. Nominalism, association of ideas, scepticism, the theory of universal tautology, the denial of logic are defended in propositions which cannot take themselves as arguments, and which as facts are arguments to contradictory principles. Their contradictory principles therefore hold sometimes at least, so that these doctrines must be false if they are put forward without restriction, and cannot be universally true, if, in Bradley's words, they "appear".

2. "I am lying"—if it be taken in isolation from all fact—is a meaningless statement. There must be some objective truth that is distorted, and unless it is provided the assertion has no significance. This proposition means either, "I am lying about X"; "I always lie," or "I have always lied". The first can be either true or false without giving rise to any problem, except where "all my assertions" is made an argument to X, in which case it is equivalent to either the second or third formulation. "I always lie" involves the same situation as with Epimenides, and the proposition is false. The supposition of its truth would involve a contradiction; the supposition of its falsity means simply that I sometimes lie and sometimes tell the truth. If what is meant is that "I have always lied" that does not involve a contradiction, for what is intended is a restricted proposition, applying to *all but* the present one. It can be true because it does not apply to all propositions; if it were false, then sometimes I lied and sometimes I did not. In short, there is nothing like a self-reflective universal liar, which is an interesting moral conclusion to derive from a logical analysis. Similarly, there cannot be a thorough scepticism held by the sceptic to be valid.

Prof. Whitehead (to whom I am also indebted for the notation) has pointed out to me that wherever a conjunction of propositions results in a *reductio ad absurdum*, there is no way of determining on logical grounds alone which of the antecedents fails, or is false (though one at least must be). Thus in the case of Epimenides we have:

4. $\{\phi p . \supset_p . \backsim p\} . \{Ep \,!\, p . \supset_p . \phi p\} . Ep \,!\, \{\phi p . \supset_p . \backsim p\}$
   (A)   (B)   (C)
   $. \supset . \backsim \{\phi p . \supset_p . \backsim p\}$
                        (D)

It is because B and C are in that case assumed to hold, that we can say that A must fail. If the truth of all these antecedents were undetermined, we should have merely the general rule: a *reductio ad absurdum* has as a necessary condition the conjunction of one or more false propositions. Transposition—

4'. $\{\phi p . \supset_p . \backsim p\} . \supset . \backsim \{\phi p . \supset_p . \backsim p\}$
    (D)   (A)
    $. \mathsf{v} . \backsim \{Ep \,!\, p . \supset_p . \phi p\} . \mathsf{v} . \backsim Ep \,!\, \{\phi p . \supset_p . \backsim p\}$
                        (B)   (C)

makes it apparent that to deny the conclusion of a *reductio ad absurdum* is to imply that at least one of the antecedents is false.

In connection with the *reductio ad absurdum* involved in the assertions, "I always lie" and "I always doubt," No. 4B reduces to the tautologies: "If I assert p, p is my assertion," and "If I doubt, the doubt is mine". In these cases, the only alternatives left are the denial of the fact of the assertion (No. 4C), or the truth of the principle itself (No. 4A).

3. Weyl's heterological-autological contradiction[1] is the result of a material fallacy of amphiboly in connection with the employment of adjectives. The simplest form of such a fallacy is due to a failure to distinguish between an adjective as substantive and an adjective as attribute. Thus if we treat both the subject and attribute in "large is small" and "small is large" as attributes united by a copula expressing identity (instead of reading it as "large is a small word," "small is a large word") we could say "whatever is small is large, and whatever is large is small". No one, I believe, since the Megarics, has been troubled by this particular confusion.

The present problem is the result of a confusion, not between substantive and adjective, but between an adjective which expresses a property, and an adjective which expresses a relation between this property and the substantive. All words can be described in terms of a property—they are long, short, beautiful, melodious, etc., words. They can be classified in accordance with these properties, giving us the class of long words, short words, etc. They can also be classified as either "autological" or "heterological," depending on whether or not the same word is at once substantive and property-adjective; the terms 'autological" and "heterological" expressing relationships between the substantive and adjective.

The autological class is made up of words, each of which expresses a property which it possesses; though all of them have unique properties. If "short" be short, and if "melodious" be melodious, they would both be members of the autological class; though in addition, "short" would be a member of the class of short words, and "melodious" would be a member of the class of melodious words.

The heterological class is made up of words, each of which expresses a property which it does not possess. If "long" be short, and if "fat" be thin, they would both be members of the heterological class; although here also "long" would be a member of the class of short words, and "fat" would be a member of the class of thin words. Though when classified according to the relationship of the adjective to the substantive, "short" would be an autological word and "long" a heterological word, they would both be members of that class which was defined in terms of the properties of words—being in this case, members of the class of short words.

Now if heterologicality were a property that a word could have, and if the word "heterological" had that property, it would be a member of the autological class, for it would then possess a property that it expressed. But it would also be a member of a class of words which had the *property* of heterologicality. This class is determined by taking the properties of words, and if it be called

---

[1] Briefly stated it is: all words which express a property they possess are autological; all words which express a property they do not possess are heterological. If 'heterological' is heterological it expresses a property it possesses and is thus autological; if it is autological, it expresses a property it does not possess and is therefore heterological. *Das Kontinuum*, p. 2.

"heterological," must be distinguished from that class which was determined not by properties, but by the relationship between properties and substantives.

If there were a property like autologicality and if "heterological" had that property,[1] it would be a member of the heterological class, for it would express a property which it did not possess. But it would also be a member of the class of words which possessed autologicality and could thus be classified.

Thus if "heterological" had the property of autologicality, it would be in the heterological class owing to the *relation* which held between the property and substantive (or between a property it possessed and the property it expressed); but it would be in the class of autological words, owing to a *property* it possessed. If it had the property of heterologicality, it would be in the autological class on the basis of the *relation*, and in the class of heterological words on the basis of *property* classification. There is no difficulty in considering something as a member of two distinct classes, owing to the employment of different methods of classification. There is no contradiction in saying: " 'heterological' expresses the property heterologicality, possesses the property autologicality, and the relation between these properties is heterological, or that it expresses and possesses the property heterologicality and the relation between them is autological." Similarly, Richard's contradiction, Berry's contradiction, and that involving the least indefinable ordinal, are resolvable by recognising that "nameable" and "indefinable" are used in two sharply distinguishable senses. They do not require a hierarchy, but a discrimination in the methods of description.

When a distinction is made between a class and its membership (the distinction between a number of numbers and a number is a particular case of this), and between a relation of objects and a relation of relations, the requirements for the solution of the other mathematical problems are provided. A class is other than its members, and a relation, like all universals, transcends any given instance or totality of instances. As they have characters of their own, universals can be described in terms of other universals, which in turn transcend them. Arguments are of a different "type" than functions, just so far as they have different logical characteristics, *i.e.* are different kinds of logical facts. The class which is an argument to a function about classes has, as argument, a different logical import than the function, and its arguments have a different import from it. This is true of all functions, restricted and unrestricted alike, for it means simply that they are discriminable from their arguments. They can, despite this difference, have characteristics in common with their arguments, and are to that extent unrestricted. Thus in the case of "the class of those classes which are identical with themselves," the class of classes can be

---

[1] 'Heterological,' in fact, has the properties of being long, polysyllabic, etc., and it is questionable whether there are properties like autologicality and heterologicality possessed by words. If there be no such properties, "heterological" is a member of the class of long words, polysyllabic words, etc. In addition it would be one of the terms related by the heterological relation, which fact would not make it have the *property* of heterologicality.

taken simply as a class, without logical embarrassment. Yet a class of classes differs from a class, and must therefore be capable of a different characterisation, and thus also be an argument to a function of a different type. With some classes, it may not be possible to consider them as arguments to their own functions, without uncovering a contradiction. In such cases (*e.g.* the class of those classes which are not members of themselves, and the relations which are connected by their contradictories), it is the difference between the function and the argument that is of moment. That *some* functions cannot take themselves as arguments does not indicate that *all* functions are restricted in scope, but simply that they are *non-restricted*. Some classes and functions are restricted and some are not. To say that all are restricted because some are is an obvious fallacy.

Whenever, as individual, a general proposition is in the class of those objects of which it treats, but cannot be considered as an argument to itself, it is either false or restricted in scope. If the second, its range of arguments must be specified. Accordingly, we can state as a *necessary* condition for the truth of a general proposition, whose scope is unspecified, that when it has a character, which is one of the characters about which it speaks, it *must* be an argument to itself. Thus if Bergson adequately described the comic, his formula should be an object of laughter, and if the theory of types is universal in application, it should be capable of being subject to itself. Conformity to this condition indicates that the unrestricted proposition is *possibly* true; not that it is necessarily true. To demonstrate that such a proposition was necessarily true, it would be essential to show that the supposition of its falsity assumes its truth. That there is danger in applying this rule can be seen from the consideration of some such proposition as: "Everything is made up of language elements". Its denial will be made up of language elements, and would seem to demonstrate that the proposition was necessarily true. Supposition of the falsity of a proposition, however, means verbal denial only in so far as the proposition applies to the realm of language. If it applies to everything, supposition of its falsity involves the positing of the objects of assertions; not the assertions. A necessary unrestricted proposition about everything can be supported only by a demonstration that the supposition of an argument for which it does not hold is self-contradictory. If the proposition has to do with grammar, meaning, logic, judgment, etc., the conditions for a necessarily true and unrestricted proposition would be: 1. the assertion of it is an argument to it; 2. any possible denial is an argument to it. That "any possible denial" rather than "any given denial" is required, is apparent from the consideration of the following propositions: "All sentences are made up of eight words," "No sentence is made up of eight words". Each of these contains eight words. It is because of the fact that we can formulate propositions such as, "It is false that every proposition must be made up of eight words," that the condition is seen not to have been met.

An unrestricted proposition applies to every member of the category, and has some aspect of itself as value. It is in some sense then a determinate in the category which it determines. If the proposition refers to some other category than the one to which it as fact, or some aspect of it as fact, belongs, it is restricted. Thus "all men are mortal" is neither man nor mortal, and as condition does not determine itself as fact. Any proposition referring to that statement would be of a different type, and would deal with its truth, falsity, constituents, historical place, logical structure, etc. Though the unrestricted propositions have no limitations, the category to which they refer may have. Epimenides' remark, for example, referred only to Cretans. As his assertion was a determinate in the category, and as his statement of the supposed conditions imposed on the members of that category was not a possible argument to the general proposition, the general proposition was seen to be false or restricted. Had he said, "All Cretans tell the truth," he would have stated an unrestricted proposition which was possibly true. It could not be said to be necessarily true unless Cretans and lie, against the evidence of history, were actually contradictories.

Accordingly, we shall say: *All true unrestricted propositions are arguments to themselves; or by transposition, those propositions which are not arguments to themselves are either restricted or false.* As this proposition can take itself as argument it is possibly true. Unless no proposition is possible which does not conform to it, it cannot be said to be necessarily true. I have not been able to demonstrate this and therefore accept it as a definition or "methodological principle of validation". The theory of types, in its most general form, may be stated as: *A proposition or function of order n, which cannot be an argument to itself, is, as fact, an argument of a proposition or function of order n+1.*

In accordance with the scheme of the criticism of the theory of types, we can describe our principle as (1) applying to all propositions, including (2) those which refer to it. (3) It is a formal implication with itself as one of its arguments. The theory of types, on the other hand, (1) does not apply to all propositions, but only to those which are restricted, (2) *may* apply to those propositions which refer to it, and (3) is a formal implication which cannot take itself as argument.

The theory of types cannot be an unrestricted proposition about all restricted propositions. As an unrestricted proposition it must take itself as argument; but its arguments are only those propositions which are *not* arguments to themselves. It cannot therefore be unrestricted without being restricted. Nor can it be a restricted proposition about all restricted propositions for it would then be one of the restricted propositions, and would have to take itself as argument—in which case it would be unrestricted. Hence it cannot be restricted without being unrestricted. Three possible solutions may be advanced. The first is that the theory of types is restricted and does not apply to *all* restricted propositions, but only to *some* of them. It is not an argument to itself but to some other proposition about restricted propositions. This in turn will have to be restricted and refer only to some propositions, and so on, giving us theories of types of various orders. The proposition made about the totality of these orders would be of a still higher order and would in turn presuppose a higher order *ad infinitum*. The theory of types thus depends on theories of types of theories of types without end. This seems probable on the ground that the theory is based on the recognition that no proposition can be made about all restricted propositions, so that it must by that very fact admit that it cannot apply to all of them. Instead, therefore, of the theory of types applying to all propositions, and determining them in various orders, it does not even apply to all of a given class of them. This interpretation would not affect unrestricted propositions, and would merely show that the determination of restricted propositions is subject to determinations without end.

The second possibility is suggested by the consideration of a proposition such as: "all truths are but partially true". If that were absolutely true, it would contradict itself, and if it were not, could apply only to some truths. Considered as referring to the necessary limitations which any finite statement must have, it would take itself as argument in so far as it was finite, thus indicating that it was absolutely true about finite propositions, and yet not absolutely true as regards all truths. By pointing out the limitations of a finite statement it indicates that there is an absolute truth in terms of which it is relatively true. On this interpretation, any condition which imposes universal limitations is unlimited in terms of what it limits, but limited in turn by some other condition. One might hold, therefore, that the theory would be unrestricted as regards restricted propositions, and restricted as regards all propositions, and would point to a higher principle which limits it.

The third possibility is to allow for "intensive" propositions which are neither restricted nor unrestricted, being incapable of any arguments. The theory of types could be viewed as such an intensive proposition, and what we have called its arguments, would merely "conform" to it. This interpretation means the downfall of a completely extensional logic, and a determination of an extensional logic as subordinate to an intensional one.

There are difficulties in each of these interpretations. The last seems to me to be best. In any of these cases, however, a restricted proposition which refers to some other than the restricted aspect of the theory would be subject to the theory and the principle we have laid down about unrestricted propositions could still hold. Those restricted propositions which refer to the restricted character of the theory would not be an argument to it on the first, would be an argument to it on the second, and would neither be nor not be an argument to it on the third solution.

To briefly summarise: The theory of types must be limited in application. Not all the problems it was designed to answer require it; another principle of greater logical import is desirable; while for the resolution of the problems in which it is itself involved, very drastic remedies are necessary. No matter how the theory fares, the possibility of the methodological principle and the possibility of other solutions for the so-called paradoxes, indicate that it is at least not as significant an instrument as it was originally thought to be.

SUPPLEMENT III

A NON-ARISTOTELIAN SYSTEM AND ITS NECESSITY FOR RIGOUR IN MATHEMATICS AND PHYSICS*

BY ALFRED KORZYBSKI

We are here dealing with a concrete mathematical problem which is not trivial, but at the same time is solvable, and I cannot imagine that any mathematician can find the courage to elude its honest solution by means of a metaphysical dogma. (549)   HERMANN WEYL

I protest against the use of infinite magnitude as something completed, which in mathematics is never permissible. Infinity is merely a *façon de parler*, the real meaning being a limit which certain ratios approach indefinitely near, while others are permitted to increase without restrictions. (74)   K. F. GAUSS

A very extensive literature shows that the problems of 'infinity' pervade human psycho-logical reactions, starting from the lowest stage of human development up to the present and that without some theory of 'infinity', modern mathematics would be impossible. Up to date, no satisfactory theory of infinity, on which all mathematicians could agree, has been produced. The results are rather bewildering because what appears to some prominent mathematicians as perfectly sound mathematics is evaluated by other equally prominent scientists as a 'mental' disease (Poincaré); or we find opinions that a large portion of mathematics is devoid of proof and has to be accepted on faith; or that some parts of mathematics must be treated as non-sense (Kronecker, Brouwer, Weyl . ,). 'There are eminent scholars on both sides and the chance of reaching an agreement within a finite period is practically excluded', says Brouwer, and certainly such a state of affairs does not allow us to have any satisfactory modern standards of proof and rigour; the last thing we should expect in mathematics.

The majority of those mathematicians who take interest in the soundness of their science seem to believe that the main difficulty centres around the validity of the 'law of excluded third' ('A is B, or not B') of the accepted, sharply two-valued, chrisippian form of A 'logic'. They disregard the fact that we are born, bred, educated, speak a language, live under conditions, institutions . , which still remain desperately A or even pre-aristotelian. If we attempt to reject one of the two-valued 'laws of thought' or postulates of the A-system, but retain A or pre-aristotelian *elementalistic* 'psychologies', 'logic', and *s.r*, no agreement in 'a finite period' can be expected, and the present mathematical chaos would continue.

---

*Paper presented before the American Mathematical Society at the New Orleans, Louisiana, Meeting of the A.A.A.S. December 28, 1931. I continue to use the abbreviations introduced in this book.

Among the more important schools we may distinguish roughly:[1]

1) The logistic school represented by Peano, Russell, and Whitehead, who accept the chrisippian, two-valued, restricted form of the *el* 'logic' and so may be called the *chrisippian school.*

2) The axiomatic school, represented by Hilbert and his followers, which may be called the *aristotelian school.*

3) The 'intuitional' school represented by Brouwer and Weyl who question the 'law of excluded third', and so may be called the *non-chrisippian school.*

4) The Polish school of: (*a*) 'intuitional' formalism with Łukasiewicz, Tarski, Leśniewski as representatives, which may be called the *non-aristotelian school.* Łukasiewicz generalized the *A* 'logic' to three-valued 'logic' which covers modality. Łukasiewicz and Tarski finally produced a general many-valued 'logic' of which the two-valued represents only a limiting case. Leśniewski produced *Protothetic,* a still more general 'logical' system, by introducing variable 'funktors', .* (*b*) The *restricted semantic* school represented by Chwistek and his pupils, which is characterized mostly by the semantic approach, and by paying special attention to the *number* of values, establishing the thesis that the older 'freedom from contradictions' depends on one-valued formulations, as discovered by Skarżeński and quoted by Chwistek. This school has already produced new foundations (still *elementalistic*) for 'logic' and mathematics, and leads to generalized arithmetics and analysis.

5) The average prevalent mathematical technician, who does not realize that he belongs to the numerically large class which may be called the 'christian science' school of mathematics, which proceeds by faith and disregards entirely any problems of the epistemological foundations of their supposed 'scientific' activities.

It should be noticed that all existing mathematical schools accept implicitly, at least, *A elementalism* and do not challenge identity, a principle which happens to be invariably false to facts and which therefore should be entirely abolished.

The above classification suggests that, in spite of great achievements in the field of mathematical foundations, no school can expect to be convincing or accepted by other schools as long as we all flounder in the *A* and *el* ambiguities which prevent any possibility of agreement. It becomes obvious also that when a *Ā* and *non-el* system is formulated it will necessitate a new paradox-free foundation for mathematics and so a new school of mathematics will arise which may be called:

6) The *general semantic, non-aristotelian, non-elementalistic* school of mathematics. It is premature to give the names of the leading pioneers in this field at present.

---

*At present Łukasiewicz and Tarski call their many-valued 'logic' non-chrisippian, but this name does not seem appropriate because these authors generalized both forms of the aristotelian 'logic' to a many-valued 'logic' of which the two-valued becomes only a limiting case. Thus it seems that their many-valued 'logic' is better described by the term *non-aristotelian,* yet still *elementalistic* 'logic'.

In a $\bar{A}$-system, the 'logical' problems of freedom from contradiction become also semantic problems of *one-valued meanings* made possible only under ∞-valued, $\bar{A}$, *non-el* general semantics, and the recognition of the $\bar{A}$ *multiordinality* of terms , . A $\bar{A}$-system introduces some fundamental innovations, such as completely rejecting identity, elementalism . , and becomes based on *m.o* structure and order, and so ultimately becomes *non-el*. The A, (3+1)-dimensional *el*, (in the main) *intensional* system becomes a four-dimensional, *non-el*, (in the main) *extensional* system. In such a system we cannot use the formulations of *elementalistic* 'logics' and 'psychologies', but must have $\bar{A}$, *non-el* general *semantics*, which when generalized become an entirely general discipline applicable to all life, as well as to *generalized* mathematics. For the above reasons I shall use the word 'logic', in its *el* sense, with quotation marks; and use the term *general semantics* for a *non-el*, $\bar{A}$ discipline corresponding to the *el*, A or $\bar{A}$ 'logics'.

Investigations show that the primitive man (and the 'mentally' ill) use *one-valued semantics* which have left more or less marked traces in all of us, reflected even in science and mathematics. The elimination of these primitive traces clears the foundation for an adult civilization, a theory of sanity, and the elimination of the scientific and mathematical paradoxes.

To assume that because a many-valued 'logic' has been produced, all the problems of mathematical infinity, irrational numbers, continuity, mathematical induction, validity of mathematical proof, mathematical existence . , have been solved, would be a mistake. The aim of the present paper is to analyse some of the fundamental complexities produced by the unconscious operation of the one-valued semantic *identification* concealed in the formulation of the 'law of identity', which have escaped notice until now, and which would make the application of a many-valued 'logic' or ∞-valued semantics and agreement impossible. Here, as in the $\bar{E}$ and $\bar{N}$ systems, only the most general formulations help us to *discriminate* between the particular cases, and so to eliminate the undesirable traces of one-valued semantics by building a $\bar{A}$-system, of which the A and pre-A represent only particular cases.

Let me recall the 'philosophical grammar' of our language which we solemnly call the 'laws of thought', as given by Jevons:[2]

1) The law of identity. Whatever is, is.
2) The law of contradiction. Nothing can both be, and not be.
3) The law of excluded third. Everything must either be, or not be.

These 'laws' have different 'philosophical' interpretations which help very little and for my purpose it is enough to emphasize that: (1) The second 'law' represents a negative statement of the first, and the third represents a corollary of the former two; namely, no third possible between two contradictories. (2) The verb 'to be', or 'is', and 'identity' play a most fundamental role in these formulations. We should not be surprised to find that the investigation of these terms may give us a long sought solution. Such an investigation is very laborious and difficult. 'The complete attempt to deal with the term *is* would go to the form and matter of everything in existence, at least, if not to the possible form and matter of all that does not exist, but might. As far as it could be done, it would give the grand Cyclopaedia, and its yearly supplement would be the history of the human race for the time', said Augustus de Morgan in his *Formal Logic*, and this opinion I found fully justified.

So I must be brief, and state but roughly, that in the Indo-european languages the verb 'to be' has at least four entirely different uses: (1) as an auxiliary verb, 'Smith is coming'; (2) as the 'is' of predication, 'the apple is red'; (3) as the 'is' of 'existence', 'I am'; (4) as the 'is' of identity, 'the apple is a fruit'. The fact that four semantically entirely different words should have one sound and spelling appears as a genuine tragedy of the race; the more so since the discrimination between their uses is not always easy.

The researches of the present writer have shown that the problems involved are very complicated and cannot be solved except by a *joint study* of mathematics, mathematical foundations, history of mathematics, 'logic', 'psychology', anthropology, psychiatry, linguistics, epistemology, physics and its history, colloidal chemistry, physiology, and neurology; this study resulting in the discovery of a general semantic mechanism underlying human behaviour, many new interrelations and formulations, culminating in a $\bar{A}$-system. This semantic mechanism appears as a general psychophysiological mechanism based on four-dimensional order, present and abused in all of us, the primitive man, the infant, the 'mentally' ill, and the genius not excluded. It gives us an extremely simple means of training our *s.r*, which can be applied even in elementary education.

The scientific problems involved are very extensive and can be dealt with only in a large volume. Here I am able to give only a very sketchy summary without empirical data, omitting niceties and technicalities.

$$(a)$$

Paris	Dresden	Warsaw
*_____	____*____	_____*

$$(b)$$

Dresden	Paris	Warsaw
*_____	___*___	_____*

If we consider an actual territory (*a*) say, Paris, Dresden, Warsaw, and build up a *map* (*b*) in which the order of these cities would be represented as Dresden, Paris, Warsaw; to travel by such a map would be misguiding, wasteful of effort , . In case of emergencies, it might be seriously harmful , . We could say that such a map was 'not true'. , or that the map had a *structure not similar* to the territory, structure to be defined in terms of relations and multi-dimensional order. We should notice that:

A) A map may have a structure similar or dissimilar to the structure of the territory.   (1)

B) Two similar structures have similar 'logical' characteristics. Thus, if in a correct map, Dresden is given as between Paris and Warsaw, a similar relation is found in the actual territory.   (2)

C) A map *is not* the territory.   (3)

D) An ideal map would contain the map of the map, the map of the map of the map . , endlessly. This characteristic was first discovered by Royce. We may call it self-reflexiveness.   (4)

Languages share with the map the above four characteristics.

A) Languages have structure, thus we may have languages of *elementalistic* structure such as 'space' *and* 'time', 'observer' *and* 'observed', 'body' *and* 'soul', 'senses' *and* 'mind', 'intellect' *and* 'emotions', 'thinking' *and* 'feeling', 'thought' *and* 'intuition'. , which allow verbal division or separation. Or we may have languages of *non-elementalistic structure* such as, 'space-time', the new quantum languages, 'time-binding', 'different order abstractions', 'semantic reactions'. , which do not involve verbal division or separation . ; also mathematical languages of 'order', 'relation', 'structure', 'function', 'variable', 'invariant', 'difference', 'addition', 'division'. , which apply to 'senses' and 'mind', that is, can be 'seen' and 'thought of', .   (5)

B) If we use languages of a structure non-similar to the world and our nervous system, our verbal predictions are not verified empirically, we cannot be 'rational' or adjusted , . We would have to copy the animals in their wasteful and painful 'trial and error' performances, as we have done all through human history. In science we would be handicapped by semantic blockages, lack of creativeness, lack of understanding, lack of vision, disturbed by inconsistencies, paradoxes , .   (6)

C) Words *are not* the things they represent.   (7)

D) Language also has self-reflexive characteristics. We use language to speak about language, which fact introduces serious verbal and semantic difficulties, solved by the theory of *multiordinality*.   (8)

The above unusually simple considerations lead to unexpectedly far-reaching consequences.

A) From (7)—it follows that the objective levels which include the events, ordinary objects, objective actions, processes, immediate feelings, 'instincts', 'ideas', *s.r* in general . , represent un-speakable levels, *are not words*.   (9)

B) From (9)—that the use of the 'is' of *identity*, as applied to objective, un-speakable levels, appears invariably structurally false to facts and must be entirely abandoned. Whatever we might *say* a happening 'is', *it is not*.   (10)

C) From (10)—*structure* appears as the only possible link between the objective, un-speakable, and the verbal levels.   (11)

D) From (11)—the only possible 'content of knowledge' becomes exclusively *structural*.   (12)

E) From (12)—the only aim of 'knowledge' and science appears as the empirical search for, and verbal formulation of, structure.   (13)

F) The only method for acquiring 'knowledge' is found in an *empirical* investigation of the potentially unknown structure of the world, ourselves included, only afterwards adjusting the structure of languages so that they would be similar, and so of maximum usefulness; instead of the delusional

reversed order of ascribing to the world the structure of an inherited primitive
language.   (14)

G) The investigation of the potentially known structure of languages in
which we predict and then verify the predictions empirically, appears as an
important method for the discovery of the structure of the world.   (15)

H) Investigations disclose that all A, *el* languages and disciplines built
on them (older 'psychologies', 'logics'. , and, based on them, economics,
sociology, politics, 'ethics'. , reflected in turn in our institutions, systems. ,)
are not structurally similar to the world and our nervous system, as they
verbally divide what empirically cannot be divided. Under such conditions
neither a higher grade civilization, nor general sanity, nor paradox-free science
and mathematics are possible. In *el* languages, our verbal predictions are not
verified empirically, and not being able to foresee we must proceed by animalistic 'trial and error'.   (16)

I) Mathematics appears as a very limited but the only language in
existence, in the main similar in structure to the world around us *and* the
nervous system.   (17)

J) From the study of mathematics, mathematical physics, and physics,
we learn, and will continue to learn, the fundamentals of *m.o* structure. It is
no mystery that all chemistry has become a branch of physics, all physics can
be made a branch of geometry, all geometry a part of analysis, and all analysis
a part of general semantics. The present work shows that the analysis *of all
human problems* of daily life or science becomes dependent on *general semantics*
which on the verbal levels becomes generalized mathematics. Thus mathematics, mathematical physics, and physics become the most important
disciplines from which we learn most about *structure*,—the only 'content of
knowledge'.   (18)

K) The older *el* 'psychologies' and 'logics' for their maximum usefulness
must be transformed into unified *non-el* psycho-logics and general semantics,
possible only after studying all forms of human behaviour, mathematics
included.   (19)

L) The study of mathematics as a form of human behaviour, appears
necessary prior to the possibility of formulating any laws of semantics.   (20)

M) The problem of mathematical foundations do not belong to mathematics but to psycho-logics which would not disregard anthropology, and
would not be vitiated by our persistence in the use of structurally inappropriate
el 'psychologies', 'logics', and an innocence of mathematics.   (21)

N) The 'intuitional' and the 'intuitional' formalist schools of mathematics
must be considered as a legitimate, yet not properly formulated, protest
against the older elementalism.   (22)

O) The general semantic school will represent the *non-el* and $\bar{A}$ school
of mathematics.   (23)

P) The present crisis of mathematics ultimately depends on the meanings
and use of a few terms such as 'all', 'there is', 'infinite'. , which solution depends
on a *non-el* theory of meanings, which ultimately can be solved by transforming

what might be called the $(3+1)$-dimensional *el*, A-system, which divides 'space' *and* 'time'. , (an attitude which is carried all through the system), into a four-dimensional *non-el*, $\bar{A}$-system (an attitude which is also carried all through the system).   (24)

Q) From (8)—it follows that statements about statements represent results of new neurological processes, that their content varies, and that we must *discriminate* and *not identify* these different meanings. In other words, only through consciousness of abstracting which represents the most general *s.r* of discrimination, or the elimination of identification, can we assign single values to words which have an essentially many-valued character. Identification confuses these many meanings into one.   (25)

R) We must differentiate between descriptive and inferential words and phrases, and never use inferential terms as descriptive, without realizing that we are doing so.   (26)

S) Certain words or phrases used to speak about languages, such as 'all statements', 'proposition about all propositions'. , lead to self-contradictions. We cannot speak about 'all' propositions without some limitations, if we proceed introducing new propositions. Even St. Paul felt the necessity for limiting the values of 'all'.* We are compelled to introduce some equivalents to the biblical 'illegitimate totalities' or the theory of types of Russell.   (27)

T) Analysis finds that certain of the most important terms we use; such as, 'yes', 'no', 'true', 'false', 'all', 'fact', 'reality', 'existence', 'definition', 'relation', 'structure', 'order', 'number', 'is', 'has', 'there is', 'variable', 'infinite', 'abstraction', 'property', 'meaning', 'value', 'love', 'hate', 'knowing', 'doubt'. , . , may apply to all verbal levels and in each particular case may have a different content or meanings and so *in general no single content or meaning*. I call such terms *multiordinal terms* (*m.o*). The definition of such terms is always given in other *m.o* terms preserving their fundamental multiordinality. In other words, a *m.o* term represents a many-valued term. If the many values are identified, or disregarded, or confused, we treat a fundamentally many-valued term as one-valued, and we must have every kind of paradox through such an identification. All known paradoxes in mathematics and life can be manufactured by the disregard of this fundamental multiordinality. Vice versa, by formulating the general semantic problem of multiordinality we gain means to discriminate between the many meanings and so assign a single meaning in a given context. A *m.o* term represents a variable in general, and becomes constant or one-valued in a given context, its value being given by that context. Here we find the main importance of the semantic fact established by Skarżeński,** that the 'logical' freedom from contradiction becomes a semantic

---

*Professor Cassius J. Keyser drew my attention to a passage in the first letter of St. Paul to the Corinthians, Chapter 15, line 27. 'For he hath put *all* things under his feet. But when he saith *all* things are put under him, it is manifest that he is *excepted*, which did put *all* things under him.' Italics are mine.

**Quoted by Chwistek in his *Neue Grundlagen der Logic und Mathematik*.

problem of one-value. But for application we must have a four-dimensional, non-el, $\bar{A}$, extensional system, based on structure . , and the complete elimination of identity. (28)

U) That the disregard of multiordinality, orders of abstractions, may lead to identification and therefore false evaluation resulting in disagreement and maladjustment. (29)

V) From (25–29)—it follows that identification or confusion of higher order abstractions must be eliminated. Because of (7, 9, 10, 25–29)—*all identification must be eliminated.* (30)

W) The elimination of identification on all levels, or a complete and unconscious discrimination between different orders of abstractions, including as a special important case the multiordinality of terms, results in general consciousness of abstracting which in turn, solves the paradoxes of life and mathematics and leads to *generalized* mathematics along the lines suggested by Chwistek. (31)

X) The realization of the inherent multiordinality of some of the most important terms we have, gives us an enormous flexibility of language. It makes the number of our words indefinitely great. When both the writer and the reader recognize this multiordinality, and look for the meaning in the context and discriminate between the orders of abstractions, indicated by the context, confusion becomes impossible. (32)

Y) The test for multiordinality is simple. We take any statement and test it to see whether a given term applies to it. Then we make a statement about this statement and again test if this term applies to the new higher order statement. If it does, the given term must be considered multiordinal, because this procedure may be repeated indefinitely. (33)

Z) *The complete elimination of identification* does not allow us to use the term 'is' of identity, and so we must use operational, functional, actional, behaviouristic . , languages, requiring new attitudes and new *s.r*, impossible without the formulation of a $\bar{A}$-system. (34)

Z_1) The *s.r* of those who produced the general theory of relativity, the unified field theory, the new quantum mechanics, the new revision of the foundation of mathematics . , depend on new $\bar{A}$, *non-el*, and non-identity, operational, actional . , attitudes. (35)

Z_2) As the $\bar{A}$-system is based on the general elimination of the 'is' of identity, or on 'is not', it is impossible to reject these premises without producing impossible data, and a theory of agreement 'in a finite period' then becomes a possibility. (36)

Z_3) The old 'unknowable' becomes abolished and limited to the simple and natural fact that the objective levels *are not* words. (37)

Observation and experience, scientific and otherwise, show that in nature we find a definite order, which establishes a *natural order;* namely, that the sub-microscopic process, called the event or the scientific object, came first; only later abstracting organisms happened and objects which represent the results of abstracting by amoebas or men, came next. In the process of evolution we find object first, label next. Descriptions first, inferences next. The above *natural order* establishes also a *natural order of evaluation*. Proper evaluation becomes the foundation for survival, *non-el s.r;* the more so since evaluation requires asymmetrical relations of 'more' or 'less'. , impossible to handle properly in an A-system. Thus the most important level is represented by the sub-microscopic processes. What the organism needs is not the three-dimensional shadow of a four-dimensional event, not the abstraction of low order produced by our nervous systems, called the object, but the sub-microscopic dynamic processes without which the desired end-results would not happen. The animal, the primitive, the infant, the ignorant man identify the two; live in a delusional world. Similarly the objective levels are more important than the verbal levels, and descriptions are more important than inferences. If we *identify* any orders while the natural order is established by the asymmetrical relation of 'more', the semantic process of evaluation is *reversed* and appears pathological in different degrees. If $a>b$ and we make them delusionally equal in value (identify), then, in the false-to-fact relation $a=b$ we have either over-evaluated the right-hand side or under-evaluated the left-hand side; in both cases reversing the natural order of evaluation. It is important to notice that by basing our *s.r* on a *natural order* of evaluation, general semantics become a generalized science of order and values, a very secure guide in life, indispensable for sanity, as experiments have shown, and include also generalized mathematics.

Another very serious mechanism of identification is found in language.

A) Thus we have only *one name*, say 'apple' for the: (*a*) un-speakable, un-eatable event or scientific process; (*b*) the un-speakable but eatable abstraction of low order, the object; (*c*) the un-speakable and un-eatable 'mental' picture, or higher order abstraction, on semantic levels; (*d*) and for a definition on verbal levels.   (38)

B) The multiordinality of terms was not discovered until 1925 and is still generally unknown. It presents a serious difficulty facilitating, perhaps even necessitating, identification unless prevented by special formulations and training. Multiordinal terms sound and look alike on all levels; experience has shown how easy it is to confuse their orders and identify the many values into one.   (39)

C) The differentiation between descriptions and inferences, and particularly between descriptive and inferential words as such, is also novel, and was, until the present $\bar{A}$-system was formulated, largely disregarded, which again led to identifications and false evaluations.   (40)

Investigations show, that in all known primitive peoples and in the 'mentally' ill, we find literal identification of different orders of abstractions, which accounts for these semantic states. Even their 'perceptions' are different from those of the so-called 'normal', 'civilized' man, because higher order abstractions are projected and identified with lower order abstractions. They identify or ascribe one value to essentially many-valued different orders of abstractions and so become impervious to contradictions with 'reality' and impervious also to higher order experience.

The infant, and the rest of us, identify a great deal because of the reasons given above. Investigations show that most of human difficulties, public, private, or *scientific* are due to this A *s.r*, which accounts for the infantile state of our commercial so-called civilization. Identification abolishes the natural order of evaluation, but so does also an unconscious assumption of an 'infinite velocity' of a process. The A trilogy involved some fanciful 'infinity' assumptions. Thus in the A-system the velocity of nerve currents, which is known to be 126 metres per second in the human nervous system, is at present assumed unconsciously as 'infinite', made evident by the elementalism of 'intellect' or 'emotions'., as something 'by themselves' and detached. In the E-system the length of a line, the space constant, and the natural unit of length were assumed 'infinite'. In the N-system the velocity of light, known to be finite, was unconsciously assumed to be 'infinite'. In the $\bar{A}$ trilogy these unjustified or meaningless 'infinities' have been eliminated. 'Infinite velocity' of a process has *no meaning*. It represents only a play upon symbols. Velocity is defined as $v = s/t$. If we assume $t = 0$ and write $v = s/0 = \infty$, this 'velocity' lacks one of the fundamental factors of its definition; namely, t, and so such an expression ceases to define anything at all and has no meaning, although it may be a symbol for a semantic disturbance. But the results of such delusional *s.r* are far reaching, no matter how mild they might be in degree. In a process propagated with 'infinite' velocity there would be no transition or delay in action, and therefore such a process would *not be ordered*. Vice versa, the disregard of order in our observations must introduce some mythological 'infinities' somewhere. So we see that the semantic process of identification is intimately connected with 'infinity' assumptions, both *abolishing order*. Training in natural order trains *s.r* away from delusional evaluation, abolishes pathological identification of different degrees and fanciful 'infinities'.

Thus we see that the problems of mathematical 'infinity' are extremely complex and involve many fundamental considerations never analysed before in connection with the semantic process of identification. Once these problems are analysed and formulated from a $\bar{A}$, *non-el*, structural point of view the problems of 'excluded third' become secondary in importance, easily managed under the creative freedom of the coveted 'consciousness of abstracting'.

Let me recall for continuity, that the mathematicians recognize at present, two kinds of 'infinities'. One with which we are familiar from our school days, symbolized by ∞, Cantor calls 'potential' infinity and defines as a *variable finite*, the misunderstanding of which introduces paradoxes even in high schools; the other, the 'actual' infinity, which introduces paradoxes in universities. All these paradoxes are due, as the present enquiry shows, to fundamental fallacies in connection with semantic processes of identification which we learn at home and in elementary schools.

The process of identification of different orders of abstractions may be due to pathological conditions, to ignorance, to 'thoughtlessness', to lack of observation, to unconscious false assumptions, to hastiness, to superficiality, to habits of speech, to the structure of language used , . In fact, under the A-system it is practically impossible to avoid it, as we can witness it in such a comparatively advanced field as mathematics. The label 'identification' is applied to the semantic process of wrong evaluation going on inside of our skins on the un-speakable objective levels, when we are not aware of the differences between different orders of abstractions. When making it conscious, we may speak of the confusion of the orders of abstractions. To make such a process conscious, we must train in the differentiation or discrimination between different orders of abstractions, and distinguish the different orders by actually learning how to order them. Such training results in general *consciousness of abstracting* which is not inborn, nor fully acquired, even in university training, but which requires *special* training. Experiments in this field are extremely encouraging; in a number of cases, pathological individuals have become 'normal' and the 'unchangeable' human nature has been actually changed. *Infantile reactions in adults are abolished*, and this training becomes a general and simple method for prevention of future semantic disturbances of false evaluation which must result in maladjustment.

To stop identification we must discriminate or differentiate to the limit between what appears always as four-dimensional, absolute individual stages of processes and situations on all levels, verbal included. Let us follow briefly such an actual performance. If we realize (7)—we accept (11–15)—and on *structural* grounds reject the elementalism of the A trilogy as expressed in its 'psychology', 'logic', the division of 'space' *and* 'time', . We accept the non-elementalism of the $\bar{A}$ trilogy as expressed in the new terms in the present work and accept also 'space-time', . The difference is very serious in all fields, when carried consistently all through the system. As we actually deal with four-dimensional dynamic processes which must be considered continually different, and with world conditions changing also continually, statements about such structural conditions, in an extensional sense, must be considered as involving variables, generating *propositional functions*, doctrinal or system-functions, funktors , . But propositional functions, which involve variables, are neither 'true' nor 'false', but ambiguous, and to have a proposition we must assign a value to the variable by *at least* permanently, in principle, assigning a date to it. We must also introduce, in principle, and as a semantic attitude, numerical subscripts to our words. Thus 'apple' in the A-system represents a *name* attached to an *intensional definition*, and space-time considerations do not enter. The term is applied to a definition which might be considered as one-valued and permanent. Now obviously such a language and s.r are *structurally non-similar* to the world and our nervous system.

If we try to identify a name for a definition, implying permanence, with the objective level which is made up of absolute individuals, and represents ever-changing processes, we must live in a delusional world in which we should expect every kind of paradoxes and psycho-logical shocks.

In a $\bar{A}$-system, for structural reasons, we must retain the general implications of the term 'apple', so we retain the word. We must make our language extensional in principle, and the name 'apple' an *individual name*, by calling it 'apple$_1$', 'apple$_2$', . The combination of letters 'a-p-p-l-e' implying similarities, the subscripts 1, 2 ., implying individual differences, which automatically prevent identification. But this is not enough. Our 'apple$_1$' represents a name applied to an object *and* a process; its *meaning* becomes only one-valued when we assign to it at least a definite date. Thus the objective 'apple$_{1 \text{ (Dec. 1, 1931.)}}$' may be a very appetizing affair, and 'apple$_{1 \text{ (Jan. 1, 1932.)}}$' an un-edible wet splash. It should be noticed that the fundamental difference between the A and $\bar{A}$ systems turns out to be a difference of semantic attitudes. The scientific facts are not changed. The 'apple' of 'Adam' or our own did not differ in essential characteristics under discussion. In both the A and $\bar{A}$ systems we actually deal, in principle, with *many-valued processes*. The important problem is to adjust the structure of our verbal processes to the structure of the world; hence a $\bar{A}$-system must be made extensional, *non-el*, four-dimensional , . Here once more, as in general semantics, the ascribing of one value (or at least limited to a small range of values in practice), in a given situation (context), eliminates paradoxes and contradictions on the older 'logical' grounds. We should notice that the multiordinal terms must be considered as names for many-valued *s.r*, depending upon the order of abstractions; hence the name *multiordinal*. Names for happenings on the objective levels apply to many-valued processes but should not be considered multiordinal. All the psycho-logics of the differential calculus, 'space-time', enter here, yet the whole field is covered semantically if we entirely abandon the 'is' of identity. Instead of training in 'allness' and 'isness'—'this *is* this', we shall train in non-allness, and non-isness—'this *is not* this', in connection with a special diagram called the Structural Differential.

Experience and experiments show that the above seems essential for sanity. It is interesting to notice that mathematics has produced a language similar in structure to the human nervous system. Roughly the central part of the brain which we call the thalamus is directly connected with the dynamic world through our 'senses' and with those semantic manifestations which we usually call 'affective', 'emotions'. , all of which manifest themselves as dynamic. The cortex which gives us the static verbal reactions and definitions, is not connected with the outside world directly but receives all impulses through the thalamus. On semantic levels the thalamus can only deal with dynamic material, the cortex with static. Obviously for the optimum working of the human nervous system, which represents a cyclic chain, where the lower centres supply the material for the higher centres and the higher centres should influence the lower, we must have means to translate the static into dynamic and the dynamic into static; a method supplied *exclusively* by mathematics.

With the above considerations we must discriminate between our semantic capacities for *infinite divisibility* of finites, and for the *generation* of infinite postulated processes which by definition *cannot be exhausted*. If we use a three-dimensional A language and apply such an 'all' to such an infinite process then we simply produce a self-contradiction. If we apply to such a semantic process a four-dimensional 'all with a date', then we have arrested, for the 'time' being, the process, or taken a static cross section of the infinite process at that date; but then we deal with a finite. Once we are constantly conscious of abstracting in different orders, these subtle differences become quite clear and the solutions of the problems of infinity follow a similar path as the older problems of the 'infinitesimal', which also was self-contradictory, unnecessary for mathematics. When treated as a *variable finite* it was satisfactory and sufficient, and has proven to be a most creative notion in 'mathematics. In the problems of the irrational, continuity., similar subtle identifications or non-discriminations of *el*, A, three-dimensional terms with $\bar{A}$, *non-el*, four-dimensional terms occur, which once eliminated, clear up not only the paradoxes, but some self-contradictory, often unconscious, postulates of some parts of mathematics.

Lack of space does not allow me to go into further details, except to suggest how some subtle *discriminations* may help to eliminate identification. In my $\bar{A}$-system the differentiation between orders of abstractions on physiological grounds, the introduction of multiordinality of terms, four-dimensional considerations., as a structural necessity for all languages, makes the theory of types unnecessary.

For a better understanding of the present work we must at least differentiate:

A) Between numerical experience and mathematics.

B) Between languages with content and languages without content.

C) Between creative building of verbal schemes which, for the sake of generality, have no content, called pure mathematics, and the application of these schemes to actual problems, with content, called applied mathematics.

D) Between the contentless mathematics and the investigation of the foundation of mathematics which represents the investigation of the *s.r* of mathematicians and belongs to a future *non-el* psycho-logics with content.

E) Between different forms of complex adjustment which we have in common with the primitive man, and even the higher animals, and reasoning which starts with conscious observations, passing to descriptions and inferences,.

F) Between the dynamic process of relating ('thinking') on the unspeakable semantic levels, and the verbal expression of 'relations'.

G) Between the use of negative terms, disagreement, contradiction; and self-contradiction. In a $\bar{A}$-system contradictions take the form of self-contradictions.

H) Between *el* 'logics' expressed in terms of 'true', 'false', and modality, and the investigations of one-, two-, three-, and ∞-valued *s.r*, which become a *general theory of values*, and which may some day include all human interests.

I) Between the inherent circularity of 'human knowledge', which must start with sets of undefined terms, and so start with some knowledge, and circular definitions or explanations which define or explain nothing.

This list of suggestions is not exhaustive, and in principle appears as inexhaustible. I selected only a few topics of immediate need.

We should also notice that because on objective levels we deal structurally with absolutely individual stages of processes and situations and by necessity we speak in higher order abstractions and generalities and use many multi-ordinal terms (without the use of which no speaking is possible), so *any positive statement* about the objective levels must be only probable in different degrees, which introduces a fundamental and entirely *general $\bar{A}$ principle of uncertainty*. Heisenberg's restricted principle in physics appears only as a special case. For structural reasons we must preserve determinism but because of (11–15)— the older two-valued determinism must be reformulated into the ∞-valued determinism of the maximum probability. The einsteinian introduction of non-elementalism in physics has resulted in the automatic elimination of some semantic blockages in the younger physicists. Some of the semantic results and triumphs of science, besides the new quantum mechanics, can be found in the latest (free from identification of the term 'time' with some objectivity) new entropy of Tolman.[3]

To sum up, we find that although the primitive man or the 'mentally' ill may have some reactions of orientation, or capacity for relating, which we have in common with the higher animals, yet these do not involve 'reasoning' in the sense defined before. Thus a boxer, football player . , does a great deal of reflexive relating and wins his match, but this cannot be considered as reasoning in the strict sense as used in a $\bar{A}$-system. If we attempt *to discuss* something with a primitive or 'mentally' ill individual and write down his processes of relating, we would have to conclude that he uses *one-valued semantics of identification* of many values into one, or a semantics of inclusion by which 'everything *is* everything else'. The 'law of contradiction', or any 'excluded' 'third', or '*n*-th', practically never appears in our sense, yet it is complicated by the use of positive and *negative* terms, to which any meanings connected with some identifications of higher orders may be ascribed. Although his prevailing semantic processes appear as a complete and *literal identification*, yet because of the general orienting and relating capacities of organisms and the character of terms used, it would not be easy or profitable to attempt an *el* formulation of his 'laws of thought'. But a semantic formulation, as given above, is very instructive and comparatively simple.

Our existing *el* 'logic', besides the two-valued type of formulation, involves many different 'philosophical' elucidations, which instead of clarifying the status of 'logic', in general, tends only to conceal the important issues involved in *non-elementalism*. The role that identification plays in a given individual appears always as a deciding factor in his adjustment. Unfortunately, at present, the sinister identification is not counteracted but fostered or even induced by the structure of the languages we use, different mythologies . , and our whole educational, economic, social . , systems.

The two-valued A, *el*, three-dimensional 'logic' does not apply to the world of events, to the objective levels . , and, for the reasons already explained, does not apply to the study of the foundations of mathematics. It applies to a large extent to contentless technical mathematics, including so-called 'formal logic' of that system.

Formalism *when free* from identification becomes a unique comparative tool in search for structure; formalism *with* identification of different orders of abstractions, a symptom of semantic disturbances, often of a morbid character. It should be realized that we may have one-, two-, three-. , many-, and ∞-valued orientations, which with the exception of one-valued, we should utilize when conditions warrant a particular use in a particular case. Thus in mathematics, for the sake of having mathematics as a standard of evaluation, *we select* a sharply two-valued orientation by which in the old language 'A is B or not B', *to allow sharp* statements that for instance, $1+1=2$. If we would deliberately postulate that $1+1$ may sometimes be equal to 2 and sometimes not equal to 2, we would have forms of representation which would apply perhaps more readily to science and life, *but* mathematics as such would be impossible, and we would be deprived of this sharp tool for evaluation.

It is interesting to notice that mathematicians, by the use of two-valued *semantics*, (not 'logic', because an *el* discipline cannot be 'lived through' at all by non-heavily pathological individuals), have produced the most important disciplines. Thus we have, for instance, the theory of 'variance' (the theory of function), the theory of invariance, the differential calculus, the 1, 2, 3, 4, and n-dimensional systems, and a host of other verbal structures similar not only to the world, but to the human nervous system. These results give us means not only to enlarge our mastery of the external world, but when generalized into a *non-el*, $\bar{A}$-system, give us the means for the mastery of the inner world, leading toward sanity.

It is amusing to discover, in the twentieth century, that the quarrels between two lovers, two mathematicians, two nations, two economic systems . , usually assumed insoluble in a 'finite period' should exhibit one mechanism— the semantic mechanism of identification—the discovery of which makes universal agreement possible, in mathematics and in life.

NOTES AND REFERENCES

In the following references the bold face numbers refer to the numbers of the literature in the bibliography; p, or, pp, indicates the page or pages; ff, indicates following pages.

In many instances the number of the page is given, but in others, when I refer to a large subject, only the number of a book or paper is indicated, and in such cases the index of the given book should be consulted.

In other cases, when no references are given, and yet the serious and educated reader may occasionally feel perplexed, may I not suggest, in this connection, that wide experience has taught me that we usually forget the structural, not entirely common, subtleties of grammar. We also often ascribe to words a very limited, personal, and *habitual* range of meanings, and, so, some purely linguistic difficulties appear as mysterious 'scientific' difficulties, which they are not. The reader, on such occasions, will be surprised to find what an enormous amount of knowledge may be found in a mature occasional perusal of a good grammar or dictionary, the neglect of which acts as a psycho-logical blockage to the understanding.

CHAPTER II

1—**25**. 2—**212**. 3—**591**, pp. 93, 94.

CHAPTER III

1—The need of International Languages, or a *Universal Language* besides mathematics is becoming increasingly urgent. At present there are several such languages, and in many large cities there are organizations, usually called International Auxiliary Language Associations, with addresses listed in telephone directories. Any of these organizations will gladly supply information about the whole of the international linguistic movement. There is also a number of books written on this subject to be found in the larger public or university libraries. Informations about the *Basic English* of Ogden as a *Universal Language,* consisting of the astonishingly small number of 850 words, which do the work of about 20,000 words, may be obtained from the Orthological Institute, 10 King's Parade, Cambridge, England (see also **376, 377**). In my opinion, the possibilities of the Basic for a scientific civilization are unlimited, *provided* the Basic is *revised* from a non-aristotelian, non-identity, point of view.

The general and serious defect of all of these languages is, that their authors have, as yet, entirely disregarded the non-aristotelian problems of non-identity, and so of *structure*, without which *general sanity,* or the elimination of *delusional worlds* is *entirely impossible.*

CHAPTER IV

1—**590**. 2—**579**, Vol. II, Part IV, *150 ff.; **455-457**. 3—**590**. 4—**457**, p. 249.

CHAPTER VII

1—**317, 318, 319**. 2—**83**.

CHAPTER VIII

1—**92**, pp. 50-52. 2—**91, 92**. 3—**92**, pp. 114, 119, 123, 242. 4—**564, 560**, pp. 16 ff., 28 ff.

CHAPTER IX

1—**487**. 2—**304, 7**, Vol. II, p. 461 ff. 3—**214**. 4—**7**, Vol. II, p. 457. 5—**214**, p. 210. 6—**7**, Vol. II, p. 458. 7—**7**, Vol. II, p. 251. 8—**7**, Vol. II, p. 634. 9—**7**, Vol. II, p. 944. 10—**310**, Chap. V. 11—**364, 365-369, 540**. 12—**9, 10, 12, 13, 196**, 210, 211, 370. 13—**257, 272, 328, 345**. 14—**7**, Vol. II, p. 917 ff., 273, 313, 416. 15—**7**, Vol. II, p. 690. 16—**7**, Vol. II, p. 961 ff. 17—**7**, Vol. II, p. 644 ff. 18—**7**, Vol. II, p. 803 ff. 19—**7**, Vol. II, p. 869 ff. 20—**363**. 21—**499, 500-504, 127, 128**. 22—**201, 533**. 23—**49**. 24—**7**, Vol. II, p. 59 ff.

CHAPTER X

1—306, p. 72 ff. 2—306, p. 72 ff. 3—306, p. 72 ff., p. 193. 4—310, p. 95 ff. 5—309, p. 197. 6—309, p. 202. 7—309, pp. 204, 211. 8—411, p. 233. 9—221, p. 139. 10—221, pp. 144 145. 11—343, 344. 12—559, p. 15. 13—242.

CHAPTER XI

1—452, 453-457. 2—82, 264, 269. 3—470. 4—176-184.

CHAPTER XII

1—220, p. 60 (5th edition). 2—222, pp. 3-5. 3—411, p. 8. 4—300, and in a personal letter. 5—220, p. 334. 6—220, p. 68. 7—411, p. 210. 8—566.

CHAPTER XIII

1—452, 453.

CHAPTER XIV

1—264. 2—203, p. 39.

CHAPTER XV

1—452, 453-457. 2—457, p. 22 ff.

CHAPTER XVI

1—449.

CHAPTER XVII

1—547, p. 102. 2—208, p. 811 (17th edition). 3—208, p. 815 (17th edition).

CHAPTER XVIII

1—455, p. 18. 2—126, pp. 14, 16. 3—74, p. 467, 468.

CHAPTER XIX

1—264. 2—430. 3—468, p. 133. 4—468, p. 141. 5—264, p. 153 ff. 6—264, p. 153 ff. 7—222, p. 166 ff. 8—289. 9—350. 10—318, 319.

CHAPTER XX

1—221, p. 277. 2—7, Vol. II, pp. 59-92, 121, 436. 3—216.

CHAPTER XXI

1—394, 395. 2—Rudyard Kipling's *Verse*. Inclusive edition, 1885-1926. New York. 3—501. 4—221, p. 197. 5—552, 553-557. 6—250, p. 194.

CHAPTER XXII

1—92, p. 112. 2—221. 3—221, pp. 86, 87, 91, Chap. XIII. 4—306, p. 48 ff. 5—306, pp. 37-40. 6—306, pp. 85, 86. 7—306, p. 61-63. 8—304, p. 45. 9—487, p. 393. 10—222, pp. 213, 214. 11—222, pp. 262, 263. 12—394, pp. 47, 388. 13—278, pp. 90-92. 14—394, p. 99.

CHAPTER XXIII

1—394, pp. 88-94. 2—394, pp. 310, 311. 3—394, pp. 310, 311. 4—394, pp. 403, 404.

CHAPTER XXIV

1—34.

CHAPTER XXV

1—All anthropology gives ample evidence. In the present very brief bibliography, consult 172, 173, 200, 298, 299, 331-336, 492.

CHAPTER XXVII

1—579 (1st edition), Vol. I, p. 40. 2—590, No. 3, 332. 3—97. 4—452, 453-457, 579, and all modern works on 'logic' and the foundations of mathematics. 5—579, pp. 63, 65 ff., Vol. I (1st edition).

CHAPTER XXVIII

1—**247**. 2—the whole of primitive and modern mythologies, consult **172, 173, 200, 298, 299, 331-336, 478, 479, 492**. The literature is very large, and cannot be given here. Consult also standard treatises on comparative religion, history of religion, and works of psychiatrists who deal with these aspects of 'mental' ills.

CHAPTER XXX

1—**241**, p. 37. 2—**241**, Fig. 2, p. 94, Fig. 3, p. 140. 3—**241**, pp. 140-142. 4—**42**, pp. 1, 19, 23, 118-120. 5—**42**, p. 23. 6—**99**, p. 897. 7—**34**, p. 230. 8—**34**, p. 180. 9—**34**, p. 233. 10—**34**, p. 242. 11—**411**, p. 237. 12—**411**, p. 10. 13—**15, 16, 286**. 14—**298**, Chap. IX. 15—**181**, Vol. IV, p. 136, **492**. 16—**70**. 17—**487**, pp. 1363, 1334. 18—**487**, pp. 1327-29, 1363. 19—**34**, p. 117. 20—**34**, p. 91. 21—**70**.

CHAPTER XXXI

1—Science. Jan. 22, 1932. 2—**56**. 3—**105, 290, 329, 413, 467, 521, 532**, the literature on this subject is very large, and I give here only examples, see also **110**.

CHAPTER XXXII

1—**72**. 2—**381, 510, 585**. 3—**220**, p. 368 (5th edition).

CHAPTER XXXIII

1—**547**, p. 104. 2—**55**, pp. 89, 174, 221. 3—**411**, p. 191.

CHAPTER XXXIV

1—**147**, p. 79. 2—Most of the above presentation follows **45**. 3—**204**, Vol. II, p. 342. 4—**45**, pp. 260, 261. 5—see, however, **472** and **508**. 6—see C. Runge, *Vector Analysis* (London, New York), p. 178. 7—**148**, p. 49. 8—**147**, pp. 86-88. 9—**45**, pp. 18, 22, 24, 25, 36. 10—**411**, pp. 24, 50, 91.

CHAPTER XXXV

1—**45**, p. 132. 2—**547**, p. 92. 3—**204**, Vol. I, pp. 247, 248. 4—**45**, pp. 158, 159. 5—**547**, p. 66. 6—**547**, p. 91. 7—**457**, pp. 21, 22.

CHAPTER XXXVIII

1—**264**, p. 327 ff., see, however, **348**. 2—**314**, p. 75 ff. 3—**147**, p. 58 ff. 4—**45**, pp. 196, 197, 213.

CHAPTER XXXIX

1—**148**, p. 23. 2—**45**, p. 232. 3—**148**, p. 136. 4—**147**, pp. 177, 178. 5—**547**, pp. 284, 285. 6—**147**, p. 178. 7—**147**, p. 149.

CHAPTER XL

1—**480**, pp. 65, 66. 2—**31**, pp. 17-19. 3—**480**, pp. 467, 468. 4—**32**, p. 38.

CHAPTER XLI

1—**142**, p. 5. 2—**204**, Vol. I, p. 79 ff. 3—**480**, pp. 194-202. 4—**32**, p. 64; **46**, p. 74. 5—**123, 39, 522**. 6—**123**, Vol. III, p. VI. 7—**146**. 8—**457**, pp. 42-44. 9—**32**, later Dirac altered his procedure, see **142**. 10—see last section of appendix 7 in the German, 4th edition of **480**, and also **481**. 11—**32**, p. 185, **63, 64, 65**. 12—**27**, p. 16 ff. 13—**32**, pp. 148, 149. 14—**32**, p. 137; **27**, pp. 25, 26. 15—**27**, pp. 64-67. 16—**27**, p. 38.

SUPPLEMENT III

1—for literature, see under respective names. 2—**249**.

BIBLIOGRAPHY

A complete bibliography for a *non-aristotelian system* would require many volumes and is, therefore, impossible here. The formulation of a *non-aristotelian system*, with the number of scientific facts known in 1933, turned out to be an extremely laborious process.

A non-aristotelian language and attitude differ considerably from the older languages and attitudes, and so a first non-aristotelian system has no literature which would deal directly with the subject. The statement 'that everything has already been said' is, unfortunately, largely true. This introduces serious complexities, because extremely few men have the genius of a Poincaré, and fully realize that the language used in making a statement plays an overwhelming role as to the consequences which eventually follow. If we even grant that 'every thing was said', I must add, 'but not so', and this *prevented the building* of a non-aristotelian system for more than two thousand years. One of the human tragedies can be found in the fact that *wise epigrams* do not work. It takes *a system* which often expresses similar notions; but they must be expressed in a *unified language* of different structure to make them workable.

In giving this extremely abbreviated, insufficient, and, perhaps, even poorly selected bibliography, I had, in the main, three aims: (1) to acknowledge some of my direct obligations; (2) to give to the future student an outline of the type of literature in existence which has bearing on my subject; and (3) to list such books and articles which give further literature.

I was particularly careful to list as few scientific periodicals as possible, because specialists in a given field do not need them; and laymen do not want them. In a number of instances, I have listed only one or two of the latest papers of an author, which give his previous titles.

Because of the lack of linguistic co-ordination, in most cases, I have had few or no opportunities to refer directly to many authors, although the titles of the books usually suggest the material needed. For further data on a given subject, the reader is referred to the respective indexes. As a rule I had to express many similar notions, but from a different angle, and in a different language.

I have prefaced the books, parts, and chapters with many important quotations, only to show that all modern science requires a fundamental non-aristotelian revision. The attentive reader will discover that, although I am in general sympathy with these quotations, yet, in many instances, I would have to express them differently.

In the case of the two volumes of *Colloidal Chemistry* edited by Alexander, which is a collection of important contributions by different authors, I would have to list about one hundred and twenty more titles, and so I mostly refer to the page without giving the name of the author or the title of his contribution; for which I apologize. In a number of instances I have utilized the material given by *Science Service,* as printed in *Science* and I indicate such references by inserting *SS* before giving the date of the issue of *Science.*

1. ACKERMANN, W. Begründung des "tertium non datur" mittels der Hilbertschen Theorie der Widerspruchsfreiheit. *Math. Ann.* B. 93, H. ½.
2. See Hilbert.
3. ADAMS, J. T. *Our Business Civilization: Some Aspects of American Culture.* New York.
4. ADLER, A. *The Practice and Theory of Individual Psychology.* London, New York.
5. *Organ Inferiority and its Psychical Compensation.* Washington.
6. ALEXANDER, J., and BRIDGES, C. B. Some Physico-Chemical Aspects of Life, Mutation, and Evolution. In *Colloid Chemistry,* edited by J. Alexander. New York.
7. ALEXANDER, J., editor. *Colloid Chemistry.* Vol. I, *Theory and Methods.* Vol. II, *Biology and Medicine.* New York.
8. ANREP, G. V. See Pavlov.

9. BABCOCK, E. B., and COLLINS, J. L. Mutations Caused by Radiations from the Earth. *SS. Science.* Aug. 2, 1929.

10. Does Natural Ionizing Radiation Control Rate of Mutation? *Proc. Nat. Acad. Sci.,* Vol. 15. 1929.

11. BACON, R. *Letter.* Translated from the Latin by T. L. Davis. Easton, Pennsylvania, London.

12. BAGG, H. J. Biological Effects of Irradiation. *SS. Science.* May 31, 1929

13. BAGG, H. J., and HALTER, C. R. Bodily Defects of X-rayed Mice. *SS. Science.* Jan. 6, 1928.

14. BALDWIN, J. M. *Thought and Things. A Study of the Development and Meaning of Thought or Genetic Logic.* 3 vol. London, New York.

15. BANCROFT, W. D., and RICHTER, G. H. The Colloid Chemistry of Insanity. *Science.* May 8, 1931.

16. BANCROFT, W. D., and RUTZLER, J. E. Reversible Coagulation in Living Tissue. VIII. *Proc. Nat. Acad. Sci.,* Nov., 1931.

17. BARON, M. A. See Gurwitch.

18. BAYLISS, W. M. *Principles of General Physiology.* New York.

19. *The Colloidal State in its Medical and Physiological Aspects.* London.

20. BECKER, O. *Mathematische Existenz.* Halle.

21. BELL, E. T. The Principle of General Relativity. See Bird, J. M., editor.

22. *Debunking Science.* Seattle.

23. *The Queen of the Sciences.* Baltimore.

24. BENT, S. *Ballyhoo the Voice of the Press.* New York.

25. BENTLEY, A. F. *Linguistic Analysis of Mathematics.* Bloomington, Indiana.

26. BERNSTEIN, B. A. Review of *Principia Mathematica,* Vol. I, 2nd edition. *Bull. Amer. Math. Soc.* Nov., 1926.

27. BIGGS, H. F. *Wave Mechanics.* London, New York.

28. BIRD, J. M., editor. *Einstein's Theories of Relativity and Gravitation.* New York.

29. BIRKHOFF, G. D. *Relativity and Modern Physics.* Cambridge, Massachusetts, London.

30. *The Origin, Nature, and Influence of Relativity.* London, New York.

31. BIRTWISTLE, G. *The Quantum Theory of the Atom.* London, New York.

32. *The New Quantum Mechanics.* London, New York.

33. BLEULER, E. *The Theory of Schizophrenic Negativism.* Washington.

34. *Textbook of Psychiatry.* New York, London.

35. BLISS, G. A. *Calculus of Variations.* Chicago, London.

36. The Calculus of Variations and the Quantum Theory. *Bull. Amer. Math. Soc.* Apr., 1932.

37. BOAS, F. *The Mind of Primitive Man.* London, New York.

38. *Anthropology and Modern Life.* New York.

39. BôCHER, M. *Introduction to Higher Algebra.* New York, London.

40. BOLTON, J. S. *The Brain in Health and Disease.* London.

41. BOLTON, L. *An Introduction to the Theory of Relativity.* London, New York.

42. BONOLA, R. *Non-Euclidean Geometry.* Chicago, London.

43. BOOLE, G. *The Mathematical Analysis of Logic.* Cambridge.

44. *Investigation of the Laws of Thought; on Which Are Founded the Mathematical Theories of Logic and Probabilities.* London, Chicago.

45. BORN, M. *Einstein's Theory of Relativity.* London, New York.

46. *Problems of Atomic Dynamics.* Cambridge, Massachusetts.

47. BOTCHARSKY, S., and FOERINGER, A. Radiation from Vitamins. *SS. Science.* July 17, 1931.

48. BOUSQUET, G. H. *Précis de sociologie d'après Vilfredo Pareto.* Paris.

49. BOVIE, W. T. The Biological Effects of Light. (*Lectures on the Biologic Aspects of Colloid and Physiologic Chemistry,* by W. T. Bovie and others. Philadelphia, London.)

50. BRAGG, SIR WILLIAM. *Concerning the Nature of Things.* London, New York.

51. BREITWIESER, J. V. *Psychological Education.* New York.

52. BRIDGES, C. B., and MORGAN, T. H. The Third Chromosome Group of Mutant Characters of *Drosophila melanogaster, Carnegie Inst. Pub.,* No. 327. 1923.
53. BRIDGES, C. B. See Alexander.
54.   See Morgan, T. H.
55. BRIDGMAN, P. W. *The Logic of Modern Physics.* New York, London.
56.   *The Physics of High Pressure.* London.
57.   Statistical Mechanics and the Second Law of Thermodynamics. Gibbs Lecture. *Bull. Amer. Math. Soc.* Apr., 1932.
58. BRIGHAM, C. C. *A Study of American Intelligence.* Princeton.
59. BRILL, A. A. *Psychanalysis: Its Theories and Practical Applications.* Philadelphia, London.
60. BRILLOUIN, L. See de Broglie.
61. BROAD, C. D. *Scientific Thought.* London, New York.
62.   *The Mind and its Place in Nature.* London, New York.
63. BROGLIE, L. DE. *Ondes et mouvements.* Paris.
64.   *An Introduction to the Study of Wave Mechanics.* London, New York.
65. BROGLIE, L. DE, and BRILLOUIN, L. *Selected Papers on Wave Mechanics.* London, Glasgow.
66. BROUWER, L. E. J. Intuitionism and Formalism. *Bull. Amer. Math. Soc.* Nov., 1913.
67.   Über die Bedeutung des *principium tertii exclusi* in der Mathematik, besonders in der Funktionentheorie. *Jour. f. Math.,* Vol. 154. 1925.
68. BRUNOT, F. *La Pensée et la langue.* Paris.
69. BRUNSCHVICG, L. *Les Étapes de la philosophie mathématique.* Paris.
70. BURROW, T. *The Social Basis of Consciousness.* London, New York.
71.   *The Structure of Insanity..* London.
72. CAJORI, F. *A History of the Conceptions of Limits and Fluxions in Great Britain from Newton to Woodhouse.* Chicago, London.
73.   *William Oughtred.* Chicago, London.
74.   *A History of Mathematics.* New York, London.
75.   *A History of Mathematical Notations,* 2 vol. Chicago, London.
76.   *A History of Physics.* New York, London.
77. CAMPBELL, N. R. *Physics; the Elements.* Cambridge.
78. CANNON, W. B. *Bodily Changes in Pain, Hunger, Fear and Rage.* New York, London.
79.   *The Wisdom of the Body.* New York.
80. CANTOR, G. *Contributions to the Founding of the Theory of Transfinite Numbers.* Chicago, London.
81. CARMICHAEL, R. D. *The Theory of Relativity.* New York, London.
82.   *The Logic of Discovery.* Chicago, London.
83. CARMICHAEL, R. D., and others. *A Debate on the Theory of Relativity.* Chicago, London.
84. CARNAP, R. *Abriss der Logistik.* Wien.
85. CARREL, A. Physiological Time. *Science.* Dec. 18, 1931.
86. CASSIRER, E. *Das Erkenntnisproblem in der Philosophie und Wissenschaft der neueren Zeit,* 3 vol. Berlin.
87.   *Substance and Function, and Einstein's Theory of Relativity.* Chicago, London.
88.   *Philosophie der Symbolischen Formen,* 4 vol. Berlin.
89. CHILD, C. M. *Senescence and Rejuvenescence.* Chicago.
90.   *Individuality in Organisms.* Chicago.
91.   *The Origin and Development of the Nervous System.* Chicago.
92.   *Physiological Foundations of Behavior.* New York.
93. CHURCH, ALONZO. On Irredundant Sets of Postulates. *Trans. Amer. Math. Soc.* July, 1925.
94.   On Irredundant Sets of Postulates. *Bull. Amer. Math. Soc.* Nov., 1926.
95.   Alternatives to Zermelo's Assumption. *Trans. Amer. Math. Soc.* Jan., 1927.
96.   On the Law of the Excluded Middle. *Bull. Amer. Math. Soc.* Jan., 1928.

97.   Review of *Principia Mathematica,* Vol. II and III, 2nd edition. *Bull. Amer. Math. Soc.* Mar., 1928.
98.   A Set of Postulates for the Foundation of Logic. *Ann. of Math.* Apr., 1932.
99.   CHURCH, ARCHIBALD, and PETERSON, F. *Nervous and Mental Diseases.* Philadelphia, London.
100. CHWISTEK, L. *Wielość Rzeczywistości.* Kraków.
101. The Theory of Constructive Types (Principles of Logic and Mathematics), 2 parts. *Ann. soc. polonaise de mathématique.* Kraków. 1924, 1925.
102. Über die Hypothesen der Mengenlehre. *Math. Zeit.* B. 25, H. 3. Berlin. 1926.
103. Une Méthode métamathématique d'analyse. *C. R. du I^ier congrès des math. des pays Slaves.* Warsaw. 1929.
104. Neue Grundlagen der Logik und Mathematik, Part I. *Math. Zeit.* B. 30, H. 5. Berlin. 1929. Part II, B. 34, H. 4. Berlin. 1932.
105. CLARK, P. *Napoleon Self-Destroyed.* New York.
106. COGHILL, G. E. *Anatomy and the Problem of Behaviour.* Cambridge.
107. The Structural Basis of the Integration of Behavior. *Proc. Nat. Acad. Sci.* Oct., 1930.
108. COHEN, E. *Physico-Chemical Metamorphosis and Some Problems in Piezo-chemistry.* New York, London.
109. COHEN, M. R. *Reason and Nature.* New York.
110. COLLINS, J. *The Doctor Looks at Love and Life.* New York.
111. COLLINS, J. L. See Babcock.
112. CONKLIN, E. G. *Heredity and Environment in the Development of Men.* Princeton.
113. *The Direction of Human Evolution.* New York.
114. COOK, W. W. Scientific Method and the Law. *Johns Hopkins Al. Mag.* Mar., 1927.
115. COOLIDGE, W. D. Cathode Rays for Drying Paint. *SS. Science.* Apr. 15, 1927.
116. COURANT, R., and HILBERT, D. *Methoden der mathematischen Physik.* B. I. Berlin.
117. COUTURAT, L. *La Logique de Leibniz, d'après documents inédits.* Paris.
118. *Les Principes des mathématiques, avec un appendice sur la philosophie des mathématiques de Kant.* Paris.
119. *The Algebra of Logic.* Chicago, London.
120. The Principles of Logic. *(Encyclopaedia of the Philosophical Sciences.* Vol. I. *Logic.* London, New York.)
121. CRILE, G. W. Autosynthetic Cells. *SS. Science.* Jan. 16, 1931.
122. CROOKSHANK, F. G. The Importance of a Theory of Signs and a Critique of Language in the Study of Medicine. (Supplement II in Ogden and Richards' *The Meaning of Meaning.* London, New York.)
123. CULLIS, C. E. *Matrices and Determinoids,* 3 vol. Cambridge.
124. CUNNINGHAM, E. *Relativity the Electron Theory and Gravitation.* London, New York.
125. CURTISS, D. R. *Analytic Functions of a Complex Variable.* Chicago, London.
126. DANTZIG, T. *Number, the Language of Science.* New York, London.
127. DARROW, C. W. Sensory, Secretory and Electrical Changes in the Skin Following Bodily Excitation. *Jour. Exp. Psych.* June, 1927.
128. The Galvanic Skin-Reflex and Finger Volume Changes. *Amer. Jour. Phys.* Mar., 1929.
129. DARROW, K. K. *Introduction to Contemporary Physics.* New York.
130. DARWIN, C. G. *The New Conception of Matter.* New York, London.
131. DAVENPORT, C. B. *Heredity in Relation to Eugenics.* New York.
132. DAVIS, H. T. *Philosophy and Modern Science.* Bloomington, Indiana.
133. DAVIS, T. L. See Bacon.
134. DEDEKIND, R. *Essays on the Theory of Numbers.* Chicago, London.
135. DÉJERINE, J. *Le Langage.* Paris.
136. DELACROIX, H. *Le Langage et la pensée.* Paris.
137. DERCUM, F. X. *An Essay on the Physiology of Mind.* Philadelphia.

138. DEWEY, J. *Human Nature and Conduct; an Introduction to Social Psychology.* New York.
139.   *Experience and Nature.* Chicago, London.
140.   *The Quest for Certainty; a Study of the Relation of Knowledge and Action.* London, Glasgow.
141. DICKSON, L. E. *Introduction to the Theory of Numbers.* Chicago, London.
142. DIRAC, P. A. M. *The Principles of Quantum Mechanics.* Oxford.
143. DRESDEN, A. Brouwer's Contributions to the Foundations of Mathematics. *Bull. Amer. Math. Soc.* Jan., 1924.
144.   Some Philosophical Aspects of Mathematics. *Bull. Amer. Math. Soc.* July, 1928.
145. DROBISCH, M. W. *Neue Darstellung der Logik.* Hamburg.
146. ECKART, C. Operator Calculus and the Solution of the Equations of Quantum Dynamics. *Phys. Rev.* Oct., 1926.
147. EDDINGTON, A. S. *Space, Time, and Gravitation.* Cambridge.
148.   *The Mathematical Theory of Relativity.* Cambridge.
149.   *The Nature of the Physical World.* Cambridge.
150. EINSTEIN, A. *Relativity.* New York.
151.   *Sidelights on Relativity.* London, New York.
152.   *The Meaning of Relativity.* Princeton.
153.   *Investigations on the Theory of the Brownian Movement.* London, New York.
154.   Zur Einheitlichen Feldtheorie. *Sitz. Preus. Akad. Wiss.* Berlin. 1929.
155.   See Lorentz.
156. EINSTEIN, A., and MAYER, W. Einheitliche Theorie von Gravitation und Elektrizität. *Sitz. Preus. Akad. Wiss.* Berlin. 1931.
157. ELLIS, H. *Studies in the Psychology of Sex,* 6 vol. Philadelphia.
158. ENRIQUES, F. The Problems of Logic. (*Encyclopaedia of the Philosophical Sciences.* Vol. I. *Logic.* London, New York.)
159.   *Problems of Science.* Chicago, London.
160. FAJANS, K., and WÜST, J. *A Textbook of Practical Physical Chemistry.* London, New York.
161. FERENCZI, S. *Further Contributions to the Theory and Technique of Psychoanalysis.* London.
162. FERENCZI, S., and RANK, O. *The Development of Psychoanalysis.* Washington.
163. FERENCZI, S., and others. *Psycho-analysis and the War Neuroses.* London.
164. FISCHER, M. H. *Fats and Fatty Degeneration.* New York.
165.   *Oedema and Nephritis.* New York.
166.   Colloid Chemistry in Biology and Medicine. (*Mayo Foundation Lectures on the Biologic Aspects of Colloid and Physiologic Chemistry.* Philadelphia, London.)
167.   Lyophilic Colloids and Protoplasmic Behavior. In *Colloid Chemistry,* Vol. II, edited by J. Alexander. New York.
168. FISCHER, M. H., and collaborators. *Soaps and Proteins.* New York.
169. FISHER, A. *The Mathematical Theory of Probabilities,* Vol. I. London, New York.
170. FOERINGER, A. See Botcharsky.
171. FRAENKEL, A. *Zehn Vorlesungen über die Grundlegung der Mengenlehre.* Leipzig, Berlin.
172. FRAZER, J. G. *Totemism and Exogamy. A Treatise on Certain Early Forms of Superstition and Society,* 4 vol. London, New York. see page xii
173.   *The Golden Bough, a Study in Magic and Religion.* London, New York.
174. FREEMAN, W. Psychochemistry. *Jour. Amer. Med. Asso.* Aug. 1, 1931.
175. FREGE, G. *Die Grundlagen der Arithmetik.* Breslau.
176. FREUD, S. *Totem and Taboo.* New York.
177.   *A General Introduction to Psychoanalysis.* New York.
178.   *Psychopathology of Everyday Life.* New York, London.
179.   *Selected Papers on Hysteria and Other Psychoneuroses.* Washington.
180.   *Beyond the Pleasure Principle.* London.

181. *Collected Papers,* 4 vol. London.
182. *The Ego and the Id.* London.
183. *The Problem of Lay-Analyses.* New York.
184. *The Future of an Illusion.* London, New York.
185. FREUNDLICH, E. *The Foundations of Einstein's Theory of Gravitation.* London, New York.
186. *The Theory of Relativity.* London, New York.
187. FREUNDLICH, H. *Colloid and Capillary Chemistry.* London, New York.
188. GANTT, W. H. Recent Work of Pavlov and his Pupils. *Arch. Neur. and Psych.* Apr., 1927.
189. *A Medical Review of Soviet Russia.* London.
190. See Kupalov.
191. See Pavlov.
192. GIBBS, W. See Wilson, E. B.
193. GOMPERZ, H. *Weltanschauungslehre,* 2 vol. Jena.
194. GOMPERZ, T. *Greek Thinkers.* London.
195. GONSETH, F. *Les Fondements des mathématiques.* Paris.
196. GOODSPEED, T. H. The Breeding of X-rayed Tobacco Plants. *SS. Science.* Jan. 6, 1928.
197. GRAVEN, P. S. A Series of Clinical Notes on Headache. *Psychoan. Rev.* July, 1924.
198. A Case of Smoke Phobia. *Psychoan. Rev.* Apr., 1925.
199. GREGORY, J. C. *A Short History of Atomism from Democritus to Bohr.* London.
200. GROOT, J. J. M. DE. *The Religious System of China.* London.
201. GURWITCH, A., and BARON, M. A. Emission of Rays by Plant Cells. *SS. Science.* June 15, 1928.
202. GUYE, C. E. *Physico-Chemical Evolution.* London.
203. HAAS, A. *The New Physics.* London, New York.
204. *Introduction to Theoretical Physics,* 2 vol. London, New York.
205. *Wave Mechanics and the New Quantum Theory.* London, New York.
206. *Quantum Chemistry. A Short Introduction in Four Non-mathematical Lectures.* London.
207. HALL, G. S. *Jesus, the Christ, in the Light of Psychology.* New York.
208. HALLIBURTON, W. D. *Handbook of Physiology.* London.
209. HALTER, C. R. See Bagg.
210. HANCE, R. T. The Results of X-ray Experiments on Warm-blooded Animals. *SS. Science.* Jan. 6, 1928.
211. HANSON, F. B., and HEYS, F. A Possible Relation between Natural (Earth) Radiation and Gene Mutations. *Science.* Jan. 10, 1930.
212. HEAD, H. *Aphasia and Kindred Disorders of Speech,* 2 vol. London, New York.
213. HEAD, H., and others. *Studies in Neurology,* 2 vol. London.
214. HEILBRUNN, L. V. *The Colloid Chemistry of Protoplasm.* Berlin.
215. HEISENBERG, W. *The Physical Principles of the Quantum Theory.* Chicago, London.
216. HELSON, H. The Tau Effect. An Example of Psychological Relativity. *Science.* May 23, 1930.
217. HENDERSON, L. J. *The Fitness of the Environment.* New York, London.
218. HERELLE, F. D'. *The Bacteriophage.* Baltimore.
219. *Immunity in Natural Infectious Disease.* Baltimore.
220. HERRICK, C. J. *An Introduction to Neurology.* Philadelphia, London.
221. *Neurological Foundations of Animal Behavior.* New York.
222. *Brains of Rats and Men.* Chicago.
223. *The Thinking Machine.* Chicago.
224. Localization of Function in the Nervous System. *Proc. Nat. Acad. Sci.* Oct., 1930.
225. HEYS, F. See Hanson.
226. HILBERT, D. *The Foundations of Geometry.* Chicago, London.
227. See Courant.

228. HILBERT, D., and ACKERMANN, W. *Grundzüge der Theoretischen Logik.* Berlin.
229. HINKLE, B. *The Re-creating of the Individual.* New York.
230. HOBSON, E. W. *The Domain of Natural Science.* Aberdeen, Cambridge.
231. HOLLANDER, B. *The Mental Symptoms of Brain Disease.* London.
232. HOLMES, S. J. *The Evolution of Animal Intelligence.* New York.
233. *Studies in Animal Behavior.* Boston.
234. HOLT, E. B. *The Freudian Wish.* New York.
235. *Animal Drive and the Learning Process.* New York.
236. HUNTER, W. S. *The Delayed Reaction in Animals and Children.* Behav. Mono., Vol. 2, No. 1.
237. HUNTINGTON, E. V. *The Continuum.* Cambridge, Massachusetts.
238. HUNTINGTON, E. V., and KLINE, J. R. Sets of Independent Postulates for Betweenness. *Trans. Amer. Math. Soc.* July, 1917.
239. JAMES, W. *The Principles of Psychology,* 2 vol. New York.
240. JEFFREYS, H. *Scientific Inference.* Cambridge.
241. JELLIFFE, S. E. *The Technique of Psychoanalysis.* Washington.
242. JELLIFFE, S. E., and WHITE, W. A. *Diseases of the Nervous System. A Text-book of Neurology and Psychiatry.* Philadelphia.
243. JENNINGS, H. S. *The Behavior of Lower Organisms.* New York.
244. *Life and Death Heredity and Evolution in Unicellular Organisms.* Boston.
245. Review of Ritter's *"The Unity of the Organism, or the Organismal Conception of Life." Phil. Rev.* Nov., 1921.
246. *Prometheus or Biology and the Advancement of Man.* London, New York.
247. *The Biological Basis of Human Nature.* New York.
248. JESPERSEN, O. *Language.* New York.
249. JEVONS, W. S. *The Elements of Logic.* New York.
250. *The Principles of Science.* London, New York.
251. JOHNSON, T. H. The Wave Length of Atoms. *SS. Science.* July 17, 1931.
252. JOHNSTONE, J. *The Mechanism of Life.* London.
253. JONES, E. *Essays in Applied Psycho-analysis.* London.
254. JØRGENSEN, J. *A Treatise of Formal Logic. Its Evolution and Main Branches, with its Relations to Mathematics and Philosophy,* 3 vol. Oxford.
255. JUNG, C. G. *Contributions to Analytical Psychology.* London.
256. *Psychological Types: or the Psychology of Individuation.* London, New York.
257. JUST, E. E. Effects of Ultra-violet Rays on Eggs of *Nereis. SS. Science.* Jan. 6, 1928.
258. KAUFMANN, F. *Das Unendliche in der Mathematik und seine Ausschaltung.* Leipzig, Vienna.
259. KEMPF, E. J. *Psychopathology.* St. Louis.
260. KENNEDY, R. J. A Refinement of the Michelson-Morley Experiment. *Proc. Nat. Acad. Sci.* Nov., 1926.
261. KENNEDY, R. J., and THORNDIKE, E. M. Experimental Establishment of the Relativity of Time. *Phys. Rev.* Nov., 1932.
262. KEYNES, J. M. *A Treatise on Probability.* London, New York.
263. KEYSER, C. J. *The Human Worth of Rigorous Thinking.* New York.
264. *Mathematical Philosophy.* New York.
265. *Thinking about Thinking.* New York.
266. *Mole Philosophy and Other Essays.* New York.
267. *The Pastures of Wonder.* New York.
268. *Humanism and Science.* New York.
269. The Nature of the Doctrinal Function and its Role in Rational Thought. *Yale Law Jour.* Mar., 1932.
270. KINDER, E. F. See Syz.
271. KLINE, J. R. See Huntington.
272. KLUGH, A. B. Ultra-violet Radiations Deadly to the Minute *Crustacea. SS. Science.* Jan. 6, 1928.
273. KNUDSON, A. Further Studies on the Antirachitic Activation of Substances by Cathode Rays. *Science.* Aug. 19, 1927.

274. KOFFKA, K. *The Growth of the Mind*. London, New York.
275. KÖHLER, W. *Gestalt Psychology*. London, New York.
276. *The Mentality of Apes*. London, New York.
277. KORMES, M. On the Functional Equation $f(x + y) = f(x) + f(y)$. *Bull. Amer. Math. Soc.* Nov., 1926.
278. KORZYBSKI, A. *Manhood of Humanity. The Science and Art of Human Engineering*. New York.
279. Fate and Freedom. *Math. Teacher*. May, 1923.
280. The Brotherhood of Doctrines. *The Builder*. Apr., 1924.
281. *Time-Binding: The General Theory*. Presented before the Internat. Math. Congress, 1924, Toronto, Can. New York.
282. *Time-Binding: The General Theory*. Second Paper. Before the Washington Soc. Nervous and Mental Diseases, June 25, 1925; and the Washington Psychopath. Soc., March 13, 1926. Washington, D. C.
283. Discussion of 'Mental Hygiene and Criminology'. *Proc. First Internat. Congress of Mental Hygiene*, Washington, D. C. May, 1930.
283a. A Non-aristotelian System and Its Necessity for Rigour in Mathematics and Physics. Presented before the Amer. Math. Soc., at New Orleans, Dec. 28, 1931. Reprinted in the present volume as Supplement III.
284. KRETSCHMER, E. *Physique and Character*. London, New York.
285. KUPALOV, P. S., and GANTT, W. H. The Relationship between the Strength of the Conditioned Stimulus and the Size of the Resulting Conditioned Reflex. *Brain*. Vol. 50, Pt. 1. 1927.
286. LANG, H. B., and PATERSON, J. A. A Preliminary Report on Functional Psychoses. *Proc. Nat. Acad. Sci.* Nov., 1931.
287. LANGDON-DAVIES, J. *Man Comes of Age*. New York.
288. LANGWORTHY, O. R., and RICHTER, C. P. The Influence of Efferent Cerebral Pathways upon the Sympathetic Nervous System. *Brain*. Vol. 53, Pt. 2. 1930.
289. LASHLEY, K. S. *Brain Mechanisms and Intelligence*. Chicago.
290. LASSWELL, H. D. *Psychopathology and Politics*. Chicago.
291. LEARNED, B. W. See Yerkes, R. M.
292. LEDUC, S. *The Mechanism of Life*. New York.
293. Solution and Life. In *Colloid Chemistry*, edited by J. Alexander. New York.
294. LENZEN, V. F. *The Nature of Physical Theory*. New York, London.
295. LEŚNIEWSKI, S. Grundzüge eines neuen Systems der Grundlagen der Mathematik. *Fund. Math.* B. 14. Warsaw. 1929.
296. Über die Grundlagen der Ontologie. *C. R. des séances de la soc. des sciences et des lettres de Varsovie*. XXIII. 1930. Classe III.
297. On the Foundations of Mathematics (in Polish). Eleven chapters in the *Przegląd Filozoficzny* (Philosophical Review). Warsaw, Poland. 1930, 1931.
298. LÉVY-BRUHL, L. *Primitive Mentality*. London, New York.
299. *How Natives Think*. London, New York.
300. LEWIS, C. I. *A Survey of Symbolic Logic*. Berkeley.
301. LEWIS, G. N. *The Anatomy of Science*. New Haven.
302. See Wilson, E. B.
303. LEWIS, N. D. C. *The Constitutional Factors in Dementia Precox*. Washington.
304. LILLIE, R. S. *Protoplasmic Action and Nervous Action*. Chicago.
305. LOBATCHEVSKI, N. *Geometrical Researches on the Theory of Parallels*. Chicago, London.
306. LOEB, J. *Comparative Physiology of the Brain and Comparative Psychology*. New York, London.
307. *Studies in General Physiology*, 2 vol. Chicago.
308. *The Dynamics of Living Matter*. New York.
309. *The Mechanistic Conception of Life*. Chicago.
310. *The Organism as a Whole*. New York, London.
311. *Forced Movements, Tropisms, and Animal Conduct*. Philadelphia, London.

312. *Proteins and the Theory of Colloidal Behaviour.* New York, London.
313. LONG, J. S. Cathode Rays for Drying Paint. *SS. Science.* Apr. 15, 1927.
314. LORENTZ, H. A., EINSTEIN, A., MINKOWSKI, H., and WEYL, H. *The Principle of Relativity.* London, New York.
315. LOTKA, A. J. *Elements of Physical Biology.* Baltimore.
316. LUCAS, K. *The Conduction of the Nervous Impulse.* London.
317. ŁUKASIEWICZ, J. The Significance and Needs of Mathematical Logic (in Polish). *Nauka Polska.* Vol. X.
318. Philosophische Bemerkungen zu mehrwertigen Systemen des Aussagenkalküls. *C. R. des séances de la soc. des sciences et des lettres de Varsovie.* XXIII. 1930. Classe III.
319. ŁUKASIEWICZ, J., and TARSKI, A. Untersuchungen über den Aussagenkalkül. *Ibid.*
320. LUMIÈRE, A. *Rôle des colloides chez les êtres vivants. Essai de biocolloidologie.* Paris.
321. MACDOUGAL, D. T. Dr. Crile's "Autosynthetic Cells." *SS. Science.* Jan. 16, 1931.
322. Reactions of Non-Living Matter. *Ibid.*
323. MACH, E. *The Science of Mechanics.* Chicago, London.
324. *Scientific Lectures.* Chicago, London.
325. *Space and Geometry.* Chicago, London.
326. *Conservation of Energy.* Chicago, London.
327. *The Analysis of Sensations.* Chicago, London.
328. MACHT, D. I. Contributions to Phytopharmacology or the Applications of Plant Physiology to Medical Problems. *Science.* Mar. 21, 1930.
329. MACLAURIN, C. *Post Mortem. Essays, Historical and Medical.* New York.
330. MAIER, H. *Die Syllogistik des Aristoteles.* Tübingen.
331. MALINOWSKI, B. *Argonauts of the Western Pacific.* London, New York.
332. The Problem of Meaning in Primitive Languages. (Supplement I in Ogden and Richards' *The Meaning of Meaning.* London, New York.)
333. *Crime and Custom in Savage Society.* London, New York.
334. *Myth in Primitive Psychology.* London, New York.
335. *Sex and Repression in Savage Society.* London, New York.
336. *The Father in Primitive Psychology.* London, New York.
337. MAST, S. O. *Light and the Behavior of Organisms.* New York.
338. MAUTHNER, F. *Beiträge zu einer Kritik der Sprache,* 3 vol. Stuttgart, Leipzig.
339. *Aristotle.* New York.
340. MAYER, W. See Einstein.
341. McCLENDON, J. F. *Physical Chemistry of Vital Phenomena.* Princeton.
342. McCLENDON, J. F., and MEDES, G. *Physical Chemistry in Biology and Medicine.* Philadelphia.
343. McCOLLUM, E. V. *The Newer Knowledge of Nutrition.* London, New York.
344. McCOLLUM, E. V., and ORENT, E. R. The Effects of Deprivation of Manganese on the Rat. *Science.* May 8, 1931.
345. McCREA, A. Ultra-violet Radiation in the Promotion of Plant Growth. *SS. Science.* Jan. 6, 1928.
346. McDOUGALL, W. *World Chaos. The Responsibility of Science.* New York.
347. MEDES, G. See McClendon.
348. MENGER, K. *Dimensionstheorie.* Leipzig, Berlin.
349. MERZ, J. T. *A History of European Thought in the Nineteenth Century.* Chicago, London.
350. MEYER, A. *Psychobiology.* New York.
351. MIKAMI, Y. See Smith, D. E.
352. MINKOWSKI, H. Space and Time. In *The Principle of Relativity* by H. A. Lorenz and others. London, New York.
353. MONAKOW, C. VON. *The Emotions, Morality and the Brain.* Washington.
354. MORGAN, A. DE. *Formal Logic or the Calculus of Inference, Necessary and Probable.* London, New York.

355. *Elementary Illustrations of the Differential and Integral Calculus.* Chicago, London.
356. MORGAN, C. L. *An Introduction to Comparative Psychology.* London.
357. *Emergent Evolution.* New York.
358. MORGAN, T. H. *The Physical Basis of Heredity.* Philadelphia, London.
359. *The Theory of the Gene.* New Haven.
360. *Experimental Embryology.* New York.
361. See Bridges.
362. MORGAN, T. H., STURTEVANT, A. H., MULLER, H. J., and BRIDGES, C. B. *The Mechanism of Mendelian Heredity.* New York.
363. MÜHL, A. M. Fundamental Personality Trends in Tuberculous Women. *Psychoan. Rev.* Oct., 1923.
364. MULLER, H. J. X-rays and Evolution. *SS. Science.* Sept. 16, 1927.
365. The Effects of X-Radiation on Genes and Chromosomes. Abstract. *Anat. Rec.,* Vol. 37, and *Science.* Jan. 27, 1928.
366. The Problem of Genic Modification. *Verhand. V. Internat. Kongresses Vererb.* B. I. 1928.
367. The Production of Mutations by X-rays. *Proc. Nat. Acad. Sci.,* Vol. 14. 1928.
368. See Morgan, T. H.
369. MULLER, H. J., and PAINTER, T. S. The Cytological Expression of Changes in Gene Alignment Produced by X-rays in *Drosophila. Amer. Naturalist.,* Vol. 63. 1929.
370. MURPHY, D. P. Dangers of X-Ray Therapy. *SS. Science.* July 19, 1929.
371. NICOD, J. *Foundations of Geometry and Induction.* London, New York.
372. NORDMANN, C. *Einstein and the Universe.* New York.
373. *The Tyranny of Time.* London, New York.
374. NORTHROP, F. S. C. *Science and First Principles.* New York, London.
375. NUNN, T. P. *Relativity and Gravitation.* London.
376. OGDEN, C. K. *The Basic Vocabulary.* London.
377. *Basic English.* London.
378. OGDEN, C. K., and RICHARDS, I. A. *The Meaning of Meaning.* London, New York.
379. ORENT, E. R. See McCollum.
380. OSBORN, H. F. *Men of the Old Stone Age.* New York.
381. OSGOOD, W. F. *A First Course in the Differential and Integral Calculus.* New York, London.
382. PACKARD, C. The Effects of Beta and Gamma Rays of Radium on Protoplasm. *Jour. Exp. Zoöl.,* Vol. 19. 1915.
383. The Susceptibility of Cells to Radium Radiations. *Biol. Bull.,* Vol. 46. 1924.
384. Biological Effects of Irradiation. *SS. Science.* May 31, 1929.
385. PAGET, SIR RICHARD. *Human Speech.* London, New York.
386. PAINTER, T. S. See Muller.
387. PARETO, V. Compiled by G. H. Bousquet. *Précis de sociologie.* Paris.
388. PARKER, G. H. *The Elementary Nervous System.* London, Philadelphia.
389. *Smell, Taste and Allied Senses in the Vertebrates.* Philadelphia.
390. PATERSON, J. A. See Lang.
391. PATON, S. *Education in War and Peace.* New York.
392. *Signs of Sanity and the Principles of Mental Hygiene.* New York.
393. *Prohibiting Minds and the Present Social and Economic Crisis.* New York.
394. PAVLOV, I. P. *Conditioned Reflexes an Investigation of the Physiological Activity of the Cerebral Cortex.* Translated and edited by G. V. Anrep. London, New York.
395. *Lectures on Conditioned Reflexes.* Translated by W. H. Gantt. New York.
396. PEANO, G. *Formulaire de mathématiques. Logique mathématique.* Turin.
397. PEARL, R. *Modes of Research in Genetics.* New York.
398. *The Biology of Death.* Philadelphia, London.
399. *Studies in Human Biology.* Baltimore.
400. *The Present Status of Eugenics.* Hanover, New Hampshire.
401. PEARSON, K. *The Grammar of Science.* London.

402. Peirce, C. S. *Chance, Love and Logic.* London, New York.
403. Peterson, F. See Church, A.
404. Piaget, J. *The Language and Thought of the Child.* London, New York.
405. *Judgment and Reasoning in the Child.* London, New York.
406. *The Child's Conception of the World.* London, New York.
407. *The Child's Conception of Physical Causality.* London, New York.
408. Piaggio, H. T. H. *Differential Equations.* London.
409. Piéron, H. *L'Evolution de la mémoire.* Paris.
410. *Principles of Experimental Psychology.* London, New York.
411. *Thought and the Brain.* London, New York.
412. Pierpont, J. Mathematical Rigor, Past and Present. *Bull. Amer. Math. Soc.* Jan., 1928.
413. Pitkin, W. B. *A Short Introduction to the History of Human Stupidity.* New York.
414. Planck, M. *The Origin and Development of the Quantum Theory. Nobel Prize Address.* Oxford.
415. *A Survey of Physics.* London, New York.
416. Plauson, H. Uses of the Cathode Rays. *SS. Science.* Aug. 24, 1928.
417. Poincaré, H. *The Foundations of Science.* New York.
418. Polakov, W. N. *Man and His Affairs.* Baltimore.
419. Prantl, C. von. *Geschichte der Logik im Abendlande,* 4 vol. Leipzig.
420. Prince, M. *The Unconscious.* New York.
421. *The Dissociation of a Personality.* New York.
422. Rabaud, E. *How Animals Find Their Way About.* London, New York.
423. Rainich, G. Y. Tensor Analysis without Coördinates. *Proc. Nat. Acad. Sci.* June, 1923.
424. Electrodynamics in the General Relativity Theory. *Proc. Nat. Acad. Sci.* Apr., 1924.
425. Twodimensional Tensor Analysis without Coördinates. *Amer. Jour. of Math.* Apr., 1924.
426. Second Note. Electrodynamics in the General Relativity Theory. *Proc. Nat. Acad. Sci.* July, 1924.
427. Electrodynamics in the General Relativity Theory. *Trans. Amer. Math. Soc.* Jan., 1925.
428. Second Paper on Tensor Analysis. *Amer. Jour. of Math.* Jan., 1925.
429. Mass in Curved Space-Time. *Proc. Nat. Acad. Sci.* Feb., 1926.
430. The Role of Groups in a Physical Theory. *Jour. Franklin Inst.* Apr., 1929.
431. Analytic Functions and Mathematical Physics. *Bull. Amer. Math. Soc.* Oct., 1931.
432. *Mathematics of Relativity. Lecture Notes.* Ann Arbor, Michigan.
433. Ramsey, F. P. The Foundations of Mathematics. *Proc. London Math. Soc.* (2). Vol. 25. 1926.
434. Rank, O. *The Trauma of Birth.* London, New York.
435. See Ferenczi.
436. Rashevsky, N. V. Reactions of Non-Living Matter. *SS. Science.* Jan. 16, 1931.
437. Drops Acting like Living Cells. *SS. Science.* May 8, 1931.
438. Reiser, O. L. *Humanistic Logic for the Mind in Action.* New York.
439. Richards, I. A. See Ogden.
440. Richter, G. H. See Bancroft.
441. Rietz, H. L. *Mathematical Statistics.* Chicago.
442. Rinaldo, J. *The Psychoanalysis of the Reformer.* New York.
443. Ritchie, A. D. *Scientific Method.* London, New York.
444. Ritter, W. E. *The Unity of the Organism; or the Organismal Conception of Life,* 2 vol. Boston.
445. Roback, A. A. *Behaviorism and Psychology.* Cambridge, Mass.
446. *The Psychology of Character.* London, New York.
447. Roback, A. A., editor. *Problems of Personality.* London, New York.
448. Robb, A. A. *The Absolute Relations of Time and Space.* Cambridge.

449. ROYCE, J. The Principles of Logic. (*Encyclopaedia of the Philosophical Sciences*. Vol. I. *Logic*. London, New York.)
450. RUARK, A. E., and UREY, H. C. *Atoms, Molecules and Quanta*. New York.
451. RUEFF, J. *From The Physical to the Social Sciences: Introduction to a Study of Economic and Ethical Theory*. Baltimore.
452. RUSSELL, B. *The Principles of Mathematics*. Cambridge.
453. *Our Knowledge of the External World*. Chicago, London.
454. *Mysticism and Logic*. New York, London.
455. *Introduction to Mathematical Philosophy*. London, New York.
456. Introduction to *Tractatus Logico-Philosophicus* by Ludwig Wittgenstein. London, New York.
457. *The Analysis of Matter*. London, New York.
458. *The A B C of Atoms*. New York.
459. See Whitehead.
460. RUTZLER, J. E. See Bancroft.
461. SANTAYANA, G. *Scepticism and Animal Faith*. New York.
462. SCHILLER, F. C. S. *Riddles of the Sphinx. A Study in the Philosophy of Humanism*. New York, London.
463. SCHLICK, M. *Allgemeine Erkenntnislehre*. Berlin.
464. *Space and Time in Contemporary Physics*. London, New York.
465. SCHRÖDINGER, E. *Wave Mechanics*. London, Glasgow.
466. *Collected Papers on Wave Mechanics*. London, Glasgow.
467. SENCOURT, R. *The Spanish Crown*. New York.
468. SHAW, J. B. *Lectures on the Philosophy of Mathematics*. Chicago, London.
469. SHEFFER, H. M. A Set of Five Independent Postulates for Boolean Algebras. *Trans. Amer. Math. Soc.* Oct., 1913.
470. *The General Theory of Notational Relativity*. Mimeographed Edition. Harvard. Cambridge, Massachusetts, and *Proc. Sixth Internat. Congress of Philos.* Harvard. Cambridge, Massachusetts.
471. Review of *Principia Mathematica*, 2nd edition, Vol. I. *Isis.* Feb., 1926.
472. SHEPPARD, W. F. *From Determinant to Tensor*. London, New York.
473. SHERRINGTON, C. S. *The Integrative Action of the Nervous System*. New York.
474. SIERPIŃSKI, W. *Leçons sur les nombres transfinis*. Paris.
475. SILBERER, H. *Problems of Mysticism*. New York.
476. SILBERSTEIN, L. *The Theory of Relativity*. London, New York.
477. SMITH, D. E., and MIKAMI, Y. *A History of Japanese Mathematics*. Chicago, London.
478. SMITH, W. B. *Der Vorchristliche Jesus*. Jena.
479. *Ecce Deus*. London.
480. SOMMERFELD, A. *Atomic Structure and Spectral Lines*. London, New York.
481. *Wave-Mechanics*. London, New York.
482. SOMMERVILLE, D. M. Y. *The Elements of Non-Euclidean Geometry*. London, Chicago.
483. SPAULDING, E. G. *The New Rationalism*. New York.
484. SPEARMAN, C. E. *The Nature of 'Intelligence' and the Principles of Cognition*. London.
485. *The Abilities of Man*. London, New York.
486. SPENGLER, O. *The Decline of the West*, 2 vol. New York.
487. STARLING, E. H. *Principles of Human Physiology*. Philadelphia, New York.
488. STEKEL, W. *Conditions of Nervous Anxiety and their Treatment*. London.
489. STEWART, SIR JAMES PURVES. *The Diagnosis of Nervous Diseases*. London.
490. STIEGLITZ, J., editor. *Chemistry in Medicine*. New York.
491. STILES, P. G. *The Nervous System and Its Conservation*. Philadelphia.
492. STORCH, A. *The Primitive Archaic Forms of Inner Experiences and Thought in Schizophrenia*. Washington.
493. STURTEVANT, A. H. See Morgan, T. H.
494. SULLIVAN, H. S. Schizophrenia: Its Conservative and Malignant Features. *Amer. Jour. Psychiat.* July, 1924.
495. The Oral Complex. *Psychoan. Rev.* Jan., 1925.

496. Peculiarity of Thought in Schizophrenia. *Amer. Jour. Psychiat.* July, 1925.
497. Research in Schizophrenia. *Amer. Jour. Psychiat.* Nov., 1929.
498. SULLIVAN, J. W. N. *Three Men Discuss Relativity.* London.
499. SYZ, H. C. Psycho-Galvanic Studies on Sixty-four Medical Students. *Brit. Jour. Psych.* July, 1926.
500. Observations on the Unreliability of Subjective Reports of Emotional Reactions. *Brit. Jour. Psych.* Oct., 1926.
501. Psychogalvanic Studies in Schizophrenia. *Arch. Neur. and Psychiat.* Dec., 1926.
502. Observations on Experimental Convulsions with Special Reference to Permeability Changes. *Amer. Jour. Psychiat.* Sept., 1927.
503. On a Social Approach to Neurotic Conditions. *Jour. Nervous and Mental Disease.* Dec., 1927.
504. SYZ, H. C., and KINDER, E. F. Electrical Skin Resistance in Normal and in Psychotic Subjects. *Arch. Neur. and Psychiat.* June, 1928.
505. TARSKI, A. Fundamentale Begriffe der Methodologie der deduktiven Wissenschaften. I. *Monatsh. f. Math. und Phys.* B. XXXVII, H. 2. 1930.
506. See Łukasiewicz.
507. THALBITZER, S. *Emotion and Insanity.* London, New York.
508. THOMAS, T. Y. *The Elementary Theory of Tensors.* London, New York.
509. THOMPSON, D'ARCY W. *On Growth and Form.* Cambridge.
510. THOMPSON, S. P. *Calculus Made Easy.* London, New York.
511. THORNDIKE, E. L. *The Original Nature of Man.* New York.
512. *Educational Psychology,* 3 vol. New York.
513. THURSTONE, L. L. *The Nature of Intelligence.* London, New York.
514. TILNEY, F. *The Brain from Ape to Man.* New York.
515. TOLMAN, R. C. *Statistical Mechanics with Applications to Physics and Chemistry.* New York.
516. On the Problem of the Entropy of the Universe as a Whole. *Phys. Rev.* June 15, 1931.
517. Nonstatic Model of Universe with Reversible Annihilation of Matter. *Phys. Rev.* Aug. 15, 1931.
518. TRAMER, M. *Technisches Schaffen Geisteskranker.* Munich, Berlin.
519. Arbeitstherapie. *Schw. Arch. f. Neur. u. Psychiat.* B. XXI, H. 2. 1927.
520. Allgemeine Psychohygiene. *Schw. ZS. f. Hygiene.* 1931.
521. TSCHUPPIK, K. *Ludendorff, The Tragedy of a Military Mind.* Boston.
522. TURNBULL, H. W. *The Theory of Determinants, Matrices, and Invariants.* London.
523. TYLER, E. B. *Primitive Culture.* London.
524. UREY, H. C. See Ruark.
525. VAIHINGER, H. *The Philosophy of 'As If'.* London, New York.
526. VASILIEV, A. V. *Space Time Motion.* London.
527. The Acquisitions and Enigmas of the Philosophy of Nature. *Proc. Sixth Internat. Congress of Philos.* Harvard. Cambridge, Mass.
528. VEBLEN, O. A System of Axioms for Geometry. *Trans. Amer. Math. Soc.* July, 1904.
529. Geometry and Physics. *Science.* Feb. 2, 1923.
530. *Analysis Situs.* Coll. Publ. Amer. Math. Soc. New York.
531. VENDRYÈS, J. *Language.* London, New York.
532. VULLIAMY, C. E., editor. *The Letters of the Tsar to the Tsaritsa,* 1914-1917. New York, London.
533. WAGNER, N. Emission of Rays by Plant Cells. *SS. Science.* June 15, 1928.
534. WASHBURN, M. F. *The Animal Mind.* New York.
535. *Movement and Mental Imagery.* New York.
536. WATSON, J. B. *Behavior.* New York.
537. *Psychology from the Standpoint of a Behaviorist.* Philadelphia.
538. WEATHERBURN, C. E. *Elementary Vector Analysis.* London.
539. *Advanced Vector Analysis.* London.
540. WEINSTEIN, A. The Production of Mutations and Rearrangements of Genes by X-rays. *Science.* Apr. 6, 1928.

541. WEISS, P. The Theory of Types. *Mind*. Vol. XXXVII, No. 147. Reprinted as Supplement II in this volume.
542.   Relativity in Logic. *Monist*. Oct., 1928.
543.   The Nature of Systems. *Monist*. Apr. and July, 1929.
544.   Two-Valued Logic. Another Approach. *Ann. Philos*. B. 2, H. 4. 1931.
545.   The Metaphysics and Logic of Classes. *Monist*. Jan., 1932.
546. WEYL, H. *Das Kontinuum. Kritische Untersuchungen über die Grundlagen der Analysis*. Leipzig.
547.   *Space Time Matter*. London, New York.
548.   Über die neue Grundlagenkrise der Mathematik. *Math. ZS*. Vol. 10. 1921.
549.   *Consistency in Mathematics*. Rice Inst. Pamphlet. Oct., 1929.
550.   *The Theory of Groups and Quantum Mechanics*. London, New York.
551.   See Lorentz.
552. WHEELER, W. M. *Ants, Their Structure, Development and Behavior*. New York.
553.   *Social Life among the Insects*. New York.
554.   *The Social Insects*. London, New York.
555.   Emergent Evolution of the Social. *Proc. Sixth Internat. Congress of Philos*. Cambridge, Mass.
556.   *Emergent Evolution and the Development of Societies*. New York.
557.   *Demons of the Dust*. New York.
558. WHITE, A. D. *A History of the Warfare of Science with Theology in Christendom*, 2 vol. New York, London.
559. WHITE, W. A. *Outlines of Psychiatry*. Washington.
560.   *Foundations of Psychiatry*. Washington.
561.   *Principles of Mental Hygiene*. New York, London.
562.   *Insanity and the Criminal Law*. New York, London.
563.   *An Introduction to the Study of the Mind*. Washington.
564.   *Essays in Psychopathology*. Washington.
565.   *The Meaning of Disease*. Baltimore.
566.   The Language of Schizophrenia. *Arch. Neur. and Psychiat*. Oct., 1926.
567.   *Medical Psychology*. Washington.
568.   A Message from Medical Psychology to General Medicine. *Milwaukee Proc. Postgrad. Med. Asso. N. Amer*. 1931.
569.   See Jelliffe.
570. WHITEHEAD, A. N. *Universal Algebra*. I. Cambridge.
571.   *An Introduction to Mathematics*. New York, London.
572.   *An Enquiry Concerning the Principles of Natural Knowledge*. Cambridge.
573.   *The Concept of Nature*. Cambridge.
574.   *The Principle of Relativity with Applications to Physical Science*. Cambridge.
575.   *Science and the Modern World*. London, New York.
576.   *Symbolism, Its Meaning and Effect*. New York, London.
577.   *The Function of Reason*. Princeton.
578.   *Process and Reality*. London, New York.
579. WHITEHEAD, A. N., and RUSSELL, B. *Principia Mathematica*, 3 vol. Cambridge.
580. WILDER, H. H. *The Pedigree of the Human Race*. New York.
581. WILLIAMS, F. E. *Adolescence. Studies in Mental Hygiene*. New York.
582. WILLIAMS, H. B. Mathematics for the Physiologist and Physician. *Math. Teacher*. Mar., 1920.
583.   Mathematics and the Biological Sciences. Gibbs Lecture. *Bull. Amer. Math. Soc*. June, 1927.
584. WILSON, E. B. Logic and the Continuum. *Bull. Amer. Math. Soc*. June, 1908.
585.   *Advanced Calculus*. New York, Boston.
586. WILSON, E. B., and GIBBS, J. W. *Vector Analysis*. New York.
587. WILSON, E. B., and LEWIS, G. N. The Space-time Manifold of Relativity. *Proc. Amer. Acad. Arts and Sci*. Nov., 1912.

588. Windelband, W. The Principles of Logic. (*Encyclopaedia of the Philoso-phical Sciences.* Vol. I. *Logic.* London, New York.)
589.   *A History of Philosophy.* New York, London.
590. Wittgenstein, L. *Tractatus Logico-Philosophicus.* London, New York.
591. Woodworth, R. S. *Psychology. A Study of Mental Life.* New York.
592. Wundt, W. *Elements of Folk Psychology.* London, New York.
593. Wüst, J. See Fajans.
594. Yealland, L. R. *Hysterical Disorders of Warfare.* London, New York.
595. Yerkes, A. W. See Yerkes, R. M.
596. Yerkes, R. M. *The Mental Life of Monkeys and Apes.* Behav. Mono. Vol. 3, No. 1. 1916.
597.   *Almost Human.* New York.
598.   The Mind of a Gorilla. *Gen. Psych. Mono.* Vol. 2. 1927. *Comp. Psych. Mono.* Vol. 5. 1928.
599. Yerkes, R. M., and Learned, B. W. *Chimpanzee Intelligence and.Its Vocal Expression.* Baltimore.
600. Yerkes, R. M., and Yerkes, A. W. *The Great Apes.* New Haven.
601. Young, J. W. *Lectures on Fundamental Concepts of Algebra and Geometry.* New York, London.
602. Young, J. W. A., editor. *Monographs on Topics of Modern Mathematics.* London, New York.
603. Zaremba, S. *La Logique des mathématiques.* Paris.

The following important items in the bibliography have either been omit-ted by inadvertence, or they appeared after the numbering of the bibliography was completed.

604. Bell, E. T. *Numerology.* Baltimore.
605. Bridges, C. B. The Genetics of Sex in Drosophila. *Sex and Internal Secre-tions.* 1932.
606.   Apparatus and Methods for Drosophila Culture. *Amer. Naturalist.* May-June, 1932.
607. Eckart, C. Application of Group Theory to the Quantum Dynamics of Monatomic Systems. *Rev. of Modern Physics.* July, 1930.
608. Eddington, A. S. *The Expanding Universe.* New York, London.
609. Gantt, W. H. *History of Russian Medicine.*
610. Hedrick, E. R. Tendencies in the Logic of Mathematics. *Science.* Apr. 7, 1933.
611. Huntington, E. V. New Sets of Independent Postulates for the Algebra of Logic, with Special Reference to Whitehead and Russell's *Principia Mathematica. Trans. Amer. Math. Soc.* Jan., 1933.
612. Klein, F. *Elementary Mathematics from an Advanced Standpoint.* English Translation. New York, Leipzig.
613. Leathem, J. G. *Elements of the Mathematical Theory of Limits.* London, Chicago.
614. Loucks, R. B. An Appraisal of Pavlov's Systematization of Behavior from the Experimental Standpoint. *Jour. Comp. Psychology.* Feb., 1933.
615. Morgan, T. H. The Rise of Genetics. *Science.* Sept., 23 and 30, 1932.
616. Smith, H. B. *Symbolic Logic.* Ann Arbor, Michigan (Edwards Bros.).
617. Tolman, R. C. Thermodynamics and Relativity. The Gibbs Lecture. *Bul. Amer. Math. Soc.* Febr., 1933.
618. Thomson, G. P. *The Wave Mechanics of Free Electrons.* New York, London.
619. Whitehead, A. N. *Adventures of Ideas.* New York, London.

SCIENCE AND SANITY

AN INTRODUCTION TO NON-ARISTOTELIAN SYSTEMS AND GENERAL SEMANTICS

BY

ALFRED KORZYBSKI

(Author of *Manhood of Humanity*)

SCIENTIFIC OPINIONS ABOUT THE FIRST EDITION, 1933:

1. ANTHROPOLOGY ... Professor B. Malinowski, University of London.
2. BIOLOGY .... Doctor C. B. Bridges, Carnegie Institution.
   Professor C. M. Child, University of Chicago.
   Professor H. S. Jennings, Johns Hopkins University.
   Professor R. Pearl, Johns Hopkins University.
3. BOTANY .... Doctor D. G. Fairchild, United States Department of Agriculture.
4. CONDITIONAL
   REFLEXES .... Doctor W. H. Gantt, Phipps Psychiatric Institute, Johns Hopkins Hospital.
5. EDUCATION .... Doctor E. L. Hardy, President States Teachers' College, San Diego, California.
   C. L. Williams, President Williams College, Berkeley, California.
6. ENTOMOLOGY ...... Professor W. M. Wheeler, Harvard University.
7. GENETICS .... See C. B. Bridges, D. G. Fairchild, H. S. Jennings.
8. OPHTHALMOLOGY .. Professor W. H. Wilmer, Johns Hopkins University.
9. MATHEMATICS ..... Professor E. T. Bell, California Institute of Technology.
10. MATHEMATICAL
    FOUNDATIONS .... Bertrand Russell, F.R.S., London, England.
    AND LOGIC.
11. MATHEMATICAL
    PHYSICS .... Professor B. F. Dostal, University of Florida.
12. NEUROLOGY .... Professor C. J. Herrick, University of Chicago.
13. PHYSICS .... Professor P. W. Bridgman, Harvard University.
    Professor R. J. Kennedy, University of Washington, Seattle, Washington.
14. PHYSIOLOGY .... Professor R. S. Lillie, University of Chicago.
    Professor H. B. Williams, Columbia University.
15. PSYCHIATRY .... Doctor P. S. Graven, Washington, D. C.
    Doctor J. A. P. Millet, New York City.
    Doctor M. Tramer, University of Bern, Switzerland.
    Doctor W. A. White, Superintendent Saint Elizabeth's Hospital, Washington, D. C.
16. SEMANTICS .... See B. Malinowski.

INTERNATIONAL		THE SCIENCE PRESS
NON-ARISTOTELIAN		LANCASTER, PA., U. S. A.
LIBRARY		DISTRIBUTORS

1. ANTHROPOLOGY.

BRONISLAW MALINOWSKI, Ph.D. (Cracow), D.Sc. (London), Professor of Social Anthropology, School of Economics, University of London.

"The functional or relational conception of matter, mind and, finally, of human culture, seems to be gradually crystallising from all attempts at scientific synthesis. Count Korzybski's work contributes to these efforts in no mean measure. I am perhaps biassed as a countryman, but to me this Polish attempt at synthesis seems to rank as one of the most important. I am of course unable to express a competent judgment on its mathematical, scientific—in the narrow sense of the word—and philosophical side. As regards however semantics and the anthropological issues discussed by Count Korzybski, I am in complete agreement with his approach. I should like to add that the approach is so new and fundamental that it will take some time for us to become completely familiar with it. For the present I should like to say that I have not yet mastered all the intricacies of Count Korzybski's system, so my appreciation must naturally be regarded as preliminary."

2. BIOLOGY.

DOCTOR CALVIN B. BRIDGES, Biologist, internationally known specialist in heredity, Carnegie Institution of Washington, in residence at California Institute of Technology, Pasadena.

"In several fields of Biological sciences the unconscious drift of thought has been for some years more and more in the direction which Count Korzybski designates as 'non-elementalistic.' Thus, the distinction, once thought fundamental, between heredity and environment, loses its force, and the organism is now redefined as the focus of both internal (hereditary) and external (environmental) activities. What is considered external rather than internal, changes with the point of view and the size of the unit (nations, man, gland, cell, nucleus, chromosome, gene) which is made the basis of the formulation. The formulation becomes relational, non-elementalistic, organism-as-a-whole. The reformulation of biological concepts is made ultimately inevitable, and is greatly hastened and aided in transition by the generalized point of view established in the non-aristotelian system of Korzybski. The elimination of 'identity' constitutes the first and most fundamental general step for such a non-aristotelian and non-elementalistic reconstruction. Following this complete reformulation and its application in science and in life, the psycho-logical and environmental conditions for man would be improved to such an extent that it is not yet possible to foresee the entire result in the new enviro-genetic manifold."

C. M. CHILD, Professor of Zoology, University of Chicago.

"I think that Count Korzybski has a view point of great interest and that his method of attack on the various problems with which he deals cannot fail to be of value."

H. S. JENNINGS, Henry Walters Professor of Zoology and Director of the Zoological Laboratory, Johns Hopkins University.

"The attempt of Count Korzybski to formulate the world and its processes, keeping in view as a guiding principle the fact that no two things are identical, seems to me of the greatest interest and value. It is something that had to be done, and it has within it the seeds of a much needed intellectual revolution."

RAYMOND PEARL, Professor of Biology, Johns Hopkins University.

"I have known and followed Count Korzybski's work for many years with the keenest interest. In this new book he makes, in my opinion, a contribution to human thought and understanding of the very first rank of importance. It states and develops a really new idea. The consequences of that idea will, in the passage of time, be far-reaching and fundamental. At long last real hope is offered of measurably freeing man from some of the dreadful consequences of his verbalistic bonds."

See also W. M. WHEELER.

784

3. BOTANY.

DOCTOR DAVID G. FAIRCHILD, Botanist, U. S. Department of Agriculture, Washington, D. C., plant explorer and pathologist.

"I am much impressed with the profundity of Korzybski's *Science and Sanity*. I find it hard to get out of the meshes of the old aristotelianism and wish that I had been able to read this book in my youth, for then I could have acquired the new language of relations.

Korzybski's masterly treatise will act as a powerful force in natural selection perhaps, when it brings into common use the non-aristotelian methods, for it will favor in most pursuits those who are capable of conceptual thought and confuse and eliminate those who want quick off hand decisions such as are usually blurted out with great show of confidence. Of one thing we may be assured: once a man grasps the general idea Korzybski is driving it, he cannot fail to look at the world of everyday language from a different standpoint.

Korzybski's criticisms are so profound that they change the very foundations upon which we have been used to depend. As I look back to my years of travel over the world I can see that I did everything I ought not to have done in the way of bad thinking (we all have done this I suppose; bad thinking must be the common plague of mankind).

These last years among West Africans in Africa and the West Indies have made me realize keenly that primitive man identifies about like the animals do and has no consciousness that he abstracts at all. Of course if we are not conscious of abstracting; in other words, if we copy primitives or animals in our nervous reactions, then I suppose any kind of maladjustment can be expected. The simple and efficient neuropsychological non-aristotelian technique which Korzybski formulates in his Structural Differential for the purpose of the elimination of identification holds promises that we may finally outgrow the infantile stage of our civilization. I wonder that Educators have not already taken up this pressing problem and made the elimination of identification and the acquiring of consciousness of abstracting, the main aims of all education.

These impressions regarding Korzybski's remarkable book come at the close of many years of travel through the world,—in savage countries, in the Orient, in South America and South Africa,—and had they only been a part of my mental training before these travels, the results of my observations could hardly have failed to have been much nearer to the actualities.

I predict a steady conversion to the point of view of this most interesting and important work."

4. CONDITIONAL REFLEXES.

DOCTOR W. HORSLEY GANTT, Phipps Psychiatric Institute, The Johns Hopkins Hospital. Formerly for five years co-worker with Professor Pavlov in Leningrad.

"I have read with great interest Count Korzybski's *Science and Sanity* and feel that it is very important for science as well as general education and progress of human thinking. It expresses a point of view and a truth that I have not seen stated previously. I was particularly interested in the chapters dealing with conditional reflexes. Korzybski dicsusses the matter with profound and accurate understanding, and the suggestions he makes are most timely and helpful to those who are working in this field. Anyone interested in the broader aspects of science I am sure will find in Korzybski's book an original and far-sighted view of the whole modern teaching of the subject."

5. EDUCATION.

DOCTOR EDWARD L. HARDY, President State Teachers College, San Diego, California.

"Count Korzybski's *Science and Sanity* should be read by all persons seriously interested in or concerned with the next necessary steps in the development of educational principles and procedures."

CORA L. WILLIAMS, Mathematician, President Williams Institute, Berkeley, California.

"What Einstein has done for the outer realm of our being, Korzybski is doing for our inner realm. It is to be hoped that some understanding person will endow a chair of Non-aristotelian General Semantics for him in the Institute of Advanced Study so that these two lines of research may go on together."

6. ENTOMOLOGY.

WILLIAM MORTON WHEELER, Professor of Entomology, Harvard University.

"Count Korzybski's work seems to me to be of great interest and value not only to the lay reader but also to the student of science and to the biologist and sociologist in particular for three reasons. First, his views give greater generality to the significance of the organism as a whole, and of structure and creative synthesis, or emergence, which are being increasingly emphasized by biologists, psychologists and sociologists working in the most diverse fields. Second, the sections of his work dealing with the intellectual vices of wishful thinking, verbalism and identification, to which we are all more or less addicted, point the way to the acquirement of mental balance and sanity. And third, his method of attaining this sanity through a non-aristotelian system and a realization of the meaning of the abstractions and symbols which we are constantly using, lays the foundation for a sound and much-needed social, commercial and political ethics."

7. GENETICS.

See CALVIN B. BRIDGES, D. G. FAIRCHILD and H. S. JENNINGS.

8. OPHTHALMOLOGY.

WILLIAM H. WILMER, M.D., Professor of Ophthalmology, Johns Hopkins University and Ophthalmologist-in-chief, Johns Hopkins Hospital.

"Count Korzybski's viewpoint is very unique, fascinating, and, I think, very logical. The induction of non-identity would cover a great many ills, mental, moral and physical. For more than a quarter of a century, I have observed the retrogression of a number of great men after a certain time; and I feel that their failure to hold their greatness has been due largely to an egocentricism. What is true of these men whom the world called great for awhile, is equally true of the masses of humanity, who have not attained to greatness. Many of these could probably have been saved by the proper attention to psychophysiology."

9. MATHEMATICS.

E. T. BELL, Professor of Mathematics, California Institute of Technology.

"I think it is obvious that Korzybski is working in a direction of the highest present importance for science and life. This is the more so as some sort of corrective seems to be needed for the well-meaning but ill-considered popular announcements by certain leading scientific men.

A little careful consideration of the recognized fundamentals of scientific and other thinking, such as Korzybski's book aims to set forth clearly, would prevent such really futile pronouncements by prophets of science and make the public more chary in swallowing every transient guess.

Korzybski, among personal contributions of his own concerning the law of identity, has succeeded incidentally in making current the fundamental revolution in mathematical and other basic thinking, which goes under the name of a non-aristotelian logic, and bringing to educated people an account of the most significant advance in abstract thought of the past millennium. The profound modifications of rational, mathematical thinking which began about thirty years ago with the work of Brouwer, have, so far as I am aware, escaped the notice of those who undertake

786

to report science and mathematics to the general public. The reader of Korzybski's book will gain an outlook on these new fields as well as an insight into the author's contributions to the problem of identity. Brouwer challenged one of the laws of Aristotle, Korzybski challenges another."

See also P. W. BRIDGMAN, B. F. DOSTAL, R. J. KENNEDY, BERTRAND RUSSELL, M. TRAMER, C. L. WILLIAMS, H. B. WILLIAMS.

10. MATHEMATICAL FOUNDATIONS AND LOGIC.

BERTRAND RUSSELL cables from London to the author:

"Your work is impressive and your erudition extraordinary. Have not had time for thorough reading but think well of parts read. Undoubtedly your theories demand serious consideration."

11. MATHEMATICAL PHYSICS.

B. F. DOSTAL, Professor of Mathematics, University of Florida.

"We still teach classical science on Mondays, Wednesdays and Fridays, and modern science on Tuesdays, Thursdays and Saturdays, as Sir William Bragg truly said. Within the bounds of the aristotelian system there appears to be no hope of ever finding the requisite unifying principle. Mathematicians have been rapidly outgrowing the old forms of so-called logic, but mathematical physicists have in general been slow to appreciate the value of these efforts or to apply these results to their own problems. Korzybski's *Science and Sanity* will be of great value to science because it contains the basis for the development of a new and wider, and more unifying, form of scientific determinism, without which the outlook for modern science would be gloomy indeed. Not only does Korzybski point out a more satisfactory, non-aristotelian, non-identity basis for a new science in general, than any hitherto employed, but he goes further in giving several promising suggestions for extensive developments and applications of the results of *modern* science, including those of the new wave and quantum mechanics. His work is bound to become a stimulus to investigators in mathematics, physics, chemistry, biology, 'psychology,' and medicine; and to educators, economists, sociologists, engineers, lawyers, and laymen as well, the majority of which still have a 'philosophy of the universe which takes one form on weekdays and another form on Sundays.' "

12. NEUROLOGY.

C. JUDSON HERRICK, Professor of Neurology, University of Chicago.

"The disturbances of mental balance and social stability now so prevalent seem to indicate a general failure to adjust our minds to our jobs. This results in futile conflict and too often in mental and social derangement. The numberless panaceas proposed fail because each attacks a single phase of a very complex situation, and generally a particular symptom rather than the cause of the trouble. Count Korzybski has diagnosed a fundamental source of confusion in thinking and in conduct and he presents a plan for radical revamping of our theory and practice that seems worthy of further trial in a wide variety of fields. His dynamic definition of *structure* in terms of relations gives promise of important applications in both science and practical affairs. It provides a generally useful symbol for experience of all sorts and a technique for recasting traditional ideas and practices more efficiently. Adjustments in terms of one dominant motive (or value) are replaced by a broader (many-valued) scheme of motivation which points the way toward personal and social sanity—a way that I believe is fundamentally correct and practicable."

See also C. M. CHILD, R. S. LILLIE, M. TRAMER, W. M. WHEELER, H. B. WILLIAMS, W. H. WILMER.

13. PHYSICS.

P. W. BRIDGMAN, Professor of Physics, Harvard University.

"Of late years the realization has been growing that the ultimate source of a large fraction of the difficulties of society, civilization, and science, is verbal in character. Among the few serious attempts to waken full self consciousness of what the situation is, and, having awaked consciousness, to provide *a technique* by which the vicious consequences of verbal habits may be avoided, I believe that of Count Korzybski must be rated as of the very first importance. I have been acquainted with his work for a number of years; not only do I believe it to be fundamentally sound, but I have always found his points of view most suggestive and stimulating, both in general and technical matters, and I have been amazed at the breadth of his interests and reading, and the diversity of the fields to which applications are made."

ROY J. KENNEDY, Professor of Physics, University of Washington, Seattle, Washington.

"Many of the impasses in which we of this lunatic world are involved are the result of verbal difficulties, and it is precisely these difficulties at which Count Korzybski's technique for the elimination of identity is chiefly aimed. He has shown a striking versatility in developing this technique which he has originated; he discusses the shortcomings of the sciences as facilely as those of religion. Whether or not the reader's sanity is improved by a careful study of the book, he cannot fail to enlarge his capacity for clear thinking. Paradoxically, athough *Science and Sanity* deals largely with the unspeakable it is suitable for discussion in the most decorous circles."

See also B. F. DOSTAL.

14. PHYSIOLOGY.

RALPH S. LILLIE, Professor of Physiology, University of Chicago.

"Count Korzybski's criticism of the present structure and usages of human society—as failing to keep pace with the advance of knowledge in the physical and biological sciences—is timely and well-founded, and is expressed with clearness, vigor and insight in this interesting book. It is certain that this knowledge, if widely diffused and acted upon, would greatly alleviate and perhaps remove many of the ills which afflict the modern world. The chief obstacle to such progress is not the lack of available knowledge, but the anachronistic survival of many mental habits and conceptions which are inconsistent with the facts of natural reality as revealed by science. These conceptions are firmly rooted in the general mind by language and custom. What is needed is a far-reaching revision of concepts, and this book points the way to such a revision. Since we are compelled by the conditions of existence to think and act in terms of symbols—concepts, words, images, formulae—it is all-important that these should conform as closely as possible to the permanent realities of life and nature. How to secure an adequate degree of such conformity and establish it by training and education is one of the most pressing problems of the time. Count Korzybski describes in detail the nature of verbal, mathematical and scientific symbolisms, and discusses clearly the biological, neurological, and other conditions which give them their representative value. He shows that misconceptions regarding the nature of language underlie many prevalent confusions and fallacies, especially the various fallacies of identification (arising mainly from verbalisms),—as when it is assumed that the application of the same label to different facts somehow renders them all alike and justifies the same action toward all. Serious consequences inevitably arise from the failures of discrimination and valuation thus resulting; and the author makes a special plea for discernment and individual treatment in the problems of human personality. It is only on this basis that many types of maladjustment can be prevented or corrected. These are only a few features in a book remarkable for its comprehensiveness, scholarship and independence."

HORATIO B. WILLIAMS, Dalton Professor of Physiology, Columbia University, and a Mathematician.

"In his *Science and Sanity* Count Korzybski undertakes to bring to the attention of his readers the importance of 'consciousness of abstracting,' or the permanent and full awareness that: (1) the object *is not* the event or the physico-chemical submicroscopic process; (2) that the symbol or label *is not* the object; and, (3) that an inference *is not* a description. Thus he is led to the formulation of a non-aristotelian system based on the complete rejection of 'identity.' To facilitate training in non-identity he proposes a simple structural diagram which he calls the *Structural Differential*. With its aid, the general *verbal* rejection of 'identity,' is translated into *ordering* which becomes a visual, kinesthetic, *neuro*-psycho-logical method to train in non-identity or discrimination, and so to eliminate the always dangerous identifications, which play such an important rôle in all maladjustments.

It has been my privilege to read a considerable part of this book in manuscript and most of it in page proof. It appears at a time when the *modus vivendi* has broken down. National and international problems demand sane thinking. Count Korzybski, by rejecting a principle invariably false to facts (identity), which in principle prevents adjustment, points the way toward better adjustment, and so perhaps toward a saner solution of our man-made difficulties.

This is distinctly not a book for superficial reading, but one that will amply repay the thoughtful reader. It should be approached not as a medium of entertainment, though it may be that, but rather in the spirit indicated by one of its passages which refers to 'the joint labors of the author and the reader.' "

See also W. H. GANTT.

15. PSYCHIATRY.

DOCTOR PHILIP S. GRAVEN, Psychiatrist, Washington, D. C.

"I have read *Science and Sanity* through completely, some parts several times, and I must admit I have never encountered a work so rich in fundamental suggestions. It clearly covers a field almost wholly neglected in our University education. From the methodological point of view, therefore, the book is indispensable to any one endeavoring to carry on sane, clear, scientific work. This of course includes Medicine and especially Medical Psychology where sane thinking about the un-sane and 'insane' is vital at all times. Statements, principles, etc., that affect one's mental attitude toward problems are of most importance. These, *Science and Sanity* provides in abundance.

In addition to the scientists being considerably aided by the use of the non-aristotelian principles (aided in carrying on sane, creative, well-tempered, rigorous thinking about their observations and experimental data), there is also another group directly affected: namely, the mentally disordered. By direct clinical application, I have found the non-aristotelian principles workable in this enormous group. My observations cover a period of about six years. I shall have a great deal to say about these observations in contributions to medical and scientific journals.

By reading the book carefully, I have derived many benefits: personal, cultural, professional, scientific. The book appeals to me as one that will supply at least a few generations of scientific workers with a means of maintaining a productive, not clogging and obstructive psycho-logical attitude towards investigations needful to human security and advancement. And what could be more urgent in this modern age than a means to attain and maintain sanity: Korzybski's Theory of Sanity already makes that a possibility."

DOCTOR JOHN A. P. MILLET, Psychiatrist, New York City.

"It seems to me that Count Korzybski in his book *Science and Sanity* calls attention to one very important difficulty in the thinking process which he so aptly sums up as 'confusion in the orders of abstraction.' We have found through our analytical work in dealing with the neuroses that many difficulties in attitude and orientation toward life are derived from unconscious identification, a situation which leads to a real difficulty in individuation and sometimes makes it permanently impossible.

From the psychiatric point of view it seems to me that this one point that Korzybski has emphasized gives the chief value to his work. I doubt whether analysts, for the most part, have ever considered the problem of identification from exactly this point of view. Korzybski's presentation should be interesting to such a group and might well lead to further experimental activities in the field of education designed to offset the dangers of such identifications.

A beginning has been made in this field already by Korzybski in his development of the 'Structural Differential.' It is too early to say what practical value this Structural Diagram may have but the principles of its development are based on the conception of 'order' in neuro-psycho-logical processes and so provides what might be described as a neuro-physiological technique for the elimination of false identification."

DOCTOR M. TRAMER, Priv.-Doz. in Psychiatry, University of Bern, Switzerland; Medical Director State Asylum, President Swiss Psychiatric Association. A Mathematician, Neurologist, formerly a co-worker with the late von Monakov, and a Psychiatrist.

"For almost three years I have had the opportunity to follow the researches of Count Korzybski. I recognized already from the formulation of 'time-binding' that his work deals with something fundamental. Further results of his researches justified this opinion. These affect not only our scientific thinking but also our daily lives. Korzybski now presents his work in a comprehensive form to the public, its study reveals that we have to deal with a structure of deeper foundations, the beginning of which have already appeared in different scientific domains, and particularly in physics. The general structure of the completed development of his thesis leads to the formulation of the urgent need for a fundamental analysis (revision) of our scientific language for which he blazes the trail. The means he selects to overcome the consequent difficulties of such an analysis, and to bring into the foreground the necessity for general structural thinking above all, are, in my opinion, results of fundamental importance."

DOCTOR WILLIAM A. WHITE, Professor of Psychiatry, George Washington University Medical School; Superintendent Saint Elizabeth's Hospital, Washington, D. C.

"Korzybski's concepts are to me very helpful, particularly his concepts of linearity, multiordinality, degrees of abstractions and finally of non-identity. I have always felt that such matters as are included in these concepts are of extreme importance to have as one's mental equipment. I am at one with Korzybski in many ways. Certainly I am sure that how we think about things is of as much importance as what we think about them. I congratulate the author upon the appearance of his book, and I think he has made a real contribution to the methodology of thinking."

16. SEMANTICS.

See B. MALINOWSKI.

INDEX

Function (al) (continued)
  mathematical theory of, 131, 133 ff.,
        259, 263, 266, 275, 281,
        286 f., 574 f., 595, 597 ff.,
        610 ff., 632, 710, 726.
    rough definition, 136.
  nervous system. *See*
        Nervous function.
    analyzing and synthesizing,
        333.
    interchangeability of, 290.
    in terms of a mathematical
        function, 631.
  physiological, 316 f., 330.
  propositional. *See* Propositional
        function.
  system. *See* System-function.
  terms, 94, 179, 379.
Funk, 118, 126.
Funktors, 748.

Galilean transformation, 654 f.
Galileo, 247, 574, 654, 730.
Gamma-rays, 237, 686, 716.
Gantt, 123, 315, 772, 774, 781.
Gauss, 86, 622, 642 ff., 725, 747.
Geddes, 49.
General method of orientation, 8.
General methods, *lxxvi*.
General semantic(s), 8, 95, 250, 265, 380,
        434, 461 f., 474, 535, 540, 545,
        752.
  as an empirical science, *xli*.
  as an extensional discipline, *xxxviii*.
  as a non-aristotelian system, *xxxviii*,
        xl f., *lxxxix*, *xci* f., *xcvii*.
  as a theory of evaluation, *xxxiv*.
  choice of name, *xxxiv*.
  confusion with semantics, *xxxiii*.
  distinction from semantics, *xxxiii* f.,
        xlii f.
  example, *xlii*.

Generalizations, *lv* ff.
  new factors, *lv* ff.
  undue, 10.
Genius(es), 30, 75, 307, 327, 485, 676.
Geodetics, 625, 647, 663.
Geometry, 151, 242, 251, 267, 575, 583,
        595, 645.
  analytical, 146, 251.
  co-ordinate, 615 ff.
  differential, 233, 575, 642, 658.
  euclidean, 27, 85 f., 95, 232, 265,
        405, 441, 611, 615, 622, 646,
        667 ff., 756.
  four-dimensional, 324, 449, 576,
        647, 658, 667 ff.
  non-euclidean, 27, 32, 64, 91, 233,
        399, 604, 642 f., 658, 669, 701.
  of paths, 151.
  projective, 265.
  Riemannian, 573, 615, 640, 646,
        667.
Gestalt psychology, 74.
Gibbs, 772, 780.
Glands.
  of internal secretion. *See* Endocrine
        and under names of glands.
  salivary, 330 ff.
Godefroy, 567.
Goitre, 523. *See also* Thyroid.
Gomperz, H., 772.
Gomperz, T., 772.
Gonads, 522.
Gonseth, 772.
Goodspeed, 772.
Goudsmit, 692 ff.
Gradient; *See* Dynamic gradient.
  excitation-transmission, 103 f.
  physiological, 101, 103, 323.
Graham, 112.
Grassmann, 618.
Graven, 185, 397, 424, 540, 772.

INDEX OF DIAGRAMS